Springer Series in Wood Science

Editor: T. E. Timell

M. H. Zimmermann
Xylem Structure and the Ascent of Sap (1983)

J. F. Siau
Transport Processes in Wood (1984)

R. R. Archer
Growth Stresses and Strains in Trees (1986)

W. E. Hillis
Heartwood and Tree Exudates (1987)

S. Carlquist
Comparative Wood Anatomy (1988)

L. W. Roberts / P. B. Gahan / R. Aloni
Vascular Differentiation and Plant Growth Regulators (1988)

C. Skaar
Wood-Water Relations (1988)

J. M. Harris
Spiral Grain and Wave Phenomena in Wood Formation (1989)

B. J. Zobel / J. P. van Buijtenen
Wood Variation (1989)

P. Hakkila
Utilization of Residual Forest Biomass (1989)

J. W. Rowe
Natural Products of Woody Plants (1989)

John W. Rowe (Ed.)

Natural Products of Woody Plants II

Chemicals Extraneous to
the Lignocellulosic Cell Wall

With 244 Figures (both volumes)

Springer-Verlag Berlin Heidelberg GmbH

John W. Rowe
USDA, Forest Service
Forest Products Laboratory
One Gifford Pinchot Drive
Madison, WI 53705-2398, USA

Series Editor:

T. E. Timell
State University of New York
College of Environmental, Science and Forestry,
Syracuse, NY 13210, USA

Cover: Transverse section of *Pinus lambertiana* wood. Courtesy of Dr. Carl de Zeeuw, SUNY College of Environmental Science and Forestry, Syracuse, New York

ISBN 978-3-642-74077-0   ISBN 978-3-642-74075-6 (eBook)
DOI 10.1007/978-3-642-74075-6

Library of Congress Cataloging-in-Publication Data
Natural products of woody plants: chemicals extraneous to the lignocellulosic cell wall / John W. Rowe, ed. p. cm. – (Springer series in wood science)
ISBN 978-3-642-74077-0
1. Wood products. 2. Plant products. 3. Wood – Chemistry. I. Rowe, John W. II. Series. TS930.N38 1989

This work is subject to copyright. All rights are reserved, whether the whole or part of the material is concerned, specifically the rights of translation, reprinting, re-use of illustrations, recitation, broadcasting, reproduction on microfilms or in other ways, and storage in data banks. Duplication of this publication or parts thereof is only permitted under the provisions of the German Copyright Law of September 9, 1965, in its version of June 24, 1985, and a copyright fee must always be paid. Violations fall under the prosecution act of the German Copyright Law.

© Springer-Verlag Berlin Heidelberg 1989
Originally published by Springer-Verlag Berlin Heidelberg New York in 1989
Softcover reprint of the hardcover 1st edition 1989

The use of registered names, trademarks, etc. in this publication does not imply, even in the absence of a specific statement, that such names are exempt from the relevant protective laws and regulations and therefore free for general use.

Typesetting: K+V Fotosatz GmbH, Beerfelden;

2131/3020-543210 – Printed on acid-free paper

Contents Volume II

Contents Volume I		XIII
8	**Isoprenoids**	691
8.1	Terpenoids	691
	SUKH DEV	
8.1.1	Introduction	691
8.1.1.1	Nomenclature	692
8.1.1.2	Biosynthesis	692
8.1.2	Occurrence in Woody Plants	695
8.1.3	Classes: Distribution and Structural Types	696
8.1.3.1	Hemiterpenoids	696
8.1.3.2	Monoterpenoids	697
8.1.3.2.1	Distribution	697
8.1.3.2.2	Structural Types	698
8.1.3.2.3	Tropolones	711
8.1.3.3	Sesquiterpenoids	711
8.1.3.3.1	Distribution	711
8.1.3.3.2	Structural Types	712
8.1.3.4	Diterpenoids	746
8.1.3.4.1	Distribution	746
8.1.3.4.2	Structural Types	748
8.1.3.5	Sesterterpenoids	766
8.1.3.6	Non-Steroidal Triterpenoids	767
8.1.3.6.1	Distribution	767
8.1.3.6.2	Structural Types	768
8.1.3.7	Tetraterpenoids: Carotenoids	785
8.1.3.8	Polyterpenoids: Polyprenols	787
8.1.3.9	Meroterpenoids	788
8.1.4	Biological Role	789
	References	791
8.2	Steroids	808
	W. R. NES	
8.2.1	Introduction	808
8.2.1.1	Meaning of the Terms "Steroid" and "Sterol"	808
8.2.1.2	Nomenclature	810
8.2.1.3	Biosynthesis of Basic Steroidal Structure	812
8.2.2	Sterols	815
8.2.2.1	Names, Structures, and Organismic Relationships	815
8.2.2.2	Identification of Sterols	818
8.2.2.3	Occurrence in Wood and Bark	819
8.2.2.4	Biosynthetic Origin of Individual Sterols	826

8.2.2.5	Function of Sterols	830
8.2.2.6	Phyletic and Phylogenetic Relationships	832
8.2.2.7	Industrial Utilization of Tree Sterols	833
8.2.3	Esters	834
8.2.4	Glycosides	835
8.2.5	Spiroketals (Saponins)	835
8.2.6	Ecdysteroids	836
8.2.6.1	Names, Structures, and Occurrence	836
8.2.6.2	Function of Plant Ecdysteroids	837
8.2.7	Cardiac Glycosides	837
	References	839

9 The Influence of Extractives on Wood Properties and Utilization ... 843

9.1 Contribution of Extractives to Wood Characteristics ... 843
H. IMAMURA

9.1.1	Introduction	843
9.1.2	Color in Wood	844
9.1.2.1	Chemical Structure and Color	844
9.1.2.2	Color of Wood	844
9.1.2.3	Pigments Occurring in Wood	851
9.1.3	Odor in Wood	851
9.1.3.1	Volatile Components	851
9.1.3.2	Fragrant Components	852
9.1.3.3	Foul-Smelling Components	853
9.1.3.4	Removal of Foul Odors	855
9.1.3.5	Insect Attractants	855
9.1.4	Physical Properties	856
9.1.4.1	Wood Density and Strength	856
9.1.4.2	Other Physical Properties	859
	References	859

9.2 Role of Wood Exudates and Extractives in Protecting Wood from Decay ... 861
J. H. HART

9.2.1	Introduction	861
9.2.1.1	Decay	861
9.2.1.2	Decay-Causing Organisms and Their Effect on Wood Structure	861
9.2.2	How Trees Defend Themselves Against Decay	863
9.2.2.1	Role of Wounds	863
9.2.2.2	Toxic Heartwood Components	867
9.2.2.2.1	Formation of Antimicrobial Compounds	867
9.2.2.2.2	Chemical Nature of Antimicrobial Compounds	868
9.2.3	Evaluation of Decay Resistance	870
9.2.3.1	Isolation and Evaluation of Compounds	870
9.2.3.2	Physiology of Decay Inhibition	872
9.2.3.3	Variation in Decay Resistance	874
9.2.3.3.1	Variation Between Tree Species	874
9.2.3.3.2	Variation Between Individuals of the Same Species	875
9.2.3.3.3	Variation Within an Individual Tree	876
	References	878

Contents

9.3	Effect of Extractives on Pulping	880
	W. E. HILLIS and M. SUMIMOTO	
9.3.1	Introduction	880
9.3.2	Pulping Processes	881
9.3.2.1	Mechanical	882
9.3.2.2	Semichemical	883
9.3.2.3	Chemical	884
9.3.3	Pulpwood Quality	885
9.3.3.1	Effect of Extractives on Pulp Yield	885
9.3.3.2	Effect of Storage	885
9.3.4	Increased Consumption of Pulping Liquors	886
9.3.5	Effect on Pulping Processes	887
9.3.5.1	Reduced Penetrability of Liquors	887
9.3.5.2	Reduced Lignin Solubility	888
9.3.6	Effect on Equipment during Pulping	888
9.3.6.1	Wear and Corrosion	888
9.3.6.2	Blockage and Deposits	889
9.3.7	Pulp Properties	895
9.3.7.1	Color Changes Arising during Pulping and Bleaching	895
9.3.7.1.1	Mechanical Pulps	895
9.3.7.1.2	Chemical Pulps	897
9.3.7.2	Speck Formation and Pitch Problems during Pulping and Bleaching	901
9.3.7.3	Wettability	910
9.3.7.4	Sticking to Press Rolls	910
9.3.8	Spent Liquor Recovery	910
9.3.8.1	Concentration and Burning Difficulties	910
9.3.8.2	Foaming during Concentration and Oxidation	912
9.3.8.3	By-Product Recovery	912
9.3.9	Observations	912
	References	914
9.4	Effect of Extractives on the Utilization of Wood	920
	T. YOSHIMOTO	
9.4.1	Introduction	920
9.4.2	Inhibition of Resin and Glue Curing by Extractives	920
9.4.2.1	Phenolic Resin Adhesives	922
9.4.2.2	Amino Resin Adhesives	923
9.4.3	Inhibition of Cement Hardening by Extractives	923
9.4.3.1	Effect of Extractives on Cement Hardening	923
9.4.3.2	Effect of Light Exposure on Wood Panels Used with Cement	924
9.4.4	Color Change by Light	925
9.4.4.1	Color Changes Upon Exposure to Light	925
9.4.4.2	Color Changes by Other Agents	927
	References	929
9.5	Health Hazards Associated with Extractives	931
	E. P. SWAN	
9.5.1	Introduction	931
9.5.2	Toxic Extractives	932

9.5.2.1		Alkaloids and Amino Acids	932
9.5.2.2		Saponins and Glycosides	933
9.5.2.3		Quinones	933
9.5.2.4		Phenolics	933
9.5.2.5		Terpenes	933
9.5.3		Allergenic Extractives	934
9.5.3.1		Quinones	935
9.5.3.2		Alkyl Phenols	935
9.5.3.3		Terpenes	936
9.5.3.4		Phenolics	937
9.5.3.5		Other Allergenic Extractives	938
9.5.4		Carcinogenic Extractives	938
9.5.4.1		Early Work	938
9.5.4.2		Hausen's Contributions	938
9.5.4.3		Tannin	939
9.5.4.4		Other Carcinogenic Extractives	941
9.5.5		Hygiene and Safety	941
9.5.6		Discussion and Further Research	943
9.5.6.1		Discussion	943
9.5.6.2		Future Research	944
References			946

10 The Utilization of Wood Extractives 953

10.1	Naval Stores		953
	D. F. ZINKEL		
	10.1.1	Introduction	953
	10.1.2	Naval Stores Sources	954
	10.1.2.1	Gum Naval Stores	955
	10.1.2.2	Wood Naval Stores	956
	10.1.2.3	Sulfate (Kraft) Naval Stores	957
	10.1.2.4	Potential New Sources	958
	10.1.3	Turpentine	959
	10.1.3.1	Pine Oil	961
	10.1.3.2	Polyterpene Resins	961
	10.1.3.3	Flavors and Fragrances	962
	10.1.3.4	Insecticides	965
	10.1.3.5	Miscellaneous Uses	966
	10.1.4	Rosin	968
	10.1.4.1	Paper Size	969
	10.1.4.2	Polymerization Emulsifiers	970
	10.1.4.3	Adhesives	970
	10.1.4.4	Inks	971
	10.1.4.5	Other Market Areas	972
	10.1.5	Fatty Acids	973
	10.1.5.1	Intermediate Chemicals	974
	10.1.5.2	Protective Coatings	974
	10.1.5.3	Other Uses	975
	10.1.6	Miscellaneous Products	975
	10.1.7	The Future for Naval Stores	977
	References		977

10.2	Gums	978
	J. N. BeMiller	
10.2.1	Introduction	978
10.2.2	Larch Arabinogalactan	979
10.2.2.1	Source	979
10.2.2.2	Production	980
10.2.2.3	Purification	980
10.2.2.4	Properties	981
10.2.2.5	Uses	982
10.2.3	Gum Arabic (Gum Acacia, Acacia Gum)	983
10.2.3.1	Source, Production, and Purification	983
10.2.3.2	Properties	983
10.2.3.3	Uses	984
10.2.4	Gum Karaya	984
10.2.5	Gum Tragacanth	985
10.2.6	Gum Ghatti	986
10.2.7	Predictions	986
	References	987
10.3	Significance of the Condensed Tannins	988
	L. J. Porter and R. W. Hemingway	
10.3.1	Introduction	988
10.3.2	Tannins as Human and Animal Nutrition Factors	988
10.3.3	Pharmacological and Physiological Properties	991
10.3.4	Tannins as Insect, Mollusc, Bacterial, and Fungal Control Factors	992
10.3.5	Leather Tannage	993
10.3.5.1	Mechanisms of Vegetable Tanning	994
10.3.5.2	Vegetable Tanning Processes	997
10.3.5.3	Treatment of Vegetable Tanning Spent Liquors	1000
10.3.6	Condensed Tannins in Wood Adhesives	1002
10.3.6.1	Wattle Tannin-Based Particleboard Adhesives	1003
10.3.6.2	Wattle Tannin-Based Plywood Adhesives	1005
10.3.6.3	Wattle Tannin-Based Laminating Adhesives	1006
10.3.6.4	Other Wattle Tannin-Based Adhesives	1007
10.3.6.5	Conifer Bark and Related Tannins as Particleboard Adhesives	1008
10.3.6.6	Conifer Bark and Related Tannins as Plywood Adhesives	1011
10.3.6.7	Conifer Bark and Related Tannins in Cold-Setting Phenolic Resins	1014
10.3.7	Specialty Polymer Applications	1016
	References	1018
10.4	Rubber, Gutta, and Chicle	1028
	F. W. Barlow	
10.4.1	Introduction	1028
10.4.2	Historical Development of Natural Rubber Production	1029
10.4.2.1	*Hevea brasiliensis*	1029
10.4.2.2	Guayule	1031
10.4.3	Natural Rubber Production Processes	1032

X   Contents

10.4.3.1	Rubber Tree Growing	1032
10.4.3.2	Latex Collection	1034
10.4.3.3	Production of Latex Concentrate	1034
10.4.3.4	Production of Dry Rubber	1035
10.4.4	Packing and Market Grades	1037
10.4.4.1	Packing	1037
10.4.4.2	Grading	1037
10.4.5	Properties	1038
10.4.5.1	Natural Rubber	1038
10.4.5.2	Modified Natural Rubber	1040
10.4.5.2.1	Deproteinized Rubber	1040
10.4.5.2.2	Depolymerized Rubber	1040
10.4.5.2.3	Peptized Rubber	1040
10.4.5.2.4	Oil Extended Natural Rubber (OENR)	1041
10.4.5.2.5	MG Rubbers	1041
10.4.5.2.6	SP Rubbers	1041
10.4.6	Natural Rubber Utilization	1041
10.4.6.1	Vulcanization	1042
10.4.6.2	Transportation Items	1043
10.4.6.3	Mechanical Rubber Goods	1044
10.4.6.4	Footwear	1044
10.4.6.5	Miscellaneous Uses	1045
10.4.6.6	Health and Safety Factors	1045
10.4.7	Natural Rubber Economy	1046
10.4.8	Gutta Percha and Balata	1048
10.4.8.1	Production	1048
10.4.8.2	Properties and Uses	1049
10.4.9	Chicle	1049
References		1050

10.5 Other Extractives and Chemical Intermediates 1051
E. P. SWAN

10.5.1	Introduction	1051
10.5.2	Conifer Extractives Utilization	1051
10.5.2.1	Balsams, Copals, Amber, and Other Products	1051
10.5.2.2	Cedar Wood Oils	1052
10.5.2.3	Non-Commercial Extractives	1053
10.5.2.3.1	Conidendrin	1053
10.5.2.3.2	Dihydroquercetin	1054
10.5.2.3.3	Juvabione and Related Insect Hormones	1054
10.5.2.3.4	Occidentalol	1054
10.5.2.3.5	Plicatic Acid	1054
10.5.2.3.6	Thujaplicins, Thujic Acid, and Methyl Esters	1055
10.5.2.3.7	Waxes	1055
10.5.3	Hardwood Extractives Utilization	1056
10.5.4	Prospects	1056
References		1058

10.6	Pharmacologically Active Metabolites	1059
	C. W. W. BEECHER, N. R. FARNSWORTH, and C. GYLLENHAAL	
10.6.1	Introduction	1059
10.6.2	Sources of Information	1059
10.6.3	Currently Used Drugs Produced in Wood	1061
10.6.3.1	Alkaloids	1061
10.6.3.1.1	Tropane	1061
10.6.3.1.2	Quinolizidine	1061
10.6.3.1.3	Phenylalanine Derivatives	1062
10.6.3.1.3.1	Simple Tyramine Derivatives	1062
10.6.3.1.3.2	Protoberberine	1062
10.6.3.1.3.3	Phthalideisoquinoline	1062
10.6.3.1.3.4	Benzo(c)phenanthridine	1062
10.6.3.1.3.5	Ipecac	1062
10.6.3.1.4	Tryptophane Derivatives	1062
10.6.3.1.4.1	Indole	1062
10.6.3.1.4.2	Quinoline	1063
10.6.3.2	Quinoids	1063
10.6.3.3	Lignans	1063
10.6.3.4	Triterpenes	1063
10.6.3.5	Pyrones	1063
10.6.3.6	Coumarins	1063
10.6.4	Potential Drugs Derived from Secondary Metabolites of Wood	1063
10.6.4.1	Alkaloids	1064
10.6.4.1.1	Tropane	1064
10.6.4.1.2	Isoquinoline	1064
10.6.4.1.3	Indole Alkaloids	1065
10.6.4.1.4	Ansamacrolides	1065
10.6.4.1.5	Quinazoline	1065
10.6.4.1.6	Pyrazine Derivatives	1066
10.6.4.2	Quinoids	1066
10.6.4.3	Lignans	1066
10.6.4.4	Diterpenes	1068
10.6.4.5	Triterpenes	1069
10.6.4.6	Flavonoids	1069
10.6.5	Summary and Conclusions	1069
	References	1120

11	The Future of Wood Extractives	1165
	H. L. HERGERT	
11.1	Introduction	1165
11.2	Requirements for Future Wood Extractives Ventures	1166
11.2.1	Low Investment Risk	1166
11.2.2	Good Sales Potential	1167
11.2.3	Inexpensive Raw Material	1168
11.2.4	Shared Capital Expense	1169
11.2.5	National Priority	1169
11.2.6	Realistic Research, Development, and Engineering	1169

XII   Contents

11.3 Prospects for Existing Extractives-Based Industries 1170
    11.3.1 Natural Rubber 1170
    11.3.2 Rosin and Terpenes from Pine 1172
    11.3.3 Carbohydrate Gums 1174
    11.3.4 Tannins 1175
11.4 Failed Wood Extractives Ventures 1176
11.5 Future Directions for Industrially Oriented Extractives Research 1178
    11.5.1 Control of Extractives Deposition 1179
    11.5.2 Manipulation of Wood Growth by Chemicals 1180
    11.5.3 New Techniques for Extractives Isolation 1182
11.6 Areas of Needed Basic Research 1184
    11.6.1 Cambial Constituents: Growth Regulators 1185
    11.6.2 Root Constituents: Role of Mycorrhizae 1186
    11.6.3 Environmental Relationships 1186
    11.6.4 Pharmacologically Active Compounds 1188
    11.6.5 Phenolic Polymers 1189
    11.6.6 Sites and Control Mechanisms of Biosynthesis 1190
11.7 Conclusions 1192
    References 1193

Index of Plant Genera and Species 1197

Organic Compounds Index 1220

Contents Volume I

Contents Volume II		V
1	**Introduction and Historical Background**	1
1.1	Historical Uses of Extractives and Exudates	1
	W. E. HILLIS	
	1.1.1 Introduction	1
	1.1.2 Major Uses of Extractives and Exudates	1
	1.1.2.1 The Use of Durable Woods	2
	1.1.2.2 Exudates	3
	1.1.2.2.1 Varnishes	3
	1.1.2.2.2 Lacquers	5
	1.1.2.2.3 Gums	5
	1.1.2.3 Tannins	5
	1.1.2.4 Dyes	6
	1.1.2.5 Perfumes	7
	1.1.2.6 Rubber	8
	1.1.2.7 Medicines	9
	1.1.3 Lessons from History	10
	References	12
1.2	Natural Products Chemistry – Past and Future	13
	K. NAKANISHI	
	1.2.1 Introduction	13
	1.2.2 Isolation and Purification	16
	1.2.3 Structure Determination	18
	1.2.4 The Future of Natural Products Science	22
	References	25
2	**Fractionation and Proof of Structure of Natural Products**	27
	J. SNYDER, R. BREUNING, F. DERGUINI, and K. NAKANISHI	
2.1	Introduction	27
2.2	Novel Techniques and Recent Developments in Fractionation and Isolation	28
	2.2.1 Countercurrent Chromatography	28
	2.2.1.1 Coil Countercurrent Chromatography	29
	2.2.1.2 Droplet Countercurrent Chromatography	36
	2.2.1.3 Rotation Locular Countercurrent Chromatography	39
	2.2.1.4 Centrifugal Partition Chromatography	44
	2.2.1.5 Comparison of Partition Chromatographic Methods	46
	2.2.2 Adsorption Chromatography	49

2.2.2.1	Ion-Pair Chromatography	50
2.2.2.2	Other New Methods of Column Chromatography	53
2.2.2.3	Supercritical Fluid Chromatography	55

2.3 Nuclear Magnetic Resonance Spectroscopy ... 60
2.3.1	Proton Nuclear Magnetic Resonance	60
2.3.1.1	Difference Decoupling	60
2.3.1.2	Difference NOE	61
2.3.1.3	Contact Shifts	64
2.3.1.4	Partial Relaxation	65
2.3.2	Carbon Nuclear Magnetic Resonance	66
2.3.2.1	J-Modulated Spin Echo	67
2.3.2.2	Insensitive Nuclei Enhanced by Polarization Transfer	68
2.3.2.3	Distortionless Enhancement by Polarization Transfer	69
2.3.2.4	Carbon-Proton Heteronuclear Coupling	70
2.3.2.5	Carbon-Proton Heteronuclear NOE	72
2.3.2.6	Deuterium Isotopic Shifts	73
2.3.3	Two-Dimensional NMR Spectroscopy	75
2.3.3.1	Two Dimensional J-Resolved Proton NMR Spectroscopy	75
2.3.3.2	Two Dimensional Correlation Spectroscopy	76
2.3.3.3	Two Dimensional-INADEQUATE (Incredible Natural Abundance Double Quantum Transfer Experiment)	83

2.4 Other Spectroscopic Techniques ... 85
2.4.1	Mass Spectrometry	85
2.4.1.1	Techniques That Enhance Sample Volatilization	86
2.4.1.2	Modern Techniques of Ionization/Desorption	88
2.4.1.3	Tandem Mass Spectrometry	91
2.4.2	Ultraviolet-Visible Spectroscopy	93
2.4.3	Infra-Red Spectroscopy	97
2.4.4	Circular Dichroism	101
2.4.4.1	The Nature of Circular Dichroism	103
2.4.4.2	The Additivity Relation in A Values	106

2.5 General Conclusions ... 109
References ... 109

3 Evolution of Natural Products ... 125
O. R. GOTTLIEB

3.1 Convergent Synthesis and the Origin of RNA-Based Life ... 125
3.2 Expansion of the Acetate, Mevalonate, and δ-Aminolevulinate Pathways in Bacteria and Algae ... 127
3.3 Expansion of the Shikimate Pathway in Terrestrial Plants ... 128
3.4 Phytochemistry and Plant Defense ... 133
3.5 Oxidation Levels of Angiospermous Micromolecules ... 137
3.6 Skeletal Specialization of Angiospermous Micromolecules ... 140
3.7 Quantification of Micromolecular Parameters ... 145
3.8 Phytochemical Gradients in Angiosperms ... 146
3.9 Future Perspectives ... 148
References ... 150

4	**Carbohydrates**		155
	J. N. BeMiller		
4.1	Introduction		155
4.2	Sucrose		156
4.3	Higher Oligosaccharides Related to Sucrose		157
4.4	Other Oligosaccharides		158
4.5	Monosaccharides		159
4.6	Alditols		159
4.7	Cyclitols		160
	4.7.1	*myo*-Inositol	160
	4.7.2	D-*chiro*-Inositol	161
	4.7.3	Quebrachitol	161
	4.7.4	D-Quercitol	162
	4.7.5	Conduritol	162
	4.7.6	Quinic Acid	162
4.8	Plant Glycosides		162
4.9	Starch		162
4.10	Extractable Polysaccharides		163
	4.10.1	Arabinogalactans	164
	4.10.1.1	Larch Arabinogalactans	165
	4.10.1.2	Other Extractable, Nonexudate Arabinogalactans	165
	4.10.2	Other Extractable Polysaccharides; The Pectic Polysaccharides	166
	4.10.3	Exudate Gums	167
	4.10.3.1	*Acacia* Gums	168
	4.10.3.2	Exudate Gums of Other Rosales Genera	168
	4.10.3.3	Gums of Combretaceae (Myrtiflorae) Genera	169
	4.10.3.4	Exudate Gums of Anacardiaceae (Sapindales)	170
	4.10.3.5	Exudate Gums of Families in the Orders Rutales, Parietales, and Malvales	170
	4.10.3.6	Exudate Gums from Other Orders	171
	4.10.3.7	Exudate Gums with Xylan Cores	172
References			172
5	**Nitrogenous Extractives**		179
5.1	Amino Acids, Proteins, Enzymes, and Nuccleic Acids		179
	D. J. Durzan		
	5.1.1	Introduction	179
	5.1.2	Composition	182
	5.1.2.1	Free and Bound Amino Acids	183
	5.1.2.2	Proteins and Enzymes	186
	5.1.2.3	Nucleic Acids and Related Products	189
	5.1.3	Factors Determining Composition	189
	5.1.3.1	Genetics	189
	5.1.3.2	Genetics × Environment	190
	5.1.3.3	Growth and Development	190

5.1.3.4	Pathology	191
5.1.3.5	Impact of Humans	192
5.1.4	Conclusion	195
References		195

5.2 The Alkaloids 200
S.-I. SAKAI, N. AIMI, E. YAMANAKA, and K. YAMAGUCHI

5.2.1	Introduction	200
5.2.2	True Alkaloids	202
5.2.2.1	Alkaloids from Ornithine	202
5.2.2.1.1	Coca Alkaloids	202
5.2.2.1.2	Elaeocarpus Alkaloids	204
5.2.2.2	Alkaloids from Lysine	205
5.2.2.2.1	Punica Alkaloids	205
5.2.2.2.2	Lythraceae Alkaloids	206
5.2.2.2.3	Securinega Alkaloids	208
5.2.2.2.4	Cytisus Alkaloids	209
5.2.2.3	Alkaloids from Anthranilic Acid	210
5.2.2.3.1	Quinoline and Furoquinoline Alkaloids	210
5.2.2.3.2	Acridone Alkaloids	211
5.2.2.3.3	Evodia Alkaloids	212
5.2.2.3.4	Carbazole Alkaloids	213
5.2.2.4	Alkaloids from Nicotinic Acid (Celastraceae Alkaloids)	213
5.2.2.5	Alkaloids from Phenylalanine and Tyrosine	214
5.2.2.5.1	Benzylisoquinoline Alkaloids	214
5.2.2.5.2	Curare Alkaloids	217
5.2.2.5.3	Sinomenine	217
5.2.2.5.4	Aporphine-type Alkaloids	218
5.2.2.5.5	Berberine	218
5.2.2.5.6	Nitidine	219
5.2.2.5.7	Erythrina Alkaloids	221
5.2.2.5.8	Cephalotaxus Alkaloids	222
5.2.2.5.9	Ipecacuanha and Alangium Alkaloids	223
5.2.2.6	Alkaloids from Tryptophan	225
5.2.2.6.1	Calycanthus Alkaloids	225
5.2.2.6.2	Picrasma (Pentaceras) and Carboline Alkaloids	226
5.2.2.6.3	Rauwolfia Alkaloids	227
5.2.2.6.4	Tabernanthe Alkaloids	229
5.2.2.6.5	Ochrosia Alkaloids	231
5.2.2.6.6	Ervatamia Alkaloids	231
5.2.2.6.7	Uncaria-Mitragyna Alkaloids	232
5.2.2.6.8	Yohimbe Alkaloids	236
5.2.2.6.9	Cinchona Alkaloids	237
5.2.2.6.10	Guettarda Alkaloids	238
5.2.2.6.11	Strychnos Alkaloids	239
5.2.2.6.12	Gelsemium Alkaloids	241
5.2.2.6.13	Gardneria Alkaloids	243
5.2.2.6.14	Camptothecins	244
5.2.3	Pseudoalkaloids	246
5.2.3.1	Alkaloids from Polyketides	246
5.2.3.1.1	Pinidine	246

5.2.3.1.2	Galbulimima Alkaloids	246
5.2.3.2	Alkaloids from Mevalonate	248
5.2.3.2.1	Spiraea Alkaloids	248
5.2.3.2.2	Erythrophleum Alkaloids	248
5.2.3.2.3	Daphniphyllum Alkaloids	249
5.2.3.2.4	Apocynaceae Steroidal Alkaloids	250
5.2.3.2.5	Buxaceae Steroidal Alkaloids	250
References		251

6 Aliphatic and Alicyclic Extractives ... 259

6.1 Simple Organic Acids ... 259
Y. OHTA

6.1.1	Introduction	259
6.1.2	Organic Acids in the TCA and Glyoxylate Cycles	259
6.1.2.1	Citric Acid	260
6.1.2.2	Aconitic Acid	261
6.1.2.3	Isocitric Acid	261
6.1.2.4	α-Ketoglutaric Acid	261
6.1.2.5	Succinic Acid	262
6.1.2.6	Fumaric Acid	262
6.1.2.7	Malic Acid	262
6.1.2.8	Oxaloacetic Acid	263
6.1.2.9	Glyoxylic Acid	263
6.1.3	Other Metabolically Important Organic Acids	263
6.1.3.1	Glycolic Acid	263
6.1.3.2	Glyceric Acid	265
6.1.3.3	Pyruvic Acid	265
6.1.3.4	Malonic Acid	265
6.1.3.5	Shikimic Acid and Quinic Acid	266
6.1.4	Organic Acids of an End-Product Nature	266
6.1.4.1	Lactic Acid	266
6.1.4.2	Oxalic Acid	268
6.1.4.3	Tartaric Acid	268
6.1.4.4	Chelidonic Acid	269
6.1.4.5	Fluoroacetic Acid	269
References		270

6.2 Complex Aliphatic and Alicyclic Extractives ... 274
Y. OHTA

6.2.1	Introduction	274
6.2.2	γ-Lactones	275
6.2.3	δ-Lactones (2-Pyrones)	280
6.2.4	Cyanogenic Glycosides and Related Compounds	281
6.2.5	Highly Oxygenated Cyclohexanes	287
6.2.6	Cyclohexane Diols	289
6.2.7	Polycyclic Compounds	291
6.2.8	Miscellaneous	292
References		294

6.3 Fats and Fatty Acids 299
D. F. ZINKEL

6.3.1	Introduction	299
6.3.2	Fats as Food Reserves.................................	299
6.3.3	Aliphatic Monocarboxylic Acids	300
6.3.3.1	Volatile Fatty Acids...................................	301
6.3.3.2	Constituent Fatty Acids of Fats	301
References	....	302

6.4 Chemistry, Biochemistry, and Function of Suberin and Associated Waxes ... 304
P. E. KOLATTUKUDY and K. E. ESPELIE

6.4.1	Introduction	304
6.4.2	Waxes	304
6.4.2.1	Analysis of Plant Waxes	304
6.4.2.2	Composition of Suberin-Associated Waxes..............	306
6.4.2.2.1	Hydrocarbons in Suberin-Associated Waxes	307
6.4.2.2.2	Wax Esters	308
6.4.2.2.3	Free Fatty Alcohols...................................	308
6.4.2.2.4	Free Fatty Acids	308
6.4.2.2.5	Polar Wax Components	309
6.4.2.2.6	Ferulic Acid Esters	309
6.4.2.2.7	Tabular Survey of Bark Wax Components	310
6.4.2.3	Biosynthesis of Wax Components	312
6.4.2.3.1	Very Long Fatty Acids	312
6.4.2.3.2	Fatty Alcohols	313
6.4.2.3.3	Wax Esters	313
6.4.2.3.4	Hydrocarbons and Derivatives	314
6.4.3	Suberin	316
6.4.3.1	Ultrastructure...	316
6.4.3.1.1	Ultrastructural Characterization	316
6.4.3.1.2	Ultrastructural Identification of Suberin in Bark	317
6.4.3.2	Chemical Composition and Structure of the Polymer ..	323
6.4.3.2.1	Composition of the Aliphatic Portion	323
6.4.3.2.2	Phenolic Composition	326
6.4.3.2.3	Structure	333
6.4.3.3	Suberin Biosynthesis...................................	337
6.4.3.3.1	Biosynthesis of the Aliphatic Monomers	337
6.4.3.3.1.1	ω-Hydroxylation of Fatty Acids.........................	337
6.4.3.3.1.2	Oxidation of ω-Hydroxy Acids........................	337
6.4.3.3.1.3	Biosynthesis of Mid-Chain Oxygenated Suberin Monomers	339
6.4.3.3.2	Biosynthesis of the Aromatic Components of Suberin ..	341
6.4.3.3.3	Biosynthesis of Suberin from Aliphatic and Aromatic Monomers	342
6.4.3.4	Function of Suberin and Associated Waxes	343
6.4.3.4.1	Prevention of Water Loss..............................	343
6.4.3.4.2	Suberization in Wound Healing	344
6.4.3.4.3	Suberization in Response to Stress.....................	344
6.4.3.4.4	Suberization as a Means of Compartmentalization	345
6.4.3.5	Regulation of Suberization	346
6.4.3.6	Enzymatic Degradation of Suberin	347
References	....	349

7 Benzenoid Extractives 369

7.1 Monoaryl Natural Products 369
O. THEANDER and L. N. LUNDGREN

7.1.1	Introduction	369
7.1.2	Simple Phenols (C$_6$)	370
7.1.3	Phenolic Acids, Salicins and Other C$_6$–C$_1$ Compounds	371
7.1.3.1	Benzoic Acids and Related Compounds	371
7.1.3.2	Salicins and Related Compounds	371
7.1.4	Acetophenones and Other C$_6$–C$_2$ Compounds	373
7.1.5	Cinnamic Acids, Coumarins and Other Phenylpropanoids (C$_6$–C$_3$)	374
7.1.5.1	Cinnamic Acids	374
7.1.5.2	Coumarins	374
7.1.5.3	Other Phenylpropanoids	383
7.1.6	Miscellaneous Monoaryl Compounds	392
	References	393

7.2 Gallic Acid Derivatives and Hydrolyzable Tannins 399
E. HASLAM

7.2.1	Introduction	399
7.2.2	Metabolism of Gallic Acid – General Observations	400
7.2.3	Biosynthesis of Gallic Acid	404
7.2.4	Metabolites of Gallic Acid	406
7.2.4.1	Simple Esters Occurrence and Detection	407
7.2.4.2	Depside Metabolites Group 2A	412
7.2.4.3	Metabolites Formed by Oxidative Coupling of Galloyl Esters Groups 2B and 2C, Ellagitannins	415
7.2.4.3.1	Hexahydroxydiphenoyl Esters	416
7.2.4.3.2	Dehydrohexahydroxydiphenoyl Esters	419
7.2.4.3.3	Group 2B Metabolites	419
7.2.4.3.4	Group 2C Metabolites	426
7.2.4.3.5	Postscript	429
7.2.5	The Interaction of Proteins with Metabolites of Gallic Acid	430
	References	433

7.3 Lignans 439
O. R. GOTTLIEB and M. YOSHIDA

7.3.1	Introduction	439
7.3.2	Nomenclature and Numbering	441
7.3.3	Chemistry	501
7.3.4	Oligomeric Lignoids	505
	References	505

7.4 Stilbenes, Conioids, and Other Polyaryl Natural Products 512
T. NORIN

7.4.1	Introduction	512
7.4.2	Stilbenes and Structurally Related Compounds	512

7.4.3	Conioids (Norlignans), Including Condensed and Structurally Related Compounds	517
7.4.4	Aucuparins and Structurally Related Biphenyls	520
7.4.5	Diarylheptanoids, Structurally Related Diarylheptanoids, and Bridged Biphenyls (Cyclophanes)	521
7.4.6	Miscellaneous Diaryl and Polyaromatic Compounds	525
7.4.7	Concluding Remarks	528
References		528

7.5 Flavonoids 533
J. B. HARBORNE

7.5.1	Introduction	533
7.5.2	Structural Types	536
7.5.2.1	Flavones and Flavonols	536
7.5.2.2	Flavonones and Flavononols	543
7.5.2.3	Chalcones and Aurones	546
7.5.2.4	Isoflavonoids and Neoflavonoids	548
7.5.3	Distribution	554
7.5.3.1	Distribution within the Plant	554
7.5.3.2	Patterns in Gymnosperm Woods	556
7.5.3.3	Patterns in Angiosperm Woods	558
7.5.3.3.1	Heartwood Flavonoids of the Anacardiaceae	559
7.5.3.3.2	Heartwood Flavonoids of the Leguminosae	562
7.5.4	Properties and Function	567
References		569

7.6 Biflavonoids and Proanthocyanidins 571
R. W. HEMINGWAY

7.6.1	Introduction	571
7.6.2	Biflavonoids	571
7.6.2.1	Structural Variations of Biflavonoids	572
7.6.2.2	Distribution of Biflavonoids	577
7.6.2.3	Significant Properties of Biflavonoids	583
7.6.3	Proanthocyanidins	584
7.6.3.1	Flavan-3-ols	586
7.6.3.1.1	Structure of Flavan-3-ols	587
7.6.3.1.2	Distribution of Flavan-3-ols	589
7.6.3.1.3	Reactions of Flavan-3-ols	594
7.6.3.2	Flavan-3,4-diols	602
7.6.3.2.1	Structure of Flavan-3,4-diols	604
7.6.3.2.2	Distribution of Flavan-3,4-diols	608
7.6.3.2.3	Reactions of Flavan-3,4-diols	608
7.6.3.3	Oligomeric Proanthocyanidins	611
7.6.3.3.1	Structure and Distribution of Oligomeric Proanthocyanidins	612
7.6.3.3.1.1	Proquibourtinidins	613
7.6.3.3.1.2	Profisetinidins	613
7.6.3.3.1.3	Prorobinetinidins	619
7.6.3.3.1.4	Proteracacidins and Promelacacidins	619
7.6.3.3.1.5	Propelargonidins	619
7.6.3.3.1.6	Procyanidins	621

7.6.3.3.1.7 Prodelphinidins		629
7.6.3.3.2	Reactions of Oligomeric Proanthocyanidins	631
References		636

7.7 Condensed Tannins ... 651
L. J. PORTER

7.7.1	Introduction	651
7.7.2	Structure and Properties	652
7.7.2.1	Isolation and Purification	652
7.7.2.2	Elucidation of the Structure of Type 1 Proanthocyanidin Polymers	653
7.7.2.3	Structure of Type 2 Proanthocyanidin Polymers	660
7.7.2.4	Molecular Weight Distribution	661
7.7.2.5	Conformation and Solution Properties	664
7.7.2.6	Complexation	667
7.7.3	Distribution in Plants	669
7.7.3.1	Chemotaxonomic and Phylogenetic Significance	669
7.7.3.2	Distribution and Structural Variations within Plants	675
7.7.4	Metabolism	676
7.7.4.1	Biosynthesis	676
7.7.4.2	Seasonal Variation and Fate in Senescent Tissues	681
7.7.5	Role in Plants	682
7.7.5.1	Resistance to Insects	682
7.7.5.2	Resistance to Decay Fungi	683
7.7.5.3	Allelopathic Relationships	684
References		685

Chapter 8
Isoprenoids

Isoprenoids – or polyprenoids, as they are sometimes called – comprise a large and important group of natural products that can be formally derived from isoprene (isopentane) units. This group consists of two major classes: terpenoids and steroids. Although terpenoids and steroids are genetically related and such a relationship was anticipated by Heilbron as early as 1926 (187) – the chemistry of the two classes has developed fairly independently. Because of this historical fact, the two groups are usually treated separately.

8.1 Terpenoids
SUKH DEV

8.1.1 Introduction

The term "Terpen" (English, "terpene") is attributed to Kekule (171, vol. 1:15–83) who coined it to describe $C_{10}H_{18}$ hydrocarbons occurring in turpentine (German, "Terpentin") oil. This term has, over the years, acquired a generic significance and is used to designate isoprene-based secondary metabolites. The term terpenoid which has come into considerable usage since 1955, is now considered to be synonymous with terpene and is the preferred generic name for this class of natural products. At present, several groups of terpenoids, classified in terms of C_{10}-units are recognized (Table 8.1.1). Within each class, members are often grouped together according to the number of carbocyclic rings – e.g., acyclic diterpenoids, monocyclic diterpenoids, etc.

However, several natural products are known that have a number of carbon atoms which is different from those shown in Table 8.1.1, but have a clear struc-

Table 8.1.1. Classification of terpenoids

Classification	No. of carbon atoms	No. of isoprene units
Hemiterpenoids	5	1
Monoterpenoids	10	2
Sesquiterpenoids	15	3
Diterpenoids	20	4
Sesterterpenoids	25	5
Triterpenoids	30	6
Tetraterpenoids	40	8
Polyterpenoids	>40	>8

tural relationship with one of these classes. These compounds, which may have fewer (nor derivatives) or more (homo derivatives) carbon atoms, apparently arise from their parent terpenoids by metabolic processes. At least one group of C_{20} terpenoids is known (see Sect. 8.1.3.6.2) that has been shown to be formed in nature by loss of 10 carbon atoms from a suitable triterpene precursor and hence is treated as decanortriterpenoids, rather than as true diterpenoids.

Terpenoids are probably the largest single group of secondary metabolites, and at present over 7500 structurally well-defined compounds embracing a bewildering but fascinating array of skeletal types are known.

8.1.1.1 Nomenclature

As is the case with other classes of natural products, a trivial name based on the biological source material is invariably assigned to a new terpenoid. Thus terpenoid compounds usually carry a name derived from the family, genus, species, or local name of the raw material from which it had been isolated. Because functionality of a new compound is readily established, a trivial name is coined to indicate the dominant functionality.

It has been suggested (212) that semisystematic names based on parent structural types be used whenever possible. The name of the parent structure is derived from the trivial name of the member first isolated. This name, when used for such derivations, besides indicating the carbon skeleton, implies, without further specification, the absolute configuration at all chiral centers of the parent. Carbon skeletons closely related to a parent structure are often named by using prefixes such as 'cyclo', 'nor', 'homo', 'seco', 'abeo' etc. with the parent name, as indicators of modification. The prefix *ent* (short for 'enantio') is used to indicate inversion at all chiral centers implied in the parent structure.

Systematic names for terpenoids, based on the general rules of nomenclature of organic compounds (324), are usually cumbersome and are seldom used unless the structures are sufficiently simple (212).

8.1.1.2 Biosynthesis

The *isoprene rule* (337) and its evolved version, the *biogenetic isoprene rule* (336, 337), have proven to be of fundamental consequence in the development of terpene chemistry. According to the biogenetic isoprene rule, which was formally enunciated by Ruzicka in 1953, terpene structures may be rationalized, or preliminary structures deduced, by accepted reaction mechanisms from hypothesized acyclic precursors such as geraniol, farnesol, geranylgeraniol, etc. Biosynthetic investigations (20, 75, 87, 100, 110, 160, 162, 179, 263) over the past three decades have fully confirmed the biogenetic isoprene rule, and have provided detailed information on several fundamental steps in isoprenoid biosynthesis.

The fortuitous discovery of mevalonic acid (3-methyl-3,5-dihydroxypentanoic acid) by Folkers and co-workers and by Tamura (110) in 1956 prompted its evaluation as a possible precursor in the biosynthesis of cholesterol. It was found that

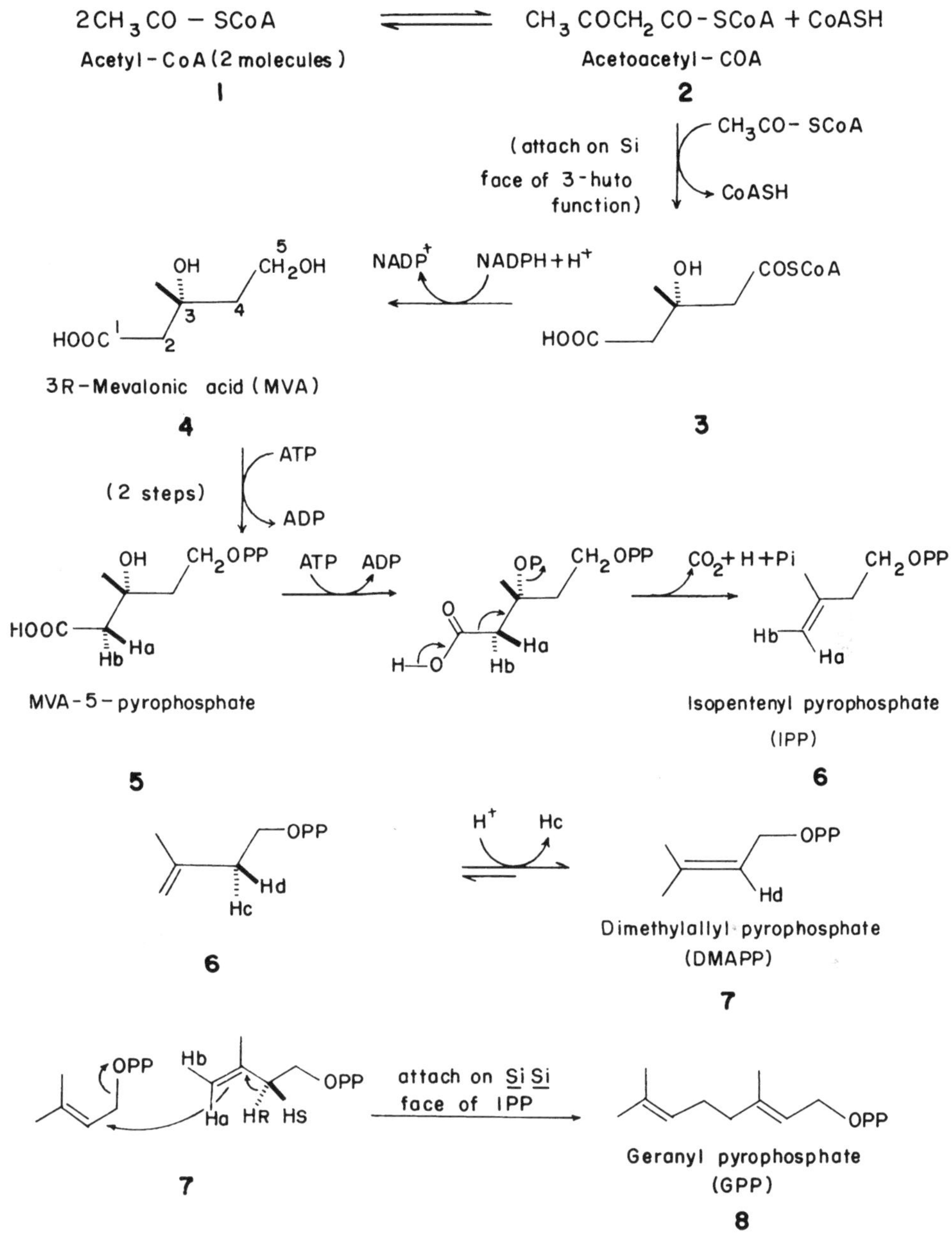

Fig. 8.1.1. Biosynthesis of geranyl pyrophosphate

one of the enantiomers was quantitatively incorporated with loss of carbon dioxide.

Lynen subsequently showed that the active C_5 unit from mevalonic acid was isopentenyl pyrophosphate (IPP, **6**) and it was at once recognized as the long-sought biological isoprene unit. Later investigations, mostly by Lynen, Cornforth,

Popjak and their collaborators, helped in clarifying the various steps involved in the biosynthesis of mevalonic acid and the acyclic terpene precursors. Figure 8.1.1 gives a simplified version of the pathway from acetyl coenzyme A to mevalonic acid, to IPP, and thence to geranyl pyrophosphate (GPP, **8**); this knowledge has been gathered essentially from studies on biosynthesis of steroids in animals and microorganisms.

It is now well established that two molecules of acetyl coenzyme A (**1**) condense to furnish acetoacetyl coenzyme A (**2**), which by an aldol-type reaction with a third molecule of acetyl coenzyme A gives β-hydroxy-β-methylglutaryl coenzyme A (**3**). This undergoes an irreversible reduction through intervention of NADPH (reduced nicotinamide-adenine dinucleotide phosphate) to R-(−)-mevalonilc acid (MVA, **4**), the building block of almost all isoprenoids. Phosphorylation of MVA by adenosine triphosphate (ATP) in two steps (two enzymes) leads to MVA-5-pyrophosphate (**5**). The latter reacts on the enzyme with ATP, generating IPP (**6**), with concomitant loss of CO_2. This elimination reaction, which had no previous analogy in biochemistry, proceeds by a concerted

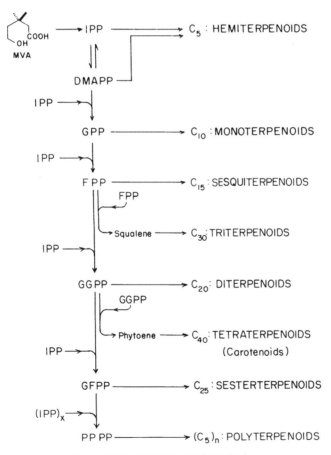

Fig. 8.1.2. General scheme of isoprenoid biosynthesis

trans-elimination. IPP is then isomerized by an enzyme-catalyzed prototropy into an equilibrium mixture with dimethylallyl pyrophosphate (DMAPP, 7), in which the latter predominates. This isomerization proceeds in a stereospecific manner: the incoming proton attacks the *re re* face of the olefinic bond in **6**, finally resulting in the loss of hydrogen Hc (in **6**). The two intermediates **6** and **7** react as shown (Fig. 8.1.1) with the intervention of prenyl transferase, generating GPP (**8**) in a stereospecific condensation (75, 113). GPP is then channelled into monoterpene biosynthesis, and/or serves as a substrate for prenylation by IPP to the sesquiterpenoid precursor farnesyl pyrophosphate (FPP; Fig. 8.1.2). Two other metabolic pathways are available to FPP: 1) tail-to-tail dimerization by the action of membrane-bound enzyme presqualene synthase to yield squalene, the precursor of all triterpenoids and steroids; and 2) chain extension to geranylgeranyl pyrophosphate (GGPP; Fig. 8.1.2). The same three options are now available to GGPP: generating diterpenoids, prenylation to geranylfarnesyl pyrophosphate (GFPP), the sesterterpenoid precursor, and/or dimerization to phytoene and thence to carotenoids (Fig. 8.1.2.). Further chain elongation of GFPP to polyprenyl pyrophosphate (PPPP, Fig. 8.1.2) is also an important reaction leading to the formation of polyprenols (392) and related products. The sequence of reactions just discussed generates only *trans* linear isoprenoids. A different prenyl-transferase is involved in production of rubber (10), which is *cis* configurated.

Significant information is now available on the regulation and compartmentation of terpenoid biosynthesis (259, 412). There is sufficient evidence to show that terpene chain elongation can take place both inside and outside the chloroplasts, whereas the final specific reactions are mostly restricted to the organelles.

8.1.2 Occurrence in Woody Plants

Terpenoids are widely distributed in higher plants. Some compounds, such as chlorophyll pigments that have a diterpene side chain or gibberellins, are essential for the growth and well-being of plants (Sect. 8.1.4) and hence, presumably, are produced by all higher green plants. Terpenoids constitute important components of many wood extractives, and are often the major constituents of extracts obtained with non-polar solvents. Essential oils (171, 275, 395, 396), latexes, and resinous exudates (377, 389) from plants are often composed mainly of terpenoids. Mono- and sesquiterpenes are the usual components of essential oils, whereas diterpenes, being less steam-volatile, are only rarely found in essential oils. Co-occurrence of diterpenoids or triterpenoids with lower terpenoids is common, though the two seldom occur together in the same tissue.

In nature, terpenoids are found to occur both in free and combined states. Most of the terpenoids that are not highly functionalized occur free. Carboxylic acids invariably occur in a free state, though occasional occurrence of methyl and other esters is known. Alcohols often occur both in the free state and as esters, in which the acid moiety is usually a simple aliphatic acid (e.g., acetic, tiglic, angelic, etc.) or less often an aromatic acid (e.g., benzoic, cinnamic), though esters derived from more exotic acids hae been encountered (143). Highly oxygenated terpenoids often occur as glycosides.

In common with many other naturally occurring organic compounds, the percentage and nature of terpenoids in a given plant may vary widely with the time of harvesting of plant material (87, 118, 171). Different parts of the same plant may contain different terpenoids (87, 154, 204).

8.1.3 Classes: Distribution and Structural Types

8.1.3.1 Hemiterpenoids

Hemiterpenes, by definition, consist of only one isopentane unit and are, thus, the simplest members of terpenoids. The basic building blocks of isoprenoids, namely IPP and DMAPP, must be universally present in green plants, being the progenitors of the phytyl side chain of chlorophyll, which is essential for biological activity (Sect. 8.1.4). These do not appear to be accumulated by plants (3), though the related prenyl alcohol (3,3-dimethylallyl alcohol, 9) and its isomer isoprene alcohol (10) have been found to occur in a few essential oils (229). Isoprene (11) itself is an important emittant of pine forests (165).

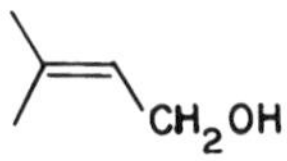

Prenyl alcohol 9

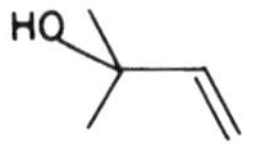

Isoprene alcohol 10

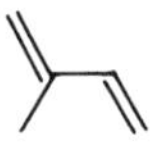

Isoprene 11

Metabolic activity of DMAPP and IPP sometimes results in prenylation of plant phenolics leading to prenyl ethers, several of which hare been isolated from woody tissues. For example, imperatorin (12), a prenylated coumarin, is present in the roots of *Imperatoria ostruthium* (228). From African sneezewood *Ptaeroxylon obliquum,* prenyletin (13) was obtained along with several other coumarins (125). Another example of a naturally occurring prenyl ether is the isoflavone, maxima-substance C (14) from roots of *Tephrosia maxima* (229). Several other ex-

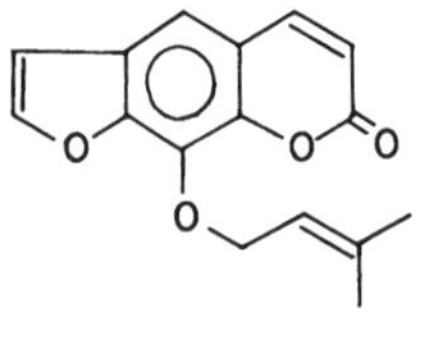

Imperatorin 12

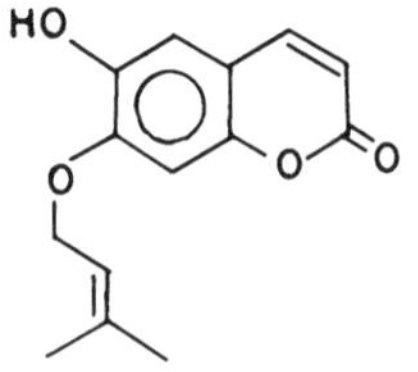

Prenyletin 13

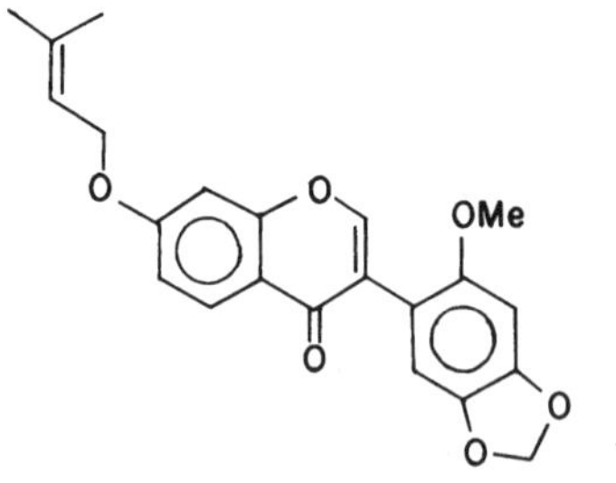

Maxima-substance – C 14

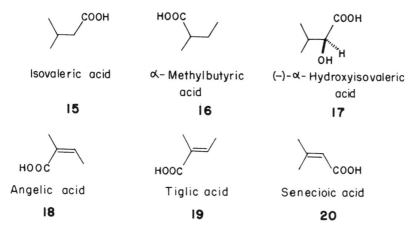

Fig. 8.1.3. Some non-mevalonoid hemiterpenoids of natural occurrence

amples may be found in the reviews by Murray (288), Ingham (209) and Grundon (170).

Another group of C_5 compounds with an isopentane framework is represented by compounds shown in Fig. 8.1.3. These acids have been found to occur, both free and as esters, in the roots of several species of the families Valerianaceae and Umbelliferae. Thus, isovaleric acid (15) and (−)-α-hydroxy-isovaleric acid (17) have been obtained from roots of *Valeriana officinalis*. Angelic acid (18) was first isolated in 1842 from the roots of *Angelica archangelica*, which also yields α-methylbutyric acid (16) (228). Tiglic acid (19) occurs in many plants including *Angelica glabra*, while roots of *Peucedanum japonicum* contain senecioic acid (20) (229). These compounds − and there are several others related to these − are of non-mevalonate origin, and have been shown to be products of amino acid metabolism (152).

8.1.3.2 Monoterpenoids

8.1.3.2.1 Distribution

Monoterpenoids are widely distributed in plant and animal kingdoms, both terrestrial and marine (379). Lists of their occurrence in higher plants have been compiled (19, 155, 175, 275, 395, 396). Table 8.1.2 lists plant families that are, by far, the most prolific in accumulating monoterpenoids. α-Pinene, cineol and limonene are the most widely distributed monoterpenoids.

Monoterpenoids apparently have a limited value in phylogeny and systematics (147) though some useful attempts in this direction have been made (87, 89 vol. 12, 155, 186).

Several monoterpenoids are known to occur in higher plants in both optically active forms, and sometimes both antipodes (partially or fully racemic compounds) have been found to occur in the same plant (379).

698   Isoprenoids

Table 8.1.2. Plant families prolific in producing monoterpenoids

Gymnospermae	Angiospermae
Cupressaceae	Monocotyledoneae
Pinaceae	Gramineae
Podocarpaceae	Zingiberaceae
Taxodiaceae	Dicotyledoneae
	Burseraceae
	Compositae
	Labiatae
	Lauraceae
	Myrtaceae
	Pittosporaceae
	Rutaceae
	Umbelliferae
	Valerianaceae

8.1.3.2.2 Structural Types

It is now fairly well-established that GPP (**8**) is the biosynthetic precursor of almost all monoterpenoids, except for a small group of so-called "irregular" types. Acyclic monoterpenes, having 2,6-dimethyloctane framework, arise directly

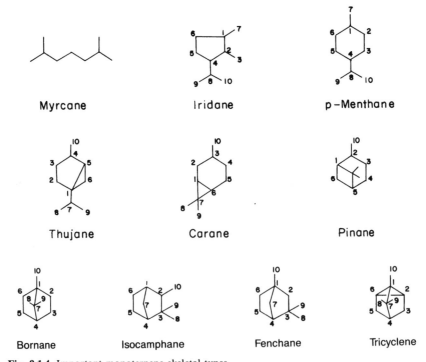

Fig. 8.1.4. Important monoterpene skeletal types

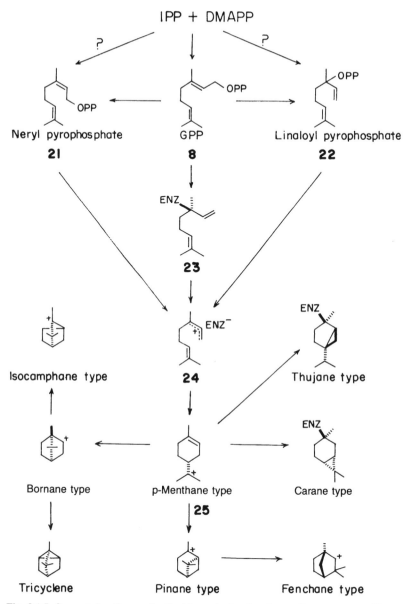

Fig. 8.1.5. Suggested pathways for the biosynthesis of major cyclic monoterpenoid types

from GPP by modifying reactions. Cyclizations of GPP as such or via neryl pyrophosphate (**21**) or via linaloyl pyrophosphate (**22**) result in cyclic structures that undergo secondary transformations to generate an extensive array of cyclic monoterpenoids. At present, over 740 monoterpenoids (only C_{10} compounds) embracing some 40 distinct skeletal types are known (379). Of these, the important ones occurring in woody tissues are shown in Fig. 8.1.4. These are also the

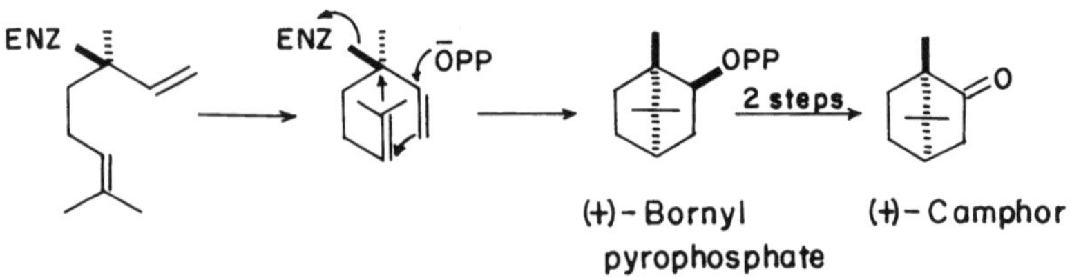

700   Isoprenoids

Fig. 8.1.6. Projected pathway for the biosynthesis of (+)-camphor

types most widely distributed in nature. Synthetic activity in the area of monoterpenoids has been reviewed (390).

Figure 8.1.5 depicts cyclization of GPP to various cyclic structure types of Fig. 8.1.4 (except iridane type). Until recently, it had been argued that these cyclizations must involve, as the first step, isomerization of GPP to neryl pyrophosphate (**21**) or linaloyl pyrophosphate (**22**), for in GPP the Δ^2 is *trans* configurated, hence, its envisaged cyclization to structure **25** is sterically prohibitive (87). However, recent studies with isolated enzyme systems, free from phosphohydrolases, have demonstrated that neryl pyrophosphate is not an obligatory intermediate and indeed GPP has been found to be the preferred substrate in the biosyntheses of several cyclic monoterpenoids (118). To rationalize these results, it has been proposed (75, 77, 119, 367) that GPP first reacts with the enzyme, with concomitant ionization of pyrophosphate (which still remains associated with the reacting complex) to furnish an enzyme-bound linaloyl derivative, such as **23** or an ion-pair like **24** (Fig. 8.1.5) with a conformation suitable for cyclization, and thus, can finally lead to various cyclic monoterpenoids. Figure 8.1.6 specificallys depicts the postulated pathway for the biosynthesis of (+)-camphor (**27**) in *Salvia officinalis* from GPP; bornyl pyrophosphate (**26**) was actually isolated and characterized (118). However, the situation is complicated: Another group of workers (370) has reported that in the biosynthesis of α-terpineol (**28**) in *Mentha spicata* and of (+)-limonene (**29**) and (−)-perillaldehyde (**30**) in *Citrus natsudaidai* and *Perilla frutescens*, linalool (**31**)and phosphate and pyrophosphate are preferred over geraniol, nerol, or their esters, both in vivo and in cell-free extracts. Thus, cyclic monoterpenoids appear to be formed from several substrates, and carbocyclases from different plant sources do not exhibit absolute substrate specificity (112).

Biosynthesis of iridoids has been studied in much detail, essentially in connection with the biosynthesis of indole alkaloids (29, 342). In the case of loganin (19)

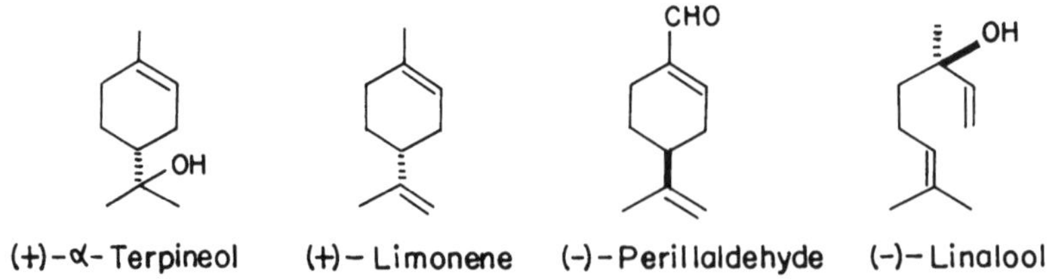

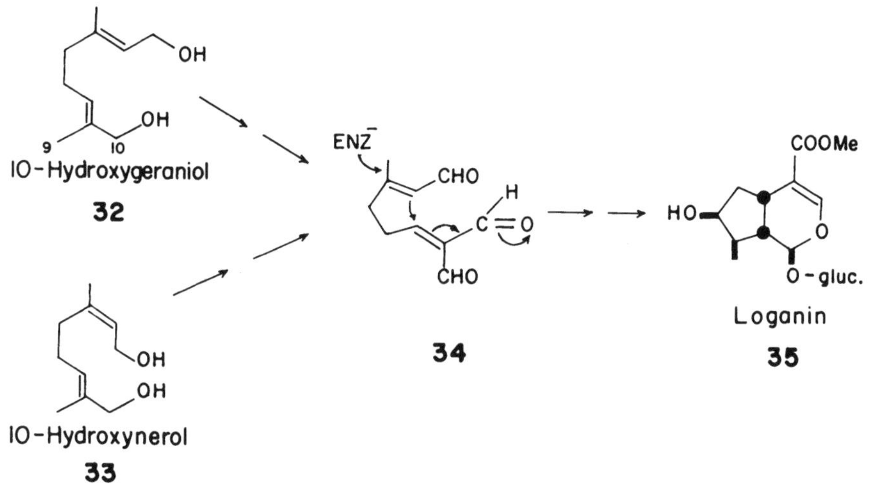

Fig. 8.1.7. Postulated biosynthesis of iridoids (loganin)

and some indole alkaloids, both 10-hydroxygeraniol (**32**) and 10-hydroxynerol (**33**) have been shown to be efficient precursors. It was also demonstrated that, at some stage in the biosynthesis, C-9 and C-10 become biosynthetically equivalent. Based on these considerations, intermediacy of a hypothetical species such as **34** (Fig. 8.1.7) has been suggested, though the exact mechanism of cyclization remains unclear (20, 142).

Evidence has been obtained that administered leucine and valine participate in the biosynthesis of certain monoterpenoids in higher plants by directly providing DMAPP, by-passing the MVA pathway (385). This provides rationalization for the long-known fact that IPP- and DMPP-derived moieties of monoterpenoids originate in vivo from different metabolic pools; MVA is essentially responsible for IPP (87).

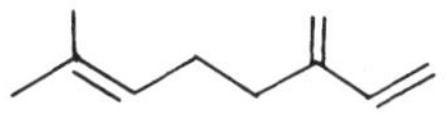

β - Myrcene   cis-β - Ocimene   trans -β- Ocimene
36   **37**   **38**

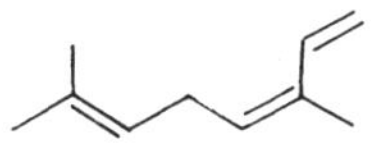

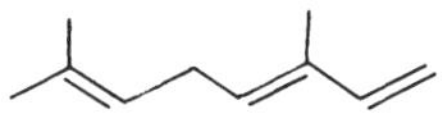

(a) *Acyclic.* A number of acyclic monoterpenoids with a 2,6-dimethyloctane framework have been isolated from extractives, oleoresins and essential oils obtained from woody tissues. Thus, β-myrcene (**36**) is present to the extent of over 60% in the essential oil from the gum-resin of *Commiphora mukul*, and in small quantities (1% – 5%) in turpentine oils from *Pinus* spp. (*P. ponderosa, P. sylvestris* and others). Myrcene is one of the few commercially important terpene hydrocarbons and is manufactured by pyrolysis of β-pinene (**90**) (Chap. 10). Ocimene of unknown stereochemistry (**37** and/or **38**) is a minor (~4%) constituent of turpentine oil from *Pinus quadrifolia* (379, 396).

702   Isoprenoids

Of the oxygenated derivatives, the alcohols and aldehydes are very widely distributed in higher plants, especially in foliage and flowers, but have a very limited distribution in the woody parts. (−)-Citronellol (**39**) is present (~4%) in the essential oil of the heartwood of *Chamaecyparis lawsoniana*. The corresponding aldehyde (but of opposite chirality), (+)-citronellal (**40**), is the major constituent (60%) of the stem oil of *Litsea citrata*. Cayenne rosewood oil, which is obtained by steam distillation of wood of *Aniba rosaeodora*, is an exceptionally rich (85%–95%) source of (−)-linalool (**31**); about 1% to 2% each of geraniol (**41**) and nerol (**42**) are also present in the oil. On the other hand, the chief component (~75%) of Mexican oil of linaloe (from the wood of *Bursera delpechiana*) is (+)-linalool, while its racemic modification is a minor constituent of the wood of *Nectandra elaiphora* (171 vol. 4, 275 vol. III c, 379, 396). Citral (**43** and/or **44**), *trans*-linalool oxide (**45**), and *cis*-linalool oxide (**46**) have been detected as minor constituents of Peruvian rosewood oil (296). From the roots of *Anthemis montana*, dehydroneryl isovalerate (**47**) has been isolated (44). The terpene alcohols and aldehydes described above are valuable fragrance raw materials (11) and are very widely used in perfumery. The bulk of these compounds is now produced commercially by synthetic methods (126).

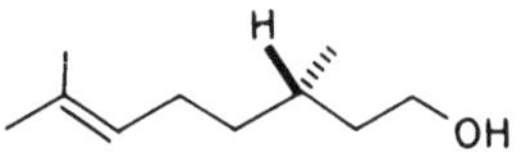

(−)- Citronellol
39

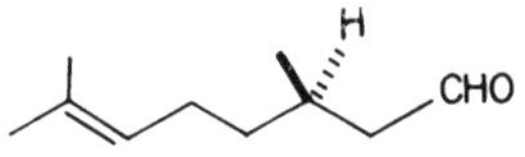

(+)- Citronellal
40

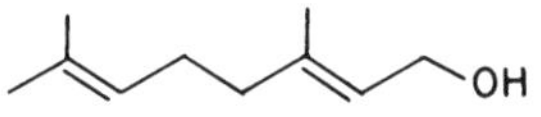

Geraniol
41

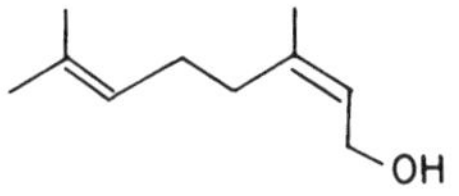

Nerol
42

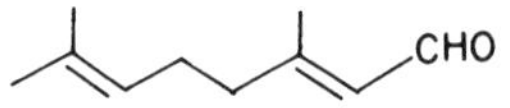

trans- Citral
43

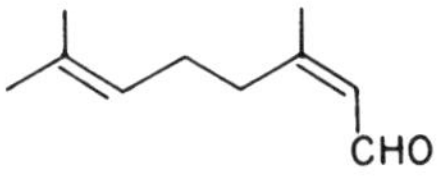

cis − Citral
44

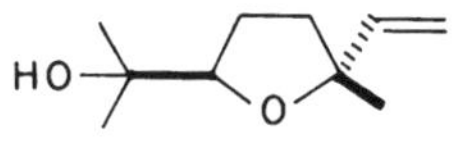

trans − Linalool oxide
45

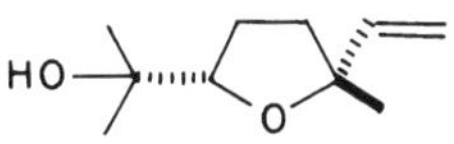

cis − Linalool oxide
46

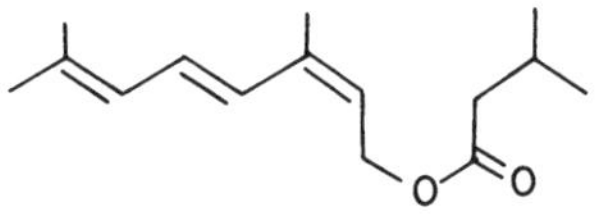

Dehydroneryl isovalerate
47

(−)-Citronellic acid (**48**) has been isolated from wood of various species belonging to the genus *Callitris* (e.g., *C. glauca*), and of *Chamaecyparis taiwanensis* and *Thujopsis dolabrata*. This acid was found to be a fungicide and a termite repellent. Dehydrogeranic acid (**49**) occurs as a geranyl ester in the essential oil from the wood of a New Caledonian conifer, *Neocallitropsis araucarioides*; it has also been isolated from the wood of *Callitris glauca* (143, 275 vol. III d).

References 275 vol. III c, d, 379; and 396 give more complete listings of acyclic monoterpenoids occurring in woody tissues.

Terpenoids 703

(−)-Citronellic acid
48

Dehydrogeranic acid
49

(b) *Monocyclic: p-Menthane and other cyclohexane derivatives.* Several menthadienes have been found to occur in woody parts of plants. The most common of these is limonene, which has been identified in a large number of turpentines, wood, and bark extractives. Thus, (−)-limonene (*ent* **29**) is the major constituesnt (75% to 95%) of turpentine oil from *Pinus pinea* (Italian), *P. pinceana*, and *P. lumholtzii*. Optically pure (+)-limonene (**29**) appears to have only a limited occurrence in woody parts, though it is the chief constituent of several citrus peel oils, which are its commercial source. The racemic modification (dipentene) is present to the extent of 12% to 13% in the oleoresin of *P. ponderosa* and in the bark oil of *Abies concolor*; in partly racemized form (with an excess of L or D), it occurs in several resin oils and turpentines. Terpinolene (**50**) has been isolated as a minor component (1% to 5%) of certain turpentine oils (e.g., *Pinus palustris, P. ponderosa* and *P. reflexa*) and heartwood oils (e.g., *Chamaecyparis lawsoniana*). Stump oil of *Pinus sylvestris* contains α-terpinene (**51**), while γ-terpinene (**53**) has been found in trace quantities in *P. eldarica*. The related β-terpinene (**52**) has not yet apparently been detected in woody tissues. (+)-α-Phellandrene (**54**) is a minor component of angelica root oil and of *Boswellia serrata* (207, 396). The major constituent (65% to 80%) of lodgepole pine (*Pinus contorta*) turpentine is (−)-β-phellandrene (**55**) (171 vol. 6, 353). *p*-Cymene (**56**) has been detected as a trace component of certain pine resins (207, 229).

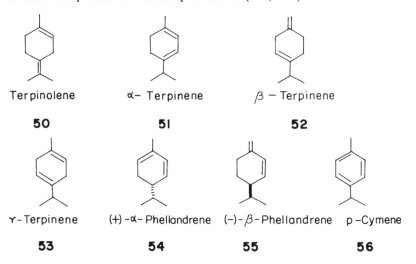

Terpinolene   α-Terpinene   β-Terpinene
50   **51**   **52**

γ-Terpinene   (+)-α-Phellandrene   (−)-β-Phellandrene   *p*-Cymene
53   **54**   **55**   **56**

Over 50 hydroxy derivatives (besides their esters and ethers) based on *p*-menthane are known to occur in nature (379). Of these, only a few have so far been encountered in woody parts of plants. The most widely distributed of these is α-terpineol, occurring in optically active and racemic forms. Bark oil of *Ocotea usambarensis* is a rich source (40%) of (−)-α-terpineol (*ent* **28**). The essential oil from the root and stump of *Chamaecyparis obtusa formosana* contains about 30% (+)-isomer. The racemic modification occurs in wood oil of *Nectandra elaiophora*. Although many esters of α-terpineol are found in nature, they have been seldom detected in woody tissues. The hydroxy derivative α-terpineol is a valuable perfumery material and is commercially produced from pinenes (Chap. 10). In many essential oils, (+)-terpineol-4 (**57**) is also found, but its occurrence in woody parts is, so far, restricted to Polish extracted turpentine oil of *Pinus sylvestris* and in the twig oil of *Juniperus occidentalis* (228, 275 vol. III b). Presence of *trans*-yabunikkeol (**58**) and *cis*-yabunikkeol (**59**) in the essential oil from the twigs of *Cinnamomum japonicum* has been reported (149). The diols, *trans*-terpin (**60**) and *trans*-sobrerol (**61**), have been found in the turpentine oil from pine stumps and in the turpentine oil of *Abies pectinata* respectively (207, 228, 379). Several phenols with *p*-menthane framework have been isolated from woody sections. Mention may be made of carvacrol (**62**) occurring in the wood oil of *Callitris quadrivalvis* (besides several other wood extractives), and thymol (**63**), which has been detected in Colombo root oil (from *Jatrorrhiza palmata*). The essential oil from the wood of *Cupressus sempervirens* contains as much as 96% carvacrol methyl ether (296). Somewhat recently, several thymol derivatives have been isolated from roots of *Inula* spp. and *Gaillardia aristata*, both of family Compositae (379). As an example, occurrence (0.11%) of 7-isobutyryloxythymohydroquinone dimethylether (64) in the roots of *Inula salicina* may be noted (6).

The compound 1,8-cineole (65) is extensively distributed in essential oils. It has been identified in many oils obtained from trunk, bark, roots, and other woody tissues; the volatile oil from *Alpinia nutans* roots contains as much as 60% of this compound. The related oxide, 1,4-cineole (66) has been found in the wood oil of *Cinnamomum kanahirai* and in the resin and wood of *Shorea maranti* (207, 296).

1,8-Cineole 1,4-Cineole
65   **66**

Only a few aldehydes and ketones of this class have so far been met within woody portions of plants. Perillaldehyde (30) has been detected in turpentine oil of *Pinus excelsa*. Cuminaldehyde (67) has been isolated from the twig oil of *Chrysothamnus nauseosa*. The presence of menthone (68) in American turpentine oil and piperitone (69) and pulegone (70) in the resin from *Shorea maranti* have been reported (207, 275 vol. III c). Racemic carvone (71) has been found to occur in the heartwood of *Bursera graveolens* to the extent of 5% (120). From the roots of *Pluchea odorata*, 5-(angeloyloxy) carvotagetone (72) has been isolated (51).

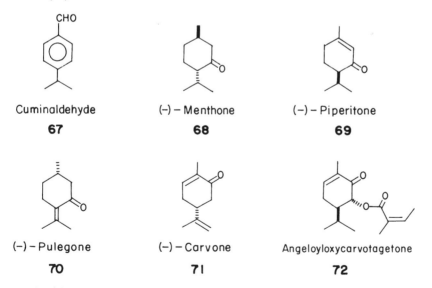

Cuminaldehyde (−)−Menthone (−)−Piperitone
67   **68**   **69**

(−)−Pulegone (−)−Carvone Angeloyloxycarvotagetone
70   **71**   **72**

Besides *p*-menthane derivatives discussed above, there are a few other isoprenoid cyclohexane types with rather limited distribution. Some of these have been found to occur in woody tissues also – e.g., **73**, an *o*-menthane derivative, and **74**, a 1,1,2,3-tetramethylcyclohexane derivative, both isolated from roots of *Piqueria trinervia* (46). Ferulol (**75**) from roots of *Selinum carvifolia* is an example of still another type (49). A more complete list will be found elsewhere (379). There are then compounds based on 1,1,3-trimethylcyclohexane, but with more than 10 carbon atoms. Strictly speaking, these compounds are not monoterpenoids, but for historical reasons have always been treated along with this class. These compounds are now considered as possibly arising from catabolism of carotenoids (Sect. 8.1.3.7) (16). Most of these are C_{13}-compounds and the only one known to occur in woody tissues is α-ionone (**76**), a minor constituent of costus (*Saussurea lappa*) root oil. Orris concrete, prepared by special methods from roots of *Iris pallida*, is valued in perfumery for its violet-like odor, attributed to the presence of irones (e.g., *cis*-α-irone, **77**) (275 vol. III c, 379).

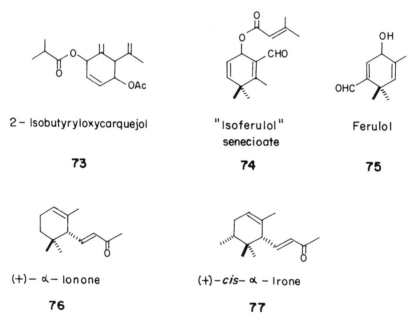

(c) *Monocyclic: Iridoids and other types*. Iridoids, derivatives of 1,2-dimethyl-3-isopropylcyclopentane, were first recognized in 1954–55, with the elucidation of structures of nepetalactone (**78**), the active principle of oil of catnip (*Nepeta cataria*). This group comprises C_{10} compounds and their nor (C_8, C_9), homo (C_{13}, C_{14}), and seco derivatives, and is the largest single class of monoterpenoids with over 270 well-defined compounds. Iridoids are widely distributed in the higher plants, especially in the dicotyledoneae. Many of these compounds occur as glycosides and have been isolated from stem, bark, roots and rhizomes, besides leaves, fruit, seeds and whole plants. As an example of occurrence of C_{10} iridoids in woody parts, isolation of (−)-valerosidate (**79**) (210), (+)-valtrate (**80**) (387), valerianine (**81**) (148), and at least seven more compounds of this class, from roots of *Valeriana officinalis*, may be cited. Picroside-I (**82**), with useful antihepatoxic activity (405), is an example of C_9-iridoid, and has been isolated from the roots of *Picrorhiza kurrooa*, a well-known *Ayurvedic* bitter tonic (235). *Plumeria rubra*

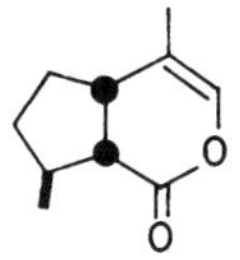

Nepetalactone
78

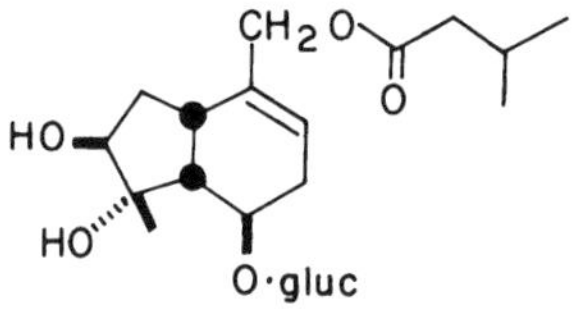

(−)−Valerosidate
79

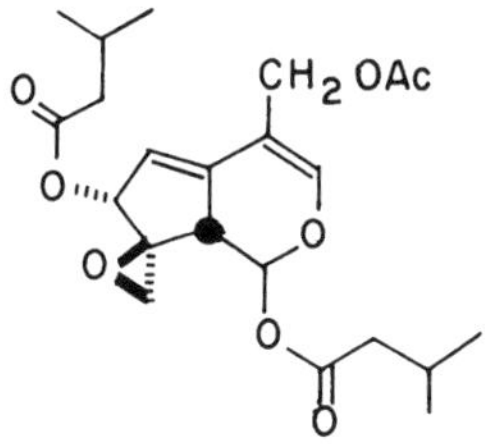

(+)−Valtrate
80

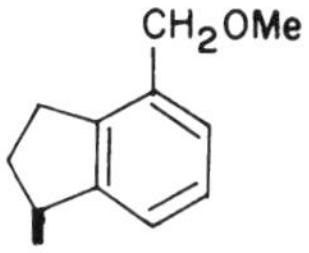

Valerianine
81

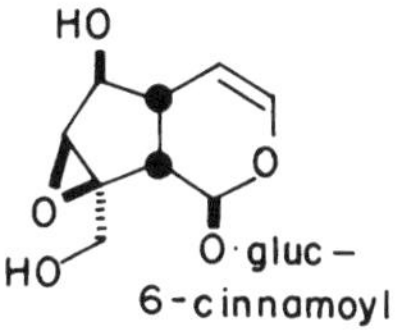

Picroside − I
82

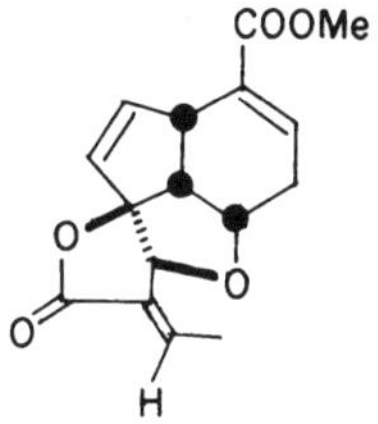

(+)−Plumericin
83

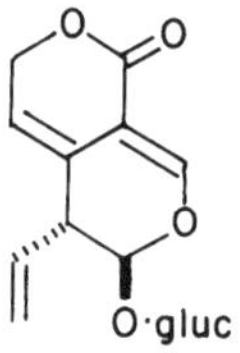

(−)−Gentiopicroside
84

elaborates a number of homo-iridoids (C_{14}), such as (+)-plumericin (**83**), isolated from its roots (2). Several secoiridoids have been isolated from roots and bark of plants – for example, (−)-gentiopicroside (**84**) from roots of *Swertia caroliniensis* (211). Reference may be made to a recent book (379) and a review article (135) for many more examples of iridoids occurring in woody portions of plants.

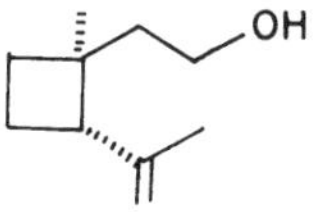

Fragranol
85

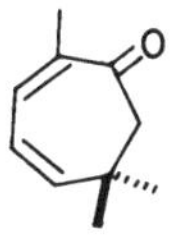

Eucarvone
86

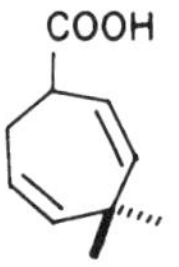

Shonanic acid
87

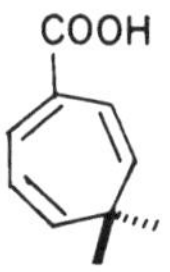

Thujic acid
88

Fragranol (**85**), a cyclobutane derivative, has been obtained from root of *Artemisia fragrans*, where several of its esters are also present (54). Eucarvone (**86**), which was isolated in 1905 (379) occurs in the essential oil from rhizomes of *Asarum sieboldii*. Two related compounds are the carboxylic acids, shonanic acid (**87**) from the wood of *Libocedrus formosana* and thujic acid (**88**) present in the heartwood of *Thuja plicata*, which also contains its methyl ester (140).

(d) *Bicyclic and tricyclic.* Bicyclic monoterpenes, though comprising a relatively small group, are important from a commercial point of view. Presently, this group consists of some 75 complounds, several of which have been isolated from woody tissues of plants. Most important of these are the pinenes, with α-pinene being the most widely distributed of all accumulating terpenoids. It is also the terpene with the largest tonnage produced commercially. Several turpentines are exceptionally rich in pinenes. For example, turpentine oil from American slash pine (*Pinus elliottii*) contains 58% to 63% (−)-α-pinene (**89**), and 28% to 33% (−)-β-pinene (**90**). Pinenes occur in nature in levo, dextro, and partly or fully racemized forms. (+)-α-Pinene often occurs in admixture with (−)-β-pinene of opposite absolute configuration, and is seldom optically pure. Lists of occurrence of pinenes in higher plants have been compiled (171 vol. 6, 207, 396). Camphene (**91**) and the related tricyclic hydrocardbon tricyclene (**92**)have also been isolated from certain wood and resin turpentine oils (207, 396). Camphene containing some tricyclene is commercially produced from pinenes (Chap. 10). The chief (55% − 65%) constituent of Indian turpentine oil is (+)-3-carene (**93**), which is obtained from *P. roxburghii* resin (80), whereas (−)-3-carene is present in *P. sylvestris* (wood and stump turpentine), and their racemate in the turpentine oil of *P. ponderosa* (396). Essential oil from gum-oleoresin of *Boswellia serrata* is a good source of (+)-α-thujene (**94**) (379).

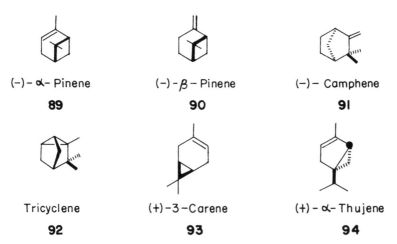

Thujone (α-thujone, **95**) and isothujone (β-thujone, **96**) have been detected (by GLC) in the essential oil from the wood and resin of *Shorea maranti* (103). Oil from the twigs of *Cupressus lusitanica* contains up to 3.5% of the interesting bicyclic ketone, umbellulone (**97**) (275 vol. III c, 379). The only oxygenated derivatives of carane-type, known to occur in woody tissues are the (+)-chamic

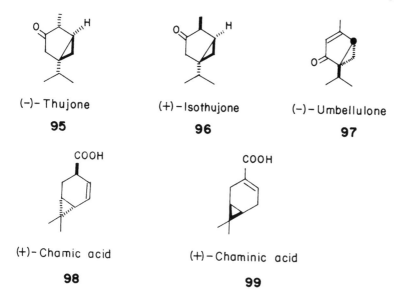

acid (98) and the (+)-chaminic acid (99) isolated from heartwood of *Chamaecyparis nootkatensis*; their antipodal character may be noted (140).

Several oxygenated derivatives based on pinane have been found to occur in woody parts. For example, myrtenol (100), myrtenal (101) and myrtenic acid (102) have been isolated from wood of *Chamaecyparis formosensis*. (−)-*trans*-Pinocarveol (103) is present in American gum turpentine. (+)-*cis*-Verbenol (104), its *trans* isomer, and (+)-*trans*-verbenone (105) are important components of frankincense from *Boswellia cartevii* (275 vol. III b, c, 296, 379). From the roots of *Paeonia albiflora*, a few highly oxygenated pinane derivatives (e.g., paeoniflorin, 106) have been obtained (223).

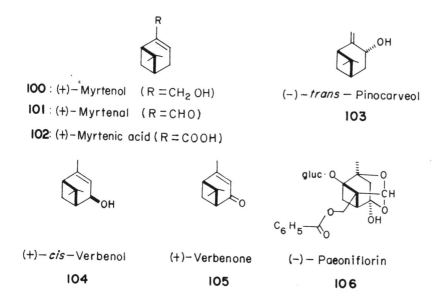

710 Isoprenoids

Borneol, both free and as esters, occurs widely in the plant kingdom, and though usually a component of foliage, it has been isolated from many heartwoods and tree exudates. The so-called "borneo-camphor," (+)-borneol (**107**), was first encountered as a crystalline deposit in the trunk crevices of the Borneo tree *Dryobalanops camphora*. (−)-Borneol ("Ngai-camphor") is the major component of the essential oil (in leaves and twigs) of *Blumea balsamifera*. The corresponding ketone, camphor, has been known in the East since antiquity. (+)-Camphor (**27**) is the chief constituent of *Cinnamomum camphora* wood oil, and is also found in many other wood oils, especially of trees of the family Lauraceae. (−)-Camphor is less common, but still occurs in many essential oils, including that from *Blumea balsamifera*. Camphor (±) is now produced commercially from pinenes (Chap. 10), though some quantity of (+)-camphor is still obtained from natural sources. The fenchane derivative, (−)-α-fenchol (**108**), is present (6%) in the essential oil of heartwood of *Chamaecyparis lawsoniana* (275 vol. III b, c).

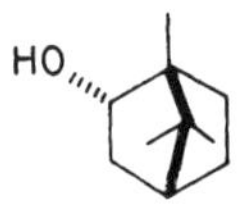

(+) − Borneol
107

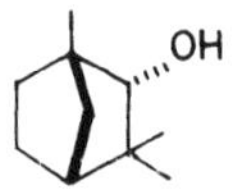

(−)-α-Fenchol
108

109 : Teresantalol (R=CH$_2$OH)
110 : Teresantalal (R = CHO)
111 : Teresantalic acid (R=COOH)

Teresantalol (**109**), the corresponding aldehyde (**110**), and acid (**111**) are minor constituents of Indian sandalwood oil, which is obtained by steam distillation of wood of *Santalum album*. These compounds are possibly degraded sesquiterpenes (379).

(e) *Nongeranyl monoterpenoids*. There is a small but significant group of monoterpenoids that apparently are not derived from GPP, and that have come to be known as "irregular" terpenes. The earliest known examples of this structurally diverse class are artemisia ketone (**112**) and chrysanthemic acid (**113**) discovered in the 1920's. Since then, other structural types have been found. These non-head-to-trail compounds have a very restricted distribution and no clear-cut example of their occurrence in woody tissues is available (379). Their biogenesis has been discussed (26, 138) and some biosynthetic evidence for involvement of chrysanthemyl alcohol (**114**) has been adduced (87).

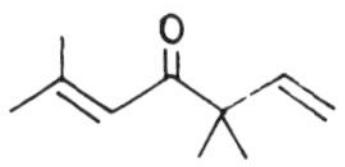

Artemisia ketone
112

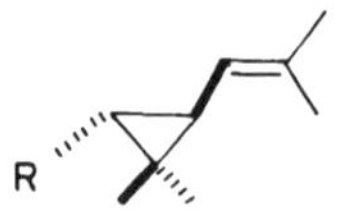

113 (+)-*trans*-Chrysanthemic acid
(R = COOH)
114 (+)-*trans*-Chrysanthemyl alcohol
(R = CH$_2$OH)

8.1.3.2.3 Tropolones

The heartwood and essential oils of trees belonging to the familys Cupressaceae contain several C_{10} compounds that exhibit strong acid character and are possibly responsible for the resistance to fungal decay of the parent wood. These compounds were later recognized as having a seven-membered enolone structure, a system named by Dewar as tropolone, and first identified by him in certain mold metabolites. The best known members of this class, occurring in wood, are the α-, β-, and γ-thujaplicins (115, 116, 117), first isolated from the heartwood of Western red cedar (*Thuja plicata*) (139, 298). Several other wood tropolones are known (379). These tropolones, as distinct from the mold tropolones, have an isopropyl group as a side chain and are considered to be of mevalonoid origin, though no tracer work has been carried out. It has been suggested (139) that these compounds arise from certain carane precursors.

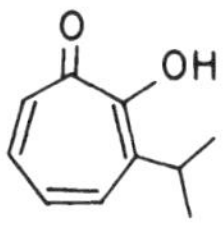

α-Thujaplicin
115

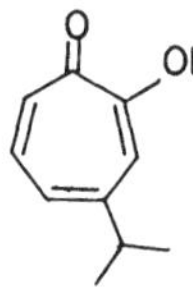

β-Thujaplicin
116

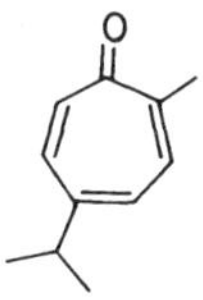

γ-Thujaplicin
117

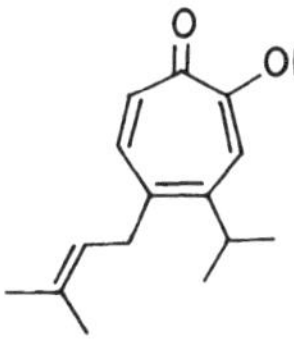

Nootkatin
118

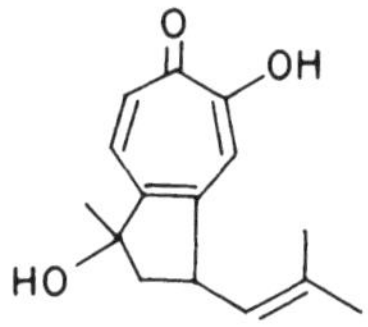

Chanootin
119

A few C_{15} tropolones, which may be considered as sesquiterpenoids (Sect. 8.1.3.3) are also known — for example, nootkatin (**118**) and chanootin (**119**) from heartwood of *Chamaecyparis nootkatensis* (Alaska-cedar) (140).

8.1.3.3 Sesquiterpenoids

8.1.3.3.1 Distribution

Lists of occurrence of sesquiterpenoids in higher plants are available (155, 207, 275 Vol. III b, c, 296, 305, 396). Table 8.1.3 gives a list of plant families known for accumulating sesquiterpenoids. Of these, Cupressaceae, Compositae, Dipterocarpaceae and Myrtaceae are specially prolific in producing these compounds. Humulene and caryophyllene are the most extensively distributed sesquiterpenes.

Table 8.1.3. Plant families prolific in producing sesquiterpenoids

Gymnospermae	Angiospermae
Cupressaceae	Monocotyledoneae
Pinaceae	Araceae
Taxodiaceae	Cyperaceae
	Zingiberaceae
	Dicotyledoneae
	Betulaceae
	Compositae
	Coriariaceae
	Dipterocarpaceae
	Meliaceae
	Myrtaceae
	Piperaceae
	Rutaceae
	Santalaceae
	Umbelliferae
	Valerianaceae
	Winteranaceae

As in the case of monoterpenoids, both antipodal forms of certain sesquiterpenoids have been found to occur in higher plants, though to a much restricted extent. Occurrence of racemic modification is rare, though a few examples are known (305). However, it may be pointed out that several sesquiterpenoids isolated from Hepaticae have been found to be enantiomeric to those from the higher plants (13, 266).

The value of sesquiterpenoids as chemotaxonomic markers has received some attention (140, 155). Much discussion has been focused on their importance in Compositae phylogeny, at the genus and species level (153, 191, 193). Sesquiterpene distribution in the family Dipterocarpaceae has been studied (37).

8.1.3.3.2 Structural Types

Presently, sesquiterpenoids (128, 237, 305, 333, 376) constitute the largest single class of terpenoids, with over 2500 well-defined (only C_{15}) compounds, embracing some 120 distinct skeletal types from acyclic to tetracyclic. Sesquiterpenoids encompass a most impressive array of carbocyclic ring systems, and this diversity and novelty have stimulated outstanding synthetic activity in this field (184). Figure 8.1.8 depicts the important sesquiterpene skeletal types most widely distributed in nature and often encountered as constituents of wood extractives.

According to the biogenetic isoprene rule, sesquiterpenes arise from a linear precursor such as farnesol. There is considerable biosynthetic evidence to support this, and farnesyl pyrophosphate is considered as the immediate biosynthetic precursor of almost all sesquiterpenoids (110, 263). Conceptually, cyclization of farnesyl pyrophosphate can proceed from either the tail end (by ionization of the pyrophosphate moiety) or the head (by a H^+ or OH^+ attack on Δ^{10}). Both

Terpenoids 713

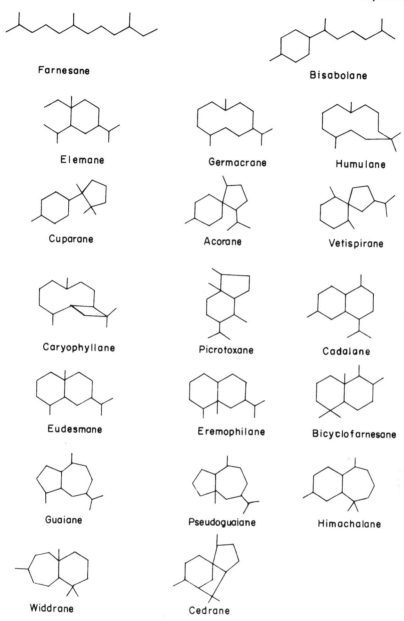

Fig. 8.1.8. Important sesquiterpene skeletal types

pathways have been exploited by nature, though the former is by far the most dominant. Starting with either *cis*- (*c*-FPP, **121**) or *trans*-farnesyl pyrophosphate (FPP, **120**), it is possible to rationalize the structures of almost all cyclic sesquiterpenoids by suitable cyclizations (310, 323, 376). Cyclization from the tail end is triggered by ionization of the pyrophosphate moiety with assistance from

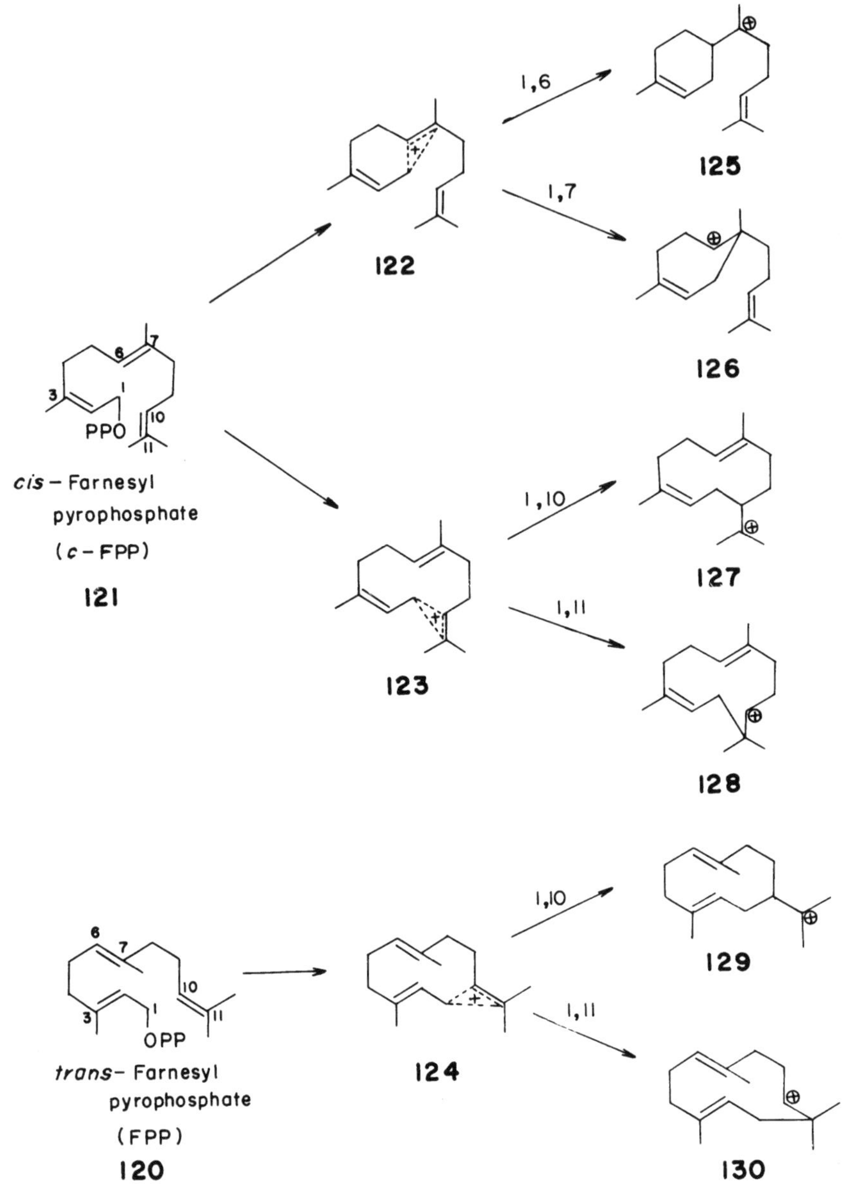

714   Isoprenoids

Fig. 8.1.9. Postulated initial cyclization of farnesyl pyrophosphate

a suitably oriented Π-electron cloud. Figure 8.1.9 depicts possible monocyclic ions (**125–130**) that would result from such a cyclization, depending on the nature of the substrate and the cyclase. Cyclization from the head, initiated by a proton or OH^+ attack, will result in C-6, C-11 bond formation generating monocyclic ion **131**. These monocyclic ions can stabilize in the usual way to give products or to undergo further cyclization/rearrangement to generate new carbocyclic systems. Nature appears to have fully exploited this simple theme of C—C bond formation to unfold an almost baffling range of structural types. Catabolism of these products leads to a variety of functionality and occasionally to newer skeletons (by ring cleavage or loss of C atoms).

An impressive amount of biosynthetic work in the area of sesquiterpenoids has been carried out and the results essentially confirm the above biogenetic theories. Ingenious tracer experiments have been devised to extract detailed stereochemical information in many cases. However, most of this work was performed on fungal metabolites, and it is not the purpose of this article to discuss details of biosynthetic studies, which will be found elsewhere (12, 19, 77, 90, 110). One basic question, though — namely the mandatory role of c-FPP in sesquiterpene biosynthesis — calls for a brief comment.

In parallel with the situation in monoterpene biosynthesis, c-FPP has been implicated in the cyclization of the substrate to systems containing six- and seven-membered rings (125, 126). Various possibilities for the generation of the required c-FPP have been discussed (75). Direct biosynthesis of c-FPP from GPP and IPP by a suitable prenyltransferase has been considered less important (261, 286). Isomerization of FPP to c-FPP has been a subject of much debate and experimental scrutiny. Isomerization through a redox system, which at one time was much favoured, is now being discounted. Strong experimental support has emerged in favour of this *trans* to *cis* isomerization, proceeding through the, intermediary of nerolidyl pyrophosphate (132). Evidence has also been presented to show that isomerization of FPP to nerolidyl pyrophosphate proceeds by a net suprafacial process involving an ion-pair intermediate, such as **133** (77). Indeed, in line with the present ideas about monoterpenoid biosynthesis (Sect. 8.1.1.2), it is suggested (77, 118) that *c*-FPP is not an obligatory substrate for generating ions **125/126** and that FPP by way of **134** or **135** can cyclize to these systems. Thus, the intermediary of nerolidyl pyrophosphate or *c*-FPP becomes a moot point.

The following discussion of sesquiterpenoids in woody tissues has been organized on the basis of initial cyclization of FPP. It is not possible to describe all such compounds, and only a few of the most important members of each dominant type will be mentioned. No complete work on the subject is available, but the reader is referred to a few comprehensive listings (89, 128, 156, 207, 228, 229).

(a) *Acyclic*. Many sesquiterpenoids with a farnesane framework are known, and several of these occur in the woody parts of higher plants. Undoubtedly, the most important of these are the alcohols, farnesol, and nerolidol. *trans*-Farnesol (**136**) occurs in many essential oils; it was first isolated from ambrette seeds (*Hibiscus abelmoschus*) in 1904, its gross structure deduced by 1913, and the stereochemistry established in 1963 (24, 74). Farnesol, present in Peru balsam oil (from *Myroxylon pereirae*), is the *trans*-isomer (294). *cis*-Farnesol (**137**) is a component of petitgrain oil, which is obtained by the steam distillation of leaves and twigs of the bitter orange tree, *Citrus aurantium* (74). (+)-Nerolidol (**138**) (402) is present to the extent of 50% to 70% in the wood oil of the Peru balsam tree; the (−)-antipode has been found in the oil from wood of *Dalbergia parviflora*, whereas the racemic modification is the chief constituent of oil from the wood of *D. sissoo* (275 vol. III b). Farnesol and nerolidol are valued in perfumery.

trans−Farnesol

136

cis−Farnesol

137

(+)−Nerolidol

138

Some of the much more restricted sesquiterpenoids occurring in woody tissues are the hydrocarbon **139** (and its 10Z-isomer) from Australian sandalwood (*Santalum spicatum*) oil (35), the nerolidol derivatives **140** and **141** from the resin of *Ocotea caparrapi* (82) and 8-angeloloxy-4,5-deshydronerolidol (**142**) in the roots of *Brickellia guatemaliensis* (48). A number of furane derivatives having the farnesane carbon skeleton have been encountered in nature and quite a few of these have been isolated from the woody parts of plants. Thus, neotorreyol (**143**) and torreyal (**144**) have been isolated from the wood of *Torreya nucifera* (338). The wood oil from *Eremophila freelingii* contains a number of novel furane derivatives, and special mention should be made of freelingyne (**145**), the first acetylenic sesquiterpenoid (269).

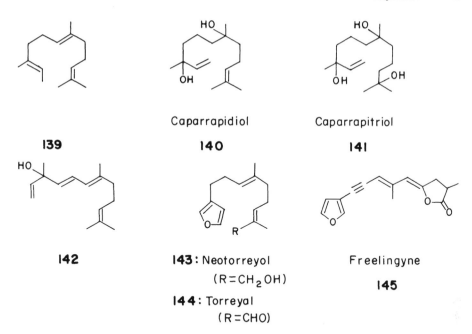

(b) *1,6-Cyclization: Bisabolanes and derived types.* Several sesquiterpenoids with the carbon skeleton of ion **125** are known. The earliest examples of this class are bisabolene (1897) and zingiberene (1900) (228). Bisabolene, for which several isomers are possible, has been characterized as a component of several essential oils from wood and resin, by formation of crystalline trihydrochloride (396). (−)-β-Bisabolene (**146**) is a constituent of the stump oil of *Chamaecyparis lawsoniana* (140) and of essential oil of wood of *Lansium anamalayanum*, where it occurs to the extent of 35% (243). Essential oil from rhizomes of *Zingiber officinale* contains (−)-zingiberene (**147**) as the chief component (396). (−)-α-Curcumene (**148**) has been isolated from turmeric oil, obtained by steam distillation of rhizomes of *Curcuma aromatica* (202, 396). Of the oxygenated derivatives, mention may be made of (−)-lanceol (**149**) from wood of *Santalum lanceolatum* (117, 275 vol. III b), nuciferal (**150**) from wood oil of *kaya* (*Torreya nucifera*) (338), *trans*-atlantone (**151**) from the essential oil of wood of *Cedrus deodara* (117, 308), and the golden-yellow quinone, (−)-perezone (**152**), from roots of Mexican *Perezia* spp. (408). (+)-Bilobanone (**153**) found in the heartwood of *Ginkgo biloba* (214, 238) and (+)-deodarone (**154**) from *Cedrus deodara* wood oil (347) are two examples of heterocyclic derivatives based on bisabolane.

The so-called "paper factor" responsible for insect juvenile hormone (JH) activity of certain American papers has been identified (1966) as the bisabolane methyl ester, (+)-juvabione (**155**), a constituent of balsam fir, *Abies balsamea* (416). As a matter of fact, the acid corresponding to **155**, todomatuic acid (**156**), had been known earlier as a component of bisulfite-treated pulp oil of *Abies sachalinensis* (399). Juvabione and related compounds possessing JH activity have been isolated from several species of the family Pinaceae, including (+)-Δ^7-dehydrotodomatuic acid (**157**) and 7-hydroxytodomatuic acid (**158**) from wood of *Cedrus deodara* (33).

718 Isoprenoids

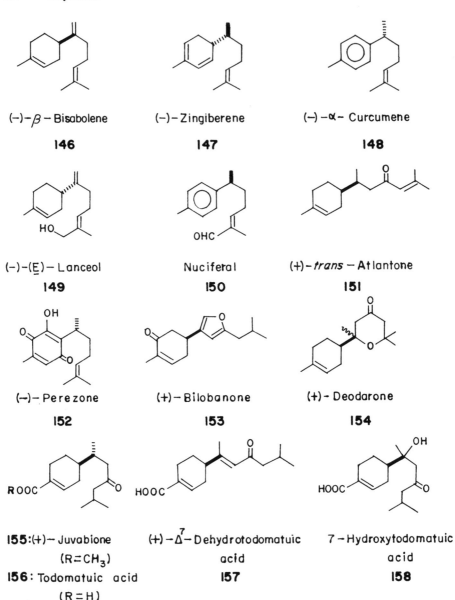

Cation **125** can be considered as the progenitor of at least eighteen different skeletal types (376) of sesquiterpenes; Fig. 8.1.10 depicts some of the types important to this discussion.

Acorone (**166**), a sesquiterpene diketone isolated from the essential oil of roots of *Acorus calamus*, was the first spirosesquiterpenoid to be characterized, with the carbon skeleton of cation **158**. At present several compounds of this class are known and an extensive review has been published (267). Hydrocarbons (e.g., γ-acoradiene ≡ α-alaskene, **167**) and alcohols (e.g., α-acorenol, **168**) directly derivable from the spiro-cation **159** have been isolated from wood of *Juniperus*

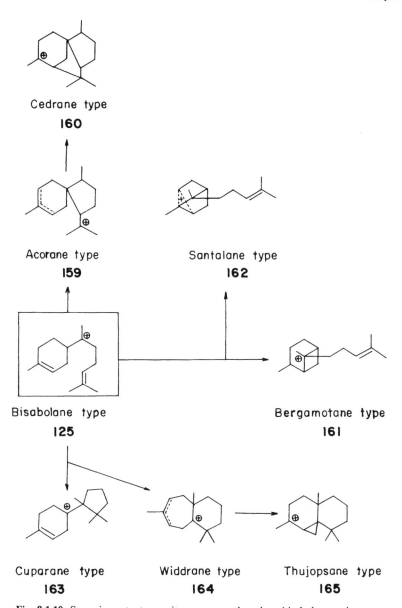

Fig. 8.1.10. Some important sesquiterpene types based on bisabolene cation

rigida (267). Acoric acid (**169**), an acid from roots of *A. calamus*, can be considered as a 9,10-seco derivative of acorone (36).

Further cyclization of the spiro-cation **159** leads to the tricyclic system **160** (Fig. 8.1.10) of which the best known examples are (−)-α-cedrene (**170**) and (+)-cedrol (**171**), compounds that are widely distributed in the cupressales woods. Both of these compounds were first isolated from cedar wood oil (from *Juniperus virginiana* in 1841 (140, 207, 229, 238). (+)-α-Cedrene has been found in vetiver

720   Isoprenoids

(+)-Acorone   (−)-γ-Acoradiene   (−)-α-Acorenol   Acoric acid
166   **167**   **168**   **169**

oil from roots of *Vetiveria zizanioides* (231). Related compounds, though with severely restricted distribution, are zizaene (**172**), prezizaene (**173**), and their derivatives, occurring in vetiver oil. These structures are biogenetically derivable from cation **159** (5, 102, 267).

(−)-α-Cedrene   (+)-Cedrol   Zizaene   Prezizaene
170   **171**   **172**   **173**

(+)-Cuparene (**174**), (−)-thujopsene (**175**) and the alcohol (+)-widdrol (**176**) are widely distributed in heartwoods of the family Cupressaceae, and often co-occur with each other (140, 229, 238). All of these types are biogenetically related as shown in Fig. 8.1.10. Biosynthesis of the cuparane type (**163**) has been investigated (110). (+)-Cuparene and the related cuparenic acid (**177**) were first isolated from the heartwood of *Chamaecyparis thyoides* in 1958 (136). Several other members of this class are known and mention may be made of cuparenones (**178, 179**) from the wood oil of "mayur pankhi" (*Thuja orientalis*) (94). Thujopsene is present to the extent of 50% to 60% in the essential oil from the wood of *Thujopsis dolabratta* and is also an important component of several species of the South African genus *Widdringtonia*. Special mention may be made of the related acid, hinokiic acid (**180**) from various *Widdringtonia* spp., and the related nor-ketone, mayurone (**181**) occurring in *Thuja orientalis* and *Thujopsis dolabratta*. Widdrol was first isolated from various *Widdringtonia* spp. (140).

Of the other skeletal types shown in Fig. 8.1.10 bergamotane type (**161**) has not, apparently, yet been isolated from woody tissues. Santalane type is represented by sesquiterpenoids from Indian sandalwood oil, which is obtained by steam distillation of wood of *Santalum album*. Important constituents are α- and β-santalene (**182, 183**), and the derived allylic alcohols **184, 185** which constitute almost 90% of the essential oil (71, 98, 238, 242). It may be noted that α-santalene was the first sesquiterpene for which the correct structure was proposed (344).

Sesquifenchene (**186**), a component of the Indian valerian root oil (*Valeriana wallichii*), is an example of another skeletal type, that can be derived from bisabolene cation **125** (168).

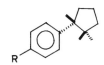

174 : (+)– Cuparene   175 : (–)–Thujopsene   (+)– Widdrol
  (R=CH₃)   (R=CH₃)   176
177 : (+)–Cuparenic acid 180 : (–)–Hinokiic acid
  (R=COOH)   (R=COOH)

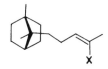   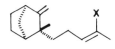   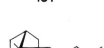

α– Cuparenone   β– Cuparenone   (+)– Mayurone
   178   179   181

182 : (+)–α–Santalene   183 : (–)–β – Santalene   Sesquifenchene
   (X=CH₃)   (X=CH₃)   186
184 : (+)–α– Santalol   185 : (–)–β – Santalol
   (X=CH₂OH)   (X=CH₂OH)

(c) *1,7-Cyclization: Daucanes.* Cation **126** resulting from 1,7-cyclization of FPP appears to be the progenitor of only one important class of sesquiterpenoids (daucane); some biosynthetic work has been carried out (110). Its first member, daucol (**187**), was isolated in 1909 from oil of carrot seeds (*Daucus carota*). Several compounds of this type have since been found in roots of several plants of the family Umbelliferae, especially those of the sub-family Apioideae. Examples are laserpitin (**188**), from roots of *Laserpitium latifolia* (201), and jaeschkeanadiol (**189**) from roots of *Ferula jaeschkeana* (363).

(d) *1,10-Cyclization: Germacranes and elemanes.* Although Ruzicka's biogenetic scheme of 1953 (336) envisaged cation **129** (Fig. 8.1.9) as a species central to the biogenesis of several different sesquiterpene types, the first member of this class, pyrethrosin (**190**), was characterized only in 1957, though it was isolated from *Chrysanthemum cinerariaefolium* as early as 1891. Soon after, another member, germacrone (**191**), readily obtainable from Bulgarian 'Zdravets' oil

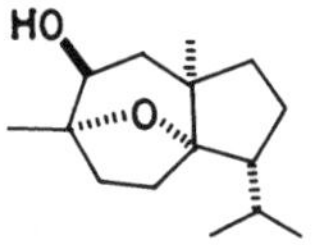

(−)−Daucol
187

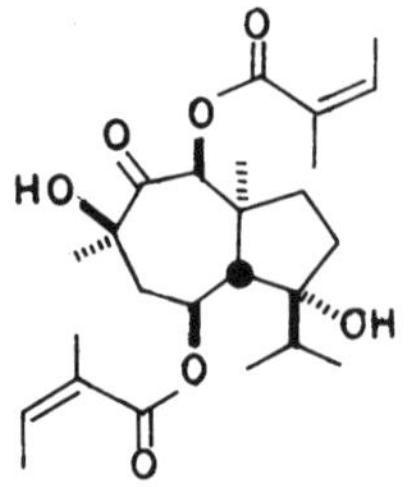

(+)−Laserpitin
188

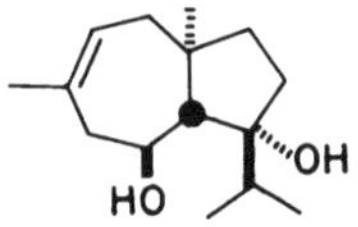

(+)−Jaeschkeanadiol
189

(from over-ground parts of *Geranium macrorhizum*) and known since 1929, was formulated. By 1970 some 65 compounds of this class had been characterized (360); now this number runs into 500 compounds, making germacranes the largest familys of sesquiterpenoids. These compounds have been chiefly obtained from higher plants, especially those of the family Compositae. Although above-ground entire plants, foliage or flowers have been mostly used in isolation work, several compounds have been obtained from roots, and a few from heartwoods and bark. A book (361) and an extensive review (146) on the subject are available.

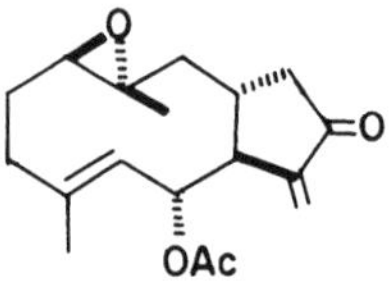

(−)−Pyrethrosin
190

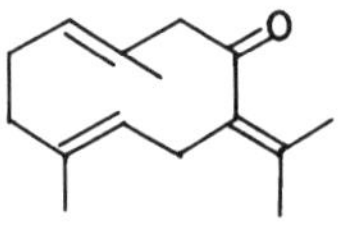

Germacrone
191

Relatively few simple derivatives based on **129** are known. Of those encountered in woody portions of plants, preisocalamendiol (**192**) from roots of *Acorus calamus* (208) and curdione (**193**) from rhizomes of *Curcuma zedoaria* (196) may be mentioned. 8β-(Angeloyloxy)ligularinon-A (**194**) isolated from roots of *Ligularia hodgsoni* is an example of a highly oxygenated germacrane (41). However, the vast majority of known germacranes are oxidatively modified to carry a lactone ring (germacranolides), often along with a variety of other oxygen functionalities. Several germacranolides have cytotoxic and antitumor activity (277). Examples of these lactones occurring in woody tissues are (+)-costunolide (**195**) from costus root (*Saussurea lappa*), (+)-linderane (**196**) from roots of *Lindera strychnifolia*, and the taraxinic acid derivative **197** from roots of *Taraxacum officinale* (146, 177).

Germacrane cation **129** with both the olefinic linkages *trans* configurated may be expected to generate products in which this stereochemical integrity is maintained. Although this is true for the majority of germacranolides and almost all other types of germacranes, an increasing number of germacrandioles having one or both the ethylenic linkage with *cis* geometry are being isolated from nature. This has led to reclassification of germacranolides into four configurationally isomeric subgroups: germacrolides (**198**), melampolides (**199**), heliangolides

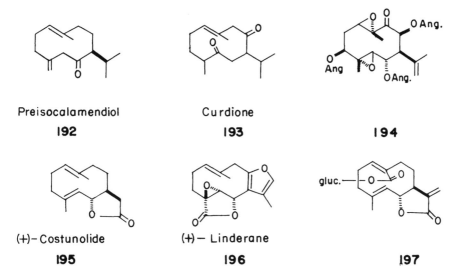

Preisocalamendiol **192**

Curdione **193**

194

(+)-Costunolide **195**

(+)-Linderane **196**

197

(**200**), and *cis,cis*-germacrolides (**201**) (146). Although the *cis*-germacrane cation **127** (Fig. 8.1.9) may be the precursor of lactones with Δ^2 (farnesol numbering) *cis* configurated, it is likely that Δ^6 *cis* configuration arises from a *trans* precursor by an isomerization process. In this connection, it may be noted that all presently known melampolides have an oxidized C-14. Isolation of *cis*-germacranolides of types **199–201** from woody parts of plants is rare, though a few examples can be cited. Thus, longipilin (**202**) has been obtained from twigs (and leaves) of *Melampodium longipilum* (343). Whereas twigs (and leaves) of *Podanthus ovatifolius* contain erioflorin (**203**) (157), roots of *Anthemis cretica* contain the *cis,cis* derivative **204** (50).

Germacranes with *trans,trans* 1,5-diene system are thermally labile and undergo Cope rearrangement to give products with elemane skeleton. Thus, costunolide (**195**) on heating gives saussurea lactone (**205**), albeit, in low yield.

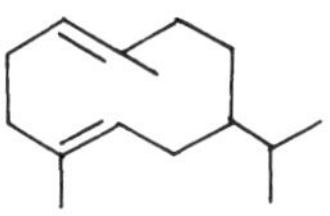

Germacrolide **198**

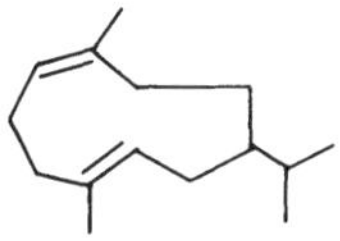

Melampolide **199**

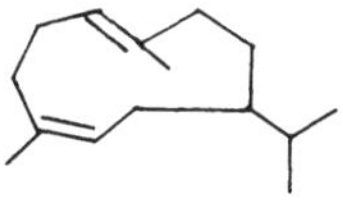

Heliangolide **200**

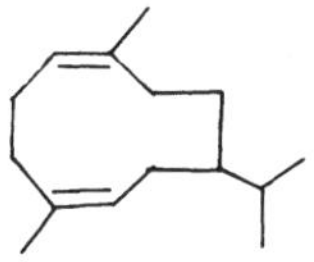

cis-cis-Germacrolide **201**

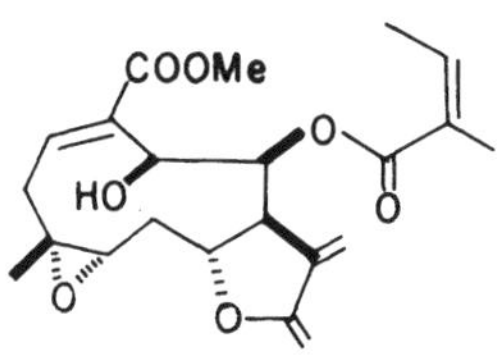

Longipilin **202**

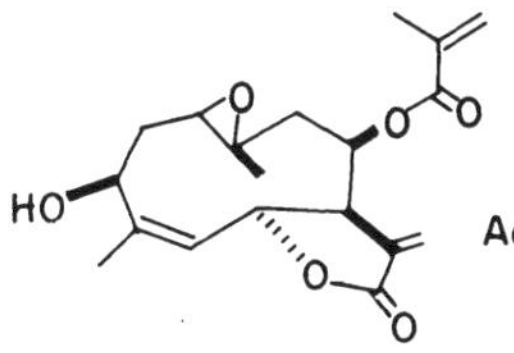

Erioflorin **203**

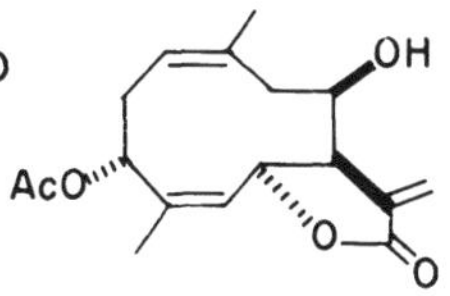

204

Lactone **205** had earlier been isolated from costus root oil, but it was later demonstrated that it is only an artifact (321). Similarly it was found that, whereas the essential oil from the leaves of *Hedycarya angustifolia* is a rich source of elemol (**206**), solvent extraction of leaves at room temperature yielded essentially hedycaryol (**207**) with only a trace of elemol (129). There are other examples (382). In view of these observations, suggestions have been made that elemanes arise in nature from suitable germacranes by a biological equivalent of Cope rearrangement, and that many natural elemanes may, in fact, be artifacts. In this connection, it is worth noting that the dilactone, vernolepin (**208**), has been isolated from *Vernonia hymenolepis* by both cold and hot solvent extraction, and thus provides an example of a true naturally occurring elemane (248).

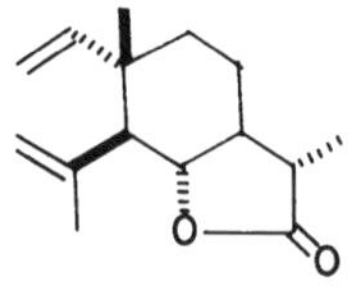

Saussurea lactone
205

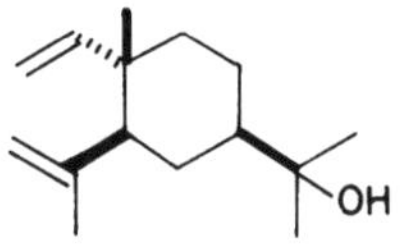

(−)−Elemol
206

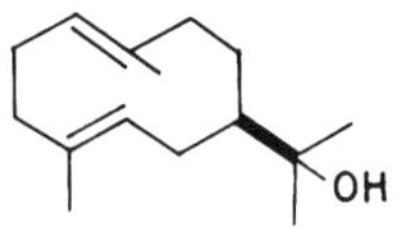

(+)−Hedycaryol
207

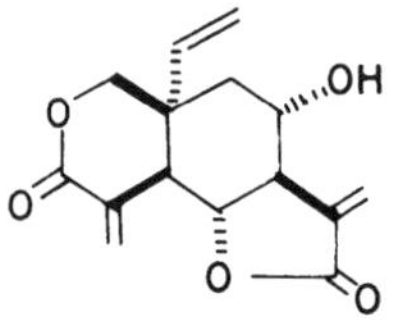

(+)−Vernolepin
208

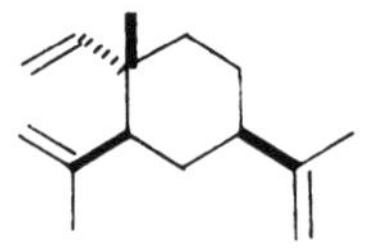

(−)−β−Elemene
209

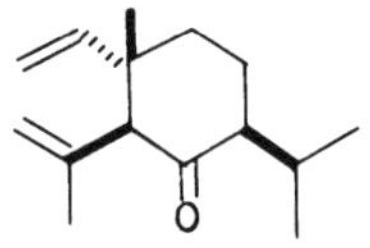

Shyobunone
210

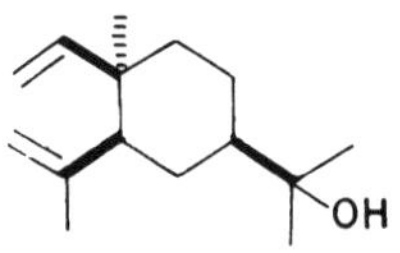

(+)−10−Epi-elemol
211

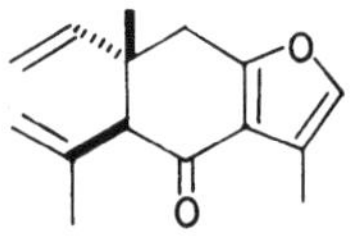

Curzerenone
212

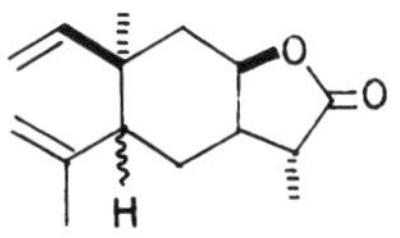

Callitrin
213

Around 60 elemanes are known at present and some of these have been isolated from woody tissues. (−)-Elemol (**206**) was first isolated in 1907 (350) from Manila elemi oil obtained from the oleoresin of *Canarium luzonicum*. (−)-β-Elemene (**209**) (238, 396), shyobunone (**210**) (419), and a few related elemanes occur in the essential oil from rhizomes of *Acorus calamus*. Resin from *Ferula* spp. (Galbanum resin) contains 10-epi-elemol (**211**) (391). The furane derivative, *curzerenone* (**212**), has been isolated from rhizomes of *Curcuma zedoaria* (195). The majority of presently known elemanes have a lactone ring and are known as elemanolides; they have been recently rewiewed (146). Callitrin (**213**) is an example of an elemanolide that has been isolated from the heartwood of *Callitris columellaris* (59).

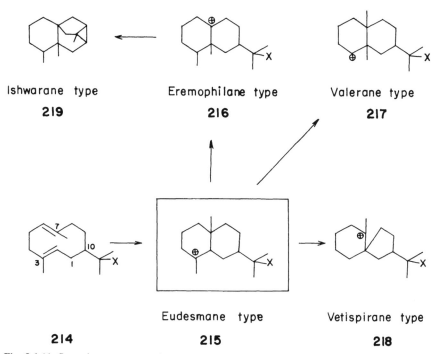

Fig. 8.1.11. Some important sesquiterpene types based on eudesmane cation

Germacrane cations **127** and **129** (Fig. 8.1.9) and/or products derived from them play a dominant role in the biosynthesis of a large variety of bicyclic and tricyclic sesquiterpenoids, which are discussed in the following subsections. Although only limited biosynthetic studies have been carried out in this area (90, 110), there is an impressive number of in vitro chemical transformations mimicking these cyclizations/rearrangements (101).

(e) *1,10-Cyclization: Eudesmanes and related types.* Markownikoff cyclization of a suitable germacrene, such as **214**, arising from the germacrane cation **129** will generate the bicyclic ion **215**, which, in principle, has been implicated in the genesis of a number of known sesquiterpene types (190). Figure 8.1.11 (X depicts OH/OPP/ENZ, etc.) depicts important types relevant to our present discussion.

The eudesmanes (selinanes) are an important family of sesquiterpenoids; at present some 300 compounds of this class are known. Classically, this group was recognized by formation of eudalene (**220**) on dehydrogenation (315). (−)-α-Santonin (**221**), the first pure sesquiterpene to be isolated from nature (in 1830) (350) belongs to this group. Like many other major sesquiterpene types, this class exhibits a variety of structural features, but a majority of these compounds have a γ-lactone ring (eudesmanolides, 146). For structure **215**, four diastereoisomers (**222–225**) are possible, and all four types have been encountered in nature. Compounds with relative stereochemistry depicted in **222** are most common, and the simple derivatives are widely distributed in the families Cupressaceae, Myrtaceae, and Taxodiaceae (140, 275 vol. III b). The essential oil from the wood of *Callitropis araucarioides* is a good source of (+)-β-eudesmol (**226**) (25, 238).

(−)-Cryptomeridiol (**227**) is a fully saturated eudesmane from heartwood of *Cryptomeria japonica* (381). From the heartwood of red sandal (*Pterocarpus santalinus*) a number of eudesmanes have been isolated, of which pterocarpdiolone (**228**) may be mentioned (246). Costic acid (**229**), a eudesmane acid, is present in costus root oil (30). Laevojunenol (**230**) from vetiver oil (*Vetiveria zizanioides*) of North Indian origin is an example of *ent*-eudesmane derivative (345). Of the many presently known eudesmanolides only a few have been isolated from woody parts of plants; (+)-dihydro-β-cyclocostunolide (**231**) was obtained from trunkwood of *Moquinea velutina* (146).

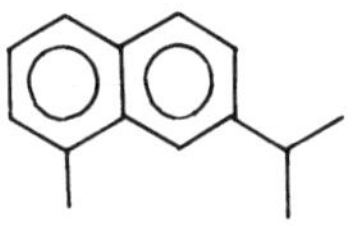

Eudalene
220

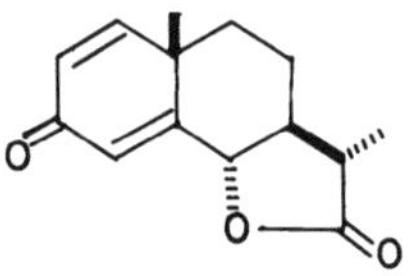

(−)-α- Santonin
221

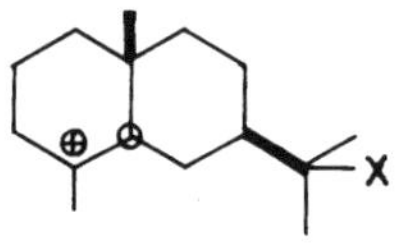

Eudesmane type
222

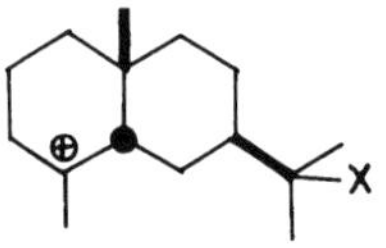

Chamaecynane type
223

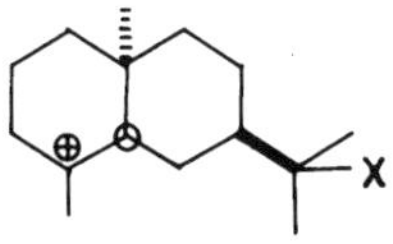

Occidentalane type
224

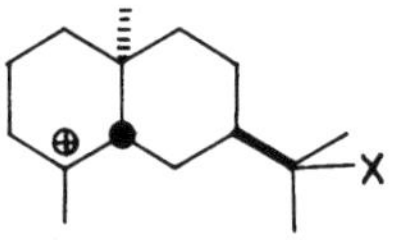

Intermedeane type
225

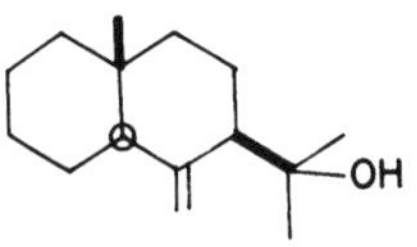

(+)- β- Eudesmol
226

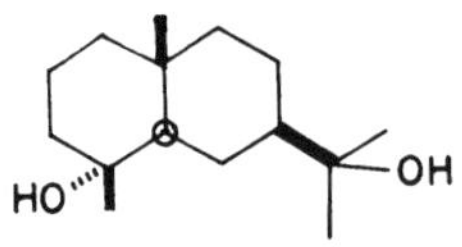

(−)- Cryptomeridiol
227

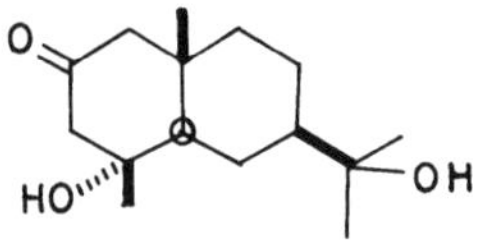

(+)- Pterocarpdiolone
228

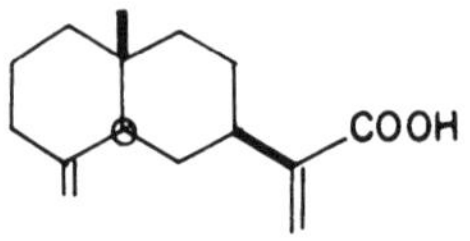

(+)- Costic acid
229

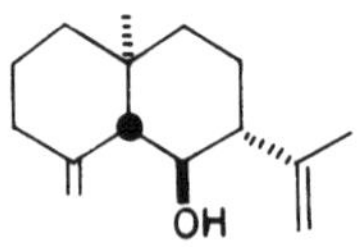

Laevojunenol
230

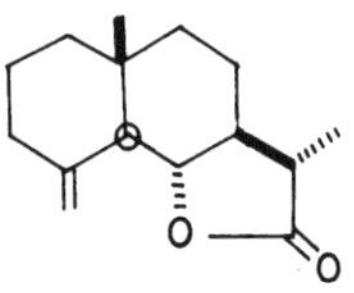

(+)- Dihydro-β- cyclo-costunolide
231

Chamaecynane type (**223**) was first encountered with the characterization of C$_{14}$ sesquiterpenes – e.g., dehydrochamaecynenol (**232**) from *Chamaecyparis formosensis* (14). Some half dozen compounds with this stereochemistry are known at present and all of these are C$_{14}$ compounds isolated from *C. formosensis*. The other *cis* fused type, **224**, was uncovered with the structure elucidation of (+)-occidentalol (**233**), isolated from the heartwood of *Thuja occidentalis* (368). The aromatic compound occidol (**234**) co-occurs with occidentalol in the *Thuja* wood. Several compounds of intermedeane type (**225**) are known at present and some of these have been isolated from woody tissues. Intermedeol (**235**) itself was first isolated from the Indian race of the grass *Bothriochloa intermedia* (95). (−)-Callitrisin (**236**) is a constituent of *Callitris columellaris* heartwood (59, 158).

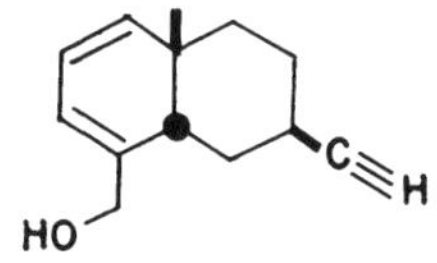

(−)- Dehydrochamaecynenol
232

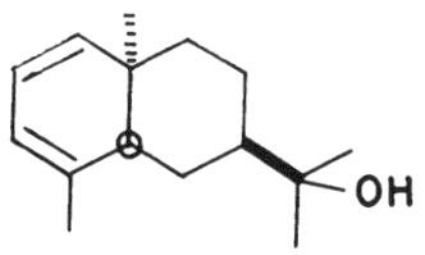

(+)-Occidentalol
233

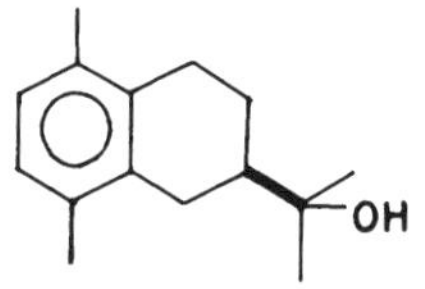

Occidol
234

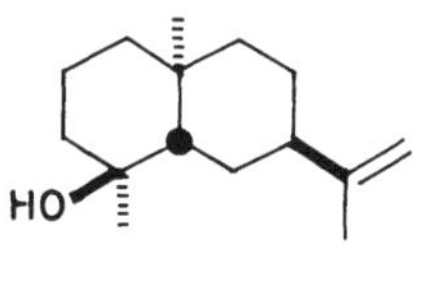

(+)- Intermedeol
235

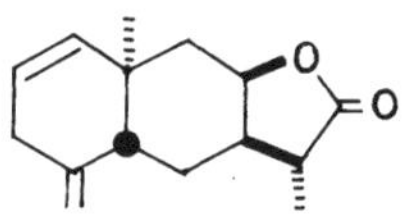

(−)- Callitrisin
236

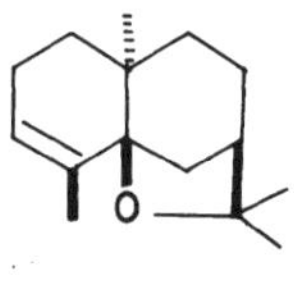

(+)- α-Agarofuran
237

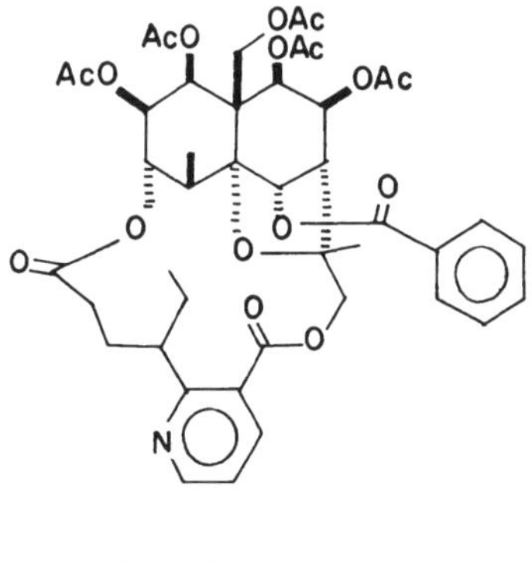

Cassinin
238

α-Agarofuran (**237**) (21) and related compounds, isolated from agarwood (fungus-infected wood of *Aquillaria agallocha*), have the stereochemistry of intermedeane type (**225**). The *Celastraceae* alkaloids (354) have a sesquiterpene core of agarofuran type. These compounds have been isolated from leaves, young shoots and seeds. However, one such alkaloid, cassinin (**238**), has been isolated from roots of *Cassine matabelica* (406).

The biogenesis of configurationally isomeric eudesmanes (**222**–**225**) has been discussed (203, 310).

Of the rearranged eudesmanes, eremophilanes (**216**) are most prolific (~150 derivatives). The subject was reviewed in 1977 (314). Two stereochemical classes, **239** and **240**, have been recognized; in principle, these are related to the eudesmane (**222**) and intermedeane types (**225**), respectively. The first member of this class to be characterized was eremophilone (**241**), which occurs in the wood oil of the Australian tree *Eremophila mitchelli*, along with other related compounds. This was the first terpene structure not consistent with the isoprene rule and Robinson (327) invoked a 1,2-methyl shift from a eudesmane precursor to rationalize this structure. (−)-Eremoligenol (**242**) is a component of roots of *Ligularia fischeri*, while isovalencenic acid (**243**) has been isolated from vetiver oil. (+)-Nootkatone (**244**) was first isolated from the heartwood of Alaska-cedar

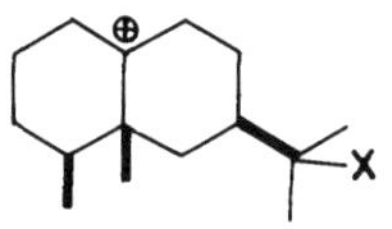

Eremophilane type
239

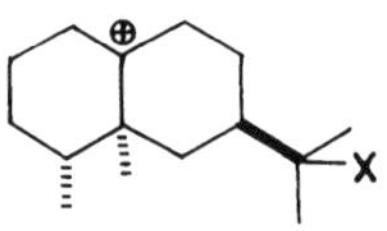

Nootkatane type
240

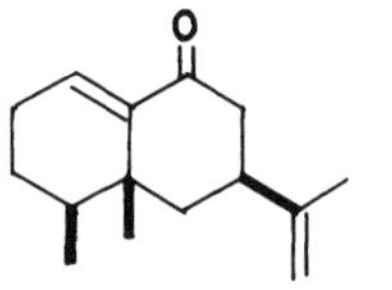

(−)-Eremophilone
241

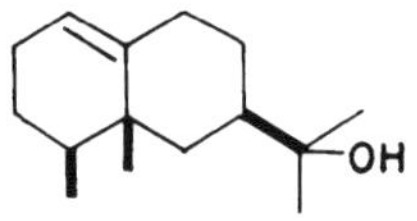

(−)-Eremoligenol
242

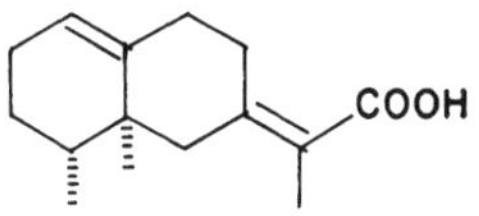

(+)-Isovalencenic acid
243

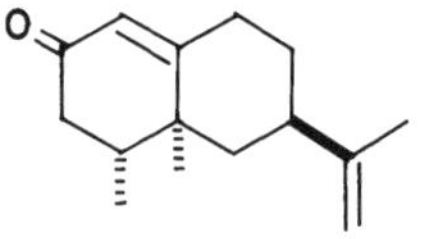

(+)-Nootkatone
244

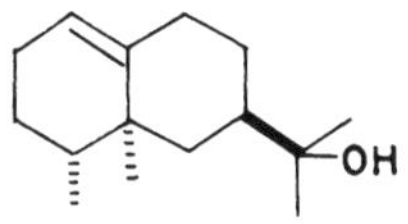

(+)-Valarianol
245

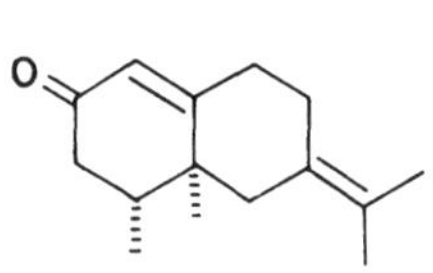

(+)-α-Vetivone
246

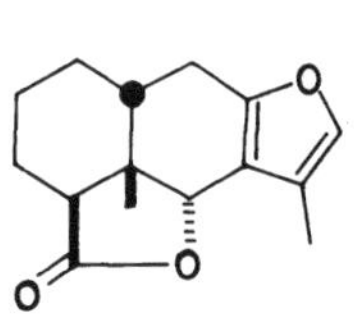

247

(*Chamaecyparis nootkatensis*) and later from grapefruit (*Citrus paradisi*) juice. It is responsible for the typical grapefruit flavor, and is commercially produced synthetically. Other examples of nootkatane type from woody parts of plants are (+)-valerianol (**245**) from roots of *Valeriana officinalis*, and α-vetivone (**246**), an important component of vetiver oil (314). Of several eremophilanolides (146), mention may be made of furanoeremophilanolide **247** from roots of *Ligularia hodgsoni*. Over 50 new furanoeremophilanes have been isolated from roots of various *Euryops* spp. (52).

What appear to be modified eremophilanes are bakkenolide-A (**248**) and related compounds from the roots of *Petasites japonica* (146, 314), and cacalol (**249**), maturinone (**250**), and related sesquiterpenoids from roots of *Cacalia* spp. (70, 221). The biogenetic relationship of these compounds to eremophilanes is strengthened by the co-occurrence of decompostin (**251**) (339) with cacalol and related compounds in *C. decomposita* (221).

(+)-Bakkenolide-A
248

Cacalol
249

Maturinone
250

Decompostin
251

From the roots of *Aristolochia indica*, the tetracyclic sesquiterpenes ishwarane (**252**) and ishwarone (**253**) have been isolated (314). A few other related compounds are known. Biogenetically, they appear to be derived from the nootkatane cation (**240**).

Valeranone (**254**) (194), a component of the roots of *Valeriana officinalis* and of the rhizomes of *Nardostachys jatamansi* was the first compound to be recognized among a minor group of rearranged eudesmanes in which C-3 methyl (farnesol numbering) has undergone a 1,2-shift. Cryptofauronol (**255**) is another related compound isolated from Japanese valerian (197).

Rearrangement of the eudesmane cation (**215**) involving C-7,C-8 bond (farnesol numbering) will generate spiro system **218** (Fig. 8.1.11). Depending on the C-10 configuration, two spiro types **256** and **257** have been recognized (267). The first compound with carbon skeleton **256/257** was agarospirol (**258**), a constituent of fungus-infected wood of *Aquillaria agallocha*. Since then, it has been shown that β-vetivone, known since 1939 as an important constituent of vetiver oil and until 1969 considered to be **259**, has the structure **260**. This led to the reformulation of hinesol (**261**), which occurs in rhizomes of *Atractylodes lancea*, and which had been earlier correlated with β-vetivone (238, 267). Other examples will be found in a review published in 1974 (267).

(f) *1,10-Cyclization: Guaianes and related types*. Anti-Markownifoff cyclization of a suitable germacradiene (**214**), can lead to bicyclic ions **262a/262b** having the hydroazulene skeleton. These ions and further important types arising from

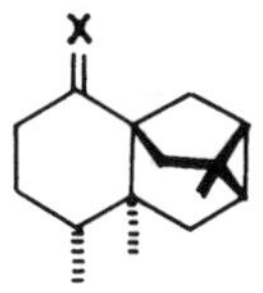

252: (−)−Ishwarone
(X=H₂)
253: (+)−Ishwarone
(X=O)

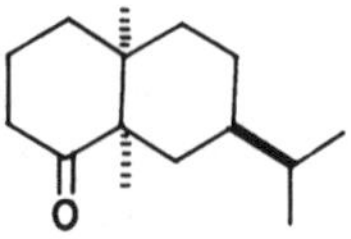

(−)−Valeranone
254

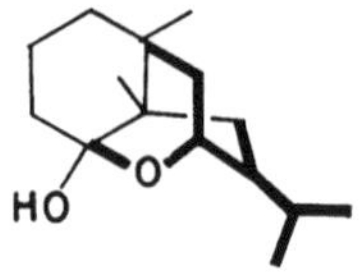

(−)−Cryptofauronol
255

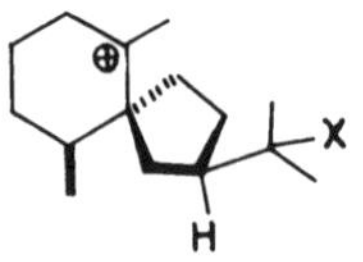

Hinesane type
256

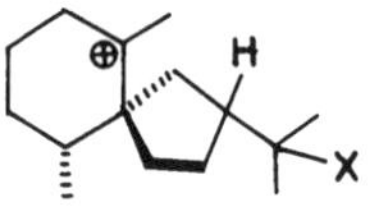

Vetispirane type
257

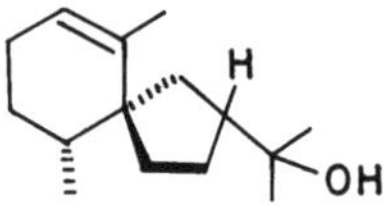

(−)−Agarospirol
258

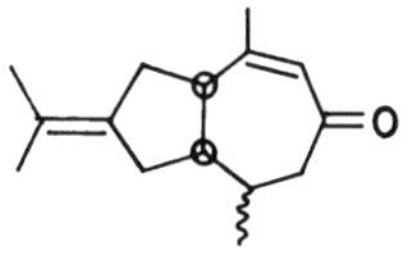

259

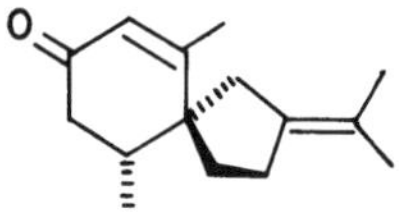

(−)−β−Vetivone
260

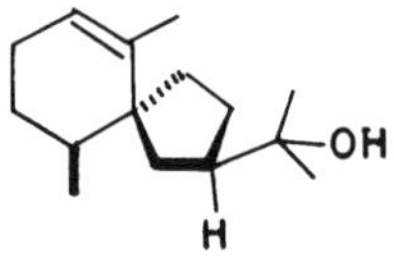

(−)−Hinesol
261

these by rearrangement/cyclization are shown in Fig. 8.1.12. Of these, the most prolific are the guaiane (**262**) and pseudoguaiane types (**263**). Guaianes with over 300 compounds are presently the second largest group of sesquiterpenoids. Some 160 pseudoguaianes are known at present. Earlier the guaiane class was recognized by dehydrogenation to S-guaiazulene (**269**) or S-chamazulene (**270**) (422) or by certain color reactions (372). A vast majority of guaianes and almost all pseudoguaianes have a γ-butenolide system (guaianolides and pseudoguaianolides); the chemistry of these lactones has been reviewed (146, 330, 361). They are most widely distributed in the Compositae. Several of these lactones have antineoplastic activity (277).

(−)-Guaiol (**271**) was first isolated in 1903 from guaiacum wood (*Bulnesia sarmienti*), and has also been obtained from wood of several *Callitris* spp. (275 vol. III b). Commercial guaiacum wood oil contains some 30% **271**, plus a number of related compounds including (−)-bulnesol (**272**), which is present to the extent of 45% (27). γ-Gurjunene (**273**) is a component of gurjun balsam of several species of *Dipterocarpus* (133). Liguloxide (**274**), earlier isolated from aerial parts

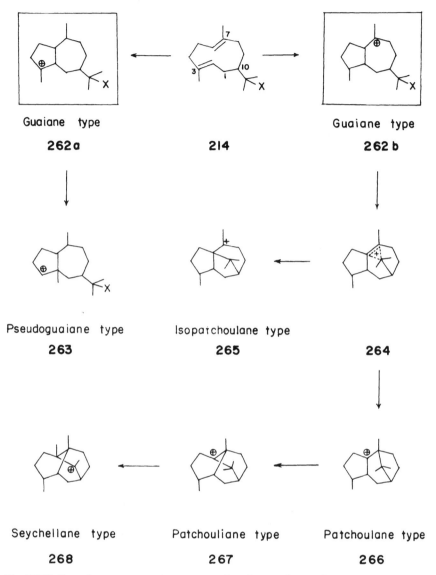

Fig. 8.1.12. Some important sesquiterpene types based on guaiane cation

of *Ligularia fischeri* (215), has been found to be a component of the Chinese drug "San-shion" from roots of *Ligularia sibirica* (199). Of the guaianolides and pseudoguaianolides, only very few have been isolated from woody tissues. Examples are dehydrocostus lactone (275) from costus root oil and collumellarin (276) from heartwood of *Callitris columellaris*. The interesting pseudoguaianolide, sulferalin (277) has been obtained from roots of *Helenium autumnale* (146).

The tricyclic sesquiterpenoids of types 265 and 266–268 are considered to arise from cation 264 resulting from a suitable olefin derived from hydroazulene

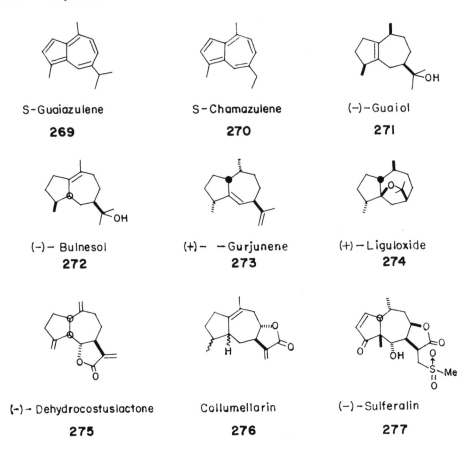

cation **262b**. These types have a very restricted distribution, and have mostly been isolated from leaf (e.g., *Pogostemon patchouli*) or tuber (e.g., *Cyperus rotundus*) essential oils, and will not be discussed further here.

(g) *1,10-Cyclization: Cadalanes and related types.* Cation **279**, arising from 1,3-hydride shlift in **127** followed by further cyclization, has been considered appropriate for genesis of a large group of sesquiterpene types, such as those depicted in Fig. 8.1.13. It is obvious that for further cyclization of ion **278** to a decalin ring system, Δ^2 (farnesol numbering) must have *cis* geometry. For this reason, it was believed that such systems require *cis*-farnesylpyrophosphate (**121**) as the first sesquiterpene precursor. In view of our earlier discussion on the subject (Sect. 8.1.3.3.2) however, the question of primary requirement of *c*-FPP becomes redundant. In fact, in the biosynthesis of γ-cadinene (**286**) by fungus imperfectus *Anthostoma avocetta*, it was demonstrated that *trans*-farnesyl pyrophosphate (**120**) is the immediate precursor (12). There is now ample biosynthetic evidence for 1,3-hydride shift of the type envisaged in going from **127/129** to the allylic cation (12, 110). Biosynthesis of gossypol (**287**), the toxic yellow pigment of the cotton plant, *Gossypium hirsutum*, is an excellent example of an alternate pathway to a cadalane type. There is sufficient evidence to show that biosynthesis of **287** involves *c*-FPP and 1,6-cyclization occurs first (110).

Terpenoids 733

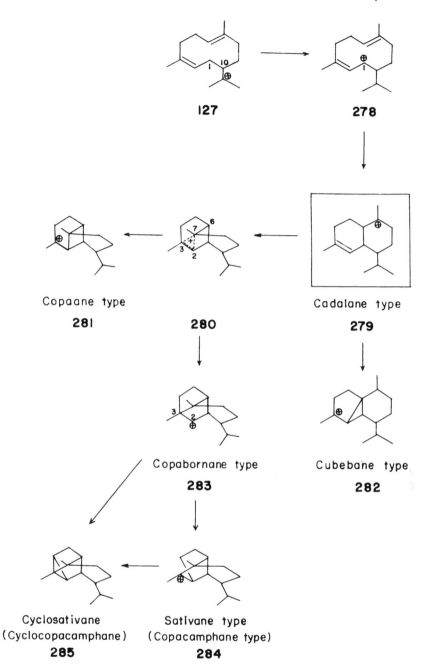

Fig. 8.1.13. Some important sesquiterpene types based on cadalane cation

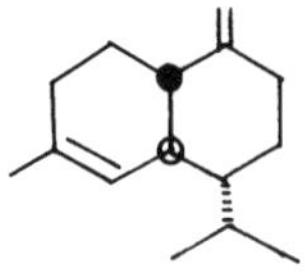

(−)-γ-Cadinene
286

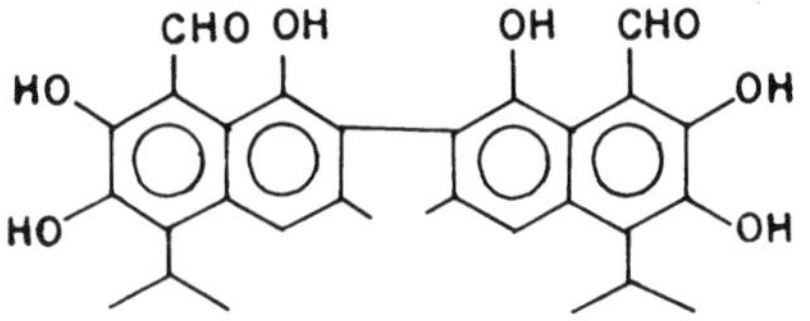

Gossypol
287

The cadalane group with carbon frame work of ion **279** is a fairly large group of bicyclic sesquiterpenoids with approximately 200 compounds. Hydrocarbons (cadinenes) and alcolols (cadinols) of this group were obtained, often as impure samples, during early investigations on various essential oils. These compounds were characterized by formation of cadalene (**288**) on dehydrogenation and often by reaction with hydrogen chloride to give a crystalline dihydrochloride (**289**) (192). At present, four configurationally isomeric types corresponding to ion **279** are recognized: cadinanes (**290**), muurolanes (**291**), amorphanes (**292**), and bulgaranes (**293**). Antipodal cadinane, muurolane, and amorphane types are known.

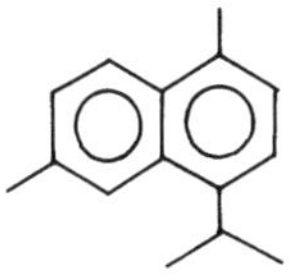

Cadalene
288

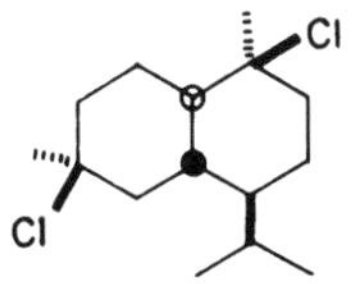

(−)-Cadinene dihydrochloride
289

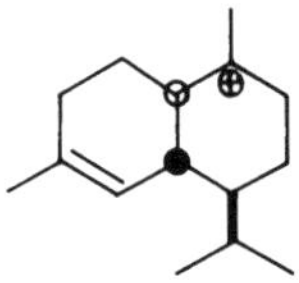

Cadinanes
290

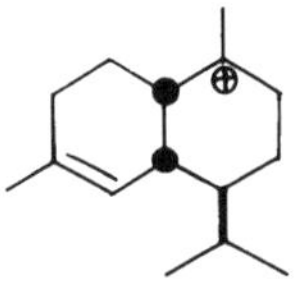

Muurolanes
291

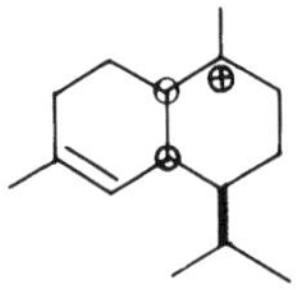

Amorphanes
292

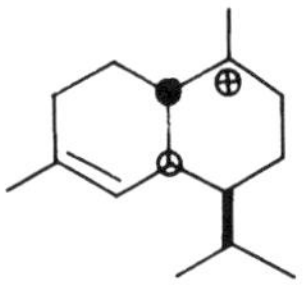

Bulgaranes
293

Of the above four types, cadinanes are most common and are fairly widely distributed in nature (396). Several of these compounds have been found in woody parts of plants. Examples are T-cadinol (**294**) and the hydroxy ketone **295** from the wood of *Taiwania cryptomerioides* (92, 255), (+)-δ-cadinene (**296**) from trunk wood of *Cedrela toona* (291), and (−)-γ-cadinene (**286**), khusinol (**297**) and several related compounds from vetiver oil of North Indian origin (222, 394). Muurolenes (e.g., ε-muurolene, **298**) were first characterized in 1966, and were isolated from wood of *Pinus sylvestris* (413). δ-Cadinol, which occurs in several species of Taxodiaceae and Cupressaceae families, was later found to have muurolane stereochemistry (**299**) (140, 291). The related epimer, **300** (T-muurolol), is a component of *Cedrela toona* wood (291). The other two types have not

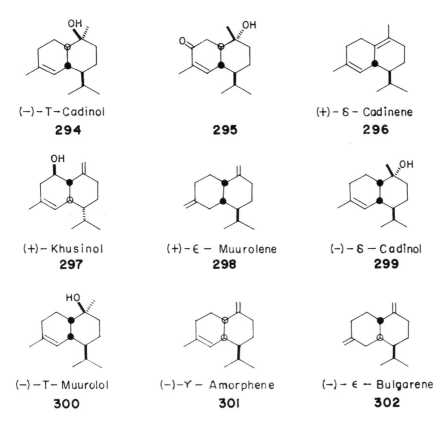

so far been found in woody tissues. However, for the sake of completeness it may be mentioned that (−)-γ-amorphene (**301**) was first isolated from essential oil of fruits of *Amorpha fruticosa* (285) and (−)-ε-bulgarene (**302**) was obtained from Bulgarian *Mentha piperita* oil (403).

Several aromatic compounds based on the cadalane skeleton are known. Thus, cadalene (**288**) itself has been found to occur in *Cedrela toona* wood, along with (−)-calamenene (**303**) and (+)-α-calacone (**304**) (291). From heartwood of *Mansonia altissima* (150) and several *Ulmus* spp. (332), a series of phenols and quinones have been isolated. Examples are mansone-G (**305**) from Chinese elm (*Ulmus parvifolia*) (91) and the violet-colored quinone mansonone-L (**306**) from *Mansonia altissima* (150). Blue Mahoe (*Hibiscus elatus*) wood of Jamaica, like many other woods, undergoes a color change on exposure to light. In an effort to understand this phenomenon, extractives of its heartwood as well as those of the related *H. tiliaceus* have been examined and several cadalanes characterized − e.g., the colorless hibiscone-C (**307**) and the purple hibiscoquinone-A (**308**) (145). Reference to gossypol (**287**), another aromatic derivative, has already been made.

Cubebane type (Fig. 8.1.13) sesquiterpenoids have a very restricted distribution and have not been, so far, found in woody tissues.

Further cyclization of cadalane ions with *cis* ring function − namely muurolane and amorphane cations (**291, 292**) − leads to tri- and tetracyclic systems

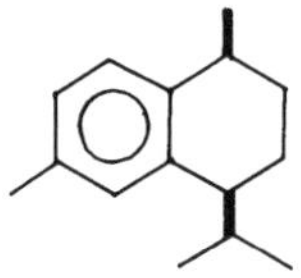

(−)-Calamenene
303

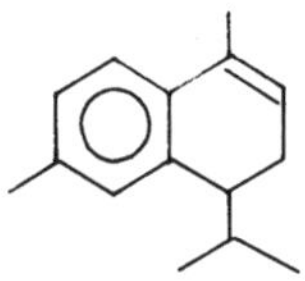

α-Calacorene
304

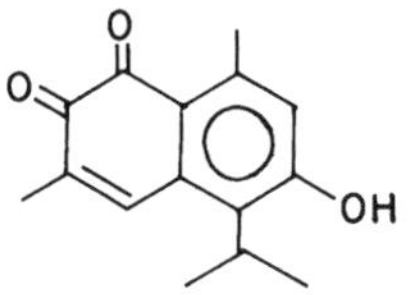

Mansone − G
305

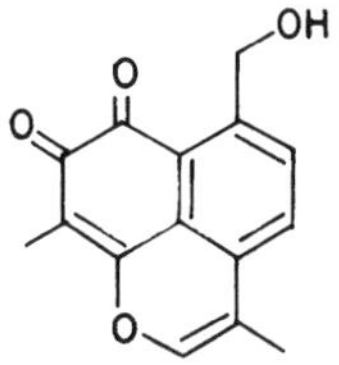

Mansonone − L
306

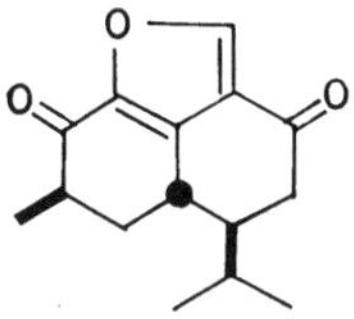

Hibiscone − C
307

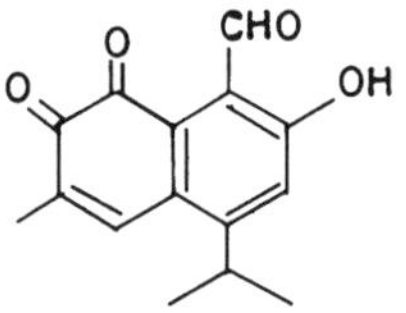

Hibiscoquinone − A
308

of types shown in Fig. 8.1.13. With the exception of copaene which is found in many essential oils, these have a severely restricted occurrence. Copaene, the oldest member of this group, was first isolated from African copaiba oil (from *Oxystigma manii*) in 1914, but correctly characterized as **309** only in 1963 (226, 270). (−)-α-Copaene has been isolated from wood of *Cedrela toona* (291) and a few other woods and resins. α-Copaene is configurationally related to the muurolane ion (291). α-Ylangene (**310**), corresponding to amorphane cation (292), was first recognized in 1965, but its stereochemical relationship with copaene could be clarified only in 1969 (301). Antipodes of both hydrocarbons are found in nature.

The secondary alcohol, copaborneol (**311**), has a structure corresponding to cation **283**; this alcohol was first isolated from wood of *Pinus sylvestris* (240).

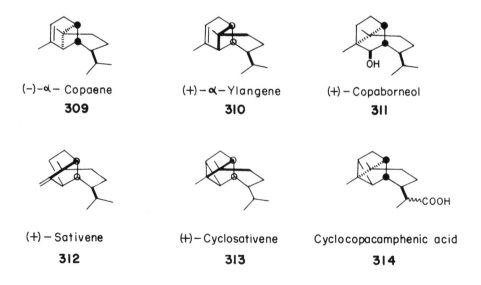

(−)-α-Copaene
309

(+)-α-Ylangene
310

(+)-Copaborneol
311

(+)-Sativene
312

(+)-Cyclosativene
313

Cyclocopacamphenic acid
314

Recently, **311** and the corresponding ketone (copacamphor) have been obtained from stems of *Espeletiopsis guacharaca* (45). From the wood of California red flir (*Abies magnifica*), sativene (**312**) and cyclosativene (**313**) – genetically derivable from copabornane cation **283** – have been isolated (352). However, it may be noted that stereochemically **312** and **313** are related to amorphane-ylangane series, whereas copaborneol is linked with the muurolane-copaane series. Some cyclocopacamphane derivatives (e.g., **314**) have been obtained from vetiver oil (232).

A number of highly oxygenated sesquiterpenoids with the picrotoxane (Fig. 8.1.8) skeleton are known. The first compound of this type to be examined was picrotoxin, from berries of *Menispermum cocculus*; it was later found to be a molecular complex of picrotoxinin (**315**) and picrotin (**316**). These structures could not be establised until 1957, although picrotoxin was isolated in 1811. At present, several compounds of this class, including nitrogen bases, are known (115, 132). They are considered to be seco-copabornanes (cleavage of C-2,C-3 bond in **283**)d and have been the subject of many biosynthetic investigations (12, 110). These results support the intermediacy of copaborneol. All these compounds have been, invariably, obtained from non-woody parts of plants and will not be discussed here further.

( )– Picrotoxinin   (–)– Picrotin
315   **316**

(h) *1,11-Cyclization: Humulanes and related types*. Deprotonation of cation **130** (Fig. 8.1.9) will generate humulene (**317**), which is known to have all *trans* geometry. Cyclization of this cation involving Δ^2 (farnesol numbering) leads to the caryophyllene cation (**318**), the precursor of caryophyllene (**319**). Structure elucidation of these compounds revealed for the first time the occurrence of medium-ring compounds in nature; germacranes, already discussed, were characterized a few years later. The past two decades have seen humulene and caryophyllene steadily emerge as progenitors of a series of new bicyclic and tricyclic sesquiterpenoids of bizarre types. Humulane-based compounds, of which 17 major types have been listed (89 vol. 12) have been invariably elaborated by mushrooms and related Basidiomycetes. Biosynthesis of these compounds has received much attention, and the intermediacy of humulene has been well established in many cases. These sesquiterpenoids will not be discussed here; an excellent review has recently been published (15). Structural types biogenetically linked with caryophyllene are also numerous, and major types have been listed (89 vol. 12). Most, if not all, of these compounds have been isolated from higher

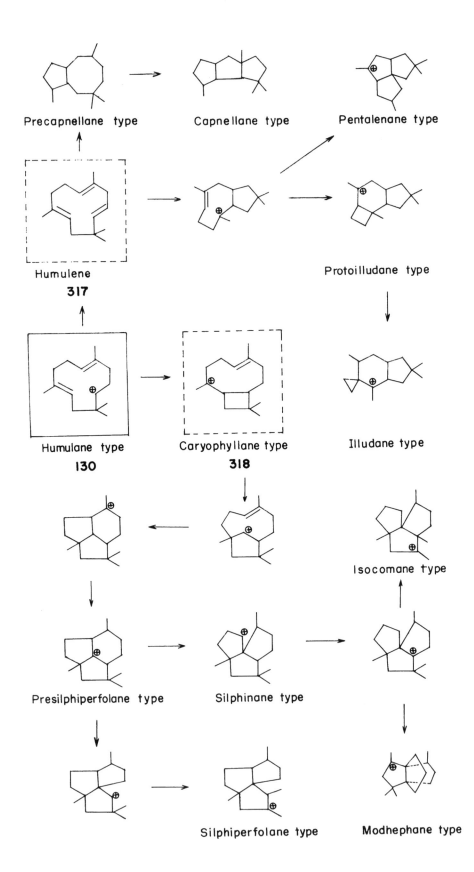

plants; little biosynthetic work in this class has been carried out. Figure 8.1.14 gives some structural types based on caryophyllene cation; a few humulene-based types have been included as illustration only.

Humulene and caryophyllene occur in the oleoresins of several species of *Dipterocarpus* (305). The essential oil from rhizomes of *Zingiber zerumbet* is an excellent source of humulene, and contains several other humulene derivatives, including zerumbone (320) and humulene oxides (e.g., humulene epoxide-II, 321) (121, 122, 371). (−)-Caryophyllene (319) is present in the essential oil from the oleoresin of *Hardwickia pinnata* to the extent of 75% (279). (+)-Caryophyllene has been isolated from the essential oil of *Libanotis transcaucasica* (207). Caryophyllene is commercially obtained as a byproduct in the isolation of eugenol from clove oil. Several oxygen derivatives of caryophyllene have been isolated from woody portions of plants, including (−)-caryophyllene oxide (322) which comprises 30% of the essential oil of oleoresin of *Dipterocarpus pilosus* (172). From the roots of *Buddleja davidii* a series of caryophyllene derivatives has been identified − e.g., buddledin-C (323) (20). In recent years caryophyllanes with *cis* ring fusion have been encountered in nature. Thus, epicaryophyllene (324) was isolated from roots of *Euryops brevipapposus* (52), whereas its epoxde 325 has been found to occur in roots of *Senecio crassissimus* (57).

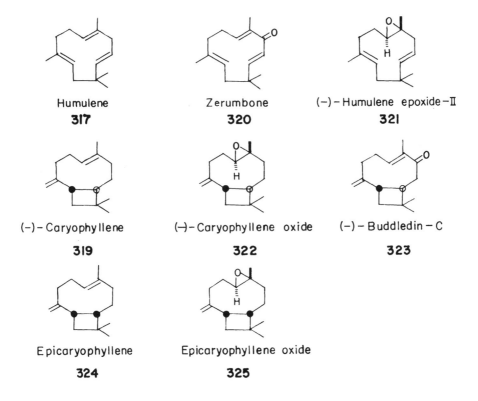

Fig. 8.1.14. Some important sesquiterpene types based on humulene cation

740   Isoprenoids

A number of tricyclic sesquiterpenoids base on caryophyllene have been isolated from woody tissues. Only a brief reference to a few of these is possible here. Caryophyllene alcohol (**326**) and clovane diol (**327**) have been obtained from oleoresin of *Hardwickia pinnata* (279). From stems of the poisonous plant *Isocoma wrightii* (rayless goldenrod) two novel sesquiterpene hydrocarbons, isocomene (**328**) and modhephene (**329**), have been isolated; **329** is the first example of a propellane occurring in nature (421, 422). These compounds have also been found to occur in the roots of *Otanthus maritimus* (386). Silphinene (**330**), silphiperfol-6-ene (**331**), and other related compounds from the roots of *Silphium perfoliatum* represent other novel types (43). Their biogenesis has been discussed (43, 421) and is summarized in Fig. 8.1.14. Biosynthesis of a related fungal metabolite has been investigated (58). These types now appear to be fairly common, and related oxygenated derivatives are being discovered (45, 53, 341). Special mention should be made of the isolation of hydroxypresilphiperfolane (**332**) from dried aerial parts of the plants *Eriophyllum staechadifolium* and *Flourensia heterolepis*, as it provides the link between **328** and **329** (Fig. 8.1.14) (56).

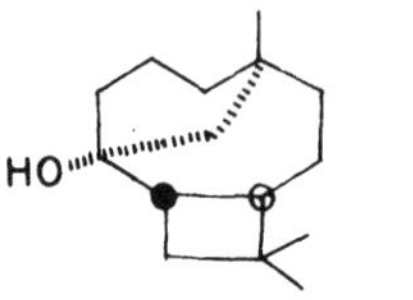

Caryophyllene alcohol
326

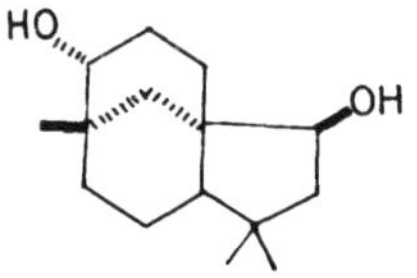

Clovane diol
327

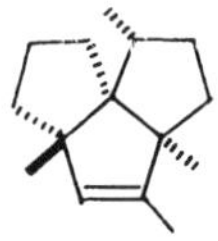

Isocomene
328

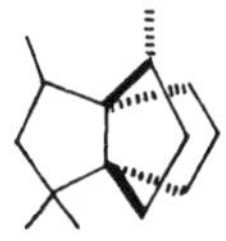

Modhephene
329

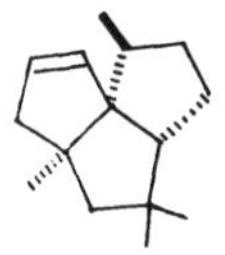

Silphinene
330

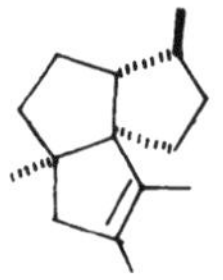

Silphiperfol-6-ene
331

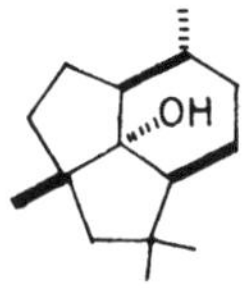

Hydroxypresilphiperfolane
332

(i) *1,11-Cyclization: Himachalenes and related types.* Cation **128** (Fig. 8.1.9) apparently behaves in the manner already outlined for **127**, the main progenitor of cadalanes and related types (Fig. 8.1.13). Thus, a 1,3-hydride shift generates ion **333**, which cyclizes to **334**, precursor of himachalenes. Further cyclization involving Δ^2 (farnesol numbering) leads to tri- and tetracyclic systems, found in nature and outlined in Fig. 8.1.15. Significant biosynthetic work in this area supports the above gross features and provides finer details. Definitive biosynthetic work has been carried out only on fungal metabolites (12, 110).

Compounds with the himachalane skeleton were first isolated from the essential oil of *Cedrus deodara*, and include α-himachalene (**341**), himachalol (**342**) and oxidohimachalene (**343**) (34, 220, 346). Many related compounds are now known to occur in nature: the keto diol ester **344** from roots of *Senecio deltoideus*

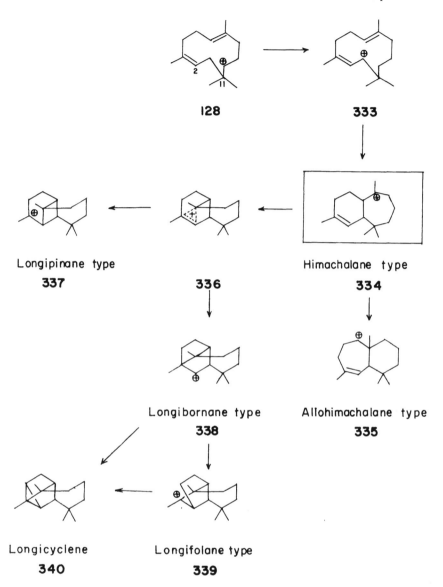

Fig. 8.1.15. Some important sesquiterpene types based on himachalene cation

and the himachalonolide **345** from roots of *Acritopappus longifolins* deserve special mention (39, 42). Allohimachalol (**346**) from the wood of *Cedrus deodara* has a rearranged himachalane skeleton (17).

(+)-α-Longipinene (**347**) arising from cation **337** was first found in wood of *Pinus sylvestris* (414). Many related compounds have since been isolated from woody tissues of higher plants (47) — for example, rastevione (**348**) from roots of *Stevia serrata* and *S. rhombifolia* (329). Longiborneol (juniperol, **349**) is widely distributed in the *Pinus* spp. but was first obtained from bark of *Juniperus*

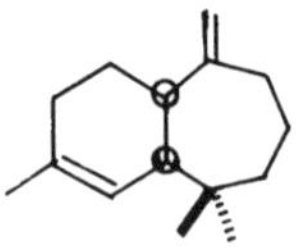

(−)-α-Himachalene
341

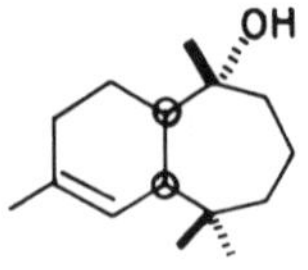

(+)-Himachalol
342

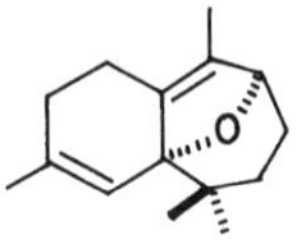

(+)-Oxidohimachalene
343

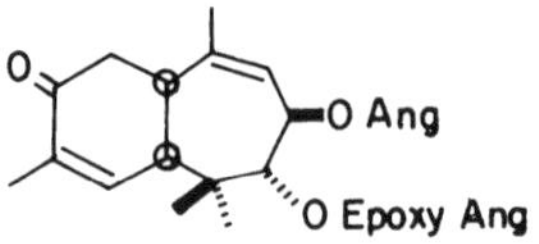

344

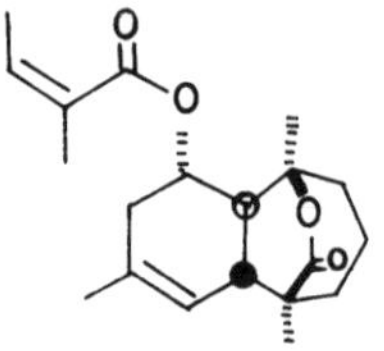

345

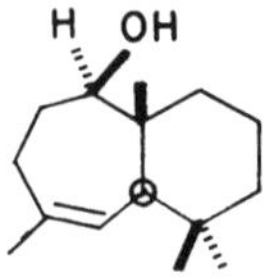

(+)-Allohimachalol
346

communis (140). (+)-Longifolene (**350**) is the major sesquiterpene constituent of Indian turpentine oil from *Pinus roxburghii*, from which it was first obtained. It appears to be ubiquitous in the genus *Pinus*. The chemistry of longifolene and its derivatives has been reviewed recently (374, 375). Another longifolane derivative, epoxide **351** from heartwood of *Juniperus conferta*, may be noted (129). Only a few other longifolanes from higher plants or molds are known. Longicyclene (**352**), the first tetracyclic sesquiterpene, is a minor component of Indian turpentine oil.

(j) *1,10-11-Cyclization: Bicyclogermacrane and derived types.* 1,3-Proton elimination of an appropriate cation **124/129/130** will result in a bicyclic diene, bicyclogermacrene (**353**), which has been considered (366) as the precursor of a

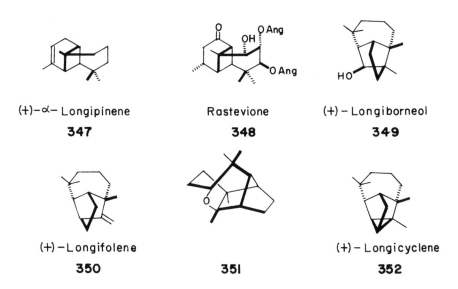

(+)-α-Longipinene
347

Rastevione
348

(+)-Longiborneol
349

(+)-Longifolene
350

351

(+)-Longicyclene
352

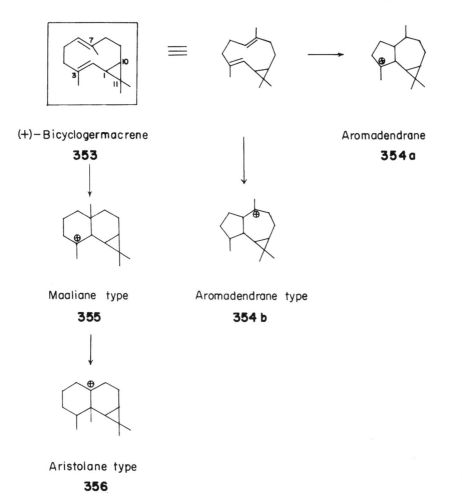

Fig. 8.1.16. Some important sesquiterpene types based on bicyclogermacrene

small group of sesquiterpenoid types (Fig. 8.1.16). This hydrocarbon was later isolated from *Citrus junos* peel oil. It is biogenetically significant that this compound occurs together with germacrane and aromadendrane types in this oil (297).

Anti-Markownikoff cyclization of **353** gives ions **354a/354b** (cf. Fig. 8.1.12), which have the carbon framework of a small group of sesquiterpenoids. The earliest known compound of this class is ledol, first isolated in 1875 from marsh tea oil, which is obtained from the leaves and flowering tips of *Ledum palustre*. Ledol has since been obtained from valerian root oil and from oil of *Aristolochia indica* roots. Its structure (**357**) was elucidated in 1969 (275 vol. IIIb). Aromadendrene (**358**) is a component of leaf oil from many *Eucalyptus* spp. and has also been isolated from twig oil of *Pseudowintera colorata* (229, 396). α-Gurjunene (**359**) occurs in gurjun balsam (*Dipterocarpus* spp.) (305).

(+)- Ledol (+)-Aromadendren (−)-α-Gurjunene
357 **358** **359**

Markownikoff cyclization of 353 is represented by maaliol (**360**), a component of Maali resin, a soft exudate from *Canarium samonense* (275 vol. IIIb). A related compound is maalioxide (**361**) from roots of *Valeriana wallichii* (305). Rearrangement of maaliane cation 355 (cf. Fig. 8.1.11), as shown in Fig.8.1.16, leads to the carbon skeleton of aristolene (**362**), which occurs in the essential oil from roots of *Nardostachys jatamansi*. (+)-Aristolene (α-ferulene) is present in the latex of *Ferula communis* (312). From roots of *Aristolochia debilis*, hydroxyketone 363 has been isolated (241).

In recent years many bicyclogermacrane and aromadendrane derivatives have been obtained from liverworts and marine organisms (13, 89 vol. 12).

(+)-Maaliol (+)-Maalioxide (−)-Aristolene Debilone
360 **361** **362** **363**

(k) *Cyclization from the head: Mono- and bi-cyclofarnesanes*. This mode represents only a minor pathway for sesquiterpene biosynthesis. Cyclization from the head is the dominant mode in diterpenoids. The first member of this class was discovered in 1954 with the characterization of γ-lactone iresin (**364**), isolated from whole plant *Iresine celosioides*. Since then, the number of compounds in this class has swelled to some 50, approximately half of these being lactones (drimanolides), which were recently reviewed (146). The simplest member is drimenol (**365**), obtained from stem bark of *Drimys winteri* which contains several other related compounds including the anhydride winterin (**366**) (7, 8). (+)-Ugandensolide (**367**) has been isolated from heartwood of *Warburgia ugandensis* (69). From the bark of this tree, a rearranged bicyclofarnesane derivative 368, having potent antifeedant (against the African army worm) and antimicrobial activity, has also been obtained (244). Several drimanolides are present in the bark of *Cinnamoma fragrans* (146). From the resin asa foetida (ex *Ferula foetida*) coumarin ethers farnesiferol A, B and C (**369–371**) have been isolated; several other similar compounds have been found to occur in various *Ferula* spp. (288). The last two compounds are monocyclofarnesane derivatives. Another example is caparrapi oxide (**372**) from *Ocotea caparrapi* wood oil (68).

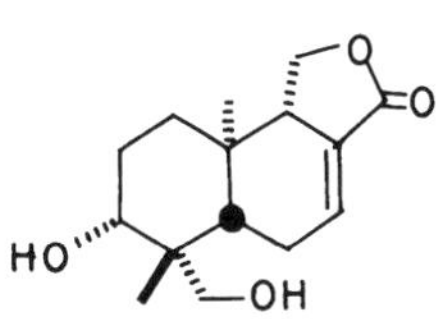

(+) − Iresin
364

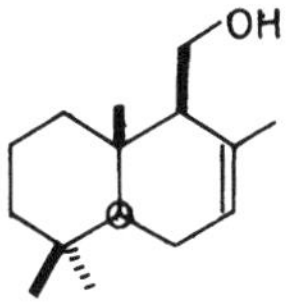

(−) − Drimenol
365

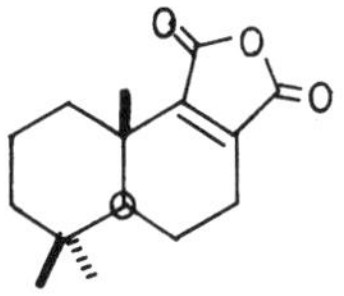

(+) − Winterin
366

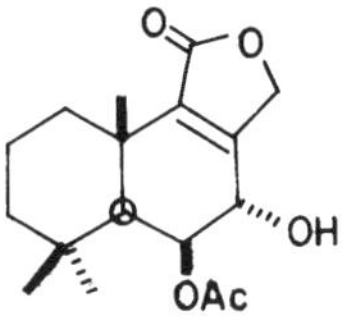

(+) − Ugandensolide
367

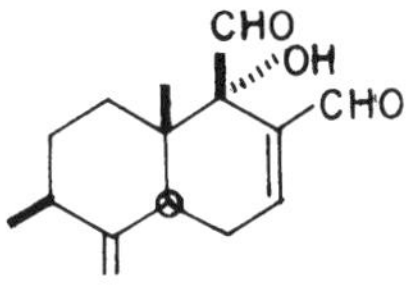

Muzigadial
368

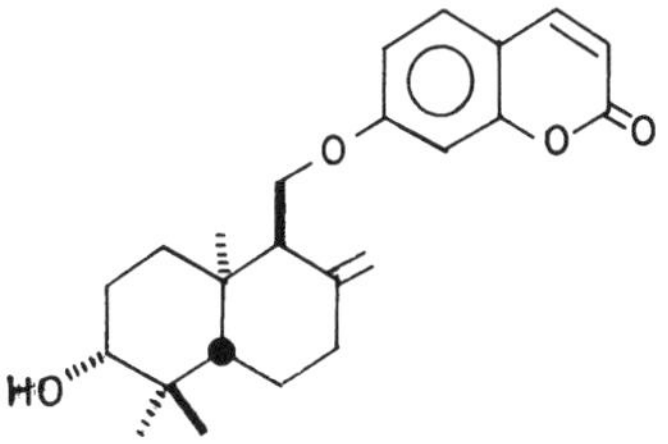

(−) − Farnesiferol − A
369

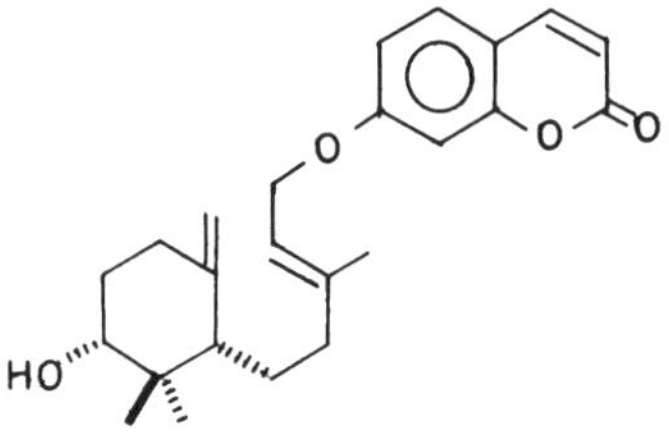

(+) − Farnesiferol − B
370

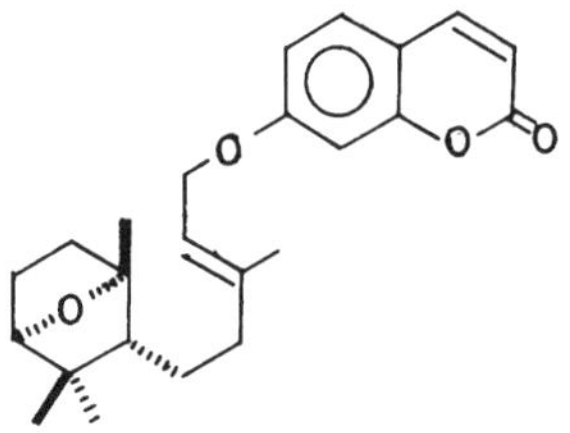

(−) − Farnsiferol − C
371

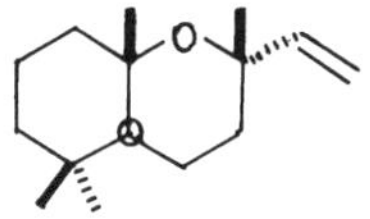

Caparrapi oxide
372

Many monocyclofarnesane derivatives have recently been identified in insect secretions and marine organisms.

Mention should be made of abscisic acid and related compounds. Abscisic acid (**373**) was first isolated from young cotton fruit as an abscission-accelerating principle. It has since been found to be ubiquitous in higher plants and has been detected in the roots of several plants. It is now recognized as a natural plant growth regulatory hormone. Although structure **373** is clearly that of a monocyclofarnesane and could arise from cyclization of FPP and further modification, there is some evidence that it may be a catabolic product of a carotenoid, such as violaxanthin. Considerable biosynthetic work has been carried out, but no clear-cut answer is available yet (110, 179, 276).

(l) *Non-farnesyl sesquiterpenoids*. there are a few sesquiterpenoids that cannot be derived from FPP or another related precursor, either directly or indirectly.

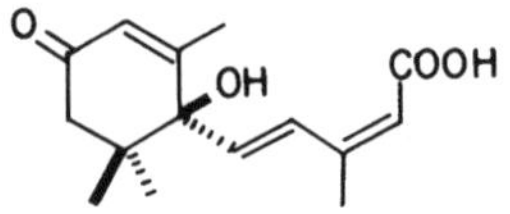

746 Isoprenoids

(+) − Abscisic acid
373

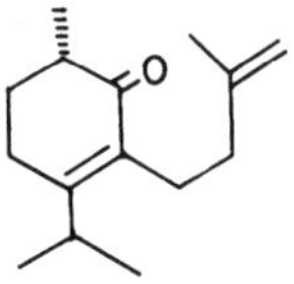

(+) − Calacone
374

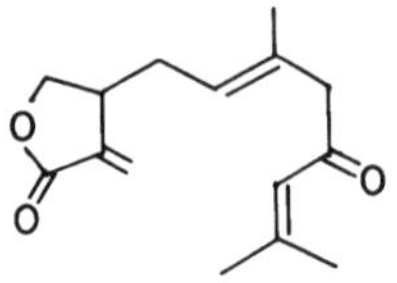

375

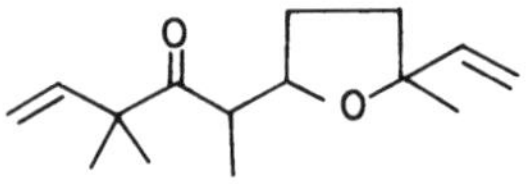

Artemone
376

These types have been encountered in higher plants, bryophytes and marine organisms. Examples from higher plants are: calacone (**374**) from essential oil of *Acorus calamus* (404), keto-lactone (**375**) from *Anthemis cotula* (55), and artemone (**376**) from "davna" oil (ex *Artemisia pallens*) (289).

8.1.3.4 Diterpenoids

8.1.3.4.1 Distribution

With the exception of the diterpene plant hormones, gibberellins, which are universal in green plants, the distribution of remaining diterpenoids is rather restricted, compared to lower terpenoids. Table 8.1.4 lists plant families known for elaborating diterpenoids. Lists of occurrence of diterpenoids at compound level are available (207, 228, 229).

Antipodes of several diterpenoids are known to occur in nature; however, this occurrence is much less frequent than in lower terpenoids. Diterpenoids

Table 8.1.4. Plant families rich in diterpenoids

Gymnospermae	Angiospermae
Araucariaceae	Dicotyledoneae
Cupressaceae	Burseraceae
Pinaceae	Cistaceae
Podocarpaceae	Euphorbiaceae
Taxodiaceae	Labiatae
	Leguminosae
	Ranunculaceae

elaborated by Hepaticae and isolated so far, invariably belong to the *ent* series (of labdanes, pimaranes, and kauranes) (13, 266).

Diterpenoids as chemosystematic markers have received some attention, especially with reference to Leguminosae phylogeny (181, 302, 317) and distribution in Cupressales (140).

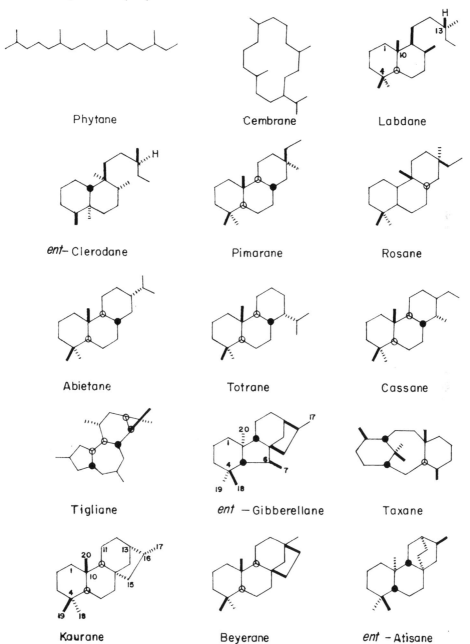

Fig. 8.1.17. Important diterpene skeletal types

8.1.3.4.2 Structural Types

Diterpenoids constitute the second largest class of terpenoids with over 2200 compounds belonging to some 130 distinct skeletal types. The discovery of new types and new members has been most dramatic during the past decade, mostly as a result of screening of marine flora and fauna (89, 144, 397). Figure 8.1.17 lists diterpene types relevant to our present discussion. An essentially complete list will be found elsewhere (378).

Geranylgeranyl pyrophosphate (GGPP, **377**; Fig. 8.1.2 is the immediate precursor of diterpenoids, as evidenced by much biosynthetic work carried out mostly on fungal metabolites (90, 178). Initiation of cyclization of GGPP can occur from either end of the molecule – tail or head – as already indicated for FPP (Sect. 8.1.3.3, under Sesquiterpenoids). Both pathways have been exploited by nature, though cyclization from the head (mostly initiation by H$^+$) is by far the more dominant mode in diterpenoids. The most important cyclization reaction of GGPP is the formation of cation **378** and generation of copalyl pyrophosphate (copalyl-PP, **379**) therefrom. These cyclization reactions proceed by a concerted, antiparallel addition mechanism, as depicted for GGPP in Fig. 8.1.18, and are considered non-stop (however, see Sect. 8.1.3.6.2).

The bicyclic cation **378** and the derived copalyl-PP (**379**) play a central role in the biosynthesis of most of the terrestrial bi-, tri-, tetra-, and pentacyclic

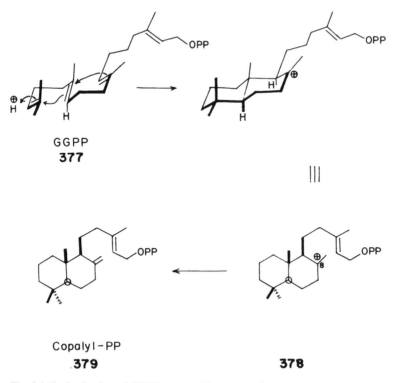

Fig. 8.1.18. Cyclization of GGPP to copalyl pyrophosphate

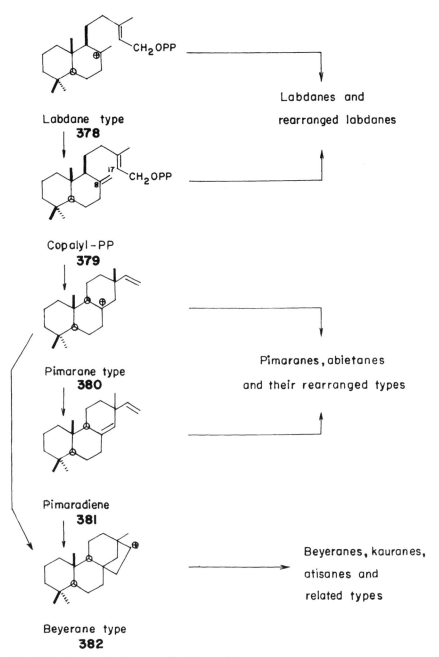

Fig. 8.1.19. Biogenesis of some cyclic diterpenoids

diterpenoids, as indicated in summary in Fig. 8.1.19. It may be noted that copalyl-PP is an obligatory intermediate in going from GGPP to these tri- and tetracyclic diterpenoids. Pimaradiene (**381**) has been implicated in certain biogenetic schemes.

750   Isoprenoids

The important diterpene types occurring in woody parts of plants are discussed in the following pages; the subject has been organized in terms of the number of alicyclic rings present in the compounds (Fig. 8.1.19) and the region of GGPP where initial cyclization is triggered. In the limited space, it will be possible to cite only a few examples from each major group; for other examples the reader is referred to a recent comprehensive work on diterpenoids (378).

(a) *Acyclic: Phytanes*. Phytol (**383**) (238) is an integral part of chlorophylls, the energy harvesting pigments of all green plants. The distribution of other phytanes, of which over 50 are presently known, in woody tissues is limited. Geranylgeraniol (**384**) occurs in the wood of *Cedrela toona* (290), whereas its allylic isomer, geranyl-linalool (**385**) has been detected in the oleoresin of *Picea abies* (233). From the stems of a Thai medicinal plant, *Croton sublyratus*, the diol **386** with potent antipeptic ulcer activity has been isolated (300).

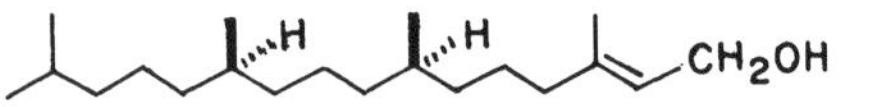

(−)-*trans*-Phytol
383

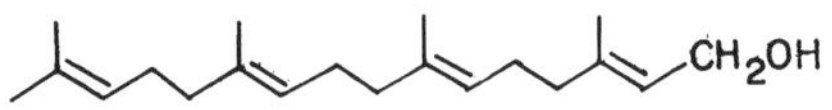

Geranylgeraniol
384

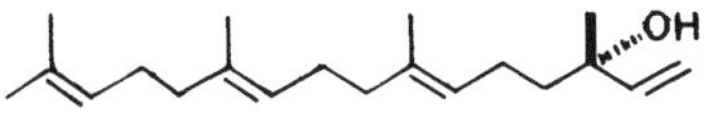

(+)-Geranyl-linalool
385

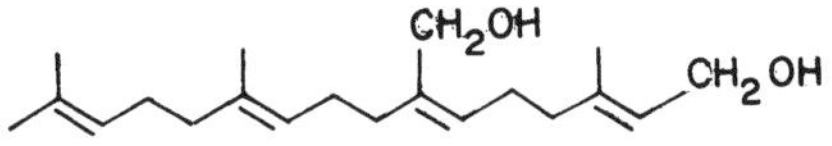

19-Hydroxygeranylgeraniol
386

(b) *Bicyclic: Labdanes, clerodanes and related types*. Labdanes (Fig. 8.1.17) presently constitute the largest diterpene group, with about 370 compounds. Labdanes are also widely distributed, albeit almost entirely restricted to higher plants, and many of these have been isolated from balsams, oleoresins, and heartwoods. Labdanes of both the normal (378) steroid absolute stereochemistry and antipodal absolute stereochemistry are known, though the former are much more preponderant. In consonance with the cyclization pathway shown in Fig. 8.1.18, labdanes invariably have the expected *trans-anti* backbone stereochemistry, and have an olefinic linkage involving C-8 or, have oxygen functionality on this carbon or, on one of its immediate neighbors.

(+)-Copalol (**387**) has been isolated from heartwood of *Dacrydium kirkii*; its antipode (*ent* **387**) has been reported from *Erythroxylon monogynum* wood (356). Its allylic isomer, (+)-manool (**388**) has been known since 1935, when it was first isolated from wood of *Dacrydium biforme* (228, 238). The heartwood of *Cupressus torulosa* contains the diol (**389**) and the related aldehyde (**390**), besides manool, and the corresponding acid, cupressic acid (**391**) has been found in the resin of *C. sempervirens* (140). A number of labdanes have been obtained from

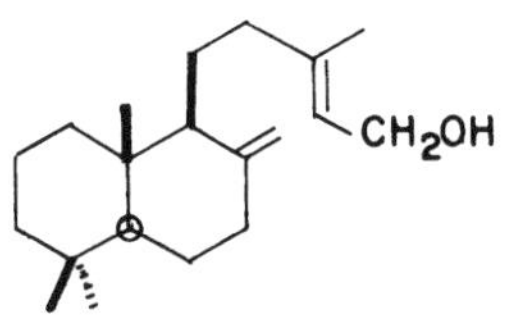

(+)− Copalol
387

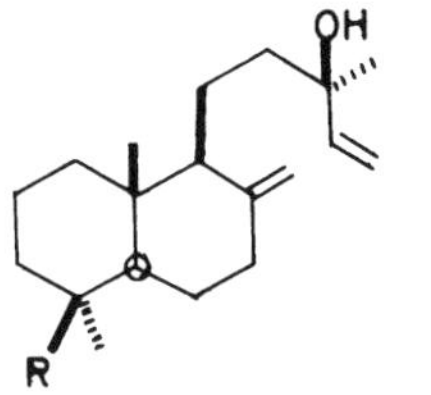

388: (+)−Manool (R=CH₃)
389: (+)−Torulosol (R=CH₂OH)
390: (+)−Torulosal (R=CHO)
391: (+)−Cupressic acid (R=COOH)

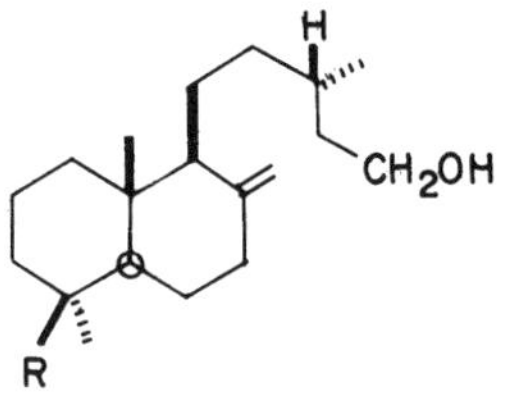

392: (+)−Imbricatadiol (R=CH₂OH)
393: (+)−Imbricatolal (R=CHO)
394: (+)−Imbricatolic acid (R=COOH)

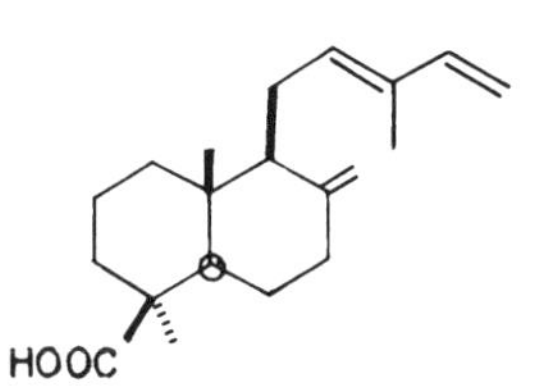

(+)-*trans*-Communic acid
395

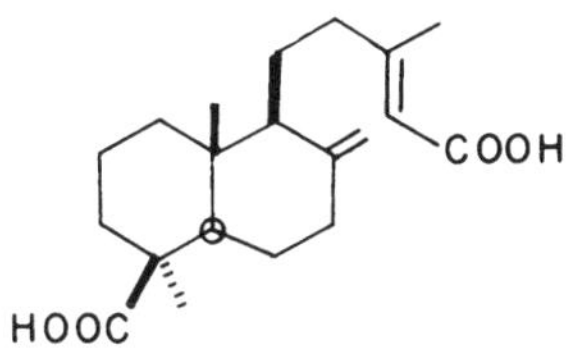

(+)−Agathic acid
396

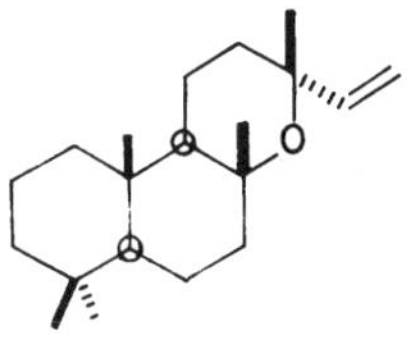

(+)−Manoyl oxide
397

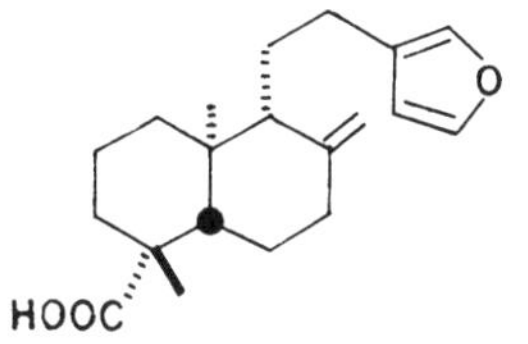

(−)−Daniellic acid
398

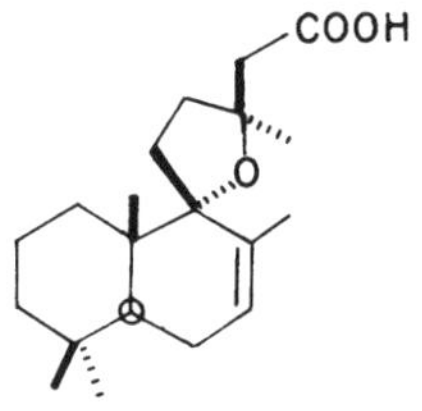

(−)−Grindellic acid
399

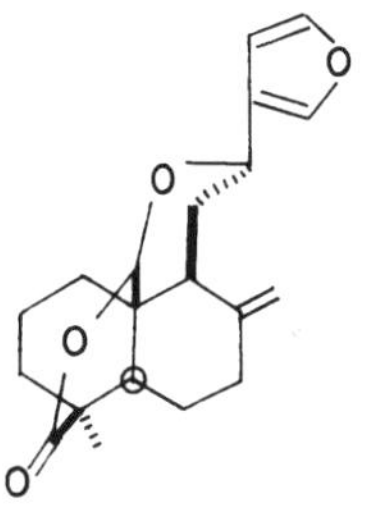

Sciadin
400

Araucaria spp., including imbricatadiol (**392**), imbricatolal (**393**), imbsricatolic acid (**394**), and a few other related compounds from *A. imbricata* (72, 73). Oleoresin from *Agathis microstachya* (Bull kauri) contains up to 27% (+)-*trans*-communic acid (elliotinic acid, **395**) and up to 40% (+)-agathic acid (**396**) (81). Labdanes with heterocyclic ring systems have been isolated from woody tissues; examples are (+)-manoyl oxide (**397**) from the heartwood of silver pine (*Dacrydium colensoi*) (200), (−)-daniellic acid (**398**) from oleoresin of *Daniellia oliveri* (175), and (−)-grindelic acid (**399**) from resin of *Grindelia robusta* (1). Several labdanolides are known, but have been mostly isolated from leaves, seeds

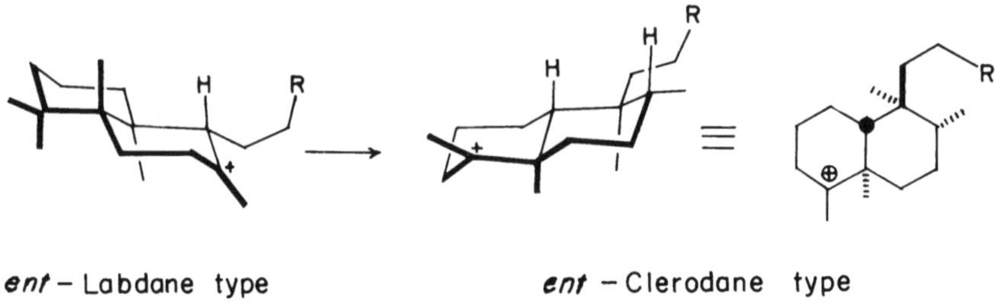

ent- Labdane type   *ent* - Clerodane type
                                    401

Fig. 8.1.20. Rearrangement of labdane cation to clerodane cation

or whole plants. Lactones – e.g., sciadin (**400**) from heartwood of Japanese umbrella pine (*Sciadopitys verticillata*) – appear to be the only examples of labdanolides from woody tissues (355, 380).

Of the rearranged labdanes, the most important are clerodanes, which are believed to arise from the labdane cation by a concerted backbone rearrangement involving sequential hydride and methyl shifts (Fig. 8.1.20) (cf. friedelanes, Sect. 8.1.3.6). Of the approximately 230 clerodanes known to date, most have the *trans-anti-trans* configuration (e.g., kolavenol, **403**) implicit in the rearrangement depicted in Fig. 8.1.20. However, quite a significant number of *cis*-clerodanes with *cis-anti* backbone (*cis-anti-trans*, *cis-anti-cis*) have been found in nature, indicating that *syn* rearrangements also occur. These cases can be accommodated by the "X-group" hypothesis (113). As expected on biogenetic grounds, almost all clerodanes have a hydrogen on C-8 and in a couple of exceptions known, this hydrogen has apparently been lost by catabolic processes. Clerodanes of both antipodal configurations are known, but *almost* all *trans*-clerodanes with well-secured absolute stereochemistry are biogenetically related to *ent*-labdanes (Fig. 8.1.20). Several genera of the families Cistaceae (*Cistus*), Compositae (*Baccharis, Solidago*), and Labiatae (*Teucrium*) have proven especially prolific in producing clerodanes. Many clerodanes have useful antimicrobial and insect antifeedant properties. Several of the classical bitter principles are now known to be clerodanes.

This class of bicyclic diterpenoids derive their generic name and chirality[1] from (−)-clerodin (**402**), the bitter principle from the leaves and twigs of *Clerodendron infortunatum* (180, 328). The oleoresin of *Hardwickia pinnata* contains several compounds of this class, including the simplest clerodane, kolavenol (**403**) (278); however, the chief constituent of the exudate is (−)-hardwickiic acid (**404**) (279). (+)-Hardwickiic acid (*ent* **404**) has been isolated from the heartwood of *Copaifera officinalis* (104). Agbanindiol-B (**405**) and related compounds from wood of *Gassweilerodendron balsamiferum* provide other examples of *trans-*

---

[1] With the reversal of absolute stereochemistry of (−)-clerodin in 1978−1979, which until then had been considered to be the mirror-image of **402**, the name *neo*-clerodane has been suggested by one group (328) for the structure and chirality shown in **401**. We feel this is unnecessary, since semisystematic names based on old, wrong absolute stereochemistry of clerodin have not been used in many cases. Furthermore, the term *neo*-clerodane does not signify its relationship with *ent*-labdane

clerodanes from woody tissues (134). (−)-Cascarillin (**406**) is the bitter principle from the trunk bark of *Croton eleuteria* (273). Examples of clerodanes from woody tissues with *cis-anti-cis*-stereochemistry are plathyterpol (**407**) from the heartwood of *Plathymenia reticulata* (234), and columbin (**408**), the bitter principle of Colombo root (*Jateorhiza palmata*) (307), which also contains several other related compounds. Compound **409** and related compounds from roots of *Solidago arguta* provide examples of clerodanes with *cis-anti-trans* geometry (272). Many C_{19} furanolactones, mostly isolated from non-woody parts of plants, are known. Crotocaudin (**410**) and a related compound from roots of *Croton caudatus* provide examples of C_{19} clerodanes (88).

Chettaphanin-I (**411**), first isolated from roots of *Andenochlaena siamensis*, represents another class (Chettaphanane) of rearranged labdanes, in which one hydride shift and one methyl shift have occurred sequentially (340). A more

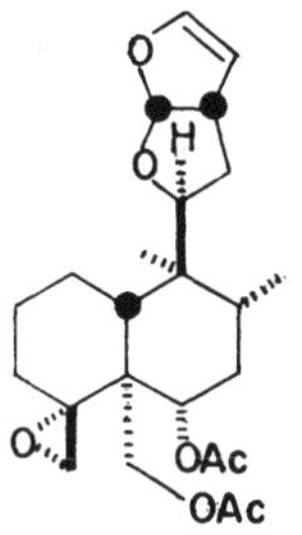

(−)−Clerodin
402

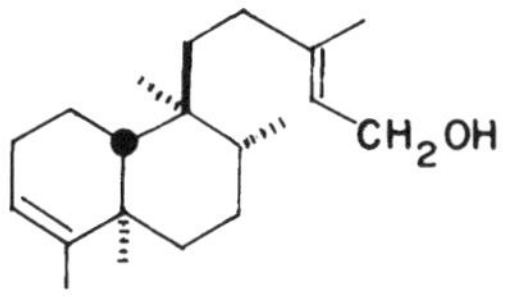

(−)−Kolavenol
403

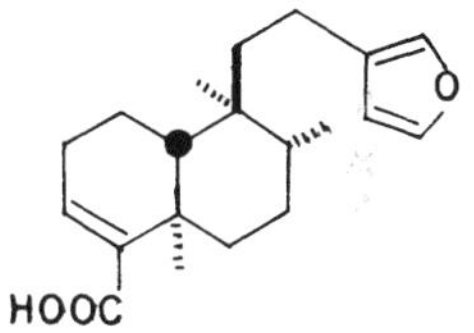

(−)−Hardwickiic acid
404

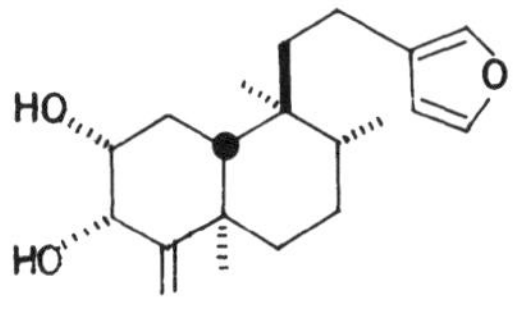

(+)−Agbanindiol−B
405

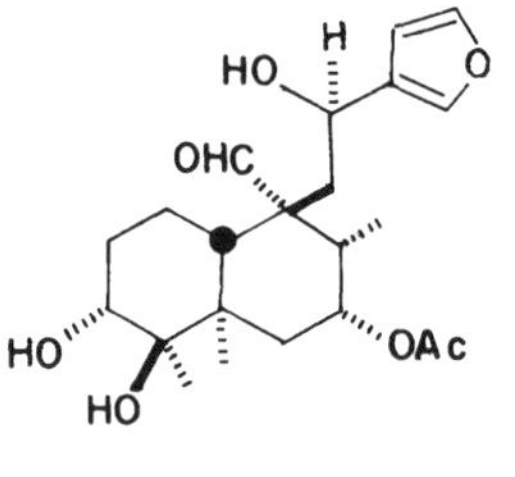

(−)−Cascarillin
406

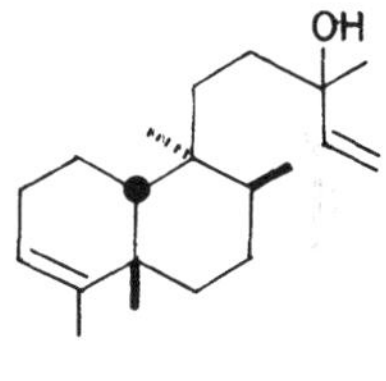

(−)−Plathyterpol
407

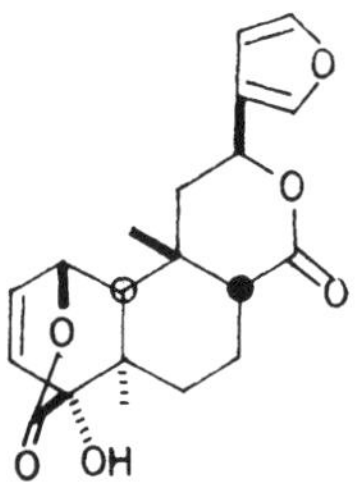

(+)−Columbin
408

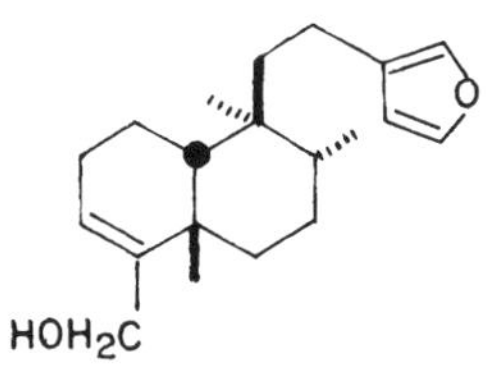

409

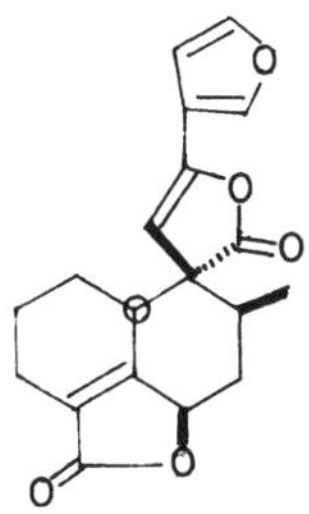

(−)−Crotocaudin
410

Chattaphanin−1
411

Diasin
412

Colensenone
413

appropriate example from woody tissues is diasin (**412**) from trunk wood of *Croton diasii* (4). Colensenone (**413**) from heartwood of *Dacrydium colensoi* is an A-nor labdane A(166).

(c) *Tricyclic: Pimaranes, abietanes and related types.* Bicyclic copalyl-PP (**379**) can cyclize further via ionization of pyrophosphate moiety and Δ^8 (17) participation, generating the tricyclic species **380** (pimarane type), in which a new asym-

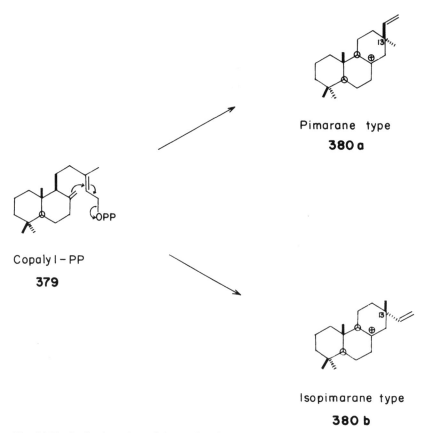

Fig. 8.1.21. Cyclization of copalyl pyrophosphate

metric center (at C-13) has been created (Fig. 8.1.21). Pimaranes with epimeric C-13 stereochemistry (cf. **380a**, *380b*) freely occur in nature. Further stereochemical diversity results from the absolute stereochemistry of the bicyclic **379**, the enantiomeric pair generating the corresponding enantiomric series of pimaranes. Although, pimaranes with "normal" (steroidal) stereochemistry predominate, several *ent*-pimaranes are known. Stereochemistry of cyclization of **379** to **380** has been elucidated for biosynthesis of certain pimaranes, and it has been demonstrated that the allylic displacement by which ring C is formed, takes place with overall *anti*-stereochemistry (76, 130). As might be expected on biogenetic considerations, pimaranes mostly have an oxygen function at C-8 or have one terminus of an olefinic linkage on this carbon. In earlier work, diterpenes with gross pimarane skeletons were characterized by dehydrogenation to pimanthrene (**414**) (350).

There are at present over 100 pimaranes known and many of these have been isolated from heartwood and wood resins. (+)-Pimaradiene (**415**) and (+)-pimarinol (**416**) occur in wood resin of *Pinus sylvestris* (141). The corresponding aldehyde (**417**) is a component of *P. elliottii* gum resin (326), and (+)-pimaric acid (**418**) has been isolated from oleoresin of *P. palustris* (182, 238). (−)-Pimaradiene

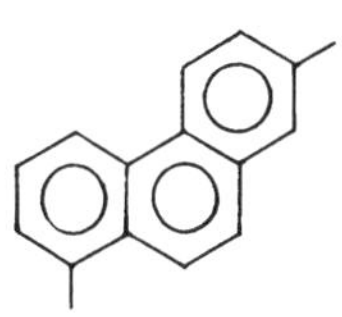

Pimanthrene
414

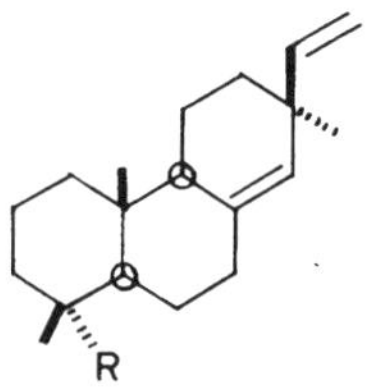

415: (+)-Pimaradiene
(R = CH$_3$)

416: (+)-Pimarinol
(R = CH$_2$OH)

417: (+)-Pimarinal
(R = CHO)

418 (+)-Pimaric acid
(R = COOH)

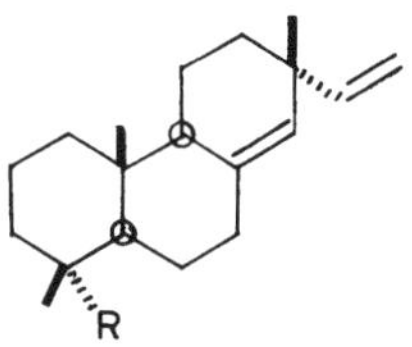

419: (−)-Sandaracopimaradiene
(R = CH$_3$)

420: (R = CH$_2$OH)

421: (−)-Sandaracopimaric acid
(R = COOH)

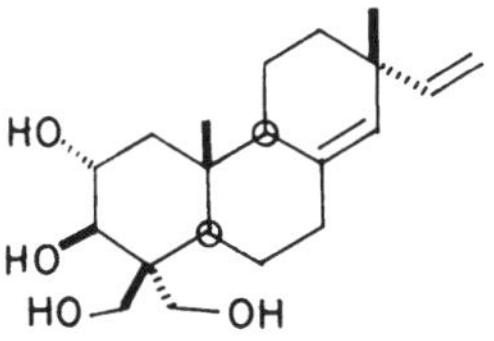

422

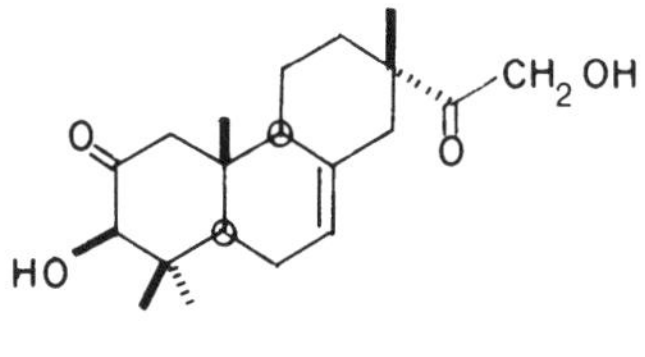

(−)-Araucarolone
423

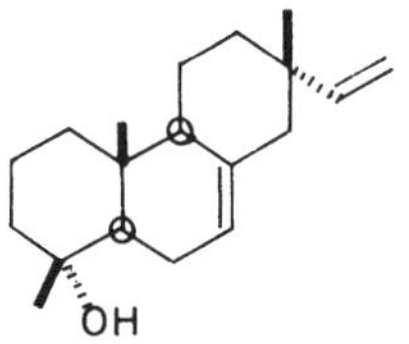

424

756 Isoprenoids

(*ent* 415) has been obtained from wood of *Erythroxylon monogynum* (224). Examples of isopimaranes are (−)-sandaracopimaradiene (419) from heartwood of *Xylia dolabriformis* (251), and an alcohol (420) and (−)-sandaracopimaric acid (421) from fossil resin of *Agathis australis* (388). In the heartwood of *Dacrydium colensoi* heavily oxygenated derivatives such as tetrol (422) have been found (80). The neutral portion of *Agathis australis* resin contains several keto alcohols such as araucarolone (423) (137). Several norpimaranes (C_{19}) in which one of the C-4 methyls has been lost are known and almost all of these have been isolated from woody tissues. As an example, mention may be made of the C_{19} alcohol 424 from heartwood of *Dacrydium bidwillii* (167).

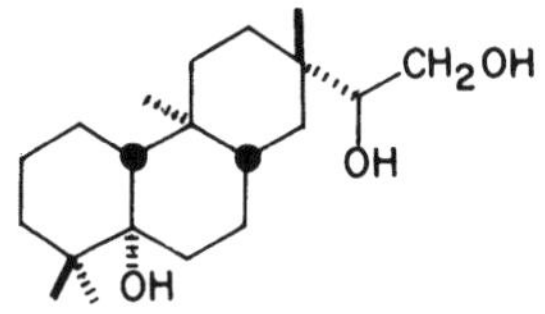

(+)-Erythroxytriol-P
425a

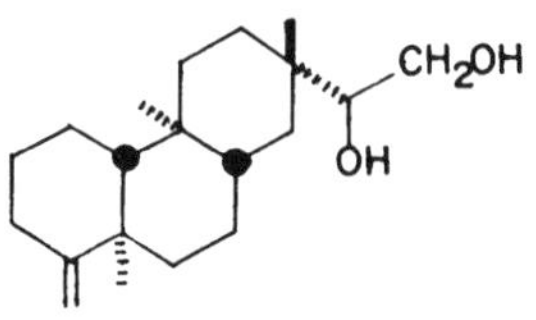

(+)-Allodevadarool
426a

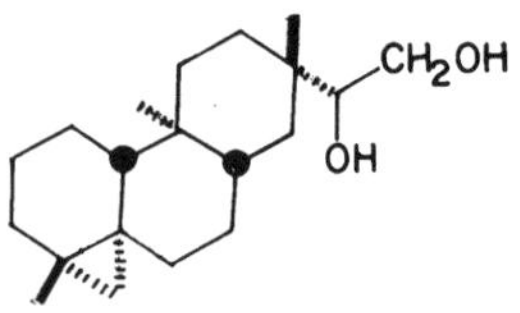

(+)-Devadarool
427a

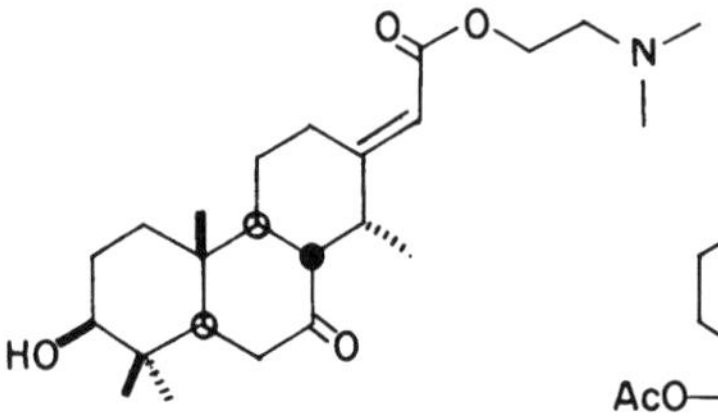

(−)-Cassaine
428a

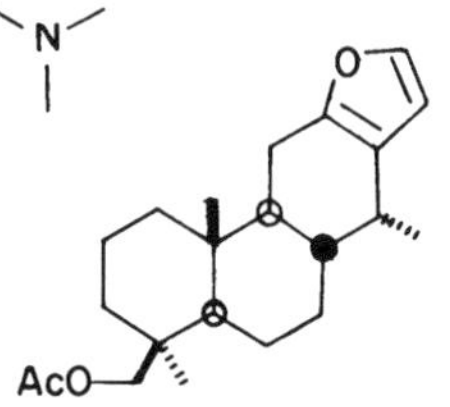

(+)-Vaucapenyl acetate
429a

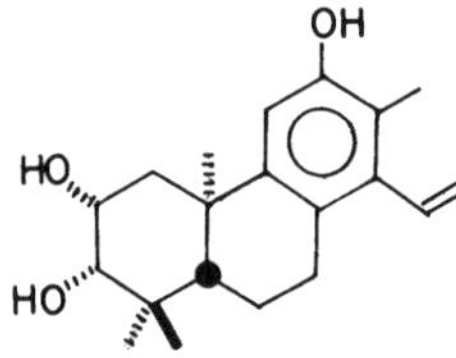

Cleistanthol
430a

A rather large variety of tricyclic diterpene types, genetically linked with pimarane, are known. Figure 8.1.22 depicts biogenesis of some of these types, which are relevant to our discussion. A hydride and a methyl shift in **380** generate the rosane type (**425**), characteristic of several fungal diterpenoids; erythroxytriol-P (**425a**) from the trunkwood of *Erythroxylon monogynum*, appears to be the only example of this rearranged pimarane found in woody tissues (106). (+)-Allodevadarool (**426a**) and devadarool (**427a**) from *E. monogynum* trunkwood are examples of products from two alternative pathways available to the fully rearranged (backbone) ion 426 (357, 358); however, very few compounds of these types are known at present. Cassanes (**429**, Fig. 8.1.17) constitute an important class of rearranged pimaranes of another type (Fig. 8.1.22), of which over 50 compounds are known at present. Cassaine (**428a**) and several related compounds, all amine esters, have been isolated from trunk-bark of *Erythrophleum* spp. (282). Voucapenyl acetate (**429a**) from heartwood of *Voucapoua macropetela* is an example of another sub-group, cassanes with a furane ring (362). Cleistanthol (**430a**), a constituent of heartwood of *Cleistanthus schlechteri*, depicts the less common isocassane type (**430**) (274).

Terpenoids 757

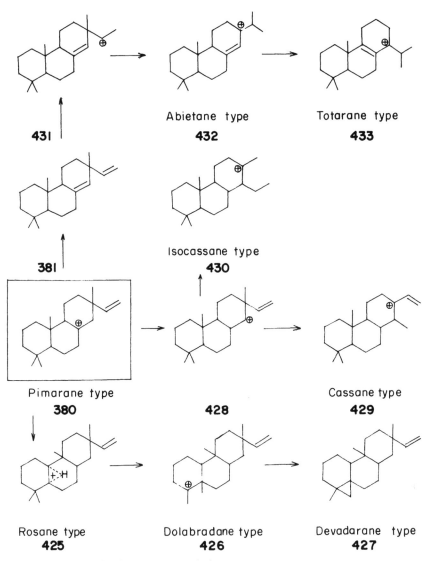

Fig. 8.1.22. Biogenesis of some rearranged pimaranes

The most important class of rearranged pimaranes is the abietane type (432), represented by abietic acid (434), the chief constituent of commercial rosin (colophony). Abietanes constitute a fairly large group of diterpenoids with more than 120 C_{20} compounds, and a variety of functionalities met within this group. In view of the stipulated intermediacy of cation 432, it is not suprising that ring C in almost 50% of the abietanes is aromatic. (−)-Abietic acid (434) has wide distribution in the genus *Pinus* (207); although pure abietic acid has been available since 1912 or so, its structure was finally settled only in 1949 (238, 350). Related compounds are (−)-abietadiene (431) from *Larix sibirica* (257) and (−)-abietinol (432) and (−)-abietinal (433) from wood resin of *Pinus sylvestris* (141).

Levopimaric acid (**435**) from oleoresin of *P. palustris* (123) and neoabietic acid (**436**) a component of *Agathis microstachya* oleoresin (**81**), are other related resin acids. Ferruginol (**437**), an abietane phenol, occurs in the resin from *Podocarpus ferrugineus*; this compound has good antimicrobial activity (60). Several polyphenols and quinones based on abietane system are known (378), but these have been isolated as pigments of leaf glands of *Plectranthus* and *Coleus* spp. A related compound is the pale orange royleanone (**438**) with cytotoxic activity, from roots of *Inula royleana* (249). (+)-Jolinolide-B (**439**), from fresh roots of *Euphorbia jolkini*, is an example of a minor group of abietanolides (400).

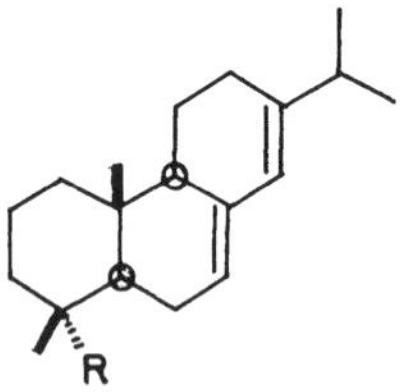

431:(−)− Abietadiene
(R=CH$_3$)

432:(−)− Abietinol
(R=CH$_2$OH)

433:(−)− Abietinal
(R=CHO)

434:(−)− Abietic acid
(R=COOH)

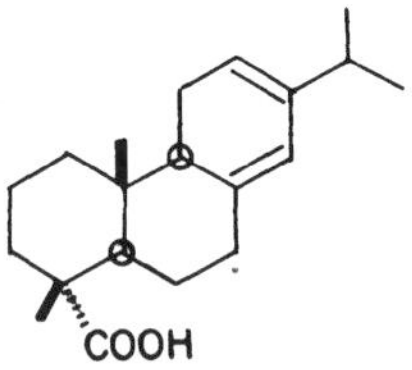

(−)−Levopimaric acid
435

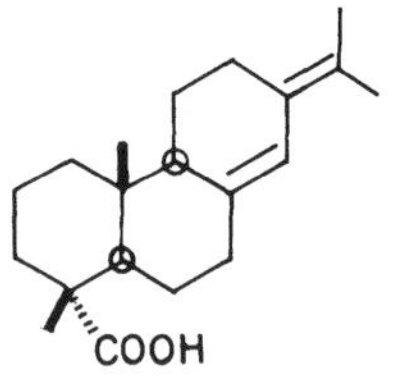

(+)−Neoabietic acid
436

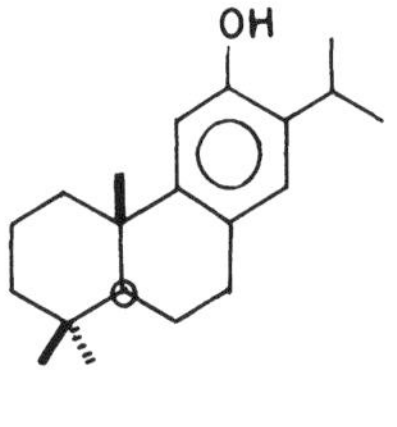

(+)−Ferruginol
437

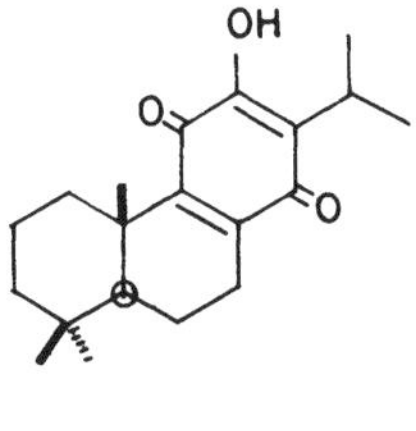

(+)− Royleanone
438

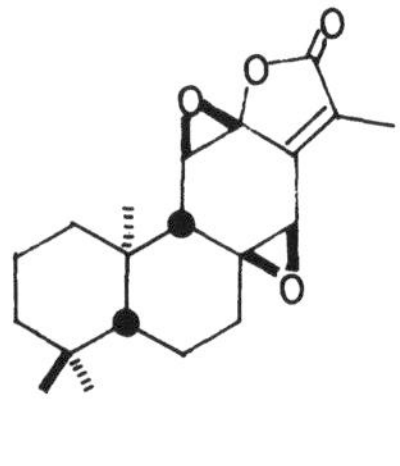

(+)−Jolkinolide− B
439

A significant number of norabietanes isolated mostly from woods, resins, and roots are known, but it will be possible here to refer only to podocarpic acid (**440**), the major constituent of the resin from *Dacrydium cupressinum* (228). Totarol (**441**), a rearranged abietane, occurs in the heartwood of *Podocarpus totara* (228) several related compounds are known, including the orange quinone, (+)-maytenoquinone (**442**), from the root-bak of *Maytenus dispermus* (268). The antileukemic compound, (−)-triptolide (**443**) from roots of *Tripterygium wilfordii*, represents a minor group of differently rearranged abietanes (247).

(+)-Podocarpic acid
440

(+)-Totarol
441

(+)-Maytenoquinone
442

(−)-Triptolide
443

(−)-Ginkgolide−B
444

Ginkgolides (e.g., ginkgolide-B, **444**), from the root-bark of *Ginkgo biloba*, have a novel carbon skeleton. Biosynthetic experiments have shown that their genesis apparently involves a seconor rearranged pimarane and C-methylation with methionine (293).

(d) *Tetracyclic: Beyeranes, kauranes, atisanes and related types.* As depicted in Fig. 8.1.19, the pimarenyl cation (**380**) cyclizes further to generate the tetracyclic

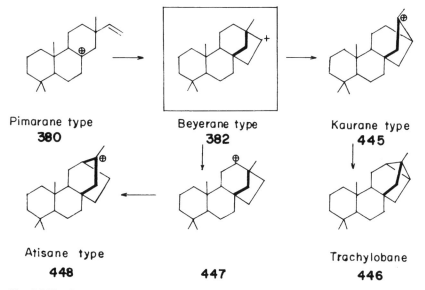

Fig. 8.1.23. Biogenesis of some tetra- und pentacyclic diterpenoids

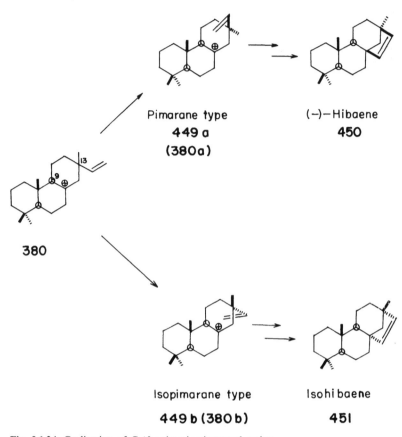

Fig. 8.1.24. Cyclization of C-13 epimeric pimarenyl cation

cation **382**, which is considered to be the progenitor of a large family of tetra- and pentacyclic diterpenoids (411). This relationship is shown in Fig. 8.1.23 where classical carbonium ions are shown for the sake of clarity. Although the scheme depicts trachylobane (**446**) as arising from the kaurane cation (**445**), its genesis is equally feasible from ions **382**, **447**, or **448** by a similar 1,3-proton elimination. It may be noted that if in cation **380** the vinyl side chain is *syn* to C-9 hydrogen, a different set of structures stereoisomeric (at C-8 and C-13/C-12) to those shown in Fig. 8.1.23 would emerge, as exemplified for hibaene and isohibaene in Fig. 8.1.24. Further stereochemical diversity results from absolute stereoisomerism. However, to date, with the exception of kaurene, kauranol, and hibaene, which are known to occur in both enantiomeric forms, all remainling tetracyclic and pentacyclic diterpenes of this class occur only as antipodes to the "normal" steroid-type series. Biosynthesis of kaurene has been investigated in some detail, essentially in connection with the biosynthesis of gibberellins.

Next to the bicyclic labdanes and clerodanes, kauranes (Fig. 8.1.17), as presently constituted are most abundant. However, these compounds have been mostly isolated from leaves, seeds, or whole plants or as fungal metabolites. (−)-Kaurene (**452**) has been obtained from the leaf oil of *Agathis australis*, while its

(+)-antipode occurs in the leaf oil of New Zealand 'matai' (*Podocarpus spicatus*) (63). (−)-Kauranol (**453**) is present as a minor constituent of wood extractive of *Erythroxylon monogynum* (224). A large number of hydroxy derivatives are known, some of which are highly oxygenated; an example from woody tissues is pentol **454** from rhizomes of *Pteris cretica* (287). Stevioside (**455**), from leaves and twigs of *Stevia rebaudiana*, is one of the sweetest natural sweeteners (239). Grandifloric acid (**456**) has been isolated from wood resin of *Espeletia grandiflora* (62); several 15-acyl derivatives of **456** are known. Many kaurenolides have been isolated, but, apparently none from woody parts of plants. A few alkaloids based on kaurane carbon framework have been isolated, mostly from bark of *Garrya* spp. − e.g., (−)-garryine (**457**) from the bark of *Garrya veatchii* (313).

(−)-Kaurene
452

(−)-Kauranol
453

454

The beyerane group (**382**) was first recognized in 1962, when beyerol cinnamate from leaves and stems of *Beyeria leschenaultii* was formulated as **458** (217). Since then, many more members of this class have been isolated from nature and, at present, about 45 such compounds are known. (+)-Hibaene (*ent* **450**) is a minor component of the wood of *Erythroxylon monogynum*, which contains several related compounds (225). (−)-Hibaene (**450**) occurs in the leaf essential oil of *Thujopsis dolabrata* (236). The atisane group of tetracyclic diterpenes arises from ion **382** after a 1,3-hydride shift and Wagner-Meerwein rearrangement to cation **448** (Fig. 8.1.23). This group derives its name from the alkaloid atisine (**459**), the major alkaloid from the roots of the Indian plant *Aconitum heterophyllum*, known since 1877; other related alkaloids based on atisane are known (313). The first non-nitrogenous member of this group was reported in 1965, when isolation and structure elucidation of atisirene (**460**) from the wood of *Erythroxylon monogynum* was described (356). Since then, other non-nitrogenous atisanes have been found. Trachylobanes (**446**) are a rather small group of some 10 compounds, first discovered in 1965 (205). The only presently known example of their occurrence in woody tissues is the isolation of (−)-trachylobane (**461**) from the essential oil of tree branches of *Araucaria araucana* (65).

Of the isomeric group of tetracyclic diterpenoids originating from the isopimarenyl cation (Fig. 8.1.24), only two types have been encountered so far. Isohibaene (**451**) from *Chamaecyparis nootkatensis* (93) has already been mentioned. The other type is represented by phyllocladene, known since 1910 (350), but correctly characterized only in 1959. (+)-Phyllocladanol (**462**) from the wood of *Cryptomeria japonica* is an appropriate example for the present discussion (64). Only a few compounds in each class are known at present, and all of these belong to the "normal" series.

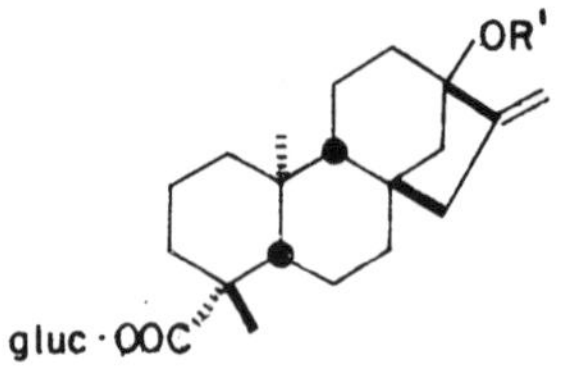

[R'= β-gluc(2-1)-β-gluc]
(−)- Stevioside
455

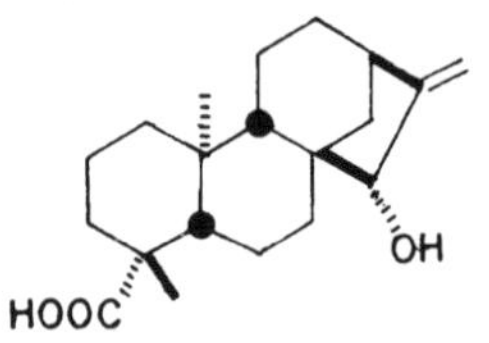

(−)- Grandifloric acid
456

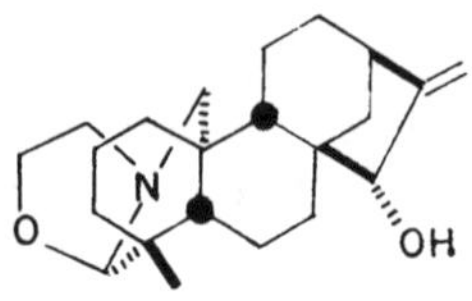

(−)- Garryine
457

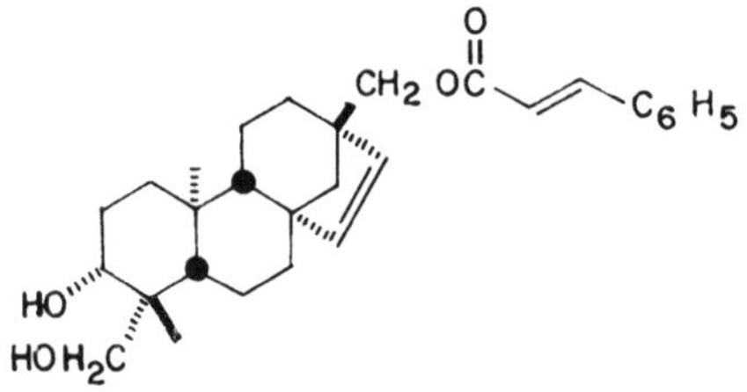

(+)- Beyerol-17-cinnamate
458

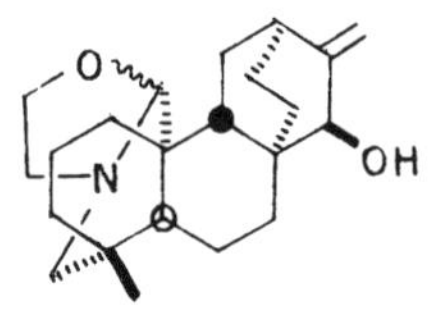

(−)- Atisine
459

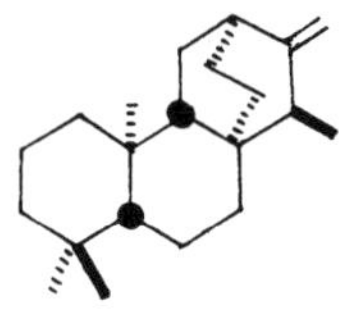

(−)- Atisiren
460

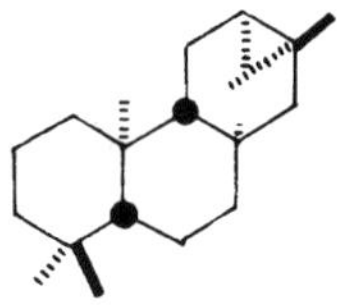

(−)- Trachylobane
461

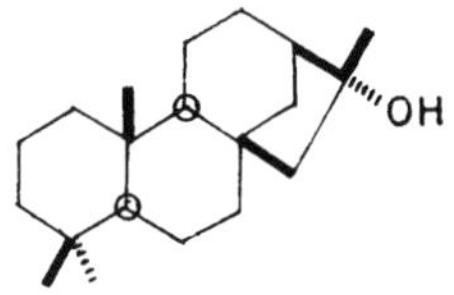

(+)- Phyllocladanol
462

On the basis of strucstural similarity, it has been postulated that certain kaurane and atisane derivatives serve as substrates for elaboration of an array of diterpene types. The reactions involved are further cyclizations and/or skeletal rearrangements, and products include complex, often highly functionalized, tetra- and pentacarbocyclic diterpenoids such as gibberellins, the grayanotoxins, and the aconite alkaloids. Because of their biological importance, only gibberellins will be discussed here. Grayanotoxins (292) have invariably been isolated only from leaves and flowers and thus fall outside the scope of the present work. Aconite alkaloids have already been covered in Chap. 5.

Gibberellins (GA) (Fig. 8.1.17) represent a most important class of biologically active diterpenoids, first discovered around 1940 as secondary metabolites of the fungus *Gibberela fujikuroi*, the causative organism for the baka-nae (foolish seedling) disease, responsible for anomalous growth of rice seedlings. Although the active factor from the causative fungus was isolated by 1940 and the name gibberellin A coined for it, it was later (1955) shown to be a mixture of three compounds GA₁ (**468**), GA₂ (**469**), and GA₃ (**470**), the last compund being identical with gibberellic acid isolated a year earlier by another group (61, 262). At present over 70 gibberellins are known, a majority of which are C_{19} compounds lacking in C-20. Gibberellins are phytohormones (Sect. 8.1.4), and are considered to occur in all green plants. Although these compounds have been isolated from fungal culture broths or from immature seeds, pods and young shoots of higher plants, their occurrence in roots of certain plants has been noted (164, 237). Gibberellic acid (**470**) is produced commercially by a fermentation process for use as plant growth regulator. Biosynthesis of gibberellins has attracted much attention, and various steps in their elaboration from *R*-mevalonic acid are now well-understood, not only in terms of chemical and stereochemical detail but often at the enzyme level. *ent*-Kaurene (**452**) has been established as an obligatory intermediate. Figure 8.1.25 depicts biosynthesis of GA₁₂-aldehyde (**467**), the key intermediate for the elaboration of various gibberellins (164, 185).

(e) *Cyclization from the tail end.* Diterpene skeletal types arising from this mode were limited until a decade ago but have rapidly proliferated since then. However, most of these types have been encountered in work on marine organ-

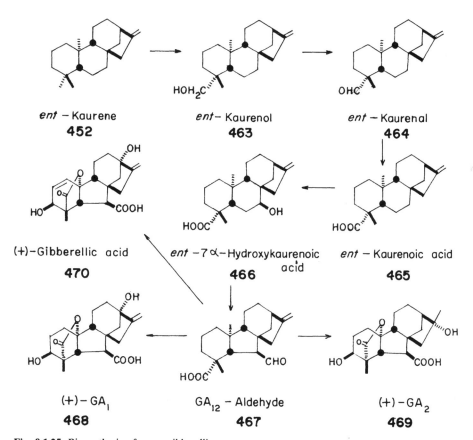

Fig. 8.1.25. Biosynthesis of some gibberellins

764   Isoprenoids

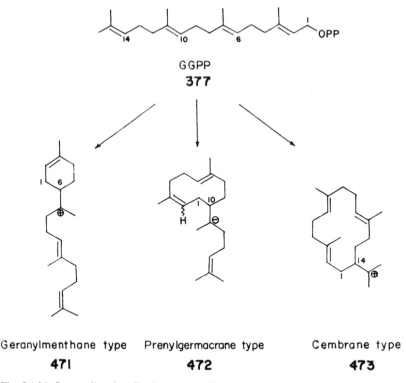

Fig. 8.1.26. Some tail-end cyclization modes of GGPP

isms (144) and hence will not be discussed here. Those relevant to our present discussion are shown in Fig. 8.1.26.

Only a few compounds have the carbon skeleton of the geranylmenthane cation (**471**), mostly isolated from genus *Helichrysum* – e.g., peroxide **474** from roots of *H. acutatum* (40). Compounds of the prenylgermacrane type (**472**) have not, so far, been isolated from higher plants; however, the bluish-red antibiotic biflorin (**475**), occurring in the roots of *Capraria biflora* (105), is an example of a type arising from further cyclization of **472** (cf. cadalanes in Sect. 8.1.3.3.2, Sesquiterpenoids).

Of the diterpenoids arising from tail end cyclization, cembranes (**473**) involving 1,14-cyclization of GGPP are most numerous. This class was first recognized in 1962, with the structure elucidation of cembrene (thunbergene, **476**), a component of *Pinus* spp. Since then, this group has grown rapidly and at present some 120 compounds of this class are known. They appear to be distributed widely, and certain exudates, tobacco, and marine invertebrates are rich sources. A comprehensive review was published in 1979 (410). As examples of cembranes isolated from woody tissues, cembrene-A (**477**) and mukulol (**478**) from gum-resin of *Commiphora mukul* (319) and incensole oxide (**479**) from *Boswellia carteri* resin (230) may be cited. Although many lactones (cembranolides) are known to occur in nature, apparently none has, so far, been isolated from woody parts of plants.

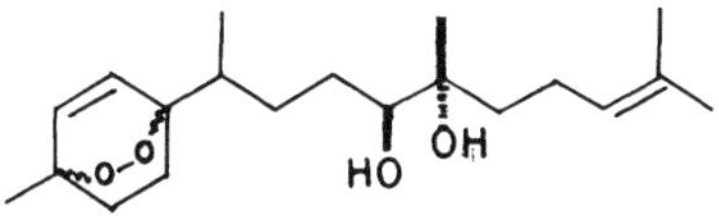

474

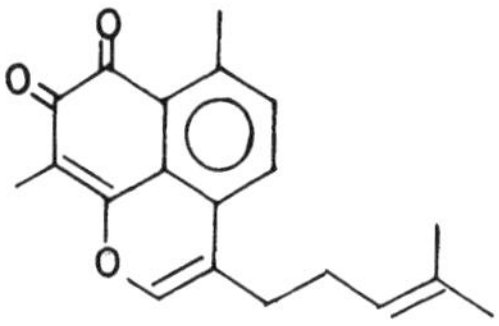

Biflorin
475

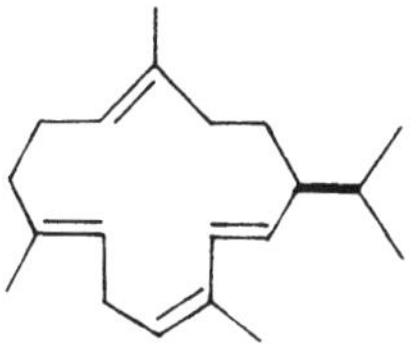

(+)-Cembrene
476

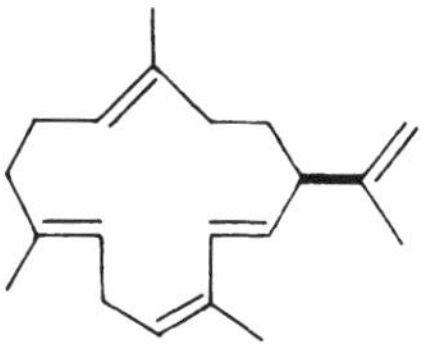

(−)-Cembrene−A
477

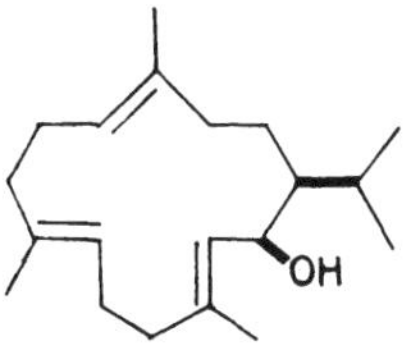

(+)-Mukulol
478

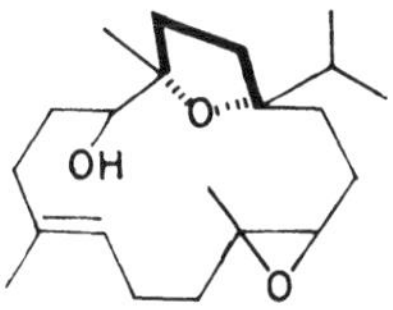

Incensole oxide
479

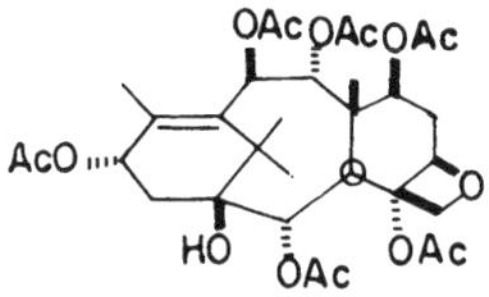

Baccatin−IV
480

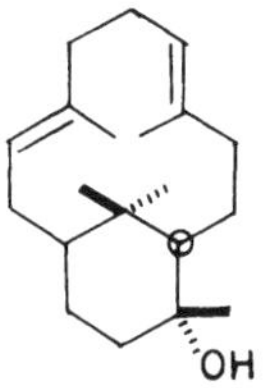

(+)-Verticillol
481

Heartwoods of certain *Taxus* spp., especially that of *T. baccata*, contain several toxic diterpenoids, such as baccatin-IV (**480**) (264). About 30 compounds are known at present, and are invariably highly functionalized. The tricyclic framework of these compounds is readily rationalized by further cyclization of cembrene cation **473** (292). Verticillol (**481**), an extractive from the wood of *Sciadopitys verticillata*, corresponds to the bicyclic stage on the pathway to the taxane type (227).

Several species of the plant families Euphorbiaceae and Thymelaeaceae especially those of the genus *Euphorbia*, produce a complex group of tri- and tetracyclic diterpenoids, of which over 100 are known. These compounds are often highly oxygenated and frequently occur esterified with a variety of fatty or aromatic acids. Many of these compounds are biologically active, being irritants and co-carcinogens. These diterpenes represent several skeletal types and their biogenetic relationship with casbene (**482**), which arises from cembrene cation (**473**) by a 1,3-deprotonation, has been postulated. Most of them have been isolated from seeds or latexes. Examples of tricyclic compounds from woody parts of plants are (+)-jolkinol-D (**483**) from roots of *Euphorbia jolkini*, (−)-jatrophatrione (**484**) from roots of *Jatropha macrorhiza*, and (+)-daphnetoxin

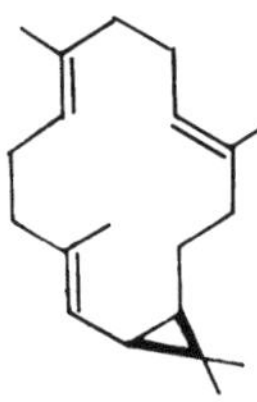

Casbene
482

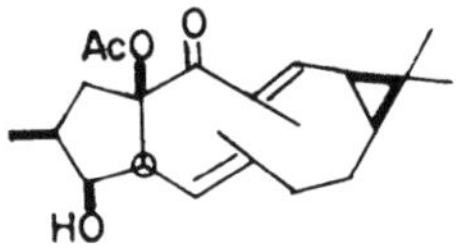

(+)-Jolkinol-D
483

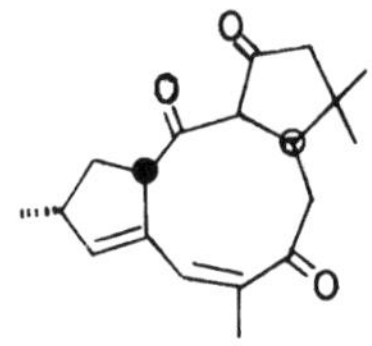

(−)-Jatrophatrione
484

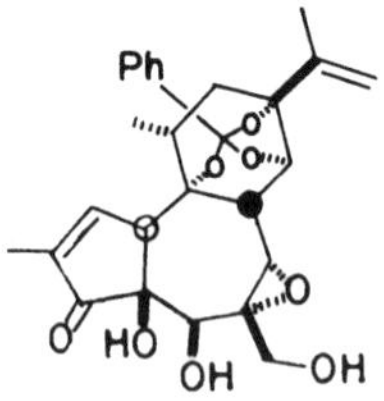

(+)-Daphnetoxin
485

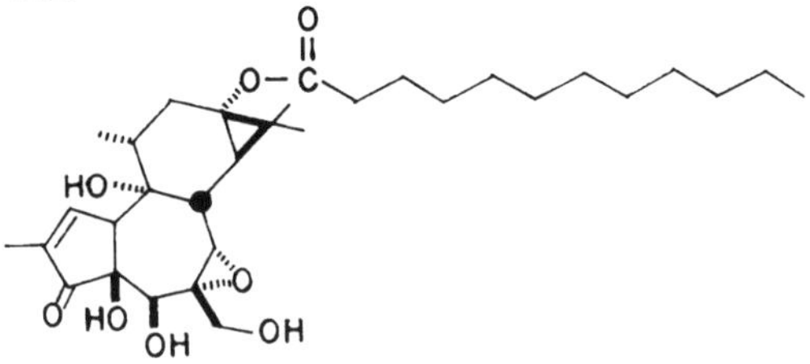

Baliospermin
486

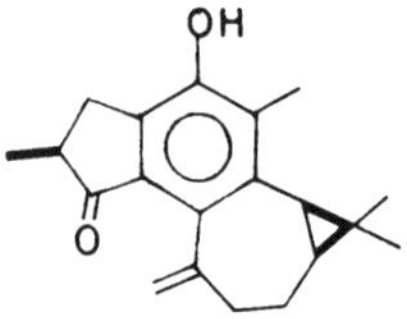

Jatropholone-A
487

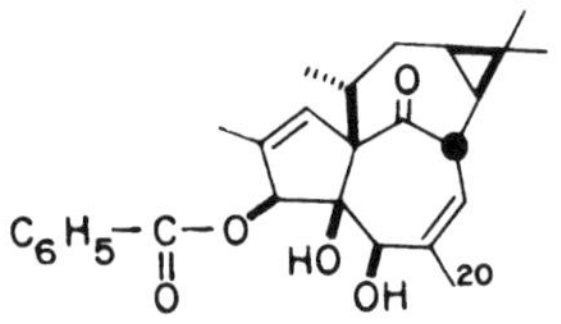

20-Deoxyingenol-3-benzoate
488

(**485**) from bark of *Daphne mezereum*. Baliospermin (**486**) from roots of *Baliospermum montanum*, jatropholone-A (**487**) from roots of *Jatropha gossypifolia*, and 20-deoxyingenol-3-benzoate (**488**) from dried roots of *Euphorbia kansui* provide examples of tetracyclic types. An excellent recent review of these diterpenes is available (143).

8.1.3.5 Sesterterpenoids

The C_{25} terpenoids, sesterterpenes, known since 1957 (299), were first characterized in 1965 with the structure elucidation of ophiobolin (**489**), a phytotoxic metabolite of the fungus *Ophiobolus miyabeanus*. This was almost immediately followed by the formulation of gascardic acid (**490**), another sesterterpenoid, from the secretion of the insect *Gascardia madagascariensis* (89 vol. 4). Since then, several other sesterterpenoids have been obtained, but invariably from insect secretions, fungi, ferns, lichens, or marine organisms. Their occurrence in higher

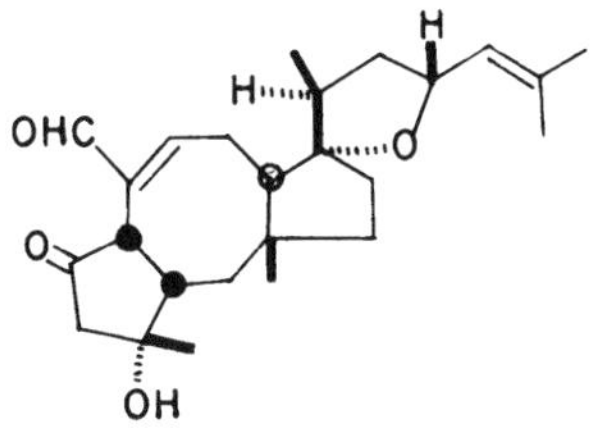

(+)-Ophiobolin
489

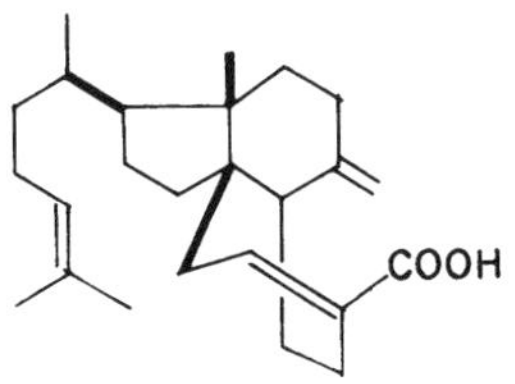

( )-Gascardic acid
490

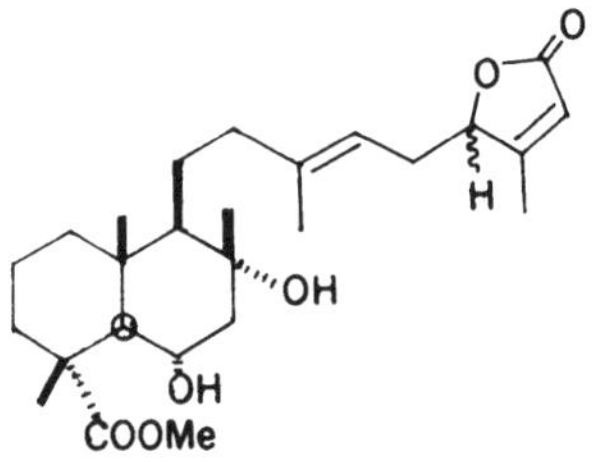

Salvileucolide methyl ester
491

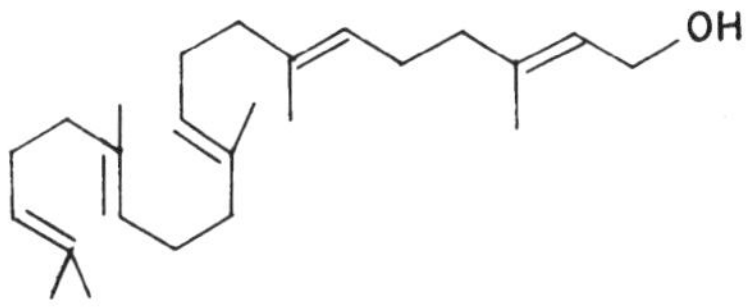

Geranylfarnesol
492

plants appears to be highly restricted; only a few compounds of this class have been isolated so far, and none from woody tissues (280, 335, 364). Salvileucolide methyl ester (**491**) from aerial parts of *Salvia hypoleuca* provides an example of a sesterterpene from higher plants.

Sesterpenoids constitute a rather small group of 45 compounds belonging to eight skeletal types. Geranylfarnesyl pyrophosphate is considered as their biosynthetic precursor, and this is supported by experimental evidence. Geranylfarnesol (**492**) itself has been isolated from the wax of the insect *Ceroplastes albolineatus*. A comprehensive review of sesterterpenes was published in 1977 (111).

8.1.3.6 Non-Steroidal Triterpenoids

8.1.3.6.1 Distribution

Of all terpenes, triterpenoids are the most distributed in nature, and are ubiquitous among the dicotyledons. Table 8.1.5 lists higher plant families especially prolific in producing triterpenoids. On the basis of isolation studies, ursolic acid, β-amyrin and friedelin have been most commonly encountered. Lists of occurrences of triterpenoids at compound level are available (86, 155, 207, 208, 229, 245, 309).

Cyclic triterpenoids occur only in "normal" configuration. The usefuness of triterpenoids in plant systematics has been assessed (155, 317). Their use as chemotaxonomic markers in Leguminosae (181) and Euphorbiaceae (306) has received special attention.

Table 8.1.5. Plant families prolific in producing triterpenoids

Gymnospermae	Angiospermae
Pinaceae	Monocotyledoneae
	Gramineae
	Dicotyledoneae
	Amarantaceae
	Apocynaceae
	Aquifoliaceae
	Araliaceae
	Betulaceae
	Burseraceae
	Cactaceae
	Caryophyllaceae
	Celastraceae
	Cucurbitaceae
	Dipterocarpaceae
	Ericaceae
	Euphorbiaceae
	Lecythidaceae
	Leguminosae
	Meliaceae
	Moraceae
	Myrtaceae
	Rutaceae
	Simaroubaceae

8.1.3.6.2 Structural Types

Presently, triterpenoids constitute the third largest family of terpenoids with over 1500 compounds (excluding nor derivatives) embracing some 40 skeletal types. Because they are biogenetically and structurally related to the biologically important steroids, triterpenoids have attracted considerable attention and, together with steroids, have provided an experimental basis for the development of principles of conformation analysis. The chemistry of triterpenoids has been reviewed several times (38, 89, 151, 176, 245, 304, 309, 351). A special review covers synthetic efforts in this area (9). Figure 8.1.27 lists triterpene skeletal types of frequent occurrence in nature or those relevant to our discussion.

According to the biogenetic isoprene rule, all-*trans*-squalene (**496**) is the immediate precursor of all cyclic triterpenoids. There is now sufficient biosynthetic evidence in support of this (22, 90, 263). Biosynthesis of squalene itself, which was believed to arise from two molecules of farnesol by tail-to-tail coupling, is now known in sufficient detail. The coupling of two FPP units is catalyzed by a membrane-bound enzyme, and occurs by way of an S_N2 displacement to generate species **493** (Fig. 8.1.28), which undergoes 1,3-deprotonation to furnish presqualene pyrophosphate (**494**), which under NADPH-starved conditions, can be isolated (318, 325). Ionization of presqualene pyrophosphate triggers a rearrangement to cyclopropylcarbinol cation **495**, which, by intervention of NADPH, furnishes squalene. This pathway,which was first demostrated for liver and yeast, has now been shown to be operative in higher plants as well (90, 162, 179).

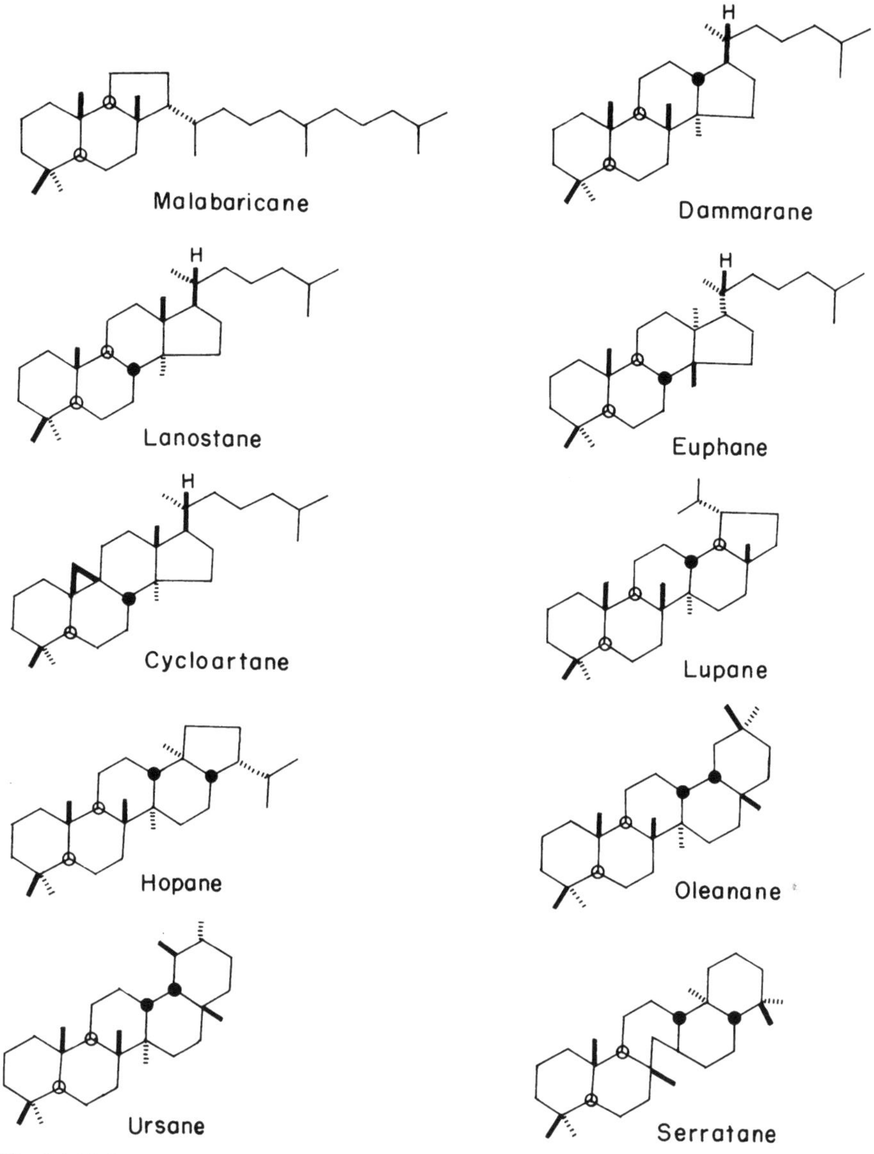

Terpenoids 769

Fig. 8.1.27. Important triterpene skeletal types

Cyclization of squalene, in the vast majority of cases, proceeds by its oxidation first to squalene 2,3-epoxide, in which the chirality at C-3 is usally *S*. The epoxidation is effected by a mixed function oxygenase requiring oxygen and NADPH. Accumulations of squalene epoxide in the presence of certain specific enzyme inhibitors has been demonstrated. Cyclization is triggered by proton attack on the epoxide. In many protozoa and ferns, direct proton-initiated cyclization of squalene to triterpenoids occurs. In some of these cases, incorporation of labelled squalene but not of labelled squalene 2,3-epoxide has been demonstrated (90).

770 Isoprenoids

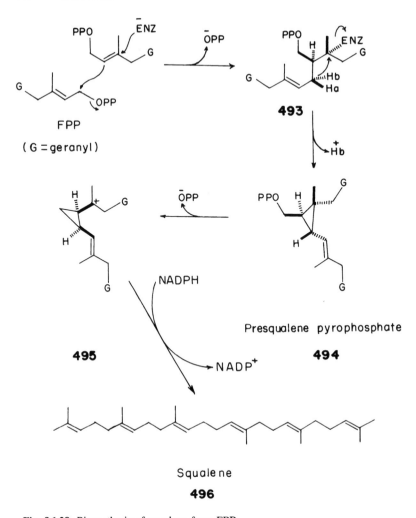

Fig. 8.1.28. Biosynthesis of squalene from FPP

Cyclic triterpenes fall into two main categories, tetracyclic and pentacyclic, and these arise from the generalized ion **497** shown in Fig. 8.1.29; two other minor cyclization pathways leading to a tricyclic and a double bicyclic system, are also shown. The outcome of the cyclization reaction is dictated by the folding of squalene on the enzyme template. The cyclization process itself is considered to proceed by a concerted, antiparallel addition mechanism and at one time was believed to be non-stop (100, 323). However, the concept of concertedness involving many centers has come under criticism (415). Some evidence has been put forward that these reactions indeed may not be non-stop, and intervention of a series of conformationally rigid carbocyclic cationic intermediates has been suggested (383). The biosynthetic evidence available so far cannot distinguish between the two situations.

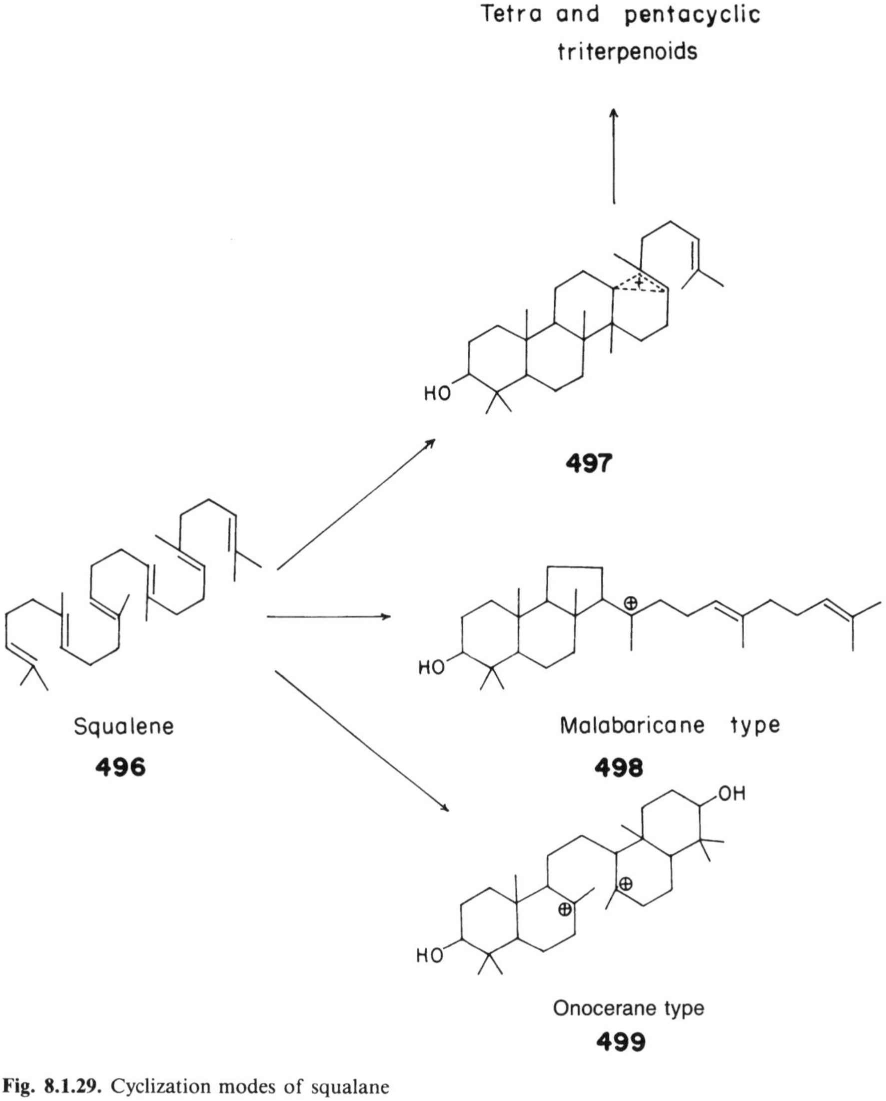

Tetra and pentacyclic triterpenoids

Fig. 8.1.29. Cyclization modes of squalane

A comprehensive rationale for structural and stereochemical outcome of squalene cyclization in terms of conformation dictated by the cyclase has been build up by the Zürich school. It provides a convenient basis for discussing various triterpene structures (263, 323). These conformations are described in terms of section-wise folding of the squalene chain into a chair (C), or boat (B) conformation or a part remaining unfolded (U). The following discussion of triterpenoids relevant to wood chemistry is based on these considerations.

(a) *C-B-C-B-U: Lanostane and related types.* This mode of cyclization leads to the lanostane, protostane, and cycloartane group. These comprise the so-called steroidal triterpenoids, and are discussed in Sect. 8.2.

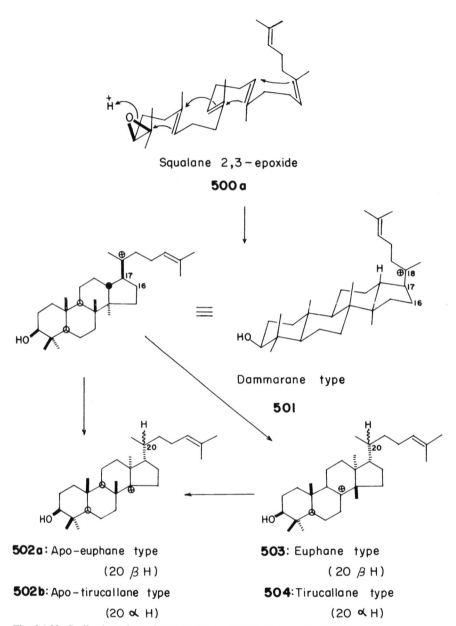

Fig. 8.1.30. Cyclization of squalane epoxide to dammarane and related types

(b) *C-C-C-B-U: Dammarane and related types.* Cyclization of squalene epoxide in C-C-C-B-U conformation (**500a**) generates a cation, that is not only the parent ion of the dammaranes (Fig. 8.1.27) but is also the precursor of a few related skeletal types, as depicted in Fig. 8.1.30.

Several dammarane-based triterpenoids have been obtained from woody tissues, especially from the resinous exudate (dammar) from trees of family

Dipterocarpaceae (173, 304). Examples are dammarenediol-I (**505**) present in commercial dammar resins and in resins of *Hopea odorata* and *Shorea vulgaris* (304), and ocotillone-II (**506**) from several Dipterocarpaceae spp. and from the wood of *Cabralea eichleriana* (37, 320, 384). From the roots of *Panax ginseng* several dammarane-based glycosides, some of them pharmacologically active, have been isolated – e.g., ginsenoside-Rg$_1$ (**507**) (349). Many more dammaranes are known, but most have been isolated from leaves, fruits, and seeds. Presently this group comprises some 80 compounds.

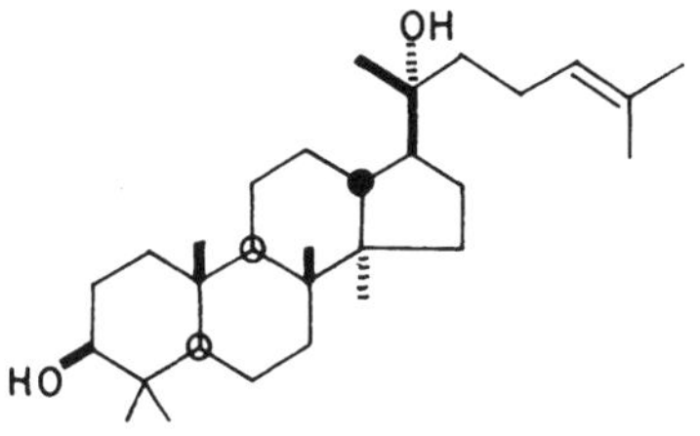

Dammarenediol – I
505

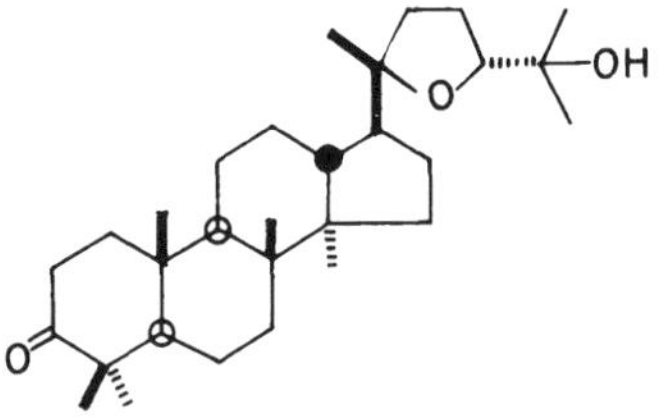

Ocotillone – II
506

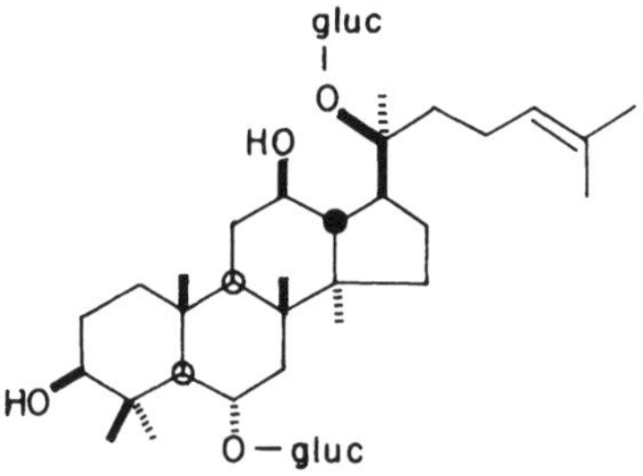

Ginsenoside – Rg$_1$
507

Euphanes (**503**) and tirucallanes (**504**) constitute an equally large group. Euphol (**503** with Δ^8) and tirucallol (**504** with Δ^8), which are epimeric at C-20 and which are the parent compounds of this group, occur in latex of several *Euphorbia* spp. (304). Gum mastic, resin from wood of *Pistacia leutiscus*, contains masticadienonic acid (**508**), which is also present along with several related compounds in the bled resin of *P. vera* (79). Manila elemi resin (from *Canarium luzonicum* and *C. commune*) contains several acids of this group, of which α-elemolic acid (**509**) with C-3αOH may be noted (116, 304). Kulinone (**510**) and related compounds occurring in the bark of *Melia azedarach* are first examples of euphanes (or tirucallanes) with oxygen functions in the D ring (97). Sapelin A (**511**) from the wood of *Entandophragma cylindricum* is an example of a heterocyclic derivative (84) and is cytotoxic (218).

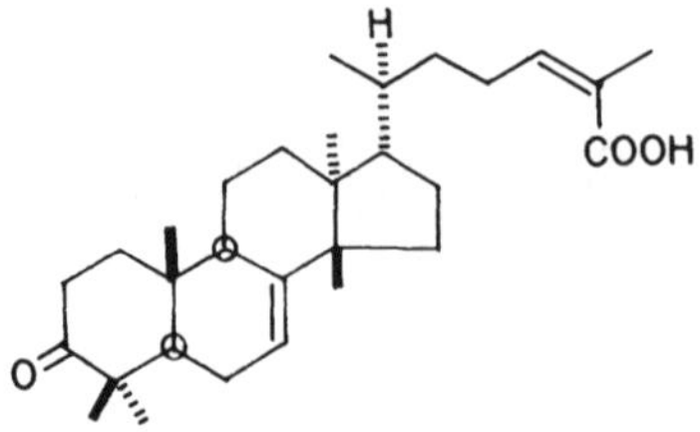

Masticadienonic acid
508

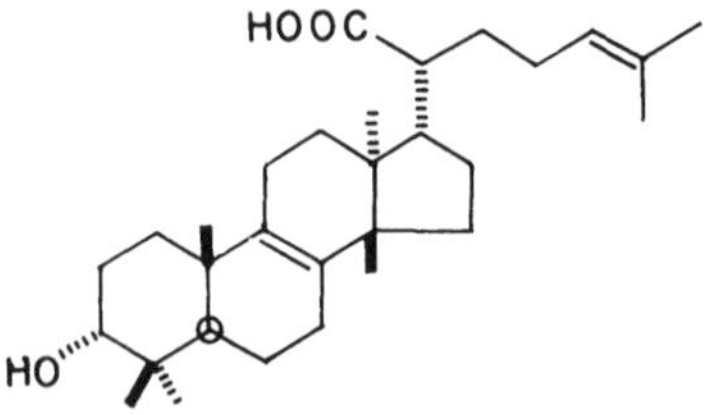

α-Elemolic acid
509

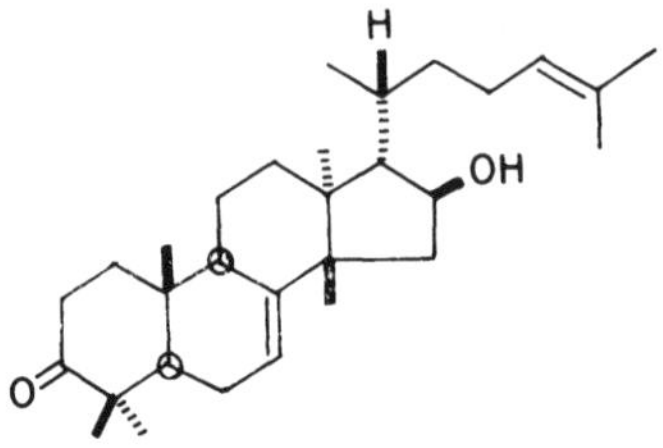

Kulinone
510

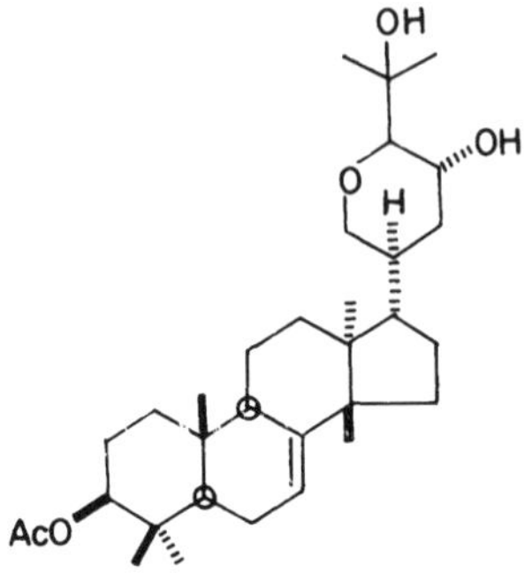

Sapelin A
511

Catabolic degradation of terpenoids is a well-recognized pathway leading to new skeletal types. With lower terpenes, this activity apparently generates a minor perturbation on the basic biosynthetic flux. However, at the tetracyclic triterpene stage, catabolism assumes a fundamental importance in that the biologically significant steroids arise from lanosterol (**512**) by this pathway. Over the years, it has become evident that nature has selected another set of related molecules (**513**) (78, 304) to fashion a flurry of degraded compounds.

These compounds fall into three main categories: *limonoids*, generally tetranor (C_{26}) compounds with the basic skeleton **514** and those derived therefrom by ring cleavage/cyclization; *quassinoids*, mainly C_{20} compounds of type **515**; and *cneoroids*, the pentanor C_{25} compounds with a framework of **516**.

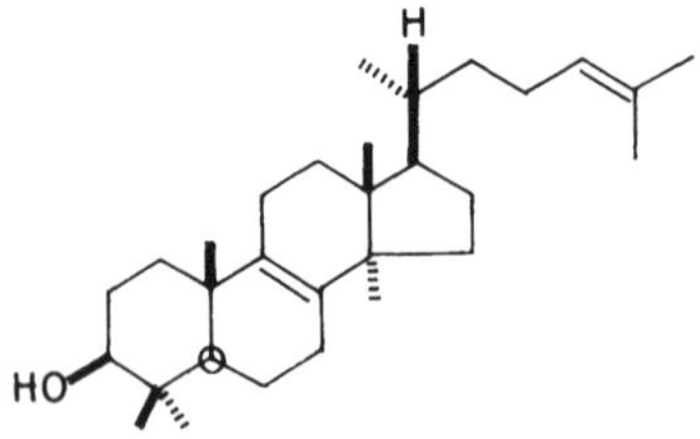

Lanosterol
512

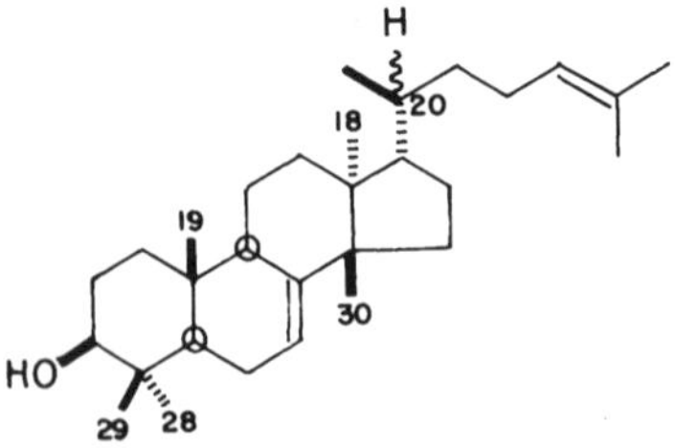

513a: Butyrospermol (20 β H)
513b: (20 S)-Butyrospermol (20α H)

Limonoids have been mostly isolated from Meliaceae and Rutaceae families, while quassinoids are typical of Simaroubaceae, and cneoroids of Cneoraceae. It may be noted that all these families belong to the order Geraniales. There is circumstantial evidence and at least some biosynthetic data (281, 316) to suggest that these compounds are degraded triterpenes. Apo-euphane/apo-tirucallane structures (502) provide an appropriate framework for generation of these types by suitable catabolic processes. Structures 502 can arise directly from C-C-C-B-U cyclization of squalene epoxide, as shown in Fig. 8.1.30. Very few triterpenes with the 502 skeleton are known. The first compound with this framework was grandifoliolenone (517), isolated from the heartwood of *Khaya grandifoliola* (108); since then, a few more compounds of this type have been isolated (107). All these compounds have a 7αOH function. Similarly, all known limonoids, quassinoids, and cneoroids have either an oxygen function at C-7 (if hydroxyl, then invariably α) or have ring B cleaved at this site. In view of these observations, it has been suggested that skeleton 502 arises from a Δ^7-euphane/tirucallane (513) by an oxidative rearrangement. There is chemical evidence to support such a transformation (281, 316). In the case of cneoranes, it may be noted that C-30 methyl of 513 becomes a part of the chain (see 516) (281). Most of these compounds are bitter and several of these have useful antifeedant activity against insects. It is not possible to discuss these compounds in any detail here, and only a brief reference to a few compounds isolated from woody tissues will be made. For more details, available reviews (89, 109, 131, 281, 316) on the subject may be consulted.

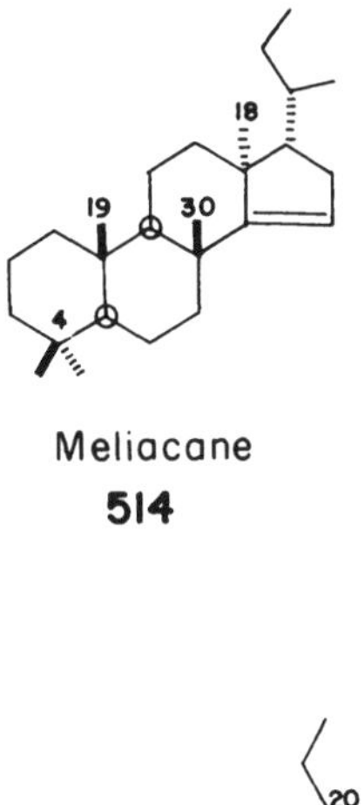

Meliacane
514

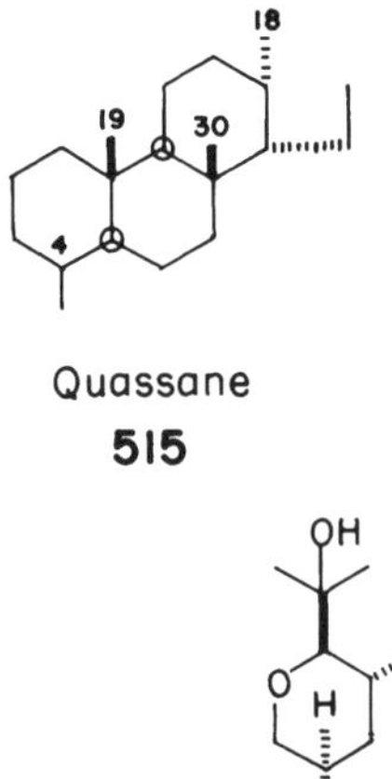

Quassane
515

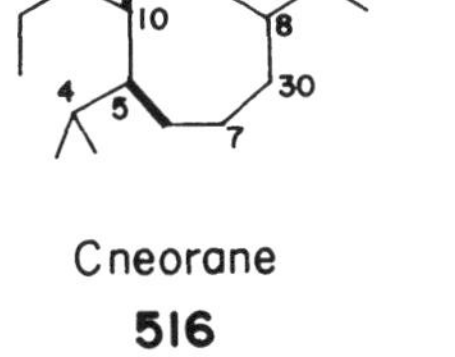

Cneorane
516

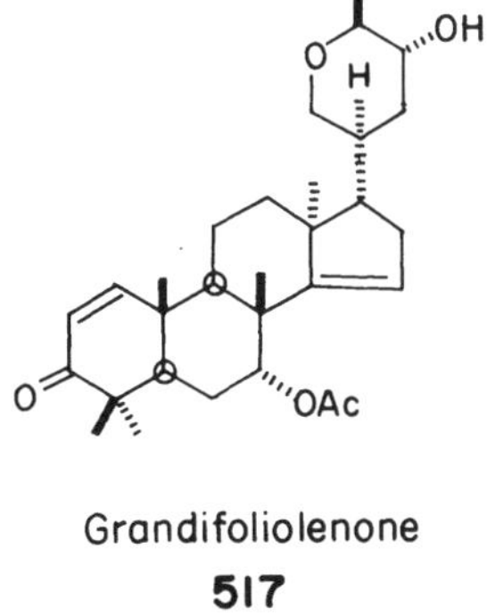

Grandifoliolenone
517

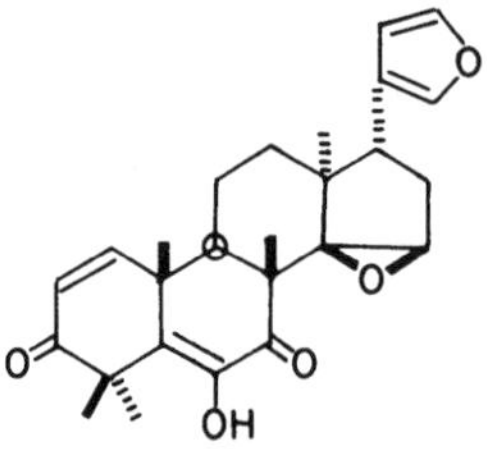

Cedrelone
518

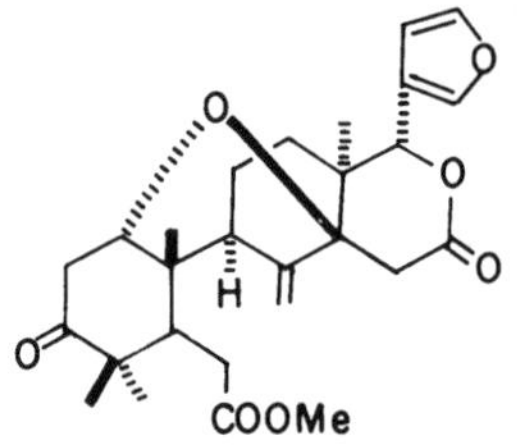

Methyl angolensate
519

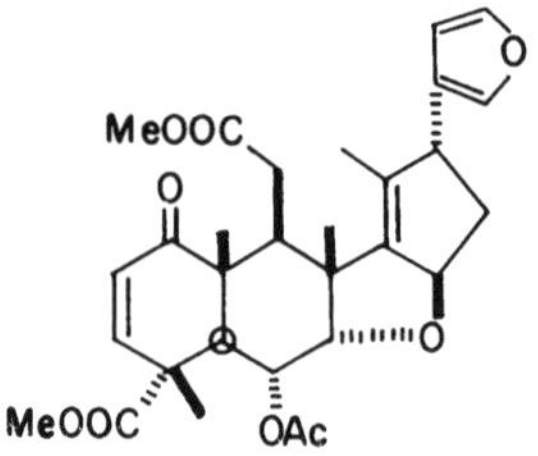

Nimbin
520

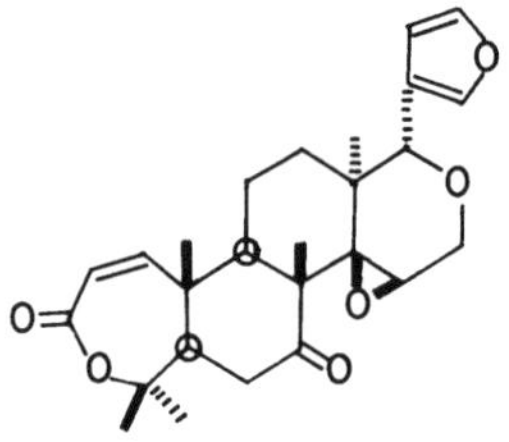

Obacunone
521

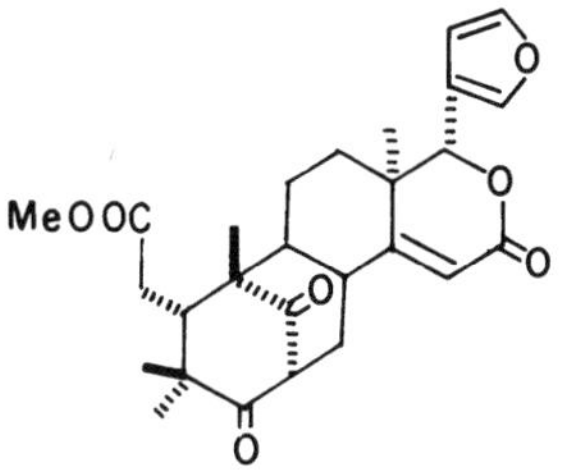

Carapin
522

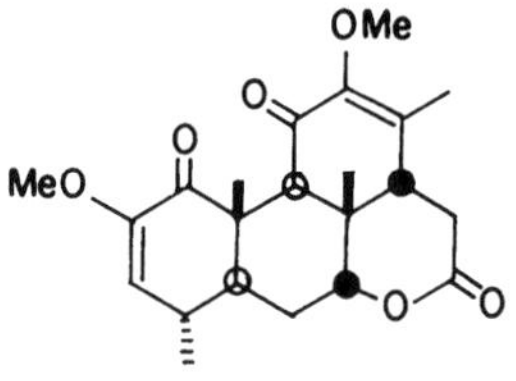

Quassin
523

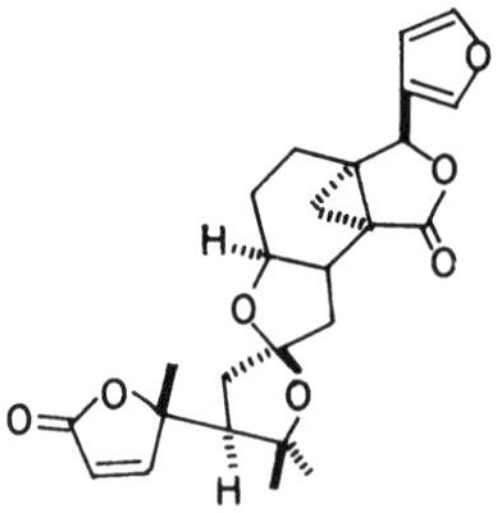

Cneorin – C
524

Cedrelone (**518**) is the major tetranortriterpenoid from heartwood of *Cedrela toona*. Methyl angolensate (**519**), a component of several West African timbers (e.g., *Khaya grandifoliola*) is an example of a B-seco limonoid. In nimbin (**520**), which is present in almost all parts of the Indian tree *Melia azedarachta*, ring C is cleaved. Obacunone (**521**) is closely related to limonin, the first tetranortriterpenoid to be characterized; abacunone was first obtained from the bark of *Phellodendron amurense*. Carapin (**522**) from heartwood of *Carapa procera* is an example of another structural variant (109, 131). Quassin (**523**), from the wood of *Quassia amara*, is typical of quassinoids (316). Cneoroids have, so far, been isolated only from the leaves and fruits of two species, *Neochamaelea pulverulenta* and *Cneorum tricoccon*; the structure of cneorin-C (**524**) from *N. pulverulenta* is shown for the sake of completeness (281).

It should be noted that a number of tetranortriterpenoids have been isolated from Cneoraceae. Protolimonoids (Δ^7-tirucallanes with highly oxygenated side-chain – e.g., **511**) have been encountered in all four families, as anticipated on the basis of biogenetic considerations.

(c) *C-C-C-B: Lupane, oleanane, ursane, and related types.* In another mode available to the tetracyclic cation **501**, expansion of the D ring is envisaged to furnish ion **525**, which can cyclize further with involvement of the end olefinic bond to generate ion **526**. This ion is the precursor to lupanes. An alternate pathway available to the lupane cation is ring-expansion to cation **527**, from which the other important pentacyclic triterpenoid types – germanicanes, taraxastanes, oleananes, ursanes and related classes – arise. These relationships are shown in Fig. 8.1.31.

Lupeol (**531**) has a wide distribution and has been isolated from the bark of many plants – e.g., *Fagara* spp. and several genera of families Apocynaceae and Leguminosae. It was first isolated from the seed peelings of *Lupinus luteus* in 1891. Bark of *Phyllanthus emblica* is reported to contain as much as 2.25% lupeol (207, 228, 229). Its C-3 epimer, epilupeol, has been isolated from exudates of several *Bersera* spp. (398). Betulin (**532**) is another important member of this class, first isolated from birch bark in 1788. Bark and wood of birches (Betulaceae) are good sources of lupeol, betulin, and related compounds; the outer cortical layer of *Betula platyphylla* has some 35% betulin. The corresponding acid, betulinic acid (**533**) is also widely distributed and is present in the bark of *Rhododendron arboreum* to the extent of 4% (207, 228). Lupeol, betulin, and betulinic acid have been shown to be active against Walker carcinoma 256 (intramuscular) tumor system (277). Presently this group of triterpenoids has some 90 compounds, and several of these occur in woody tissues.

Germanicane and taraxastane are minor groups. Morolic acid (**534**) from heartwood of *Mora excelsa* and ψ-taraxasterol (**535**) from roots of *Taraxacum officinale* are examples of these types present in woody parts of plants (228).

Oleananes constitute the largest class of triterpenoids, with more than 375 compounds. The most widely occurring compounds of this class are β-amyrin (**536**) and oleanolic acid (**537**). β-Amyrin has been known since 1890 as a constituent of Manila elemi resin, in which it co-occurs with α-amyrin (vide infra). Oleanolic acid was first obtained from leaves of the olive tree (*Olea europaea*); it occurs in the bark of *Olearia paniculata* to the extent of 2% (207, 228). A large number of glycosides based on oleanolic acid are known (23, 85). 3-Epioleanolic acid has been obtained from balsam of *Liquidambar orientalis* (206). Hederagenlin (23-hydroxyoleanolic acid) occurs in the wood of *Archas sapota*, and is the genin of several glycosides (85, 183). Arjunolic acid (2α,23-dihydroxy-

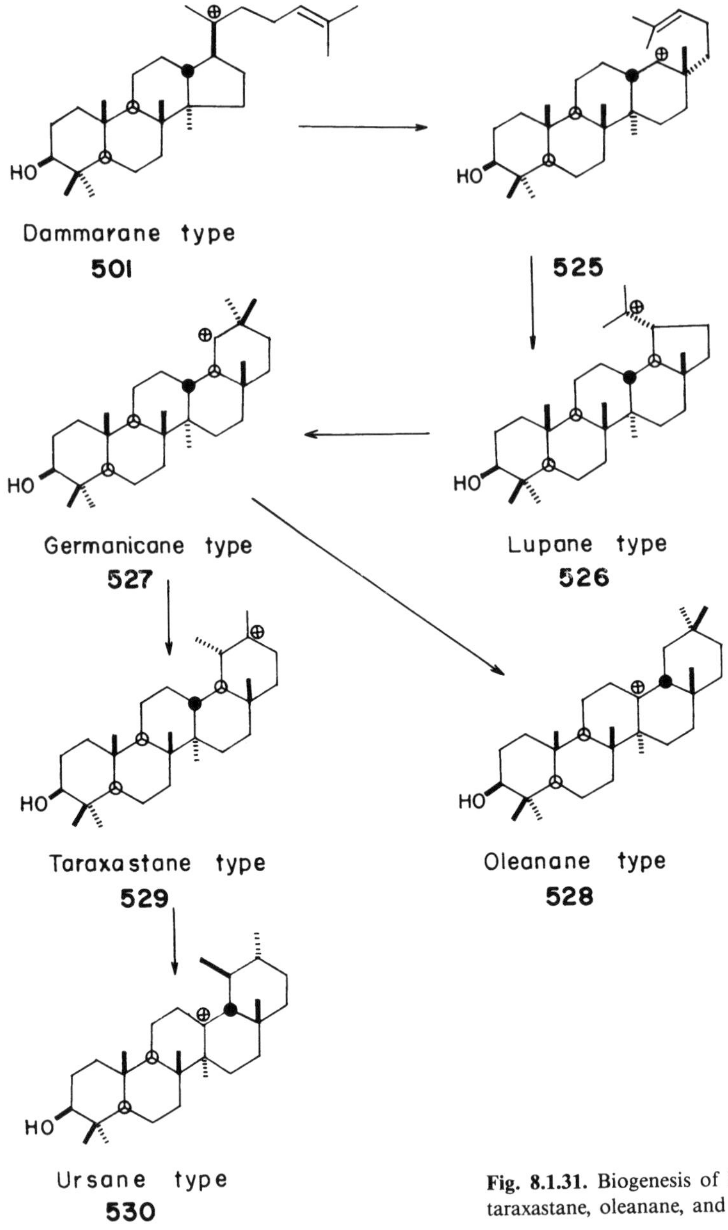

778   Isoprenoids

Dammarane type
501

525

Germanicane type
527

Lupane type
526

Taraxastane type
529

Oleanane type
528

Ursane type
530

Fig. 8.1.31. Biogenesis of lupane, germanicane, taraxastane, oleanane, and ursane types

oleanolic acid) has been isolated in a yield of 2.6% from the wood of *Terminalia arjuna* (228). Several oleananes have been obtained from licorice, the root of *Glycyrrhiza glabra*, the most important among these being glycyrrhizin, the principal flavoring amd medicinally important component of licorice. Glycyrrhizin consists of the calcium and potassium salts of glycyrrhizic acid, the diglucopyranosiduronic acid of the aglycone glycyrrhetic acid (**538**) (229, 348).

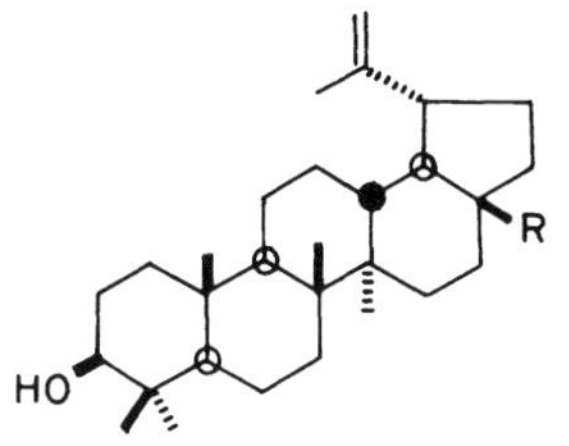

531: Lupeol (R=CH$_3$)
532: Betulin (R=CH$_2$OH)
533: Betulinic acid (R=COOH)

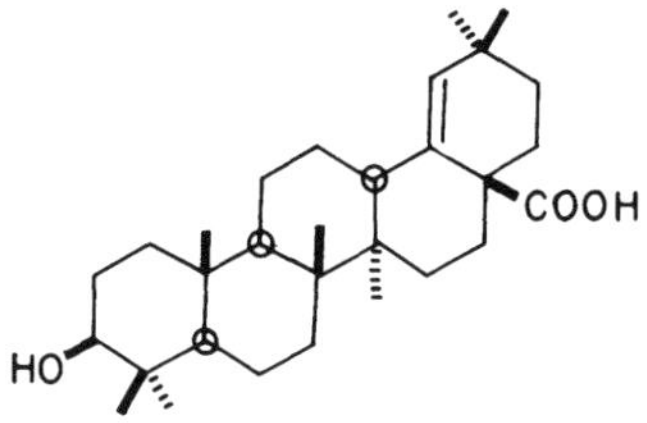

Morolic acid
534

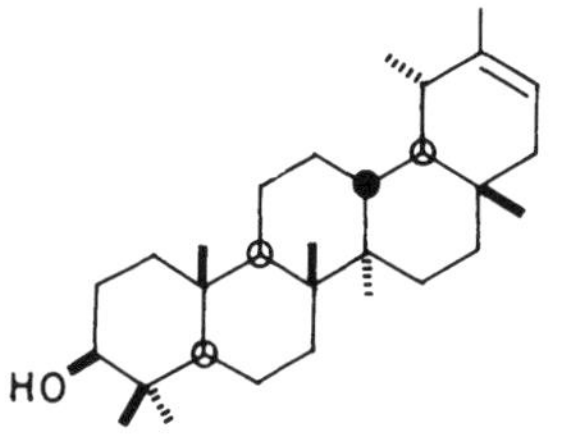

ψ-Taraxasterol
535

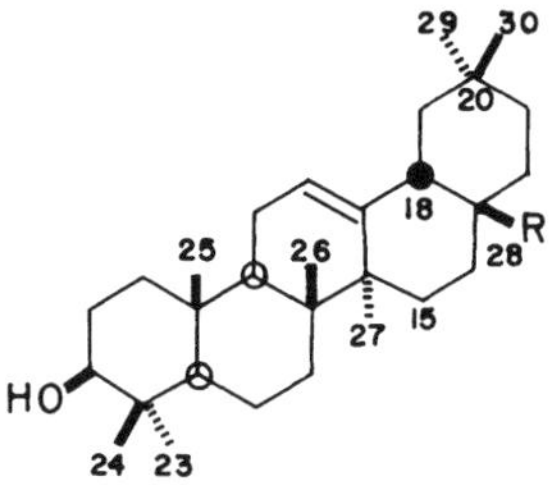

536: β-Amyrin (R=CH$_3$)
537: Oleanolic acid (R=COOH)

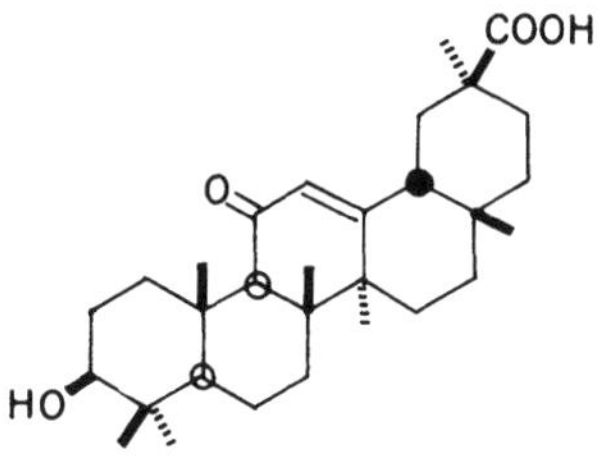

β-Glycyrrhetic acid
538

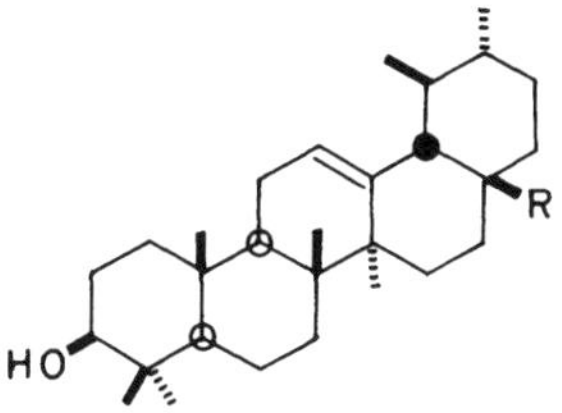

539: α-Amyrin (R=CH$_3$)
540: Ursolic acid (R=COOH)

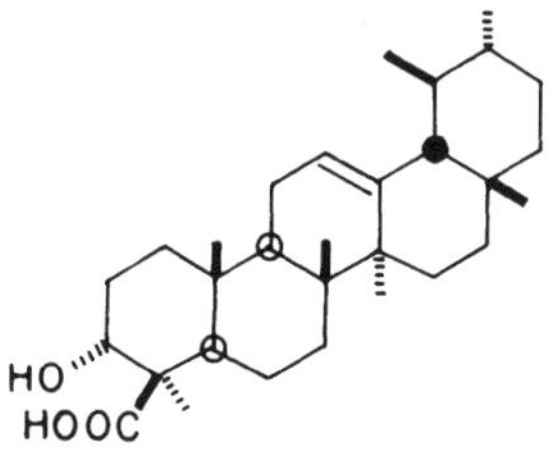

β-Boswellic acid
541

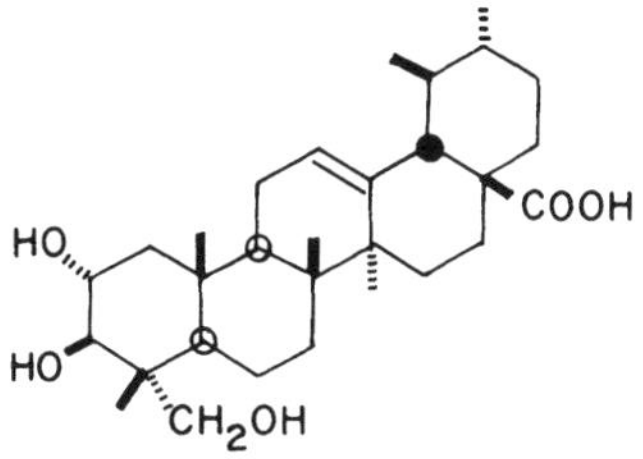

Asiatic acid
542

About one hundred ursanes have been obtained from nature so far. Most important of these is ursolic acid (**540**), first isolated in 1854 from leaves of *Arctostaphylos uva-ursi*. It is widely distributed in nature and has been mostly obtained from leaves and fruits, but has also been isolated from the bark of certain trees — e.g., *Olearia paniculata* and *Betula pendula* (207, 228). As already mentioned, α-amyrin (**539**) occurs together with β-amyrin in Manila elemi resin. α-Amyrenone (the ketone corresponding to **539**) occurs in the heartwood of *Diospyros ebenum* (174). 11β-Hydroxy-α-amyrin has been isolated from black dammar, the resinous exudate of *Canarium strictum* (198). Other important members are β-boswellic acid (**541**) from the oleoresin from *Boswellia* spp., and asiatic acid (**542**) first isolated from the leaves of *Centella asiatica* but also present in dammar resin (109, 228). 11-Oxoasiatic acid is a component of the resin from

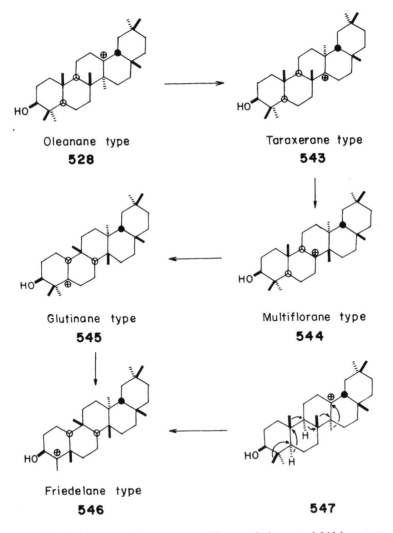

Fig. 8.1.32. Biogenesis of taraxerane, multiflorane, glutinane, and fridelane types

Dryobalanops aromatica (96). Several ursane-based glycosides are known (23, 85).

That cation **528** can rearrange furtlher to give new skeletal types was first recognized with the structure elucidation of friedelin (**548**), the major triterpene constituent of cork (the outer bark of *Quercus suber*). Structure **548** arises from cation **528** by a series of methyl and hydride shifts as depicted in **547** (Fig. 8.1.32). Such sequential backbone rearrangements have come to be known as "friedo" rearrangements. Since all adjacent methyls and hydrogens involved in this rearrangement have a *trans/anti* relationship, the rearrangement has been considered as concerted. However, as already pointed out, this concept of concertedness over many centers is now considered less realistic (415).

Besides friedelin (**548**) and the related cerin (2β-hydroxyfriedelin), some 70 friedelanes, mostly with a 3-oxo function, are known. Cerin co-occurs with friedelin in cork. From the bark of *Siphonodon australe*, many friedelanes have

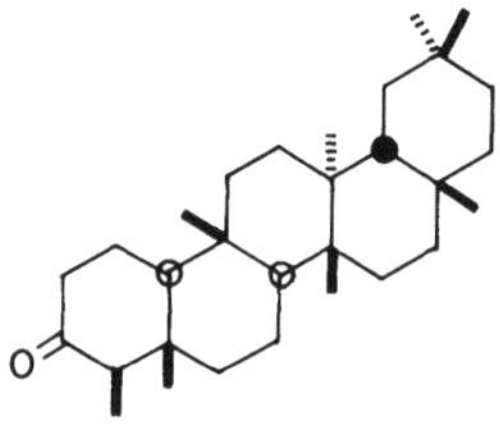

Friedelin
548

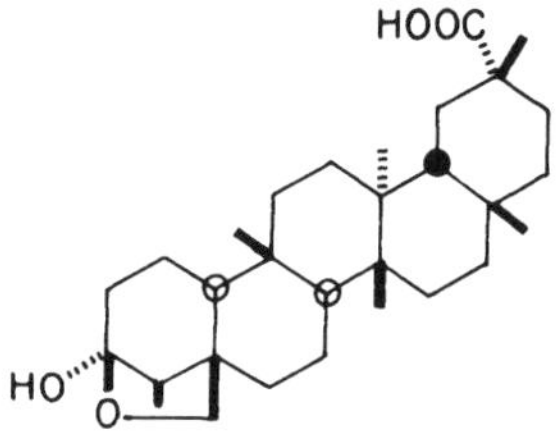

Salaspermic acid
549

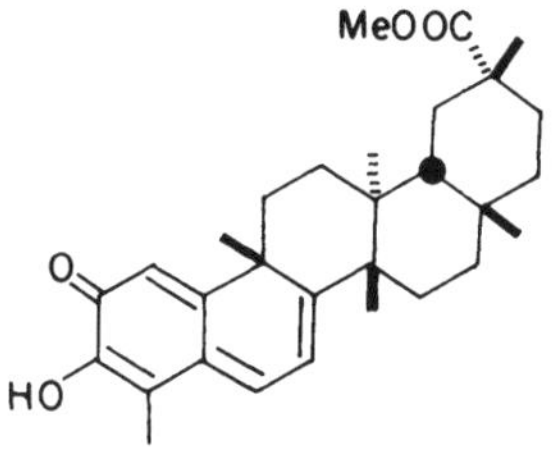

Pristimerin
550

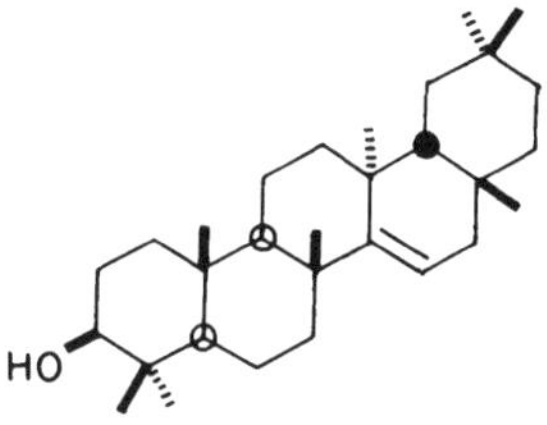

Taraxerol
551

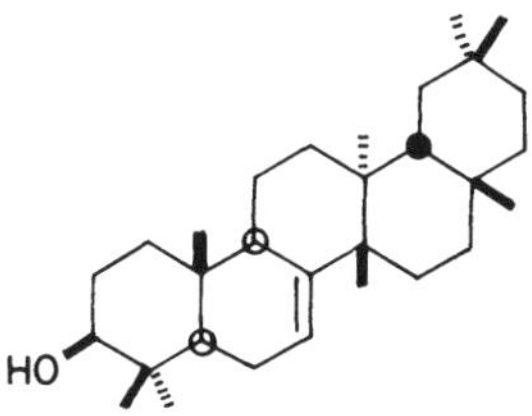

Multiflorenol
552

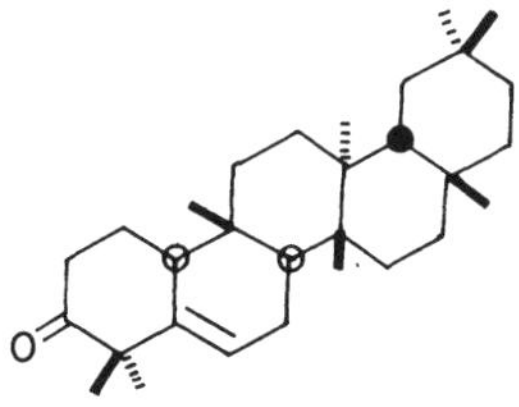

Alnusenone
553

782   Isoprenoids

been isolated; as examples, the occurrence of 21α- and 21β-hydroxyfriedelin may be noted (99). The interesting salaspermic acid (549) has been obtained from wood of *Salacia macrosperma* (401). A number of cytotoxic quinone methides based on the norfriedelane framework are known. Pristimerin (550) from roots of *Pristimera indica*, a potent, but toxic, antitumor agent, is an example of this group (277). Natural occurrence of friedelanes has been reviewed (86). Taraxeranes, multifloranes, and glutinanes (Fig. 8.1.32) constitute only minor groups. Taraxerol (551) from the roots of *Taraxacum officinale*, multiflorenol (552) from bark of *Gelonium multiflorum*, and alnusenone (553) from black alder (*Alnus glutinosa*) bark provide examples of their occurrence in woody tissues (228, 229).

Similar rearrangements in the ursane series (cf. cation 530) are known but appear to be of minor consequence in elaborating new compounds; no more than a dozen triterpenoids belonging to all these types (cf. Fig. 8.1.32) have been isolated so far. Of these, only baueranes (corresponding to the multiflorane type) (Fig. 8.1.32) have been encountered in woody tissues. Thus, the simplest member, bauerenol (554), has been found to occur in the barks of *Acronychia baueri* and *Ilex crenata* (229). Phyllanthol (555), from the root-bark of *Phyllanthus engleri*, represents an alternative fate for the isoursane cation (ursane equivalent of cation 543) (109, 228).

Bauerenol
554

Phyllanthol
555

(d) *C-C-C-C-C: Hopane and related types.* Cyclization of squalene in an all-chair conformation (Fig. 8.1.33) leads to ion 556, the progenitor of the hopane class of triterpenoids. Ion 556 is set to undergo "friedo" rearrangement to varying extents ultimately yielding cation 560. The initial two 1,2-hydrogen migrations are envisaged to proceed in a *syn* fashion. Compounds belonging to all the rearranged ions shown in Fig. 8.1.33 have been met with in nature. These compounds have often been isolated from ferns, lichens, grasses, and microorganisms, but only occasionally from higher plants (32, 303). A noteworthy characteristic of this family is the frequent absence of the usual oxygen function at C-3.

Although some 50 compounds with hopane skeleton (556) have been obtained so far from nature, their occurrence in woody tissues is quite rare. Hydroxyhopanone (561) is a component of dammar from *Hopea micrantha* (229). Moretenol (562), from bark of *Sapium sebiferum*, is an example of a C-21αH hopane; apparently this compound results from C-C-C-C-B folding of squalene. To date, none of the rearranged hopane types depicted in Fig. 8.1.33 has been reported to occur in woody parts of plants.

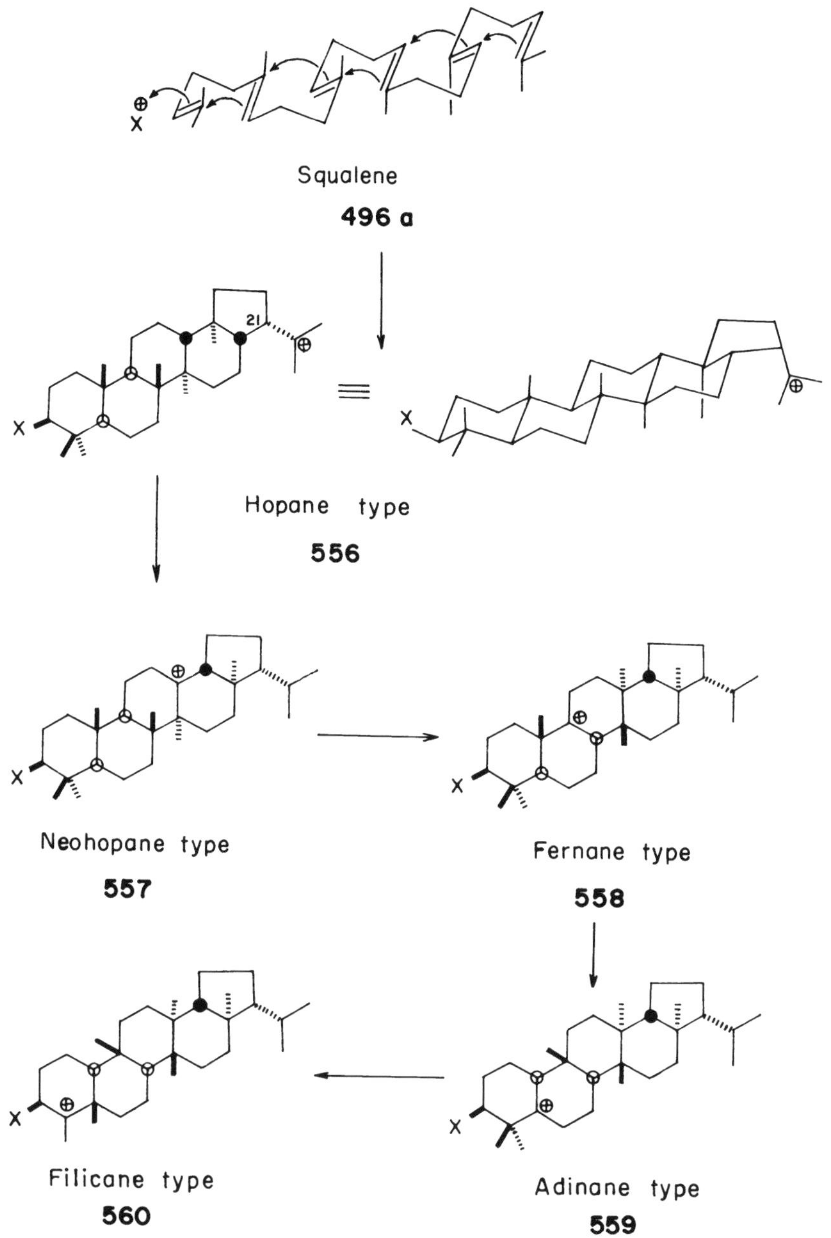

Terpenoids 783

Squalene
496 a

Hopane type
556

Neohopane type
557

Fernane type
558

Filicane type
560

Adinane type
559

Fig. 8.1.33. Cyclization of squalene to hopane and related types

In recent years, hopanes and modified hopanes (nor and homo derivatives), collectively termed hopanoids, have come to be recognized as ubiquitous in sedimentary organic matter of varied origin (303).

(e) *C-C-C-U-U: Malabaricanes.* Malabaricol (**563**) and related compounds, first discovered in 1967 as components of the trunk exudate of the tree *Ailanthus*

Hydroxyhopanone
561

Moretenol
562

malabarica, represent a squalene 2,3-epoxide folding so that cyclization terminates after the closing of the third ring (Fig. 8.1.34). It may be noted that the third ring closes in a Markownikoff fashion (312, 373, 377). This mode of cyclization appears to be a minor pathway in the biogenesis of triterpenoids, and no other compounds of this class except those obtained from *A. malabarica* are known at present. Recently, related compounds with 8α-Me,9β-H stereochemistry

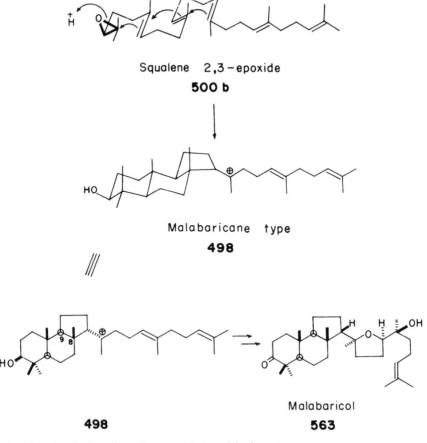

Fig. 8.1.34. Cyclization of squalene epoxide to malabaricane type

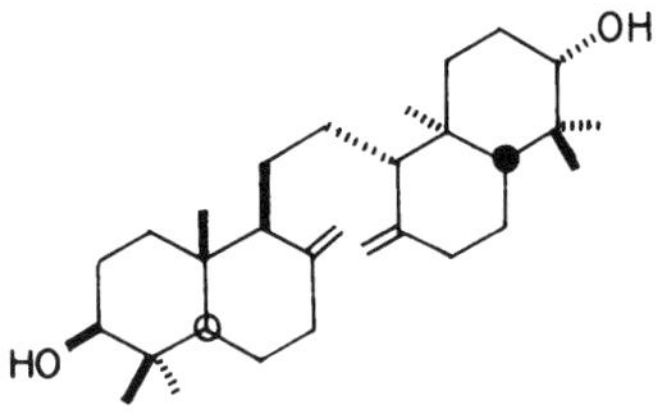

α - Onocerin
564

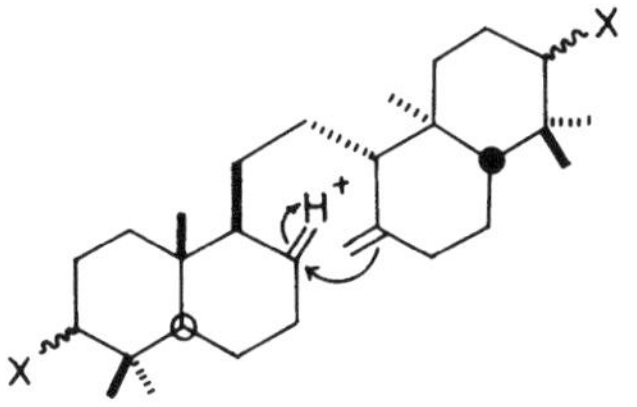

565

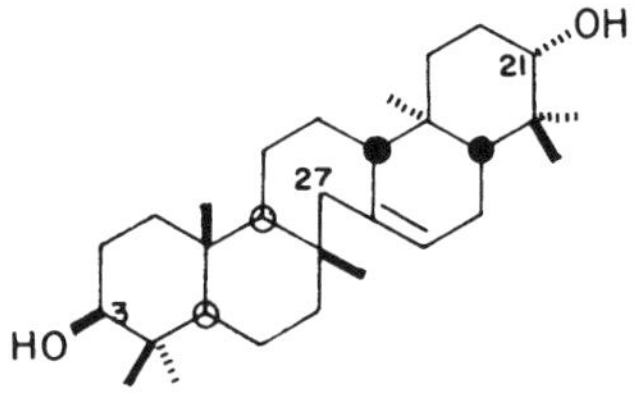

Serratenediol
566

(iso-malabaricane) have been encountered as metabolites of sponges of genus *Stelletta* (271).

(f) *Cyclization of squalene from both ends.* α-Onocerin (**564**), known since 1855 as a constituent of roots of *Ononis spinosa*, provides an example of a small group of bis-bicyclic triterpenoids arising from simultaneous cyclization from both ends of squalene (Fig. 8.1.29) (109). However, this mode of cyclization appears to be central in the biogenesis of a fair-sized group of pentacyclic triterpenoids, serratanes (Fig. 8.1.27). These compounds are believed to be formed from onocerin-type precursors by a cyclization mode depicted in **565**. Presently, some 40 compounds of this class are known. Most of these have been obtained from *Lycopodium* spp. (mosses), though a few have been isolated from bark of certain *Pinus* and *Picea* spp. Serratenediol (**566**), first isolated from *Lycopodium serratum*, has also been isolated from bark of several *Pinus* spp. (213, 331). 21-Episerratenediol occurs in the bark of *Picea sitchensis* (250).

8.1.3.7 Tetraterpenoids: Carotenoids

Unlike the terpenoid families discussed so far, tetraterpenoids consist of only one class, namely carotenoids, the yellow and red pigments of plants. At present, about 500 different carotenoids are known. A vast majority of these are C_{40} compounds, though compounds with fewer than 40 (nor- and apocarotenoids – e.g., C_{30}) and more than 40 (homocarotenoids – e.g., C_{45}, C_{50}) carbon atoms have been isolated. The occurrence, chemistry, and biochemistry of carotenoids have been covered in a number of reviews and books (67, 124, 163, 216, 254, 282, 365).

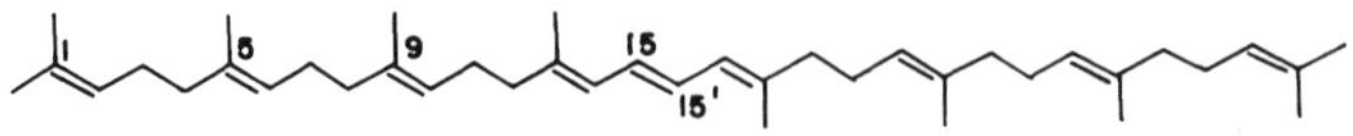

567: Phytoene

↓ −2H

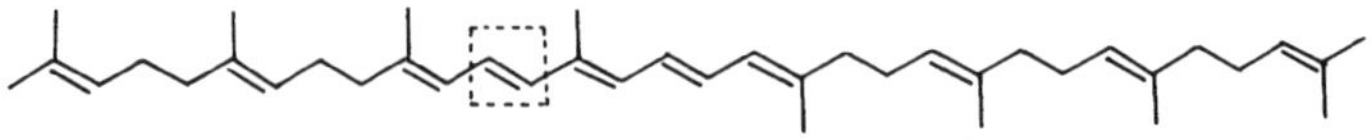

568: Phytofluene

↓ −2H

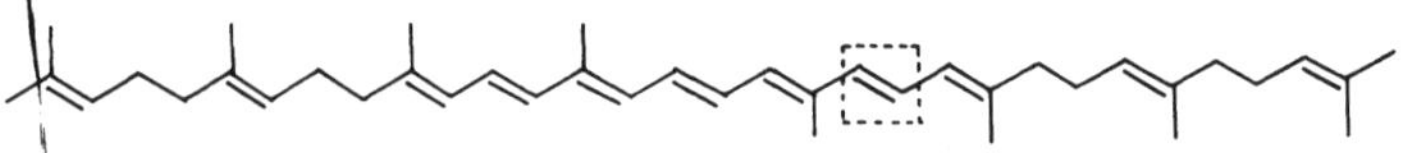

569: ζ - Carotene

or

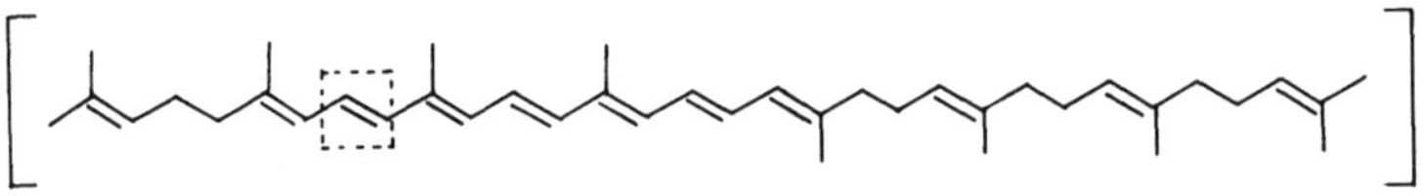

570: 7,8,11,12-Tetrahydrolycopene

↓ −2H

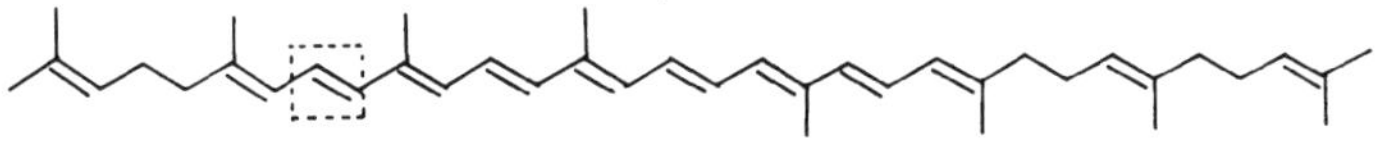

571: Neurosporene

↓ −2H

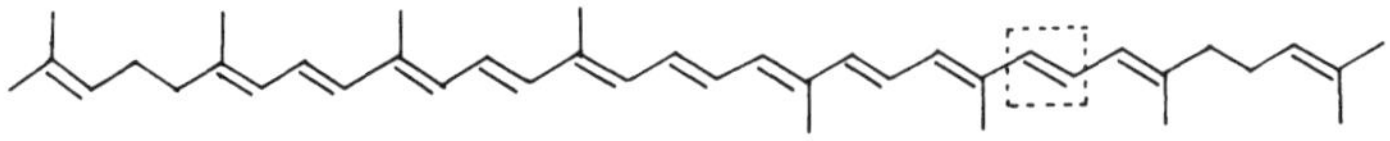

572: Lycopene

Fig. 8.1.35. Biosynthesis of lycopene (box shows the new double bond introduced)

Phytoene (**567**) is the first C_{40} precursor of carotenoids, and arises by tail-to-tail coupling of two GGPP units via prephytoene pyrophosphate in a sequence of reactions exactly parallel to those already outlined for the biosynthesis of squalene. The phytene produced can be either all-*trans* (**567**) or with $\Delta^{15(15')}$ *cis* configurated. Either of these isomers then undergoes step-wise desaturation to furnish phytofluene, all-*trans*-ζ-carotene (or its isomer 7,8,11,12-tetrahydrolycopene), neurosporene and finally lycopene (Fig. 8.1.35). Both neurosporene (**571**) and lycopene (**572**) have been implicated in the biogenesis of other carotenoids by reactions such as hydrogenation, dehydrogenation, isomerization, cyclization, hydration, oxygenation, chain homologation or cleavage (66, 162, 163). Oxygenated carotenoids are often called xanthophylls, whereas the hydrocarbons are designated carotenes.

Carotenoids are essential components of all photosynthetic tissues. They have also been isolated from non-photosynthetic organs of higher plants, such as flowers, fruits, and seeds. However, they do not seem to occur in woody tissues, and hence will not be discussed any further.

8.1.3.8 Polyterpenoids: Polyprenols

Rubber, gutta, and chicle are 1,4-polyisoprenes of different molecular weights and geometries. These products are of great commercial significance and are discussed in some detail in Chap. 10.

The linear homologous isoprenoid alcohol of maximum size discussed so far is the C_{25} geranylfarnesol (**492**), the pyrophosphate of which is the biogenetic precursor of sesterterpenoids. However, during the past two decades, it has become evident that such acyclic polyisoprenoid primary alcohols of longer lengths are fairly widely distributed in the higher plants, algae, fungi, mammals, and marine invertebrates. This class of isoprenoids has come to be designated as *polyprenols*. Polyprenols with chain lengths varying from 6 to 24 isoprene units have been isolated. These compounds may have all-*trans*- or mixed geometry. In some cases, a few isoprene units occur as saturated. All-*trans*-polyprenols have been found only in higher plants and are less common than those with mixed stereochemistry (188, 189, 392).

A short-hand convention to describe structures of polyprenols has been suggested (392). Abbreviations ω (ω-isoprene residue), T (*trans*-isoprene residue), C (*cis*-isoprene residue), and S (saturated isoprene residue) are used to designate a structure. Thus, the all-*trans*-nonaprenol, solanesol, the first polyprenol to be isolated from nature, can be represented as ω-T-T-[T]$_5$-T-OH.

Polyprenols of higher plants have mostly been isolated from leaves; solanesol (C_{45}) was first obtained in 1956 from tobacco leaves. However, their occurrence in certain woody tissues has been demonstrated. Silver birch (*Betula verrucosa*) wood has been shown to contain fatty acid esters of betulaprenols -6 to -9; the major polyprenol, betulaprenol-7 has the structure ω-T-T-C-[C]$_2$-C-OH (**573**); the polyprenols from the leaves were free, and the major component had 11 isoprene units (188, 256).

Betulaprenol – 7
573

All *trans*-polyprenols are believed to be formed from GFPP by the chain extension process already discussed (Sect. 8.1.1.2), followed by hydrolysis by appropriate phosphatases. There is evidence to show that synthesis of all-*trans*-polyprenol pyrophosphates occurs in mitochondrial membranes and chloroplasts. Biosynthetic studies have shown that *cis,trans*-polyprenols are synthesized from FPP or GGPP by *cis*-prenylation by IIP (189, 369, 392).

8.1.3.9 Meroterpenoids

The term meroterpenoids has been suggested (114) and is adopted here to cover compounds with "part-terpene" carbon-framework, arising from mixed biogenesis. A number of indole alkaloids (28, 342), certain plant phenolics anld oxygen heterocycles (170, 207, 228, 229, 288), chlorophyll (164), certain quinones and chromanols, and some other compounds come under this category. Of these, examples of such indole alkaloids and oxygen heterocycles will be found in Chaps. 5 and 7. Of the remaining, only a brief reference to terpene quinones and chromanols will be made here.

Terpene quinones and chromanols (334, 392, 393, 417) have a wide distribution in higher plants, bryophytes, mosses, algae, bacteria, and animals. From the higher plants, they have been isolated/detected mostly in leaves, fruits, tubers and seeds (207, 229, 396). All these compounds have an aromatic moiety that has been alkylated by a polyprenyl chain. Biosynthetically, they fall into three main types: plastoquinones and plastochromanols (e.g., plastoquinone-9 and α-tocopherol) derived from 2,5-dihydroxyphenylacetic acid (homogentisic acid); phylloquinones (e.g., vitamin K_1) derived from 2-succinoylbenzoic acid; and ubiquinones (e.g., ubiquinone-10) based on 4-hydroxybenzoic acid. Thus, all these compounds arise from a shikimic acid – mevalonic acid mixed pathway. These conclusions are based on much biosynthetic evidence obtained from feeding experiments using labelled precursors. These experiments also showed that plastoquinones are synthesized de novo in the chloroplast, whereas ubiquinones are extrachloroplastidic compounds elaborated in the mitochondrial membrane (334, 392, 393).

α-Tocotrienolquinone (**578**) isolated from latex of *Hevea brasiliensis* is an example of occurrence of such compounds in woody parts of plants (260, 392).

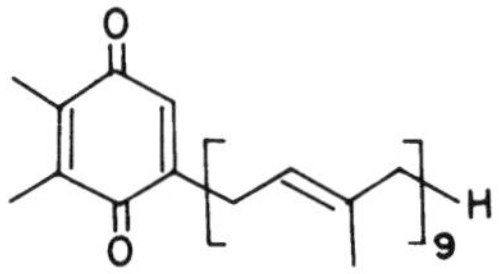

Plastoquinone – 9
574

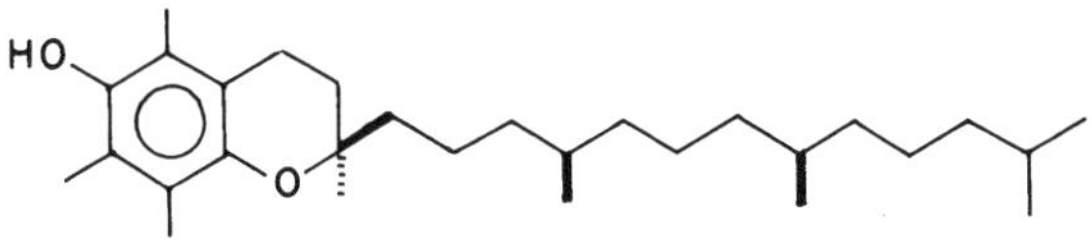

α – Tocopherol
575

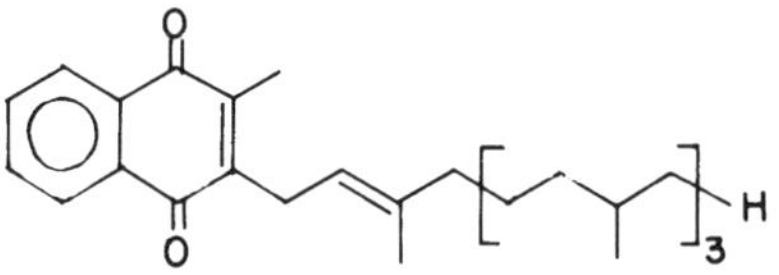

Vitamin K$_1$
576

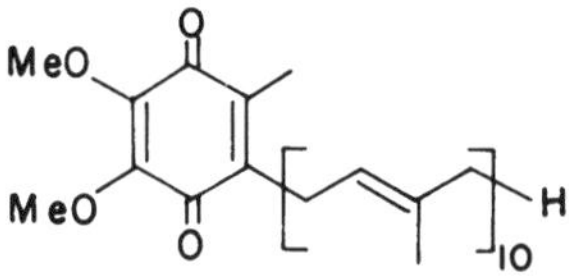

Ubiquinone – 10
577

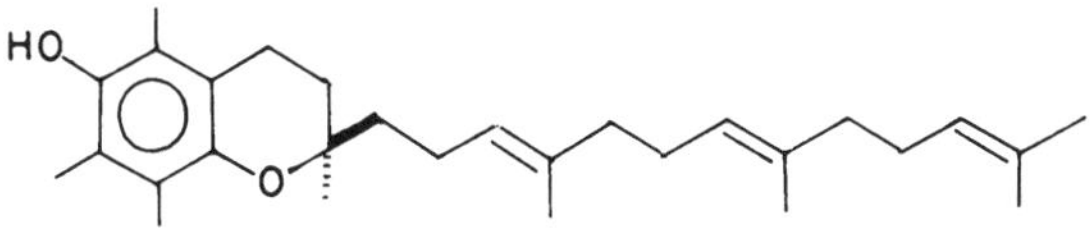

α – Tocotrienol
578

8.1.4 Biological Role

Certain groups or members of the terpenoid family have a well-recognized role in plant physiology and plant biochemistry. The lipophilic phytyl side chain in chlorophyll is essential for its biological activity. Carotenoids are accessory energy harvesting pigments associated with photosynthesis; they also protect chlorophylls against photo-oxidative degradation. Plastoquinones are also involved in the photosynthetic process; plastoquinone-9 **(574)** is an obligatory electron carrier in the photosynthetic electron transport system. Polyprenyl phosphates act as coenzymes in protein and oligosaccharide glycosylation (159, 161). Gibberellins have a clear role as endogenous plant growth regulators; they promote cell elongation, induce parthenocarpy, and induce new RNA and protein synthesis. Abscisic acid **(373)** and cytokinins with C_5-terpene side chain (e.g., zeatin, **579**) are other natural plant hormones; abscisic acid inhibits growth and promotes abscission and senescence, while cytokinins induce cell division (127, 161, 253, 262).

Zeatin
579

The above represent only a minute fraction of known terpenoids. Questions arise as to why their numbers are so great and why there is such a bewildering variety. Do these products have any biological significance for the host organism? These questions are part of the general question concerning the role of so-called secondary metabolites. Although the older view that these compounds are just metabolic waste products is still held in some quarters (284), this view does not appear to be realistic, and is certainly not true in the case of terpenoids, especially in view of the biological role of several terpenoids already discussed above. Evidence has also accumulated indicating that monoterpenoids (and to some extent sesquiterpenoids) are metabolically active (87, 258). However, it would be prudent to realise that it is unrealistic to expect any unique role for a class as a whole, composed as they are of such diverse types and invariably with restricted distribution.

Several possible factors and needs have been cited to account for the multiplicity of structural types and numbers. Of these, one might consider random synthesis by the organism to arrive at a useful compound for a certain physiological function, and to provide an alternative strategy to switching off metabolic pathways temporarily (159, 317).

During the past few decades, it has become clear that a large number of terpenoids are involved in ecological interactions: insect-insect, insect-plant,

plant-plant, plant-herbivore (18, 252, 265, 311, 322, 356, 407, 409, 418). Furthermore, plants are subject to mechanical injury, attack by predators, microorganism infection and climatic and nutritional stresses. In this fight for survival, it appears that plants and other organisms have come to exploit their primary metabolic pools to engineer compounds and mixtures (for good physical characteristics) as defense and attack chemicals. This can be seen in the contemporary scene: de novo synthesis of compounds (phytoalexins) by plants under conditions of stress (31, 169). Since ecological interactions are a continuing phenomemon, it would be reasonable to say that a vast majority of these compounds are essentially *stress compounds*, evolutionarily retained because of continuous environmental stress. Furthermore, different ecological situations would invoke different responses at the genus, species, and even at individual level, thus compounding further the numbers and types.

References

1. Adinolfi M, Laonigro G, Parrilli M, Mangoni L 1976 The synthesis of grindelic acid. Gazz Chim Ital 106:625–631
2. Albers-Schönberg G, Schmid H 1961 Über die Struktur von Plumericin, Isoplumericin, β-Dihydroplumericin und der β-Dihydroplumericinsäure. Helv Chim Acta 44:1447–1473
3. Allen K G, Banthorpe D V, Charlwood B V, Ekundayo O, Mann J 1976 Metabolic pools associated with monoterpene biosynthesis in higher plants. Phytochemistry 15:101–107
4. Alvarenga M A, Gottlieb H E, Gottlieb O R, Magalhaes M T, da Silva V O 1978 Diasin, a diterpene from *Croton diasii*. Phytochemistry 17:1773–1776
5. Andersen N H, Falcone M S 1971 Prezizaene and biogenesis of zizaene. Chem Ind 62–63
6. Anthonsen T, Kjosen B 1971 New thymol derivatives from *Inula salicina* L. Acta Chem Scand 25:390–392
7. Appel H H, Bond R P M, Overton K H 1963 Sesquiterpenoids-III. The constitution and stereochemistry of valdivolide, fuegin, winterin and futronolide. Tetrahedron 19:635–641
8. Appel H H, Brooks C J W, Overton K H 1959 The constitution and stereochemistry of drimenol, a novel bicyclic sesquiterpenoid. J Chem Soc 3322–3332
9. ApSimon J W, Hooper J W 1973 The synthesis of triterpenes. In: ApSimon J W (ed) The total Sunthesis of natural products, vol 2. Wiley-Interscience New York, 559–640
10. Archer B L 1980 Polyisoprene. In: Bell E A, Charlwood B V (eds) Secondary plant products. Springer Heidelberg, 309–327
11. Arctander S 1969 Perfume and flavor chemicals, vol 1–2. S Arctander Montclair NJ
12. Arigoni D 1975 Stereochemical aspects of sesquiterpene biosynthesis. Pure Appl Chem 41:219–245
13. Asakawa Y 1982 Chemical constituents of the Hepaticae. Fortschr Chem Org Naturst 42:1–285
14. Asao T, Ibe S, Takase K, Cheng Y S, Nozoe T 1968 The structure of two new acetylenic nor-sequiterpenoid, dehydrochamaecynenol and dehydrochamaecynenal, isolated from *Chamaecyparis formosensis* Matsm. Tetrahedron Lett 3639–3642
15. Ayer W A, Browne L M 1981 Terpenoid metabolites of mushrooms and related Basidiomycetes. Tetrahedron 37:2199–2248
16. Ayers J E, Fishwick M J, Land G, Swain T 1964 Off-flavor of dehydrated carrot stored in oxygen. Nature 203:81–82
17. Bajaj A G, Sukh Dev, Tagle B, Telser J, Clary J 190 The stereochemistry of allohimachalol. Tetrahedron Lett 325–326
18. Baker R, Herbert R H 1884 Insect pheromones and related natural products. Nat Prod Rep 1:299–318
19. Banthorpe D V, Charlwood B V 1980 The terpenoids. In: Bell E A, Charlwood B V (eds) Secondary plant products. Springer Heidelberg, 185–220
20. Banthorpe D V, Charlwood B V, Francis M J O 1972 The biosynthesis of monoterpenes. Chem Rev 72:115–155
21. Barrett H C, Büchi G 1967 Stereochemistry and synthesis of α-agarofuran. J Am Chem Soc 89:5665–5667
22. Barton D H R, Jarman T R, Watson K C, Widdowson D A, Boar R B, Damps K 1975 Investigations on the biosynthesis of steroids and terpenoids. Part XII. Biosynthesis of 3β-hydroxy-triterpenoids and -steroids from (3S)-2,3-epoxy-2,3-dihydrosqualene. J Chem Soc Perkin Trans I 1134–1138
23. Basu N, Rastogi R P 1967 Triterpenoid saponins and sapogenins. Phytochemistry 6:1249–1270
24. Bates R B, Gale D M, Gruner B J 1963 Stereisomeric farnesols. J Org Chem 28:1086–1089
25. Bates R B, Hendrickson E K 1962 γ-Eudesmol from *Callitropsis araucarioides*. Chem Ind 1759–1760
26. Bates R B, Paknikar S K 1965 Biogenesis of some monoterpenoids not derived from a geranyl precursor. Tetrahedron Lett 1453–1455
27. Bates R B, Slagel R C 1962 β-Bulnesene, α-guaiene, β-patchoulene, and guaioxide in essential oils. Chem Ind 1715–1716

28 Battersby A R 1970 Biosynthesis of terpenoid alkaloids. In: Goodwin T W (ed) Natural substances formed biologically from mevalonic acid. Academic Press New York, 157–168
29 Battersby A R, Thompson M, Glüsenkamp K-H, Tietze L-F 1981 Untersuchungen zur Biogenese der Indolalkaloide. Synthese und Verfütterung radioaktiv markierter Monoterpenaldehyde. Chem Ber 114:3430–3438
30 Bawdekar A S, Kelkar G R 1965 Terpenoids. LXVIII. Structure and absolute configuration of costic acid, a new sesquiterpene acid from costus root oil. Tetrahedron 21:1521–1528
31 Bell A A 1981 Biochemical mechanisms of disease resistance. Ann Rev Plant Physiol 32:21–81
32 Berti G, Bottari F 1968 Constituents of ferns. Progr Phytochem 1:589–685
33 Bhan P, Pande B S, Soman R, Damodaran N P, Sukh Dev 1984 Products active on Arthropod-V. Insect juvenile mimics (Part 5): Sesquiterpene acids having JH activity from wood of *Cedrus deodara* Loud. Tetrahedron 40:2961–2965
34 Bhan P, Sukh Dev, Bass L S, Tagle B, Clardy J 1982 The stereochemistry of himachalol. J Chem Res(S) 344–345
35 Birch A J, Chamberlain K B, Moore B P, Powell V H 1970 Termite attractants in *Santalum spicatum*. Aust J Chem 23:2337–2341
36 Birch A J, Hochstein F A, Quartey J A K, Turnbull J P 1964 Structure and some reactions of acoric acid. J Chem Soc 2923–2931
37 Bisset N G, Diaz M A, Ehret C, Ourisson G, Palmade M, Patil F, Pesnelle P, Streith J 1966. Études chimio-taxonomiques dans la Famille des Dipterocarpaceés. II. Constituants du genre *Dipterocarpus* Gaertn. F. Essai de classification chimio-taxonomique. Phytochemistry 5:865–880
38 Boar R B 1984 Triterpenoids. Nat Prod Rep 1:53–65
39 Bohlmann F, Abraham W R 1978 Neue Sesquiterpene und Acetylenverbindungen aus *Cineraria*-arten. Phytochemistry 17:1629–1635
40 Bohlmann F, Abraham W-R 1979 Neue Diterpen aus *Helichrysum acutatum*. Phytochemistry 18: 1754–1756
41 Bohlmann F, Ehlers D, Zdero C, Grenz M 1977 Über Inhaltsstoffe der Gattung *Ligularia*. Chem Ber 110:2640–2648
42 Bohlmann F, Gupta R K, Robinson H, King R M 1981 Labdane derivatives and a himachalanolide from *Acritopappus longifolius*. Phytochemistry 20:275–279
43 Bohlmann F, Jakupovic J 1980 Neue sesquiterpen-Kohlenwasserstoffe mit anomale Kohlenstoffgerüst aus *Silbhium*-arten. Phytochemistry 19:259–265
44 Bohlmann F, Kapteyn H 1963 Isolierung von Dehydronerolisovalerat aus *Anthemis montana* L. Tetrahedron Lett 2065–2066
45 Bohlmann F, Suding H, Cuatrecasas J, Robinson H, King R M 1980 Tricyclic sesquiterpenes and further diterpenes from *Esplethiopsis* species. Tetrahedron 19:2399–2403
46 Bohlmann F, Suwita A 1978 New Terpene derivatives from *Piqueria trinervia*. Phytochemistry 17: 560–561
47 Bohlmann F, Suwita A, Natu A A, Czerson H, Suwita A 1977 Über weitere α-Longipinen-Derivate aus Compositen. Chem Ber 110:3572–3581
48 Bohlmann F, Zdero C 1969 Neue Sesquiterpene aus *Brickellia guatemaliensis*. Tetrahedron Lett 5109–5110
49 Bohlmann F, Zdero C 1969 Über Terpenderivate aus *Ferula hispania*. Chem Ber 102:2211–2215
50 Bohlmann F, Zdero C 1975 Über neue Inhaltsstoffe der Gattung *Anthemis*. Chem Ber 108: 1902–1910
51 Bohlmann F, Zdero C 1976 Über neue Inhaltsstoffe aus *Pluchea odorata* Cass. Chem Ber 109: 2653–2656
52 Bohlmann F, Zdero C 1978 Neue Furanoeremophilane und andere Sesquiterpene aus Vertretern der Gattung *Euryops*. Phytochemistry 17:1135–1153
53 Bohlmann F, Zdero C, Bohlmann R, King R M, Robinson H 1980 Neue Sesquiterpene aus *Liabum*-arten. Phytochemistry 19:579–582
54 Bohlmann F, Zdero C, Faass U 1973 Über die Inhaltsstoffe von *Artemisia fragrans* Willd. Chem Ber 106:2904–2909

55 Bohlmann F, Zdero C, Grenz M 1969 Über ein neues Sesquiterpen aus *Anthemis cotula* L. Tetrahedron Lett 2417–2418
56 Bohlmann F, Zdero C, Jakupovic J, Robinson H, King R M 1981 Eriolanolides, eudesmanolides and a rearranged sesquiterpene from Eriphyllum species. Phytochemistry 20:2239–2244
57 Bohlmann F, Ziesche J 1981 Sesquiterpenes from the three *Senecio* species. Phytochemistry 20: 469–472
58 Bradshaw A P W, Hanson J R 1981 Stable isotope studies of the biosynthesis of the sesquiterpenoid dihydrobotrydiol. J Chem Soc Chem Commun 1169–1170
59 Brecknell D J, Carman R M 1978 Callitrin, callitrisin, dihydrocallitrisin, columellarin and dihydrocolumellarin, new sesquiterpene lactones from the heartwood of *Callitris columellaris*. Tetrahedron Lett 73–76
60 Bredenberg J B 1957 Ferruginol and Δ-dehydroferruginol. Acta Chem Scand II:932–935
61 Brian P W, Grove J F, Mac Millan J 1960 The gibberellins. Fortschr Chem Org Naturst 18: 350–433
62 Brieskorn C H, Pohlmann E 1969 Kauradien-[9(11), 16]-säure-(19) und 15α-Acetoxy-kauren-(16)-säure-(19). Chem Ber 102:2621–2628
63 Briggs L H, Cain B F, Cambie R C, Davis B R, Rutledge P S, Wilmshurst J K 1963 Diterpenes. Part VII. Kaurene. J Chem Soc 1345–1355
64 Briggs L H, Cambie R C, Rutledge P S 1963 Diterpenes. Part VIII. Reaction of epoxides in the (−)-kaurene and phyllocladene series, and a direct correlation of the diterpenes. J Chem Soc 5374–5383
65 Briggs L H, White G W 1975 Constituents of the essential oil of *Araucaria araucana*. Tetrahedron 31:1311–1314
66 Britton G 1979 Carotenoid biosynthesis and vitamin A. In: Barton D, Ollis W D (eds) Comprehensive organic chemistry, vol 5. Pergamon Press Oxford 1025–1042
67 Britton G, Goodwin T W (eds) 1982 Carotenoid chemistry and biochemistry. Pergamon Press Oxford, 395 pp
68 Brooks C J W, Campbell M M 1969 Caparrapi oxide, a sesquiterpenoid from caparrapi oil. Phytochemistry 8:215–218
69 Brooks C J W, Draffan G H 1969 Sesquiterpenoids of Warburgia species. II. Ugandensolide and ugandensidial (cinnamodial). Tetrahedron 25:2887–2898
70 Brown P M, Thomson R H 1969 Nasturally occurring quinones. Part XVI. Structure of maturinone and related compounds. J Chem Soc (C) 1184–1186
71 Brunke E-J, Hammerschmidt F-J, Struwe H 1980 (+)-Epi-β-santalol, Isolierung aus Sandelholzl und Partialsynthese aus (+)-α-Santalol. Tetrahedron Lett 21:2405–2408
72 Bruns K 1968 Diterpene. V. Über die C_{13}-Konfiguration der Diterpene aus *Araucaria imbricata* Pavon (Araucariaceae). Tetrahedron 24:3417–3423
73 Bruns K 1970 Zur Absoluten Konfiguration einiger Labdanderivate am C-atom 13. Tetrahedron Lett 3263–3264
74 Burrell J W K, Garwood R F, Jackman L M, Oskay E, Weedon B C L 1966 Carotenoids and related compounds. Part XIV. Stereochemistry and synthesis of geraniol, nerol, farnesol and phytol. J Chem Soc(C) 2144–2154
75 Cane D E 1980 The stereochemistry of allylic pyrophosphate metabolism. Tetrahedron 36: 1109–1159
76 Cane D E, Hasler H, Materna J, Cagnoli-Bellavita N, Ceccerelli P, Madruzza G F 1981 ^{1}H NMR determination of the stereochemistry of an allylic displacement in the biosynthesis of virescenol-B. J Chem Soc Chem Commun 280–282
77 Cane D E, Iyengar R, Shiao M-S 1981 Cyclonderodiol biosynthesis and the enzymatic conversion of farnesyl to nerolidyl pyrophosphate. J Am Chem Soc 103:914–931
78 Caputo R, Mangoni L, Monaco P, Palumbo G 1979 Triterpenes from the galls of *Pistacia palestina*. Phytochemistry 18:896–898
79 Caputo R, Mangoni L, Monaco P, Palumbo G, Aynehchi Y, Bagheri M 1978 Triterpenes from the bled resin of *Pistacia vera*. Phytochemistry 17:815–817

80 Carman R M, Corbett R E, Grant P K, McGrath M J A, Munro M H G 1966 The diterpenes of *Dacrydium colensoi*. Part V. Tetrahedron Lett 3173–3179
81 Carman R M, Marty R A 1966 Diterpenoids. IX. *Agathis microstachya* oleoresin. Aust J Chem 19:2403–2406
82 del Castillo J B, Brooks C J W, Campbell M M 1966 Caparrapidiol and caparrapitriol: Two new axyclic sesquiterpene alcohols. Tetrahedron Lett 3731–3736
83 Chadha Y R 1982 Indian pine resin – its chemical composition and utilization. In: Atal C K, Kapur B M (eds) Cultivation and utilization of aromatic plants. Regional Research Laboratory CSIR Jammu-Tawi, 337–347
84 Chan W R, Taylor D R Yee T 1970 Triterpenoids from *Entandrophragma cylindricum* Sprague. Part I. Structures of sapelins A and B. J Chem Soc (C) 311–314
85 Chandel R S, Rastogi R P 1980 Triterpenoid saponins and sapogenins: 1973-1978. Phytochemistry 19:1889–1908
86 Chandler R F, Hooper S N 1979 Friedelin and associated triterpenoids. Phytochemistry 18: 711–724
87 Charlwood B V, Banthorpe D V 1978 The biosynthesis of monoterpenes. Progr Phytochem 5: 65–125
88 Chatterjee A, Banerjee A, Bohlmann F 1977 Crotocaudin; a rearranged labdane type nor-diterpene from *Croton daudatus* Geisel. Tetrahedron 33:2407–2414
89 Chemical Society of London 1971–1983 Specialist periodical reports: Terpenoids and steroids, vol 1–12. Chemical Society London.
90 Chemical Society of London 1972–1980 Specialist periodical reports: Biosynthesis, vol. 1–6. Chemical Society London
91 Chen F-C, Lin Y-M, Chen A-H 1972 Sesquiterpenes from the heartwood of Chinese elm. Phytochemistry 11:1190–1191
92 Cheng Y S, Kuo Y H, Lin Y T 1967 Extractive components from wood of *Taiwania cryptomerioides* Hayata: the structures of T-cadinol and T-muurolol. J Chem Soc Chem Commun 565–566
93 Cheng Y S, von Rudloff E 1970 The volatile oil of the leaves of *Chamaecyparis nootkatensis*. Phytochemistry 9:2517–2527
94 Chetty G L, Sukh Dev 1964 Ketones from "mayur pankhi": some new cuparene-based sesquiterpenoids. Tetrahedron Lett 73–77
95 Chetty G L, Zalkow V B, Zalkow L H 1968 The synthesis and absolute configuration of juniper camphor and selin-11-en-4α-ol. The structure of intermedeol. Tetrahedron Lett 3223–3225
96 Cheung H T, Tokes L 1968 Oxygenated derivatives of asiatic acid from *Dryobalanops aromatica*. Tetrahedron Lett 4363–4366
97 Chiang C-K, Chang F C 1973 Tetracyclic triterpenoids from *Melia azedarach* L. III. Tetrahedron 29:1911–1929
98 Christenson P A, Secord N, Wills B J 1981 Identification of trans-β-santalol and epi-cis-β-santalol in East Indian sandalwood oil. Phytochemistry 20:1139–1141
99 Clarke B J, Courtney J L, Stern W 1970 Triterpenes of the friedelane series. VII. The structure of friedelan-x-one. Aust J Chem 23:1651–1654
100 Clayton R B 1965 Biosynthesis of sterols, steroids and terpenoids. Quart Rev 19:168–230
101 Coates R M 1976 Biogenetic-type rearrangements of terpenes. Fortschr Chem Org Naturst 33: 73–230
102 Coates R M, Sowerby R L 1972 Stereoselective total synthesis of (±)-zizaene. J Am Chem Soc 94:5386–5396
103 Cocker W, McMurry T B H, Sainsbury D M, Stanley L E 1964 Extractives from wood. VI. Extractives from *Shorea maranti*. Perfum Essent Oil Rec 55:442–445
104 Cocker W, Moore A L, Pratt A C 1965 Dextrorotatory hardwickiic acid, an extractive of *Copaifera officinalis*. Tetrahedron Lett 1983–1985
105 Comin J, Goncalves de Lima O, Grant H N, Jackman L M, Keller-Schierlein W, Prelog V 1963 Über die Konstitution des Biflorins, eines o-Chinons der Diterpen-Reihe. Helv Chim Acta 46: 409–415

106 Connolly J D, Gunn D M, McCrindle R, Murray R D H, Overton K H 1967 Constituents of *Erythroxylon monogynum* Roxb. Part III. Erythroxytriols P and Q. J Chem Soc(C) 668–674
107 Connolly J D, Labbe C, Rycroft D S, Taylor D A H 1979 Tetranortriterpenoids and related compounds. Part 22. New apo-tirucallol derivatives and tetranortriterpenoids from the wood and seeds of *Chisocheton paniculatus* (Meliaceae). J Chem Soc Perkin Trans I 2959–2964
108 Connolly J D, McCrindle R 1971 Tetranortriterpenoids and related substances. Part XIII. The constitution of grandifoliolenone, an apo-tirucallol derivative from *Khaya grandifoliola* (Meliaceae). J Chem Soc(C) 1715–1718
109 Connolly J D, Overton K H 1972 The triterpenoids. In: Newman A A (ed) Chemistry of terpenes and terpenoids. Academic Press New York, 207–287
110 Cordell G A 1976 Biosynthesis of sesquiterpenes. Chem Rev 76:425–460
111 Cordell G A 1977 The sesterterpenes – a rare group of natural products. Progr Phytochem 4: 209–256
112 Cori M O 1983 Enzymic aspects of the biosynthesis of monoterpenes in plants. Phytochemistry 22:331–341
113 Cornforth J W 1968a Olefin alkylation in biosynthesis. Angew Chem Int Ed 7:903–911
114 Cornforth J W 1968b Terpenoid biosynthesis. Chem Brit 102–106
115 Coscia C J 1969 Picrotoxin. In: Taylor W I, Battersby A R (eds) Cyclopentanoid terpene derivatives. Marcel Dekker New York, 147–201
116 Cotterrell G P, Halsall T G, Wriglesworth M J 1970 The chemistry of triterpenes and related compounds. Part XLVII. Clarification of the nature of the tetracyclic triterpene acids of elemi resin. J Chem Soc(C) 739–743
117 Crawford R J, Erman W F, Broaddus C D 1971 Metalation of limonene. A novel method for the synthesis of bisabolane sesquiterpenes. J Am Chem Soc 94:4298–4306
118 Croteau R 1980 The biosynthesis of terpene compounds. Perfumer and Flavorist (VIII Int Congress Essent Oils issue) 35–59
119 Croteau R, Karp F 1979 Biosynthesis of monoterpenes: preliminary characterization of bornyl pyrophosphate synthetase from sage (*Salvia officinalis*) and demonstration that geranyl pyrophosphate is the preferred substrate for cyclization. Arch Biochem Biophys 198:512–522
120 Crowley K J 1964 Some terpenic constituents of *Bursera graveolens*. J Chem Soc 4254–4256
121 Damodaran N P, Sukh Dev 1968a Studies in sesquiterpenes. XXXVII. Sesquiterpenoids from the essential oil of *Zingiber zerumbet* Smith. Tetrahedron 24:4113–4122
122 Damodaran N P, Sukh Dev 1968b Studies in sesquiterpenes. XXXVIII. Structure of humulene epoxide-I and humulene epoxide-II. Tetrahedron 24:4123–4132
123 Dauben W G, Coates R M 1963 Determination of the configuration of C-9 in levopimaric acid. J Org Chem 28:1698–1699
124 Davies B H 1976 Carotenoids. In: Goodwin T W (ed) Chemistry and biochemistry of plant pigments, vol II. Academic Press New York, 38–165
125 Dean F M, Parton B, Somvichien N, Taylor D A H 1967 The coumarins of *Ptaeroxylon obliquum*. Tetrahedron Lett 2147–2151
126 Derfer J M, Derfer M M 1983 Terpenoids. In: Kirk-Othmer encyclopedia of chemical technology. Wiley-Interscience New York 22:709–762
127 Devlin R M 1969 Plant physiology. Van Nostrand Reinhold New York, 446 pp
128 Devon T K, Scott A I 1972 Handbook of naturally occurring compounds, vol II. Academic Press New York, 576 pp
129 Doi K, Shibuya T, Matsuo T, Miki S 1971 Longifol-7(15)-en-5β-ol and longifolan-3α,7α-oxide: new sesquiterpenes from *Juniperus conferta* Parl. Tetrahedron Lett 4003–4006
130 Drengler K A, Coates R M 1980 Stereochemistry of the S_N' cyclization in the biosynthesis of *ent*-sandaracopomaradiene with enzyme extracts from seedlings of *Ricinus communis* L. J Chem Soc Chem Commun 856–857
131 Dreyer D L 1968 Limonoid bitter principles. Fortschr Chem Org Naturst 26:190–244
132 Edwards O E 1969 Sesquiterpene alkaloids. In: Taylor W I, Battersby A R (eds) Cyclopentanoid terpene derivatives. Marcel Dekker New York, 357–408

133 Ehret C, Ourisson G 1969 Le γ-gurjunene, structure et configuration, isomerisation de l'α-gurjunene. Tetrahedron 25:1785–1799

134 Ekong D E U, Okogun J L 1969 West African timbers. Part XXV. Diterpenoids of *Gossweilerodendron balsamiferum* Harms. J Chem Soc(C) 2153–2156

135 El-Naggar L J, Beal J L 1980 Iridoids. A review. J Nat Prod 43:649–707

136 Enzell C, Erdtman H 1958 The chemistry of natural order Cupressales. XXI. Cuparene and cuparenic acid, two sesquiterpenic compounds with a new carbon skeleton. Tetrahedron 4: 361–368

137 Enzell C R, Thomas B R 1965 The chemistry of the order Araucariales. 3. Structure and configuration of Araucarolone and some related compounds from *Agathis australis*. Acta Chem Scand 19:1875–1896

138 Epstein W W, Poulter C D 1973 A survey of some irregular monoterpenes and their biogenetic analogies to presqualene alcohol. Phytochemistry 12:737–747

139 Erdtman H 1952 Chemistry of some heartwood constituents of conifers and their physiological and taxonomic significance. Progr Organ Chem 1:22–63

140 Erdtman H, Norin T 1966 The chemistry of order Cupressales. Fortschr Chem Org Naturst 24: 206–287

141 Erdtman H, Westfelt L 1963 The neutral diterpenes from pine wood resin. Acta Chem Scand 17: 1826–1827

142 Escher S, Loew P, Arigoni D 1970 The role of hydroxygeraniol and hydroxynerol in the biosynthesis of loganin and indole alkaloids. J Chem Soc Chem Commun 823–825

143 Evans F J, Taylor S E 1983 Pro-inflammatory, tumour-promoting and anti-tumour diterpenes of plant families Euphorbiaceae and Thymelaeaceae. Fortschr Chem Org Naturst 44:1–99

144 Fenical W 1978 Diterpenoids. In: Scheuer P J (ed) Marine natural products, vol II. Academic Press New York, 173–245

145 Ferreira M A, King T J, Ali S, Thomson R H 1980 Naturally occurring quinones. Part 27. Sesquiterpenoid quinones and related compounds from *Hibiscus elatus*: crystal structure of hibiscone C (gmelofuran). J Chem Soc Perkin Trans I 249–256

146 Fischer N H, Olivier E J, Fischer H D 1979 The biogenesis and chemistry of sesquiterpene lactones. Fortschr Chem Org Naturst 38:47–390

147 Forsen K V, Schantz M 1973 Chemotypen von *Chrysanthemum vulgare* (L.) Bernh. In: Bendz G, Santesson J (eds) Chemistry in botanical classification. Academic Press New York, 145–152

148 Frank B, Petersen U, Hüper F 1970 Valerianine, a tertiary monoterpene alkaloid from *Valerian*. Angew Chem Int Ed 9:891

149 Fujita Y, Fujita S, Yoshikawa H 1970 trans- and cis-Yabunikkeol, new monoterpene alcohols isolated from the essential oils of *Cinnamomum japonicum* Sieb. Bull Chem Soc Jpn 43:1599

150 Galeffi C, Monache E M, Casinovi C G, Marini-Bettolo G B 1969 A new quinone from the heartwood of *Mansonia altissima* Chev: mansonone-L. Tetrahedron Lett 3583–3584

151 Gascoigne R M 1955 The tetracyclic triterpenes. Quart Rev 9:328–361

152 Geissman T A, Crout D H G 1969 Organic chemistry of secondary plant metabolism. Freeman Cooper San Francisco, 360–370

153 Geissman T A, Irwin M A 1973 Chemical constitution and botanical affinity. In: Bendz G, Santesson J (eds) Chemistry in botanical classification. Academic Press New York, 135–143

154 Ghisalberti E L, Jefferies P R, Proudfoot G M 1981 The chemistry of *Eremophila* spp. XV. New acyclic diterpenes from *Eremophila* spp. Aust J Chem 34:1491–1499

155 Gibbs R D 1974 Chemotaxonomy of flowering plants, vol II. McGill-Queen's University Press Montreal, 770–872

156 Glasby J S 1982 Encyclopaedia of the terpenoids. John Wiley New York, 2646 pp

157 Gnecco S, Poyser J P, Silva M, Sammes P G, Tyler T W 1973 Sesquiterpene lactones from *Podanthus ovatifolius*. Phytochemistry 12:2469–2477

158 Godfrey J D, Schultz A G 1979 The total synthesis of dl-dihydrocallitrisin. Tetrahedron Lett 3241–3244

159 Goodwin T W 1967 The biological significance of terpenes in plants. In: Pridham J B (ed) Terpenoids in plants. Academic Press New York, 1–23

160 Goodwin T W (ed) 1970 Natural substances formed biologically from mevalonic acid. Academic Press New York, 186 pp
161 Goodwin T W 1978 The biochemical functions of terpenoids in plants. R Soc London, 161pp
162 Goodwin T W 1979 Biosynthesis of terpenoids. Ann Rev Plant Physiol 30:369–404
163 Goodwin T W 1980 The biochemistry of the carotenoids, vol I. Chapman and Hall London, 377 pp
164 Goodwin T W, Mercer E I 1983 Introduction to plant biochemistry. Pergamon Press Oxford, 677 pp
165 Graedel T E 1979 Terpenoids in the atmosphere. Rev Geophysics Space Physics 17:937–947
166 Grant P K, Hill N R 1964 Synthesis of colensan-1-one. Aust J Chem 17:66–69
167 Grant P K, Huntrakul C, Sheppard D R J 1967 Diterpenes of *Dacrydium bidwillii*. Aust J Chem 20:969–972
168 Grieco P A, Masaki Y 1975 Synthesis of the *Valeriana wallichii* hydrocarbon sesquifenchene. A route to specifically functionalized 7,7-disubstituted bicyclo[2,2,1]heptane derivatives. J Org Chem 40:150–151
169 Gross V D 1977 Phytoalexine und verwandte Pflanzenstoffe. Fortschr Chem Org Naturst 34: 187–247
170 Grundon M F 1978 The biosynthesis of aromatic hemiterpenes. Tetrahedron 34:143–161
171 Guenther E 1948–1952 The essential oils, vol 1–6. Vam Nostrand Reinhold New York
172 Gupta A S, Sukh Dev 1971a Studies in sesquiterpene. XLVI. Sesquiterpenes from the oleoresin of *Dipterocarpus pilosus*: humulene epoxide-III, caryophyllenol-I and caryophyllenol-II. Tetrahedron 27:635–644
173 Gupta A S, Sukh Dev 1971b Higher terpenoids. I. Triterpenoids from the oleoresin of *Dipterocarpus pilosus*: hollongdione and dipterocarpolic acid. Tetrahedron 27:823–834
174 Gupta R K, Mahadevan V 1967 Chemical examination of the heartwood of *Diospyros ebenum*. Indian J Pharm 29:289–291
175 Haeuser J, Lombard R, Lederer F, Ourisson G 1961 Isolement et structure d'un nouveau diterpene: l'acide daniellique. Stereochemie de l'acide daniellique. Tetrahedron 12:205–214
176 Halsall T G, Aplin R T 1964 A pattern of development in the chemistry of pentacyclic triterpenes. Fortschr Chem Org Naturst 22:153–202
177 Hänsel R, Kartarahardja M, Huang J-T, Bohlmann F 1980 Sesquiterpenlacton-β-D-glucopyranoside sowie ein neues Eudesmanolid aus *Taraxacum officinale*. Phytochemistry 19:857–861
178 Hanson J R 1971 The biosynthesis of the diterpenoids. Fortschr Chem Org Naturst 29:395–416
179 Hanson J R 1979 Terpenoid biosynthesis. In: Barton D, Ollis W D (eds) Comprehensive organic chemistry, vol 5. Pergamon Press Oxford, 989–1023
180 Harada N, Uda H 1978 Absolute stereochemistries of 3-epicaryoptin, caryoptin, and clerodin as determined by chiroptical methods. J Am Chem Soc 100:8022–8024
181 Harborne J B 1971 Terpenoids and other low molecular weight substances of systematic interest in the Leguminosae. In: Harborne J B, Boulter D, Turner B L (eds) Chemotaxonomy of the Leguminosae. Academic Press New York, 257–283
182 Harris G C, Sanderson T F 1948 Resin acids. III. The isolation of dextropimaric acid and a new pimaric-type acid, isodextropimaric acid. J Am Chem Soc 70:2079–2081
183 Hart N K, Lamberton J A, Triffett A C K 1973 Triterpenoids of *Achras sapota* (Sapotaceae). Aust J Chem 26:1827–1829
184 Heathcock C H, Graham S L, Pirrung M C, Plavac F, White C T 1983 The total synthesis of sesquiterpenes, 1970–1979. In: ApSimon J (ed) SThe total synthesis of natural products, vol 5. Wiley-Interscience New York, 550 pp
185 Hedden P, MacMillan J, Phinney B O 1978 The metabolism of the gibberellins. Ann Rev Plant Physiol 29:149–192
186 Hegnauer R 1969 Chemical evidence for the classification of some plant taxa. In: Harborne J B, Swain T (eds) Perspectives in phytochemistry. Academic Press New York, 121–138
187 Heilbron I M, Kamm E D, Owens W M 1926 Unsaponifiable matter from the oils of elasmobranch fish. I. Contribution to the study of the constitution of squalene (spinacene). J Chem Soc 1630–1644

188 Hemming F W 1967 Polyisoprenoid alcohols (prenols). In: Pridham J B (ed) Terpenoids in plants. Academic Press New York, 223–239
189 Hemming F W 1970 Polyprenols. In: Goodwin T W (ed) Natural substances formed biologically from mevalonic acid. Academic Press New York, 105–117
190 Herout V 1971 Biochemistry of sesquiterpenoids. In: Goodwin T W (ed) Aspects of terpenoid chemistry and biochemistry. Academic Press New York, 53–94
191 Herout V, Sorm F 1969 Chemotaxonomy of the sesquiterpenoids of the compositae. In: Harborne J B, Swain T (eds) Perspectives in phytochemistry. Academic Press New York, 139–165
192 Herout V, Sykora V 1958 The chemistry of cadinenes and cadinols. Tetrahedron 4:246–255
193 Herz W 1973 Pseudoguaianolides in Compositae. In: Bendz G, Santesson J (eds) Chemistry in botanical classification. Academic Press New York, 153–172
194 Hikino H, Hikino Y, Takeshita Y, Meguro K, Takemoto T 1965 Structure and absolute configuration of valeranone. Chem Pharm Bull 13:1408
195 Hikino H, Konno C, Agatsuma K, Takemoto T, Horibe I, Tori K, Ueyama M, Takeda K 1975 Sesquiterpenoids. Part XLVII. Structure, configuration, conformation, and thermal rearrangement of furanodienone, isofuranodienone, curzerenone, epicurzerenone, and pyrocurzerenone, sesquiterpenoids of *Curcuma zedoaria*. J Chem Soc Perkin Trans I 478–484
196 Hikino H, Sakurai Y, Takahashi H, Takemoto T 1967 Structure of curdione. Chem Pharm Bull 15:1390–1394
197 Hikino H, Takeshita Y, Hikino Y, Takemoto T 1966 Structure and absolute configuration of fauronyl acetate and cryptofauronol. Chem Pharm Bull 14:735–741
198 Hinge V K, Wagh A D, Paknikar S K, Bhattacharyya S C 1965 Terpenoids. LXXI. Constituents of Indian black dammar resin. Tetrahedron 21:3197–3203
199 Hirota H, Tanahashi Y, Takahashi T 1980 Structure of a guaioxide-type sesquiterpene from "Sanshion". Bull Chem Soc Jpn 53:785–788
200 Hodges R, Reed R I 1960 The stereochemistry of manoyl oxide. Tetrahedron 10:71–75
201 Holub M, Tax J, Sedmera P, Sorm F 1970 On terpenes. CCVII. Contribution to the stereochemistry of substances with daucane skeleton and the determination of absolute configuration of laserpitin. Collect Czech Chem Commun 35:3597–3609
202 Honwad V K, Rao A S 1965 Terpenoids LXIX. Absolute configuration of (−)-α-curcumene. Tetrahedron 21:2593–2604
203 Horkmann A G 1968 A hypothesis for the formation of cis-fused eudesmane-type sesquiterpenes. Tetrahedron Lett 5785–5786
204 Hugel G, Lods L, Mellor J M, Theobald D W, Ourisson G 1965 Diterpenes de Trachylobium. Bull Soc Chim Fr 2882–2887
205 Hugel G, Lods L, Mellor J M, Theobald D W, Ourisson G 1965 Diterpenes de *Trachylobium*. II. Structure des diterpenes tetra- et pentacycliques de *Trachylobium*. Bull Soc Chim Fr 2888–2894
206 Huneck S 1963 Triterpene – IV. Die Triterpensäuren des Balsams von *Liquidambar orientalis* Miller. Tetrahedron 19:479–482
207 Hürlimann H, Cherbuliez E 1981 Konstitution und Vorkommen der organischen Pflanzenstoffe. Suppl 2, Part 1. Birkhäuser Basel, 939 pp
208 Iguchi M, Nishiyama A 1970 Preisocalamendiol, a plausible precursor of isocalamendiol. Tetrahedron Lett 855–857
209 Ingham J L 1983 Naturally occurring isoflavonoids. Fortschr Chem Org Naturst 43:1–265
210 Inouye H, Ueda S, Uesato S, Shingu T, Thies P W 1974 Die absolute Konfiguration von Valerosidatum und von Dihydrovaltratum. Tetrahedron 30:2317–2325
211 Inouye H, Yoshida T, Nakamura Y, Tobita S 1968 Die Stereochemie einiger Secoiridoidglucoside und die Revision der Struktur des Gentiopicrosids. Tetrahedron Lett 4429–4432
212 International Union of Pure and Applied Chemistry 1976 Nomenclature of organic chemistry: Section F – Natural products and related compounds. Information Bull Int Union Pure Appl Chem, 22 pp
213 Inubushi Y, Sano T, Tsuda Y 1964 Serratenediol: a new skeletal triterpenoid containing a seven membered ring. Tetrahedron Lett 1303–1310
214 Irie H, Kimura H, Otani N, Ueda K, Uyeo S 1967 The revised structure of bilobanone. J Chem Soc Chem Commun 678–679

215 Ishi H, Tozyo T, Nakamura M, Minato H 1970 Studies on sesquiterpenoids. XIX. Structure and absolute configuration of liguloxide, liguloxidol and liguloxidol acetate. Tetrahedron 26: 2911–2918
216 Isler O (ed) 1971 Carotenoids. Birkhäuser Basel, 932 pp
217 Jefferies P R, Rosich R S, White D E 1963 The absolute configuration of beyerol. Tetrahedron Lett 1793–1799
218 Jolad S D, Wiedhopf R M, Cole J R 1977 Cytotoc agents from *Bursera klugii* (Burseraceae) I: Isolation of sapelins A and B. J Pharm Sci 66:889–890
219 Jones R V H, Sutherland M D 1968 Hdycaryol, the precursor of elemol. J Chem Soc Chem Commun 1229–1230
220 Joseph T C, Sukh Dev 1968 Studies in sesquiterpenes. XXXI. The absolute stereochemistry of himachalenes. Tetrahedron 24:3841–3852
221 Joseph-Nathan P, Negrete Ma C, Gonzales Ma P 1970 Studies in *Cacalia* species. Tetrahedron 9:1623–1628
222 Kalsi P S, Gupta B C, Chahal S, Mehta Y K, Wadia M S 1979 Structure of isokhusinoloxide. A new sesquiterpene epoxy alcohol from vetiver oil. Bull Soc Chim Fr Part 2, 599–600
223 Kaneda M, Iitaka Y, Shibata S 1972 Chemical studies on the oriental plant drugs. XXXIII. The absolute structures of paeoniflorin, albiflorin, oxypaeoniflorin and benzoylpaeoniflorin isolated from Chinese paeony root. Tetrahedron 28:4309–4317
224 Kapadi A H, Soman R, Sobti R R, Sukh Dev 1983 Higher isoprenoids: Part XIV. Diterpenoids of *Erythroxylon monogynum* Roxb. (Part 1): Introduction, isolation and biogenetic considerations. Indian J Chem 22B:964–969
225 Kapadi A H, Sukh Dev 1964 The diterpenoids of *Erythroxyion monogynum*. III. furtsher constituents, the absolute stereochemistry of monogynol, and hydroxymonogynol. Tetrahedron Lett 2751–2757
226 Kapadia V H, Nagasampagi B A, Naik V G, Sukh Dev 1965 Studies in sesquiterpenes. XXII. Structure of mustakone and copaene. Tetrahedron 21:607–618
227 Karlsson B, Pilotti A M, Söderholm A C, Norin T, Sundin S, Sumimoto M 1978 Structure and absolute configuration of verticillol, a macrocyclic diterpene alcohol from the wood of *Sciadopitys verticillata* Sieb. et Zucc. (Taxodiaceae). Tetrahedron 34:2349–2354
228 Karrer W 1958 Konstitution und Vorkommen der organischen Pflanzenstoffe. Birkhäuser Basel, 1207 pp
229 Karrer W, Cherbuliez E, Eugster C H 1977 Konstitution und Vorkommen der organischen Pflanzenstoffe. Suppl 1. Birkhäuser Basel, 1038 pp
230 Kato T, Yen C C, Kobayashi T, Kitahara Y 1976 Cyclization of polyenes. XXI. Synthesis of d1-incensole. Chem Lett 1191–1192
231 Kaiser R, Naegeli P 1972 Biogenetically significant components in vetiver oil. Tetrahedron Lett 2009–2012
232 Kido F, Sakuma R, Uda H, Yoshishi A 1969 Minor acidic constituents of vetiver oil. Part II (1). Cyclocopacamphenic and epicyclocopacamphenic acids. Tetrahedron Lett 3169–3172
233 Kimland B, Norin T 1967 The oleoresin of Norwegian spruce, *Picea abies* (L.) Karst. Isolation of (−)-geranyllinalool. Acta Chem Scand 21:825–826
234 King T J, Rodrigo S, Wallwork S C 1969 The stereochemistry of plathyterpol, a diterpene with A: B-*cis*-ring fusion. J Chem Soc Chem Commun 683
235 Kitagawa I, Hino K, Nishimura T, Iwata E, Yoshioka I 1971 On the constituents of *Picrorhiza kurrooa* (1). The structure of picroside I, a bitter principle of the subterranean part. Chem Pharm Bull 19:2534–2544
236 Kitahara Y, Yoshikoshi A 1964 The structure of hibaene. Tetrahedron Lett 1771–1774
237 Klein E, Rojahu W 1970 Die Konfigurationen der Sesquiterpenoide − Literaturreport. Dragaco Holzminden Germany, 71 pp
238 Klyne W, Buckingham J 1974 Atlas of stereochemistry, absolute configurations of organic molecules. Chapman and Hall London, 311 pp
239 Kohda H, Kasai R, Yamasaki K, Murakami K, Tanaka O 1976 New sweet diterpenes glucosides from *Stevia rebaudiana*. Phytochemistry 15:981–983

240 Kolbe-Haugwitz M, Westfelt L 1970 Copaborneol, constitution and synthesis. Acra Chem Scand 24:1623–1630
241 Krepinsky J, Jommi G, Samek Z, Sorm F 1970 On terpenes. CCIV. The sesquiterpenic constituents of *Aristolochia debilis* Sieb. et Zucc. The structure of debilone. Collect Czech Chem Commun 35:745–748
242 Kretschmar H C, Erman W F 1970 The total synthesis and geometric configuration of d1-β-santalol. Tetrahedron Lett 41–44
243 Krishnappa S, Sukh Dev 1973 Sesquiterpenes from *Lansium anamalayanum*. Tetrahedron 12: 823–825
244 Kubo I, Miura I, Pettei M J, Lee Y W, Pilkiewicz F, Nakanishi K 1977 Muzigadial and warburganal, potsent antifungal, antiyeast, and African army worm antifeedant agents. Tetrahedron Lett 4553–4556
245 Kulshreshtha M J, Kulshreshtha D K, Rastogi R P 1972 The triterpenoids. Phytochemistry 11: 2369–2381
246 Kumar N, Ravindranath B, Seshadri T R 1974 Terpenoids of *Pterocarpus santalinus* heartwood. Phytochemistry 13:633–636
247 Kupchan S M, Court W A, Dailey R G, Gilmore C J, Bryan R F 1972 Triptolide and tripdiolide, novel antileukemic diterpenoid triepoxides from *Tripterygium wilfordii*. J Am Soc 94: 7194–7195
248 Kupchan S M, Hemingway R J, Werner D, Karim A 1969 Vernolepin, a novel sesquiterpene dilactone tumor inhibitor from *Vernonia hymenolepis* A Rich. J Org Chem 34:3903–3908
249 Kupchan S M, Karim A, Marcks C 1969 Tumor inhibitors. XLVIII. Taxodione and taxodone, two novel diterpenoid quinone methide tumor inhibitors from *Taxodium distichum*. J Org Chem 34:3912–3918
250 Kutney J P, Rogers I H 1968 Novel triterpenes from Sitka spruce (*Picea sitchensis* [Bong.] Carr). Tetrahedron Lett 761–766
251 Laidlaw R A, Morgan J W W 1963 The diterpenes of *Xylia dolabriformis*. J Chem Soc 644–650
252 Leonhardt B A, Beroza M (eds) 1982 Insect pheromone technology: Chemistry and applications. Am Chem Soc Symp Series 190:260 pp
253 Letham D S, Palni L M S 1983 The biosynthesis and metabolism of cytokinins. Ann Rev Plant Physiol 34:163–197
254 Liaaen-Jensen S 1980 Stereochemistry of naturally occurring carotenoids. Fortschr Chem Org Naturst 39:123–172
255 Lin Y T, Cheng Y S, Kuo Y H 1968 Extractive components from the wood of *Taiwania cryptomerioides* Hayata: a new sesquiterpene keto alcohol, cadinane-3-ene-9α-ol-2-one. Tetrahedron Lett 3881–3882
256 Lindgren B O 1965 Homologous aliphatic C_{30}–C_{45} terpenols in birch wood. Acta Chem Scand 19:1317–1326
257 Lisina A I, Pentegova V A 1965 Abietadiene from the rosin of the Siberian larch (*Larix sibirica*). Izvest Sibir Otdel Akad Nauk SSSR Ser Khim Nauk 96–100
258 Loomis W D 1967 Biosynthesis and metabolism of monoterpenoids. In: Pridham J B (ed) Terpenoids in plants. Academic Press New York, 59–82
259 Luckner M, Diettrich B, Lerbs W 1980 Cellular compartmentation and channelling of secondary metabolism in microorganisms and higher plants. Progr Phytochem 6:103–142
260 Machlin L J 1980 Vitamin E, a comprehensive treatise. Marcel Dekker New York, 660 pp
261 Mackie H, Overton K H 1977 Hydrolysis and isomerization of trans,trans-farnesyl diphosphate by *Andrographis* tissue culture enzymes. Eur J Biochem 77:101–106
262 MacMillan J, Pryce R J 1973 The gibberellins. In: Miller L P (ed) Phytochemistry, vol III. Van Nostrand New York, 283–326
263 Manitto P 1981 Biosynthesis of natural products. Ellis Horwood Chichester, 213–345
264 de Marcano D P D C, Halsall T G 1975 Structures of some taxane diterpenoids, baccatins-III, -IV, -VI, and -VII and 1-dehydroxybaccatin-IV, possessing an oxetan ring. J Chem Soc Chem Commun 365–366
265 Marini-Bettolo G H (ed) 1977 Natural products and the protection of plants. Elsevier Amsterdam, 846 pp

266 Markham K R, Porter L J 1978 Chemical constituents of the bryophytes. Progr Phytochem 5:181–272
267 Marshall J A, Brady St F, Andersen NH 1974 The chemistry of spiro[4,5]decane sesquiterpenes. Fortschr Chem Org Naturst 31:283–376
268 Martin J D 1973 New diterpenoids extractives of *Maytenus dispermus*. Tetrahedron 29:2553–2559
269 Massy-Westropp R A, Reynolds G D, Spotswood T M 1966 Freelingyne, an acetylenic sesquiterpenoid. Tetrahedron Lett 1939–1946
270 Mayo P, Williams R E, Büchi G, Feairheller S H 1965 The absolute stereostructure of copaene. Tetrahedron 21:619–627
271 McCabe T, Clardy J, Minale L, Pizza C, Zollo F, Riccio R 1982 A triterpenoid pigment with the isomalabaricane skeleton from the marine sponge *Stelletta* sp. Tetrahedron Lett 23:3307–3310
272 McCrindle R, Nakamura E, Anderson A B 1976 Constituents of *Solidago* species. Part VII. Constitution and stereochemistry of the *cis*-clerodanes from *Solidago arguta* Ait. and of related diterpenoids. J Chem Soc Perkin Trans I 1590–1597
273 McEachan C E, McPhail A T, Sim G A 1966 The structure of cascarillin: X-ray analysis of deacetylcascarillin acetal iodoacetate. J Chem Soc(B) 633–640
274 McGarry E J, Pegel K H, Phillips L, Waight E S 1971 The constitution of the aromatic diterpene cleistanthol. J Chem Soc(C) 904–909
275 Merkel D 1962–1966 E Gildemeister/Fr Hoffmann. Die Ätherischen Öle, vol IIIb–IIId. Akademie Berlin
276 Milborrow B V 1971 Abscisic acid. In: Goodwin T W (ed) Aspects of terpenoid chemistry amd biochemistry. Academic Press New York, 137–151
277 Misra R, Pandey R 1981 Cytotoxic and antitumor terpenoids. In: Aszalos A (ed) Antitumor compounds of natural origin, vol II. CRC Press Boca Raton, 145–192
278 Misra R, Pandey R C, Sukh Dev 1978 Higher isoprenoids. X. Diterpenoids from the oleoresin of *Hardwickia pinnata*. Part 3: Kolavenol, kolavelool and a nor diterpene hydrocarbon. Tetrahedron 35:985–987
279 Misra R, Pandey R C, Sukh Dev 1979 Higher terpenoids. VIII. Diterpenoids from the oleoresin of *Hardwickia pinnata*. Part I: hardwickiic acid. Tetrahedron 35:2301–2310
280 Mondon A, Callsen H, Epe B 1975 Zur Kenntnis der Bitterstoffe aus Cneoraceen, III. Tetrahedron Lett 703–706
281 Mondon A, Epe B 1983 Bitter principles of Cneoraceae. Fortschr Chem Org Naturst 44:101–187
282 Morin R B 1968 Erythrophleum alkaloids. Alkaloids 10:287–303
283 Moss G P, Weedon B C L 1976 Chemistry of carotenoids. In: Goodwin T W (ed) Chemistry and biochemistry of plant pigments, vol I. Academic Press New York, 149–224
284 Mothes K 1976 Secondary plant substances as materials for chemical high quality breeding in higher plants. In: Wallace J W, Mansell R L (eds) Recent advances in phytochemistry, vol 10. Biochemical interaction between plants and insects. Plenum Press New Jork, 385–405
285 Motl O, Romanuk M, Herout V 1966 On terpenes. CLXXVIII. Composition of the oil from *Amorpha fruticosa* L. fruits. Structure of (−)-γ-amorphene. Collect Czech Chem Commun 31:2025–2033
286 Muller B, Tamm C 1975 The mode of incorporation of (3R)-[(5R)-5-^{3}H]-mevalonate into verrucarol. Helv Chim Acta 58:483–488
287 Murakami T, Satake T, Tezuka M, Tanaka K, Tanaka F, Chen C M 1974 Chemische Untersuchungen der Inhaltsstoffe von *Pteris cretica* L. Chem Pharm Bull Jpn 22:1686–1689
288 Murray R D H 1978 Naturally occurring plant coumarins. Fortschr Chem Org Naturst 35:199–429
289 Naegli P, Klimes J, Weber G 1970 Structure and synsthesis of artemone. Tetrahedron Lett 5021–5024
290 Nagasampagi B A, Yankov L, Sukh Dev 1967 Isolation and characterisation of geranylgeraniol. Tetrahedron Lett 189–192
291 Nagasampagi B A, Yankov L, Sukh Dev 1968 Sesquiterpenoids from the wood of *Cedrela toona* Roxb.: partial synthesis of T-muurolol, T-cadinol and cubenol; structures of δ-cadinene and δ-cadinol. Tetrahedron Lett 1913–1918
292 Nakanishi K 1974 Diterpenes. In: Nakanishi K, Goto T, Ito S, Natori S, Nozoe S (eds) Natural products chemistry, vol 1. Academic Press New York, 185–312

293 Nakanishi K, Habaguchi K 1971 Biosynthesis of ginkgolide B, its diterpenoid nature, and origin of the *tert*-butyl group. J Am Chem Soc 93:3546–3547
294 Naves Y-R 1960 Sur la presence de geraniol, de nerol, de linalol, de farnesols et de nerolidols dans les huiles essentielles. CR Acad Sci 251:900–902
295 Nayak U R, Sukh Dev 1968 Studies in sesquiterpenes. XXXV. Longicyclene, the first tetracyclic sesquiterpene. Tetrahedron 24:4099–4104
296 Nigam I C, Levi L 1963 Essential oils and their constituents. XIX. Detection of new trace constituents in oil of rosewood. Perfum Essent Oil Rec 54:814–816
297 Nishimura K 1969 A new sesquiterpene, bicyclogermacrene. Tetrahedron Lett 3097–3100
298 Nozoe T 1956 Natural tropolones and some related troponoids. Fortschr Chem Org Naturst 13: 232–301
299 Nozoe S, Morisaki M, Tsuda K, Itaka Y, Takahashi N, Tamura S, Ishibashi K, Shirasaka M 1965 The structure of ophiobolin, a C_{25} terpenoid having a novel skeleton. J Am Chem Soc 87: 4968–4970
300 Ogiso A, Kitazawa E, Kurabayashi M, Sato A, Takahashi S, Noguchi H, Kuwano H, Kobayashi S, Mishima H 1978 Isolation and structure of antipeptic ulcer diterpene from Thai medicinal plant. Chem Pharm Bull 26:3117–3123
301 Ohta Y, Hirose Y 1969 Stereochemistry of (+)-α-ylangene. Tetrahedron Lett 1601–1604
302 Ourisson G 1973 Some aspects of the distribution of diterpenes in plants. In: Bendz G, Santesson J (eds) Chemistry in botanical classification. Academic Press New York, 129–134
303 Ourisson G, Albrecht P, Rohmer M 1979 The hopanoids: plaeochemistry and biochemistry of a group of natural products. Pure Appl Chem 51:709–729
304 Ourisson G, Crabbe P, Roding O 1961 Tetracyclic triterpenes. Holden-Day San Francisco, 237 pp
305 Ourisson G, Munavalli S, Ehret C 1966 Data relative to sesquiterpenoids. Pergamon Press Oxford, 70 pp
306 Ourisson G, Rohmer M, Anton R 1979 From terpenes to sterols: Macroevolution and microevolution. In: Swain T, Waller G R (eds) Topics in the biochemistry of natural products. Plenum Press New York London, 131–162
307 Overton K H, Weir N G, Wylie A 1966 The stereochemistry of the Colombo root bitter principles. J Chem Soc(C) 1482–1490
308 Pande B S, Krishnappa S, Bisarya S C, Sukh Dev 1971 Studies in sesquiterpenes. XLVII. cis- and trans-Atlantones from *Cedrus deodara* Loud. Tetrahedron 27:841–844
309 Pant P, Rastogi R P 1979 The triterpenoids. Phytochemistry 18:1095–1108
310 Parker W, Roberts J S, Ramage R 1967 Sesquiterpene biogenesis. Quart Rev 21:331–363
311 Pasteels J M, Gregoire J-C, Rowell-Rahier M 1983 The chemical ecology of defence in arthropods. Ann Rev Entomol 28:263–289
312 Paton W F, Paul I C, Bajaj A G, Sukh Dev 1979 The structure of malabaricol. Tetrahedron Lett 4153–4154
313 Pelletier S W, Keith L H 1970 The diterpene alkaloids. Alkaloids 12:1–206
314 Pinder A R 1977 The chemistry of the eremophilane and related sesquiterpenes. Fortschr Chem Org Naturst 34:81–186
315 Plattner P A 1948 Dehydrogenation with sulfur, selenium, and platinum metals. In: Newer methods of preparative organic chemistry. Wiley-Interscience New York, 21–59
316 Polansky J 1973 Quassinoid bitter principles. Fortschr Chem Org Naturst 30:101–150
317 Ponsinet G, Ourisson G, Oehlschlager A C 1968 Systematic aspects of the distribution of di- and triterpenes. In: Malory T J, Alson R E (eds) Recent advances in phytochemistry, vol I. Proc. 6th Ann Symp. Phytochem Soc N Am. Plenum Press New York, 271–302
318 Popjak G, Edmond J, Wong S-M 1973 Absolute configuration of presqualene alcohol. J Am Chem Soc 95:2713–2714
319 Prasad R S, Sukh Dev 1976 Chemistry of α-yurvedic crude drugs. IV. Guggulu (resin from *Commiphora mukul*). 4. Absolute stereochemistry of mukulol. Tetrahedron 32:1437–1441
320 Rao M M, Meshulam H, Zelnik R, Lavie D 1975 *Cabralea eichleriana* D C (Meliaceae). I. Structure and stereochemistry of wood extractives. Tetrahedron 31:333–339
321 Rao A S, Paul A, Sadgopal, Bhattacharyya S C 1961 Terpenoids. XXV. Structure of saussurea lactone. Tetrahedron 13:319–323
322 Rice E L 1974 Allelopathy. Academic Press New York, 368 pp

323 Richards J H, Hendrickson J B 1964 The biosynthesis of steroids, terpenes and acetogenins. Benjamin New York Amsterdam, 416 pp
324 Rigaudy J, Klesney S P (eds) 1979 Nomenclature of organic chemistry. Int Union Pure Appl Chem. Pergamon Press Oxford, 559 pp
325 Rilling H C 1966 A new intermediate in the biosynthesis of squalene. J Biol Chem 241: 3233–3236
326 Roberts E M, Lawrence R V 1956 The occurrence of dextropimarinal and isodextropimarinal in commercial gum rosin. J Am Chem Soc 78:4087–4089
327 Robinson R 1939 Cited in Penfold A R, Simonsen J L Constitution of eremophilone, hydroxyeremophilone and hydroxydiydroeremophilone. III. J Chem Soc 87–89
328 Rogers D, Ünal G G, Williams D J, Steven V L, Joshi B S, Ravindranath K R 1979 The crystal structure of 3-epicaryoptin and the reversal of the currently accepted absolute configuration of clerodin. J Chem Soc Chem Commun 97–99
329 Roman L U, del Rio R E, Hernandez J D, Joseph-Nathan P, Zabel V, Watson W H 1981 Structure, chemistry and stereochemistry of rastevione, a sesquiterpenoid from the genus *Stevia*. Tetrahedron 37:2769–2778
330 Romo J, Romo de Vivar A 1967 The pseudoguaianolides. Fortschr Chem Org Naturst 25:90–130
331 Rowe J W 1964 Triterpenes of pine barks: Identity of pinusenediol and serratenediol. Tetrahedron Lett 2347–2353
332 Rowe J W, Conner A H 1979 Extractives in Eastern hardwoods – a review. US For Serv Gen Tech Rep FPL-18, 67 pp
333 Rücker G 1973 Sesquiterpenes. Angew Chem Int Ed 12:793–806
334 Rudney H 1970 The biosynthesis of terpenoid quinones. In: Goodwin T W (ed) Natural substances formed biologically from mevalonic acid. Academic Press New York, 89–103
335 Rustaiyan A, Niknejad A, Nazarians L, Jakupovic J, Bohlmann F 1982 Sesterterpenes from *Salvia hypoleuca*. Phytochemistry 21:1812–1813
336 Ruzicka L 1953 The isoprene rule and the biogenesis of terpenic compounds. Experientia 9: 357–367
337 Ruzicka L 1959 History of isoprene rule. J Chem Soc 341–360
338 Sakai T, Nishimura K, Hirose Y 1965 The structure and stereochemistry of four new sesquiterpenes isolated from the wood oil of "Kaya" (*Torreya nucifera*). Bull Chem Soc Jpn 38: 381–387
339 Samek Z, Harmatha J, Novotony L, Sorm F 1969 On terpenes. CCII. Absolute configuration of adenostylone, neoadenostylone and isoadenostylone from *Adenostyles alliariae* (Gouan) Kern, and of decompositin from *Cacalia decomposita* A. Gray. Collect Czech Chem Commun 34: 2792–2808
340 Sato A, Kurabayashi M, Nagahori H, Ogiso A, Mishima H 1970 Chettaphanin-I, a novel furanoditerpenoid. Tetrahedron Lett 1095–1098
341 Schmitz R, Frahm A W, Kating H 1980 Ein neues Sesquiterpenketon aus *Arnica*-arten. Phytochsemistry 19:1477–1480
342 Scott A I 1970 Biosynthesis of the indole alkaloids. Acct Chem Res 3:151–157
343 Seaman F, Fischer N H 1978 Longipilin, a new melampolide from *Melampodium longipilum*. Phytochemistry 17:2131–2132
344 Semmler F W 1910 Constituents of ethereal oils. Constitutions of the α-santalol and of the α-santalene series, and of sesquiterpene alcohols and of sesquiterpenes. Ber Deutsch Chem Ges 40: 1893–1898
345 Shaligram A M, Rao A S, Bhattacharyya S C 1962 Terpenoids. XXXII. Absolutse configuration of junenol and laevojunenol and synthesis of junenol from costunolide. Tetrahedron 18:969–977
346 Shankaranarayan R, Bisarya S C, Sukh Dev 1977 Studies in sesquiterpenes. LIV. Oxidohimachalene, a novel sesquiterpenoid from the wood of *Cedrus deodara* Loud. Tetrahedron 33: 1207–1210
347 Shankaranarayan R, Krishnappa S, Bisarya S C, Sukh Dev 1977 Studies in sesquiterpenes – LIII. Deodarone and atlantolone, new sesquiterpenoids from wood of *Cedrus deodara* Loud. Tetrahedron 33:1201–1205
348 Shibata S 1977 Saponins with biological and pharmacological activity. In: Wagner H, Wolff P (eds) New natural products and plant drugs with pharmacological, biological or therapeutical activity. Springer Heidelberg, 177–195

349 Shibata S 1981 Chinese drug constituents: isolation of the biologically active principles. In: Natori S, Ikekawa N, Suzuki M (eds) Advances in natural products chemistry. Kodansha Tokyo, 398–416
350 Simonsen J L, Barton D H R 1951 The terpenes, vol III. Cambridge University Press Cambridge, 579 pp
351 Simonsen J, Ross W C J 1957 The terpenes, vol IV–V. Cambridge University Press Cambridge
352 Smedman L, Zavarin E 1968 Cyclosativene – a tetracyclic sesquiterpene. Tetrahedron Lett 3833–3835
353 Smith R H 1964 The monoterpenes of lodgepole pine oleoresin. Phytochemistry 3:259–262
354 Smith R M 1977 The Celastraceae alkaloids. Alkaloids 16:215–248
355 Smith C R, Madrigal R V, Weisleder D, Mikolajczak K L, Highet R J 1976 Potamogetonin, a new furanoid diterpene. Structural assignment by carbon-13 and proton magnetic resonance. J Org Chem 41:593–596
356 Soman R, Kapadi A H, Sobti R R, Sukh Dev 1983 Higher isoprenoids. Part XVIII. Diterpenes of *Erythroxylon monogynum* Roxb. (Part 5): minor constituents. Indian J Chem 22B:989–992
357 Soman R, Sukh Dev 1983a Higher isoprenoids: Part XVI. Diterpenoids from *Erythroxylon monogynum* Roxb. (Part 3): ($\pm$)-Devadarool and ($-$)-hydroxydevadarool. Indian J Chem 22B: 978–983
358 Soman R, Sukh Dev 1983b Higher isoprenoids: Part XVII. Diterpenoids from *Erythroxylon monogynum* Roxb. (Part 4): Absolute stereostructure of (+)-allodevadarool. Indian J Chem 22B: 984–988
359 Sondheimer E, Simeone J B (eds) 1970 Chemical ecology. Academic Press New York, 336 pp
360 Sorm F 1971 Sesquiterpenes with ten-membered carbon rings. A review. J Agr Food Chem 19: 1081–1087
361 Sorm F, Dolejs L 1965 Guaianolides and germacranolides. Hermann Paris, 153 pp
362 Spencer T A, Smith R A J, Storm D L, Villarica R M 1971 Total synthesis of (+)-methyl vinhaticoate and (+)-methyl vauacapenate. J Am Chem Soc 93:4856–4864
363 Sriraman M C, Nagasampagi B A, Pandey R C, Sukh Dev 1973 Studies in sesquiterpenes. XLIX. Sesquiterpenes from *Ferula jaeschkeana* Vatke (Part 1): jaeschkeanadiol – structure, stereochemistry. Tetrahedron 29:985–991
364 Stipanovic R D, Bell A A, O'Brien D H, Lukefahr M J 1978 Heliocide H$_1$. A new insecticidal C$_{25}$ terpenoid from cotton. J Agr Food Chem 26:115–118
365 Straub O 1976 Key to carotenoids. Lists of natural carotenoids. Birkhäuser Basel, 160 pp
366 Streith J, Pesnelle P, Ourisson G 1963 Le β-gurjunene. Identification avec le calarene. Bull Soc Chim Fr 518–522
367 Suckling C J 1984 Reactive intermediates in enzyme-catalyzed reactions. Chem Soc Rev 13: 97–129
368 Suga T, Imamura K, Shishibori T, von Rudloff E 1972 The stereochemistry of (+)-occidentalol. Further support for the revised structure. Bull Chem Soc Jpn 45:3502–3504
369 Suga T, Shishibori T 1980 Structure and biosynthesis of cleomeprenols from the leaves of *Cleome spinosa*. J Chem Soc Perkin Trans I 2098–2104
370 Suga T, Shishibori T, Morinaka H 1980 Preferential participation of linaloyl pyrophosphate rather than neryl pyrophosphate in biosynthesis of cyclic monoterpenoids in higher plants. J Chem Soc Chem Commun 167–168
371 Sukh Dev 1960 Studies in sesquiterpenes. XVI. Zerumbone, a monocyclic sesquiterpene ketone. Tetrahedron 8:171–180
372 Sukh Dev 1962 Azulenes from natural precursors. In: Gore T S, Joshi B S, Sunthankar S V, Tilak B D (eds) Recent progress in the chemistry of natural and synthetic colouring matters and related fields. Academic Press New York, 59–76
373 Sukh Dev 1979 Biogenetic concepts in terpene structure elucidation. Pure Appl Chem 51: 837–856
374 Sukh Dev 1981 Aspects of longifolene chemistry. An example of another facet of natural products chemistry. Acct Chem Res 14:82–88
375 Sukh Dev 1981 The chemistry of longifolene and its derivatives. Fortschr Chem Org Naturst 40: 49–104
376 Sukh Dev 1982 Sesquiterpenoidi. Enciclopedia della Chimica Uses, vol IX. Edizioni Scientifiche Firenze, 554–556

377 Sukh Dev 1983 Chemistry of resinous exudates of some Indian trees. Proc Indian Nat Sci Acad 49(A):359–385
378 Sukh Dev, Misra R 1985 Handbook of terpenoids: Diterpenoids, vol 1–4. CRC Press Boca Raton
379 Sukh Dev, Narula A P S, Yadav J S 1982 Handbook of terpenoids: monoterpenoids, vol 1–2. CRC Press Boca Raton
380 Sumimoto M 1963 Heartwood constituents of *Sciadopitys verticillata* Sieb. et Zucc. I. The constitustion of sciadin. Tetrahedron 19:643–655
381 Sumimoto M, Ito H, Hirai H, Wada K 1963 Cryptomeridiol, the direct precursor of the eudesmane series. Chem Ind 780–781
382 Takeda K 1974 Stereospecific Cope rearrangement of the germacrane-type sesquiterpenes. Tetrahedron 30:1525–1534
383 van Tamelen E E, James D R 1977 Overall mechanism of terpenoid terminal epoxide polycyclizations. J Am Chem Soc 99:950–952
384 Tanaka O, Yahara S 1978 Dammarane saponins of leaves of *Panax pseudo-ginseng* subsp. *himalaicus*. Phytochemistry 17:1353–1358
385 Tange K 1981 The biosynthesis of monoterpenoids in higher plants. The biosynthetic pathway leading to the monoterpenoids from amino acids with a carbon-skeleton similar to mevalonic acid. Bull Chem Soc Jpn 54:2763–2769
386 Teresa J P, Feliciano A S, Barrero A F, Medarde M, Tome F 1981 Sesquiterpene hydrocarbons from the roots of *Otanthus maritimus*. Phytochemistry 20:166–167
387 Thies P W 1968 Die Konstitution der Valepostriate. Mitteilung über die Wirkstoffe des Baldrians. Tetrahedron 24:313–347
388 Thomas B R 1966 The chemistry of the order Araucariales. 4. The bled resins of *Agathis australis*. Acta Chem Scand 20:1074–1081
389 Thomas B R s1970 Midern and fossil plant resins. In: Harborne J B (ed) Phytochemical phylogeny. Academic Press New York, 59–80
390 Thomas A F, Bessiere Y 1981 The synthesis of monoterpenes, 1971–1979. In: ApSimon J (ed) The total synthesis of natural products, vol 4. Wiley-Interscience New York, 451–591
391 Thomas A F, Ozainne M 1978 New sesquiterpene alcohols from *Galbanum* resin: the occurrence of C(10)-epi-sesquiterpenoids. Helv Chim Acta 61:2874–2880
392 Threlfall D R 1980 Polyprenols and terpenoid quinones and chromanols. In: Bell E A, Charlwood B V (eds) Secondary plant products. Springer Heidelberg, 289–308
393 Threlfall D R, Whistance G R 1971 Biosynthesis of isoprenoid quinones and chromanols. In: Goodwin T W (ed) Aspects of terpenoid chemistry and biochemistry. Academic Press New York, 357–404
394 Trivedi G K, Chakravarti KK, Bhattacharyya S C 1971 Isolation & characterization of antipodal (−)-γ-cadinene, (−)-δ-cadinol and khusimol from North Indian vetiver oil. Indian J Chem 9: 1049–1051
395 Treibs W (ed) 1956 E Gildemeister/Fr Hoffmann Die Ätherischen Öle, vol 1–2. Akademie Berlin
396 Treibs W, Merkel D 1960 E Gildemeister/Fr Hoffmann Die Ätherischen Öle, vol IIIa. Akademie Berlin, 628 pp
397 Tursch B, Braekman J C, Daloze D, Kaisin M 1978 Terpenoids and coelenterates. In: Scheuer P J (ed) Marine natural products, vol II. Academic Press New York, 247–296
398 Tursch B, Tursch E 1961 Triterpenes of the latex of *Bursera*. Bull Soc Chim Belges 70:585–591
399 Tutihasi R, Hanazawa T 1940 Ketonic acid isolated from sulfite turpentine oil. J Chem Soc Jpn 61:1041–1047
400 Uemura D, Katayama C, Hirata Y 1977 Crystal and molecular structure of jolkinolide B, a novel oxidolactone diterpene. Tetrahedron Lett 283–284
401 Viswanathan N I 1979 Salaspermic acid, a new triterpene acid from *Salacia macrosperma* Wight. J Chem Soc Perkin Trans I 349–352
402 Vlad R, Soucek M 1962 On terpenes. CXXXVII. Absolute configuration of nerolidol. Collect Czech Chem Commun 27:1726–1729
403 Vlahov R, Holub M, Herout V 1967 On terpenes. CLXXXV. The structure of two hydrocarbons of cadalene type isolated from *Mentha piperita* oil of Bulgarian origin. Collect Czech Chem Commun 32:822–829

404 Vrkoc J, Herout V, Sorm F 1961 On terpenes. CXXIII. On the structure of calacone, a new sesquiterpenic ketone from the sweet-flag oil (*Acorus calamus* L.). Collect Czech Chem Commun 26:1343–1349
405 Wagner H 1981 Plant constituents with antihepatotoxic activity In: Beal J L, Reinhard E (eds) Natural products as medicinal agents. Hippokrates Stuttgart, 217–242
406 Wagner H, Brüning R, Lotter H, Jones A 1977 Die Struktur von Cassinin, einem neuen Sesquiterpenalkaloid aus *Cassine matabelica* Les. Tetrahedron Lett 125–128
407 Wallace J W, Mansell R L (eds) 1975 Biochemical interaction between plants and insects. Recent advances in phytochemistry, vol 10. Biochemical interaction between plants and insects. Plenum Press New York, 425 pp
408 Walls F, Padilla J, Joseph-Nathan P, Giral F, Escobar M, Romo J 1966 Studies in perezone derivatives, structures of the pipitzols and perezinone. Tetrahedron 22:2387–2399
409 Wegler R (ed) 1980 Chemie der Pflanzenschutz- und Schädlingsbekämpfungsmittel, vol 6. Springer Heidelberg, 512 pp
410 Weinheimer A J, Chang C W J, Matson J A 1979 Naturally occurring cembranes. Fortschr Chem Org Naturst 36:285–387
411 Wenkert E 1955 Structural and biogenetic relationships in the diterpene series. Chem Ind 282–284
412 West A W, Dudley M W, Dueber M T 1979 Regulation of terpenoid biosynthesis in higher plants. In: Swain T, Walker G R (eds) Recent advances in phytochemistry, vol 13. The biochemistry of natural products. Plenum Press New York, 163–198
413 Westfelt L 1966 α,γ, and ε-Muurolene, major sesquiterpenes of the wood of *Pinus silvestris* L. and of Swedish sulphate turpentine. Acta Chem Scand 20:2852–2864
414 Westfelt L 1967 The structure of α-longipinene, a minor sesquiterpene of the wood of *Pinus silvestris* L. and of Swedish sulphate turpentine. Acta Chem Scand 21:159–162
415 Whitlock H W, Olson A H 1970 The concertedness, or lack thereof, of a multiple carbonium ion rearrangement. J Am Chem Soc 92:5383–5388
416 Williams C M 1970 Hormonal interactions between plants and insects. In: Sondheimer E, Simeone J B (eds) Chemical ecology. Academic Press New York, 103–132
417 Wiss O, Gloor U 1970 Nature and distribution of terpene quinones. In: Goodwin T W (ed) Natural substances formed biologically from mevalonic acid. Academic Press New York, 79–87
418 Wood D L, Silverstein R M, Nakajima M (eds) 1970 Control of insect behaviour by natural products. Academic Press New York, 345 pp
419 Yamamura S, Iguchi M, Nishiyama A, Niwa M, Koyama H, Hirata Y 1971 Sesquiterpenes from *Axorus calamus* L. Tetrahedron 27:5419–5431
420 Yoshida T, Nobuhara J, Fujii N, Okunda T 1978 Studies on constituents of Buddleja species. II. Buddledin C, D and E, new sesquiterpenes from *Buddleja davidii* Franch. Chem Pharm Bull 26:2543–2549
421 Zalkow L H, Harris R N, Van Derveer D 1978 Modhephene: a sesquiterpenoid carbocyclic [3,3,3] propellane. X-ray crystal structure of the corresponding diol. J Chem Soc Chem Commun 420–421
422 Zalkow L H, Harris R N, Van Derveer D, Bertrand J A 1977 Isocomene: a novel sesquiterpene from *Isocoma wrightii*. X-ray crystal structure of the corresponding diol. J Chem Soc Chem Commun 456–457

8.2 Steroids

W. R. NES

8.2.1 Introduction

The amount of definitive work that has been done on steroids of bark or the woody parts of dicotyledons is unfortunately small in comparison to the substantial investigations of terpenoids. This would suggest that there might be little to say about the relationship of steroids to the trunks of the great forest trees were it not for the fact that those studies that have been done correlate quite well with the more extensive research with seeds, seed oils, fruits, stems, and leaves (28, 29, 33, 67, 69–72). That is, the major steroids found in tree trunks appears to be of the same kind as found in other parts of the tree or in other plants related botanically. This is not surprising, for, in eukaryotes generally, steroids act as consituents of membranes (66, 71) (cf., Sect. 8.2.2.4). Of special importance is the so-called "bulk membrane" role, which is thought to utilize most of the steroid in a cell. In the bulk membrane role the steroid acts nonmetabolically in an architectural way that is thought to influence physical properties such as fluidity. The membranous steroid is nearly always a free, monohydroxylic sterol (cf., Sect. 8.2.1.1), but much smaller amounts of the fatty acid esters and glycosides of the sterols are also sometimes associated with membranes. Most of the esters, however, are found elsewhere in the cell – e.g., in lipid droplets in the cytosol – and they serve functions that need more study but that probably include acting as a reverse or store of fatty acid and sterol. The other steroids (alkaloids, saponins, ecdysones, etc.) found in plants are not well defined in terms of function but in many cases probably serve to control insects, disease, and other environmental aspects of the tree's life.

8.2.1.1 Meaning of the Terms "Steroid" and "Sterol"

The term "steroid" ("oid" being a suffix implying similarity) was coined by Fieser and Fieser (21) to mean a compound that (as with cholesterol (**1**), its epimeric dihydro derivatives, cholestanol (**2**) and coprostanol (**3**), and the bile acids – e.g., cholic acid (4)) has a tetracyclic ring system (the "nucleus") of the sort found in 1,2-cyclopentanoperhydrophenanthrene (5). It is obvious that this definition is neither stereochemically limiting nor is it precise concerning the number and position of carbon atoms other than those directly in the four rings. The failure to specify details about the number and arrangement of extracyclic carbon atoms has been very useful in allowing different but closely related compounds to be grouped together, but the lack of definition of the nuclear stereochemistry has been confusing. For instance, there is a whole group of triterpenoids, called euphoids (67, 71), that empirically have the same tetracyclic skeleton as cholesterol but differ in the stereochemistry of the junction of the five- to the six-membered ring as well as in the junction of the acyclic side chain to the tetracyclic nucleus (77). If we adopt the usual numbering system for the annular carbon atoms (6) and assign the letters A through D to the rings as shown in 6, the euphoids represent stereoisomers of the steroids in which carbon atoms 13, 14, and 17 have

6

been inverted. That is, with respect to the portion of the molecule at the junction of the side chain to ring D and the junction of ring D to ring C, the euphoids are formal antipodes of the steroids (63). This difference is apparently very important biologically. The euphoids are never to our knowledge metabolized so that the analogs of membranous steroids are obtained, and the few plants — e.g., in the genus *Euphorbia* — that make euphoids also make sterols (97) presumably for functional reasons. The author and most other investigators have therefore not included euphoids within the family of steroids. The Fieser and Fieser definition has recently been extended to accommodate this along with other distinctions by including both stereochemistry and biosynthesis (67, 71). Steroids are taken to be the tetracycles derived biosynthetically (Fig. 8.2.1) from a folding pattern of the C_{30}-precursor, squalene oxide, which mimics and then by electron redistribution actually gives the *trans-syn-trans-anti-trans-anti* stereochemistry in the tetracyclic C_{30}-product, whereas euphoids are formed through a folding pattern similar to the more familiar all-*trans-anti* stereochemistry (67, 71). This bifurcation is shown by **7a** to **8** to **9** and **7** to **10** to **11** where **7a** is squalene oxide and **8** and **10** are the enzyme-bound *trans-syn-trans-anti-trans-syn* and all-*trans-anti* intermediate states, which, by elimination of enzyme (Enz) and proton with 1,2-migrations of methyls and H-atoms (cf., Sect. 8.5.1.3), give C_{30}-steroids — e.g., lanosterol (**9**) — and the isomeric C_{30}-euphoids — e.g., euphols (**11a**, 20*R*) and tirucallol (**11b**, 20*S*). The euphoids are, as already mentioned, generally regarded as tetracyclic triterpenoids and share an all-*trans-anti* transition from squalene oxide with their pentacyclic relatives — e.g., the amyrins (71). Removal of the methyl groups at C-4 and C-14 of the C_{30}-steroids — e.g., lanosterol — with some other changes gives the common sterols found in membranes. It is this

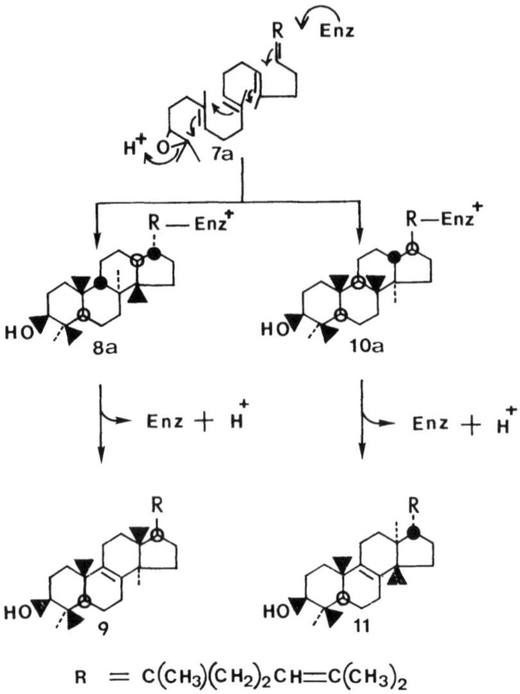

Fig. 8.2.1. Biosynthetic origin of steroids and tetracyclic triterpenoids from squalene oxide

kind of metabolism, especially trisdemethylation, that does not seem to occur with the euphoids and other triterpenoids. Euphoids are metabolically active, however, and proceed in four families of Rutales by extensive oxidations to various liminoids and quassinoids (19).

The term "sterol", a subcategory of "steroid", is derived from the word "cholesterol" and implies that, as with cholesterol, the compound is hydroxylic without being highly polar; thus, sterols are lipids. Cholic acid (**4**), on the other hand, is much too polar to be considered a sterol, while cholestanol, 22-dehydrocholesterol, and such compounds as the 5,6-epoxy, Δ^7,4,4-dimethyl, and 25,26,27-trisnor derivatives of cholesterol all have general lipid characteristics and are considered to be sterols even though they differ in the degree and kind of unsaturation, the number of carbon atoms, and the number of oxygen atoms. These differences, however, will change the lipid properties in detail and therefore will alter the precise biological function of the compounds. As many as one to two thousand sterols may exist naturally as biosynthetic intermediates, various functional end products, and their simple metabolites. The term stenol means an unsaturated sterol; stanol implies a saturation, and stenone and stanone imply the corresponding ketones with the carbonyl group at C-3 unless otherwise noted.

8.2.1.2 Nomenclature

Except in very special cases, the stereochemistry of plant steroids at C-3, C-5, C-8, C-9, C-10, C-13, C-14, C-17, and C-20 is always the same when these positions are chiral. The normal stereochemistry is shown by cholestanol (**2a**). In the steroid field it is customary to omit a full notation of the stereochemistry at these positions. Cholestanol (**2a**), for instance, is usually shown simply by formula **2b** implying **2a**. However, if one wishes to emphasize that it is in the A/B-*trans* instead of the A/B-*cis* series as in coprostanol (**3**) or bile acids (**4**), the H-atom at C-5 can be shown with an interrupted line attaching it to C-5 (**2c**). This simplification will be used in this chapter – i.e., unless indicated, the stereochemistry is the same as in **2a**. However, a solid line joining an atom to one of the carbon atoms of the four rings (steroid nucleus) always implies the β-configuration, meaning that the atom attached to the annular carbon is (either axially or equatorially) on the same side (β-face) of the nucleus as are C-18 and C-19 (the so-called angular methyls). An interrupted line joining an atom to a carbon atom in the nucleus implies the attached atom is on the side (α-face) opposite that containing the angular methyls. The substituent is then said to be α-oriented. When the molecule is viewed as in **1–11** with the angular methyls (C-18 and C-19 or just C-19 in the euphoid case) pointing toward the observer with C-3 to the left, this is referred to as the "usual view" of the molecule, and the β- and α-faces then become the front and rear of the molecule, respectively.

Three naming systems are in use for steroids: trivial names (cholesterol, cholestanol, etc.), semisystematic names (use of trivial name such as cholesterol as a parent), and truly systematic names with commonly accepted saturated hydrocarbons as parents. The most often encountered steroids are derivatives of the parent, 5α-cholestane (**2** lacking the HO-group) where "5α" refers to the presence of an α-H-atom on C-5. Thus, cholestanol (**2**) is systematically 5α-cholestan-3β-ol and cholesterol is cholest-5-en-3β-ol. Cholestanol could also be described semisystematically as 5α-dihydrocholesterol.

The stereochemical nomenclature of the side chain is an extension by the author (67, 71) of one devised by Fieser and Fieser (22). The side chain attached to C-17 of ring D is arbitrarily taken to have the numbering system and conformation given by **12a** (simplified to **12b**). In this stereocondition, assumed for purposes of nomenclature, C-22 is *trans*-oriented to C-13 around the 17(20)-bond and all of the carbon atoms of the side chain are in the fully staggered arrangement (C-23 *trans* to C-21, etc.) with C-20, C-21, C-22, C-23, C-24, C-25, and C-26 lying in a plane. The H-atoms, C-27, and any substituent(s) then lie in front or in back of the latter plane. Those that lie in back – e.g., C-28 at C-24 in **13** – are given the designation β, and those in front – e.g., C-28 at C-24 in **14a** are called α. It should be noted that for historical and not for logical reasons this convention is the reverse of the one used for the nucleus.

In the crystal structure of steroids deduced from X-ray diffraction, C-22 is actually *trans*-oriented to C-13 as assumed for side chain nomenclature. However,

C-23 is *trans*-oriented to C-17, not to C-21. The preferred conformation about the 17(20)-bond is dependent on the configuration at C-20 and C-13 (22), because the H-atom on C-20 is preferably adjacent to C-18 (22).

It is perhaps also worth mentioning here that, although α-, β-, and γ-sitosterols were originally thought to be isomers, this is now known not to be true (67). The α- and γ-sitosterols were found to be mixtures, and only β-sitosterol proved to be a single entity. Therefore, the "β" should now be deleted (67), and 24α-ethylcholesterol is properly called just "sitosterol", not "β-sitosterol."

8.2.1.3 Biosynthesis of Basic Steroidal Structure

As discussed in detail elsewhere (33, 67, 71), steroids originate biosynthetically in the polymerization of acetyl coenzyme A through the acetoacetyl C_4-stage to the branched C_6-compound, 3-hydroxy-3-methylglutaric acid as the mono-coenzyme A derivative called, in short, hydroxymethylglutaryl CoA (HMG-CoA). Reduction of the HMG-CoA at the thiol ester grouping by two moles of NADPH leads to the corresponding C_6-alcohol, (R)-mevalonic acid (MVA), which, after three phosphorylations with ATP by loss of CO_2 and the elements of water (actually as phosphoric acid after phosphation), becomes the C_5-compound, 3-methylbut-3-en-1-yl pyrophosphate (Δ^3-isopentenyl pyrophosphate, 15). The latter in part isomerizes to the Δ^2-isomer (16, also known as dimethylacrylic acid). Polymerization of the C_5-pyrophosphates then proceeds, starting with the Δ^2-isomer (16) by successive additions of the tail end (Δ^3) of the Δ^3-isomer (15) to the head end (phosphorylated) of the Δ^2-monomer (16) and then to the head end of the succeeding polymers – e.g., geranyl pyrophosphate (17). These polymers, which can become very large (as in rubber), are the backbone of the whole immense family of isopentenoids (isoprenoids), which includes the steroids (sterols, hormones, bile alcohols and acids, alkaloids, sapogenins, and other types), carotenoids, various quinone cofactors and vitamins, and the many kinds of terpene.

The pathway to steroids and triterpenoids rather than other kinds of isopentenoids is determined at the trimeric stage of C_3-polymerization when the all-*trans*-product (farnesyl pyrophosphate, 18) undergoes a multistep reductive (NADPH) dimerization with elimination of the pyrophosphate groups giving the C_{30}-compound, squalene (19), as shown in Fig. 8.2.2. One of the two terminal double bonds of this molecule is then epoxidized (O_2, NADPH) giving the (S)-enantiomer of 2,3-oxidosqualene (7b). the cyclization of 7b in photosynthetic systems proceeds by a slight modification of the process shown in Sect. 8.2.1.1 by 7a via 8 to 9. Instead of elimination of a proton from C-9 of 8 giving lanosterol (9), which occurs in fungi and animals, elimination of an H-atom comes from C-19 in organisms such as trees, which are phylogenetically photosynthetic. The C-19 carbon then attacks C-9 with migration of the H-atom on C-9 to C-8 and formation of a three-membered ring (9β,19-cyclo grouping). The product of this sequence of events is cycloartenol (20), which is the first steroid formed in the photosynthetic pathway (33, 67, 71, 72) leading to functional end products such as cholesterol or sitosterol. The cycloartenol pathway is dependent not on whether

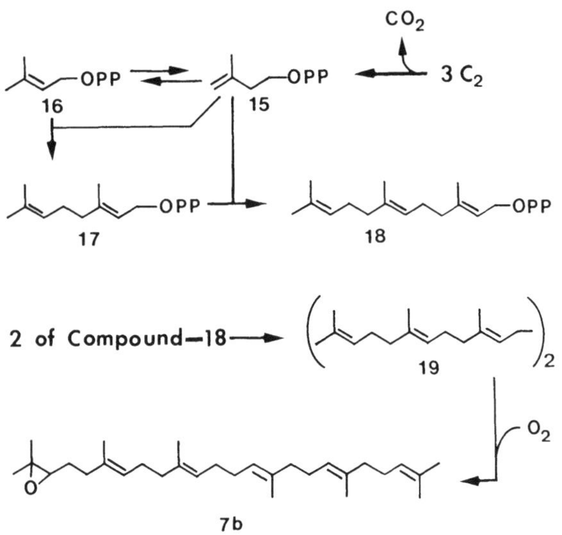

Fig. 8.2.2. Biosynthetic origin of squalene oxide

functional photosynthesis exists but rather on whether the organism is in the photosynthetic lineage (32, 67, 71). For instance, nonphotosynthetic angiosperms use the cycloartenol route (62) even though they have no operational chloroplasts and are parasitic. Similarly, nonphotosynthetic colorless tissue (endosperm of *Pinus pinea* seeds) of a photosynthetic gymnosperm uses the cycloartenol route (32). The cyclization of squalene oxide to cycloartenol (Fig. 8.2.3) is believed (67, 71) to involve five main elements: (a) assumption of an appropriate conformation (**7c**) controlled by binding to the cyclase; (b) attack of a proton on the oxygen atom in ring A with simultaneous attack of a pair of electrons from the enzyme onto C-20, giving **8b**; (c) rotation around the 17(20)-bond, giving **8c**; (d) elimination of the enzyme and its pair of electrons from C-20 with a series of *trans*-migrations (9β-H to C-8, 8α-CH$_3$ to C-14, 14β-CH$_3$ to C-13, 13α-H to C-17, and 17β-H to C-20) together with reentry of the enzyme at C-9, giving **8d**; and (e) elimination of the enzyme from C-9 and *trans*-attack of C-19 on C-9 with loss of a proton from C-19, giving cycloartenol (**20**). The formation of lanosterol (**9**) would proceed by the same set of *trans*-migrations from **9c** except that the enzyme would remove the 9β-H (producing the Δ^8-bond) instead of attacking the α-side of C-9 forcing the H to migrate. In some cases – e.g., the genus *Euphorbia* (13) – cycloartenol (**20**) seems to be isomerized (9β,19-cyclo to Δ^8) to lanosterol (**9**) as the very next step, but in most cases it appears that the 9β,19-cyclo grouping is opened only after other reactions have occurred – e.g., removal of a nuclear methyl group and metabolism of the Δ^{24}-bond by reduction or alkylation. Thus, lanosterol itself (**9**) is not usually an intermediate in living systems that are phylogenetically photosynthetic. Whereas the occurrence of lanosterol is rare in higher plants, cycloartenol is found frequently (2).

The conversion of cycloartenol (**20**) (which appears to act only as an intermediate) to functional end products involves oxidative (O$_2$, NADPH) removal of the three methyl groups at C-4 and C-14, addition (frequently) of one or two car-

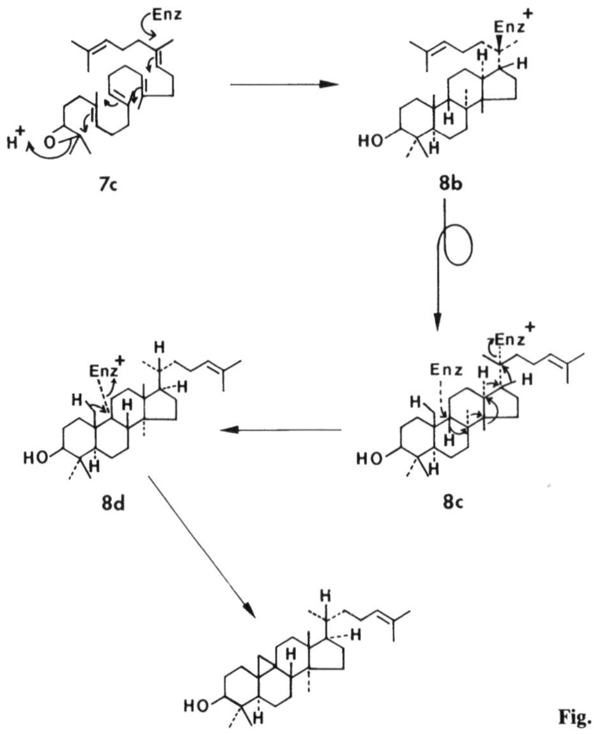

Fig. 8.2.3. Biosynthesis of cycloartenol from squalene oxide

bon atoms at C-24 of the side chain, and manipulation of the unsaturation. Reduction (by reduced pyridine nucleotide) of double bonds can occur – e.g., of Δ^{24} and Δ^{14} – as well as introduction of double bonds. In the latter case direct aerobic dehydrogenation can lead to a double bond – e.g., Δ^{22} – a particular double bond also can arise by isomerization – e.g., Δ^8 to Δ^7 – or in the course of oxidative removal of a methyl group as in the introduction of Δ^{14} when the 14α-methyl is lost. The number of possible sequences for these reactions is large and incompletely investigated (65, 71). The mechanisms of a few of the individual reactions have been studied, however, in some detail (71). For a consideration of the biosynthetic origin of individual end products, see Sect. 8.2.2.3.

All of the sterol in a photosynthetic plant is derived by biosynthesis de novo in that plant, but neither the contribution of the various tissues to the total biosynthetic pool nor the role of transport (41, 64) is well investigated. It is not known, for instance, how much of the sterol of wood or bark actually is biosynthesized in the stem itself or is biosynthesized in the leaves and then carried to the stem.

8.2.2 Sterols

8.2.2.1 Names, Structures, and Organismic Relationships

Sterols in general can be divided into two categories: (a) the end products that accumulate, becoming the principal sterols on a quantitative basis, which are called the dominant sterols (66, 67, 70, 71), and (b) their biosynthetic intermediates that are usually found in very much smaller concentrations, if at all. For the most part we will consider here only the dominant sterols. These substances are thought to be the main functional end-products of the sterol pathway.

In vascular plants (tracheophytes), a category that includes trees, the sterols found in significant amounts are all formal derivatives of cholestanol (**2c**) − that is, each of the dominant sterols has the 3β-hydroxy group, the all-*trans-anti* nuclear stereochemistry at positions that are saturated, no 9β,19-cyclo group, and a side chain that has minimally the same number and skeletal arrangement of carbon atoms as in cholesterol. They differ from each other (a) by the position and number of double bonds, (b) by the presence or absence of an additional methyl or ethyl group at C-24, and (c) by the configuration of C-24.

The most common end products have either a Δ^5- or, less commonly, a Δ^7-bond in the nucleus, a 24α- or a 24β-methyl or ethyl group in the side chain, and either no double bond or a Δ^{22}- and/or in some cases a $\Delta^{25(27)}$-bond in the side chain. With some expansion − e.g., to include $\Delta^{5,7}$-dienic compounds − the structures of the sterols of fungi, mosses, ferns, lycopods, and limnic and terrestrial algae are much the same as found in flowering plants, although in the lower organisms the 24β-alkylsterols are more frequent in occurrence than in the higher plants (67, 71, 72). In marine organisms there is a greater disparity as a result of addition of carbon atoms not only to C-24 but also the several other of the side chain positions, causing elongation and more branching; shortening of the side chain as well as skeletal alterations in the nucleus also occur (18). In vertebrate animals quite the reverse is true: instead of variety there is a strong restriction in the types of sterols present except in pathological conditions. No biosynthetic addition of carbon atoms to the side chain occurs, and incorporation of exogenous 24-alkylsterols is discriminated against by both absorption and metabolism. Furthermore, no double bonds are added to the side chain of vertebrate sterols, and no skeletal alterations to the nucleus occur. In fact, one sterol alone (cholesterol, **1**) usually makes up nearly all (98% or so) of the sterol present. It is accompanied principally by the corresponding 5α-stanol (cholestanol, **2c**), the Δ^7-analog (lathosterol, **21**), the $\Delta^{5,7}$-dienol (7-dehydrocholesterol, **22**), and (especially in fetuses) the $\Delta^{5,24}$-derivative (desmosterol, **23**). Contrastingly, all tracheophytes that have been examined contain 24-alkylsterols, which, in the great majority of cases, are the principal ones present.

Among the tracheophytes, most of the plants examined belong to the "main line" in that they have the same type of dominant sterols frequently comprised by the homologous Δ^5-series, cholesterol (**1**) and its 24-methyl and 24-ethyl derivatives together with some of the E-Δ^{22}-analogs of the 24α-ethyl and 24β-methyl compounds. The Δ^{22}-24-methylsterols seem to be virtually absent from higher plants. All of the 24-ethylsterols of main line plants have the α-configuration at C-24 (by the sequence rule: *R* without the Δ^{22}-bond, *S* with it). In the 24-methyl fraction (lacking Δ^{22}) both epimers occur, but 24α-methyl is dominant. Although there can be substantial variation in the relative amounts of these various sterols, in most of the mature plants the monoenic 24α-ethylsterol (24α-ethylcholest-5-en-3β-ol, sitosterol, **24**) is the major component of the mixture followed in order of amount by the 24α-methyl (24α-methylcholest-5-en-3β-ol, campesterol, **14** in the Δ^5-series) and 24β-methyl (24β-methylcholest-5-en-3β-ol, dihydrobrassicasterol, **13** in the Δ^5-series) analogues. The α-to-β ratio for the 24-methylsterols is usually about 2:1, at least in the Δ^5-series. Measurements of the ratio for Δ^7-sterols have not been reported, because when present they are very minor components. Cholesterol (**1**) is frequently present only as a trace constituent but in a few cases – e.g., *Solanum dulcamara* and *Euphorbia pulcherrima* (13) – it can be the major sterol. Similarly, in some cases – e.g., peel of the banana (*Musa sapientum*) (1) – the Δ^{24}-24α-ethylsterol (24α-ethylcholesta-5,22E-dien-3β-ol, stigmasterol, **25**) can become the principal component, although usually it is only a minor one (traces to 20%). Similarly, the Δ^{22}-sterol with a 24β-methyl group (24β-methylcholesta-5,22E-dien-3β-ol, brassicasterol, **26**), while usually not present in more than traces, can rise (at the expense of dihydrobrassicasterol) to significant levels in seeds of the Cruciferae – e.g., *Brassica rapa* (rapeseed) – where it is a 10% component (1). The Δ^5-sterols are sometimes accompanied by smaller amounts of the biosynthetic intermediate with a Z-24-ethylidene group (24-ethylcholesta-5,24(28)Z-dien-3β-ol, isofucosterol also known as Δ^5-avenasterol, **27**) and by other sterols, the most common of which is the Δ^7-analog of sitosterol (24α-ethylcholest-7-en-3β-ol, dihydrospinasterol, also known as schottenol and as 24α-ethyllathosterol, **28**). The Δ^7-24-methylsterol is also found occasionally as a trace component but, as already alluded to, the configuration at C-24 is uncertain. It may mimic that found with the Δ^5-analogue (α with lesser amounts of β), but this is far from a necessary conclusion. The Δ^5-24-ethyl component (all α) does not mimic the Δ^5-24-methyl component (α and β) configurationally, and the ratio of Δ^7-24-ethylsterol to Δ^7-24-methylsterol does not mimic what is found with Δ^5-sterols. Cholestanol (2c) and other stanols are also (but rarely) found in traces except, as we shall see, in wood tissue.

Main line plants (those having sterols with 24-alkyl groups mostly of the α-configuration) have been designated as being of category I, and two subcategories exist: I-A, having mostly sterols with a Δ^5-bond, and I-B, having sterols mostly with a Δ^7-bond (6). In the previous paragraph only I-A plants were considered. They range (67, 69–72) throughout the Tracheophyta from the ferns – e.g., *Polystichum acrostichoides* (the "Christmas fern") in the Polypodiaceae – through the gymnosperms – e.g., *Ginkgo biloba* in the Ginkoaceae and *Pinus pinea* in the Pinaceae – to both the lowest – e.g., *Liriodendron tulipifera* (the tulip tree) in the Magnoliaceae – and the highest – e.g., *Pisum sativum* (the pea) and *Phaseolus vulgaris* (the bean) in the Fabaceae – of the angiosperms. There is, however, a small group of plants that are I-B. They include representatives of the Chenopodiaceae (leaves of the spinach plant, *Spinacia oleracea*), the Compositae (roots of *Aster scaber*, although many other Compositae are I-A), the Theaceae (seed oils of species of *Camellia* and *Thea*), the Sapotaceae (seeds of *Butyrospermum parkii*), the Fabaceae (seed oil of *Medicago sativa*, alfalfa), and mature tissue (not seeds) of the Cucurbitaceae. In each of the latter cases, 22-dihydrospinasterol (24α-ethylcholesta-7-en-3β-ol, 28) and spinasterol (24α-ethylcholesta-7,22E-dien-3β-ol, 29) are the major components of the sterol mixture (33, 67, 70–72). Minor sterols are avenasterol (24-ethylcholesta-7,24(28)Z-dien-3β-ol, 30) and 24-methylenelathosterol (24-methylcholesta-7,24(28)-dien-3β-ol). In rarer cases – e.g., *Cucurbita maxima* seeds (28) – the 24α- and 24β-alkylsterols both make large contributions of the composition of the sterol mixture.

In addition to the main line plants, representatives of two families (*Kalanchoe diagremontiana* in the family Crassulaceae and several species of the genus *Clerodendrum* in the family Verbenaceae) have a markedly divergent sterol pattern (33, 67, 70–72, 85). They contain only 24β-ethylsterols with Δ^5- and $\Delta^{25(27)}$-double bonds – viz., 24β-ethylcholesta-5,25(27)-dien-3β-ol (clerosterol, also known as 25(27)-dehydroclionasterol, 30) and its 22E-dehydro derivative, 24β-ethylcholesta-5,22E,25(27)-trien-3β-ol (22E-dehydroclerosterol, also known as 25(27)-dehydroporiferasterol, 31). Plants of this sort in which the 24β-alkylsterols are dominant are called Category II, or, more specifically for the present case (Δ^5-sterols), they are called Category II-A. In between categories I and II are all examined genera and species of the family Cucurbitaceae. These plants are a mixture varying from mostly II-B in many of the seeds to virtually all I-B when mature (28, 29, 33, 67, 70–72). For instance, 75% of the seed sterols of *Cucurbita pepo* (pumpkin) are the 24β-ethylsterols, 25(27)-dehydrochondrillasterol (24β-ethylcholesta-7,22E,25(27)-trien-3β-ol, 32) and its 22-dihydro derivative (24β-ethylcholesta-7,25(27)-dien-3β-ol, 33) together with 25% of the 24α-ethylsterol, spinasterol (29). However, in the mature plant, spinasterol (42%) and 22-dihydrospinasterol (**28**) (36%) become the major sterols with avenasterol (**34**) (16%), 24ξ-methyllathosterol (0.5%), lathosterol (cholest-7-en-3β-ol, **21**, 0.2%), and 5% of an incompletely identified sterol comprising the remainder of the mixture (5). Few if any 24β-ethylsterols are present in the mature plant, and the principal minor sterol (avenasterol) is a biosynthetic intermediate to the major 24α-ethylsterols (33, 67, 71). The ontogeny of the Cucurbitaceae plants is thought to represent an evolutionary recapitulation with the seed stage reflecting early phylogenesis in the algae that have 24-β-alkylsterols (67, 71, 72). The seeds of some cucurbits also contain various Δ^5-sterols as minor components. Most thoroughly investigated is *Cucurbita maxima* (squash) (**29**). In keeping with the Category II designation (106) derived from the dominant (Δ^7) sterols (**28**), the principal Δ^5-compound (**29**) is the 24β-methylsterol, codisterol (24β-methylcholesta-5,25(27)-dien-3β-ol, **35**) along with two 24β-ethylsterols (Clerosterol, **30** and 25(27)-dehydroporiferasterol, **31**) as well as the usual main line sterols.

[Structures 31, 32, 33, 34 with side chains; labeled Δ^5-N, Δ^7-N, Δ^7-N, Δ^7-N respectively]

[Structures 35, 36, 37 with side chains; labeled Δ^5-N, $\Delta^{5,7}$-N, Δ^5-N respectively]   N=Nucleus

8.2.2.2 Identification of Sterols

Sterols usually occur in mixtures that can often be quite complex, as indicated in the previous section. Thus, melting point and other classic data on the total sterol fraction is not very informative. However, by chromatographic and spectral means, individual components can be separated from one another, their structures proven, and their amounts determined. A single gas-liquid or reversed phase liquid chromatogram is usually far more helpful than a melting point, infrared spectrum, or optical rotation. Yet, one chromatogram still does not constitute adequate identification, since no chromatographic system will separate all sterols. For a key to the literature on methods of identification and quantitation, see the recent review by this author (68). Two recent papers (28, 29) on the resolution of the sterols in *Cucurbita maxima* seeds into no fewer than 14 individual components of defined structures might also be consulted as a current example of structural analysis and quantitative assay by various chromatographic (TLC, GLC, HPLC) and spectral procedures (^{1}H-NMR, MS). Configurational assignments at C-24 are of special importance, and it is worth noting that they cannot be made by HPLC or packed column GLC. Clear determination of the configuration can only be obtained from ^{1}H-NMR at 100 to 500 MHz, ^{13}C-NMR, or in certain cases by capillary GLC. The NMR spectra can also be used, of course, for still other structural assignments.

An illustration of the determination of configuration is as follows. Two configurational standards among others have been established. They are stigmasterol

(24α-ethylcholesta-5,22E-dien-3β-ol, **25**) and ergosterol (24β-methylcholesta-5,7,22E-trien-3β-ol, **36**). Stigmasterol was ozonolytically degraded at the Δ^{22}-bond (24) to yield the C_7-fragment consisting of C-23 to C-29. This had only a single chiral center (C-24), the configuration of which (α in steroid nomenclature) was determined (106) from its optical rotation by comparison with material of known configuration ultimately determined by X-ray diffraction. The configuration of natural sitosterol was demonstrated unequivocally later (69) by showing that its ^{1}H-NMR spectrum is the same as that of a sample derived by reduction of the Δ^{22}-bond of stigmasterol and different from a sample of the epimeric clionasterol (24β-ethylcholesta-5-en-3β-ol, **37**) derived from nature.

In the 24-methyl case, Tsuda (106) attempted the same sort of series of reactions and comparisons as he had done with stigmasterol. Unfortunately, the results were ambiguous, though suggesting ergosterol to be 24β. The suggestive assignment was recently confirmed as true (2) as follows. 24ζ-Methylcholesterol (ε signifies the configuration is unknown) from equilibrium crystallization of the 24-methylsterol component of soybeans and 24ζ-methylcholesterol prepared from yeast ergosterol were shown to have distinctly different ^{1}H-NMR spectra. The doublet at C-21 was upfield in the epimer from soybeans compared to the sample derived from ergosterol. Since, in the sitosterol-clionasterol epimeric pair, the C-21 doublet was upfield in the sitosterol (C-24α), the soybean 24-methylcholesterol must also be α (campesterol) and the one originating in ergosterol must be β (22-dihydrobrassicasterol). We were also able to show that the same ^{1}H-NMR shift occurs when yeast ergosterol itself is compared with 24-epiergosterol, which was derived by biological (*Tetrahymena pyriformis*) dehydrogenation of the soybean 24α-methylcholesterol. 24α-Methylcholesterol can also be obtained from the seeds of *Brassica* species without equilibrium crystallization, which requires a large sample. This is a good source of a small sample. We have found the ^{1}H-NMR spectrum to be identical with that of the sample from soybean except for very weak signals from the presence of a few percent of the β-epimer. These results are unpublished.

The spectra of the two 24-methyl epimers are, incidentally, also quite different in regions other than where the C-21 doublet resonates. In particular, the methyl (C-28) on C-24 is adjacent to C-25, which contains two further methyl groups. Inversion of C-24, therefore, causes marked differences in the signals from these methyls, and the differences are of diagnostic as well as of quantitative value (28, 29, 69, 70). In one of the epimers, a branch of one of the doublets occurs at the position of a minimum in the spectrum of the other epimer, and the intensity of the branch is especially useful for quantitation if no other sterol is present with a signal there. To avoid contamination in this region, the 24-methylsterols can be obtained as a separate fraction by reversed phase chromatography on lipophilic Sephadex or by preparative HPLC.

8.2.2.3 Occurrence in Wood and Bark

Although none of the investigations of the sterols of bark and wood have utilized techniques such as ^{1}H-NMR, ^{13}C-NMR, or appropriate capillary GLC, which can distinguish epimers at C-24 (23), is has been shown in Sect. 8.2.2.1 that the pattern of sterols in main line plants is quite distinct. Thus, the principal sterols usually are the C_{29}-(sitosterol with or without stigmasterol, MW 414 and 412, respectively) and C_{28}-(24-methylcholesterol(s), MW 400) Δ^5-sterols with frequent traces of cholesterol (MW 386). Since the molecular weights can be obtained by mass spectrometry and since analogous Δ^5 and Δ^7-sterols move at different rates in GLC (23), the pattern of Δ^5-C_{29}- and Δ^5-C_{28}-sterols can frequently be distinguished by a combination of MS and GLC from which a presumption of configuration can be made. In several laboratories appropriate MS and GLC have been done with sterols from woody parts of trees and in all such cases the pattern observed is consistent with designations for the plants as main line, mostly of Category I-A but a few of Category I-B. No trees of Category II have been reported (yet?). No sterols have been found in bark or wood that do not occur in other parts (leaves, seeds, fruits, flowers, etc.) of tracheophytes. Thus, with respect to sterol structure, bark and wood do not constitute tissue specialization. The thing that is remarkable is that heartwood, which is dead material, actually contains sterols. It is conceivable that these substances might have been removed to living parts of the tree, as lipid phosphorus (37, 38) and storage carbohydrates seem to be. Instead, the total sterol content (free and ester) actually rises by an order of magnitude as cross-sections of spruce tree trunks (*Picea abies*) are traversed from the outside (sapwood) to the middle (heartwood) (37, 38). The free sterol to lipid phosphorus ratio is even more impressive; it increases no less than 400-fold from ~0.5 to ~200 (37), largely as a result of the loss of lipid phosphorus (37, 38). Höll and Goller (37) suggest that the higher levels of sterols in the heartwood (compared to the living tissue of the same tree) reflect an increased level of sterol during youth, from which, of course, the heartwood is derived. In support of this idea, they claim that 15-year-old spruce trees that contain no heartwood have sterols in amounts similar to those in the heartwood of trees four decades old (37). Geuns (30, 31) and Karunen and Ekman (49) have also found the youngest tissue of mung beans and *Sphagnum* moss, respectively, to have the most sterol.

Fischer et al. (23) have reported an especially good study (Table 8.2.1) of the structure and amounts of sterols from the wood of various conifers.

All of the 14 species studied appeared to be of the main line. Although neither NMR nor capillary GLC was used, the principal sterols had molecular weights by MS and retention times in packed column GLC that corresponded to 24-ethyl- and 24-methylcholesterol, presumably sitosterol (**24**) and a mixture of campesterol (**14**) in Δ^5-3β-ol series) and dihydrobrassicasterol (**13** in Δ^5-3β-ol series). In one case (the tree-of-life cypress, also known as the Port-Orford-Cedar, *Chamaecyparis lawsoniana*) the 24-ethylidenol, presumably stigmasterol (**25**) was also present. Fischer et al. (23) also identified 5α-sitostanol (**38**) in all of the cases using mass spectroscopy and, for tall oil, also using GLC on QF-1, which separates Δ^5- (faster) from 5α-Δ^0-(slower) sterols (68). With the other common liquid phases for GLC, the Δ^0- and Δ^5-sterols do not separate. The Δ^0-analog of one or both of the epimeric 24-methylcholesterols (referred to in the paper as "campestanol", **39**, which of course is only one of the two possibilities, the other being 24β-methyl-5α-cholestanol, **40**) was also found by MS in all of the firs, spruces, and hemlocks, and in some of the larches (in *Larix decidua* and *L. lep-*

Table 8.2.1. Sterol composition of wood from conifers[a] (23)

Tree name		Sterol[b] in mixture, percent			
Botanical	Common	Campesterol and dihydrobrassicasterol	24-Methylstanols	Sitosterol	5α-Sitostanol
Abies alba	Silver fir	29	13	38	20
Chamaecyparis lawsoniana[c]	Port-Orford-cedar	14		65	11
Larix decidua	European larch	23	9	44	24
L. leptolepis	Japanese larch	25	8	45	22
L. potanini	Chinese larch	33		57	10
Picea abies	Norway spruce	27	13	52	17
P. sitchensis	Sitka spruce	14		75	11
Pinus banksiana	Jack pine	13		71	16
P. nigra	Black pine	11		71	18
P. strobus	E. white pine	16		69	15
P. sylvestris	Scots pine	13		69	18
Pseudotsuga menziesii	Douglas-fir	30	5	51	14
Tsuga canadensis	Eastern hemlock	32	17	32	19
T. mertensiana	Mountain hemlock	24	9	46	21

[a] From M$^+$ in mass spectroscopy
[b] For a discussion of structural assignments, see text
[c] Also has 10% stigmasterol

tolepis but not in *L. potanini*). The 24-methylstanol was not found in the pines or in *Chamaecyparis lawsoniana*. In the two species of *Larix*, interestingly, the 24-methylstanol was present only in the wood and not in the bark. This, together with the relatively high content of stanols (Table 8.2.1) in wood compared to other tissues (37, 71), suggests that reduction of the Δ^5-bond (or maybe of the Δ^7-precursor) plays some important role in wood. Although no cholesterol was reported by Fischer et al. (23) for any of the trees, Höll and Goller (37) found 1% to 2% of this Δ^5-24-desalkylsterol by GLC in spruce (*Picea abies*) along with about the same amount of desmosterol (68). Stigmasterol (2), as already mentioned, was found only in the tree-of-life cypress (*C. lawsoniana*). The absence of this dienol in the genus *Pinus* bad been noticed earlier by Rowe (92). In the leaves of the ancient gymnosperm, *Ginkgo biloba*, stigmasterol similarly is at best a very minor (1% or so) component of the sterol mixture containing sitosterol with lesser amounts of campesterol, 22-dihydrobrassicasterol, and cholesterol.

Plants in the other conifer family (Taxaceae) seem to be of the main line just as the Pinaceae are. The heartwood of species of the tribe Taxeae (*Taxus baccata*, *T. cuspidata*, *T. floridana*, and *T. brevifolia*) are reported (20) to contain sitosterol (**25**) and campesterol (**14** in the Δ^5-3β-ol series) (presumably with its C-24 epimer). The other tribe (Podocarpae) of this family, which includes the important genus *Podocarpus*, may therefore also be main line. The leaves and bark of *Podocarpus lambertius* actually yielded an incompletely identified sample of sitosterol (7).

Main line Δ^5-sterols (including 12% of stigmasterol) have also been reported (107) to be present in the wood of *Gmelina arborea*, which, as with the trees studied by Fischer et al. (23), seemed to contain some sitostanol (**38**, 7.5% of the sterol mixture). Sitostanol similarly has been reported (74) in the wood of *Alnus japonica* along with sitosterol, and, together with stigmasterol and sitosterol, sitostanol was found in the wood of European beech (*Fagus sylvatica*) (86). Heartwood of another tree in the Fagaceae (the white oak, *Quercus alba*) also contains sitostanol along with sitosterol, stigmasterol, campesterol, and presumably the latter's 24β-methyl analog (9).

A number of other trees among the angiosperms have yielded sterols of the main line but without mention of stanols. A typical main line distribution (sitosterol, **25**, with smaller amounts of cholesterol, **1**, campesterol, **14**. **14**, in the Δ^5-3β-ol series probably with its epimer, **13**, at C-24, and stigmasterol, **25**) has been found along with the intermediates 24-methylenelophenol (lophenol being 4α-methyllathosterol – i.e., the 4α-methyl derivative of **21**) and 24Z-ethylidenelophenol (citrostadienol, 4α-methyl-Z-ethylidene derivative of **21**) in heartwood of the rock elm (*Ulmus thomasii*), and without 24-methylenelophenol in the slippery elm (*U. rubra*) (93) The 24-ethyl- and 24-methyl-Δ^5-sterols (**21, 14**, and presumably **13**) were also isolated from the American elm (*U. americana*) (93). In the Lauraceae, the wood of *Neolitsea sericea* from Japan similarly appears to represent the main line (51), since GLC on SE-30 indicated the presence of sitosterol (**24**), stigmasterol (**25**), and campesterol (**14** in the Δ^5-3β-ol series, presumably mixed with the epimer, **13**, at C-24). This plant is noteworthy in that the 24-methylsterol is listed as the major component. The sterols (sitosterol, stigmasterol, and campesterol) were found by GLC to be in a ratio of 31:3:68, respectively (51). Another member of the Lauraceae family (*Nectandra polita*, which grows in the Andes) is thought to contain sitosterol but adequate information is lacking (103). A "sitosterol" has also been reported in wood of the Rutaceae (*Melicope octandra*) (46), the Fagaceae in *Quercus robur* from Poland (110), the Betulaceae in birches (*Betula* sp.) (78, 79, 98), the Bignoniaceae in *Tecomella undulata* (47) and species in the genera *Haplophragma*, *Heterophragma*, and *Millingtonia* (45, 48), the Moraceae in *Ficus bengalensis* (102), the Salicaceae in the quaking aspen (*Populus tremuloides*) (81), and the Lauraceae in *Sassafras albidum* (36).

In their work on the sterols of wood zones (sapwood to heartwood) Höll and Goller (37) found the inner heartwood of spruces contained about 3 µmoles of total sterol per gram of dry weight, with lesser amounts in other zones. This translates into about 1 mg/g or 0.1% of inner heartwood of which about 67% was free sterol and 33% ester. Fischer et al. (23) found similar though somewhat smaller amounts for the total (free and esterified) sterol based on the whole of the woody part. They expressed their numbers as kg/ton and found 0.66 for the pines and 0.64 for the firs, spruces, and hemlocks. The average (0.65 kg/t) represents 0.72 mg/g or 0.07% of the wood. Fischer et al. (23) found substantially less (averaging about 0.025% or 0.25 mg/g) in the larches (0.22 kg/t) and tree-of-life cypress (0.25 kg/t). A still lower value (0.01% or 0.1 mg/g) was observed by Komae and Hayashi (51) who examined the wood sterols of *Neolitsea sericea* (Lauraceae). It is not clear, however, whether the various values are exactly comparable to one another. Höll and Goller (37) extracted wood powder with refluxing acetone in a Soxhlet apparatus and then separated the free sterols from the esters on a column of silica gel. Fischer et al. (23) also used wood powder but say the solvent was methylene chloride without giving time or temperature. The extract was saponified prior to chromatography yielding total sterol. Komae and Hayashi (51) extracted wood chips with diethyl ether at room temperature for 10 days and then directly chromatographed, giving them only the free sterols. The ester fraction was not examined, but its addition would probably only raise the amount of sterol by about 50% based on the observations of Höll and Goller (37) with spruce trees. Also not clear is the amount of water in the wood samples extracted. Except for Höll and Goller who lyophilized slices of freshly felled tree trunks the various authors do not mention what, if any, drying was given to the wood. Despite these ambiguities, one probably has to assume tentatively that the studies actually were comparable, giving us a range of about 0.01% to 0.1% for the total sterol in the wood of conifers and angiosperms. Almost precisely the same range has been found in leaves. The amounts of free and esterified leaf sterols from a fern, the Ginkgo, a laurel, spinach, and a cucurbit all were found to be between 0.01% and 0.06% of the fresh, wet weight when obtained with refluxing acetone in a Soxhlet apparatus (70). This of course would rise substantially when placed on a dry weight basis, so it would appear the wood sterol may be present in a smaller concentration than in leaves, relative to materials other than water. Water content, incidentally, probably ought to be considered as one of the possible sources of differences observed from tree to tree – e.g., wood from spruce and larch.

Spinasterol (29), the Δ^7-analogue of stigmasterol (25), is thought to represent the sterol of the bark of some trees belonging to the leguminous Mimosaceae family (*Pithecellobium saman*, also known as *Inga saman*, and *Samanea saman*, from India) (73) and the Indian evergreen *Mimusops littoralis* (17). In the Mimosaceae there are also three species of the genus *Albizzia* – *A. amara*, *A. julibrissin* (called the "silk tree" and widely distributed in eastern Asia including China, Japan and Korea), and *A. procera* – in which spinasterol is believed to occur (105). More commonly, however, bark – like wood, leaves, and fruit – contains Δ^5-sterols of the main line.

Clotofiski's paper on the European beech (*Fagus sylvatica*) in 1942 (12) appears to be the earliest report of sterols in bark, but the compounds were not identified. One of the compounds the bark yielded (12) had an empirical formula of $C_{29}H_{48}O$. This is the formula for sitosterol and other more recent evidence for sitosterol in the wood of this tree has actually been presented (86), indicating that this plant is of the main line. Clearly of the main line from the sterol pattern in GC/MS is the Indian legume, *Saraca indica*. From 2 kg of bark, 200 mg (0.01%) of crystalline sterol was obtained and was resolved by GLC into material with the rates of movement of sitosterol (75.9%), stigmasterol (22.1%), and campesterol (1.5%) or its C-24 epimer (5). The mass spectra taken on the sterols as they emerged in the effluent from GLC agreed with these assignments. Similarly, the assigned composition (63% stigmasterol, 31% sitosterol, 5% campesterol, and 1% cholesterol) in the bark of *Erythrina suberosa*, an Indian plant, seems to be main line (100). The large amount, however, of the dienol is reminiscent of Category I-B plants ($\Delta^{7,22}$, 24α) in which the dienol, spinasterol ($\Delta^{7,22}$-24α-ethyl), is often the major component followed by 22-dihydrospinasterol (the Δ^{7}-analogue of sitosterol), 24ζ-methyllathosterol, and lathosterol itself.

Qualitatively the inner and outer barks seem to be much the same, though different quantitatively. The outer bark (89%) of loblolly pine (*Pinus contorta*) is thought to contain sitosterol, **24** (0.060% of the oven-dried bark solids), as well as campesterol (**14** in the Δ^{5}-3β-ol series, presumably with its C-24 epimer, 0.004%), whereas the inner bark (11%) has the same sterols but in much higher concentration based on an oven-dried sample (0.29% and 0.04%, respectively) (80). Reconstitution of the whole dried bark then by calculation gives 0.085% 24-ethylsterol and 0.008% 24-methylsterol. On a fresh weight basis the data would be somewhat different, because the outer bark is 14.9% water and inner is 46.8%. Rowe and his associates (13) made an extensive study of the constituents (more than 70 of them) in a benzene extract of western white pine bark (*Pinus monticola*). The extract amounted to 3.2% of the oven-dried bark and, of the extract, 1% was free sterol and 1.5% steryl ester. The free sterols were 74% sitosterol, 13% stigmasterol, 13% campesterol (presumably mixed with dihydrobrassicasterol), and a trace of cholesterol. The ester fraction was 97% sitosterol and 3% the 24-methyl analogue. Similarly, the bark of *Pinus radiata* contains an 85/15 mixture of Δ^{5}-24-ethyl- and 24-methylsterols by GLC and MS and amounts to 0.050% of the air-dried bark with 5% residual moisture. *Cunninghamia*, which belongs to the Taxodiaceae, also is main line (10). The principal sterol was sitosterol (**24**) which was obtained in crystalline form. The mother liquor gave smaller amounts of cholesterol (**1**) and stigmasterol (**25**) by GLC (10). The crystalline sitosterol (identified by m.p., IR, and NMR, but not by GLC) probably had a small, undetected quantity of the 24-methyl epimers along with it. A representative of the Podocarpaceae, namely *Podocarpus lambertii*, which grows in Brazil, has also been studied (7). Unfortunately, "sitosterol" isolated chromatographically from a benzene extract was identified only by m.p. It amounted to about 0.015% of the dried bark.

A "sitosterol" is also thought to be present in bark of many other plants including *Marsdenia volubilis* (108), *Eugenia wallichii* (56), *Ficus racemosa* (6), *Dalbergia latifolia* (16), *Lophopetalum rigidum* (96), *Zantoxylum conspersipunctatum* (Rutaceae) (50), *Ceanothus velutinus* (35), *Delonix regia* (94), *Flindersia laevicarpa* (82), *Melicope octandra* (Rutaceae) (25), *Dichrostachys cinerea* (46), *Larix decidua* (Pinaceae) (76), *Fluggen microcarpa* (Euphorbiaceae) (89), the Bengalese tree, *Skimmia wallichii* (Rutaceae) (89), *Guazuma tomentosa* (Sterculeaceae), a deciduous tree native to tropical America that has been introduced into India (3), *Gymnosporia montana*, also known as *Maytenus senegalensis*, in the family Celastraceae (43), *Heterophragma adenophylum* and *Millingtonia hortensis* in the Bignoniaceae (45), as well as in *Populus tremuloides* bark (39), the bark of *Abies magnifica* (4), and the bark of *Amaroria soulameoides*, the sole member of a monotypic genus of the Simaroubaceae found in the Fiji islands (42). Curiously, *Harpullia pendula* bark is reported to contain stigmasterol (**25**) without mention of sitosterol (**24**) (11).

3-Ketones corresponding to the sterols of wood and bark were isolated as early as 1963 by Lavie and Kaye (54), who obtained sitost-4-en-3-one (24α-ethylcholest-4-en-3-one, **41**). Rowe (92) suggested such compounds may be auto-oxidation artifacts, but Norin and Winell (75) think they may have their origin in oxidation by contaminating microorganisms. Joshi et al. (44) described the mass and ^{1}H-NMR spectra (at 60 MHz) of sitost-4-en-3-one (**41**) from heartwood of *Tabebuia rosea*, which also contains sitosterol (44). Campello et al. (7) extensively characterized the 5α-hydroxy-6-keto derivative (**42**) of sitosterol from a *Podocarpus lambertii* extract. Conner et al. (13) found both sitost-4-en-3-one and campest-4-en-3-one (24α-methylcholest-4-en-3-one, **43**, presumably with its C-24 epimer) in extracts of western white pine (*Pinus monticola*) along with the dehydrogenation products (sitost-4,6-dien-3-one, **44**, campest-4,6-dien-3-one, **45**, probably with the C-24 epimer, **46**, as well as cholest-4,6-dien-3-one, **47**). They also further broadened the subject by finding the 7-ketone dehydrated at C-3 (sitost-3,5-dien-7-one, **48**). The sitostenone and campestenone were also obtained by Weston (109) from bark of *Pinus radiata*, and at about the same time Nair and

Chang (58) isolated the 6β-hydroxy derivatives of these ketones – viz., 6β-hydroxysitost-4-en-3-one (**49**) and its 24-methyl analog (6β-hydroxycampest-4-en-3-one, **50**, presumably with the C-24 epimer, **51**), – in extracts of *Melia azedarach* bark. Of perhaps special interest was the identification of the saturated 5,6-dihydro ketone, sitostanone (24α-ethyl-5α-cholestan-3-one, the 3-keto derivative of **38**) in extracts of *Alnus japonica* wood (74), which contained the corresponding 3β-alcohol, sitostanol (**38**). Similarly, extracts of a spinasterol-containing bark from *Samanea saman* (73) have yielded the related $\Delta^{7,22}$-3-ketone (**54**) called "spinasterone" (**52**) (24α-ethyl-5α-cholesta-7,22E-dien-3-one).

From the bark of the Indian Kurchi plant (*Holarrhena antidisenterica*) material was isolated by column chromatography, which presumably represented the entire or at least one major sterol fraction (59). This fraction was thought to consist of a single component and to be one of the Δ^{23}-isomers of stigmasterol as shown by (**53**) (the 24-ethylcholesta-5,23-dien-3β-ol in which C-22 is *trans*-oriented to C-25). Further proof of structure and homogeneity, however, would be worthwhile for this interesting and unusual sterol.

8.2.2.4 Biosynthetic Origin of Individual Sterols

Cholesterol and its 24-methyl and 24-ethyl derivatives do not appear to be interconverted in higher plants. These and other of the dominant sterols arise by partly separate pathways from cycloartenol. While understood in principle (33, 67, 71), the pathways still require further investigation in terms of precise sequences and the factors that control them (33, 65, 6, 71).

The best evidence at this time is that the pathway leading to cholesterol diverges early from that ending in the 24-alkyl derivatives. Introduction of the alkyl groups at C-24, which never occurs prior to cyclization of squalene oxide to cycloartenol, utilizes the π-electrons of the Δ^{24}-bond in an electrophilic attack on C-24 (8) by the positively charged methyl sulfonium moiety of *S*-adenosylmethionine. By contrast, in the formation of 24-desalkyl sterols – e.g., cholesterol (**1**) – the Δ^{24}-bond is reduced by pyridine nucleotide, thereby preventing methyl sulfonium ion attack on it. Based on the structures of natural sterols, this seems to occur quite early in the pathway (at the C_{30}-stage). 24(25)-Dihydrocycloartenol (cycloartanol) has been isolated from a number of higher plants, and in *Euphorbia pulcherrima* (Euphorbiaceae), which contains large amounts of cholesterol (97), the presence of 24(25)-dihydrolanosterol (**54**) is well documented (97). The latter is thought to arise from cycloartenol via opening of the 9β,19-cyclo ring and reduction of Δ^{24}. Thus, from various other higher plants (33, 67, 71) have been isolated the 24(25)-dihydro compounds with 29 carbon atoms, 4-monodesmethylcycloartanol (**55**) and the analog 4-monodesmethyl-24(25)-dihydrolanosterol (**56**) in which the three-membered ring has been opened. Similarly, the C_{28}-stanol, 4,4-bisdesmethylcycloartanol (pollinastanol, **57**) is known representing not only reduction of Δ^{24}, opening of the three-membered ring, and removal of one methyl group at C-4 but also removal of the other methyl group at C-4. In addition, the C_{28}-Δ^{7}-compound with Δ^{24} reduced and a methyl at C-4 and at C-14 both removed occurs naturally (lophenol, 4α-methylcholest-7-en-3β-ol, **58**). A po-

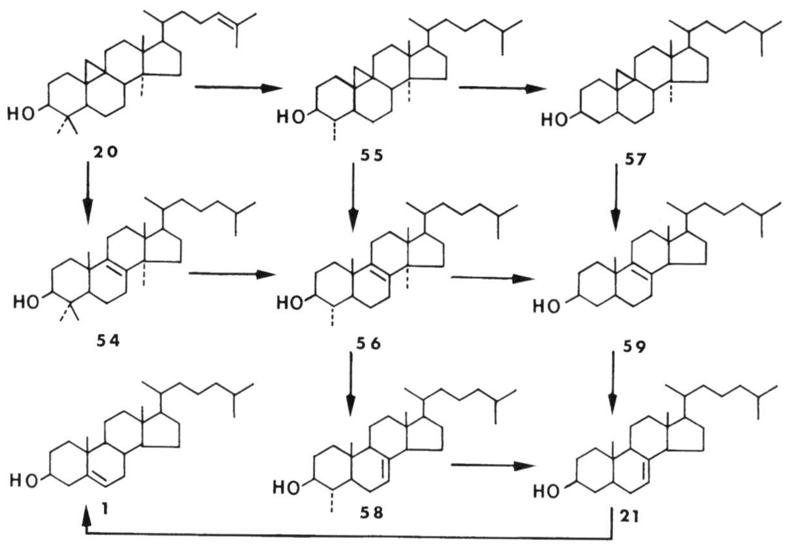

Fig. 8.2.4. Some possible alternative sequences to cholesterol

tential C_{27}-Δ^7-intermediate, which in some species is probably an end product, also occurs in higher plants. It is the Δ^7-analogue of cholesterol (also a C_{27}-sterol) called lathosterol (cholest-7-en-3β-ol, **21**). The Δ^7-sterols are expected isomerization products of the Δ^8-analogs – e.g., 24(25)-dihydrozymosterol (**59**) and its 4α-methyl derivative (**56**). These various sterols cannot be put on a single route. Instead, they represent alternative sequence (Fig. 8.2.4) that operate depending on taxonomy, tissue, and perhaps other variables. The fundamentals of the process of conversion to cholesterol are (a) reduction of the Δ^{24}-bond, (b) opening of the tree-membered ring to give a Δ^8-bond, (c) oxidative removal of the methyl groups at C-4 and C-14, and (d) isomerization of Δ^8 to Δ^7. This gives us lathosterol (**21**) via the intermediates mentioned (Fig. 8.2.4). Lathosterol by analogy to the mammalian process would probably then proceed (d) by dehydrogenation at C-5(6) and reduction at C-7(8) through the $\Delta^{5,7}$-derivative (**22**, 7-dehydrocholesterol) to cholesterol. Pollinastanol (**57**) has actually been experimentally converted to cholesterol in angiosperm tisssue. It should be noted, however, that the sequences shown in Fig. 8.2.4 are not the only ones possible. For instance, the Δ^{24}-bond could be reduced as the last step. Desmosterol (**23**), isolated from red algae, then becomes the penultimate sterol in the sequence instead of 7-dehydrocholesterol.

The 24-alkyl route is more complicated: It not only requires introduction of two C_1-groups, the 24-C_2-groups being biosynthesized stepwise (86) rather than being performed, but there is also the chiral problem at C-24. How and at what stage is the configuration determined? Some uncertainties still surround this question. The problem of the configuration of the 24β-methyl group may be the simplest. The configuration is thought to be established immediately on C_1-attack from the β-side of C-24 with elimination of a proton from C-27. In one

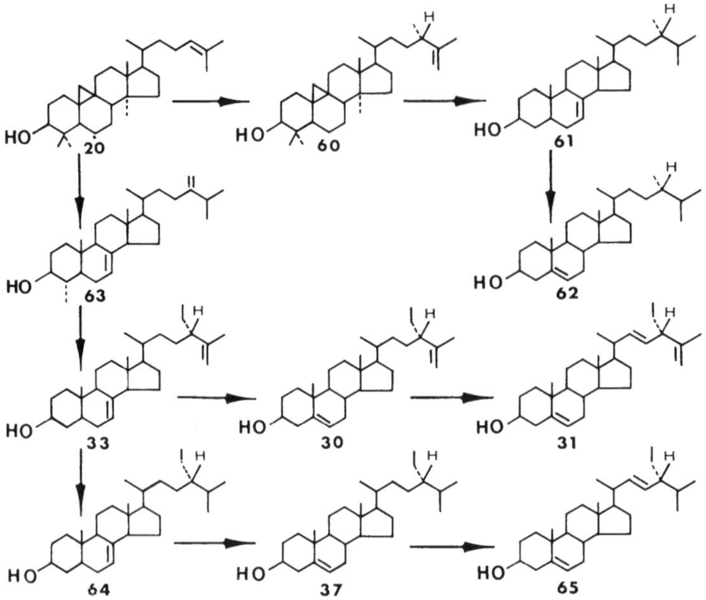

Fig. 8.2.5. Reasonable biosynthetic sequences to 24β-alkylsterols

sequence this clearly can occur right away at the cycloartenol stage owing to the natural occurrence of cyclolaudenol (24β,4α,4β,14α-tetramethyl-9β,19-cyclocholesta-25(27)-en-3β-ol, **60**) (Fig. 8.2.5.). This sterol, by the principles outlined for cholesterol biosynthesis, would proceed together with reduction of the $\Delta^{25(27)}$-bond to 24β-methyllathosterol (**13**) and then to the common constituent of functional sterols, 24β-methylcholesterol (**13** in the Δ^5-3β-ol series or **62** (Fig. 8.2.5) 22(23)-dihydrobrassicasterol). On the other hand, attack of a C$_1$-group on C-24 with elimination of a proton from the incoming methyl group and migration of the H-atom from C-24 to C-25 would give a 24-methylenesterol (possessing a $\Delta^{24(28)}$-bond). An example is the widely occurring 24-methylenecycloartanol (**66**, Fig. 8.2.6), which is believed to be a metabolite of cycloartenol (**20**) and to act as a major intermediate to 24-alkylsterols in many photosynthetic systems. Other intermediates with a 24-methylene group known to occur naturally are cycloeucalenol (4-monodesmethyl-24-methylenecycloartanol, **67**), obtusifoliols (4α,14-dimethyl-24-methylene-cholest-8-en-3β-ol, **68**), and gramisterol (4α-methyl-24-methylenelathosterol, **63**). A second C$_1$-transfer to the β-side of the newly added olefinic carbon atom (C-28) with migration of the H-atom on C-25 back to C-24 on the α-side then gives the side chain of the 24β-ethyl-$\Delta^{25(27)}$-sterols of Category II plants, which presumably lead in turn to their 25(27)-dihydro derivatives – e.g., the naturally occurring clionasterol (**37**) and chondrillasterol (Δ^{22E}-derivative of **64**), which are the 24β-ethyl epimers of sitosterol (**24**) and spinasterol (**28**), respectively.

The 24β-alkyl route is summarized in Fig. 8.2.5. It shoud be remembered, however, that there remain a number of uncertainties about the sequencing of

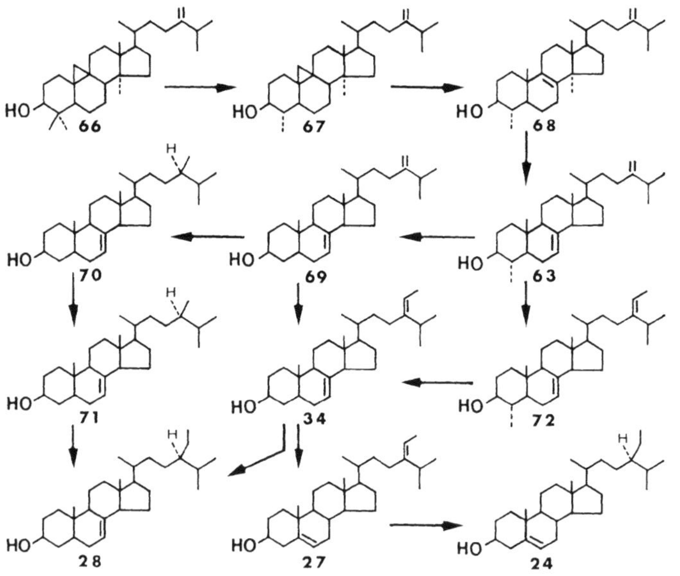

Fig. 8.2.6. Resonable biosynthetic sequences to 24α-alkylsterols

events. The basic thing to note is that there is a bifurcation in which C_1-transfer producing a 24-C_1-group with introduction of the $\Delta^{25(27)}$-bond leads to 24β-methylsterols, while C_1-transfer with formation of the 24-methylene group leads by a second C_1-transfer to 24β-ethylsterols. The 24-methylenesterols – e.g., 24-methylenecholesterol – can be end products in special cases such as pollen. They can also lead to 24α-methyl and 24α-ethylsterols as shown in Fig. 8.2.6. The latter figure also illustrates how 24-methylenecycloartanol (**66**), cycloeucalenol (**67**), and obtusifoliol (**68**) can be intermediates to gramisterol (**63**), which could be an intermediate (as suggested by Figs. 8.2.5 and 8.2.6) to 24α-methyl- as well as to the epimeric 24α- and 24β-ethylsterols. Reasons exist for not wanting to place the 24-methylenesterols on the pathway to the 24β-methylsterols in higher plants, although it remains a possibility. In fungi there is little or no doubt that they are on the pathway.

The 24α-alkyl route (Fig. 8.2.6) also is not entirely clear, but it seems to involve the $\Delta^{24(28)}$ sterols at both the 24-C_1- and 24-C_2-levels. As is seen in Fig. 8.2.6, the 24-methylenesterols are believed to lead to the 24α-methylsterols and the 24Z-ethylidene analogs to the 24α-ethylsterols. Currently it is thought that the Δ^{23}- or $\Delta^{24(25)}$-sterols might be intermediates by isomerization of the $\Delta^{24(28)}$-bond (**55**). Why this should be so, however, is far from clear, as is the reason for the ubiquitous configuration (Z) of the 24-ethylidene group (as in **27** and **34**, Fig. 8.2.6) in higher plants. In brown algae the configuration of the $\Delta^{24(28)}$-bond is E, but for reasons equally unclear the E compound in the Δ^5-series (fucosterol) fails to proceed to a 24-ethylsterol in this genus (*Fucus*). Commonly encountered 24Z-ethylidenesterols are isofucosterol (**27**), avenasterol (**34**), poriferasterol (**65**), and citrostadienol (**72**).

The origin of the Δ^{22}-sterols (commonly of the E configuration in higher plants as well as in the most well studied of the algae having Δ^{22}-sterols), which include inter alia stigmasterol (**25**), spinasterol (**29**), 25(27)-dehydroporiferasterol (**31**), poriferasterol (**65**), 25(27)-dehydrochondrillasterol (**32**), and brassicasterol (**26**), is, in terms of sequence, still something of a mystery. That is, at what point in the pathway, despite what is shown in Figs. 8.2.5 and 8.2.6, is the Δ^{22}-bond actually introduced? For instance, does stigmasterol (**25**) arise from sitosterol (**24**), as some work suggests, or does sitosterol arise from stigmasterol with stigmasterol being a metabolite of 22Z-dehydroisofucosterol? Or, maybe there is an interconversion. Only future work will resolve this question — which could be dependent on species, and/or tissue, and/or development, or environment. At least in terms of the final result — i.e., the ratio of Δ^{22} to 22(23)-dihydro products — development definitely is involved. The brassicasterol/22(23)-dihydrobrassicasterol ratio in the genus *Brassica* is strongly influenced by ontogeny. In the seed, the Δ^{22}-compound dominates, but in the mature plant, the dihydro derivative does. The Δ^{22}/dihydro ratio in the $\Delta^{25(27)}$-24β-ethyl series also depends on species within a genus.

A problem similar to the one with the Δ^{22}-bond exists with a number of other differences in plant sterols. Thus, we have the Δ^{5}/Δ^{0} ratio to explain as well as the 24α-methyl/24β-methyl and 24-methyl/24-ethyl ratios not to mention the curious fact that 24-ethylsterols of main line plants are always of the α-configuration although the 24-methyl analogs are not. In view of the statistics of occurrence, such as that sitosterol is usually the major sterol in main line plants, it would appear that the ratios are rather tightly controlled. This tight control probably constitutes a fine tuning of the function(s) that the sterol pool plays.

8.2.2.5 Function of Sterols

Sterols are among the oldest known biochemicals, cholesterol having been discovered in human gallstones in 1815, ergosterol in the ergot fungus infecting rye in 1889, and stigmasterol in the Calabar bean (*Physostigma venenosum*) in 1906. Yet, it is only very recently that the wide distribution and function of these materials has become evident. All eukaryotic cells with but a few exceptions are now known to contain sterols (71). While sterols frequently are precursors to other materials, notably to hormones such as cortisol, progesterone, estradiol, and testosterone in vertebrate animals, brassinolide in higher plants, and ecdysones in both invertebrate animals and higher plants, this precursor role in the well examined cases is quantitatively minor (71). Hormones are chemical messengers, agents that switch processes on and off, and are required in very small amounts. Consequently, they are usually not stored and have minute circulating concentrations. The best evidence is in mammals. Cortisol, for instance, is present in human peripheral blood plasma at a level of about 10 µg/dl (dl = deciliter, the common unit for blood values), whereas cholesterol, from which it is derived (in the adrenal gland), is present in blood at a concentration (about 160 mg/dl), which is four orders of magnitude higher. Similarly, vanishingly small concentrations of hormones are found in the tissues where they are biosynthesized. Thus, from adrenal glands, cortisol has been isolated in a yield of only about 1 mg/kg of tissue. Cholesterol by comparison is present at a level of about 50 g/kg, or 50 000 times more. Although for reasons that are obscure, adrenal tissue is exceptionally rich in sterol. A whole human being still has as much as 4.5 g/kg, or 0.45% of wet weight. It is therefore obvious that sterols do not function primarily as precursors.

With the development of techniques beginning in the 1960s to isolate and purify not only organelles but membranes themselves and to analyze the constituents, it has become clear from a wealth of data [for a key to the literature, see Nes (66, 71)] that the free sterol of cells in vertebrates, fungi, higher plants, protozoa, mycoplasmas, and other organisms resides primarily in membranes. There is also growing evidence to implicate the plasma membrane as the primary repository of sterol [see, for instance, Lange and Ramos (53)].

That the sterol present in a cell is a necessary part of its life has been demonstrated with animal, yeast, and mycoplasma cells, which are auxotrophic for sterol (71) by virtue of mutation [see, for instance, Ramgopal and Bloch (87)], a natural genetic deficiency [see, for example, Lala et al. (52)], or an experimental block in the sterol pathway [see, for instance, Pinto and Nes (84)]. A careful study of how much sterol is required and of the influence of structure on the ability of sterols to support the life of the cell has revealed that sterol plays at least two roles in a given cell. In one of them, the major part of the free sterol is utilized in what has been called the "bulk role." It is thought the "bulk role" involves interdigitation of the sterol into the phospholipid matrix of each of the monolayers that are subunits of the bilayered lipid leaflet of membranes. The overall length [see ref. (71) for a key to the literature] of a 20R-sterol that allows C-22 to be *trans*-oriented to C-13 (12) is, from the hydroxyl group to the end of the side chain, nearly exactly the same as the thickness of the monolayer (~12 nm) (15, 40). The polar (hydroxyl group) end of the sterol is believed to lie toward the polar end of the phospholipid, which in turn lies on the aqueous side of each of the monolayers. The reason for inclusion of a sterol in a membrane is almost certainly to modulate the physical properties of the membrane, for the sterol is not chemically changed – i.e., metabolized – as it plays its role in the membrane. This has been demonstrated in insects, yeast, and mycoplasmas that extend over a wide area of biology. Furthermore, it has been shown that the presence of a sterol alters the permeability and rigidity (usually referred to in the converse as fluidity) of membranes. Modulation of the fluidity is thought to involve dampening of the conformational freedom of the alkyl chains of the fatty acid moieties of the phospholipids. Triterpenoids may also have an analogous role (60) in some cases.

In addition to the "bulk role" in membranes, sterol plays still other role(s) that are not yet well defined. However, there can be no doubt that the other role(s) exist(s). In yeast (83, 84, 87, 91), which makes ergosterol (**73**, 24β-methylcholesta-5,7,22E-trien-3β-ol), the minimum amount of ergosterol necessary for viability

73

can be replaced by cholesterol or other sterol laking a 24β-methyl group, but only to the extent of about 90% or so. Viable cells will form only when some of the sterol has a 24β-methyl group (83, 84, 91). Similar findings using lanosterol (66) and cholesterol (71) have been found with the prokaryotic *Mycoplasma capricolum* (14) in that some but not all of the cholesterol necessary for growth can be replaced by lanosterol. This quantitatively minor but nevertheless essential role of sterol has been called "regulatory" (83), "sparking" (91), and "synergistic" (14), while the quantitatively major role is known as the "bulk" role. Ergosterol can play both roles in yeast as cholesterol can in *M. capricolum* and presumably in animals. However, in most higher plants, where a more complicated array of sterols is present, as is also the case in some fungi – e.g., *Gibberella funjikuroi* (61) – a single sterol does not appear to be able to play all of the roles. However, there are some exceptional tracheophytes with virtually a single sterol – e.g., species of *Clerodendrum* (85) and cottonseed (71). Trees, on the other hand, as discussed in Sect. 8.2.2.3, usually have a sterol mixture, and the different sterols presumably reflect different roles.

8.2.2.6 Phyletic and Phylogenetic Relationships

The fact that sterols seem to be nearly ubiquitous constituents of higher plants has prompted attempts (24, 28, 29, 61, 66, 67, 70–72, 85) to use these compounds as chemotaxonomic and evolutionary markers. Two main conclusions have been arrived at. Living systems that are phylgenetically photosynthetic utilized the cycloartenol pathway, whereas, in those that are nonphotosynthetic, lanosterol (**9**) is formed in place of cycloartenol (**20**) as the cyclization product of squalene oxide (Figs. 8.2.1 and 8.2.3). Secondly, as the evolutionary ladder is climbed, there is a statistical shift from 24β-alkylsterols to 24α-alkylsterols in plants. Both the ancient gymnosperm *Ginkgo biloba* and the angiosperm *Liriodendron tulipifera* in the primitive Magnoliales are in the main line (mostly 24α-alkylsterols, Category I) as are the majority of higher angiosperms and gymnosperms. Algae, on the other hand, are primarily of Category II (mostly 24α-alkylsterols), while mosses are intermediate. This leaves the origin of the tracheophytes of Category II uncertain. Did they have a separate parallel origin from the line to which the ginkgos and magnolias belong, or did they diverge at some point? Similarly, the origin of the Δ^7-producers (Category B) is uncertain. This problem has been discussed at some length elsewhere (28, 29, 70–72), and a detailed consideration is beyond the scope of the present section. The problem is complicated, because there is no unanimity of opinion of how many genera are related to one another especially from an evolutionary point of view and because within a given taxonomic group, whether it be genus or order, there are known differences in sterol composition. However, as might be expected, the more closely two plants are related by classic botanical parameters, the more likely they are to have a similar sterol pattern. The simplest difference in pattern, and one which can vary not only between species but even between parts of the same plant, is the relative amount of sterols. Thus, two plants or plant parts may both contain sitosterol and stigmasterol but the proportions of the two sterols may be different. A difference in the biosynthetic steps operating – e.g., producing an inversion of the configuration at C-24 – constitutes a much more substantial change and one that would not be expected at the species level. Real additions and deletions of steps seem tentatively to occur, primarily or perhaps only at the genus and higher levels, but it remains somewhat unclear what the rules are that govern why certain steps are present or absent. Very difficult and as yet unanswered questions arise as to the extent to which structural genes may be present for enzymes for all the major pathways and the degree to which history, development, and environment influence expression of these genes. Nor are there any empirical relationships relating sterol composition to habit of growth, geography, etc. Thus, chemotaxonomic use of sterols is not yet a clear and simple technique and should be approached with caution.

8.2.2.7 Industrial Utilization of Tree Sterols

At present, the principal use of sterols from plants is as starting materials in the production of natural and artificial hormones of mammals. It is possible to oxidize sterols with CrO_3 and other reagents nonselectively (protecting double bonds in some way such as by bromination) and to obtain 20-carboxylic acids and 17-ketones – e.g., 17β-carboxyandrost-5-en-3β-ol and the corresponding 17-ketone from, say, cholesterol or sitosterol – that can be used as a source of the steroid nucleus. The yield, however, is very poor, since many other products of degradation are also obtained. This has led to the use of sterols, such as stigmasterol (25) with a Δ^{22}-bond, which permits a more direct attack on the side chain. For instance, oxidative cleavage – e.g., by ozonolyis – leads to the 22-aldehyde, which, by enolization and a second oxidation, gives the 20-ketone. Thus, a Δ^5-sterol can yield the C_{21}-compound, 3β-hydroxypregn-5-en-20-one (pregnenolone) or by additional manipulation the hormone progesterone (pregn-4-en-3,20-dione). The C_{21}-steroids are used industrially for production of still other hormonally active materials such as the corticoid, cortisol, also known as hydrocortisone (77). Unfortunately, the major sterols of trees do not have a Δ^{22}-bond, but the development of biological methods has circumvented this difficulty at least to some extent (Fig. 8.2.7). The use of *Mycobacterium* NRRL B-7683, for instance, yields the C_{19}-steroid, androst-1,4-dien-3,17-dione (74) in a 52% yield from sitosterol (24) (23). This can be used to produce androgens such as testosterone (75). The C_{22}-steroid, 20-carboxypregn-1,4-dien-3-one (76) can also be obtained microbiologically from sitosterol, and the carboxylic acid then by Curtius degradation can be converted to the 20-amine and on to progesterone [Fischer et al. (23) and references therein].

Sterols are found in the tall oil that results from the kraft (sulfate) process of making wood pulp. Tall soap (~3% sterol by dry weight, or 50% of the nonsaponifiable fraction) yields tall oil (90 kg per ton of cellulose) by acidification with dilute sulfuric acid and then tall pitch (~20% of the tall soap) as a residue after distillation. Sterols tend to be retained and concentrated to ~10% in the tall pitch. However, losses are encountered between tall soap and tall pitch. Sterols are obtained from these fractions by extraction with organic solvents. A simple example depends on an extraction with warm alcohol from which the crude product

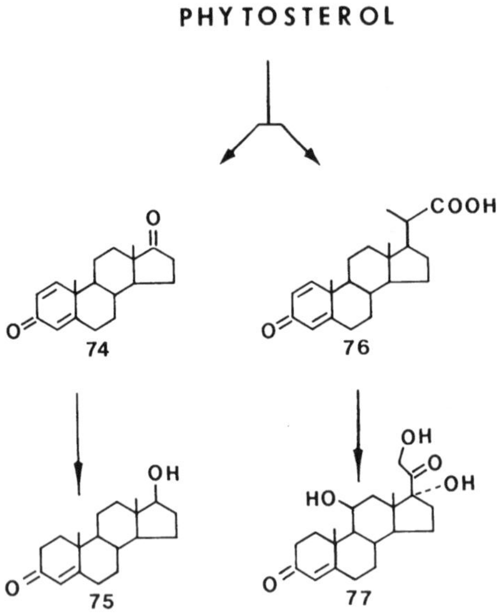

Fig. 8.2.7. Industrial degradation of sterols by microorganisms

is obtained on cooling. The crude material is dissolved in benzene and filtered. The filtrate then yields a mixture on evaporation that is about 50% sterol. Further purification is made chromatographically. For a key to the literature, see Fischer et al. (23).

8.2.3 Esters

While sapwood contains very little sterol in the ester form (traces to 0.15 moles/g, dry weight basis), the amount rises strongly to about 1.1 moles/g in the inner heartwood (37, 38). Höll and Pieczonka (38) found the acid component of the esters in both cases was comprised of a number of fatty acids. In terms of the usual nomenclature (chain length: number of double bonds), the fatty acid composition of the steryl esters in spruce heartwood was found to be 12:0 (10%), 14:0 (12%), 16:0 (20%), 16:1 (12%), 18:0 (14%), 18:1 (16%), 18:2 (5%), and unidentified (11%). The composition varied only slightly from this in sapwood. In both sapwood and heartwood of the spruce, the ester was a minor form of the total sterol. The ester to free sterol ratio was about 1:2 in the inner heartwood (37). Sitosterol and other sterols have also been found to exist partly in the ester form in the heartwood of angiosperms – e.g., the slippery elm (*Ulmus rubra*) (24). Similarly, the sterols in the bark of the western white pine (*Pinus monticola*) are partly (60%) esterified (13). In both heartwood and bark the steryl composition of the ester fraction approximately reflected the composition of the free sterol fraction in that sitosterol was the major component followed by its 24-methyl analogs (13, 37). A similar situation has been observed with non-woody angiosperms – e.g., *Zea mays* (71).

8.2.4 Glycosides

The heartwoods of the American trees, *Betula alleghaniensis* and *B. lenta* (Betulaceae), are reported to contain sitosterol β-D-glucoside (78) (98). Similarly, from the heartwood of *Melia birmanica* (Meliaceae) this glucoside was isolated (0.016% of shavings) and very well characterized (88). There is also a report of its presence in heartwood of the legume *Piptadenia macrocarpa* (57).

The bark of the legumes *Samanea saman* (also called *Pithecellobium saman*) (73), growing in India, and of a Korean strain of *Albizzia julibrissin* (105), which is widespread in Eastern Asia, each contains both spinasterol and its β-D-glucoside. Glucosides and their 6'-fatty acyl derivatives (where prime refers to the sugar) are well described in other, non-woody tracheophytes (71). Their function is unknown.

8.2.5 Spiroketals (Saponins)

A number of steroids with an unusal type of oxygenated side chain occur in angiosperms as glycosides having several sugar residues. They frequently have the abilities to form a lather in water and have been used by various peoples in place of soap, hence the name "saponins". After hydrolytic removal of the sugars, the remaining aglycone is called a sapogenin. These sapogenins are spiroketals of 22-keto-26 (or 27),16β-dihydroxysteroids. That is, C-16 and C-26(or 27) are both linked to C-22 by an oxygen bridge as a result formally of addition of one of the hydroxyl groups to the 2-carbonyl group, giving a hemiketal, followed by bonding of the other hydroxyl group to C-22 with loss of water yielding a full ketal. Two new rings (a five-membered *E* and a six-membered *F*) have been formed that are at right angles to one another by virtue of having one carbon atom in common (spiro arrangement). An important example is diosgenin (79), which structurally is the spiroketal of cholesterol. Two new chiral centers are produced at C-22 and C-25 by spiroketal formation. While the configuration at C-22 is the same in other sapogenins as in diosgenin, the one at C-25 is variable. The configuration (C-25) found in diosgenin is generically called "iso". Inversion of C-25 leads to the "neo" series. Thus, neodiosgenin occurs naturally and has its own trivial name (yamogenin). Other variations of sapogenins depend on the number, kind, and position of oxygen functions. Digitogenin (the aglycone of digitonin), for instance, is in the iso-series and has additional 2α- and 15β-hydroxyl groups. It is, therefore, 2α,15β-dihydroxydiosgenin. Sapogenins also vary in terms of the dou-

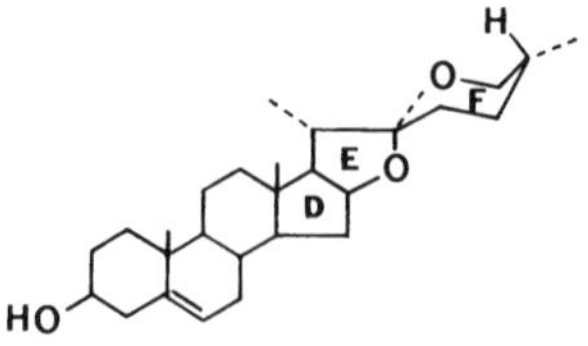

836   Isoprenoids

79

ble bond structure in ring B, and when the Δ^5-bond is missing, both the 5α- and 5β-series exist naturally.

Many tens of sapogenins have been found in plants. Several have been isolated from bark. Both diosgenin and yamogenin occur as glycosides in the bark of the forest tree *Balanites wilsoniana* (Balanitaceae) (101). One has been characterized as the monoglucoside, diosgenin β-D-glucoside (101). *Balanites aegyptica, B. orbicularis,* and *B. pedicellaris* also contain saponins [see ref. (101) for a key to the literature]. From *Balanites wilsoniana* the 3,5-dienes corresponding to diosgenin and yamogenin were also isolated (101), perhaps as artifacts. Diosgenin is also believed to be present in the root bark of *Solanum macrocarpum* after acid hydrolysis (99). The bark of *Dracaena draco* has also yielded (after acid hydrolysis) diosgenin, 1β-hydroxyyamogenin (neoruscogenin), and a new sapogenin called dracogenin with a methylene group in place of the methyl group on C-25 and hydroxyl groups at 1β, 23, and 24 (34).

8.2.6 Ecdysteroids

8.2.6.1 Names, Structures, and Occurrence

Ecdysteroids were first isolated from insects where they act as hormones for (among other things) induction of molting (ecdysis). For a key to the literature see the reviews by Rees (90) and by Svoboda (104). Some 14 ecdysteroids are known to be present in insects. Two of the compounds [ecdysone, also known as α-ecdysone, and 20α-hydroxyecdysone (**80**), also known simply as 20-hydroxyecdysone or as β-ecdysone, crusdecdysone, and ecdysterone] have most often been isolated, and 20α-hydroxylation is thought to be required for hormonal activity. That is, ecdysone may be a prohormone and 20-hydroxyecdysterone the hormone that actually produces results. Other structural features frequently of importance are a Δ^7-bond, a 6-keto group, 2β-, 14α-, and 22α-hydroxyl groups in a cholesterol or 24-alkylcholesterol side chain, and a *cis* fusion of the A/B-ring junction.

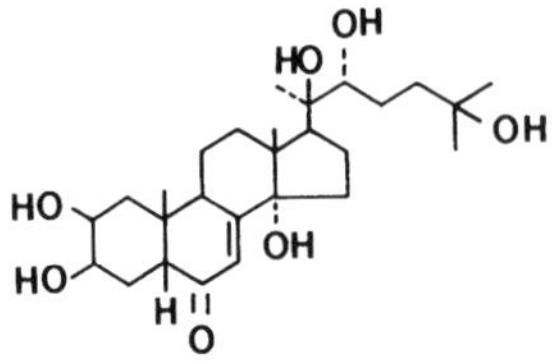

80

All of these features are present in and illustrated by 20-hydroxyecdysone (**80**), which has, in addition, a 25-hydroxyl group. The other ecdysteroids, regardless of their source, usually vary by additions of still other hydroxyl groups, by deletions of one or more of those in 20-hydroxyecdysone, or by additions of an alkyl group at C-24.

Ecdysteroids have also been isolated from plants. In fact twice as many have been isolated from plants as from insects (90), and the range of plants containing material with ecdysteroid activity varies from ferns – e.g., *Polypodium vulgare* and *Blechnum niponicum* – through the gymnosperms – e.g., the genera *Podocarpus* and *Taxus* – to both the monocots and the dicots among the angiosperms. Some correlations exist between structure and genera. The ecdysteroids also exist in a variety of plant parts. In *Polypodium vulgare* they represent 2% of the rhizomes. In *Podocarpus macrophyllus* they can be isolated from the leaves, and from *P. elatus* they are present in the bark (26). The bark ecdysteroids are makisterone A (also known as podecdysone D, one of the stereoisomers of 24-methyl-20-hydroxyecdysterone), the latter's 24-(presumably α)-ethyl analog, makisterone C (also known as lemmasterone and podecdysone A), a podecdysone B, which is the $\Delta^{8,14}$-diene corresponding to 20-hydroxyecdysone. The diene could be either a biosynthetic precursor of 20-hydroxyecdysone or, as suggested by Galbraith et al. (26), a metabolite of it. 20-Hydroxyecdysone is also present in *P. elatus*, as it is in a variety of other plants. *P. elatus* bark also contains 26 (or 27),20α-dihydroxyecdysone (podecdysone C) (27). The 5β-hydroxy derivative (dacrysterone) of makisterone A also occurs in the bark of *Dacrydium intermedium* (95).

8.2.6.2 Function of Plant Ecdysteroids

Exactly the same ecdysteroids have in some cases been isolated from both plants and insects, although biosynthesis from sterols demonstrably occurs in insects (in the prothoracic gland) as well as in plants. An example of the same structures in plants and animals is the presence of podecdysone C in bark of the *Podocarpus elatus* tree (27) as well as in the insect *Manduca sexta* (the tobacco hornworm) (104). Similarly, makisterone A is the major ecdysteroid in eggs of the large milkweed bug, *Oncopeltus fasciatus* (104), and it also occurs in at least two species of coniferous trees in the *Podocarpus* genus (26, 90). Since ecdysteroids can penetrate the insect cuticle when dissolved in other organic materials, it seems possible, even likely, that insect control constitutes the function of these materials in plants, but definitive proof has not yet been forthcoming.

8.2.7 Cardiac Glycosides

A number of glycosides with the sugar(s) attached to C-3 of a steroid have a strongly modified and shortened side chain that is present as a five- or six-membered unsaturated lactone ring. The former are called cardenolides and the latter bufadienolides. About two dozen sugars including D-glucose have been isolated from these glycosides, and many of them are unusual in that they are 6-deoxy or 2,6-bisdeoxy aldohexoses often with methylated hydroxyl groups as in the case of D-thevetose (the 3-methyl ether of 6-deoxy-D-glucose), its C-4 epimer D-digitalose, D-digitoxose (2,6-bisdeoxy-D-allose), and D-bovinose (2,6-bisdeoxy-galactose). L-sugars – e.g., L-rhamnose – also are found. These glycosides occur in various parts, including the bark, of a number of plants, especially in the families Apocynaceae, Liliaceae, Moraceae, Ranunculaceae, and Scrophulariaceae. In sufficient dose, which still can be quite small (<0.1 mg for a mouse), the compounds can have a severe effect on the heart muscle, stopping it and causing death. In lesser doses, they can have a beneficial stimulation of heart contractions. The cardiac properties of these substances have led them to be used both as medicines (as in the case of digitalis) and as poisons. For a detailed discussion of their structure and occurrence, see Fieser and Fieser (22). For a key to the more current literature see Yamauchi et al. (111) and Abe and Yamauchi (1).

81

Digitoxigenin (**81**) (genin = aglycone, the steroidal moiety of the glycoside) is a typical cardenolide. As in most of such steroids, digitoxigenin has a *cis* fusion of rings A and B and a β-oriented hydroxyl group at C-14 producing a *cis* fusion of the C/D-juncture also. The steroidal aglycones of the cardiac glycosides thus have a geometry in which the tetracyclic ring system is strongly bent backwards, in contrast to the three-dimensional structure of the membranous sterols, which have a flat character due to ring junctions that are *trans* or *anti* or that have double bonds at the juncture. Digitoxigenin occurs in several glycosidic forms such as thevetin and digitoxin. The latter is the tridigitoxoside, whereas the former is a glucosyl-glucosyl-thevetoside. Acetylated sugar moieties are sometimes present too, as in cerberin, which is the 21-*O*-acetyl-β-L-thevetoside of digitoxigenin. The other genins usually vary from digitoxigenin in the position and number of hydroxyl or other oxygen functions. Gitoxigenin, for instance, is a 16β-hydroxydigitoxigenin, and sarmentogenin is 11α-hydroxydigitoxigenin.

The bark of *Cerbera manghas* (Apoxynaceae) contains five cardiac glycosides (1) with two genins represented. One of the genins is digitoxigenin and the other is 7β,8β-epoxydigitoxigenin (tanghinin). In addition, the epimer of tanghinin at C-17 with a 17β-H (17β-tanghinin) is present and believed to be natural. All of the glycosides have L-thevetose linked with an equatorial oxygen at the anomeric carbon to C-3β of the steroid. The sugar moieties then differ in whether or not there is a second sugar (D-glucose of gentiobiose) linked β(1→4) to the L-thevetose.

Both the stem bark and the root bark of *Nerium odorum* (Apocynaceae) contain cardiac glycosides (111 and references therein). The root bark alone (111) contains no fewer than 15 glycosides, involving three genins, digitoxigenin, the latter's 5α-epimer (uzarigenin), in which quite unusually the A/B-junction is *trans*, and the 16β-acetoxy derivative (oleandrigenin) of digitoxigenin. The β-D-diginoside, the β-D-digitaloside, the β-D-glucosyl-β-D-diginoside and the β-D-glucosyl-(1→4)-β-D-digitaloside of each of the genins was found in addition to the β-gentiobiosyl-β-D-diginoside of uzarigenin and the β-gentiobiosyl-(1→4)-β-D-digitaloside of digitoxigenin and oleandrigenin. Interestingly, the stem bark of this tree growing in Japan contains two glycosides (the digitaloside and glucosyl-digitaloside of uzarigenin) that do not seem to be present in the stem bark of the same species growing in India (111).

References

1. Abe F, Yamauchi T 1977 Studies on *Cerbera*. I. Cardiac glycosides in the seeds, bark, and leaves of *Cerbera manghas* L. Chem Pharm Bull 25:2744–2748
2. Adler J H, Young M, Nes W R 1977 Determination of the absolute configuration at C-20 and C-24 of ergosterol in Ascomycetes and Basidiomycetes by NMR spectroscopy. Lipids 12:364–366
3. Anjaneyulu A S R, Suryanarayana M 1977 Chemical examination of the bark of *Guazuma tomentosa* Kinth. Planta Med 32:247–248
4. Becker E S, Kurth E F 1958 The chemical nature of extractives from the bark of the red fir. Tappi 41:380–400
5. Behari M, Andhiwal C K, Ballantine J A 1977 Phytosterols from the bark of *Saraca indica* Linn. Indian J Chem 15B:765–766
6. Bhatt K, Agrawal Y K 1973 Chemical investigation of the trunkbark from *Ficus racemosa*. J Indian Chem Soc 50:611
7. Campello J D P, Fonseca S F, Chang C-J, Wenkert E 1975 Terpenes of *Podocarpus lambertius*. Phytochemistry 14:243–248
8. Castle M, Blondin G A, Nes W R 1963 Evidence for the origin of the ethyl group of β-sitosterol. J Am Chem Soc 85:3306–3308
9. Chen C-L 1970 Constituents of *Quercus alba*. Phytochemistry 9:1149
10. Cheng Y S, Tsai M D 1972 Terpenes and sterols of Cunninghamia konishii. Phytochemistry 11:2108–2109
11. Cherry R F, Khong P W, Lewis K G 1977 Chemical constituents of *Harpullia pendula*. II. Further constituents of the bark and leaves. Aust J Chem 30:1397–1400
12. Clotofiski E, Herr W 1942 Beechnut bark (*Fagus silvatica*). III. Chem Ber 75B:237–243
13. Conner A H, Nagasampagi B A, Rowe J W 1980 Terpenoid and other extractives of western white pine bark. Phytochemistry 19:1121–1131
14. Dahl J S, Dahl C E, Bloch K 1980 Sterols in membranes: Growth characteristics and membrane properties of *Mycoplasma capricolum* cultured on cholesterol and lanosterol. Biochemistry 19:1467–1472
15. Dalton A J, Haguenau F (eds) 1968 The membranes. Academic Press New York, 223 pp
16. Dhingra V K, Mukerjee S K, Saroja T, Seshadri T R 1971 Chemical investigation of the bark and sapwood of *Dalbergia latifolia*. Phytochemistry 10:2551
17. Dixit B S, Srivastava S N 1977 Sapogenins and other constituents of the bark of *Minusops littoralis* Kurz. Indian J Pharm 39:85
18. Djerasssi C 1981 Recent studies in the marine sterol field. Pure Appl Chem 53:873–890
19. Dreyer D L 1984 Biogenetic relationships of degraded triterpenes in the Rutales. In: Nes W D, Fuller G, Tsai L-S (eds) Isopentenoids in plants. Biochemistry and function. Marcel Dekker New York, 596 pp
20. Erdtman H, Tsumo K 1969 *Taxus* heartwood constituents. Phytochemistry 8:931–932
21. Fieser L F, Fieser M 1949 Natural products related to phenanthrene. Reinhold New York, 704 pp
22. Fieser L F, Fieser M 1959 Steroids. Reinhold New York, 945 pp

23 Fischer F, Koch H, Borchers B, Höntsch R, Pruzina K-D 1981 Gewinnung und Verwertung von Phytosterolen aus Holz. Pharmazie 36:456–462
24 Fracheboud M, Rowe J W, Scott R W, Fanega S M, Buhl A J, Todo J K 1968 New sesquiterpenes from yellow-wood of slippery elm. For Prod J 18:37–40
25 Free A J, Read R W, Ritchie E, Taylor W C 1976 Some extractives of *Melicope octandra* (Rutaceae). Aust J Chem 29:695–697
26 Galbraith M N, Horn D H S, Middleton E J, Hackney R J 1969 Structure of podecdysone B, a new phytoecdysone. J Chem Soc Chem Commun 402–403
27 Galbraith M N, Horn D H S, Middleton E J, Kaplanis J N, Thompson M J 1973 Structure of podecdysone C, a steroid with moulting hormone activity from the bark of *Podocarpus elatus* R Br. Experientia 29:782
28 Garg V K, Nes W R 1984 Studies on the C-24 configurations of Δ^7-sterols in the seeds of *Cucurbita maxima*. Phytochemistry 23:2919–2923
29 Garg V K, Nes W R 1984 Codisterol and other Δ^5-sterols in the seeds of *Cucurbita maxima*. Phytochemistry 23:2925–2929
30 Geuns J M C 1973 Variations in sterol composition in etiolated mung bean seedlings. Phytochemistry 12:103–106
31 Geuns J M C 1975 Regulation of sterol biosynthesis in etiolated mung bean hypocotyl sections. Phytochemistry 14:975–978
32 Gibbons G F, Goad L J, Goodwin T W, Nes W R 1971 Concerning the role of lanosterol and cycloartenol in steroid biosynthesis. J Biol Chem 246:3967–3976
33 Goad L J, Goodwin T W 1972 The biosynthesis of plant sterols. In: Reinhold L, Liwschitz Y (eds) Progress in Phytochemistry, vol 3. Interscience New York, 113–198
34 Gonzalez A G, Freire R, Garcia-Estrada M G, Salazar J A, Suarez E 1972 Nuevas fuentes naturales de sapogeninas esteroidales – XV. Dracogenia, nueva sapogenina espirostanica de la *Dracaena*. Rev Latinoam Quim 3:8–18
35 Graig A R, Das K C, Farmer W J, Lin Y Y, Woo W R, Weinstein B 1971 Constituents of the leaves and root bark of *Ceanothus velutinus*. Phytochemistry 10:908
36 Hoke M, Hansel R 1972 A new investigation on sassafras root. Arch Pharm 305:33–39
37 Höll W, Goller I 1982 Free sterols and steryl esters in the trunkwood of *Picea abies* (L.) Karst. Z Pflanzenphysiol 106:409–418
38 Höll W, Pieczonka K 1978 Lipids in the sap- and heartwood of *Picea abies* (L) Karst. Z. Pflanzenphysiol 87:191–198
39 Hossfeld R L, Hunter W T 1958 The petroleum ether extract of aspen bark. Tappi 41:359–362
40 Huang C, Mason J T 1978 Geometric packing constraints in egg phosphatidylcholine vesicles. Proc Nat Acad Sci USA 75:308–310
41 Heupel R C, Nes W D 1984 Evidence for differences in sterol biosynthesis and derivatization in *Sorghum*. J Nat Prod 47:292–299
42 Jewers K, Ross M S F 1973 Fatty esters and sterols from the bark of *Amaroria soulamenoides*. Phytochemistry 12:956–957
43 Joshi K C, Bansal R K, Patni R 1978 Chemical constituents of *Gymnosporia montana* and *Euonymus pendulus*. Planta Med 34:211–214
44 Joshi K C, Bansal R K, Singh P 1974 Mass and NMR spectral studies of sitost-4-en-3-one from *Tabebuia rosea* DC. Indian J Chem 12:903–904
45 Joshi K C, Prakash L, Singh P 1973 Chemical investigations of the barks of *Tecomella undulata*, *Heterophragma adenophylum*, and *Millingtonia hortensis* (Bignoniaceae). J Indian Chem Soc 50:561–562
46 Joshi K C, Sharma T 1974 Triterpenoids and some other constituents from *Dichrostachys cinera*. Phytochemistry 13:2010–2011
47 Joshi K C, Singh L B 1974 Quinonoid and other constituents from the heartwood of *Tecomella undulata*. Phytochemistry 13:663–664
48 Joshi K C, Singh P, Pardasani R T, Singh G 1979 Quinones and other constituents from *Haplophragma adenophylum*. Planta Med 37:60–63
49 Karunen P, Ekman R 1981 Senescence-related changes in the composition of free and esterified sterols and alcohols in *Sphagnum fuscum*. Z Pflanzenphysiol 104:319–330
50 Knajniak E R, Ritchie E, Taylor W C 1973 The chemical constituents of Australian *Zanthoxylum* species. Part 6: A further examination of the constituents of the bark of *Zantoxylum consperspunctatum* Rutaceae. Aust J Chem 20:687–689

51 Komae H, Hayashi N 1971 Palmitone and phytosterols from *Neolitsea sericea*. Phytochemistry 10:1953–1954
52 Lala A K, Buttke T M, Bloch K 1979 On the role of the sterol hydroxyl group in membranes. J Biol Chem 254:10582–10585
53 Lange Y, Ramos B V 1983 Analysis of the distribution of cholesterol in the intact cell. J Biol Chem 258:15130–15134
54 Lavie D, Kaye I 1963 Isolation of β-sitostenone from *Quassia amara*. J Chem Soc 5001–5002
55 McKean M L, Nes W R 1976 Evidence for separate intermediates in the biosynthesis of 24α- and 24β-alkylsterols in tracheophytes. Phytochemistry 16:683–686
56 Mujamdar S G, Thakur S 1968 Chemical investigations of the stem bark of *Eugenia fracticosa* and *Eugenia wallichii*. Part 3. J Indian Chem Soc 45:785–790
57 Miyauchi Y, Yoshimoto T, Minami K 1976 Extractives from heartwood of *Piptadenia* sp. Mokuzai Gakkaishi 22:47–50
58 Nair M G, Chang F C 1973 6β-Hydroxy-4-stigmasten-3-one and 6β-hydroxy-4-campesten-3-one. Phytochemistry 12:903–906
59 Narayanan C R, Naik D G 1981 A new triterpene and steroid from Indian Karchi bark. Indian J Chemistry 20B:62–63
60 Nes W D, Heftmann E 1981 A comparison of steroids and triterpenoids as membrane components. J Nat Prod 44:377–400
61 Nes W D, Heupel R C, Le P H 1974 A comparison of sterol biosynthesis in fungi and tracheophytes and its phylogenetic and functional implications. In: Siegenthaler P A, Eichenberger W (eds) Structure, function, and metabolism of plant lipids. Elsevier Amsterdam, 207–216
62 Nes W D, Patterson G W, Southhall M A, Stanley J L 1979 Sterols and fatty acids of *Epifagus virginiana*, a nonphotosynthetic angiosperm. Lipids 14:274–276
63 Nes W D, Wong Y W, Benson M, Landrey J R, Nes W R 1984 Rotational isomerism about the 17(20)-bond of steroids and euphoids as shown by the crystal structures of euphol and tirucallol. Proc Nat Acad Sci USA 81:5896–5900
64 Nes W D, Yaniv Z, Heftmann E 1982 Translocation of cholesterol from leaves to ripening fruits of *Solanum khasianum*. Phytochemistry 21:581–583
65 Nes W R 1971 Regulation of the sequencing in sterol biosynthesis. Lipids 6:219–224
66 Nes W R 1974 The role of sterols in membranes. Lipids 9:596–612
67 Nes W R 1977 The biochemistry of plant steroids. In: Paoletti R, Kritchevski D (eds) Advances in lipid research, vol 15. Academic Press New York, 233–324
68 Nes W R 1985 A comparison of methods for the identification of sterols. In: Law J H, Rilling H C (eds) Steroids and isoprenoids. A volume of methods in enzymology, vol III. Academic Press New York, 3–37
69 Nes W R, Krevitz K, Behzadan K 1976 Configuration at C-24 of 24-methyl- and 24-ethylcholesterol in tracheophytes. Lipids 11:118–126
70 Nes W R, Krevitz K, Joseph J M, Nes W D, Harris B, Gibbons G F, Patterson G W 1977 The phylogenetic distribution of sterols in tracheophytes. Lipids 12:511–527
71 Nes W R, McKean M L 1977 Biochemistry of steroids and other isopentenoids. University Park Press Baltimore, 690 pp
72 Nes W R, Nes W D 1980 Lipids in evolution. Plenum Press New York, 244 pp
73 Nigam S K, Misra G, Mitra C R 1971 Constituents of *Samanea saman* bark. Phytochemistry 10:1954–1955
74 Nomura M, Tokoroyama T, Kubota T 1981 Bisarylheptanoids and other constituents from wood of *Alnus japonica*. Phytochemistry 20:1097–1104
75 Norin T, Winell B 1972 Extractives from the bark of Scots Pine, *Pinus silvestris* L. Acta Chem Scand 26:2297–2304
76 Norin T, Winell B 1974 Neutral constituents of *Larix decidua* bark. Phytochemistry 13:1290–1292
77 Ourisson G, Crabbe P, Rodig O R 1964 Tetracyclic triterpenes. Holden-Day San Francisco, 237 pp
78 Passonen P 1964 Extraction of birchwood. Suom Khemistil 37B:142
79 Passonen P 1965 Extraction of birchwood. Suom Khemistil 38:169–170
80 Pearl I A, Buchanen M A 1976 A study of the inner and outer barks of loblolly pine. Tappi 59:136–139
81 Pearl I A, Harrocks J A 1961 Neutral materials from the benzene extractives of *Populus tremuloides*. J Org Chem 26:1578–1583

82 Picker K, Ritchie E, Taylor W C 1976 The chemical constituents of Australian *Flindersia* species. XVI: An examination of the bark and leaves of *F. laevicarpa*. Aust J Chem 29:2023–2026
83 Pinto W J, Lozano R, Sekula B C, Nes W R 1983 Stereochemically distinct roles for sterol in *Saccharomyces cerevisiae*. Biochem Biophys Res Commun 112:47–54
84 Pinto W J, Nes W R 1983 Stereochemical specificity for sterols in *Saccharomyces cerevisiae*. J Biol Chem 258:4472–4476
85 Pinto W J, Nes W R 1985 24β-Ethylsterols, *n*-alkanes and *n*-alkanols of *Clerodendrum splendens*. Phytochemistry 24:1095–1098
86 Pišová M, Souček M 1973 Triterpenes and sterols from *Fagus sylvatica*. Phytochemistry 12:2068
87 Rangopal M, Bloch K 1983 Sterol synergism in yeast. Proc Nat Acad Sci USA 80:712–715
88 Rao M M, Krishna E M, Gupta P S, Singh P P 1979 Constituents of *Melia birmanica*. Indian J Chem 17B:177–178
89 Ray T K, Misra D R, Khastgir H N 1975 Phytosterols in Euphorbiaceae and Rutaceae. Phytochemistry 14:1876–1877
90 Rees H H 1971 Ecdysones. In: Goodwin T W (ed) Aspects of terpenoid chemistry and biochemistry. Academic Press London, 182–222
91 Rodriguez R J, Taylor F R, Parks L W 1982 A requirement for ergosterol to permit growth of yeast sterol auxotrophs on cholestanol. Biochem Biophys Res Commun 106:435–441
92 Rowe J W 1975 Sterols of pine bark. Phytochemistry 4:1–10
93 Rowe J W, Seikel M K, Roy D N, Jorgensen E 1972 Chemotaxonomy of *Ulmus*. Phytochemistry 11:2513–2517
94 Roy S, Sengupta P 1968 Chemical investigation of the bark of *Delonix regia*. J Indian Chem Soc 45:464–465
95 Russell B, Fraser J G 1973 Insect moulting hormones: Dacrysterone, a new phytoecdysone from *Dacrydium intermedium*. Aust J Chem 26:1805–1807
96 Sainsbury M, Webb B 1972 Hydrocarbons and terpenoids from the bark of *Lophopetalum rigidum*. Phytochemistry 11:3541
97 Sekula B C, Nes W R 1980 The identification of cholesterol and other steroids in *Euphorbia pulcherrima*. Phytochemistry 19:1509–1512
98 Seshadri T R, Vedantham N C 1971 Chemical examination of the barks and heartwoods of *Betula* species of American origin. Phytochemistry 10:897–898
99 Shabana M M, Mirhom Y W, Hilal S H 1978 Screening of steroids in certain Egyptian plants. III. Steroidal constituents of *Solanum macrocarpum*. Egypt J Pharm Sci 19:1–4
100 Singh H, Chawla A S, Rowe J W, Todo J K 1970 Waxes and sterols of *Erythrina suberosa* bark. Phytochemistry 9:1673–1675
101 Sofowora E A, Hardman R 1973 Steroids, phthalyl esters, and hydrocarbons from *Balanites wilsoniana* stem bark. Phytochemistry 12:403–406
102 Subramanian S K, Nair A G R 1970 Sterols and flavonols of *Ficus bengalensis*. Phytochemistry 9:2583–2584
103 Suarez J, Bonilla J, De Diaz M P, Achenbach H 1983 Dehydrodieugenols from *Nectandra polita*. Phytochemistry 22:609–610
104 Svoboda J A 1984 Insect steroids: Metabolism and function. In: Nes W D, Fuller G, Tsai L-S (eds) Isopentenoids in plants: Biochemistry and function. Marcel Dekker New York, 367–400
105 Tovivich P, Woo W S, Chamsuksai P 1981 Flavonoid and steroid constituents in the stembark of *Albizzia julibrissin* Durazz. Proc First Int Conf Chem Biotechnol Biologically Active Natural Products. Am Chem Soc Washington DC, 322–346
106 Tsuda K, Hayazu R, Kishida Y 1959 Absolute configuration of 24-ethyl of stigmasterol. Chem Ind (London) 1411–1412
107 Unkonen K 1982 Nonvolatile dichloromethane extractives of *Gmelina arborea*. Tappi 65:71
108 Venkata R D, Venkata R E 1969 Constituents of the bark of *Marsdenia volubilis*. Phytochemistry 8:1609
109 Weston R J 1963 Neutral extractives from *Pinus radiata* bark. Aust J Chem 26:2729–2734
110 Wrzeciono U 1965 Triterpenes and plant sterols from the wood of *Quercus robur*. Rocz Chem 39:943–946
111 Yamauchi T, Takahashi M, Abe F 1976 Cardiac glycosides of the root bark of *Nerium odorum*. Phytochemistry 15:1275–1278

Chapter 9

The Influence of Extractives on Wood Properties and Utilization

An important aspect of wood extractives is the effect they have on wood properties and wood utilization. Lignocellulose is essentially colorless, so the attractive colors of woods such as cherry, ebony, and mahogany are due to the extractives. The density, strength, elasticity, permeability, and hygroscopicity of wood are greatly affected by the extent that extractives fill the lumens and, occasionally, the cell wall. Flammability is very much a function of resin content, and the odor of wood is due entirely to volatile extractives or volatile metabolites of microbial action. Some wood extractives are valuable drugs such as quinine, while others are allergenic or actually highly toxic; in some cultures, toxic extractives have proven useful in hunting and fishing. The way in which wood can be pulped is a function of how the extractives react with the pulping liquors. Some extractives inhibit the setting of concrete, which affects the use of wood in concrete forms and concrete-excelsior combinations. Some extractives will accelerate the corrosion of metal in contact with wood. The paintability of wood can be adversely affected by extractives, resulting in poor adhesion or staining. Of considerable significance is the role of extractives and exudates in protecting wood, both in the living tree and in wood products, from insect and microbiological attack and decay.

In the sections that follow in this chapter, these various effects are discussed in greater detail.

9.1 Contribution of Extractives to Wood Characteristics

H. IMAMURA

9.1.1 Introduction

The major components of wood are cellulose, hemicelluloses, and lignin, and many of the properties of wood are a function of this lignocellulosic network. Although extractives are a minor component, often constituting less than 10% of the wood, they contribute disproportionately to the characteristics of wood. It is extractives that give wood its color, its odor, and, to some extent, its physical properties. Extractives can have a significant influence on how wood is used.

9.1.2 Color in Wood

9.1.2.1 Chemical Structure and Color

Of the solar radiation hitting the earth, that with wavelengths between approximately 380 and 750 nm penetrates the atmosphere most readily. The human eye is not sensitive to all wavelengths, but is limited to a wavelength sensitivity from around 400 to 700 nm. This range of visible light is of fundamental importance in maintaining life. Substances that have the special property of absorbing all or a part of visible light are the coloring matter or pigments. The atomic groups necessary for absorbing some of the wavelengths of visible light are chromophores and auxophores. In organic compounds, the chromophores include functional groups that are easily polarized such as $-N=N-$, $>C=C<$, $>C=O$, $-N{\overset{\nearrow O}{\underset{\searrow O}{}}}$, $-N=O$, and $>C=S$, and the auxophores include functional radicals that contain atoms possessing a lone pair of electrons, such as $-NH_2$, $-NHR$, $-NR_2$, $-OH$, and $-OR$. Furthermore, for color to be present, the chromophores must take a conjugated arrangement in the molecule. Most pigments isolated from deep-colored wood contain conjugated chromophores in their molecules (Fig. 9.1.1).

9.1.2.2 Color of Wood

Color is one of the most distinctive properties of wood, and is of considerable importance in woods that are used for decorative purposes (Sect. 9.4.4). Color does not depend upon the main structural components (i.e., cellulose, hemicellulose, and lignin, all of which are colorless or nearly so), but rather upon the minor components. Many wood species are classified for taxonomic purposes on the basis of the color exhibited by their heartwood; these colors range from almost white in a few woods, through varying shades of yellow, red, and brown, to black in ebony (*Diospyros* spp.). The wide range of colors encountered is due to natural coloring matter present in the wood (16, 19). Commonly used woods, however, are generally light or dull-tinted, and terms such as "brownish" and "pale strawbuff" have proved adequate to distinguish many species. The color range of 100 species of wood widely used as construction materials in Japan is shown by the LAB system in Fig. 9.1.2 (23). The values on the lightness index are distributed in a wide range from 20 to 80, but the bulk of the values on the chromaticity indices (a and b) are between 10 and 18. These indicate, that the majority of woods are fairly reddish yellow in color.

The colors of living organs such as flowers and fruits are generally brilliant, offering an advantage in pollination and seed dispersal, and their pigment composition is comparatively simple. Many bright pigments – e.g., anthocyanins, carotenoids, and betanins – have been isolated and identified from both types of tissue. However, the coloring matter and other constituents of dead tissues such as heartwood and outer bark are difficult to isolate and characterize because, as these tissues die, the enzymatic functions in the cell become disordered, and the

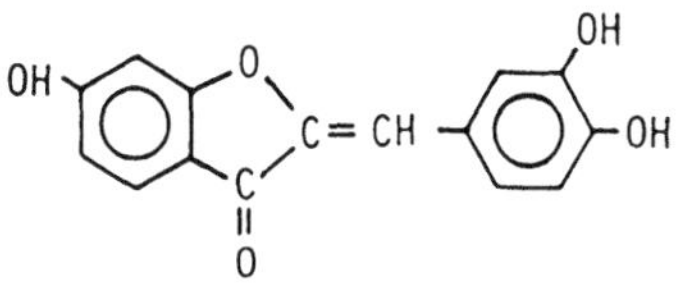

1 sulfuretin (orange yellow, Rhus spp.)(17)

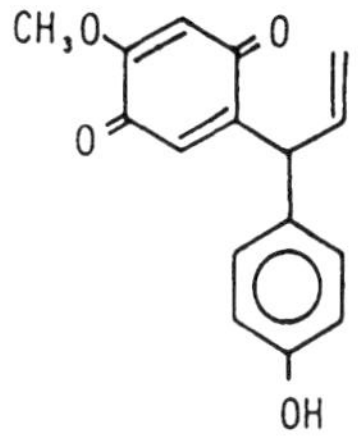

2 4-hydroxydalbergione (orange, Dalbergia and Machaerium spp.)(7)

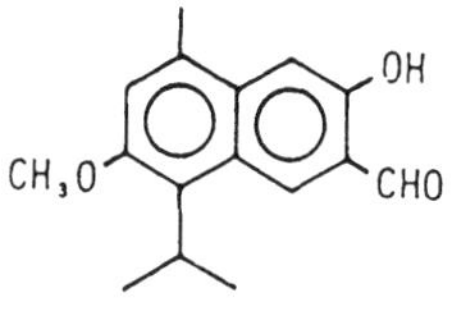

3 2-formyl-3-hydroxy-8-isopropyl-7-methoxy-5-methylnaphthalene (orange, Ulmus spp.)(10)

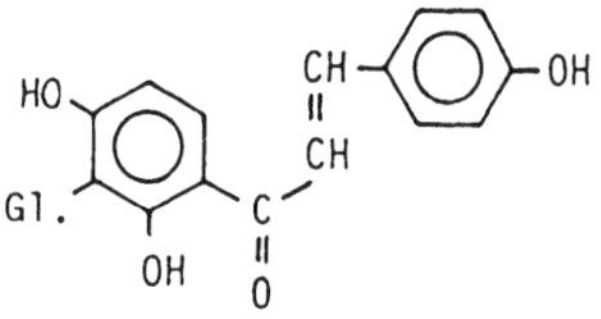

4 isoliquiritigenin-3-C-glucoside (deep yellow, Cladrastis platycarpa)(25)

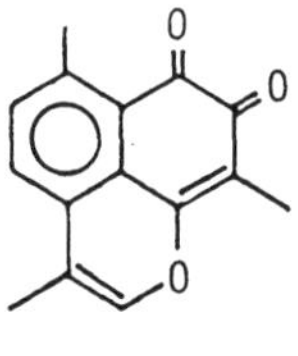

5 mansonone F (purple, Mansonia altissima)(31)

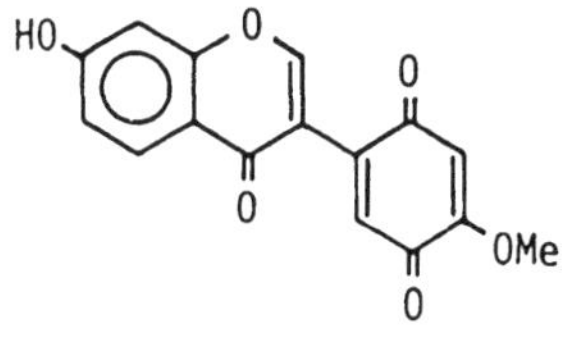

6 bowdichinone (gold brown, Bowdichia nitida)(6)

Fig. 9.1.1. Several types of coloring matter isolated from woods

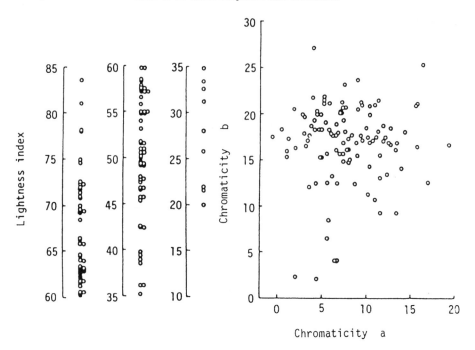

Fig. 9.1.2. Color distribution of various woods

cell contents – pigments, in particular – undergo oxidation and polymerization.

Although colored extracts may be separated from most woods by extraction with suitably neutral solvents, it is generally impossible to remove all the coloring matter in this way. This fact means that a considerable part of the coloring matter in wood may be present in macromolecular (insoluble) form, or so firmly bound to the skeleton components – lignin or polysaccharides – that it is not extractable by means of neutral solvents without destroying the wood structure. Their chemistry is frequently very complex, and few of them have been investigated or elucidated in detail.

A few woods such as rosewood (*Dalbergia, Machaerium*), ebony (*Diospyros*), cocobolo (*Dalbergia*), padauk (*Pterocarpus*) have splendidly colored heartwoods peculiar to the species and have been used for furniture and fancy goods since ancient times. Also, the heartwood extracts of several species such as logwood (*Haematoxylon compechianus*), sappan wood (*Caesalpinia sappan*), and old fustic (*Chlorophora tinctoria*) have been employed as dyestuff for fibrous materials, and logwood is presently important in producing a black dye. The colored extractives isolated from these heartwoods have been investigated since about a century ago and have been shown to possess special constitutions related to phenolic and quinonoid structures (Table 9.1.1).

A better knowledge of the character of pigments and coloring matter would be useful in suggesting ways to avoid certain troublesome discolorations that occur on some woods.

Table 9.1.1

Species	Pigment and structure	Reference
Brazilwood (*Caesalpinia braziliensis*) sappanwood (*C. sappan*)	**7a** brazilin (R=H), **8a** brazilein (R=H)	3
cocus wood (*Brya ebenus*)	**9** bryaquinone	8
ebony (*Diospyros* spp.)	**10** mamegaki-quinone, **11** macassar II	5, 3

Table 9.1.1 (continued)

Species	Pigment and structure	Reference
logwood (*Haematoxylon campechianum*)	**7b** haematoxylin (R=OH), **8b** haematein (R=OH)	3
mansonia (*Mansonia altissima*)	**12** mansonone C, **5** mansonone F	31
old fustic (*Chlorophora tinctoria*)	**13** maclurin, **14** 1,3,6,7-tetrahydroxyxanthone	12, 21
padauk (*Pterocarpus* spp.)	**15a** santalin A (R=H), **15b** santalin B (R=Me)	13

Table 9.1.1 (continued)

Species	Pigment and structure	Reference
rosewood, cocobolo (*Dalbergia* spp.)	**16a** (S)-4-methoxydalbergione (R=H) **16b** (S)-4'-hydroxy-4'-methoxy-dalbergione (R=OH), **16c** (S)-4,4'-dimethoxydalbergione (R=OMe)	7
slippery elm (*Ulmus rubra*)	**17a** 7-hydroxycadalenal (R=H), **17b** 7-hydroxy-3-methoxycadalenal (R=OMe)	10
teak (*Tectona grandis*)	**18** tectoquinone **19** lapachol	27

Table 9.1.1 (continued)

Species	Pigment and structure	Reference
yellow wood (*Cladrastis* spp.)	**20** C-glucosylisoliquiritigenin	25
young fustic (*Rhus cotinus*)	**21** fisetin **1** sulfuretin	20

9.1.2.3 Pigments Occurring in Wood

The majority of the plant pigments hitherto isolated from woods in pure form are the flavonoids, such as flavone, flavanone, isoflavone, isoflavanone, pterocarpane, and chalcone derivatives, along with the phenols stilbene and xanthone, which are almost white or yellow in color. Other flavonoids such as aurone and neoflavone derivatives (orange pigments) have a limited distribution and sometimes occur as wood extractives in species of the Anacardiaceae and Leguminosae (4, 7). The presence of the typical anthocyanin derivatives of flowers and fruits is extremely rare in wood, but their leuco-compounds such as flavan-3-ol and flavan-3,4-diol have been found to occur in the wood of a considerable number of tree species (14). Besides anthocyanidin, the few other deep-colored (red, purple, or blue) pigments from woods usually possess a quinone structure.

9.1.3 Odor in Wood

9.1.3.1 Volatile Components

An odor is detected by the olfactory organs when the molecules of volatile substances reach the olfactory cells and stimulate them. The sensitivity of olfaction varies widely among individuals, and is also affected by environmental conditions. Whereas color intensity can be quantified, it is difficult to assign numerical values to odors.

Wood odors can be classified into two groups according to the formation process of the odor source (33). In the first group, the odor is produced during normal tree metabolism, and in the second group, the odor is formed as a by-product of decomposition of some wood components by parasitic microorganisms. In the first group, most conifer woods and some deciduous woods have a pleasant odor, but a few species emit foul odors. In the second group, several species of tropical woods occasionally form foul-smelling components as artifacts, and consequently those components are sometimes troublesome to the wood industries. These odor substances can be isolated by steam distillation or extraction by solvents, and some of them can be used as raw materials for incense, medicine, fungicides, insecticides, etc.

The odor compounds in wood are usually in a liquid state at room temperature. They can be classified by chemical structure into mono- or sesquiterpenes (e.g., α-pinene, camphor, cedrene), aromatic compounds (e.g., methyl salicylate, safrol), nitrogenous compounds (e.g., pyridine, 3-methylpyrrolidine), and sulfur compounds (e.g., allylsulfide) (35). The latter two groups are rarely formed except as bad-smelling components. The content of these compounds in wood is quite small compared with the major components such as cellulose and lignin, and they are irregularly distributed among species.

9.1.3.2 Fragrant Components

Fresh wood after cutting has a characteristic odor that diminishes continuously over time. However, some species retain a fragrance for a long time, which gives them an additional value. The odor constituents of wood are formed as secondary metabolites by species-specific biosynthetic pathways, like the coloring matter described in the previous section. Therefore the odor of wood can sometimes be used for the identification of wood species.

The fragrant components of the major conifer woods are summarized in Table 9.1.2 (11, 18). Since larger amounts of these essential oils are formed when the living tree is wounded, it is presumed that they protect the tree and preserve it from decay, repel insects, act as a waterproofing, etc. Some of these components emitted from the leaves into the atmosphere act as phytocides, the so-called "green showers" that are coming into popular concern. The odors of the hardwoods are more varied than those of conifer woods. Some species are used for

Table 9.1.2. Main odor constituents of some conifer woods

Genus and species	Odor compounds
Abies spp.	bornyl acetate, camphene, α- & β-phellandrene, α- & β-pinene
Agathis alba	camphene, limonene, α- & β-pinene
Araucaria spp.	caryophyllene, elemol, α-, β- & γ-eudesmol, guaiol, humulene, limonene, myrcene, α- & β-pinene, terpinolene
Cedrus deodara	allohimachalol, centdarol, himadarol, isocentdarol
Chamaecyparis spp.	benihiol, bornyl acetate, camphene, carvacrol, o- & p-cresol, limonene, myrcene, α- & β-pinene, α-terpineol
Cryptomeria japonica	δ-cadinene, δ-cadinol, calacorene, calamenene, copaene, β-eudesmol
Cunninghamia spp.	α-cadinol, caryophyllene, α- & β-cedrene, cedrol, β-selinene, α-terpineol
Cupressus spp.	borneol, bornyl acetate, carvacrol, carvacrol methylether, cedrol, humulene, thujopsene
Juniperus spp.	γ-calacorene, cedrene, cedrenol, cedrol, β-elemene, farnesene, humulene, β-selinene, thujopsene
Larix spp.	bornyl acetate, dipentene, limonene, α- & β-pinene
Libocedrus decurrens	carvacrol, p-methoxycarvacrol, p-methoxythymol
Metasequoia glyptostroboides	α-cadinol, calacorene, calamenene, α- & β-pinene, terpineol, terpinyl acetate
Picea spp.	bornyl acetate, camphene, limonene, α- & β-pinene
Pinus spp.	δ-cadinol, calamenene, camphene, limonene, longifolene, myrcene, β-phellandrene, α- & β-pinene, γ-terpinene, terpinolene
Pseudotsuga menziesii	bornyl acetate, bornyl alcohol, camphene, 3- & 4-carene, p-cymene, 3,8-menthediene, α- & β-phellandrene, α- & β-pinene, α-terpinene, terpinolene
Sciadopitsy verticillata	cedrene, cedrol, α-pinene
Sequoia sempervirens	limonene, myrcene, α- & β-pinene
Thuja spp.	α-, β- & γ-eudesmol, methyl thujate, nezukone, occidentalol, occidol
Thujopsis dolabrata	cedrol, p-cresol, cuparene, elemenal, thujopsene, widdrol
Torreya nucifera	δ-cadinol, dendrolasin, o-methoxycinnamic aldehyde, nuciferal, nuciferol

Table 9.1.3. Main odor constituents of some deciduous woods

Genus and species	Odor compound
Anisoptera spp.	caryophyllene, copaene, humulene
Aquilaria spp.	α- & β-agarofuran, agarospirol, dihydroagarofuran
Canarium spp.	canarone, β-elemene, elemol, junenol, limonene, α-phellandrene, pinene
Cedrela spp.	aromadendrene, cadalene, δ-cadinol, α-calacorene, calamenene, copaene, farnesane
Celtis reticulosa	skatole[a]
Cinnamomum spp.	camphene, camphor, cineol, cinnamic aldehyde, eugenol, linalool, α-phellandrene, safrol
Cotylelobium spp.	caryophyllene, copaene, β-elemene, humulene
Daniellia spp.	caryophyllene, humulene
Dipterocarpus spp.	calarene, caryophyllene, copaene, α- & γ-gurjunene, humulene
Doona spp.	caryophyllene, copaene, β-elemene, humulene
Dryobalanops spp.	borneol, bornyl acetate, camphene, p-cymene, limonene, α- & β-pinene, terpineol
Hernandia peltata	perillaldehyde
Lindera umbellata	geraniol, limonene, linalool, α-pinene
Liquidambar orientalis	cinnamyl cinnamate, vanillin
Magnolia kobus	anethole, cineol, citral, eugenol, methylchavicol
Ocotea pretiosa	eugenol methylether, 1-nitro-2-phenylethane[a], skatole[a]
Octomeles sumatrana	3-methylpyrrolidine[a], pyridine[a]
Populus spp.	benzylalcohol, p-ethylphenol, methyl benzoate, phenol, β-phenylethanol
Santalum album	α- & β-santalene, α- & β-santalol, santenol, santenone
Shorea spp.	β- & γ-cadinene, caryophyllene, 1,4- & 1,8-cineol, citronellal, α-gurjunene, isopulegol, myrcene, β-terpineol
Styrax benzoin	benzoic acid, vanillin

[a] Foul-smelling compound

the preparation of soap and perfume. Some tropical woods, such as sandalwood (*Santalum* spp.) and styrax (*Liquidambar* spp.), have been used as incense woods since ancient times. The fragrant components of some deciduous woods are shown in Table 9.1.3 (11, 18).

9.1.3.3 Foul-Smelling Components

Several species with foul odors cause problems in wood industries, such as lumber sawing and the manufacture of plywood and laminated wood. A few species contain these components in fresh wood – for example, pyridine and 3-methylpyrrolidine in elima wood (*Octomeles sumatrana*) and sulfides in Bawang hutan (*Scordocarpus borneensis*) (35).

Odor problems are almost always caused by woods colonized by microorganisms. The bad odor is usually formed during wood storage in the log pond by fermentation of available carbohydrates. The released odors not only cause worker discomfort and reduce worker efficiency in the sawmill or the drykiln, but also cause a public nuisance in air pollution drifting beyond the mill. These foul-smelling substances, which vary among wood species but generally have an odor of fer-

Table 9.1.4. Lower fatty acids in foul-smelling woods

Common name	Genus and species	Fatty acid content[a]						
		acetic	propionic	isobutyric	n-butyric	isovaleric	n-valeric	caproic
Amberoi	*Pterocymbium beccarii*	+++	++	+	++	+		
Antiaris	*Antiaris africana*	+	++		+++		++	+
Celtis	*Celtis* spp.	++	+	+		+		
Elima	*Octomeles sumatrana*	+++	+		++	+	+	++
Lauan	*Shorea* spp.	+++	++	++	++	+	++	+
Meranti[b]	*Shorea* spp.	+++	++	+	++	++	++	
Poplar[c]	*Populus* spp.	+	++	+	+++	++	+	+
Ramin	*Gonystylus* spp.						+++	
Silkwood	*Planchonella* spp.	+++	+	+	++	+		+++

[a] +++ = relatively large amount; ++ = moderate amount; + = small amount
[b] Injured by insects
[c] Living tree

menting compost or feces, have been determined by gas chromatography to be mainly C_4 and C_5 fatty acids (butyric and valeric acids) (33). Examples of temperate species producing such odors are the oaks (*Quercus* spp.) and the false heartwood of beeches (*Fagus* spp.). Among the tropical species are melanti (*Shorea* spp.), jelutong (*Dyera* spp.), and ramin (*Gonystylus bancanus*) (33).

The foul odor of ramin results from metabolic products of microorganisms growing on carbohydrates and proteins in the wood (1). The odorous components of ramin, extracted with ether, are *n*-butyric acid (C_4), isovaleric acid (C_5), caproic acid (C_6), and caprilic acid (C_8); ramin also contains the nitrogenous compound skatole (2). Other species also known to generate foul-smelling fatty acids via microorganisms include *Antialis* sp., *Celtis* sp., *Octomeles sumatrana*, *Planchonella* sp., *Pterocymbium beccarii*, and *Spondias* sp. All contain butyric and valeric acid, to which the human nose is sensitive (butyric acid – 0.00056 µl/l; valeric acid – 0.00062 µl/l) (32). *Ceiba samauma* (lupna) and *Virola* sp. (banak) release caproic acid as well as butyric acid (37). Because ramin and other woods exhibiting a foul odor contain relatively small amounts of phenolic compounds – especially tannin – in comparison with normal woods, they are good media for microorganisms.

Table 9.1.4 lists wood species likely to generate bad smells from the lower fatty acids (32). Since even a small amount of formic acid or acetic acid in the group of acids described in Table 9.1.4 has a tendency to corrode metal, it is desirable to handle these species carefully.

9.1.3.4 Removal of Foul Odors

When wood with a foul odor is dried, it seems to be almost odorless. However, in many cases, the wood regenerates its odor under conditions of adequate moisture. The complete removal of the source of foul odor is very difficult because the source is hydrophilic fatty acids formed inside the wood structure and therefore tightly held by water and carbohydrates in the wood. Treatments such as spraying with dilute ammonia solution, applying calcium carbonate, and heating ($\geq 100\,°C$, several hours) can decrease the malodorous substances merely on the outer part of the wood. Under adequate moisture conditions, however, foul-smelling acids move to the surface from inside the wood by diffusion have a tendency to re-emit odor. Because the threshold value of butyric and valeric acids is extremely low (0.0005 to 0.0006 ppm), it is preferred that the residual quantity of these compounds should be as low as possible (33). Therefore, wood species likely to emit a foul odor are used for wood products that are generally carefully controlled throughout the process from timber felling to product manufacturing.

9.1.3.5 Insect Attractants

Although wood is one of the most durable of natural organic materials, it is liable to be attacked under appropriate circumstances by certain biological agents, of which the most important are insects, bacteria, and fungi.

Insects can be roughly classified from the standpoint of food habits. All insects, especially phytophagous ones, have a clear selection of hosts in their feeding habits. There are many insects (not only termites) that attack wood, causing serious deterioration. Among these insects, wood-boring beetles are oligophagous and show pronounced host selection. Wood extractives, especially volatile constituents, play a more important role than do the morphological features of wood in attracting wood-boring beetles to their host (28).

The life cycle and feeding habits of *Monochamus alternatus*, a kind of longhorn beetle and carrier of the timber nematode, *Bursaphelenchus lignicelus*, which causes serious withering of Japanese pine trees, have recently been investigated. It has been determined that fragrant compounds such as minor amounts of ethanol, acetone, and monoterpenes (α-pinene, β-pinene, camphene, β-myrcene, limonene, terpinolene, etc.) are strong attractants (15). In the temperate zone, a pinhole borer (*Cryphalus fulvus*), which causes serious damage to both standing and cut logs of Pinaceae, is also strongly attracted by the fragrant components in wood such as α-pinene; some kinds of fragrant components are recognized as active pheromones (29, 30).

Interestingly, while the volatile monoterpene and sesquiterpene derivatives have an attractant function toward beetles, they show a repellent action toward the termites that cause much destruction of wood in tropical and subtropical zones (26).

9.1.4 Physical Properties

The effects of extractives on physical properties of wood have not been clarified because this question involves chemistry (especially organic chemistry), physics, and structural dynamics. However, it is well known empirically that wood species containing large amounts of extractives have better durability, dimensional stability, and plasticization; for these reasons, extractives-rich woods have been used for construction and fancy goods since ancient times.

9.1.4.1 Wood Density and Strength

Since wood is strong, durable, and noncombustible compared to other materials of biological origin, and is also renewable, a large amount of it is used for construction materials. In that case, the density of wood, which is closely related to many mechanical properties, is important as an indicator of wood characteristics.

Wood extractives are located mainly in the cell lumen. This means that they fill vacant space in the wood, decreasing the porosity and increasing the density or specific gravity. Even with in the same species, the extractives contents and densities vary to some extent with the individual tree, as prescribed by the tree age, region, living environment, inherited quality, etc. Average values (Table 9.1.5) (9, 24) indicate that the relationship between the extractives content and density is generally positive.

Table 9.1.5. Extractives contents and densities of various woods

Genus and species	Extractives content (%)		Density
	EtOH-C$_6$H$_6$	hot H$_2$O	
Gymnospermae			
Abies firma	1.5	1.0 – 3.0	0.41
A. sachalinensis	2.6 – 4.5	1.3 – 4.2	0.42
Chamaecyparis nootkatensis	–	3.6	0.51
C. obtusa	1.2 – 4.1	2.1 – 4.8	0.41
Cryptomeria japonica	1.3 – 5.0	1.3 – 3.0	0.38
Larix gmelinii	3.0	11.4	0.58
L. leptolepis	1.8 – 5.5	3.9 – 20.1	0.53
Picea abies	1.4	3.3	0.45 – 0.50
P. glauca	2.8	2.9	0.41 – 0.47
P. jezoensis	1.8 – 4.0	1.8 – 5.5	0.43
Pinus densiflora	1.9 – 3.9	1.6 – 3.6	0.53
P. koraiensis	6.0	5.4	0.45
P. sylvestris	2.4	4.6	0.47 – 0.77
P. thunbergii	1.8 – 2.8	2.3 – 4.3	0.57
Pseudotsuga menziesii	2.7	6.5	0.55
Tsuga heterophylla	3.5	3.3	0.46
T. sieboldii	2.1 – 2.8	1.7 – 3.1	0.51
Angiospermae			
Acer spp.	1.4 – 3.0	2.2 – 5.2	0.61 – 0.67
Betula spp.	0.5 – 7.3	1.5 – 7.5	0.60 – 0.94
Cassia sp.	21.7	27.6	0.83 – 1.17
Dalbergia sp.	33.5	23.2	0.86 – 1.20
Diospyros sp.	34.7	18.2	0.95 – 1.21
Dipterocarpus spp.	1.9 – 4.0	2.3 – 5.7	0.75
Dryobalanops sp.	3.4	8.5	0.70
Dyera sp.	4.5	8.9	0.75
Fagus crenata	0.6 – 3.8	1.5 – 3.6	0.63
Milletia spp.	11.4 – 15.8	–	0.78 – 1.06
Ochroma lagopus	2.9	1.4	0.17
Palaquium sp.	2.1	2.7	0.64
Parashorea malaanonan	3.1	2.5	0.58
Pentacme contorta	3.7 – 9.2	2.6 – 7.5	0.58
Populus maximowiczii	1.8 – 3.6	2.9 – 4.5	0.38
Quercus acuta	2.4 – 6.3	2.6 – 5.4	0.92
Q. crispula	0.6 – 1.0	3.3 – 8.0	0.67
Shorea negrosensis	1.3 – 2.7	1.0 – 3.8	0.53
Swietenia sp.	11.3	11.7	0.55
Tectona grandis	6.5	11.1	0.69
Tilia japonica	6.0	2.9 – 4.1	0.48

Extractives also slightly enhance several strength properties of wood. Crushing and tensile strengths are to some extent proportional to the extractives content (Table 9.1.6), in deciduous woods in particular (9, 22). Some correlation between the other strength or physical properties (e.g., bending and shearing strengths, abrasion resistance, plasticization, etc.) and the minor constituents of wood is anticipated, but this field has yet to be studied adequately.

Table 9.1.6. Extractives contents and strengths of various woods

Genus and species	Extractives content (%) hot H$_2$O	Extractives content (%) EtOH-C$_6$H$_6$	Strength (kg/cm^2) Crushing	Strength (kg/cm^2) Tensile[a]
Gymnospermae				
Abies procera	2.3	2.7	390	15[b]
A. sachalinensis	2.6	2.6	330	1100
Chamaecyparis lawsoniana	3.5	3.1	455	28[b]
C. nootkatensis	3.1	2.6[c]	444	25[b]
C. obtusa	4.2	5.1	440	1200
C. pisifera	7.4	9.4	330	800
Cryptomeria japonica	3.1	2.6	350	900
Larix leptolepis	9.5	3.2	450	850
Picea jezoensis	3.6	1.3	350	1200
Pinus densiflora	3.9	4.1	450	1400
P. ponderosa	5.1	8.5[c]	370	28[b]
P. sylvestris	0.6	1.6[c]	290	950
P. thunbergii	3.0	3.3	450	1400
Pseudotsuga menziesii	5.6	4.4	522	24[b]
Sciadopitys verticillata	6.6	11.0	350	–
Sequoia sempervirens	9.9	1.1[c]	432	17[b]
Taxus cuspidata	11.1	11.5	400	850
Thuja plicata		10.2[d]	353	15[b]
T. standishii	10.8	8.8	300	800
Thujopsis dolabrata	3.8	4.2	400	1050
Tsuga heterophylla	0.4	1.6	437	22[b].
T. sieboldii	4.1	3.0	450	1100
Angiospermae				
Acer mono	3.7	1.6	450	1350
Betula maximowicziana	1.9	0.9	430	1400
Calophyllum vexans	2.0	1.8	450	862
Canarium indicum	4.2	1.4	342	1309
Castanopsis cuspidata	2.7	3.7	450	1100
Celtis kajewskii	4.6	1.6	494	1324
Cinnamomum camphora	4.5	2.3	400	1100
Cotylelobium sp.	10.7	13.8	804	2240
Dipterocarpus sp.	5.2	5.2	534	1457
Distylium racemosum	5.4	1.9	650	1700
Eucalyptus deglupta	1.7	1.2	493	1262
Fagus crenata	2.6	1.0	450	1350
Hopea pierrei	10.6	11.8	695	2616
Juglans sieboldiana	7.2	3.9	420	1200
Octomeles sumatrana	1.5	1.9	268	609
Planchonella thysoidea	1.0	2.1	407	1648
Pterocymbium beccarii	4.0	1.3	313	1047
Quercus acuta	8.6	3.6	550	1500
Q. crispula	6.5	0.8	450	1200
Shorea sp.	9.8	11.3	762	1847
Spondias dulcis	2.9	2.1	288	850
Tilia japonica	2.7	3.7	350	700
Vatica sp.	12.8	11.6	678	1729

[a] Parallel to the grain
[b] Perpendicular to the grain
[c] Diethylether extract
[d] Whole extract

9.1.4.2 Other Physical Properties

Durability and dimensional stability are valuable physical properties of wood that are enhanced by the extractives content. The popularity of these characteristics has led to the development of wood-polymer composites (WPC$_s$). WPC$_s$ plastic monomers or oligomers are applied to wood under pressure and polymerized within the wood component. These are being developed for use in a wide array of goods on the market. Woods such as ebony (*Diospyros* spp.) and rosewood (*Dalbergia* spp.) containing large amounts of extractives could be considered as naturally occurring WPC$_s$.

Extractives have various effects on other physical properties. Extractives with special characteristics, such as quinones, seriously affect the adhesive and finishing qualities of the wood (9.4.2). The nonpolar extractives with a lower oxygen content such as terpenoids, oils, fats, and waxes, affect the hydroscopicity, and permeability. This causes trouble in adhesion and finishing by the inhibition of the wetting of the wood. On the other hand, since these nonpolar extractives have high caloric value, they increase the flammability of wood and make these woods valuable as fuel.

The roles of these various wood properties and their influences in the utilization of wood are discussed in detail in the *Wood Handbook* (34).

References

1. Abe Z, Minami K 1976 Ill smell from the wood of *Gonystylus bancanus* (Miq.) Kurz. I. Mokuzai Gakkaishi 22:119–120
2. Abe Z, Minami K 1976 Ill smell from the wood of *Gonystylus bancanus* (Miq.) Kurz. II. Mokuzai Gakkaishi 22:121–122
3. Bentley K W 1960 The chemistry of natural products. IV. The natural pigments. Interscience New York, 74–92
4. Bohm A B 1975 Chalcones, aurones and dihydrochalcones. In: Harborne J B, Mabry T J, Mabry H (eds) The flavonoids. Chapman and Hall London, 466–492
5. Brown A G, Lovie J C, Thomson R H 1965 Ebenaceae extractives. 1. Naphthalene derivatives from Macassar ebony. J Chem Soc 2355–2361
6. Brown P M, Thomson R H 1974 Über die Inhaltsstoffe von *Bowdichia nitida*. Erstmalige Isolierung eines Isoflavonchinons. Justus Liebigs Ann Chem 8:1295–1300
7. Donnelly D M X 1975 Neoflavonoids. In: Harborne J B, Mabry T J, Mabry H (eds) The flavonoids. Chapman and Hall London, 843–848
8. Ferreira M A, Moir M, Thomson R H 1975 Pterocarpenequinones (6H-benzofuro[3,2-C] [1] benzopyranquinones) from *Brya ebenus*. J Chem Soc Perkin Trans I:1113–1115
9. Forestry and Forest Products Institute of Japan 1982 (Handbook for wood industry.) Maruzen Tokyo, 3–25, 186–197
10. Fracheboud M, Rowe J W, Scott R W, Fanega S M, Buhl A J, Toda J K 1968 New sesquiterpenes from yellow wood of slippery elm. For Prod J 18(2):37–40
11. Glasby J S 1982 Encyclopaedia of the terpenoids. Wiley-Interscience New York, 2646
12. Gottlieb O R, de Lima R A, Mendes P H, Magalhães M T 1975 Constituents of Brazilian Moraceae. Phytochemistry 14:1674–1675
13. Gurudutt K N, Seshadri T R 1974 Constitution of the santalin pigments A and B. Phytochemistry 13:2845–2847
14. Haslam E 1975 Natural proanthocyanidins. In: Harborne J B, Mabry T J, Mabry H (eds) The flavonoids. Chapman and Hall London, 510–526

15 Ikeda T, Oda K, Yamane A, Enda N 1980 Volatiles from pine logs as the attractant for the Japanese pine sawyer *Monochamus alternatus*. J Jpn For Soc 62:150–152
16 Imamura H 1980 [Color and artificial coloration of woods.] In: Hayashi K (ed) [Plant pigments. an introduction to research and experiments]. Yokendo Tokyo, 314–328
17 Imamura H, Kurosu H, Takahashi T 1967 The heartwood constituents of *Rhus javanica*. Mokuzai Gakkaishi 13:295–299
18 Imamura H, Yasue M 1983 [Wood extractives.] In: Imamura H, Okamoto H, Goto T, Yasue M, Yokota T, Yoshimoto T (eds) [Chemistry of wood utilization.] Kyoritsu Shuppan Tokyo, 324–399
19 Kai Y 1975 [Color of wood.] Mokuzai Kogyo 30:291–294, 345–349
20 King H G C, White T 1961 The colouring matter of *Rhus cotinus* wood (young fustic). J Chem Soc 3538–3539
21 Laidlaw R A, Smith G A 1959 Heartwood extractives of some timbers of the family Moraceae. Chem Ind 1604–1605
22 Murakami K 1982 [Chemical compositions of woods.] In: Society of Materials Science, Japan (ed) [Dictionary for wood technology.] Kogyo Shuppan Tokyo, 775–786
23 Nakamura F 1982 [Textbook for the meeting of wood technological association of Japan]. Wood Technol Assoc Jpn, Hokkaido Branch, 6–15
24 Nishida K 1946 [Chemical industry of wood.] Asakura Shoten Tokyo, 40–171
25 Ohashi H, Goto M, Imamura H 1977 A C-glucosylchalcone from the wood of *Cladrastis platycarpa*. Phytochemistry 16:1106–1107
26 Saeki I, Sumimoto M, Kondo T 1971 The role of essential oil in resistance of coniferous woods to termite attack. Holzforschung 25:57–60
27 Sandermann W, Simatupang M H 1966 Zur Chemie und Biochemie des Teakholzes. Holz Roh-Werkst 24:190–204
28 Sumimoto M 1983 [Insect attacks and wood components.] In: Imamura H, Okamoto H, Goto T, Yasue M, Yokota T, Yoshimoto T (eds) [Chemistry of wood utilization.] Kyoritsu Shuppan Tokyo, 135–141
29 Sumimoto M, Kondo T, Kamiyama T 1974 Attractants for the Scolytid beetle, *Cryphalus fulvus*. J Insect Physiol 20:2071–2077
30 Sumimoto M, Suzuki T, Shiraga M, Kondo T 1975 Further attractants for the Scolytid beetle, *Taenoglyptes fulvus*. J Insect Physiol 21:1803–1806
31 Tanaka N, Yasue M, Imamura H 1966 The quinonoid pigments of *Mansonia altissima* wood. Tetrahedron Lett 24:2767–2773
32 Terashima N 1975 [Ill smelling components in wood.] Mokuzai Kogyo 30:508–510
33 Terashima N 1983 [Odor of wood and metal corrosion.] In: Imamura H, Okamoto H, Goto T, Yasue M, Yokota T, Yoshimoto T (eds) [Chemistry of wood utilization.] Kyoritsu Shuppan Tokyo, 120–136
34 US Department of Agriculture 1987 Wood Handbook: Wood as an engineering material.US Dep Agr Agr Handb 72:466 pp
35 Yasuda S, Ito T, Kadota O, Terashima N 1979 Ill-smelling components of elima, *Octomeles sumatrana* Mokuzai Gakkaishi 25:549–551
36 Yoshihira K, Tezuka M, Takahashi C, Natori S 1971 Four new naphthoquinone derivatives from *Diospyros* spp. Chem Pharm Bull 19:851–854
37 Zinkel D F, Ward J C, Kukachka B F 1969 Odor problems of plywoods. For Prod J 19(12):60

9.2 The Role of Wood Exudates and Extractives in Protecting Wood from Decay

J. H. HART

9.2.1 Introduction

9.2.1.1 Decay

Ever since humans began to use wood, its tendency to decay has been noticed but not understood. The decay of heartwood in standing trees is a common phenomenon in nature. Some have regarded it as just another expected and unavoidable consequence of old age. However, all of us have observed that some trees are decayed when only 80 years old while others live for centuries, sound to the core. There are obvious differences between species in their resistance to decay. But why? Even within a species, decay varies greatly between stands and even between individuals of the same age. Why is this so?

A clear understanding of decay has been acquired only recently. According to ancient Greek and Roman ideas, all matter consisted of four elements: earth, air, fire, and water. The speed with which a certain kind of wood decayed depended upon the relative proportions of these elements in its composition. Thus, woods that contained a large proportion of fire decayed rapidly while those that contained more earth were durable. Throughout most of recorded history mankind realized there was an association between fungi and decay, but the prevailing theory was that decay caused the fungus. It was not until the 1870's when Robert Hartig published his classic work (26) on wood decay that the correct relationship was understood.

Hawley and co-workers (27) were the first to demonstrate that the unusual decay resistance of heartwood, as compared with sapwood, results from the presence of small amounts of extractives that are toxic to fungi and other wood-attacking organisms. Extractives are defined as those nonstructural constituents of wood that can be extracted with neutral organic solvents or water. These components (waxes, fatty acids, organic acids, resins, terpenes, tannins, and other benzenoid compounds) were found in the heartwood and were considered to be responsible for the relatively greater durability of the heartwood (both in the standing tree and in wood in service) than the sapwood. This chapter will discuss the role of extractives primarily as preformed antimicrobial agents and will concentrate on findings reported since the landmark review of Scheffer and Cowling (45). The role of post-infection inhibitory compounds (phytoalexins) will only briefly be covered and the reader is referred to other sources (20).

9.2.1.2 Decay-Causing Organisms and Their Effect on Wood Structure

The vast majority of wood-rot fungi belong to the class Basidiomycetes. They produce fruiting bodies, referred to as "conks" or sporophores, which are easily seen.

Fig. 9.2.1. A typical conk or fruiting body of a decay fungus

The conks develop on decaying trees or other woody material and have a lower surface composed of many pores (Fig. 9.2.1) or, more rarely, gills. The pores are lined with basidia upon which four basidiospores are produced.

A single conk releases millions of basidiospores each day into the external environment, where they are randomly distributed by the wind. Infection takes place when the basidiospores come in contact with a woody substrate, provided that the wood moisture is above the fiber saturation level, there are no antifungal inhibitors, the temperature is moderate, oxygen is available, and the pH is slightly acidic. The basidiospore germinates to form hyphae that grow in the lumens of the wood cells. The hyphae secrete extracellular enzymes, which allow the fungus to begin to digest or decay the wood. A film of free water apparently must be present in the lumens to provide a medium for diffusion of the fungal enzymes and of the solubilized degraded products of the wood. It is in this film of free moisture that some wood extractives act to inhibit the extracellular fungal enzymes and hence reduce the rate of decay.

Cellulose, hemicelluloses and lignin account for about 95% of the chemical constituents of the woody cell. Cellulose is a long-chain polymer of glucose units held together with β-1,4 glycosidic bonds. Lignin is a complex polymer of phenylpropane units with many cross linkages. Hemicelluloses consist of relatively short heterogenous polymers of glucose, galactose, mannose, arabinose, xylose, and uronic acid units linked by glycosidic bonds between the sugar residues. At least some covalent bonds exist between lignin and hemicelluloses. A large part of the remainder can be considered extractives.

Basidiomycetous wood-decay fungi produce cellulose-degrading enzymes of two principal types. Endocellulases hydrolytically split the long cellulose chain randomly, thereby producing many short cellulose chain fragments of varying length. Exocellulases systematically degrade cellulose by splitting off two glucose units at a time from one end of the chain. Some wood decay fungi produce enzymes capable of splitting the lignin polymer and oxidizing the phenylpropane molecule. The enzymes are released from growing hyphae into the film of free water in the cell lumen. The degradation products that are soluble diffuse back to the fungus and are used for fungal respiration and growth. Hence, decay-causing microorganisms return CO_2 and water to the atmosphere and help complete the carbon cycle.

Three major types of decay are recognized: white rot, brown rot, and soft rot. Wood-decay fungi that enzymatically digest cellulose, hemicelluloses, and lignin are termed white-rot fungi, whereas fungi that digest only the wood polysaccharides – leaving the lignin – are called brown rotters. Brown-rotted wood usually is darker in color than normal and has a crumbly texture. White-rotters, which use the lignin, change the wood to a lighter color and the decayed wood is often stringy. Soft-rot fungi utilize both lignin and wood polysaccharides and usually confine their effects to the surface of a piece of wood. Soft rot occurs only in wood that is saturated with water.

9.2.2 How Trees Defend Themselves Against Decay

9.2.2.1 Role of Wounds

What defense mechanisms have living trees developed to prevent decay fungi from degrading their wood? The first line of defense is an unbroken layer of bark and sapwood. Bark is somewhat analogous to skin in humans, for we rarely develop infections unless we are wounded; but in the case of trees, we must consider not only bark but sapwood as well. Decay fungi penetrate trees primarily through wounds that reach the sapwood and kill some of the sapwood cells. Killed sapwood is highly susceptible to decay, whereas living sapwood is extremely resistant. An explanation of this phenomenon is lacking at present, but obviously the vital nature of the living sapwood cells is paramount.

Certainly any organism that lives for more than a very short time will be injured sooner or later. Since trees are large and long-lived, they are wounded many times during their lifetime. Trees are uniquely well adapted for life on earth. In many ways they are superior organisms. They live longer and grow taller and larger than any other living thing. Hence they must have developed effective means to repair wounds and to protect open wounds from invasion by fungi.

A tremendous variety of agents wound trees. Trees are immobile so they are usually wounded by external moving agents. However, spontaneous wounds occurring naturally at branch stubs are very common; they form a bridge of dead tissue from the external environment to the heartwood, which can be an avenue for infection by decay fungi.

Trees are frequently wounded, but only rarely develop decay. Their second line of defense against decay fungi is the response of the injured tissue in such a manner as to restrict the entry of invading microorganisms. Most wounds are either never invaded by pathogenic organisms or, if invasion does occur, the pathogen is restricted to the area of disrupted cells.

The response strategy of the tree is twofold: first, to close the wound with new callus tissue, thus re-establishing that unbroken layer of living tissue and, second, to develop a "reaction layer" in the uninjured sapwood tissue adjacent to the damaged cells. The rate of wound closure and the production of the reaction zone are dependent not only upon the size of the wound but also upon the growth rate or vigor of the host.

What are these changes occurring in the reaction zone that prevent most wounds from becoming extensively colonized by potential pathogens? There are two general types of responses: cytological and chemical. Starch grains disappear, and gums and tyloses develop. Gradually the living cells die and their contents impregnate all tissue in the zone. There is an increase in pH, inorganic elements, and antimicrobial compounds. Many of these antimicrobial compounds are extractives and hence they will be briefly discussed here, but the reader is referred to other sources for an in-depth treatment of decay resistance in living trees (37, 49, 50).

Death of the parenchyma in the sapwood of woody plants normally results in a marked increase in susceptibility of that tissue to decay. However, certain types of wounds apparently stimulate in adjacent cells the synthesis and accumulation of compounds such as lignans, stilbenes, resin acids, or tannins. The reaction zones surrounding wounds in some species show an accumulation of compounds inhibitory to decay fungi; in other species the discolored wood surrounding the wounds is no more decay-resistant than the sapwood of that particular species. The ability of the sapwood to form discolored wood containing toxic components may be linked with an ability to produce heartwood resistant to decay fungi.

There can be differences in the composition of extractives formed in normal heartwood as the result of internal stimuli associated with aging and in the reaction zone where the stimulus is wounding. The ratio of pinosylvin to pinosylvin monomethyl ether differed significantly between the heartwood and the reaction zone surrounding injured sapwood (20). The reaction zones of Norway spruce (*Picea abies*) and loblolly pine (*Pinus taeda*) were characterized by the accumulation of normal heartwood phenols, but the most fungitoxic extractives (hydroxymatairesinol and pinosylvin) increased 3 to 20 times their concentration in heartwood (48). Thus there appears to be a disproportionate mobilization of normal antifungal heartwood constituents in the reaction zone of these two species, produced in response to fungal invasion. Phenols not normally present in heartwood may accumulate in injured sapwood. Isoolivil and scopoletin accumulated in the reaction zones of *Prunus jamasakura* and *P. domestica*, respectively, after fungal attack (50). Neither of these compounds was detectable in sound heartwood.

The best documented evidence for an extractive (pinosylvins) functioning as a phytoalexin (an antimicrobial compound produced as an active response of the plant to the pathogen) is the reaction zones produced in pine in response to invasion by *Heterobasidion annosum*, the *Sirex-Amylostereum* complex, and the *Dendroctonus*-stain fungi complex (20). As with other phytoalexins, the accumulation of sitilbenes in pine sapwood is a nonspecific response to injury. In some cases the wounding agent causes rapid death of the invaded tissue and no stilbene forms. In the sapwood of some trees, however, a reaction zone develops between the healthy tissue and the invaded tissue. As long as the cells die slowly under conditions favorable for cellular metabolism, pinosylvins may be produced. *Pinus resinosa, P. taeda*, and *P. nigra* all synthesized pinosylvins in response to *Heterobasidion annosum* attack.

The white-rot fungus, *Amylostereum areolatum*, is injected into the sapwood of *Pinus radiata* during oviposition by the wood wasp, *Sirex noctilio*. Death of the tree generally is ascribed to the combined attack of the fungus and the wasp. Vigorous trees can generally resist attack, while suppressed trees cannot. In trees that recover from attack, pinosylvins develop in the reaction zones surrounding oviposition wounds. Histological studies indicate that the fungus begins to penetrate the stilbene-impregnated edges and then stops growing. In trees that do not resist attack, stilbene accumulation was not detected.

Resinosis − the abnormal flow of resin − is considered the most important factor contributing to the resistance of lodgepole pine to bark beetles (*Dendroctonus*) and blue-stain fungi, but stilbenes also may be involved. Wound-response tissue from resistant trees attacked by *Dendroctonus ponderosae* and its associated microorganisms contained approximately twice the level of pinosylvins 5 weeks after attack, as did similar tissue from successfully colonized trees.

While the stilbenes may not be an important cause of disease resistance in this situation, they are probably important in determining which fungi act as primary invaders of wounds of lodgepole pine and thus which fungi are important in bark beetle ecology. Pinosylvin is considerably less toxic to blue-stain fungi and yeasts associated with *D. ponderosae* than it is to decay fungi. The ability of these organisms to reduce the formation of pinosylvins and to grow at concentrations of pinosylvins inhibitory to other fungi may explain their role as pioneer invaders. The stain fungi and yeasts decompose the inhibitory substances, thereby allowing the less tolerant decay fungi to become established. A similar situation has been reported for wounds of *Acer rubrum* and *Liquidambar styraciflua* and during the decay of dead coniferous wood.

Oleoresin (a solution of resin acids in a volatile terpene) frequently accumulates at wounds or sites of infection in various coniferous species. These substances might account for resistance of the wood to bacterial, fungal, or insect invasion. Data on the effectiveness of these substances in preventing or reducing invasion by microorganisms are conflicting, due in part to differences in techniques used to bioassay these complex materials. Some workers have employed bioassay systems in which the volatile materials have been allowed to escape, while other workers have used closed systems.

The volatile components of oleoresin (monoterpenes) have been found to inhibit the growth of wood-inhabiting fungi (9, 11, 14−16, 39, 51, 53). In other cases monoterpenes stimulated fungal growth (15, 16). Both α- and β-pinene have been reported to be toxic to *Heterobasidion annosum* and some other decay fungi (11, 14, 53), whereas in other studies these compounds either stimulated or failed to reduce the growth of decay or stain fungi (15, 16, 31).

Data on the fungitoxic effects of resin acids and oleoresin are just as conflicting. Wood with a high oleoresin content has an increased decay resistance (1, 13, 25, 46) and inhibits colonization by blue-stain fungi (32, 51). Gibbs (18, 19) has reviewed the literature that supports the hypothesis that oleoresin is important in the resistance of certain conifers to *Heterobasidion annosum*. The ability of a pine to mobilize resin largely determines, its resistance to *Heterobasidion annosum* and perhaps to many other diseases, although the mechanisms of resistance are not known. Prior (39) attributed decreased fungal growth in living pines affected by resinosis to mechanical blocking of tracheids rather than to fungitoxicity of the oleoresin. *Heterobasidion annosum* penetrated resin-impregnated xylem very slowly (7), but an inverse correlation between resin exudation and subsequent infection has also been reported (29). In addition, resin acids themselves have been reported to be both nontoxic (7) and toxic to *Heterobasidion annosum* (39, 47). Resin accumulation was shown to act as a barrier to infection of white spruce roots by four heart-rot fungi, but not by two others (57). Oleoresin may act as a nontoxic water-proofing layer that prevents penetration of the wood by the fungus.

Hence, there are some results that support the hypothesis that oleoresin is chemically toxic to various wood-rotting fungi and other results that indicate that it is nontoxic but acts as a mechanical barrier. Other findings indicate that oleoresin may not play any role in disease resistance. The only way to unravel the problem is to use pure compounds and determine if various pathogenic fungi are able to penetrate impregnated tissue as quickly as they do normal tissue.

The last groups of compounds in wood reported to act as phytoalexins are the lignans (see Chap. 7.3). Shain and Hillis (48) found that the major response of the inner sapwood of Norway spruce (*Picea abies*) to *Heterobasidion annosum* was a 15-fold increase in the concentration of hydroxymatairesinol in a narrow band around the affected tissue. They concluded that a high hydroxymatairesinol concentration in association with the alkalinity of the reaction zone contributed to the resistance of this tissue to *Heterobasidion annosum*. Neither hydroxymatairesinol nor conidendrin was present in detectable quantities in the sapwood of *Picea abies*. However, hydroxymatairesinol contents of sapwood, injured sapwood, and heartwood of *Picea glauca* were similar, whereas injured sapwood was much more resistant to decay than was normal sapwood or heartwood (25). Therefore hydroxymatairesinol apparently was not important in the increased resistance of injured sapwood to decay.

Parasite damage and mechanically-inflicted injury often initiate the same response in plants. However, the lignans formed in response to injury may differ substantially from one zone of injury to another even within the same tree (8, 25). The lignans produced may or may not be similar quantitatively or qualitatively to those found in normal sapwood or heartwood.

In summary, it is not yet possible to separate cause and effect phenomena in experiments correlating expression of resistance with phytoalexin accumulation. Accumulation of antimicrobial extractives is probably a consequence of, and perhaps incidental to, the actual resistance mechanism. There is no evidence that potential pathogens come in contact with these toxic compounds in the early phases of pathogenesis. Available evidence suggests that resistant trees retard the advance of the pathogen by an unknown mechanism that allows the formation of a reaction zone. Following the expression of resistance, the living cells in the sapwood die slowly, allowing the synthesis and accumulation of compounds, some of which may have antimicrobial properties. This accumulation may create an environment that, along with other factors, limits further fungal growth in otherwise susceptible tissue. Tolerance to these compounds may bestow a competitive advantage on an organism, allowing it to colonize the injured tissue preferentially to another organism that is intolerant of such compounds, but tolerance to toxic extractives is probably not a prerequisite for pathogenicity.

9.2.2.2 Toxic Heartwood Components

The first line of defense against invasion by microorganisms is an unbroken layer of bark and sapwood. The second line of defense is the development of a reaction zone in the tissue adjacent to damaged cells. But what happens when the wound is so large or deep that no reaction zone can form between the heartwood and the outside world? What happens in those situations where the fungus can successfully overcome the best defense mechanisms the living cells of the host can offer? This brings us to the third line of defense: fungus-inhibitive compounds formed during the sapwood-heartwood transformation.

Inside the sapwood, most trees contain heartwood, a core of dark, dead tissue. The formation of heartwood is a unique process in nature, in which the death of a large portion of an organism is beneficial to that organism. The tree needs the central core of the trunk only for support, and the core is just as strong dead as alive. An obvious advantage to the tree is that no energy is needed to support dead tissue or to keep it functional. However, dead tissue cannot actively fight infection. The advantage of having a strong, inert central skeleton is so great that many trees have developed mechanisms to ensure that this relatively defenseless heartwood does not become infected under normal conditions.

Toxic extractable substances deposited during the formation of heartwood are the principal source of decay resistance of heartwood (45). This conclusion is based on the following observations: 1) Extracts from durable heartwood are much more toxic to decay fungi than those from the sapwood of the same tree; 2) extracts from nondurable heartwood have little inhibitory effect on decay fungi; 3) extraction of the antimicrobial compounds with various solvents decreases the decay resistance of the wood from which the compounds were obtained; 4) the toxicity of extracts obtained from heartwood of various species corresponds broadly with the decay resistance of those species; and 5) generally, but not always, nondurable wood impregnated with extractives from durable heartwood increases in decay resistance (24, 45).

9.2.2.2.1 Formation of Antimicrobial Compounds

The reader is referred to earlier chapters (particularly Chaps. 3 and 7) for detailed information on the biosynthesis of heartwood extractives. Only a brief summary of our knowledge is included here. Heartwood extractives are formed in situ at the heartwood periphery from translocated carbohydrate or lipid substrates. The composition of the substances formed is under genetic control and the amount formed is significantly influenced by the physiological conditions impinging on the parenchyma at the time of formation. In some species the transition zone between sapwood and heartwood has an enhanced metabolic activity. Ethylene has been associated with this activity, and in some species it may play a key role. Its formation could be triggered by water stress (28).

When heartwood is formed, the vacuoles containing the extractives rupture, and the vacuolar components remain in the cell lumens or they may diffuse into the cell walls, where they probably combine with other wall components. Stilbenes may be bound to lignins, which partly explains why they cannot be extracted from wood with ether alone, although they are soluble in ether. Knowledge of the precise location of extractives in natural and in impregnated wood would help to understand their role in decay resistance.

9.2.2.2.2 Chemical Nature of Antimicrobial Compounds

The various compounds responsible for the decay resistance of heartwood, their structures, their biosynthesis, and their chemotaxonomy are discussed in Chaps. 7 and 8. Most of the important antimicrobial compounds are phenolics, although the terpenoids may also be involved. Often the major portion of the phenolic extractives are polymeric materials (MW > 500) that cannot be identified. These materials may be very important in increasing the durability of heartwood but little work has been reported.

The extractives of related species are often similar and thus are useful for chemotaxonomy. For example, the Cupressaceae are the sole source of tropolones. On the other hand, the chemical nature of extractives may differ distinctly between closely related species or the same fungitoxic material may occur in unrelated plant families.

The types of compounds known to be important in decay resistance are listed in Table 9.2.1. Each group of compounds will be discussed briefly. More detailed information has previously been published (45).

The simple phenolic acids and aldehydes are almost always present in heartwood extractives, at least in small amounts. Rudman (42) tested a number of benzene derivatives (MW = 100–200) against four species of decay fungi, but none displayed a high degree of inhibition. Gallic and ellagic acid, two of the more common compounds identified in heartwood extractives, were usually nontoxic (22, 40), although the former was inhibitory to some decay fungi (22, 42). Their generally low molecular weight may be responsible for their lack of toxicity.

Several ellagitannins isolated from *Eucalyptus* and *Quercus* are moderately fungitoxic against some decay fungi (22, 23). They apparently act to "tan" extracellular fungal enzymes, and the durability of these heartwoods results from relatively high levels ($\approx 2\%$) of these compounds. Gallotannins have not been demonstrated to be fungitoxic.

Four lignans (conidendrin, isoolivil, pinoresinol, and mataresinol) were tested against several decay fungi (42, 43) and were generally nontoxic. *Lentinus lepideus* was inhibited by 1.0% of pinoresinol or mataresinol but was unaffected by the other compounds. Lignans often are produced by *Picea* or *Pinus* in response to

Table 9.2.1. Chemical classification of compounds known to possess anti-microbial activity and to occur in the heartwood of decay-resistant species

Groups of extractives	Classes of compounds	Examples of each class
Benzenoid	Phenolic acids and phenolpropanes	Gallic and ferulic acids
	Hydrolyzable tannins	Ellagitannins
	Lignans	Hydroxymatairesinol
	Stilbenes	Pinosylvin
	Flavonoids	Robinetin
Terpenoids	Oleoresin	
	monoterpenes	α-pinene
	diterpene (resin acids)	Abietic acid
	Tropolones	β-thujaplicin
	Other related compounds	Ferruginol

a pathological condition (48) (see Sect. 9.2.2.1). *Tsuga heterophylla* is rich in hydroxymatairesinol, which is more water soluble and more toxic to *Heterobasidion annosum* than matairesinol (48). Although not studied to date, the production of hydroxymatairesinol by *Tsuga heterophylla* in response to invasion by *Heterobasidion annosum* may play a role in helping this species to resist attack.

Stilbenes occur in a number of plant families and have long been considered important in heartwood durability and in the resistance of sapwood to its pathogens (24). Wood containing stilbenes frequently decomposes slowly when exposed to various rot fungi. When tested on a nutrient agar substrate, stilbenes possess high fungitoxic properties. However, when bioassayed in a woody substrate, the toxicity of stilbenes is reduced 90% to 99%. This enigma is still unresolved.

The very low toxicity associated with flavonoids (of which 30 were tested using wood as a substrate) indicates that this group of compounds is not of great importance in determining decay resistance (42, 43). Robinetin, a constituent of the heartwood of *Robinia pseudoacacia*, was fungitoxic. The heartwood of *Robinia pseudoacacia* has exceptionally high decay resistance, apparently attributable to the high concentration (2%) of robinetin and possibly dihydrorobinetin (5.3%) present. Taxifolin, dihydrorobinetin, and quercetin have sometimes (30) been credited with fungitoxic properties but more recent results suggest otherwise (10, 42).

The possible role of monoterpenes and diterpenes (resin acids) in decay and disease resistance is discussed in Sect. 9.2.2.1. That the presence of resin retards attack by decay fungi is evidenced by the durability of pine stumps and resinous knots. Whether this protection is afforded by the waterproofing nature of these compounds or as metabolic inhibitors is in question. The abnormal flow of resin (resinosis) is stimulated by wounds or pathogen attack. These materials may serve as a natural wound dressing to repel bark beetles and inhibit the growth of decay fungi. The role of the terpenes as protective agents deserves further study.

The most clearcut, unambiguous results have been reported for the tropolones. They are clearly the most fungitoxic group of extractives known (5, 42). Tropolones (thujaplicins) have only been isolated from members of the Cupressaceae (*Cupressus, Chamaecyparis, Juniperus, Libocedrus,* and *Thuja*), all of which contain very decay-resistant heartwood. Comprehensive studies (4, 42, 52) showed that the thujaplicins are highly toxic to a wide variety of wood-destroying fungi. βThujaplicin and nootkatin are the most commonly occurring tropolones. The total amount of tropolones present in the heartwood is usually exceedingly small (often less than 0.1%) but some species of *Thuja* and *Cupressus* contain 1% or 2% (17). These minute quantities of tropolones, perhaps acting synergistically with other compounds, confer a high degree of decay resistance on heartwoods in which they occur, being comparable with those of commercial fungicides used in wood preservation.

Several other terpenoids have fungistatic properties. The phenolic diterpenes ferruginol and totarol, obtained from members of the genus *Podocarpus*, are inhibitory to some decay fungi but not to others (43). Hydrothymoquinone and *p*-methoxythymol, phenolic compounds with a *p*-cymene structure, isolated from *Libocedrus* heartwood, had broad-spectrum inhibitory effects on decay fungi (5).

The naphthopyran, lapachonon, was found to approach the tropolones in its toxicity to decay fungi (42). Lapachonon occurs in the heartwood of the decay resistant genera *Paratecoma* and *Tabebuia*. Little additional work has been conducted on this compound and it appears to be a fruitful area for additional studies.

9.2.3 Evaluation of Decay Resistance

9.2.3.1 Isolation and Evaluation of Compounds

Procedures used in the extraction, isolation, and identification of extractives have been outlined in earlier chapters. Generally, the wood is air-dried and then ground to a powder to facilitate extraction. The ground wood is placed in a Soxhlet apparatus or in a column and treated sequentially with solvents of increasing polarity or with a single solvent (usually methanol or acetone). Which system and which solvents are used depends on the chemical nature of the compounds being removed. For example, stilbenes, although soluble in ether, cannot be extracted directly by ether but they are extractable with acetone or methanol. Once the extract is obtained, it may be used as a crude extract or it may be further divided into fractions containing compounds of a similar chemical nature. If possible, pure compounds identified as to structures should be used in the bioassays.

Before decay resistance can reasonably be attributed to a particular substance, wood containing the material should be significantly resistant to decay, and removal of the substance(s) from the durable wood should render the extracted wood susceptible to decay. A number of tests have been developed to determine the rate at which certain fungi are capable of decaying wood (2, 3, 38). Either wood blocks or wood meal may be used; the latter is sometimes mixed with agar or a nutrient solution. Amounts of decay are measured by determining the loss in weight, by the quantity of glucosamine formed, by respirometry (oxygen uptake or CO_2 evolution), or by changes in natural frequency. Good correlations have been obtained between field resistance and resistance measured in the laboratory by these methods.

The effects of crude extracts or purified compounds on spore germination, respiration, radial growth of mycelium on solid media, accumulation of fungal mycelium in liquid culture, or rates of decay of woody substrates have been used as bioassay tests. Each type of test has its advantages and its limitations, and no one type of test is superior to all others. Researchers should be aware of the limiting factors for each method. Conclusions about antifungal activity should rely on data from as many types of bioassays as possible. In any bioassay, toxicity will depend on the compound being tested, its concentration, and the conditions under which the test is performed. The apparent sensitivity of a fungus to a compound can be markedly affected by the type of bioassay conditions employed. For example, the antifungal property of most stilbenes is decreased by 90% to 99% when bioassayed with a woody substrate rather than with agar alone. Hence all in situ events are not duplicated in any bioassay method, notwithstanding the presence or absence of wood as a substrate. In all bioassays, the concentration, or at least the presence, of the substance being tested must be determined at the end of the experiment.

The most common method of evaluating the toxicity of extracts to wood-destroying fungi has been to determine the effect of the extractive on the development of pure cultures growing in or on a nutrient solution or agar. An extractable antimicrobial substance should be inhibitory in agar or nutrient solution and the degree of inhibition should be quantitatively related to the concentration of the substance. Measurements of radial growth of mycelium on a solid medium or of fungal dry matter in liquid cultures are commonly made. Such assays are quick and simple but, more important, they permit the bioassaying of very small quantities of the test compound. Nutrient solution or agar bioassays are particularly valuable for comparative studies.

However, basing conclusions regarding the causes of decay resistance on tests utilizing agar or nutrient solutions is unreliable (24). Comparative tests on wood do not always agree with results obtained from nutrient media. The induction and synthesis of cellulolytic and ligninolytic enzymes by wood-decaying fungi differ depending on the substrate. The use of a substrate containing a directly available carbon source (such as agar supplemented with soluble sugars) reduces or delays the production of cellulase and laccase (34). An additional drawback to the use of nutrient media is that the extractives must possess some water solubility. Another problem that is unique to assays on agar is that the mycelium frequently grows superficially on the agar surface with minimal contact with the test chemical. Hence, higher amounts (generally twice) of the toxic extractive are required in agar tests to achieve the same degree of inhibition obtained in nutrient solutions.

If the contribution of extractives or of a single component of heartwood to overall decay resistance is being tested, the preferred substrate is wood (45). When incorporated into a decay-susceptible woody substrate at a concentration equal to that in the resistant wood, the test compound should impart an equivalent level of decay resistance. Usually a certain amount of decay resistance is lost in this process — i.e., the impregnated wood is never quite as resistant as the original piece of wood. The toxicities of ellagitannins, tropolones, and some alcohols in unextracted heartwood and their toxicities when impregnated into decay-susceptible wood at their original concentrations are highly correlated (23). However, when stilbenes are impregnated into solid wood or sawdust, their inhibition is reduced approximately to 1/100 of original (24).

While the use of a woody substrate may approach the in situ situation more closely than does agar, the use of wood certainly does not duplicate the situation in a tree. Wood has a number of significant drawbacks. No data exist to show that naturally and artificially impregnated woods contain extractives in the same location or in the same chemical form. Binding between woody components and extractives may differ in naturally and artificially impregnated wood. The amount of test material required to impregnate wood blocks is large, although the sawdust-disk technique and the wood meal-respiratory technique require much smaller amounts. A long period of exposure to decay fungi, normally 8 or more weeks, is required to complete the test. Finally, a uniform concentration is hard to achieve, at least in solid wood blocks (24).

Once the extractives have been impregnated into the woody substrate and exposed to the test fungus, weight loss is the standard measure to quantify the rate of decay (23, 38). A relatively quick and simple procedure to quantify the toxic effect of extractives on decay fungi — one that requires relatively small amounts of the compound being studied — is respirometry (55). Another novel method of evaluating extractives is to measure changes in the frequency of free vibration of an elastic body (6, 56). Since certain strength properties of decaying wood diminish more rapidly than weight, it is theoretically possible to determine the effectiveness of an extractive more rapidly using strength loss rather than weight loss. Significant decreases in natural frequency of free vibration were observed after 7 days' exposure to a decay fungus.

In summary, it should be emphasized that wood extractives are a complex mixture of toxic and nontoxic components interacting with one another and with fungi during decay. With any in vitro testing, the succession of organisms that occur during the natural decay process does not occur and hence the results could differ widely from natural behavior (50). In addition, evidence is accumulating that decay resistance is a multifunctional phenomenon and it may be impossible to confer on one substance sole responsibility for resistance of the entire heartwood. Very likely the compounds act synergistically. Testing compounds singly, even in wood, may yield very misleading results.

9.2.3.2 Physiology of Decay Inhibition

The wide diversity of decay fungi and of the chemical nature of heartwood extractives suggests that inhibition results from a variety of mechanisms. Two basic mechanisms are thought to occur: 1) the extractive is absorbed by the fungus and directly disrupts its metabolism; or 2) the extractive inactivates extracellular fungal enzymes, thereby indirectly starving the fungus. It should be remembered that heartwood is biologically inert, and that after its initial formation during the sapwood-heartwood transformation, no more extractives are produced. The fungi that live in heartwood that contains fungitoxic compounds must be tolerant of the fungitoxic compounds residing therein, be able to detoxify them, or follow other organisms that detoxify the compounds.

Both stilbenes and tropolones act as uncoupling agents that inhibit oxidative phosphorylation, the main source of energy in decay fungi (45). Free hydroxyl groups are essential if phenolic compounds are to act as uncoupling toxins. Pinosylvin monomethyl ether loses its toxicity after oxidation by fungal oxidases (36). In addition, pinosylvin monomethyl ether has been shown to inactivate enzymes containing SH groups in their active sites (35). Cellulase, xylanase, and pectinase are all extracellular enzymes that are activated by sulfhydryl groups and hence are sitilbene-sensitive. When exposed to sublethal concentrations of pinosylvin, some decay fungi produce abnormal clamp connections, which suggests that pinosylvin may have an effect on mitosis (20).

The strongly inhibitory influence of tannins and some other extractives is based on their ability to bind with extracellular enzymes rather than their toxicity to decay fungi per se (22, 36). Ellagitannins isolated from *Quercus alba* heartwood interact with the fungal proteins to form tannin-protein complexes. Tween 80 or polyvinyl pyrrolidone are capable of splitting these complexes and, when added to media containing ellagitannins, eliminated the fungistatic effects. In addition, conditions that favor a rapid synthesis of extracellular enzymes reduce the fungal toxicity of the extractives present (20). Once an excess of binding protein occurs, the fungus can begin growth, as the inert heartwood is incapable of synthesizing more extractives.

Wood extractives do not always control the growth of a broad spectrum of fungal species but may be quite specific. The extracellular enzymes of brown-rot and white-rot fungi have been shown to differ in their mode of action and molecular size (58). The different extracellular enzymes of the two groups of fungi may require different receptor sites – i.e., they are stereospecific. The binding capacity of different tannin-protein mixtures varies depending on the molecular size and chemical structure of the compounds involved. A minimum molecular weight for inhibitory activity of enzymes by tannins is about 500 (59). This may explain why ellagitannins (minimum molecular weight 481) are fungistatic, whereas gallic and ellagic acid, with lower molecular weights, are not toxic (22). It also may explain the higher toxicity of resveratrol oligomeric polymers compared to resveratrol itself (33).

Toxicity may be modified by the presence of free hydroxyl groups. Rudman (41) suggested that vicinal trihydroxy groups are important factors in determining the toxicity of heartwood phenolics. Methylation or acetylation of hydroxylated stilbenes completely destroys or greatly reduces their antifungal activity (20). Pinosylvin or resveratrol are both more toxic than *cis-*, *trans-* or *p-*hydroxy stilbene. However, *cis-*stilbene has consistently greater toxicity than *p-*hydroxy stilbene. Hence, free hydroxyls apparently are not a prerequisite for toxicity, and the location of the OH groups may be of greater importance than the number of such groups, as the formation of enzyme-phenol complexes is stereospecific. Compounds containing hydroxyl groups would generally have higher water solubility and therefore possibly a greater potential for interaction with decay organisms.

If binding to specific enzymes or proteins occurs, a specific molecular shape may be responsible for antifungal activity. The *trans*-isomer of stilbene is planar, while the *cis*-isomer is aplanar; the latter is consistently more toxic than the former (20). The difference in toxicity of ellagitannins to white- and brown-rot fungi is probably due to a structurally critical interaction between the tannin molecule and the different extracellular enzymes of the two groups of fungi (22). Difference in toxicity between two geometric isomers or between extracellular enzymes of different molecular size suggests that the toxicity depends on a receptor site onto which only molecules of a certain conformation can bind.

Hence the molecular size (of both phenolics and enzymes), geometric isomerism, number and location of free hydroxyl groups, plus factors yet unknown determine whether a compound will be antimicrobial or not. While these factors are important, to date no results are available that allow general rules to be established.

9.2.3.3 Variation in Decay Resistance

9.2.3.3.1 Variation Between Tree Species

Durability of heartwood is dependent on the kinds and concentration of extractives present, their chemical stability, and their resistance to microbial inactivation. Many tree species (e.g., *Maclura pomifera* or *Robinia pseudoacacia*) contain only one or two toxic compounds, often in very small quantities, yet the wood is extremely durable (45, 54). In other species a number of similar toxic compounds may occur together (*Libocedrus decurrens* (5)), or toxic compounds of a very different chemical nature may be present in the heartwood of a single species (stilbenes and ellagitannins in *Eucalyptus sideroxylon* (23). In either case, the combination of compounds acts additively or perhaps even synergistically to produce an inhibitory environment.

The comparative decay resistance of the more common North American woods (Table 9.2.2) has frequently been published (37, 44, 45). This list has not been revised since 1961 and new information would change the decay rating of several species (e.g., *Juglans nigra* from resistant to moderately resistant (21)). Other lists (Table 9.2.3) are available that give the relative durability of species not native to North America (44, 45).

Like all botanical characteristics, the amount and type of heartwood extractives are genetically determined (see Table 9.2.1 for a list of the compounds known to be responsible). Several species (*Maclura pomifera, Robinia pseudoacacia, Sequoia sempervirens, Eucalyptus sideroxylon, Tectona grandis,* and *Thuja* spp.) possess heartwoods that are nearly as durable as wood treated with preservatives even when used under conditions favorable to decay. Fence posts of black locust or osage orange in eastern North America and barn foundations of red ironbark in southeastern Australia have lasted over a century. In general, trees with a highly

Table 9.2.2. Grouping of some trees native to North America according to approximate relative heartwood decay resistance

Resistant or very resistant	Moderately resistant	Slightly or nonresistant
Baldcypress (old growth)	Baldcypress (young growth)	Alder
Catalpa	Douglas-fir	Ashes
Cedars	Honeylocust	Aspens
Cherry, black	Larch, western	Basswood
Chestnut	Oak, swamp chestnut	Beech
Cypress, Arizona	Pine, eastern white	Birches
Junipers	Southern pine:	Buckeye
Locust, black[a]	longleaf	Butternut
Mesquite	slash	Cottonwood
Mulberry, red[a]	Tamarack	Elms
Oak:		Hackberry
bur		Hemlocks
chestnut		Hickories
gambel		Magnolia
oregon white		Maples
post		Oaks (red and black species)
white		Pines (other than long-leaf,
Osage organe[a]		slash, and eastern white)
Redwood		Poplars
Sassafras		Spruces
Walnut, black		Sweetgum
Yew, pacific[a]		True firs (western and eastern)
		Willows
		Yellow-poplar

[a] These woods have exceptionally high decay resistance

decay resistant heartwood are also remarkably resistant to a wide range of diseases and insects.

As mentioned, the concentration of the extractives may be as important as their chemical nature or more so. The nondurable heartwood of *Eucalyptus regnans* contains the same toxic components (tannins) present in the extractives from the durable heartwood of *E. microcorys* and *E. triantha*, but the concentration in *E. regnans* is much lower than in the latter two species (41). *Quercus alba* and probably many other oaks in the subgenus *Lepidobalanus* contain ellagitannins in their heartwoods in amounts adequate to suppress decay by most fungi. Most members of the subgenus *Erythrobalanus* have non-durable heartwood containing ellagitannins only in trace amounts (22).

9.2.3.3.2 Variation Between Individuals of the Same Species

Individual trees of the same species may vary considerably in decay resistance (45). In a rangewide study of the decay resistance of *Juglans nigra*, individual trees from the same site varied significantly in decay resistance (21). This variation is attributed mainly to genetic variability. Similarities in heartwood decay resistance between trees collected at different sites suggest that site is not a major fac-

Table 9.2.3. Grouping of some woods not native to North America according to approximate relative heartwood decay resistance

Resistant or very resistant	Moderately resistant	Slightly or nonresistant
Angelique	Andiroba[a]	Balsa
Apamate	Apitong[a]	Banak
Brazilian rosewood	Avodire	Cativo
Caribbean pine	Capirona	Ceiba
Courbaril	European walnut	Jelutong
Encino	Gola	Limba
Goncalo alves	Khaya	Lupuna
Greenheart	Laurel	Mahogany,
Guijo	Mahogany	Philippine:
Iroko	Philipine	Mayapis
Jarrah	Almon	White lauan
Kapur	Bagtikan	Obeche
Karri	Red lauan	Parana pine
Kokrodua (Afrormosia)	Tanguile	Ramin
Lapacho	Ocote pine	Sande
Lignumvitae	Palosapis	Virola
Mahagony, American		
Meranti[a]		
Peroba de campos		
Primavera		
Santa Maria		
Spanish-cedar		
Teak		

[a] More than one species included, some of which may vary in resistance from that indicated

tor in determining decay resistance. Since the formation of extractives is under genetic control, it is reasonable that individuals would vary in the quantity and type of compounds formed and hence in decay resistance. *Pinus sylvestris* growing side by side can differ greatly in pinosylvin content, whereas pines from widely separated areas and from different altitudes can have similar amounts of pinosylvin (12). Hence, differences in resistance among trees of the same species may be as large as among species, which helps to explain the variation in durability of untreated wood of the same species.

9.2.3.3.3 Variation Within an Individual Tree

Several generalities concerning decay resistance within an individual tree can be made (45) but exceptions do occur (21). Generally, decay resistance is greatest in the outer (youngest) heartwood and decreases from there to the pith. Decay resistance of the outer heartwood decreases progressively from the base of the tree upward, while the reverse is true for the inner heartwood. However, an extensive study of the resistance of *Juglans nigra* heartwood to decay revealed no significant gradients in decay resistance with radial position (21). Nor did heartwood from the top, middle, and base of the trees vary significantly in decay resistance when exposed to *Coriolus versicolor*, but heartwood from the tops of some trees

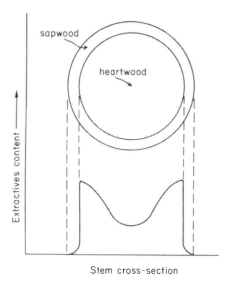

Fig. 9.2.2. Distribution of extractives across the stem

Fig. 9.2.3. A marvel of nature, the hollow tree stump

was more decay resistant than that from the bottom when it was bioassayed with *Poria placenta*. Hence, the fungus used to determine differences in decay resistance may be as significant as the inherent differences within the tree.

Outer heartwood is often more decay resistant because it contains the greatest amount of extractives (Fig. 9.2.2). The amount and toxicity of the extractives decrease radially from outer heartwood toward the pith. This phenomenon has been explained in terms of polymerization of the toxic phenolics to less toxic materials as the wood ages (5). Gradual detoxification also could be achieved by hydrolysis in trees with acidic heartwood — e.g., ellagitannin into ellagic acid and glucose (22). Once decay begins, additional degradation could occur via fungal enzymes. Since the heartwood is dead, there would be no replacement of inhibitory materials to offset losses. The heartwood would gradually become more decay-susceptible as the fungistatic compounds were destroyed by enzymatic detoxification, polymerization, or hydrolysis. These factors help explain what we have all observed in nature — hollow trees (Fig. 9.2.3).

References

1. Amburgey T L, Carter F L, Roberts D R 1978 Resistance of wood from paraquat-treated southern pine trees to termites and the fungus *Gloeophyllum trabeum*. Wood Sci 10:187–192
2. American Society for Testing and Materials 1980 Standard method of testing wood preservatives by laboratory soil-block cultures. Standard D 1413-61. ASTM Philadelphia
3. American Society for Testing and Materials 1980 Standard method for accelerated laboratory test of natural decay resistance of wood. Standard D 2017-63. ASTM Philadelphia
4. Anderson A B, Scheffer T C, Duncan C G 1962 On the decay retardant properties of some tropolones. Science 137:859–860
5. Anderson A B, Scheffer T C, Duncan C G 1963 The chemistry of decay resistance and its decrease with heartwood aging in incense cedar. Holzforschung 17:1–5
6. Bariska M, Osusky A, Bosshard H H 1983 Änderung der Mechanischen Eigenschaften von Holz nach Abbau durch Basidiomyceten. Holz Roh-Werkst 40:241–245
7. Bega R V, Tarry J 1966 Influence of pine root oleoresins on *Fomes annosus*. Phytopathology 56:870 (Abstr)
8. Chen C, Chang H, Cowling E B, Hsu C H, Gates R P 1976 Aporphine alkaloids and lignans formed in response to injury of sapwood in *Liriodendron tulipifera*. Phytochemistry 15:1161–1167
9. Cobb F W, Krstic M, Zavarin E, Barber H W 1968 Inhibitory effects of volatile oleoresin components on *Fomes annosus* and four *Ceratocystis* species. Phytopathology 58:1327–1335
10. Cserjesi A J 1969 Toxicity and biodegradation of dihydroquercetin. Can J Microbiol 15:1137–1140
11. De Groot R C 1972 Growth of wood-inhabiting fungi in saturated atmospheres of monoterpenoids. Mycologia 65:863–870
12. Erdtman H, Frank A, Lindstedt G 1951 Constituents of pine heartwood. XXVII. The content of pinosylvin phenols in Swedish pines. Svensk Papperstidn 54:275–279
13. Eslyn W E, Wolter K E 1981 Decay resistance of red pine wood chips enriched with oleoresin. Phytopathology 71:1248–1251
14. Flodin K 1979 Effects of monoterpenes on *Fomes annosus* (Fr.) Cooke and its phenol oxidase activity. Eur J For Pathol 9:1–6
15. Flodin K, Fries N 1978 Studies on volatile compounds from *Pinus silvestris* and their effect on wood-decomposing fungi. II. Effects of some volatile compounds on fungal growth. Eur J For Pathol 8:300–310
16. Fries N 1973 The growth-promoting activity of terpenoids on wood-decomposing fungi. Eur J For Pathol 3:169–180
17. Gardner J A F 1962 The tropolones. In: Hillis W E (ed) Wood extractives. Academic Press New York, 317–330
18. Gibbs J N 1970 The role of resin in the resistance of conifers to *Fomes annosus*. In: Toussoun T A (ed) Root diseases and soil-borne pathogens. University California Press Berkeley, 161–163
19. Gibbs J N 1972 Tolerance of *Fomes annosus* isolates to pine oleoresins and pinosylvins. Eur J For Pathol 12:147–151
20. Hart J H 1981 Role of phytostilbenes in decay and disease resistance. Ann Rev Phytopathol 19:437–458

21 Hart J H 1982 Variation in inherent decay resistance of black walnut. U S For Serv Tech Rep NC-74, 221 pp
22 Hart J H, Hillis W E 1972 Inhibition of wood-rotting fungi by ellagitannins in the heartwood of *Quercus alba*. Phytopathology 62:620–626
23 Hart J H, Hillis W E 1974 Inhibition of wood-rotting fungi by stilbenes and other polyphenols in *Eucalyptus sideroxylon*. Phytopathology 64:939–948
24 Hart J H, Shrimpton D M 1979 Role of stilbenes in resistance of wood to decay. Phytopathology 69:1138–1143
25 Hart J H, Wardell J F, Hemingway R W 1975 Formation of oleoresin and lignans in sapwood of white spruce in response to wounding. Phytopathology 65:412–417
26 Hartig R 1878 Die Zersetzungserscheinungen des Holzes. Berlin, 127 pp
27 Hawley L F, Fleck L C, Richards C A 1924 The relation between durability and chemical composition in wood. Ind Eng Chem 16:699–700
28 Hillis W E 1977 Secondary changes in wood. In: Loewus F F, Runeckles V C (eds) Recent advances in phytochemistry, vol 11. Structure, biosynthesis, and degradation of wood. Plenum Press New York, 247–309
29 Hodges C S 1969 Relative susceptibility of loblolly, longleaf, and slash pine roots to infection by Fomes annosus. Phytopathology 59:1031 (Abstr)
30 Kennedy R W, Wilson J W 1956 Variation in taxifolin content of a Douglas-fir stem exhibiting target ring. For Prod J 6:230–231
31 Keyes C R 1969 Effect of pinene on mycelial growth of representative wood-inhabiting fungi. Phytopathology 59:400
32 Kidd F, Reid C P P 1979 Stimulation of resinosis and apparent inhibition of blue stain development in ponderosa pine by paraquat. Forest Sci 25:569–575
33 Langcake P, Pryce R J 1977 A new class of phytoalexins from grapevines. Experientia 33:151–152
34 Loman A A 1970 Bioassays of fungi isolated from *Pinus contorta* var. *latifolia* with pinosylvin, pinosylvin monomethyl-ester, pinobanksin, and pinocembrin. Can J Bot 48:1303–1308
35 Lyr H 1961 Hemmungsanalische Untersuchungen an einigen Ektoenzymen Holzzerstörender Pilze. Enzymologia 23:231–248
36 Lyr H 1962 Detoxification of heartwood toxins and chlorophenols by higher fungi. Nature 194:289–290
37 Manion P D 1981 Tree disease concepts. Prentice-Hall Englewood Cliffs, 399 pp
38 McNabb H S Jr 1958 Procedures for laboratory studies on wood decay resistance. Proc Iowa Acad Sci 65:150–159
39 Prior C 1976 Resistance by Corsican pine to attack by *Heterobasidion annosum*. Ann Bot 40:261–279
40 Rudman P 1959 The causes of natural durability in timber. III. Some conspicuous phenolic components in the heartwood of eucalypts and their relationship to decay resistance. Holzforschung 13:112–115
41 Rudman P 1962 The causes of natural durability in timber. VIII. The causes of decay resistance in tallowwood, white mahogany and mountain ash. Holzforschung 16:56–61
42 Rudman P 1963 The causes of natural durability in timber. XI. Some tests on the fungi toxicity of wood extractives and related compounds. Holzforschung 17:54–57
43 Rudman P 1965 The causes of natural durability in timber. XVIII. Further notes on the fungi toxicity of wood extractives. Holzforschung 19:57–58
44 Scheffer T C 1973 Microbiological degradation and the causal organisms. In: Nicholas D D (ed) Wood deterioration and its prevention by preservative treatments. Syracuse University Press Syracuse, 31–106
45 Scheffer T C, Cowling E B 1966 Natural resistance of wood to microbial deterioration. Ann Rev Phytopathol 4:147–170
46 Shain L 1967 Resistance of sapwood in stems of loblolly pine to infection by *Fomes annosus*. Phytopathology 57:1034–1045
47 Shain L 1971 The response of sapwood of Norway spruce to infection by *Fomes annosus*. Phytopathology 61:301–307
48 Shain L, Hillis W E 1971 Phenolic extractives in Norway spruce and their effects on *Fomes annosus*. Phytopathology 61:841–845
49 Shigo A L 1985 Compartmentalization of decay in trees. Sci Am 252:96–103

50 Shigo A L, Hillis W E 1973 Heartwood, discolored wood and microorganisms in living trees. Ann Rev Phytopathol 11:197–222
51 Shrimpton D M, Whitney H 1968 Inhibition of growth of blue stain fungi by wood extractives. Can J Bot 46:727–761
52 Southam C M, Ehrlich J 1943 Effects of extracts of western red-cedar heartwood on certain wood-decaying fungi in culture. Phytopathology 33:517–524
53 von Schuck H J 1977 Die Wirkung von Monoterpenen auf das Mycelwachstum von *Fomes annosus* (Fr.) Cooke. Eur J For Path 7:374–384
54 Wang S, Hart J H 1983 Heartwood extractives of *Maclura pomifera* and their role in decay resistance. Wood Fiber Sci 15:290–301
55 Wang S, Hart J H, Behr E A 1980 Procedure for evaluating the effect of heartwood extractives on decay resistance. For Prod J 30(1):55–56
56 Wang S, Suchsland O, Hart J H 1980 Dynamic test for evaluating decay in wood. For Prod J 30:35–37
57 Whitney R D, Denyer W B G 1969 Resin as a barrier to infection of white spruce by heart-rotting fungi. For Sci 15:266–267
58 Wilcox W W 1973 Degradation in relation to wood structure. In: Nicholas D D (ed) Wood deterioration and its prevention by preservative treatments. Syracuse University Press Syracuse, 107–148
59 Williams A H 1963 Enzyme inhibition by phenolic compounds. In: Pridham J B (ed) Enzyme chemistry of phenolic compounds. Pergamon New York, 87–95

9.3 Effect of Extractives on Pulping

W. E. HILLIS and M. SUMIMOTO

9.3.1 Introduction

Paper used for writing, printing, and drawing has been made from fibers from various sources for centuries. With the increasing importance of paper in the past two centuries, the supply of raw materials then used became inadequate for the need, and attention was given to alternative supplies. Wood gained importance as a source of fiber in 1844 when grinding wood to pulp was introduced and subsequently when chemical processes were also used. Now, globally, wood supplies more than 94% of pulp fiber and in countries such as the United States, the amount is higher than 98%.

Paper and board products are of major importance, with 193 million tons having been produced in 1985. Annual rates of consumption per person vary from 284 kg in 1985 in the United States, to 9 kg in China, 2.5 kg in India, and smaller amounts in some tropical countries (144). In order to meet the growing demands for paper of increasing quality and in greater yields, a variety of chemical pulping processes has been developed, together with improved bleaching methods. Furthermore, new, less suitable pulpwoods are being used when pulpwoods of the desired quality are not available in sufficient quantities to meet needs. With these changing processes and resources, the presence of extractives has been revealed in different ways.

Extractives affecting pulp and paper manufacture can occur in all the anatomical elements of wood, resin canals, cell walls and, as well, in the pockets and veins in the wood that have resulted from injury to the living tree. Both the nature and amounts of extractives vary from botanical family to family, between species, between sapwood and heartwood, between injured and undamaged wood, and sometimes between the anatomical features. Different classes of extractives have different chemical effects during and after pulping; some result in colored specks in the pulp, others cause discoloration, and still others result in sticky deposits in the pulp and/or paper mill or in the paper sheet. The viscous lumps that accumulate on equipment and the specks in paper, which appear oily, or which soften or melt on heating (see Sect. 9.3.7.2) are referred to as pitch; they contain considerable amounts of wood resin (110).

Almost all thin-walled parenchyma cells, both horizontal and vertical, contain extractives, and in some heartwoods they may be completely filled. In heartwoods containing large amounts of extractives the thin-walled vessels may also contain extractives. The vessels are larger, sometimes much larger, in size than the parenchyma and may contain a single compound (82). In some species with very large amounts of extractives (e.g., *Eucalyptus marginata* (72)), the fibers may contain extractives in their lumens. In normal heartwood, the fiber walls, which are thick-walled in comparison with the other elements, also contain polyphenolic extractives. In addition, some species contain specialized cells forming latex and resins (35), and the conifers contain resin canals with extractives different in composition from those in the parenchyma of the same species. The resin canals are completely ringed by thin-walled epithelial parenchyma cells exuding resin into the canal, which, in the living tree, is under pressure so that resin exudes when a canal is ruptured.

During pulping, the resin canals, if present, readily rupture and the bulk of the resin that is not within a cell wall is released directly into the pulping liquor; this can cause pitch problems at later stages. Also, resin, gum, or kino pockets and veins are opened up during chipping or pulping to release their contents. The thin-walled vessels may be ruptured during pulping, beating, etc. The parenchyma tend to remain as groups of cells, so that their contents may be squeezed out during beating or in similar situations. Alternatively the parenchyma may continue to contain their extractives so that they appear as colored specks in the pulp or paper.

In the past, when softwoods were the source of most pulpwood, the major problems caused by extractives were those of pitch, largely resulting from the resin and fatty acids and unsaponifiable materials. Next came the problems caused by the discoloration of pulp due to the presence of polyphenols in hardwoods, notably eucalypts. Now with the increasing use of relatively unfamiliar tropical hardwoods, some of which contain large amounts of extractives, new problems are arising due to polyphenols and a group of resins different from those previously encountered. Differences in their behavior include difficulties in their removal by bleaching and in their removal from pulp mill equipment (e.g., 99). New methods need to be devised to minimize their influence and overcome these difficulties.

9.3.2 Pulping Processes

The major difference between the various processes for the manufacture of pulp for paper and paperboard is the method used in the first step of pulping. Pulp is produced with the application of mechanical, chemical, or heat energy, or a combination of all three, to the wood. To a large extent the types of energy used determine the yield and the properties of the pulp. The amount of mechanical energy required to pulpwood and to separate fibers can be reduced by the increasing use of chemicals. However, the 92% or greater yield of mechanical pulp from a conifer of low extractives content can be reduced to 60% or less by the use of chemical pulping. At the same time some strength properties will be improved and the ease of bleaching to a higher degree of brightness (or whiteness) is increased. The high proportion of the structural components of wood retained in high-yield pulps also contains significant proportions of the extractives originally present, particularly if they are water-insoluble and solid. Not only will they affect the color but also the surface properties of the fibers and resultant paper.

9.3.2.1 Mechanical Pulps

Mechanical pulps contain substantially all the lignin originally present in the wood. Considerable mechanical treatment is required to separate and prepare the fibers from wood (135).

Fibers and fiber bundles are ripped or pulled off billets of wet wood pressed against the peripheral surface of large ($\approx$ 2-m-diameter), wet, rotating, abrasive "stones" in the presence of large volumes of water to form a slurry of groundwood in yields of about 92% or more. Groundwood is used in the manufacture of newsprint and other related products. The yield of groundwood and other mechanical pulps is directly affected by the amount of hot-water-soluble extractives present. Refiner mechanical pulps with similar properties are produced by processing wood chips between parallel metal discs ($\leqslant$ 160 cm diameter) with fluted surfaces positioned closely together. Some of these attrition mills or disc refiners have only one rotating disc, but in the majority the discs are rotated rapidly ($\leqslant$ 1600 rpm) in opposite directions, and the chips are moved by a screw-feed mechanism into the center of the refiner. On their way to the periphery the chips are reduced to fibers. There is a certain amount of shattering of the fibers in these processes. During the grinding or refining processes, resin is released from the parenchyma and ray cells, resin canals, and fibers. (Resin canals are present only in softwoods and some *Shorea* spp. (20)). This high temperature and shear in the pulping process reduce the resin to colloidal-size particles and disperses them into the pulp slurry.

Increasing amounts of mechanical pulps are being produced by the thermomechanical pulp (TMP) process. The principal difference between this process and simple refiner pulping lies in the chips being preheated to 120°–125°C with steam before they are passed under pressure through the refiner. Both steam pressure ($\leqslant$ 700 kPa) and temperature (100° to 170°C) in the refining stages soften first the hemicelluloses and then the lignins (76) to facilitate fiber separation; they also have a pronounced effect on the resultant pulp properties. Together with high yields, these pulps generally have a slightly lower brightness but a significantly higher strength than those of refiner groundwood and stone-groundwood pulps. However, the TMP process requires about 25% more energy to produce pulps than does the stone-groundwood process.

The pulp for hardboards and similar panel products can be prepared by passing chips through refiners, or by heating them with water under elevated temperatures and pressures and suddenly releasing the chips to atmospheric pressure.

Unlike softwoods, young *Eucalyptus* spp. and most other hardwoods do not readily yield acceptable pulps by grinding or by refiner-mechanical or thermomechanical pulping. However, when the billets from young eucalypt trees are impregnated with dilute alkali by the Groundwood Impregnated Billet (GIB) process before stone-grinding, they give pulps at lower energy cost, but of equal or superior strength than (*Picea* spp.) spruce pulps (135). Furthermore, chemithermomechanical pulps from hardwoods, which unlike TMP, are of acceptable quality (68, 71).

9.3.2.2 Semichemical Pulps

Semichemical or chemimechanical pulping procedures involve a mild chemical treatment aimed at weakening the lignin-carbohydrate complex of the fiber bond. The mechanical stage achieves the fiber separation to produce the pulp, which is followed by refining to develop improved properties. The yields of pulp obtained are between 65% and 90% depending on the extent of chemical treatment and on the wood species.

When hardwood chips are given a short pretreatment (0.5 h at 60°C) with dilute sodium hydroxide, the secondary wall swells. The lignin network limits the swelling so that the stress resulting from the swelling differential results in the shedding of the highly lignified middle lamella and the primary wall and the S1 layer of the secondary wall of the fiber in its passage through the disc refiner. The pulps from this cold soda semichemical process have higher strengths than mechanical pulps and are obtained with lower energy requirements. Shive-free pulps can be obtained with little fiber fracture and with yields of 60% to 80%. Pulps with higher brightness can be achieved by using sodium sulfite.

Pretreatment of chips with sodium hydroxide, sodium sulfite, or other chemicals, combined with preheating to 120°C before being passed under pressure through a refiner, will give chemithermomechanical pulps (CTMP) (68, 70, 71). These CTMP pulps require less energy for production than TMP pulps (37, 69). If the pulp from these processes is to be used in newsprint, magazine grades, and light-colored carton boards, then the processes are limited to pulpwoods, which, among other properties, can be brightened with small amounts of low-cost bleaching agents such as calcium hypochlorite.

During the chemical stage of semichemical pulping various wood constituents are removed to a greater or lesser extent, but far less than in the chemical processes. In the neutral sulfite semichemical (NSSC) process, the highly lignified middle lamella is partially dissolved and the fiber bond thereby weakened. The hemicelluloses in hardwoods are high in pentosans, which are largely resistant to NSSC pulping. The high yields of resultant pulp containing large amounts of pentosans as well as lignins give products with a high degree of crush resistance. In the NSSC process, the wood chips are cooked (e.g., 2 h at 175°C) under pressure with a solution of sodium sulfite (e.g., 15% Na_2SO_3) containing small quantities of an alkali (e.g., 3.5% Na$_2$CO$_3$) to neutralize the acids formed during the digestion of the wood and to maintain neutral or slightly alkaline conditions. Pulps are produced in about 65% to 85% yields and are suitable for use in corrugating mediums, linerboards, and bag and wrapping papers. The addition of anthraquinones and related compounds (0.01% to 0.3% on dry wood basis) increases the effectiveness of NSSC pulping liquors (49). The digestion temperatures needed are reduced to 120°–150°C, and shorter times can be used to produce pulps of improved strength. The NSSC process has been largely restricted to hardwoods because softwoods require long cooking times to produce suitable pulps, and they exhibit much lower crush resistance, making them less suitable for corrugating medium. However, the additions of 0.1% anthraquinone (dry wood basis) significantly increases the delignification rate during NSSC pulping of *Pinus radiata*, and results in a pulp with improved strengths (29).

9.3.2.3 Chemical Pulps

Digestion of wood chips under pressure and temperatures around 150°C with various chemicals reduces the amount of mechanical energy needed to fiberize the chips. The chemicals also remove varying amounts of lignins and hemicelluloses so that subsequent processes can improve strength, as well as color and brightness.

The formerly important sulfite process utilized a mixture of sulfurous acid and calcium, sodium, magnesium, or ammonium bisulfite. The process worked well with genera such as *Picea*, *Tsuga*, and *Abies*. However, the need to recover cooking chemicals and process heat, the need to reduce effluent problems, and the inability to pulp highly resinous pine species has reduced its importance. The process is still employed, but mainly with ammonium or magnesium bisulfites for the production of pulp for fine grades of paper. Its use is now only about 10% that of kraft (sulfate) pulping. However, it is still used for specific purposes and a recent large expansion of a sulfite pulp mill was made on the basis that magnesium-base bisulfite pulping gave a stronger, brighter pulp at a higher yield with substantial saving of energy than ammonium-base sulfite pulping. Also, the pulp had better forming properties than kraft pulp for the same purpose, the pulp brightness was higher than that of kraft, bleaching was not required with the pulpwood used, and the capital costs of the plant were lower (47).

In the kraft pulping process, chips are heated with a sodium hydroxide (15% to 30% based on wood) liquor containing sodium sulfide at 160% to 170°C for 2 to 3 hours. The yield of pulp is low and unlikely to exceed 55% with softwoods. The widespread use of the kraft process is partly due to the fact that it produces the strongest pulps available and that the high proportion of lignin removed in the process enables the pulps from a number of woods to be readily bleached. Also the fibers are flexible. During the digestion period, terpenes are distilled if pine chips are being pulped. The spent cooking liquor is concentrated and burned to raise steam and recover the chemicals; sodium sulfate is added (hence the original name of sulfate pulping process) as a make-up chemical before burning. Massive capital equipment is required for the recovery of the inexpensive chemicals used for pulping. However, the burning of the spent liquor can provide a net source of energy. There are few residue disposal problems although the control of the malodorous volatile reduced sulfur compounds is difficult and expensive. In addition to the advantages mentioned, the kraft pulping has few important parameters for its operation and these can be varied without affecting the quality of the endproduct greatly. Also and more importantly, a wide range of wood species can be pulped by the process.

The soda process uses 18% to 20% sodium hydroxide (based on wood weight) at a temperature of 170° to 180°C for 2 to 3 hours during cooking. This process attacks the hemicelluloses more readily than the kraft process and produces easily bleached pulps in lower yield than kraft pulps, but they are softer and better for certain types of printing papers. The soda process does not produce an odor, but is restricted to use with hardwoods. The spent liquor can be recovered by a wet oxidation system (97). Recently, the pulping process has been improved with the addition of 0.1% anthraquinone (wood basis) to the alkali. Higher yields and higher strengths of pulp are obtained, notably with softwoods (152). With *Pseudotsuga menziesii*, the soda/anthraquinone pulping process removes middle lamella lignin more extensively than either the kraft or soda pulping procedures and results in fiber liberation with less lignin removal (151). The addition of small amounts of ethylenediamine to soda/anthraquinone cooking liquor gives pulps with better tearing strength than when it is absent (96).

9.3.3 Pulpwood Quality

9.3.3.1 Effect of Extractives on Pulp Yield

Naturally, the pulp yield is reduced in relation to the amount of non-structural material present in the wood. If the extractives content is high and the lignin content is normal, the pulp yield will be correspondingly lower. The basic density of the wood and its extractives content are the two most important parameters for predicting the yield and quality of kraft pulp from eucalypts (57, 67). If the extractives − such as terpenes − are inert to the pulping conditions, there may be no other effect. However, most extractives react with pulping chemicals, and occasionally with the cell wall components, so as to affect further pulp yield and quality (see Sect. 9.3.4).

9.3.3.2 Effect of Storage

The increasing production and transport of wood chips rather than pulp logs as a source of pulp, the increasing proportions of sapwood of different species in these chips, and the storage of living parenchyma in the sapwood under aerobic conditions can lead to the formation of abnormal extractives. In addition, these conditions can result in abnormal discoloration of polyphenols, the oxidation products of which are much darker than the original substances. These changes, however, are likely to be less severe than the microbial deterioration of the wood chips. In turn, the storage of chips rather than pulp logs is more likely to result in discoloration, even when contamination with metal ions is excluded.

The color changes that occur in a species are not always uniform. Normally the groundwood billets from logs of *Eucalyptus regnans* and *E. delegatensis* are cream in color. However, logs from variants of these species when cut open become pink or red within a few minutes and after a few days become dull brown and darker in color than the normal billets. The changes could be due to the presence of proanthocyanidins.

Changes of the physical and chemical composition of chips of aspen wood (*Populus tremula*) during storage depend primarily on the storage time. If the purpose of storage is to reduce pitch problems, it should be done in self-heating piles for 1 to 1.5 months. Changes in the wood color depend on the felling time, evidently because of changes in enzyme activity and oxidation of the extractives (107). To reduce the pitch problems caused by kraft pulping of softwoods, the chips should be stored for more than 10 to 15 weeks before cooking and the level of calcium ions in the process water kept at a minimum (23) (Sect. 9.3.6.2).

9.3.4 Increased Consumption of Pulping Liquors

The alkali in alkaline pulping processes is consumed by reaction with the lignins, by degradation of the carbohydrates, and by reaction with organic acids and the extractives present in the wood. The amount of alkali consumed by the lignins and carbohydrates is largely dependent on the conditions of the pulping procedure. On the other hand, acidic and polyphenolic extractives react rapidly with the sodium hydroxide so that the amount of alkali available for delignification is reduced. There is a limit to the amount of extra alkali that can be added to alkaline processes to handle large amounts of extractives and be economically and efficiently recovered.

Conifers and some hardwoods containing significant amounts of resinous components are best pulped by the kraft process. The free resin acids and fatty acids are neutralized by the alkali, their esters are saponified, and the alkali is lost in the tall-oil recovery system.

The hot-water solubles content or the hot dilute alkali (0.1 N NaOH) solubles content have been used as indexes of the amount of alkali-consuming polyphenols present. The values are only comparative, as the materials removed contain not only polyphenols but also organic acids and the simple carbohydrates. Nevertheless, useful information is obtained. In a study to determine the cooking conditions that would yield pulps of specified lignin content from pale-colored eucalypts, a relationship was found between hot-water solubles and the amount of alkali consumed (31):

alkali requirements (%) = $0.504 \times$ hot-water solubles (%) $- 16.77 \times \log_{10}$ permanganate number $+ 35.62$.

Furthermore, it was found that the relationship between hot-water solubles and alkali requirements for a specified lignin content in the pulp was less accurate. From this behavior there were indications of a rapid retardation of delignification when the hot-water solubles exceeded 7% to 8% (see also 54). A similar conclusion was reached with the red-colored *Eucalyptus camaldulensis* (21).

The effect of including bark in pulpwood varies according to the species. The bark content in the chips of *Pseudotsuga menziesii*, for semichemical pulps, has less effect on pulp yield and strength than other variables (28). On the other hand, the alkali consumption in kraft cooking increases with increased properties of bark in the wood chips of *Pinus*, *Betula*, and *Picea* species. The effect is particularly marked for stemwood bark and results in a higher Kappa number (lignin content) and impaired pulp drainage (59).

9.3.5 Effect on Pulping Processes

9.3.5.1 Reduced Penetrability of Liquors

The first step in the chemical pulping of wood is the movement of the cooking chemicals into it by one of two processes: first, diffusion of the dissolved chemicals into the water already in the wood under the influence of a concentration gradient; second, movement of the liquor itself into the wood voids under the influence of hydrostatic pressure. This latter penetration is controlled by the operating conditions and the penetrability of the wood. Penetrability is partly controlled by the diameter of the pits and this is characteristic of the genus to which the species belongs. Also, at conditions close to neutrality, the heartwood is less readily penetrated than sapwood owing to the closure of the pits in some cases, incrustations in the pit chamber, or the covering of the chamber by extractives in other cases. The extractives accumulate in the elements that are used first for penetration and, in hardwoods, tyloses reduce the penetrability of the vessels.

The penetration pathway of pulping liquors into hardwoods is first through the vessels, and then through the pits to the ray cells, vertical parenchyma, and the fibers. Penetration of softwoods follows a similar course through the open ends of the tracheids and then to adjacent cells through the pits. The rays facilitate lateral penetration, and the pits provide channels to the lumen of adjacent cells. The liquor then diffuses from the lumen of the fibers through the secondary wall to the highly lignified middle lamella, which is dissolved, and fiber separation commences (183). Pulping liquors dissolve and react with the extractives during their passage through the wood. Consequently, a weakened and chromogenic liquor initially reaches the fiber wall to penetrate and commence delignification. Its effect will depend on the amount and the nature of extractives with which it has previously reacted. However, the effect of extractives would depend partly on the amount present. In the case of an European pine (probably *Pinus sylvestris*), the different levels of extractives content in sapwood and heartwood did not influence kraft pulping significantly, although density differences did influence liquor impregnation (142).

9.3.5.2 Reduced Lignin Solubility

Polyphenols in extractives can condense with lignin when wood is heated to high temperatures (25). Polyphenols may combine with lignin in kraft pulping to result in larger fractions of higher molecular weights, as in the case of *Eucalyptus marginata* (130).

The inhibition of acid pulping of pines containing the stilbenes pinosylvin and pinosylvin monomethyl ether has long been established (46, 60). The resorcinol moiety of the stilbenes condenses with the lignins to form insoluble products under the acid conditions of the normal sulfite process or in the living tree (60). The problem is reduced by a two-stage sulfite process in which the first stage is kept at pH 4 or above to minimize the lignin-phenol condensation. This stage is followed by a more acidic stage to sulfonate the less readily sulfonated groups in the lignin so that it may be dissolved (54).

Similar difficulties would be encountered using the acid sulfite process on other woods containing stilbenes or polyphenols such as proanthocyanidins containing resorcinol of phloroglucinol moieties. For example, dihydroquercetin has been identified as the cause of inhibition of calcium base sulfite pulping of *Pseudotsuga menziesii* (138).

9.3.6 Effect on Equipment During Pulping

9.3.6.1 Wear and Corrosion

The wear and corrosion of pulp mill equipment (including chipper knives) is due not only to mechanical abrasion, but also to the acidity of the wood and to the chelating materials in the secondary components.

The acetyl groups present in the galactoglucomannans of the softwood and in the xylans of the hardwoods are slowly and autocatalytically converted into acetic acid within the living tree. The acidity of the heartwood is therefore often high in older trees, especially in hardwoods. Wood in the middle of old, living eucalypt trees has been reported to have a pH value as low as 2.6 (C. M. Stewart, personal communication, 160), although values of pH 3–4 are more common. Storage of undried hardwood and softwood chips or wood at temperatures up to 70 °C (and above) also increases acidity (50, 62, 134).

Under conditions conducive to electrochemical action, such as with undried wood, the major part of the total wear of cutting tools is due to corrosion (79, 106). The extent of the electrochemical action can be influenced by the amount and type of tool binder material. Nickel is a better binder for cemented tungsten carbide than cobalt, and iron is an unsuitable binder (106).

It has been claimed (106, 182) that the higher acidity of incense cedar (*Calocedrus decurrens* ≡ *Libocedrus decurrens*) relative to ponderosa pine (*Pinus ponderosa*) is responsible for the more rapid wear of high-speed tools. While acidity would be a contributing cause, the presence of compounds capable of chelating with metals would probably be a greater contributor to the extent of corrosion. Tropolones, which chelate with a range of metals (53) are present in incense cedar but not in ponderosa pine. The tropolones in western red cedar (*Thuja plicata*) preferentially remove cobalt from cemented carbide tools (86). The vicinal trihydroxy phenolic moieties in the ellagitannins and proanthocyanidins of eucalypts rapidly etch and corrode iron cutting-tools and iron equipment (79). Corrosion during cutting can be reduced by the application of a low voltage difference across the cutting face (5, 87, 101, 102, 182).

The volatility of acids and particularly of tropolones (such as the thujaplicins in *Thuja plicata*) results in corrosion of the dome of digesters and upper parts of pulping equipment (54, 95). The practice of steaming chips of some species before the addition of pulping liquor in order to improve penetration could lead to increased corrosion of equipment. As has been pointed out above, the chelating ability of vicinal di- and tri-hydroxy phenolic groupings accelerates corrosion. Chelates of tropolones are unstable in alkali whereas some of those of phenols (such as dihydroquercetin in *Pseudotsuga menziesii*) are stable in hot alkali. Corrosion of kraft digesters has been encountered with *Pseudotsuga menziesii* (95) and particularly with certain eucalypts. With species containing gallic acid derivatives, such as some eucalypts, severe loss of color and brightness occurs with iron equipment so that stainless steel must be used for all pulping equipment (54).

9.3.6.2 Blockage and Deposits

A variety of blockages and deposits are observed in different parts of pulp mills and are of a granular to viscous, sticky nature. The deposition of small amounts of these substances on equipment can be handled, but the use of certain pulping procedures with some species can lead to rapid accumulations that will cover the surfaces of various parts of the equipment of both the pulp and paper mill, as well as becoming entrained in the pulp. Some of these deposits are due to the addition of de-foaming agents, coating components, and so on, but only those caused by wood extractives are considered here.

Wood resin is the water-insoluble material that is soluble in neutral, non-polar organic solvents. Large amounts of it pass through mill systems that utilize softwoods and some of the hardwoods. This resin can cause pitch problems, particularly with sulfite pulping and the manufacture of unbleached sulfite pulps and to a lesser extent with mechanical and chemi-mechanical pulping. Problems are less frequently encountered with kraft pulping and usually they are less severe and of a different type.

The accumulation of very small amounts of resin can result in blockages causing a shutdown of operations. These blockages have long been a serious problem in the industry and are responsible for reduced levels of production, higher equipment maintenance costs, higher operating costs, and an increased incidence of quality defects. Various pitch control measures are taken to reduce the level of resin in the system, to stabilize the hydrophobic colloidal system, or to fix the pitch onto the fibers at the time of their release (8, 11, 18, 30, 112).

In the woods of *Pinus*, *Picea*, and *Larix* genera and *Pseudotsuga menziesii*, the resin is contained in resin canals (which are absent from hardwoods, *Tsuga*, and *Abies* spp.) and, as well, in the parenchyma cells. Depending on whether the parenchyma cells are in the sapwood or heartwood, the resin composition differs and it differs in both cases from that of the canal resin (110).

During pulping or refining, the canals and parenchyma cells can be ruptured to release and disperse spherical particles of wood resin. These particles, which can be small in size, then coalesce into larger droplets of pitch which deposit on the surface of fibers or equipment or remain suspended to be discharged in effluent or wash water. Resin acids may not appear in deposits from mechanical pulps in the same proportion as they appear in wood because they are slightly soluble in aqueous systems, are less prone to the formation of calcium salts, and have higher melting points than fatty acids (42). Most of the problems caused by pitch from mechanical pulp are found in the paper mill rather than in the pulp mill.

The chemical methods to minimize pitch problems aim to control the colloidal resin by flocculating the particles with alum, by adsorption of pitch droplets onto the surface of finely-divided talc, or by dispersing the colloidal resin in suspension so that it is less likely to deposit. One possible way to minimize the potential for mechanical pulps to cause pitch problems is to reduce the resin in the pulp by washing in the pulp mill.

When the movement of a free-flowing suspension of pulp containing colloidal resin or pitch is abruptly changed by a stirrer or by encountering a wall of a container, the particles stick to give a tacky deposit at those points of hydrodynamic shear. The deposits build up also on the edges and slots on suction boxes and can accumulate dirt. Subsequently the deposits separate from metallic surfaces as lumps of various sizes to come to rest on pulp webs or equipment. In addition, because the density of wood resin is less than that of water, it can float to the surface of stored pulp suspensions (or of "white" water), form a "bath-tub ring" on the walls of the containing vessel, and break off at intervals. Reducing entrained air bubbles in the stock water reduces the tendency of resin to float and hence reduces pitch problems. Pitch can accumulate on grinder stone surfaces and the problems are accentuated by beating. Pitch can block centricleaners, deckers, and screens, and paper machine wires and felts. The accumulation of pitch from the bisulfite pulp of *Pinus radiata* was more rapid on polyvinylchloride plastic than on metal (181). Pitch flecks can occur randomly in paper and affect its quality and strength (see Sect. 9.3.7.2). The pitch problem is worst in non-kraft mills when fresh wood is pulped and when there are more glycerides and stearyl esters and fewer fatty and resin acids present. As would be expected the problems are worst when the concentration of colloidal pitch is high, and this situation arises more readily with some softwoods than others. The problems occur to a lesser extent with some hardwood genera such as *Betula*, *Populus*, and *Tilia* (30, 110).

Various published methods for measuring depositable pitch rely on the amounts collected on stirrers, beaker walls, etc., but are often unreliable and difficult to interpret. The most commonly used and most reliable method is that of Allen (7) who developed a colloidal pitch measurement technique using a hemacytometer to measure pitch particle concentration. It is a valuable method for monitoring pitch levels on paper machines but does not quantitatively measure them. Pitch levels have been determined by solvent extraction (103) and by adsorption onto XAD-2 resin (149) but these are not entirely satisfactory. A rapid method has been developed for the determination of the amount of extractives absorbed onto talc by an oxidative chromic acid technique (58, 91). Most pitch deposition occurs where there is a drop in zeta potential, which is a measure of the charge on the colloidal particles. Since unlike charges repel, the closer the zeta potential is to zero the more the particles are likely to agglomerate. The greater the ionic charge is on the ions added the greater are the drop in zeta potential and the rise in depositing tendency. Trivalent ions, such as those of aluminum, would have about 50 times the depositing power of divalent ions, such as those of calcium or magnesium, and they, in turn, about 50 times that of monovalent sodium (181).

The terpenes present in softwood canal resins readily undergo condensation and conversion reactions. Under the conditions of sulfite pulping they are converted to a number of compounds that increase the stickiness of the resin or pitch (14, 30, 110). With kraft pulp mills the pitch is considerably less sticky than in non-kraft mills. The deposits are frequently gray or brown to black and are found mainly in the brownstock screen room. The differences in characteristics of the pitch are due to the variable degree of saponification of the components during the alkali pulping. The brownstock pitch contains about 20% wood resin mainly composed of the neutral and unsaponifiable material. Woods with relatively large amounts of the latter, such as aspen (*Populus* spp.), are more likely to have pitch problems than other species (13). In addition, the production of the carbonate ions during the decomposition of certain hemicelluloses under kraft pulping conditions reacts with calcium ions to produce calcium carbonate particles (6). Consequently, the pitch deposits contain colloidal hydrolyzed wood resin, calcium carbonate, and the calcium soaps of fatty and resin acids. Kraft lignin depresses the deposition of calcium soaps by tying up calcium ions, thereby inhibiting the formation of calcium soaps, and by dispersing them (43, 44, 94). One method of control is to ensure that the residual effective alkali concentration in the black liquor remains sufficiently high. Generally, pitch problems are less severe in kraft pulping systems than in other systems.

Wood resin contains a mixture of relatively non-polar, oleophilic, water-insoluble, and low-molecular weight components which, when dispersed in water, tend to coagulate (30). The composition of this resin can influence its "plasticity" and "stickiness" or "tackiness" and tendency to form pitch. A resin with a low plasticity will be less likely to coagulate into globules than a soft resin. Sticky resins will tend to hinder coagulated globules from being redispersed by physical forces, and will adhere to the surfaces of papermaking equipment or fibers.

The four main components of wood resin from most species are the resin acids, free fatty acids, combined fatty acids (esters), and unsaponifiable sterols, alcohols, terpenes, and waxes. (The resins from tropical hardwoods (e.g., *Shorea* spp.) can be mainly terpenoid and steroid compounds (e.g., 88, 89, 156, 157, 184, 187) and the cause of color reversion and resin spots (156, 157) (see Sect. 9.3.7.2).) As mentioned, the composition of canal and parenchyma resins differ, and in living *Picea* and *Pinus* spp. woods, levopimaric acid predominates in the canal resin over abietic, neoabietic, and palustric acids. The resin acids are the most brittle or hardest, whereas the others range between the fluid and solid state. Consequently, variations in composition will affect the plasticity or stickiness of the resultant pitch (30). Deposits at different points in a paper mill processing thermochemical pulp made from *Pinus radiata* had very different ash, extractives, and resin acid contents. The fatty acids and particularly the resin acids reacted rapidly when heated to polymerize to new compounds of higher molecular weight (84). After sulfite pulping, the pitch-forming potential was affected most by the content of "liquid fatty acid" and least by the unsaponifiable material (55). On the other hand, other workers have found the ratio of components to be important and abietic acid to contribute significantly to pitch difficulties (113, 159).

Because of the wide range of problems, there is no universal panacea for controlling pitch problems. Pitch control agents must be applied before agglomeration occurs. The addition of alum (aluminum sulfate) is the most commonly used pitch control agent with the addition of talc being the next. Sequestrants such as sodium hexametaphosphate are used to remove calcium and other metallic ions (33, 34). Dispersants and surfactants act on the surface of colloidal particles to make the surface more hydrophilic and thus stabilize the colloidal suspension (40). The addition of highly charged cationic polymers to pulp fibers increases the adsorption of the anionic resin acids. Alteration of the pH can have an effect depending on the nature of the pulp. Reduction of pH from 10 to 2 for a *Pinus radiata* thermochemical pulp can cause a linear reduction in pitch particle count, but for a bisulfite pulp the effect was negligible (181). Sudden reductions in pH, temperature, and zeta potential can accentuate pitch problems (85, 181). Sometimes removal with organic solvents is practiced, or the use of mechanical means, or the addition of polyethylene glycol (137). Blockages or deposits on surfaces can be removed with organic solvents, by mechanical means, by steaming, or by the use of solutions of alkali and wetting agents. The occurrence of pitch problems has increased with the increasing practice of recirculating water from various parts of paper machines.

The ionizable carboxyl groups in the resin ionize completely in alkaline solution to stabilize the pitch dispersion. However, in the neutral conditions in paper mills, the multivalent metal ions, such as calcium or magnesium, when inadequately washed from the pulp, precipitate the resin and fatty acids as soaps that are difficult to dissolve and remove. Within the normal pH range in the paper mill system, aluminum soaps are charged positively and partly stabilize the dispersion and partly enable the soap particles to be drawn to the negatively charged fiber and thus removed from the system with the paper. Care is needed with the addition of alum, which decreases the pH value and increases corrosion, lowers the strength of the paper, and affects other paper properties. Nevertheless acid conditions of pH 4.5–5.5 can considerably reduce the pitch problem encountered at pH 6–8 and, for that purpose, sulfuric acid alone has been added.

Although the addition of dispersants can prevent the coagulation of colloidal pitch particles, a major difficulty remains in their efficient removal. Alum has long been used for pitch control and retention of pitch on pulp fibers, for pH control, and for drainage of the pulp, although it can have undesirable side effects such as flocculation of the alum and precipitation of aluminum hydroxide above pH 5. In his study of the role of alum in the control of pitch problems in groundwood from *Pinus radiata*, Johnsson (85) followed the view of Back (17) that aluminum ions prevent the pitch from forming deposits by attaching themselves to the negatively charged pitch particles. The whole positively charged complex is then fixed to the negatively charged cellulose fibers and removed with the paper. In his discussion of pitch deposition in newsprint mills utilizing *Picea* and *Abies* species and its control, Allen (9) proposed a different mechanism. He believed that alum flocculates the colloidal pitch by hydrolyzing to form polymeric aluminum hydroxysulfates, which are insoluble and form gelatinous flocs that entrap resin particles and/or chemically bond to them. These flocs are then attached to the fibers or are entrapped by them.

The effectiveness of talc in removing pitch droplets is due to its high hydrophobicity and thus high potential for attracting pitch. Similarly, the hydrophobic end of surfactants is attracted to the hydrophobic pitch, the surfactant forming a protective hydrophilic layer around the droplet. The latter prevents agglomeration of the pitch particles and they can be washed out on a filter. It is important that talc is added before the droplets agglomerate to a particular size; otherwise the talc is less effective.

The mode of talc clarification differs from that of alum, and laboratory assessments of its effectiveness have been reported (58, 83, 91, 154). Small changes in the fatty acid/resin acid proportions can result in a marked change in the amount of pitch deposited (83). Consequently the determination of pitch particle concentration (7) may be inadequate for predicting pitch problems when the type of pulp, the wood species, and age are fluctuating. Saponified fatty acids are more strongly adsorbed on talc surfaces than saponified resin acids (83). With a newsprint furnish of predominantly *Pinus radiata*, the fatty acids are adsorbed with much greater efficiency than fatty acid esters and resin acids (58). A more detailed understanding of the adsorptions obtained by the application of size-exclusion and reverse-phase-high-performance-liquid chromatographic techniques has shown that dehydroabietic acid was adsorbed more readily than the pimarane type acids, and that in some situations the abietane type acids were not adsorbed at all (150). There was no support for the theory that talc adsorbs pitch particles by a "blotting paper" effect. Rather, pitch components, such as resin acids, are adsorbed onto talc at the molecular level. A significant decrease in the amount of resin acids adsorbed occurs at high pH values (58, 149). A wide range of effectiveness has been found between various talcs, and particle size is most important in this context (58).

In addition to species, growth rate, age of the tree, and location of the material within the tree all have significant effects on resin content and in turn the properties of the pitch. Storage of wood can reduce pitch problems considerably depending on the storage conditions. Except in kraft pulping when it is sometimes necessary to maintain a suitably high resin content to assist tall-oil recovery, it is often best to reduce resin content by appropriate storage before pulping. Reduction occurs more slowly during water storage than by storage on land but much more quickly during storage of wood chips (12). The proportion of free fatty acids increases and deresinification of the pulps can proceed further than with the presence of esters (e.g., 15). Heating *Pinus radiata* chips in air at about 70 °C for 7 days resulted in a 50% reduction in resin acids and fatty acids (65). There are reports that pitch problems resulting from the extractives in *Pinus radiata* are worst when the pulpwood is harvested at the onset of the rainy season and with the increase of temperature in the spring season. Recently developed techniques have been used (116) to analyze the polymeric components in the pitch deposits obtained from kraft pulp mills using mixed hardwoods. The large proportion of high molecular mass substances was found to originate from unsaturated fatty acids or alcohols. It was concluded that long-range storage of woods containing large amounts of these substances could be unfavorable for pulping because of the formation of polymers that can cause pitch deposits.

Deresinification of the raw material on storage and during pulping depends on the nature of both the resin and the wood (10). Pine wood has a high proportion of saponifiable resin components in relation to the unsaponifiable material so that the removal of the latter is facilitated. Furthermore, the parenchyma cells have large pits that enable diffusion of the resin. Spruce (*Picea* spp.) resin contains a significant proportion of saponifiable material, but the relatively smaller pits hinder the diffusion of the resin so that the pulp is more difficult to deresinate than pine pulp. Birch (*Betula* spp.) sulfite pulp has an unfavorable resin composition and also small pits in the parenchyma; deresinification of fresh birchwood in particular is therefore difficult to achieve (13). Recently, the resin content of spruce (*Picea abies*) pulp has been reduced economically by passing the sulfite pulp through a screw defibrator after the addition of sodium hydroxide (16). With an understanding of the nature of the resinous extractives, it may be possible to select the cheapest of possible methods for the reduction of pitch troubles. Channelling the most troublesome wood to a kraft pulping process or to proper storage and seasoning of the wood could be the most economical methods available (8).

Kraft pulping lowers the intensity of the pitch problems where they may be of two types — the organic pitch type composed of colloidal wood extractives or the inorganic pitch from suspended material. With kraft pulping the glycerol esters are completely saponified and the fatty and resin acids dissolved; the sterol and terpenoid esters are the main components of the extractives remaining in the pulp. If calcium ions are present, calcium salts of the fatty and resin acids may be formed, and, if they do, the viscosity of the pitch deposits increases (167). Calcium ions are released rapidly from the wood in the early stages of kraft pulping and later from calcium carbonate particles. These adhere to pitch deposits that can contain over 50% inorganic material (1).

It has been found that pitch problems and papermill slime can be interrelated. A large proportion of the mass of pitch deposits can be made up of microorganisms. The latter attach themselves to various surfaces in the papermill, where they metabolize the pitch and agglomerate and attract further amounts of resinous materials.

Increasing attention is being given to the use of tropical hardwood species as a source of pulp. Some of these potential pulpwood species contain resins of a nature different from those encountered in temperate zone species. *Shorea* spp. and some other members of the family Dipterocarpaceae contain resins composed largely of terpenes (24), and some dipterocarps leave a white deposit on pulp disintegrators after kraft pulping (38). Kraft pulps from a number of peat swamp species and dipterocarps of Sarawak left black spots on the bars of a phosphorbronze PF1 mill during laboratory beating. The genera included *Calophyllum, Cratoxylon, Dyera, Garcinia, Goniothalmus, Litsea, Parartocarpus,* and *Tristania*

(38, 92, 93). Methods to control pitch problems in kraft mills pulping tropical hardwoods have been summarized (127).

When woods, such as eucalypt, containing ellagitannins are pulped by the soda or kraft processes, distinctively colored green or yellow-green deposits accumulate at all stages of the process. Heartwoods from some eucalypt species, particularly from mature and over-mature trees, contain large amounts of ellagitannins (54), and it is desirable that chips from such material be maintained in a minor proportion in the raw material. When the amounts of deposits are high, particles may be entrained in the pulp and hence increase bleach requirements. Under pulping conditions, the insoluble ellagic acid present in the wood or resulting from hydrolysis of the ellagitannins forms nearly insoluble complexes with inorganic ions such as magnesium, calcium, sodium, and aluminum (54, 66, 90, 111, 143). The optimum conditions for the formation of the magnesium-ellagic acid complex appear to be pH 9 to 12 and 110° to 150°C (19), which are similar to those in pulping processes. The methyl ellagic acids present in some eucalypts may be even more insoluble (74, 75, 78, 130). The deposits contain largely (about 60%) ellagic acid and adhere strongly to many parts of the plant (78). During the pulping of over-mature eucalypts, deposits have accumulated and blocked discharge pipes of counter-current soda and NSSC cooking systems as well as the tubes of heat exchangers and of evaporators. In an NSSC mill pulping low-quality eucalypts, the deposits on a heat exchanger accumulated at the rate of 1 mm/day. Apart from blocking the flow of liquor, the deposits seriously affect heat transfer rates (19, 54). Furthermore, a green deposit formed on other steel equipment such as flat screens contacted by spent liquor, necessitating frequent cleaning. The stability of the ellagic acid-magnesium complex decreases sharply in the pH 8 to 10 range, and it is interesting that deposits on equipment can be removed with sodium hypochlorite solutions at pH 8.5 to 10 within 10 min. (100) or with strong nitric acid. The rate of decomposition of ellagic acid is inversely proportional to a fractional power of hydrogen ion concentration (143).

9.3.7 Pulp Properties

9.3.7.1 Color Changes Arising During Pulping and Bleaching

9.3.7.1.1 Mechanical Pulps

The hemicellulose and lignin fractions contribute to the discoloration of pulp during its preparation and storage, but that contribution is small compared with the loss of brightness caused by extractives (brightness is a term used in the paper industry to refer to the amount of light reflected by paper). Some of these changes occur when wood surfaces are exposed to various conditions, but mostly they are due to the pulping conditions and the nature of the pulp so that changes are accelerated.

The rate of darkening of some pulp suspensions from different species was found to be approximately in the order of *Pinus, Fagus, Betula, Alnus, Picea, Abies, Larix, Populus, Acer,* and *Tilia* (e.g., 81). Samples of newsprint (made from *Eucalyptus* spp.) exposed to sunlight darken much more rapidly than unexposed samples and those stored in the presence of oxygen and moisture darken more rapidly than those stored in their absence (179). The rate of change can be associated with the amount and nature of the extractives present. Stilbenes (such as those present in pine and a range of hardwoods) darken in sunlight (e.g., 108). The influence of metal ions in this darkening is not known. However, with 3,3'-dimethoxy-4,4'-dihydroxystilbene synthesized from lignin degradation products during bisulfite pulping of *Pinus radiata*, trace amounts of ferric and copper ions accelerate the formation of a red color that becomes brown on exposure to light (80). The importance of polyphenolic extractives, particularly those with *ortho*-dihydroxy groups, in relation to color development has been shown with refiner mechanical pulp from *Tsuga heterophylla* (22). Polyphenols containing vicinal trihydroxy moieties, when absorbed on paper and exposed to sunlight, darken more readily than those containing *ortho*-dihydroxy groups. In addition some classes are less light-stable than others and the prodelphinidins darken more rapidly than gallotannins and ellagitannins. Discoloration was particularly noticeable in newsprint from eucalypt groundwood that contained a significant proportion of the ellagitannins originally present in the pulpwood (179); developments in pulping and bleaching, however, have reduced the problem.

With the preparation of groundwood and newsprint, the presence of metallic ions, such as iron (in particular), aluminum, and copper, must be controlled if the wood or pulp contains polyphenolic extracts. The presence of organic acids, which can be present in significant quantities in hardwoods, can release these ions from equipment and also affect the extent of discoloration. With dilute solutions, the most intense color was formed by ellagitannins with iron at pH 4.5 (179). When stainless steel equipment is used, the particles arising from the erosion of that equipment and much of that originating from other sources do not have a severe effect on brightness. However, as soon as the ferric ion is introduced in impurities, or when iron particles are reduced to soluble ferrous ions during bleaching of the pulp and its subsequent aerial oxidation, dark ferric-polyphenol complexes are formed. Accordingly it is necessary to wash out the soluble ferrous ion at pH 2.5 immediately after bleaching.

Bleaching of groundwood involves the use of low-cost chemicals such as hypochlorites, sodium or zinc dithionites, sodium bisulfite, or peroxides. Chip groundwood produced from *Pseudotsuga menziesii* (Douglas-fir) is darker and more difficult to bleach than refiner groundwood obtained from the more commonly used *Tsuga heterophylla* in the Pacific Northwest regions. *Pseudotsuga menziesii* sapwood yields a relatively light-colored groundwood that is readily bleached with zinc dithionite. The color and bleachability of groundwood from heartwood is inferior to that from sapwood and varies according to the wood source. Quercetin, a major component of the heartwood extractives of Douglas-fir, is highly detrimental to the color, but dihydroquercetin has little effect on bleachability (27, 28). Polyphenolic extractives with vicinal phenolic hydroxyls, such as catechin, also play an important role in discoloration of refiner mechanical pulp from western hemlock (*Tsuga heterophylla*) (22). The yellow color produced during processing of sodium bisulfite and refiner groundwood pulps from *Pinus banksiana* (jack pine) heartwood is caused by the chelation of the flavanone pinobanksin with aluminum ions.

When groundwood is produced from a mixture of three eucalypt species – *Eucalyptus regnans*, *E. delegatensis*, and *E. obliqua* – under alkaline conditions, it is first washed to free it from alkali-extracted material. It can be brightened by bleaching with sodium hydrosulfite (dithionite) at 60 °C for 90 min. While this hydrosulfite brightening is cheap, the color stability of the product is not good, and reversion occurs on standing (136, 179). Hydrogen peroxide is an effective bleaching agent under the same conditions, and if iron and manganese are present above certain levels a stabilizer (sodium silicate) must be added to achieve a stable color (51), and relatively large amounts of sodium hydroxide must be added to optimize bleaching.

9.3.7.1.2 Chemical Pulps

As pointed out in Sect. 9.3.5.1, the movement of the pulping liquor in hardwoods takes place most readily through those wood elements that can contain most extractives. Consequently, the extractives in the vessels and parenchyma have an influence on pulping out of proportion to their amount. This is noticeable in the eucalypts, where the amount and the nature of the extractives have an adverse effect on the color of the pulp, greatly exceeding that of lignin (73). Similar effects have been found with *Pinus banksiana* (148).

In order to utilize the free alkali in black liquor, the latter was used instead of water to dilute white or fresh liquor in the soda pulping of *Eucalyptus delegatensis* ($\equiv$ *E. gigantea*). However, a pulp of inferior color was obtained. The time of contact of the pulp with the black liquor during pulping was a factor influencing bleach demand. The colored materials were highly resistant to oxidation and were not removed either by washing with water or by alkali (39). It was proposed that compounds containing the gallic acid moiety had been oxidized to quinones, and these compounds in the black liquor adhered strongly to cellulose in the pulp (54). The greater amount of dark-colored pulps obtained by pretreating wood chips of *Betula mandshurica* with black liquor has been attributed to interaction of a low-molecular weight fraction of alkali lignin having a phenolic character (185). Lee et al. (89) made a comprehensive study of several *Shorea* spp. of the Dipterocarpaceae, as well as some other tropical species, and associated the presence of extractives with the low brightness of the unbleached and bleached kraft pulps obtained from them. Color reversion of bleached kraft pulp from *Dipterocarpus* spp. and *Shorea laevifolia* has been attributed to residual fatty alcohols, triterpenoids, and steroids in the pulp (156, 157). The brightness of bleached kraft pulp from *Shorea negrosensis* is lowered by neutral extractives (41) – e.g., fatty alcohols, triterpenoids, and steroid alcohols, and the esters of the latter two with fatty acids (88).

Depending on the species, eucalypt woods contain extractives belonging mainly to the ellagitannin or gallotannin class or to the proanthocyanidin type when the major components are usually prodelphinidins. In addition, kino veins or pockets, which are frequently present, contain proanthocyanidins in a concentrated form which are released to the pulping liquor early in the cooking process. At room temperature, the components of eucalypt extractives give vivid colors with differing intensities on mixing with alkali (at pH 9.5 to 10) or NSSC pulping liquors at pH 9.5 to 10 or with kraft pulping liquors (73, 153). In air, the colors are much denser and darker and can be irreversibly retained by the cellulose (73, 158); the color changes occur much more rapidly at higher temperatures. Under the same alkaline conditions the color formed by the same weight of material is most intense with proanthocyanidin polymers (as in kinos), followed by the strong colors of ellagitannins and the weaker colors of gallic acid, catechin, ellagic acid, stilbenes, and milled wood lignin (73). A study on polyphenols (73), such as those found in the extractives of *Eucalyptus* spp., showed that they were primarily responsible for the dark color of alkali pulps, and not lignin as previously thought. When oxidized in alkaline solution by gaseous oxygen under pressure, at adequate temperatures and for appropriate times, polyphenolic extractives changed to light-colored products (64). Attempts to reduce color formation by kraft pulping with the addition of oxygen under pressure succeeded only for the surface layers of the chips. The reduced solubility of oxygen in kraft pulping liquor at elevated temperatures prevented adequate penetration of oxygen. Preliminary impregnation of the chips from unspecified species with reducing and oxidizing agents before kraft pulping has been tried. The agents were sodium borohydride, sodium sulfite, sodium chlorite, and sodium peroxide. Sodium borohydride and hydrazine were the most effective in reducing color (109).

Chromogenic properties also differ within a class. Prodelphinidins with their pyrogallol moieties contribute more than procyanidins with catechol moieties, yet

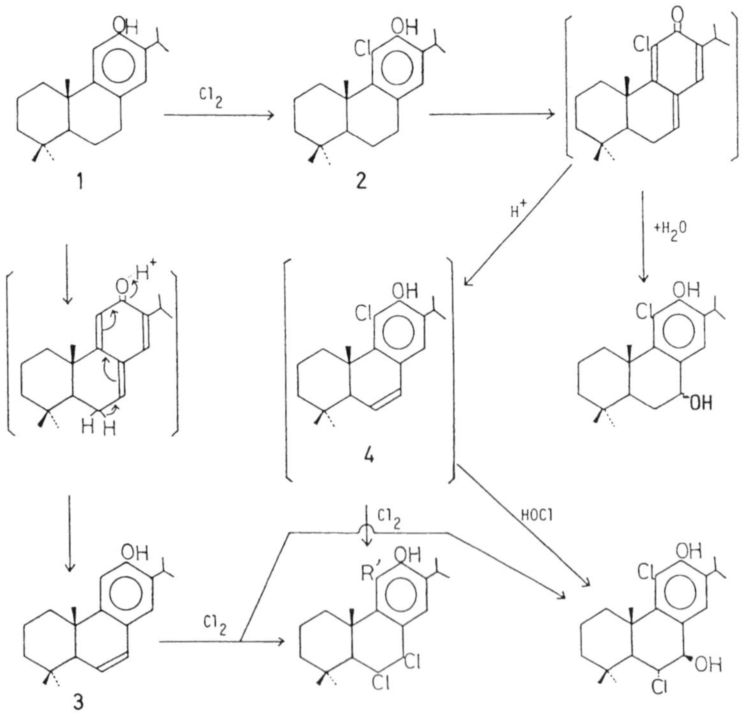

Fig. 9.3.1. Chlorination reaction of ferruginol (1)

catechin (catechol moiety) forms much less color than proanthocyanidins. The components of the ellagitannins also differ in their intensity of color formation in the presence of alkali. The extractives contents of seven eucalypts have been correlated with the brightness of the kraft pulps prepared from them (26). Water extraction before pulping improved the brightness of the pulp and reduced the chlorine demand, but the darkest woods, which contained the largest amounts of polymerized tannins, responded least to extraction. In addition to these variations in behavior, the oxidation products of prodelphinidin polymers are more strongly retained by the cellulose. As the kinos, which result in the strong colors retained by the pulps, are mostly formed as a result of injury to eucalypt trees, the value of good silvicultural practices can be seen.

The aim of bleaching of groundwood (mentioned in the previous section) is to reduce color while retaining lignin. On the other hand, bleaching of chemical pulp involves lignin removal, and multi-stage bleaching has long been adopted by the bleach plants for this purpose. One of the most conventional processes employed in the sulfate mill is the sequence of $CE_1D_1E_2D_2$, where C = chlorine, E = aqueous NaOH, and D = chlorine dioxide.

The careful removal of lignin by bleaching does not always remove the extractives, however. For example, chlorination of unsaturated fatty acids increases the difficulty of saponification on subsequent treatments. In addition the resin becomes more hydrophobic. If the resin acids are chlorinated, their solubility is decreased and their ability to solubilize neutral resin substances such as sterols is reduced. On the other hand, bleaching with chlorine dioxide produces components that are easily saponified with alkali, and the free acids can solubilize some of the other components (13).

A sheet of fully bleached sulfate pulp of *Cryptomeria japonica*, one of the most common plantation species in Japan, develops a slight yellow color. Ferruginol, a diterpenoid cryptophenol of the wood extractives, was shown to be the major cause for the low brightness and intense color reversion. Only about 36% of the ferruginol in the wood is dissolved in the cooking liquor, while the remainder is finely dispersed in the pulp and the discoloration of ferruginol proceeds during bleaching (4). It is noteworthy that ferruginol and its derivatives are distributed among a rather diverse group of conifer species.

In a more detailed investigation, xanthoperol was found to be more harmful than ferruginol in both lowering the brightness and intensifying the color reversion. In *Cryptomeria japonica*, however, xanthoperol is only a minor component and therefore has less effect than ferruginol, the major component.

At the chlorination stage (C), ferruginol (**1**) is converted to 11-chloroferruginol (**2**), dehydroferruginol (**3**), and 11-chlorodehydroferruginol (**4**), followed by the gradual formation of a number of hydroxylated and chlorinated products at C_7 and C_6 (Fig. 9.3.1). The chlorination rate at C_{11}, ortho to the cryptophenolic hydroxyl, is extremely low due to the steric hindrance against the hydrogen atom at C_{11} and the hydroxyl group at C_{12} (120).

When the chlorinated mixture is treated with chlorine dioxide (D_1), a more complex mixture involving the yellow substances – e.g., 11-chlorodehydrosugiol (**5**), xanthoperol (**6**), and 11-chloroxanthoperol (**7**) – is produced (Fig. 9.3.2). Under the conditions, the reaction rates of oxidation leading to the colored sub-

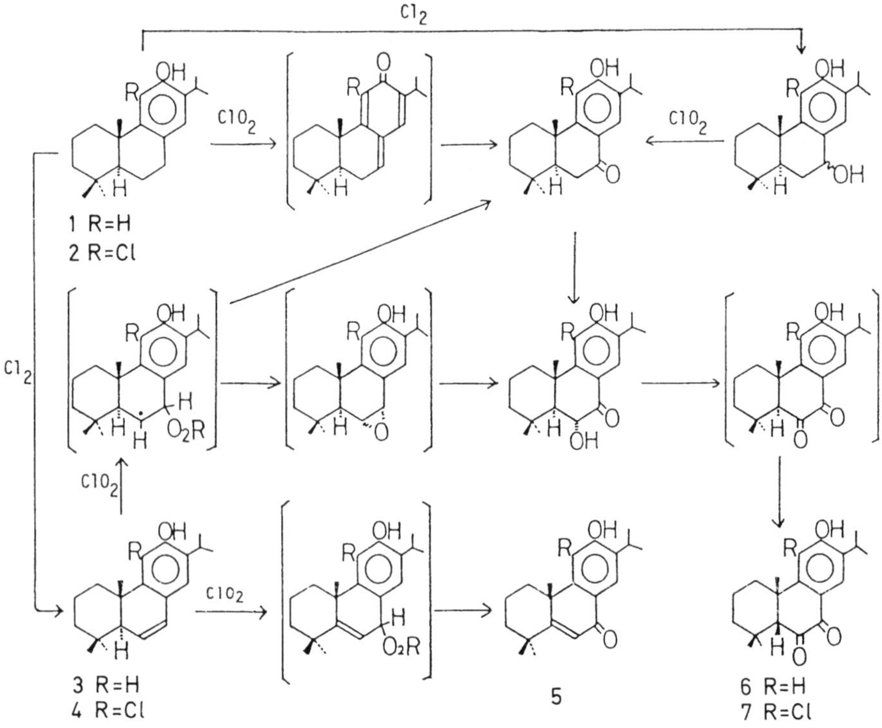

Fig. 9.3.2. Plausible reaction pathways when treating ferruginol with Cl_2 followed by ClO_2

stances are so low that the resulting mixture contains the complex products of various stages of oxidation.

When the repeated treatment with chlorine dioxide (D_2) is performed, a slight reduction in the total amount of the yellow substances is observed. However, most of the coloring materials cannot be removed even after the five-stage bleaching, which involves the repeated alkaline extraction stages (E_1 and E_2) (121).

The difficulties mentioned above can be progressively overcome when an increased ratio of chips of *Pinus* spp. is mixed with those of *Cryptomeria japonica* in advance of cooking (122). Apparently, the resin acids from pine chips behave as a surfactant and dissolve some of the neutral resins as well as the cryptophenols out of *Cryptomeria japonica* chips into the aqueous media (18).

Based on the result mentioned above, a moderate improvement in brightness and color reversion of the bleached pulp can also be attained by using additional amounts of chlorine dioxide at the bleaching stage D_2. Even better results were obtained by adding hydrogen peroxide alone or together with a minute amount of phase transfer catalysts at the stage E_2. The best results, however, were furnished with an oxygen-alkali treatment at the first stage of a multi-stage bleaching (CEDED) (105). The significant effect of the treatment with oxygen-alkali on reducing the extractives content from unbleached spruce (*Picea* spp.) sulfite pulps was reported earlier by several workers (98, 117–119, 145–147). The same was true in the reduction of neutral resin in red lauan (*Shorea* spp.) sulfate pulps (61). Although oxygen bleaching in paper mills has now been introduced in many countries, it is not yet very common.

9.3.7.2 Speck Formation and Pitch Problems During Pulping and Bleaching

A few tropical species produce particular types of extractives that form specks that may resist pulping and subsequent bleaching operations, so that it is desirable to exclude some species from the pulping of mixed tropical hardwoods. The latex of rubber wood (*Hevea brasiliensis*) and the extractives of *Nauclea undulata* (which have similar infrared spectra) produce specks in the pulp and resultant paper. Kraft pulp from the latter species contained over 13% chloroform solubles (70) and soda pulps behaved similarly. Some other tropical species produce resin specks that vary in size, number, and intensity according to the particular species (70). The specks remain or are changed in color after bleaching and the chemical nature of the cause can vary. Some of the species producing specks belong to the *Intsia*, *Hopea*, *Shorea*, *Vatica*, *Maniltoa*, *Artocarpus*, *Neonauclea*, and *Cryptocarya* genera. Attention was drawn in 1967 to the significance of neutral extractives as the cause of pitch problems in bleached kraft pulp from *Shorea* spp. (184).

Two methods for the semi-quantitative measurement of the resin specks have been reported. One method is based on the total number of specks and the other on the determination of the total speck area per pulp sheet of a certain weight. If the speck sizes are similar, the former is a quick and convenient way when the specks are counted under irradiation with ultraviolet light (114). A recent report dealing with pulps from trees of the Dipterocarpaceae adopted this method (186). If the speck sizes are different, the second method of measuring with a stereoscopic microscope ($\times 10$) gives much better results, although it is time-consuming (177). With dark-colored pulps, the accurate measurement of a variety and large number of speck sizes is difficult. The specks can be classified by size into three grades of <0.01, 0.01 to 0.1, >0.1 mm^2. The resin-speck areas in the bleached kraft pulps of several Dipterocarpaceae woods ranged from 0.1 to 250 mm^2/g (89).

Bleaching pulps with either chlorine dioxide (D) alone or with chlorine (C)-alkaline extraction (E)-chlorine dioxide (D) successively can affect the speck area. As mentioned in the previous section, chlorination makes the resins not only more hydrophobic but also causes them to aggregate more. In some cases, this results in a total speck area of 1.5 to 2.0 times higher than that before chlorination. In contrast to this, chlorine dioxide usually results in a lower speck area (2). The pitch problems of the kraft pulp of *Dipterocarpus* sp. can be considerably reduced if the pulp is oxidatively bleached at the first stage. If chlorination is employed, the chlorination product of the terpenoid hydroxydammarenone-II (the main component of the extractives in the pulp) adversely affects the appearance of resin spots and color reversion (157).

Speck formation in kraft (sulfate) pulps can be classified into at least two types. One affords sticky specks with a slight color, relatively low melting point, and moderate solubility in organic solvents. This type of speck often can be seen in both unbleached and bleached sulfate pulps prepared from woods of the Dipterocarpaceae. The other type gives solid specks with a more intense color, high melting point, and high solubility in organic solvents. The latter type appears only in bleached kraft pulps, but not in unbleached ones prepared from woods such as sepetir-paya (*Pseudosindora palustris*, Leguminosae) and rengas (*Gluta* and *Melanorrhorea* spp., Anacardiaceae) (176).

In an investigation of the first or sticky type of resin specks, woods having resin canals of either different length or volume were cooked to compare the speck area in the resulting pulp. The presence of the resins in long vertical resin canals of the wood chips was shown to play an important role in the formation of resin specks of larger size (3). Further investigation of the woods of 24 species of the Dipterocarpaceae grown in Sarawak confirmed the significant role of the natural distribution of resins in the wood. Taking into consideration the average diameter size of vertical resin canals and the manner in which the canals were filled with resins and their distribution in the wood, the incidence of resin (the proportion of canal volume filled with resin in different concentrations) could be classified into four grades that were in good agreement with the order of speck area (89). The relationship between the amounts of bleached pulp extractives and the total speck areas can also be well explained primarily in terms of incidence of resin in the vertical resin canals. This was further supported by a consideration of the percentage of the speck area for each incidence grade in relation to the total speck area.

However, the distribution of resin canals even in the same wood species can be quite different and, furthermore, highly dependent on the diameter size of the wood discs from which the chips were produced (Lee et al. 1987, unpublished data), the reason for the great differences in resin specks found between the results of Logan et al. (92, 93) and those of Lee et al. (89). This is important not only from the biological point of view but also from a practical one – e.g., when Dipterocarpaceae woods are utilized for paper making.

The importance of morphological distribution of extractives has also been shown with *Pseudosindora palustris* when the number and areas of the larger-sized specks increased with an increase in the concentration of apotracheal parenchyma (175). In this species, these parenchyma cells, which existed contiguously in the axial direction in the wood, were considerably larger than the ray parenchyma cells. Moreover, the largest amounts of phenolic compounds that make the greatest contribution to speck area, and the largest amounts of the neutral compounds that have the highest phenol retainability, are found in the apotracheal parenchyma rather than in the ray parenchyma.

Early work (114, 184) suggested that the liquid neutral extractives in *Shorea* species belonging to the Dipterocarpaceae family were highly responsible for the pitch problems in which calcium ions may have played an important role (180). The resin specks first form during kraft pulping, and their total area increases proportionally as the cooking temperature rises from 130° to 170°C (3). The acidic and phenolic fractions the extractives function as agglomeration aids, whereas the neutral constituents act as core substances in the agglomeration of resin (3). The canal resins in *Shorea* spp. contain mainly the neutral constituents.

Another factor responsible for unusual speck formation in pulping is the presence of a compound with a high molecular weight, such as β-resene in the resin

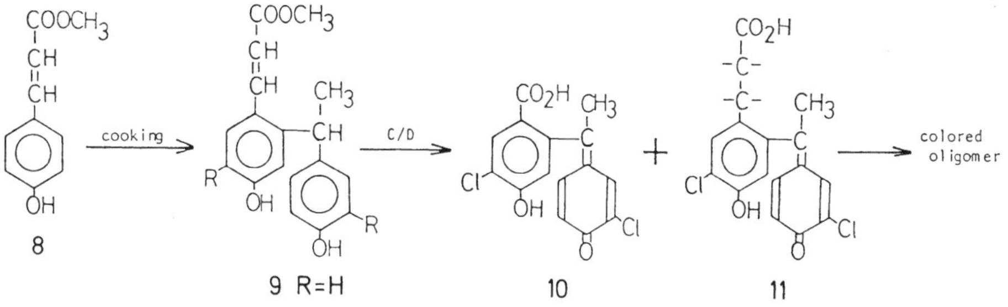

Fig. 9.3.3. Conversion of methyl p-coumarate (**8**) to the colored oligomer

canal (2). Our knowledge of the distribution of β-resene in the woods of Dipterocarpaceae and also its chemical or physicochemical characters is extremely limited. Its molecular weight was suggested earlier to be less than 2000 (104). Recently the presence of two kinds of β-resene with much higher molecular weights has been found. β-Resenes have high melting points and are soluble in benzene or chloroform but not in alcohol. Very recently pyrolysis and NMR studies of β-resene showed that it was made up of a cadalene type unit. This may be the first example of a naturally occurring sesquiterpene polymer, though the mode of conjunction among monomers is yet uncertain (155). These properties indicate that the remarkable speck formation attributed to β-resene accompanied by some other extractives is apparently accelerated with the aid of hydrophobic bond formation between the β-resene molecules. The occurrence of this type of aggregation must be compared with the other type of aggregation of phenols, which is discussed below.

As mentioned, the colored specks from the woods of *Pseudosindora palustris* and rengas (*Gluta* and *Melanorrhorea* spp.) can be found only in bleached sulfate pulps, but not in unbleached ones. Furthermore, the specks appear only when the woods are cooked and subsequently bleached, and do not appear without both procedures. Chemical conversions of extractives to colored specks should therefore involve first a conversion during cooking and then a second one during bleaching (177). Based on this assumption, methyl p-coumarate (**8**, Fig. 9.3.3) was first isolated as one of the major compounds responsible for the specks. The conversion product during cooking is a dimer (**9**) having a diphenyl methane structure accompanied by demethoxycarbonylation (168). When the chlorination product of the dimer (**9**) was treated with chlorine dioxide, the quinone methide intermediates (**10**) and (**11**) with yellow color were obtained (Fig. 9.3.3). Further treatment finally produced the yellow oligomer with the same chromophore at the terminal unit (169).

However, it is unusual that dimerization of methyl p-coumarate accompanied by demethoxycarbonylation proceeds without the compound being dissolved in the alkaline cooking media at 170 °C. This is also the case for the alkaline extraction stages (E_1 and E_2) in five-stage bleaching. Conversion in both cooking and bleaching should therefore occur only in the interior of a neutral and hydrophobic membrane. This was fully supported by a model experiment using lauryl alcohol as a neutral phenol carrier although it has a lower phenol retainability than β-sitosteryl glucoside and other extractives in unbleached kraft pulp (117, 176). Without the carrier, speck area was almost absent. Furthermore, addition of a minute amount of a phase-transfer catalyst, chosen from a range of quaternary ammonium salts (178), to the stage E_2 conspicuously reduced the speck area. The reduction was accelerated when hydrogen peroxide together with the catalyst was added at the same stage (170). It was later shown (175) that the basic fraction of the phase transfer catalyst is, first, to transport the oxidant anions through the hydrophobic neutral membrane so that they can decompose or remove the phenols of *Pseudosindora palustris* and, second, to break down the resin particles and thus enlarge their surface area.

Based on the results mentioned above, an investigation using lauryl alcohol as phenol carrier was undertaken to discover the active phenols remaining in the wood of *Pseudosindora palustris*. Besides methyl *p*-coumarate, five phenols – methyl ferulate, butein (**12**) (Fig. 9.3.4), pseudosindorin, sulfuretin, and rengasin – were isolated as those responsible for the colored specks. Two cinnamates, two chalcones, and two aurones all have a common skeleton of phenyl-α,β-unsaturated-γ-carbonyl or methoxycarbonyl group in the molecules (171).

During cooking, butein (**12**) converts to butin (**13**) and a dimerization product having a dihydrofuranoflavanone type of structure. The latter two are the major intermediates from butein to colored specks (172). On chlorination at 20°C, butin (**13**) affords 5-chlorobutein (**14**), a ring-opening product of 5'-chlorobutin (**15**) in highly acidic media. By the treatment of 5-chlorobutein with chlorine dioxide, a ring-closure to 5'-chlorobutin again proceeds, first in the weakly acid media at 60°C and then its oxidation to the corresponding *ortho*-quinone (**16**) follows. An acid-catalyzed oxidoreductive condensation of the *ortho*-quinone (**16**) occurs via the dimer (**17**) to give the final oligomer (**18**), the speck substance (Fig. 9.3.4) (173).

Rengasin (**19**, Fig. 9.4.5) is not only one of the active phenols in sepetir-paya (*Pseudosindora palustris*) wood but also the sole component responsible for the colored specks in bleached rengas (*Gluta* and *Melanorrhoea* spp.) pulps. As with the other specks, these can also be reduced by application of phase-transfer catalysts at the alkaline extraction stage E_2 (123).

On cooking sulfuretin (**20**), as a model of rengasin (**19**), in water at 170°C, a novel type of compound (**21**) was produced in only 0.3% yield, but in 8.2% yield when 6-hydroxycoumaranone and protocatechuic aldehyde were cooked under similar conditions. This proves the presence of an equilibrium (Fig. 9.3.5). The compound (**21**) is one of the postulated intermediates for the colored specks (124).

When either rengasin alone or both 4-methoxy-6-hydroxycoumaranone and protocatechuic aldehyde were cooked as above, the corresponding product (**22**) was obtained in 0.2% and 6.3% yields, respectively. Isolation of **22** in a 0.0012% yield from the rengas pulp but not from the wood, confirmed the occurrence of the equilibrium involving the true intermediate (**22**) to the final specks (118). The respective yields of the intermediates (**9, 13, 22**) produced by cooking methyl *p*-coumarate, butein, and rengasin in water is not so different from the actual yield of the three isolated from unbleached pulps of sepetir-paya and rengas. This suggests that aggregation tends to occur among the same species of phenolic molecules even at elevated temperatures (163).

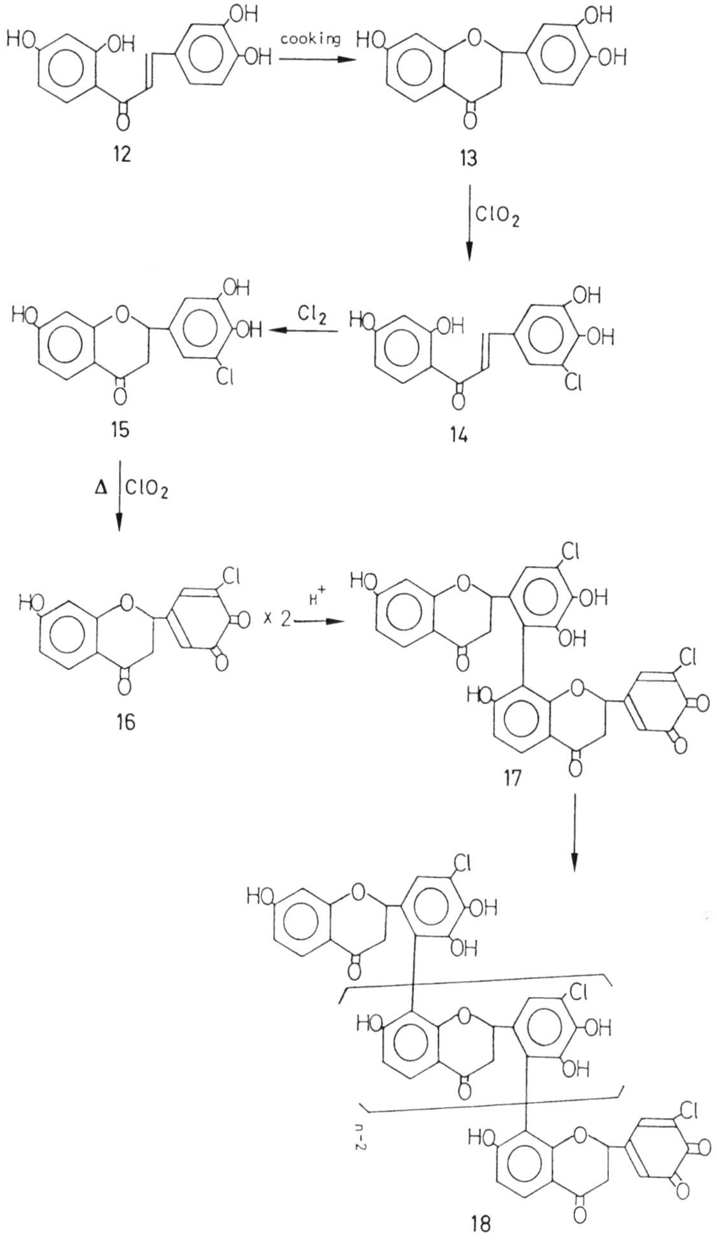

Fig. 9.3.4. Conversion of butein (12) to the speck substance (18)

Equilibrium in the highly acidic media during chlorination of sulfuretin also involved **23** (Fig. 9.3.5). Treatment of **23** with chlorine dioxide affords *ortho*-quinone **(24)** (Fig. 9.3.6), which is further subjected to successive acid-catalyzed condensation via the dimer **(25)** to the colored oligomer **(26)**, the final speck substance (Fig. 9.3.6) (119).

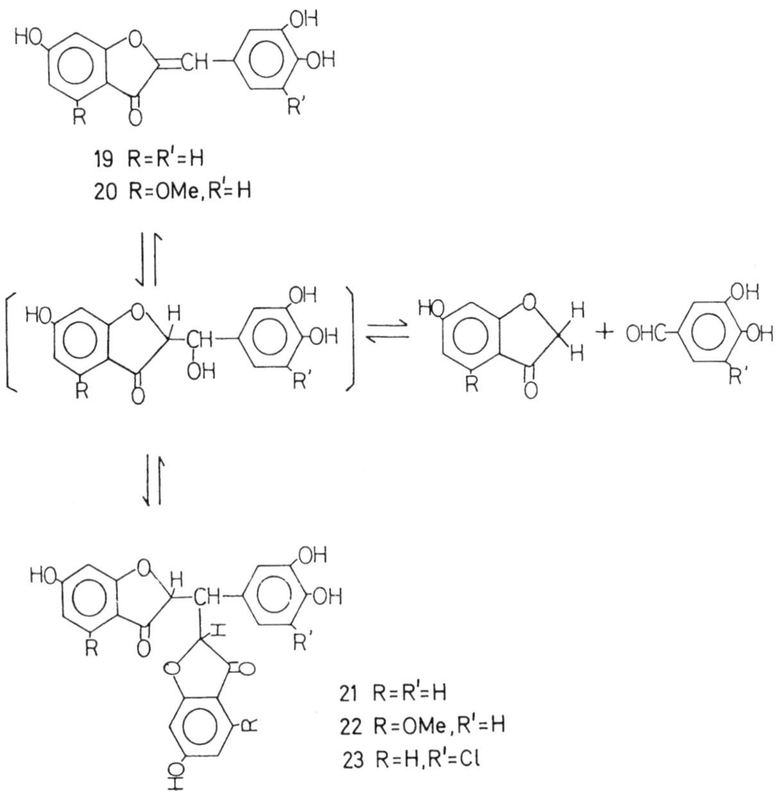

Fig. 9.3.5. Equilibria between aurones (19, 20) and coumarane adducts (21, 22, 23)

Similar to the other phenols responsible for colored specks, the aggregation and the subsequent conversion of rengasin and its derivatives proceeds inside a neutral carrier. The formation, retention, and dispersion of an aggregate may be highly affected by the various forces that occur throughout the whole process from cooking to bleaching. A neutral carrier abundant in polar constituents shows a high capability to retain the aggregate of rengasin, a relatively hydrophilic phenol. On the other hand, the aggregate of methyl *p*-coumarate, a more hydrophobic phenol, can be retained at a higher level than that of rengasin by any neutral carrier almost without regard to the nature of the constituents. Addition of a small amount of fatty acid to a neutral carrier results in a remarkable increase in the speck area due to rengasin, but much less so in the case of methyl *p*-coumarate. In the former, aggregation is probably promoted and stabilized by the formation of a kind of reversed micelle. Aggregation of the latter may be more efficiently supported by the neutral carrier itself rather than by a reversed micelle, indicating that the phenolic molecules of the same species aggregate mostly with the aid of intermolecular hydrogen-bond formation (117).

Based on the above results, the neutral extractives of sepetir-paya wood (*Pseudosindora palustris*) were divided into five fractions: 1) *n*- and *iso*-alkanes,

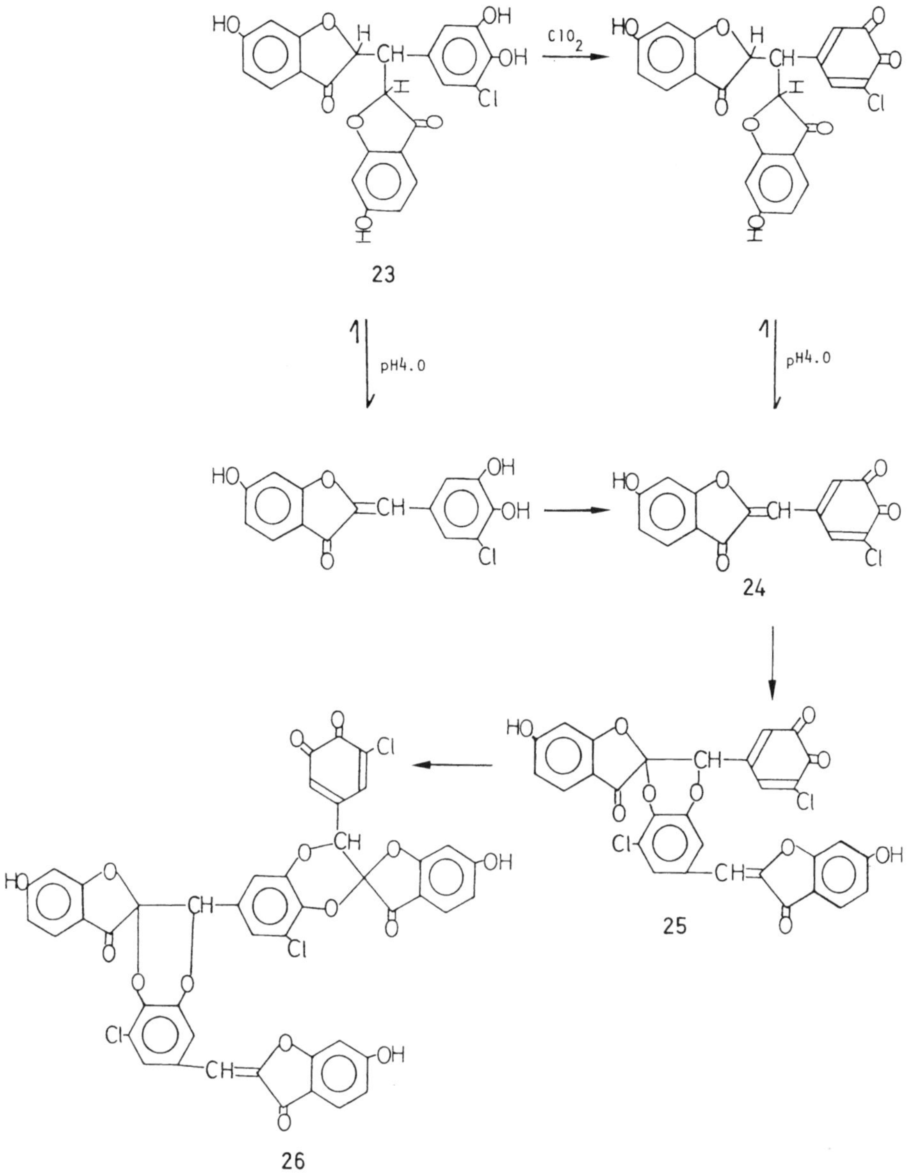

Fig. 9.3.6. Conversion of the adduct (23) to the speck substance (26)

2) *n*-alkanols, 3) β-sitosterol, 4) 1,10-dihydroxy-9-alkanones accompanied by some triterpenes, and 5) β-sitosteryl glucoside. The order of phenol retainability of the individual fraction corresponds to the order of hydrophile-lipophile balance (HLB) value. Among them, fraction 5 – i.e., β-sitosteryl glucoside – with the highest HLB value, melting point, and the least solubility in water and the usual organic solvents, shows an extremely high phenolic retainability. This is one

of the reasons for the high value of the total speck area in bleached pulps of sepetir-paya (174).

However, there is a difference in the speck area between untreated woods and extractives-free woods to which the extractives were added before cooking. After bleaching, the untreated woods show larger speck areas involving more specks with larger sizes than the latter woods. This is true in both sepetir-paya and rengas pulps. The differences, therefore, probably originate from the differences in the natural and artificial distribution of the resins. Accordingly, the differences may be closely related to the specks formed by the canal resin of red meranti (*Shorea negrosensis*) as described above. In fact, the apotracheal parenchyma tissue in sepetir-paya woods is found to be filled with resins. Cooking and bleaching of chips rich in this tissue reveal a much larger speck area involving more specks of larger sizes compared with those of the standard. In addition, the apotracheal resin evidently contains more phenols and β-sitosteryl glucoside than do the ray parenchyma resins (164).

When accompanied by a neutral carrier, as mentioned above, some phenols — e.g., methyl *p*-coumarate, butein, sulfuretin, and rengasin — are subjected to aggregation followed by chemical conversions during cooking and bleaching; they finally produce the colored specks in bleaching pulps. The aggregation of the phenols seemed to be primarily accelerated with the aid of intermolecular hydrogen-bond formation. On the other hand, neither ferruginol, a diterpenoid cryptophenol, nor its weakly phenolic derivatives formed any specks visible to the naked eye even after multi stage-bleaching. Instead, the slightly colored derivatives were finely dispersed and they lowered the brightness of the bleached pulps, as mentioned in the previous section. The capability of a phenol either to aggregate or to disperse may be highly dependent on its intrinsic nature, which will finally determine the type of pitch problem that appears in pulps.

To elucidate this, 13 various phenols, with lauryl alcohol as a neutral carrier, were dispersed into disintegrated filter paper and then cooked under sulfate (kraft) pulping conditions. The residual percent (w/w) ("R") of a phenol in the pulps was determined, and the speck areas ("S") of the aggregates, as revealed by dyeing with phosphomolybdic acid, were also determined and compared. If either the pKa value of a phenol became lower or its HLB value higher, there was a tendency for its R to decrease. When the S/R ratio was obtained, an almost inverse relationship was found, as expected. It has been observed under sulfate conditions that the lower the pKa value and the higher the HLB value of a phenol, the greater is the ability of the phenolic molecules to aggregate. However, a stereochemical environment around the phenolic hydroxyl group is an additional and important factor in aggregation. The reason for the absence of specks in the cases of ferruginol and its derivatives may be ascribed to the unusually high contribution of the steric hindrance in addition to the relatively low HLB value. Hydrogen-bond formation through the 12-hydroxy group of ferruginol and its relatives is prevented not only by the 13-isopropyl group but also by an interaction between 11-H and 1β-H (115).

If the pKa value of a phenol is low enough, the molecules should aggregate easily. In spite of this capability, many of the aggregates formed under sulfate conditions are generally dissolved owing to the low residual percent (R). However, it is noteworthy that R is highly variable depending on the quality and quantity of the concomitant neutrals and acids, as already mentioned.

In another type of speck formation, a number of ink-repellent specks appear on sheets of Japanese hand-made paper (washi) prepared from pokasa bark (*Broussonetia papyrifera*), one of the raw materials imported from Thailand. Conventionally, washi is prepared by cooking the chips for 2 hours with 20% sodium hydroxide at boiling point under atmospheric pressure followed by bleaching with sodium hypochlorite. Specks on the sheet are not visible to the naked eye but appear as white spots when Chinese ink is applied. Ink-repellent specks of this type create difficulties in use. Both the ethyl acetate solubles of the extractives and inorganics (especially calcium) present in the tissue were shown to be major factors responsible for speck formation (161). The acid fraction of the ethyl acetate extractives caused the greatest increase in turbidity in the presence of calcium ions even though it was the smallest one. The acidic, phenolic, and neutral fractions of the bark extractives exhibited a remarkable synergistic effect, with the phenolic fraction playing a certain role in speck formation. The triterpenes and steroids were the major organic components and mainly responsible for the ink-repellant specks (162).

Addition of small amounts of anionic polyacrylamide, with low molecular weights, to the bleaching liquor strongly trap calcium and metallic ions and drastically reduce speck area. Moreover, drying temperature and aging time greatly influence speck formation, and formation of the ink-repellent specks is completed by interactions between the extractives and inorganic material at the drying stage (161). Although calcium ions were the major inorganic contributors to the increase in turbidity caused by the extractives of the bark of *Broussonetia papyrifera*, aluminum and magnesium also interacted strongly (162). The wood extractives of *Pinus densiflora* and *Cryptomeria japonica* caused a much greater increase in turbidity with aluminum and magnesium than with calcium ions (162).

Pitch deposition in some equipment, as discussed in Sect. 9.3.6.2, may not always be related directly to speck formation in pulp. The significance of the neutrals in pitch deposition has been pointed out (114) and, as already mentioned, the components of speck resin in pulp have been recently confirmed to originate from the canal resin, to which the origin of deposited resin in the cooking equipment is also ascribed (2). This is the case for *Vatica* species. These results strongly suggest that pitch deposition — i.e., speck formation — in pulp is closely related to that found in digestion equipment. However, pitch deposition on equipment at the later stages may not be so simple. For example, the significance of washing temperature, participation of minerals (especially calcium), and the tackiness of resin has been emphasized (180).

Pitch deposition at the microscopic level in groundwood, sulfite, and sulfate pulps from some coniferous woods shows a close relation to the observations mentioned above. The aggregates on fiber surfaces were comprised to a large extent of inorganic salts of calcium, presumably $CaCO_3$. Much of the $CaCO_3$ precipitation and, hence, the formation of aggregates takes place inside the chips during a sulfate cook. Thus the $CaCO_3$ aggregates appear to play a central role in kraft pitch deposition by serving as fillers, with resinous pitch binding them together and to the metal surface (6). Methods for measuring calcium soap deposition have been reported (43–46).

9.3.7.3 Wettability

Pitch deposition resulting from the presence of wood resins occurs during the preparation of paper from mechanical and sulfite pulps in addition to pitch deposition developing during the pulping processes. In the press nip, resin in the ray cells can be transferred from the paper web to the felts where it plugs the pores of the felt or deposits on it subsequently during the felt dewatering. This type of problem is frequently the most troublesome in newsprint mills.

9.3.7.4 Sticking to Press Rolls

A roll that is wetted by contact with a moist pulp web containing a suspension of resin or pitch globules will partly or wholly dry during the rest of a revolution. Drying will concentrate the colloidal suspension on the surface of the granite roll so that it becomes sticky, as experienced by mills with pitch problems (8, 52). Sticking to Yankee dryers has also been reported (11). Sticking to the rolls is particularly troublesome with high-speed paper machines.

Eucalypt woods do not contain resinous extractives, but adhesion of kraft eucalypt pulps to the press can occur, and this has been overcome by overcooking the wood. The problem can be prevented more cheaply by adding dilute solutions of sodium hexametaphosphate to the press roll or some other compounds that would form sparingly dissociated compounds with iron and aluminum ions (48).

The tropical species *Dillenia pulchella* and *Tetramerista glabra* produce very sticky pulps that are difficult to remove from the laboratory sheet machine (92, 93).

The recycling of the process water from a hardboard mill utilizing hardwoods has resulted in sticking of the boards at the press (36). In some cases the cause may be the thin layer of hemicelluloses, which under wet conditions and with heat from the rolls have passed the glass transition point to become plastic so that the paper web from semichemical pulps sticks to the rolls. The addition of a mixture of solvents, dispersants, and emulsifiers to NSSC pulps can control deposit formation at the wet end of the machine (56).

9.3.8 Spent Liquor Recovery

9.3.8.1 Concentration and Burning Difficulties

Spent or black pulping liquors contain 10% to 20% solids and must be concentrated to a solids content exceeding 50% in order to maintain burning in recovery furnaces. The liquor from the acid sulfite process requires equipment that will resist corrosion and be unaffected by scaling, but extractives in the liquor do not affect concentration or burning.

Black liquor from kraft or soda mills pulping hardwoods may present problems during its evaporation and combustion by conventional methods. These problems have been encountered with some eucalypt species, particularly with wood from trees of old age, and special measures may be needed to overcome the problems caused by extractives. Apart from the formation of deposits and scale, which reduces heat transfer in the evaporators, the extractives increase the viscosity (so that the evaporation process is affected) and reduce the swelling properties when incinerated, resulting in incomplete combustion (130). These difficulties with black liquor from the soda process can be avoided with the use of a wet oxidation process conducted under high pressure (97).

Oxidation of polyphenols containing catechol and particularly pyrogallol moieties, such as the gallotannins and ellagitannins found in eucalypts during alkali pulping, result in viscous liquors on concentration or evaporation of the liquor (78, 131). High viscosity liquors move more slowly through the evaporator tubes than low viscosity liquors. In some cases the viscosity becomes so high that concentration cannot be continued owing to the blockage of the evaporator tubes.

The viscosity of the black liquor depends on the organic matter content of the total solids as well as the total solids as such. Ellagic acid has a much greater effect on viscosity than a number of other variables (130, 133). An empirical formula has been derived to express the relationship between the viscosity of the kraft black liquor from pulping trials of eucalypts and mixed tropical hardwoods and the organic matter content (130). Large differences in viscosity and the swelling volume of evaporated kraft liquor have been found between tropical hardwoods and regrowth eucalypts.

Usually a black liquor with high viscosity has a low swelling volume when burned and an empirical relationship enables the prediction of the swelling volume from the viscosity of a solution of 40% total solids content at 40°C (130). However, the two properties are different, for viscosity depends largely on the configuration of molecules in solution whereas swelling volume is influenced considerably by the properties of the black liquor solids and their thermal decomposition products (125, 126, 133).

Some studies have resulted in procedures to reduce viscosity and improve the combustion properties of eucalypt black liquor. These include oxidation of hot liquors and high pressure-wet air oxidation, pyrolysis under controlled conditions, digesting initially below 120°C to allow decomposition of the kino of eucalypts, and aqueous washing of the chips (e.g., 63, 64, 97, 128–130, 132).

The presence of magnesium ions increases the stability of ellagic acid under cooking conditions (19, 54). The particles of the granular magnesium-ellagic acid complex that pass through the system into the concentrated black liquors presented to the recovery furnace are still able to affect burning. Apparently, the colloidal properties of the liquor are changed so that the desirable swelling is reduced, and complete burning or combustion does not occur (19). As a result, the excessive carbon in the ash interferes with the efficient recovery of the alkaline salts. Black liquor from the pulping of pines, and also fuel oil have been added to eucalypt black liquor to achieve satisfactory burning.

Black liquor from the kraft pulping of mixtures of Papua New Guinea hardwoods has poor swelling properties and sometimes high viscosities that could lead to burning difficulties in commercial recovery furnaces (130, 139–141).

9.3.8.2 Foaming During Concentration and Oxidation

The resin in the wood of resinous species is dissolved during pulping by alkaline processes with the formation of sodium salts of fatty and resin acids. The presence of these soaps in black liquors facilitates foaming, thereby reducing the efficiency of pulp washing and liquor concentration operations in the pulp mill. Removal of the soap for the production of tall oil reduces these problems. Foaming is encountered from time to time with other species but the causes are not known.

9.3.8.3 By-Product Recovery

The recovery of terpenes and fatty and resin acids from the kraft pulping of resinous species can improve the economy of the operations. During the pulping of *Pinus* spp., the gases released from the digester are condensed and the turpentine decanted from the aqueous distillate. The removal of the sodium salts of fatty and resin acids from the black liquor followed by acidification give tall oil, which is refined into a number of commercially important products. These aspects will be dealt with elsewhere in this volume (Sect. 10.1).

9.3.9 Observations

When paper was prepared from wood pulp 250 years ago, the ample supplies of wood from many species enabled a choice of the most suitable raw material. The number of species used for wood pulp increased to meet the increasing demand for paper. With the use of additional species, it was necessary to determine, and to control, the causes of problems hindering the development of improved methods of pulping and paper making. At the beginning, a single class of extractives was found to be responsible for a specific problem, such as wood resins causing pitch problems with softwood pulps, or stilbenes hindering the removal of lignin during acid pulping of *Pinus* species. Later, when mature and overmature *Eucalyptus* species were examined as raw material, not only did morphological problems, such as the short fiber length of these hardwoods, have to be overcome but extractives were found to be responsible for a range of new problems. The polyphenols in the extractives resulted in low pulp yields, color problems with the pulp, and difficulties in processing the black liquor from pulping. These problems were reduced to varying levels, but are not encountered to the same extent by today's industry now that the former resource is being replaced by younger-aged material. In the most recent studies with another group of hardwoods, namely the Dipterocarpaceae of tropical southeast Asia, two or more classes of compounds act synergistically to form colored specks in the pulp that are difficult to remove by the usual pulping procedures. Thus, during the history of pulp and paper making from wood, the problems caused by extractives have become increasingly complex and intricate.

The demand for both quality and quantity of pulp and paper products will increase. In the short term, more raw material will be obtained from secondary tropical species and pulpwoods of decreasing quality. Past events indicate problems caused by extractives will continue to be encountered and will probably be of new types. With the development of new pulping technologies aimed at higher yields and quality of pulp, the influence of extractives may show in ways not previously encountered. Further knowledge of the behavior of extractives on storage of pulpwood and wood chips will assist the control of these problems. Greater understanding of the morphology of pulpwood with regard to the location of various classes of extractives will be necessary, so that the processes will not accentuate the problems. The increasing demand for high quality and speciality papers will be met by pulpwood from plantations of the progeny of selectively bred trees. Some of these plantations will be fast-grown and harvested after a few years. Just as the different wood quality of the increasing proportions of juvenile wood in these fast-grown trees requires re-examination, so also will the nature of the extractives. These trees will contain a small proportion, if any, of heartwood and hence a lower content of extractives. However, the ratio of the components in the wood may change from that normally encountered in older trees of the same species. This situation has been experienced with young *Eucalyptus globulus* trees used for pulpwood. Whereas the polyphenols are of overwhelming importance in the wood from mature trees, they and most other components of the extractives are present in smaller amounts in the younger trees (166). However, kraft pulping of pulpwood from the latter yielded an unbleached pulp containing a resin of mainly unsaponifiables material (β-sitosterol) (165). The resin, mainly located in the parenchyma cells, is difficult to remove and causes problems when the pulp is used in the viscose process (32).

The increasing need for recirculation and conservation of water in parts of pulp and paper mills and the increasingly stringent controls of the composition of effluents to meet environmental protection requirements demand a greater understanding of the behavior of small quantities of extractives and their derivatives formed during various processes.

A greater understanding is needed of the formation of extractives in various anatomical elements and in the heartwood of the living tree, as well as their formation as a response to injury. This knowledge should assist the development of optimum silvicultural practices for the production of the most suitable pulpwood. Whereas the absence of extractives would give a more desirably uniform raw material for pulp and paper manufacture, the formation of extractives is an inherent function of the tree (77). The maintenance and development of the understanding of the effects of extractives during the pulping and paper making processes will always be necessary if paper in sufficient amounts and of required quality are to be provided for future generations.

References

1. Abrams E, Nelson J R 1977 A systems approach to control of kraft mill deposits. Tappi 60:109–112
2. Agatsuma S, Tachibana S, Sumimoto M 1979 Mechanisms for resin aggregation in cooking and bleaching of Dipterocarpaceae plants. Proc Res Meet Jpn Tappi 45:13–19
3. Agatsuma S, Tachibana S, Sumimoto M 1986 Studies on pitch problems caused by pulping and bleaching of tropical wood. Part 17. Mokuzai Gakkaishi 32:34–43
4. Akimoto H, Sumitomo M 1980 Extractives from temperate wood species in pulping and papermaking. Part I. Mokuzai Gakkaishi 26:347–357
5. Alekseev A V 1957 [The influence of electrical phenomena which arise in the cutting of wood on the wear of the tool.] Derevoobrab Promst 6(8):15–16
6. Allen L H 1975 Pitch in wood pulps. Pulp Pap Mag Can 76:T139–T146
7. Allen L H 1977 Pitch particle concentration: an important parameter in pitch problems. CPPA Trans Tech Sect 3(2):32–40
8. Allen L H 1980 Mechanisms and control of pitch deposition in newsprint mills. Tappi 63(2):81–87
9. Allen L H 1981 Pitch control – optimization of alum usage in newsprint mills. Pulp Pap Mag Can 82:T397–T404
10. Andersson A 1969 Alterations in rosin composition during storage of wood. Svensk Papperstidn 72:304–311
11. Andersson B, Wåhlén S 1969 Pitch control in paper mills. Svensk Papperstidn 72:537–543
12. Assarsson A 1969 Reactions during chip storage and how to control them. Pulp Pap Mag Can 70:T323–T328
13. Assarsson A 1969 Release of resins from sulfite pulps. Svensk Papperstidn 72:380–385
14. Assarsson A 1969 Deresinification during processing of sulfite pulp. Svensk Papperstidn 72:403–410
15. Assarsson A, Åkerlund G 1967 Studies on wood resin, especially the changes during seasoning of the wood. Svensk Papperstidn 70:205–212
16. Assarsson A, Lindahl A, Lindqvist B, Ostman H 1981 Controlling the resin content of pulp. Ind Pulp Pap 36(3):8–12
17. Back E 1956 Pitch control by combined alkali and aluminium sulfate addition. Some principles in controlling pitch problems. Svensk Papperstidn 59:319–325
18. Back E 1969 Mechanism of rosin deposition. Svensk Papperstidn 72:226–231
19. Baklien A 1960 The effect of extractives on black liquor from eucalypt pulping. Appita 14:5–17
20. Bamber R K 1976 Tylosoids in the resin canals of the heartwood of some species of Shorea. Holzforschung 30:59–62
21. Barbadillo P 1967 Summary of Spanish experiments on the pulping of eucalypts. Appita 21:27–40
22. Barton G M 1973 The significance of western hemlock phenolic extractives in groundwood pulping. Tappi 56(5):115–118
23. Benda V 1978 Problems with harmful pitch in kraft pulp mills. Papir Celul 33(11):217–220
24. Bisset N G, Chavanel V, Lantz J P, Wolff R E 1971 Constituants sesquiterpéniques et terpeniques des resins de genre *Shorea*. Phytochemistry 10:2451–2463
25. Bland D E, Menshun M 1965 Combination of ellagic acid with lignin during high temperature methanol extraction of *Eucalyptus* wood. Holzforschung 19:33–36
26. Bowman D F, Nelson P F 1965 Effect of extractives on the color of eucalypt pulps. Appita 19:8–13
27. Bublitz W J, Fang Y P 1977 Douglas-fir green liquor semichemical pulp: How suitable is it for corrugating medium? Tappi 60(4):90–93
28. Bublitz W J, Meng T Y 1974 Effect of certain flavonoids on the bleaching of Douglas-fir refiner groundwood. Pulp Pap Mag Can 75(3):85–90 (T91-6)
29. Cameron D W, Farrington A, Nelson P F, Raverty W D, Samuel E L, Vanderhoek N 1981 The effect of addition of quinonoid and hydroquinonoid derivatives in NSSC pulping. Appita 35:307–315
30. Cohen W E 1962 The influence of resins on paper manufacture. In: Hillis W E (ed) Wood extractives. Academic Press New York London, 421–450

31 Cohen W E, Mackney A W 1951 Influence of wood extractives on soda and sulfite pulping. Appita 5:315–335
32 Croon I 1968 Comments on the effect of pulp extractives in the viscose process. Svensk Papperstidn 71:143
33 Croon I 1969 Additives for control of resin in pulp mills. Svensk Papperstidn 72:583–589
34 Croon I, Ponton S 1969 Additives for control of resins in pulps. Svensk Papperstidn 72:498–508
35 Dadswell H E, Hillis W E 1962 Wood. In: Hillis W E (ed) Wood extractives. Academic Press New York, 3–55
36 Dallons V 1979 In-plant pollution control in the hardboard industry. For Prod J 29(10):70–74
37 Danielson O, Hartler N, Ryrberg G 1981 Energy reductions in thermomechanical pulping. Appita 34:306–308
38 Davies G W, Logan A F 1984 Observations on Sarawak peat swamp species in relation to pitch deposition during kraft pulping. Holzforschung 38:25–30
39 Dean J C, Saul C M, Turner C H 1956 Role of added black liquor in alkaline pulping. Appita 10:84–98
40 Dickens J H 1986 Dispersants for pitch control. Tappi 69(12):55–58
41 Djamal S, Kojima Y, Kayama T 1984 Bleaching of red lauan chemical pulps with ozone and hydrogen peroxide. Jpn Tappi 38:1136–1143
42 Dorris G M, Dovek M, Allen L H 1983 Analysis of metal soaps in kraft mill brownstock pitch deposits. J Pulp Paper Sci 9(1):TR 1–7
43 Douek M, Allen L H 1978 Kraft mill pitch problems. Tappi 61(7):47–51
44 Douek M, Allen L H 1980 Calcium soap deposition in kraft mill brownstock systems. Pulp Pap Mag Can 81(11):T317–T322
45 Douek M, Allen L H 1983 A laboratory test for measuring calcium soap deposition from solutions of tall oils. Tappi 66(2):105–106
46 Erdtman H 1949 Compounds inhibiting the sulfite cook. Tappi 32:302–305
47 Evans J C W 1982 Sticking with sulfite. Fraser (Paper Inc) goes against trend with kraft. Pulp Pap 56(3):68–71
48 Farmer C T 1949 Sticking at the press: its nature and remedy. Proc Appita 3:105–120
49 Farrington A, Henderson V T, Nelson P F 1977 Alkaline pulping process. Australian patent 9.939/77 (May 2, 1977); Canadian patent 1 087 892 (Oct 21, 1980)
50 Feist W C, Hajny G J, Springer E L 1973 Effect of storing green wood at elevated temperatures. Tappi 56(8):91–95
51 Flowers A G, Banham P W 1985 The bleaching of eucalypt groundwood with hydrogen peroxide. Appita 38:127–131
52 Fuxelius K 1967 Adhesion of paper webs to M.G. cylinders during dry creping. Svensk Papperstidn 70:164–168
53 Gardner J A F 1962 The tropolones. In: Hillis W E (ed) Wood extractives. Academic Press New York, 317–330
54 Gardner J A F, Hillis W E 1962 Extractives and the pulping of wood. In: Hillis W E (ed) Wood extractives. Academic Press New York, 367–403
55 Glover G F, Purdie A J C 1957 The pitch tendency of pulps. Brit Pap Board Makers Assoc Proc Tech Sect 38:1–10
56 Granehult L, Wenzl D, Ellow A 1981 Chemical additives cut operating costs. Pulp Pap Int 23(5):80–82
57 Hall M, Hansen N W, Rudra A B 1973 The effect of species, age and wood characteristics on eucalypt kraft pulp quality. Appita 26:348–354
58 Hamilton K A, Lloyd J A 1984 Measuring the effectiveness of talc for pitch control. Appita 37:733–740
59 Hartler N, Lindström C, Stade Y 1977 Influence of stem, stump, root and branch bark in kraft processing of pine, spruce and birch. Svensk Papperstidn 80:121–124
60 Hata K, Sogo M 1952 The difficulty of sulfite digestion of pine heartwood. J Jpn For Soc 35:74–77
61 Hatanaka S, Sakai K, Kondo T 1973 Studies on oxygen-alkali treatment of wood. Part 1. J Jpn Tappi 27:296–303
62 Hatton J V 1970 Precise studies on the effect of outside storage on fiber yield. Tappi 53:627–638

63 Hemingway R W, Hillis W E 1971 Behavior of ellagitannins, gallic acid and ellagic acid under alkaline conditions. Tappi 54:933–936
64 Hemingway R W, Hillis W E 1972 Alkaline oxidation of eucalypt extractives. Appita 25:445–454
65 Hemingway R W, Nelson P J, Hillis W E 1971 Rapid oxidation of the fats and resins in *Pinus radiata* chips for pitch control. Tappi 54:95–98
66 Hewitt D G, Nelson P F 1965 Ellagic acid and the pulping of eucalypts. Part I. Some aspects of the chemistry of ellagic acid. Holzforschung 19:97–101
67 Higgins H G 1984 Pulp and paper. In: Hillis W E, Brown A G (eds) Eucalypts for wood production.. CSIRO, Academic Press New York, 290–316
68 Higgins H G, Garland C P, Puri V 1977 Thermomechanical and chemithermomechanical pulps from eucalypts and other hardwoods. Appita 30:415–423
69 Higgins H G, Irvine G M, Puri V, Wardrop A B 1978 Conditions for obtaining optimum properties of radiata and slash pine thermomechanical and chemithermomechanical pulps. Appita 32:23–33
70 Higgins H G, Phillips F H, Logan A F, Balodis V 1973 Pulping of tropical hardwoods: individual and mixed species, wood and paper properties, resource assessment. CSIRO Aust Div Appl Chem, Technol Pap 70, 22 pp
71 Higgins H G, Puri V, Garland C 1978 The effect of chemical pretreatments in chip refining. Appita 32:187–200
72 Hillis W E 1956 Leucoanthocyanins as the possible precursors of extractives in woody tissues. Aust J Biol Sci 9:263–280
73 Hillis W E 1969 The contribution of polyphenolic wood extractives to pulp colour. Appita 23:89–101
74 Hillis W E 1972 Properties of eucalypt woods of importance to the pulp and paper industry. Appita 26:113–122
75 Hillis W E 1975 The quality of eucalypt woods and the effect of extractives. J Jpn Tappi 29:341–346
76 Hillis W E 1984 High temperature and chemical effects on wood quality. Part 1. Wood Sci Technol 18:281–293
77 Hillis W E 1987 Heartwood and tree exudates. Springer Heidelberg, 268 pp
78 Hillis W E, Carle A 1959 The influence of extractives on eucalypt pulping and papermaking. Appita 13:74–83
79 Hillis W E, McKenzie W M 1964 Chemical attack as a factor in the wear of woodworking cutters. For Prod J 14:310–312
80 Hillis W E, Nelson P F, Zadow G 1966 The cause of discoloration of *Pinus radiata* bisulphite pulp. Appita 19:111–114
81 Hillis W E, Swain T 1962 Extractives in groundwood and newsprint. In: Hillis W E (ed) Wood extractives. Academic Press New York, 405–419
82 Hillis W E, Yazaki Y 1973 Polyphenols of *Intsia* heartwoods. Phytochemistry 12:2491–2495
83 Hughes D A 1977 Method for determining the pitch adsorption characteristics of mineral powders. Tappi 60(7):144–146
84 Jewell I S, Richardson D E 1987 Pitch chemistry. Appita Preprint A 14 18 pp
85 Johnsson S A 1968 The role of alum in pitch control in groundwood and newsprint manufacture. Appita 22:25–29
86 Kirbach E, Chow S 1976 Chemical wear of tungsten carbide cutting tools by western red cedar. For Prod J 26(3):44–48
87 Kivimaa E 1952 Was ist die Abstumpfung der Holzbearbeitungswerkzeuge? Holz Roh-Werkst 10:425–428
88 Kojima Y, Djamal S, Kayama T 1985 Isolation of 24-methylenecycloartanol and its related compounds from red lauan. Mokuzai Gakkaishi 31:312–315
89 Lee H K, Lim N, Sumimoto M 1987 Pitch problems in the pulping and bleaching of Sarawak hardwoods. Part 1. Appita 40:351–358
90 Lees G J, Nelson P F 1967 An examination of hot water extracts of *Eucalyptus sieberi*. Appita 20:113–117
91 Lloyd J A, Stratton L M 1986 The determination of extractives and pitch by chromic acid oxidation. Appita 39:287–288
92 Logan A F, Balodis V, Tan Y K, Phillips F H 1984 Kraft pulping properties of individual species from Sarawak forest resources. Part I. Peat swamp species. Malaysian For 47:3–27

93 Logan A F, Balodis V, Tan Y K, Phillips F H 1984 Kraft pulping properties of individual species from Sarawak forest resources. Part 2. Mixed dipterocarp forest species. Malaysian For 47:87–115

94 Louden L 1981 Pitch, problems and control. Bibliographic Ser 268 Suppl 1. Inst Pap Chem Appleton WI, 65 pp

95 MacLean H, Gardner J A F 1953 Heartwood extractives in digester corrosion. Pulp Pap Mag Can 54(12):125–130

96 MacLeod J M, Kubes G J, Fleming B I, Bolker H I 1981 Sodaanthraquinone-ethylenediamine pulping. Tappi 64(6):77–80

97 Maddern K N 1980 Wet air oxidation recovery at Burnie. Appita 34:130–134

98 Makkonen H, Pitkänen M, Nikki M 1973 Oxygen bleaching of a sulphite pulp of easy-bleach grade, with subsequent peroxide bleaching. Pap Puu 55:947–958

99 Mayer B K 1971 Practical experiences in the use of mixed tropical hardwoods for the manufacture of pulp and paper. Expert Group Meet Pulp Pap Ref ID/NG, 102/33, UNIDO Vienna

100 McKenzie A W, Pearson A J, Stephens J R 1968 The removal of the deposits of metal-ellagic acid complexes. Appita 21:19

101 McKenzie W M, Hillis W E 1965 Evidence of chemical acceleration of wear in cutting plant materials. Wear 8:238–243

102 McKenzie W M, McCombe B M 1968 Corrosive wear of veneer knives. For Prod J 18(3):45–46

103 McMahon D H 1980 Analysis of low levels of fatty and resin acids in kraft mill process streams. Tappi 63(9):101–103

104 Mills J S, Werner A E A 1955 The chemistry of dammar resin. J Chem Soc 1955:3121–3140

105 Miura K, Sumimoto M 1984 Extractives from temperate wood species in pulping and paper making. Part 6. J Jpn Tappi 38:352–360

106 Mohan G D, Klamecki B E 1982 The susceptibility of wood-cutting tools to corrosive wear. Wear 74:85–92

107 Molotkov L K, Ospishcheva M N, Smirnova L A, Grigor'eva E D, Makova L I, Goryunova R V 1975 [Preparation of hardwood to be used in the manufacture of dissolving pulp. (3). Effect of felling time on changes of aspenwood during chip storage.] Sb Tr Vses Nauch. Issled Inst Tsellyul. Bumazh Prom 67:14–25

108 Morgan J W W, Orsler R J 1968 The chemistry of color changes in wood. Part 1. Holzforschung 22:11–16

109 Mroz W, Surewicz W 1977 Modification of the kraft pulping process aimed at improving the color of the pulp. Przeglad Pap 33:227–280

110 Mutton D B 1962 Wood resins. In: Hillis W E (ed) Wood extractives. Academic Press New York, 331–363

111 Nelson P F 1965 Ellagic acid and the pulping of eucalypts. Part II. Complexes of ellagic acid with cations and solvents. Holzforschung 19:102–105

112 Nelson P J, Hemingway R W 1971 Resin in bisulphite pulp from *Pinus radiata* wood and its relation to pitch troubles. Tappi 54:968–971

113 Nishida K, Kuroki K, Ono T 1957 Pitch troubles, cause and prevention. J Jpn Tappi 11:437–440, 544–547

114 Nishida C, Tanaka M, Kondo T 1964 Studies on hardwood resin. Part 3. J Jpn Tappi 18:429–434

115 Ohtani Y, Miura K, Sumimoto M 1983 Extractives from temperate wood species in pulping and papermaking. Part 5. J Jpn Tappi 37:1102–1108

116 Ohtani Y, Shigemoto T, Okagawa 1986 Chemical aspects of pitch deposits in kraft pulping of hardwood in Japanese mills. Appita 39:301–306

117 Ohtani Y, Sumimoto M 1981 Studies on pitch troubles caused by pulping and bleaching of tropical woods. Part 10. Mokuzai Gakkaishi 27:302–310

118 Ohtani Y, Shutoh K, Sumimoto M 1982 Studies on pitch troubles caused by pulping and bleaching of tropical woods. Part 12. Mokuzai Gakkaishi 28:59–66

119 Ohtani Y, Sumimoto M 1982 Studies on pitch troubles caused by pulping and bleaching of tropical woods. Part 14. Acta Chem Scand B 36:613–621

120 Ohtani Y, Sumimoto M 1983 Extractives from temperate wood species in pulping and papermaking. Part 2. J Jpn Tappi 37:829–833

121 Ohtani Y, Sumimoto M 1983 Extractives from temperate wood species in pulping and papermaking. Part 3. J Jpn Tappi 37:921–931

122 Ohtani Y, Sumimoto M 1983 Extractives from temperate wood species in pulping and papermaking. Part 4. J Jpn Tappi 37:1011–1017
123 Ohtani Y, Tachibana S, Sumimoto M 1980 Studies on pitch troubles caused by pulping and bleaching of tropical woods. Part 7. Mokuzai Gakkaishi 26:534–542
124 Ohtani Y, Tachibana S, Sumimoto M 1980 Studies on pitch troubles caused by pulping and bleaching of tropical woods. Part 9. Mokuzai Gakkaishi 26:796–805
125 Okayama T, Oye R 1977 Properties of eucalyptus black liquor during evaporation and combustion. Part 5. Polymer properties of organic matter in eucalyptus black liquor. Mokuzai Gakkaishi 23:162–167
126 Okayama T, Oye R 1977 Properties of eucalyptus black liquor during evaporation and combustion. Part 6. Pyrolysis of eucalyptus black liquor. J Jpn Tappi 31:404–456
127 Olm L 1984 Pitch problems and their control in kraft mills using hardwoods from temperate and tropical zones: a literature survey. Appita 37:449–483
128 Oye R, Hato N, Mizuno T 1973 Extractives in eucalyptus wood and their influence on the properties of black liquor during evaporation and combustion. J Jpn Tappi 27:71–79
129 Oye R, Hato N, Mizuno T 1973 Properties of eucalyptus black liquor in evaporation and combustion. Part 3. Partial wet combustion of eucalyptus black liquor. J Jpn Tappi 27(8):394–398
130 Oye R, Langfors N G, Phillips F H, Higgins H G 1977 The properties of kraft black liquors from various eucalypts and mixed tropical hardwoods. Appita 31:33–40
131 Oye R, Mizuno T, Hato N 1971 Influence of extractives on Eucalyptus pulping. Timber Bull Eur 23 (Suppl 5):24–48
132 Oye R, Mizuno T, Hato N 1972 The influence of the extractives in *Eucalyptus* woods on the viscosity of concentrated black liquors. J Jpn Tappi 26:599–608
133 Oye R, Okayama T 1974 Properties of eucalyptus black liquor in evaporation and combustion. Part 4. Pyrolysis of eucalyptus black liquor. J Jpn Tappi 28:507–514
134 Packman D F 1960 The acidity of wood. Holzforschung 14:178–183
135 Pearson A J 1979 Mechanical pulp: its status and future. Appita 32:261–266
136 Pearson A J, Coopes I H 1964 Hydrosulphite brightening of eucalypt groundwood made in the presence of caustic soda. Appita 18:26–32
137 Pelton R H, Allen L H, Nugent H M 1982 Additives for increased retention and pitch control in paper manufacture. Can Patent CA 1 137710 (Cl D21H3/18)
138 Pew J C 1949 Douglas-fir heartwood flavonone: its properties and influence on sulphite pulping. Tappi 37:39–41
139 Phillips F H, Logan A F 1976 Papua New Guinea hardwoods: future source of raw material for pulping and papermaking. Appita 30:29–40
140 Phillips F H, Logan A F 1977 The pulping and papermaking potential of tropical hardwoods. Part IV. CSIRO Aust Div Chem Technol Tech Pap 8
141 Phillips F H, Logan A F, Langfors N G 1979 The pulping and papermaking potential of tropical hardwoods. Part VI. CSIRO Aust Div Chem Technol Tech Pap 10
142 Poller S 1971 Influence of heart- and sapwood in the sulfate pulping of pinewood. Zellst Pap 20:331–338
143 Press R E, Hardcastle D 1969 Some physico-chemical properties of ellagic acid. J Appl Chem 19:247–251
144 Pulp and Paper International 1986 Annual review. World trends and trade. Pulp Pap Int 28(8):90
145 Pusa R, Virkola N E 1974a The effect of oxygen delignification on the extractives of spruce sulphite pulps. Part 1. Pap Puu 56:599–614
146 Pusa R, Virkola N E 1974b The effect of oxygen delignification on the extractives of spruce sulphite pulps. Part 2. Pap Puu 56:687–695
147 Pusa R, Virkola N E 1975 The effect of oxygen delignification on the extractives of spruce sulphite pulps. Part 3. Pap Puu 57:43–55
148 Redmond W A, Coffey B B, Shastri S, Manchester D F 1971 Nonstructural chromophoric substances in jack pine wood and pulps. Pulp Pap Mag Can 72(1):85–92 (T15–22)
149 Richardson D E, Bloom H 1982 Analysis of resin acids in untreated and biologically treated thermo-mechanical pulp effluent. Appita 35:477–482
150 Richardson D E, Stanborough M S 1987 The measurement of pitch adsorption on talc. Appita Preprint A 15, 23 pp

151 Saka S, Thomas R J 1982 Fiber surface structure and fiber liberation in soda-anthraquinone kraft and soda pulps as determined by conventional electron microscopy. Wood Fiber 14:144–158
152 Saul C M 1979 Chemical pulp: its status and future. Appita 32:345–350
153 Seikel M K, Hillis W E 1970 Hydrolysable tannins of *Eucalyptus delegatensis* wood. Phytochemistry 9:1115–1128
154 Shelton R L 1985 Using talc to control pitch problems in paper and pulp mills. Paper Trade J 169(8):48–49
155 Shigemoto T, Ohtani Y, Okagawa A, Sumimoto M 1987 NMR study of β-resene. Cellul Chem Technol 21:249–254
156 Shimada K 1969 Studies on pitch trouble in tropical wood pulp. Part 1. Mokuzai Gakkaishi 15:126–130
157 Shimada K, Kayama T 1970 Studies on the pitch trouble in tropical wood pulp. Part 2. Mokuzai Gakkaishi 16:388–393
158 Somerville J L, Pearson A J 1958 Application of the cold soda process in the use of eucalypts for newsprint manufacture. Appita 12:57–72
159 Starostenko N P, Nepenin N N, Leshchenko IG 1957 [Influence of resin composition on pitch troubles in pulp.] Bumazh Prom 32:2–5
160 Stewart C M, Kottek J F, Dadswell H E, Watson A J 1961 The process of fiber separation. Part III. Tappi 44:798–813
161 Su Y C, Tanaka H, Sumimoto M 1984 Pitch problems in making Japanese paper. Part 1. Mokuzai Gakkaishi 30:490–500
162 Su Y C, Tachibana S, Sumimoto M 1986 Pitch problems in making Japanese paper. Part 2. Mokuzai Gakkaishi 32:190–202
163 Sumimoto M, Tachibana S, Ohtani Y 1981 A new type of pitch trouble, its formation and exclusion. Proc Int Symp Wood Pulp Chem, Chem Biochem Wood-Based Proc Prod 2:150–155
164 Sumimoto M, Tachibana S, Ohtani Y 1983 Behaviors of the phenolic wood extractives in pulping and bleaching. Proc Int Symp Wood Pulp Chem, Chem Struct Wood 1:115–119
165 Swan B 1967 Extractives of unbleached and bleached prehydrolysis kraft pulp from *Eucalyptus globulus*. Svensk Papperstidn 70:616–619
166 Swan B, Åkerblom I-S 1967 Wood extractives from *Eucalyptus globulus*. Svensk Papperstidn 70:239–244
167 Swanson J W, Cordingley R H 1956 Surface chemical studies on pitch. Part 1. Tappi 39:684–690
168 Tachibana S, Sumimoto M 1978 Studies on pitch troubles caused by pulping and bleaching of tropical woods. Part 3. Mokuzai Gakkaishi 24:575–582
169 Tachibana S, Sumimoto M 1979 Studies on pitch troubles caused by pulping and bleaching of tropical woods. Part 4. Mokuzai Gakkaishi 25:636–643
170 Tachibana S, Sumimoto M 1979 Studies on pitch troubles caused by pulping and bleaching of tropical woods. Part 5. Mokuzai Gakkaishi 25:726–742
171 Tachibana S, Sumimoto M 1980 Studies on pitch troubles caused by pulping and bleaching of tropical woods. Part 6. Holzforschung 34:131–137
172 Tachibana S, Sumimoto M 1981 Studies on pitch troubles caused by pulping and bleaching of tropical woods. Part 8. Holzforschung 35:71–80
173 Tachibana S, Sumimoto M 1982 Studies on pitch troubles caused by pulping and bleaching of tropical woods. Part 11. Mokuzai Gakkaishi 28:45–58
174 Tachibana S, Sumimoto M 1982 Studies on pitch troubles caused by pulping and bleaching of tropical woods. Part 13. Mokuzai Gakkaishi 28:452–462
175 Tachibana S, Sumimoto M 1985 Studies on pitch problems. Part 15. Mokuzai Gakkaishi 31:375–382
176 Tachibana S, Sumimoto M, Kondo T 1976 Studies on pitch troubles caused by pulping and bleaching of tropical woods. Part 1. Mokuzai Gakkaishi 22:34–39
177 Tachibana S, Sumimoto M, Kondo T 1976 Studies on pitch troubles caused by pulping and bleaching of tropical woods. Part 2. Mokuzai Gakkaishi 22:258–263
178 Tanaka T, Tachibana S, Sumimoto M 1986 Studies on pitch problems caused by pulping and bleaching of tropical woods. Part 18. Mokuzai Gakkaishi 32:103–109
179 Tardif J W 1959 Improving the brightness of eucalypt groundwood. Appita 13:58–73
180 Toyota K, Kondo T 1969 Studies on hardwood resin. Part 7. J Jpn Tappi 23:450–456

181 Trafford J 1987 Pitch investigations with *Pinus radiata* bisulphite and thermochemical pulps. Appita Preprint No 13, 19 pp
182 Tsai G S C, Klamecki B E 1980 Separation of abrasive and electrochemical tool wear mechanisms in wood cutting. Wood Sci 12:236–242
183 Wardrop A B, Davies G W 1961 Morphological factors relating to the penetration of liquids into wood. Holzforschung 15:129–141
184 Yaga S, Sumimoto M, Kondo T 1967 Studies on hardwood resin. Part 5. J Jpn Tappi 21:749–754
185 Yakimovets T L, Kossoi A S, Chudakov M I 1976 [Lignin sorption by wood during treatment of chips with black liquor.] Khim Drev (Riga) 1:64–69
186 Yatagai M, Takahashi T 1980 Tropical wood extractives' effects on durability, paint curing time, and pulp sheet resin spotting. Wood Sci 12:176–182
187 Yoshinaga Y, Sumimoto M, Kondo T 1969 Studies on hardwood resin. Part 6. J Jpn Tappi 23:324–330

9.4 Effect of Extractives on the Utilization of Wood

T. YOSHIMOTO

9.4.1 Introduction

Heartwood – industrially the most important component of wood – contains many diverse extractives. These extractives often affect wood utilization and some have been known to be injurious to the health of woodworkers (44) (see Sect. 9.5). When a wooden surface is painted or glued, extractives may dissolve into the paint or glue and retard normal setting. Another effect is the change in the setting of cement in contact with wooden forms. Further, when wood surfaces are irradiated by light or colonized by microorganisms, extractives may change the wood color and cause new compounds to be synthesized. These effects of extractives are described in this section.

9.4.2 Inhibition of Resin and Glue Curing by Extractives

Resins are an essential component of many paints and glues. When wood is painted or glued, extractives sometimes inhibit resin curing. The inhibition often

Table 9.4.1. Extractives detrimental to paintability

Genus and species	Toxic extractives	References
Cryptomeria japonica	ferruginol (1)	8
Dryobalanops sp.	vanillic acid (2)	11
Larix leptolepis	taxifolin (3)	25
Mansonia altissima	mansonone-F, (4), -H, (5)	46
Taiwania cryptomerioides	taiwanin E (6)	17
Taxus mairei	isotaxiresinol (7)	18
Tectona grandis	tectol (8), dehydroxylapacol (9)	33
Zelkova serrata	mansonone-G methyl ether (10)	47

occurs when unsaturated esters are used in the resin. Because curing of the resin involves radical polymerization of monomers, such as methyl methacrylate, the polymerization process is stopped when certain extractives enter into this reaction.

Many wood species are known to contain inhibitory extractives (Table 9.4.1). Phenolic compounds and benzoquinones are common substances shown to be inhibitory to polymerization of vinyl monomers (27). Wood rich in *n*-hexane extracts also is inclined to inhibit polymerization (45). Ether-soluble extractives can enter paint resin and inhibit the curing (9). The extent of this inhibitory effect is dependent on the specific extractives. The most inhibitory compound among a group examined was pyrogallol (26).

Reactions between some natural compounds and inhibitory extractives are inevitable. Some substances, such as resin acids, are not inhibitory themselves but have an inhibitory effect in the presence of other extractives (19). Some effects on inhibition by extractives depend on the type of resin monomer. Methyl methacrylate and polystyrene are commonly used for paint resin. The former is damaged more than the latter by the addition of isotaxiresinol (7) and taxifolin (3) (26). Moreover, different initiators of polymerization reactions may influence the inhibitory effect of extractives (18).

Extractives show a greater inhibitory effect at higher concentrations. If large amounts of extractives are concentrated in one location, such as at a knot, paint film at that location is damaged. For example, the ether extract of *Pinus ponderosa* knots is 28.5% of the ovendried weight of the wood. Reaction temperature also has a significant influence on inhibitory effect; the effects are greater at 60°C than at 40°C.

The extent of the inhibition of polymerization is observed by measurement of gel-time, curing time, maximum temperature of reaction, and dilatometric change. In the process of gluing wood, extractives interact with the glue at two stages: The first is during glue penetration into the wood; the second is during the polymerization of the glue. These interactions are dependent on specific extractives. The effect of extractives is mainly evaluated by the decrease of gluing power and by changes in gelation time on mixing extractives into the glue.

None of the extractives that prevent gluing have been characterized yet, although the extractives of some wood species have been studied in relation to gluing problems.

9.4.2.1 Phenolic Resin Adhesives

An inhibitory effect of *n*-hexane and ether extracts on gluing is shown in *Dryobalanops* sp. (kapur). Some compounds in these extracts retard penetration of the resin into wood surfaces. The presence of 1% ferulic acid in the ether extract also inhibits setting of the resin (11). An alkaline extract of kapur shows an inhibiting effect. An ether-soluble part of the alkaline extract causes gluing problems, as carboxylic acids lower the pH to make the resin acidic (1).

Dipterocarpus sp. (apitong) contain many ether- and benzene-soluble extractives that retard resin penetration into wood (13). A similar effect is found in *Shorea robusta* (sal). In this case, the problem is avoided by ether extraction of the wood; water extraction, by contrast, affords no improvement in gluing properties (12).

Hydrolyzable tannins play important roles in gluing problems with *Eucalyptus regnans* and *E. maculata*. Tannins can have a number of effects on a phenolic resin adhesive, some of them affecting bond quality in opposite directions. For example, tannins may physically block resin penetration into the wood. For these and other species, reactions with formaldehyde and methylol groups might retard cure by depleting those moieties or might enhance the rate of effective network formation; the former would increase resin penetration and the latter decrease it. The acidity of the tannins may reduce the resin pH, thereby decreasing resin solubility and retarding both cure rate and penetration. Moreover, because phenolic resin cure rate catalysis goes through a minimum at intermediate pH, the acidity change could either retard or enhance cure rate.

9.4.2.2 Amino Resin Adhesives

Normal reaction of amino resins proceeds under acidic conditions. The reaction demands hydrochloric acid as a catalyst, which is produced according to the following reaction:

$$4\,NH_4Cl + 6\,CH_2O \rightarrow (CH_2)_6N_4 + 4\,HCl + 6\,H_2O$$

In kapur, oak, and other wood species this reaction is retarded. Hydrochloric acid formation is delayed by the presence of a 1% hot-water extract of kapur in the resin. This extract contains tannin and lignin-like substances (2). The pH can be lowered by ethanol extract in *Quercus alba* and *Q. falcata*. This changes the gel time by 40%, but in turn this causes other problems (38).

In *Tectona grandis* (teak) extractives are responsible for three kinds of gluing problems (14): 1) Alcohol- and benzene-soluble compounds (11.5% of dry wood) hinder glue penetration into wood; 2) formation of hydrochloric acid is delayed by hot-water-soluble compounds (the inhibitors originate in Ca, Mg, and K salts of acidic sugars in the methanol insolubles of the hot-water extract); and 3) the solidified glue is hydrolyzed by the methanol-soluble fraction of the hot-water extracts.

9.4.3 Inhibition of Cement Hardening by Extractives

The normal mechanism of cement hydration is altered in the presence of wood extractives. There are many cases in which wood is used in contact with concrete. For example, cement is mixed in or molded with wooden forms, or cement and wood are mixed together for board production. Cement yields calcium ions when mixed with water and exhibits a pH of 13. The alkaline solution dissolves extractives from wood into the slurry. The resulting substances have a low solubility and they precipitate with CaO. CaO is one of the most important chemical compounds in concrete, and its loss retards the normal hydration reaction (35). These occurrences are dependent not only on the kinds of extractive, but also on whether the exposed wood surface is a transvers or a longitudinal section.

9.4.3.1 Effect of Extractives on Cement Hardening

Many substances are known to be inhibitory to cement hardening – for example, sucrose, galactose, fructose, glucose, lactose, dextrin, glycogen, mannitol, tannic acid, and succinic acid. Most of these are sugars and common wood extractives. Sucrose may be one of the most effective. Cement hardening proceeds very slowly if it contains as little as 0.025% sucrose. When 0.25% sucrose is present, the expected reaction mechanism in the hydration of the cement does not occur and the resulting block of concrete has very little bending strength (39).

Many species of wood used as chips in wood-cement composites inhibit normal hardening of cement (31). Measurable amounts of glucose and other common sugars are present in the hot-water extracts of these woods and have been shown to be inhibitory (32, 37). The effect of sugars and tannin on the inhibition of cement setting can be ascertained. Hot-water extracts from wood meal are developed on thin layer chromatography (TLC) and then sprayed with a cement slurry instead of chemical reagents to detect inhibitory substances (5).

The extent of the disturbance of cement setting by wood is evaluated by observing the heat generated in a mixture of cement, water, and wood meal. The extent of the disturbance varies with the wood species. For example, *Picea abies* has little effect, but *Larix decidua*, which is rich in arabinogalactan, causes a dramatic change, reflecting the presence of inhibitory extractives in the wood. Using this method, Sandermann (34) showed certain species to have an inhibitory effect on cement setting: *Larix decidua* and *Acer* sp., which contain major amounts of tannins and sugars, and *Pseudotsuga menziesii* and *Sequoia sempervirens*, which contain large amounts of tannins. The heat generated in a mixture of cement, wood, and water does not change if the wood is first completely extracted with hot water (24). The resin content of wood does not contribute to this inhibition (34). An inhibitory component of manngashinoro (*Shorea* sp.) was found to be a water-soluble glycoside (43).

Cement slurry is capable of dissolving significant quantities of extractives, but only for about 30 minutes; after this period only a negligible amount of extractive is dissolved. The amount of extractive dissolved into the cement slurry is greater from wood cross-sections than from longitudinal sections (48). In wooden concrete forms, cement slurry is in contact only with the longitudinal section of wood. But in wood-cement board production, both longitudinal and cross-sections of wood are in contact with the cement. Therefore, wood used in wood-cement boards is more prone to have its extractives dissolved into the cement slurry than wood used for concrete forms. It follows that fewer wood species can be used for the boards than for the forms. *Thuja plicata* (western cedar), *Anisoptera* sp. (mersawa), and *Shorea* sp. (merapi) have been shown to be unsuitable for use as concrete forms (50).

9.4.3.2 Effect of Light Exposure on Wood Panels Used with Cement

When a wooden form exposed to sunlight for more than 3 days is used for cement molding, the cement adjacent to the surface in contact with the exposed form sets abnormally (20). Without exposure to sunlight, almost all wood species used for concrete forms have little effect on cement setting; *Thuja plicata*, *Anisoptera*, and *Shorea* are exceptions. As exposure to light irradiation increases, cement setting is disturbed more and more. Reducing sugars have been shown to occur in greater concentration in hot-water extract from light-irradiated wood than from wood not exposed to light (52). Photo-produced carbohydrates soluble in cement slurry are polysaccharides of xylose (55). The molecular weight of the degraded polysaccharides from *Tilia japonica* is reduced to about 5000 (51). Decomposition of xylan proceeds both in the solid and in the suspended state by light irradiation (28).

9.4.4 Color Change by Light

The original color of wood is mostly due to its extractives (Sect. 9.1.2). Most woods change color by the actions of various factors such as light, metals, water, and fungi. Distinct changes are observed in a few kinds of woods, which are consequently limited to fewer uses. The extent of the color change is expressed by differences in absorbance or reflection of light. The extent and practical value of color differences include a consideration of the sensitivity of the human eye to specific wavelengths.

9.4.4.1 Color Changes Upon Exposure to Light

Almost all woods show color changes on surfaces exposed to sunlight. After irradiation, woods generally become tan. The nature and rate of the color change by light vary according to the species, and many woods turn to a dull, less attractive shade. It has been shown with many woods that the ultraviolet part of the spectrum is mainly responsible for changes brought about by exposure to light, but visible light is involved in some cases. Some extractives, such as pinosylvin found in pine (*Pinus* spp.), change color in response to irradiation. Specific extractives in woods show distinct color changes (Table 9.4.2). These extractives change color in solution or on TLC plates upon light irradiation. Few of the chemical structures of the coloring substances produced by light irradiation have been characterized. Much effort to demonstrate the relation between functional groups, double bonds, or substituents, and color change has been made. A hydroxy group at position 7 of flavan-4-β-ol has been associated with color change. The chemical change following light absorption is hypothesized in Fig. 9.4.1 (6, 30).

The color change in leucorobinetinidin (**11**) and anhydroleucofisetinidin (**13**) are included in this mechanism (6, 22). A hydroxy group at position 6 of an

Table 9.4.2. Extractives showing abnormal color change

Genus and species	Extractives	Yields (%)[a] %[a]	References
Acacia mearnsii	leucorobinetinidin (**11**)	0.9	30
Afrormosia elata	3,3',4,5'-tetrahydroxystilbene (**12**)	0.3	23
Baikiaea plurijuga	anhydroleucofisetinidin (**13**)	–	22
Chlorophora rigita	2,3',4,5'-tetrahydroxystilbene (**14**), chlorophorin (**15**)	2.0	23
C. excelsa	2,3',4,5'-tetrahydroxystilbene (**14**), chlorophorin (**15**)	4.0	23
Gluta or *Melanorrhoea* sp.	rengasin (**16**)	0.4	53
Sequoia sp.	sequirin C (**17**)	–	3

[a] On basis of ovendried wood

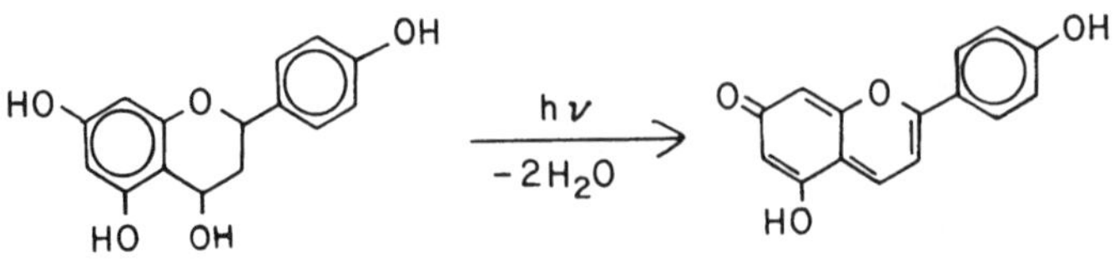

flavan 4β-ol

Fig. 9.4.1. Tentative mechanism of color change of flavan 4β-ol

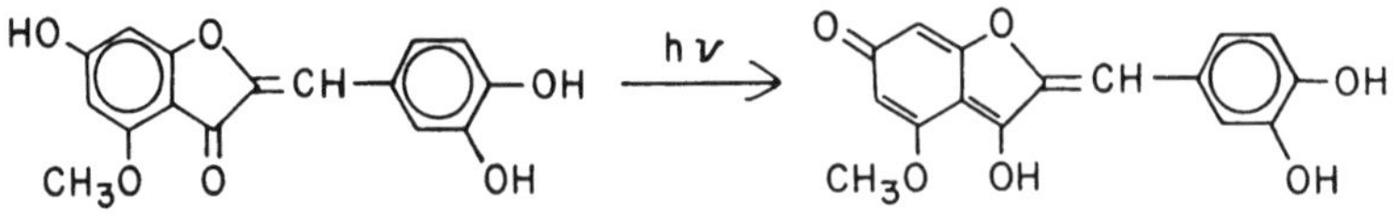

rengasin

Fig. 9.4.2. Tentative mechanism of color change of rengasin

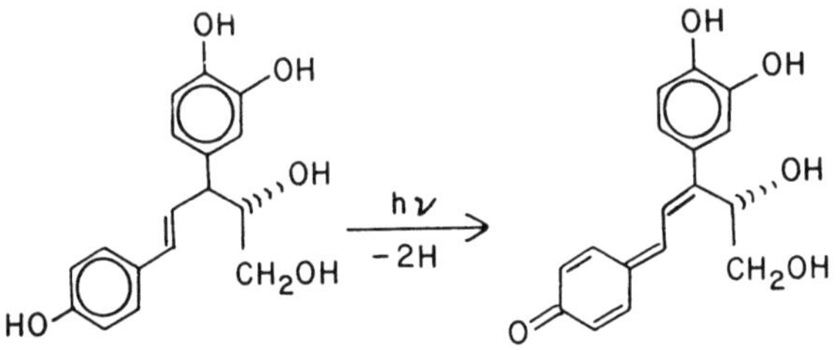

sequirin – C

Fig. 9.4.3. Tentative mechanism of color change of sequirin C

aurone is considered to play a role similar to that of the hydroxy group at position 7 of flavan-4-β-ol. The chemical change of rengasin (**16**) proceeds perhaps as shown in Fig. 9.4.2. UV and visible spectra of irradiated rengasin support the hypothesis of this mechanism (53). Some norlignans, such as sequirin C (**17**), hinokiresinol (**18**), and agatharesinol (**19**), readily change color after light irradiation (40). The chemical structures of photo-produced substances are not shown, but the chemical reaction is presumed to proceed as in the scheme shown in Fig. 9.4.3. The norlignans have a chemical structure similar to flavan 4β-ol, including a phenol conjugated with a double bond and a nonconjugated phenol. Isoflavene (**23**) on the surface of *Milettia* sp. was found to easily cause a change in color, as in the scheme shown in Fig. 9.4.4 (21).

A few woods that contain light-sensitive compounds, such as the hydroxylated stilbene derivatives as a minor component, turn to a deeper shade by the oxidative polymerization of the extractive (10, 23). It is well known that stilbenes easily isomerize after light absorption. Stilbenes isolated from woods are often reported to form colored compounds upon exposure to light, as in the case of

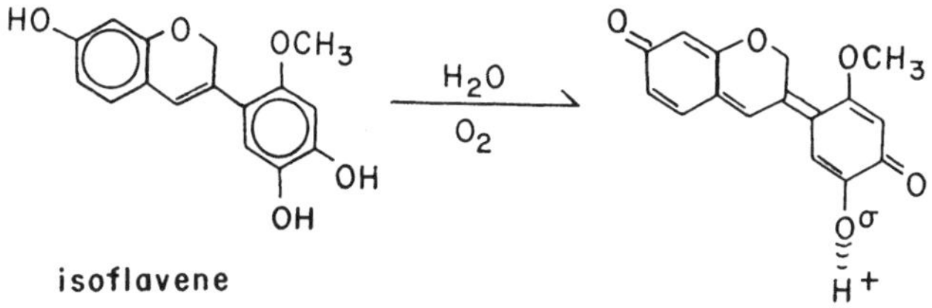

Fig. 9.4.4. Tentative mechanism of color change of isoflavene

3,3′,4,5′-tetrahydroxy stilbene (**12**). *m*-Hydroxy groups of the B-ring are necessary for the color change of stilbenes (23). UV and visible spectra of irradiated stilbenes without hydroxyl groups showed no production of a coloring substance (23). Not all woods containing the above compounds always cause distinct color changes. For a significant color change to be observed, the wood should contain over 0.5% extractives, and the original color of the wood should preferably be nearly white.

The extent of the color change caused by irradiation varies. The maximum change is more than five times the minimum change observed on a series of sampled wood species (54). It is a reasonable assumption that the extent of the color change of wood depends on specific wood constituents, most of which have not been characterized except for those in Table 9.4.2. In addition, two experiments support the hypothesis that the presence of specific wood constituents corresponds to a distinct color change. One study shows wood to change color when irradiated by visible light. Among 75 wood species tested, 22 species demonstrated this phenomenon (36). The second study shows that the amount of photo-produced radicals is greater in wood that changes color more easily than in noncoloring wood (49). The difference in photo-produced radicals clearly reflects irradiation by visible light (440 nm). The results of these two studies provide some understanding of the presence and role of specific constituents that correspond to distinct color changes, although specific compounds have not been isolated from all species of wood known to change color in response to visible-light irradiation.

9.4.4.2 Color Change by Other Agents

Woods that contain water-soluble coloring matter may give rise to color problems if they are used in wet situations; the coloring matter is carried to the wood surface by the water, and leaves watermarks and stains upon evaporation. These problems are generally found in several species of the Fagaceae, which contain tannin and other phenols such as flavonol glycosides. Woods with water-soluble phenols containing galloyl groups, such as tannins, are particularly prone to troublesome discoloration. The best known are the blue-black stains due to the formation of complex compounds when the galloyl residues containing phenols come into contact with iron or iron compounds, particularly under damp conditions (16). Many woods contain some water-soluble phenolic compounds, but a few of them – notably *Quercus* spp., *Castanea* spp., *Juglans* spp. – are well known for

928 Influence of Extractives on Wood Properties and Utilization

their susceptibility to iron compound discoloration (41). Conifer woods that contain troponids, peculiar to the Cupressaceae, are often stained red or green when they come into contact with iron or copper compounds (42). This is due to the formation of colored tropolone metal salts.

The so-called blue stain, which occurs commonly in the sapwood of *Pinus* spp. and several other timbers, is due to the invasion of wood by fungi with hyphae that are colored by melanin-like pigments rather than wood extractives (15). The sap-staining fungi are a number of molds, of which *Ophiostoma* spp. and *Leptographium* spp. occur worldwide. The resulting discoloration affects the appearance of the wood, but the strength properties are not impaired (7). Accumulated extractives in contact with ferro-compounds may result in coloration of paint films. Pyrocatechol causes such a transformation in *Sequoia* sp. (9) and β-thujaplicin and phenolics in *Chamaecyparis nootkatensis* react in a similar manner (4).

Finally, wood constituents corresponding to color change are specific, but not always extractable. Several wood species that were exhaustively extracted continued to exhibit a degree of color change similar to the unextracted samples (54). Unexplained is the fact that some extractives in white woods seem to be light-absorbent (54).

References

1 Abe I, Akimoto N 1976 [The inhibitory effect of kapur wood extractives on the curing reaction of the resol.] Mokuzai Gakkaishi 22:191–196
2 Akaike Y, Nakagami T, Yokota T 1974 The inhibitory effect of kapur wood extracts on the gelation of the urea resin adhesive. Mokuzai Gakkaishi 20:224–229
3 Balogh B, Anderson A B 1965 Isolation of sequirins, a new phenolic compound from the coast redwood (*Sequoia sempervirens*). Phytochemistry 4:565–576
4 Barton G M 1976 A review of yellow cedar (*Chamaecyparis nootkatensis* (D. Don)) extractives and their importance to utilization. Wood Fiber 8:172–176
5 Broeker F W, Simatupang M H 1973 [Thin layer chromatographic detection of substances which disturb the hardening of cement.] Zement-Kalk-Gips 245–247
6 Drew S E, Roux D G 1964 Chromophoric properties of flavan 4 β-ols, flavan 3,4-diols and condensed tannins. Biochem J 92:559–564
7 Farmer R H 1967 Fungal decay of wood. In: Chemistry in the utilization of wood. Pergamon Press New York London, 128–137
8 Funakoshi H, Nobashi K, Yokota T 1978 The inhibitory extractives in sugi heartwoods for the vinyl polymerization. Mokuzai Gakkaishi 24:141–145
9 Gardner J A F 1965 Extractive chemistry of wood and its influence on finishing. Official Digest 698–706
10 Imamura H 1970 [Phenolic constituents of wood and woodworking.] Mokuzai Kogyo 25:201–205
11 Imamura H, Takahashi T, Yasue M, Yagishita M, Karasawa H, Kawamura J 1970 [Effect of wood extractives on gluing and coating of kapur wood.] Bull Gov For Exp Sta 232:65–96
12 Jain N C, Gupta R C, Chauhan B R S 1974 Effect of extractives on gluing of *Shorea robusta* (sal). Holzforsch Holzverwert 26:129–130
13 Jordan D L, Wellons J D 1977 Wettability of *Dipterocarpus* veneer. Wood Sci 10:22–27
14 Kanazawa H, Hakagami T, Nobashi K, Yokota T 1978 [Study on the gluing the wood. XI. The effects of teak wood extractives on the curing reaction and the hydrolysis rate of the urea resin adhesive.] Mokuzai Gakkaishi 24:55–59

15. Kitamura Y, Kondo T 1958 Chemical study on blue-stained pine wood. On the pigment obtained from mycelia of blue-stained fungi. Mokuzai Gakkaishi 4:51–55
16. Kondo T, Ito H, Suda M 1956 On the heartwood components of *Platycarya strobilacea*. J Agr Chem Soc Jpn 30:281–283
17. Lee C L, Hirose Y, Nakatsuka T 1974 [The inhibitory effect of *Taiwania cryptomerioides* heartwood extractives on the curing of the unsaturated polyester resin.] Mokuzai Gakkaishi 20:558–563
18. Lee C L, Hirose Y, Nakatsuka T 1975 [The inhibitory effect of *Taxus mairei* S.Y. Hu heartwood extractives on the curing of unsaturated polyester resin.] Mokuzai Gakkaishi 21:249–256
19. Mibayashi S, Nakatsuka K, Yokota T 1978 Effect of phenolic compounds and resin acids on the polymerization of styrene in the coexistence of inhibitors. Bull Kyoto Univ For 50:209–215
20. Minami K, Kondo M, Yoshimoto T 1967 Imperfect hardening of cement caused by moulding with wood board exposed to sunlight. Mokuzai Gakkaishi 14:91–95
21. Mitsunaga T, Kondo R, Imamura H 1987 The chemistry of the color of wood. IV. The phenolic constituents to the coloration of murasakitagayasa (*Milettia* sp.) heartwood. Mokuzai Gakkaishi 33:239–245
22. Morgan L W W, Orsler R J 1967 Rhodesian teak tannin. Phytochemistry 6:1007–1012
23. Morgan J W W, Orsler R J 1968 The chemistry of color changes in wood. Holzforschung 22:11–16
24. Moslemi A A, Garcia J F, Hofstrand A D 1983 Effect of various treatment and additives on wood-portland cement-water systems. Wood Fiber Sci 15:164–176
25. Nobashi K, Yokota T 1975 The inhibitory effects of extractives prepared from some coniferous wood on the polymerization of methyl methacrylate. Mokuzai Gakkaishi 21:315–322
26. Nobashi K, Yokota T 1976 The effects of wood extractives on the reaction products of vinyl polymerization. Mokuzai Gakkaishi 22:466–472
27. Nobashi K, Shimada T, Yokota T 1977 Inhibition of the vinyl polymerization with extractives from keyaki wood. Mokuzai Gakkaishi 23:670–675
28. Peng J-Y, Minami K, Yoshimoto T 1976 Photolysis of xylan. Mokuzai Gakkaishi 22:401–411
29. Plomley K F, Hillis W E, Hirst K 1976 The influence of wood extractives on the glue-wood bond. 1. The effect of kind and amount of commercial tannins and crude wood extracts on phenolic bonding. Holzforschung 30:14–19
30. Roux D G, Drew S E 1965 Structural factors associated with the redness induced in certain condensed tannins by sunlight or heat. Chem Ind 1442–1446
31. Sandermann W, von Dehn U 1951 Einfluss chemischer Faktoren auf die Festigkeitseigenschaften Zementgebundener Holzwolleplatten. Holz Roh-Werkst 14:97–101
32. Sandermann W, Brendel M 1956 Die "zementvergiftende" Wirkung von Holzinhaltsstoffen und ihre Abhängigkeit von der chemischen Konstitution. Holz Roh- Werkst 14:307–313
33. Sandermann W, Dietrichs H-H, Puth M 1960 Über die Trocknungsinhibierung von Lackanstrichen auf Handelshölzern. Holz Roh-Werkst 18:63–75
34. Sandermann W, Kohler R 1964 Über eine kurze Eignungsprüfung von Hölzern für zementgebundene Werkstoffe. Holzforschung 18:53–59
35. Sandermann W, Preusser H-J, Schweers W 1960 Über die Wirkung von Holzinhaltsstoffen auf den Abbindevorgang bei zementgebundenen Holzwerkstoffen. Holzforschung 14:79–77
36. Sandermann B, Schlumbom F 1962 Über die Wirkung gefilterten ultravioletten Lichtes auf Holz 2. Änderung von Farbwert und Farbempfindung an Holzoberflächen. Holz Roh-Werkst 20:285–291
37. Simatupang M H 1986 Abbaureaktion von Glucose, Cellobiose und Holz unter dem Einfluss von Portland-Zementmörtel. Holzforschung 40:149–155
38. Slay J R, Short P H, Wright D C 1980 Catalytic effects of extractives from pressure-refined fiber on gel time of urea-formaldehyde resin. For Prod J 30(3):22–23
39. Suzuki S, Nishi H 1959 [Effect of sugars and several organic compounds in hydration of cement.] J Jpn Cement Eng Assoc 160–170
40. Takahashi K 1981 Heartwood phenols and their significance to color in *Cryptomeria japonica* D. Don. Mokuzai Gakkaishi 27:654–657
41. Takenami K 1964 Sensitivities of various wood species for the dyeing effect with iron. Mokuzai Gakkaishi 10:22–29
42. Takenami K 1964 Sensitivities of various wood species for the dyeing effect with copper and chrome. Mokuzai Gakkaishi 10:30–35

43 Yasuda S, Niwa J, Terashima N, Ota K, Tachi M 1986 [Inhibitory component of manngashimoro wood in cement hardening.] Mokuzai Gakkaishi 32:748–751
44 Yasue M, Ogiyama K, Abe T 1975 [Basic data on chemical constitutents of imported wood and injury to health occurring in the course of woodworking.]. Nat Res Sci Technol Jpn Publ 28:100–104
45 Yatagai M, Takahashi T 1980 Tropical wood extractives' effects on durability, paint curing time, and pulp sheet resin spotting. Wood Sci 13:176–182
46 Yokota T, Nakagami T, Ito S, Tsujimoto N 1972 The inhibitory substance for the polymerization of styrene. Mokuzai Gakkaishi 18:307–314
47 Yokota T, Noguchi A, Nobashi K 1976 The inhibitory effects of extractives prepared from keyaki heartwood on the vinyl polymerization. Mokuzai Gakkaishi 22:632–637
48 Yoshimoto T 1978 [A simple method for selecting wood suitable for a wood-cement board.] Mokuzai Kogyo 33:18–20
49 Yoshimoto T 1983 [Photodegradation of wood.] In: Imamura H, Okamoto H, Goto T, Yasue M, Yokota T, Yoshimoto T (eds) [Chemistry in wood utilization.] Kyoritsu Shuppan Tokyo, 7–22
50 Yoshimoto T, Minami K 1975b [Toxic activity of 50 tropical woods to cement hardening.] Mokuzai Kogyo 30:23–26
51 Yoshimoto T, Minami K 1976 Imperfect hardening of cement caused by moulding with wood board exposed to sunlight. (4). Mokuzai Gakkaishi 22:376–379
52 Yoshimoto T, Minami K, Kondo M 1967 Imperfect hardening of cement caused with reducing substances produced in wooden moulding board exposed to sunlight. Mokuzai Gakkaishi 13:96–101
53 Yoshimoto T, Samejima M 1977 [Rengas wood extractes relating to light induced reddening.] Mokuzai Gakkaishi 23:601–604
54 Yoshimoto T, Shibata A, Minami K 1975 [Evaluation of the contribution of extractives to the overall color change by light irradiation on tropical woods.] Mokuzai Gakkaishi 21:381–386
55 Yoshimoto T, Uchida M, Minami K 1971 Imperfect hardening of cement caused with reducing substances produced in wooden moulding board exposed to sunlight. (2). Mokuzai Gakkaishi 17:22–27

9.5 Health Hazards Associated with Extractives

E. P. SWAN

9.5.1 Introduction

> "with juice of cursed hebona in a vial, and in the porches of mine ear did pour the leprous distillment; whose effect holds such an enmity with the blood of man..."
>
> (Hamlet, Act I, Sc. IV.)

The story told by the dead king, of being murdered by his brother using hebona juice, leaves an indelible picture in the mind. Perhaps the reason for this is that the story was so well and succinctly told, or it may have been that the use of a natural product as a weapon made the story more horrible. The latter explanation may account for the success of the modern murder mystery. We are used to its victims being dispatched with a variety of natural poisons: from curare-tipped darts to tetrodotoxin-poisoned meals. Compounds such as curare and tetrodotoxin are toxic natural products of wide diversity and have many curious structures, chemically speaking. However, these compounds, which are extracted from leaves or puffer fish, are not included in the present survey. More commonplace examples of poisoning woods such as poison oak, poison ivy, and sumac (57) are known only too well. Thus, toxic (dermatological) effects are known to be caused by extractives present in the wood and bark of species much closer to home.

The toxic effect of wood extractives has sometimes been fatal; however, the toxins have generally been discovered after some instance of distress in workers or explorers. Such toxic effects were generally the first indication of the toxicity of particular classes of compounds that were extracted. The first examples were application route, whose action was across the skin – that is, transdermally. Senear (168) in 1933 reported on a case of dermatitis caused by exposure to *Taxodium* sp. Also, he surveyed the field and described 143 species of wood that were troublesome. For 54 years, his review has been a classic in this field.

A second application route was discovered when asthma or bronchial reactions were noted by many sufferers. Thus, the offending substances were acting through inhalation. A third method of introducing the toxins was by direct injection: the effects of curare and other poisons (68) have been described. Finally, all three possible application routes of intoxication (or a combination of them) may be necessary to cause the development of various cancers (121).

The natural consequence of these four toxic effects has been the study of their occurrence in woodworkers and others. The study of the toxic extractives in worker safety and industrial hygiene was essential to understanding the causes of disease and to preventing the effects from developing. Several early reviews in the field will be covered, but the book by Hausen (75) is the most recent, and best, comprehensive study.

The final section in this chapter traces the history of developments in this field and looks at future developments, research recommendations, and methods for designing such studies on the secondary metabolites of woody plants.

9.5.2 Toxic Extractives

9.5.2.1 Alkaloids and Amino Acids

There are very few alkaloids present in wood (75, 152), but they are, of course, found extensively in plants (143, 187, 188). There were many instances of toxic alkaloids in the ancient literature. The most infamous example was the use of coniine from poison hemlock, *Conium maculatum*, to kill the greatest philosopher of all, Socrates. The hemlock woods of commerce (*Tsuga* spp.) are not related to this species, and they do not contain coniine. Nowadays, more poisoning of livestock, rather than people, occurs as a consequence of alkaloid ingestion. Keeler (93) recently reviewed the many plant toxins and their effects on livestock. Various alkaloids from *Magnolia* spp. have been shown (152) to possess cytotoxic and antibacterial effects. There are several toxic non-protein amino acids, but only mimosine has been extracted from wood. The effects of this compound have been reported by Keeler (93). The comparative toxicity of some non-protein amino acids has been reviewed (70, 151).

9.5.2.2 Saponins and Glycosides

There are very few toxic carbohydrates; except for carrageenan-produced edema (47), the gums, pectins, and syrups appear to be beneficient. The carbohydrate derivatives, the glycosides, are quite the reverse. They are generally divided into cardiac glycosides, saponins, and cyanogenic glycosides (152), and act on the heart, blood, and brain, respectively. The relationship between the biological activity and the carbohydrate moiety in cardiac glycosides has been established (4). Most glycosides have been isolated from foliage, but poisoning from the wood of *Nerium oleander* has been reported (121). A hypoglycemic principle was isolated from *Ficus glomerata* (11), and Sanduja et al. (159) isolated cardenolides from *Cryptostegia madagascariensis*. West et al. isolated saponins from *Ilex opaca* (185).

9.5.2.3 Quinones

The quinones are generally skin irritants. Some compounds have both toxic and dermatological actions. For example, juglone was known to be toxic as well as to cause dermatitis (170). Other examples were reviewed most recently by Hausen (75). Jones et al. (89) showed that 2,6-dimethoxybenzoquinone, a cytotoxic constituent of the woody stems of *Tibouchina pulchra*, has an LD$_{50}$ of 2.5 µg/ml in KB cell culture.

9.5.2.4 Phenolics

A few examples are found of the toxicity of some of the phenolics to livestock. The phenol ether, tremetone, from *Eupatorium rugosum*, caused toxic effects in livestock as did the isoflavones, genistein, and coumestrol (93).

MacRae and Towers (106) reviewed the biological activities of the lignans and discussed those that are toxic to fungi, insects, and vertebrates, as well as those lignans that are known to have anti-tumor, anti-mitotic, and anti-viral activities. The most studied lignan thus far is podophyllotoxin because of its anti-cancer activity. However, podophyllotoxin is itself toxic (144).

9.5.2.5 Terpenes

The pharmacology and percutaneous absorption of common monoterpenes were studied by Schafer and Schafer (160). Among the monoterpenes, the ketones appeared to be more toxic than related compounds. In particular, camphor has been used medically since ancient times. The abuse of camphor and its toxic effects, especially in children, continued to be a problem (7, 69, 87, 88, 95). The toxicology of some borneol isomers (139) and of synthetic camphor (146) has been studied.

Another monoterpene ketone, thujone, was more toxic than camphor. Arctander (6) stated that some authorities considered thujone to be the most toxic of all common essential oil components. He considered thujone to be very toxic, with a paralyzing effect on the human central nervous system. Its presence in *Artemesia* spp., whose oil was added to various liqueurs, was responsible for the banning of the one known as "absinthe". The toxicity data are given in Food and Cosmetics Toxicology (60), which quotes an oral LD_{50} value (in rats) of 830 mg/kg. However, the data for subcutaneous injection (in mice) was even lower: Rice and Wilson (147) showed that the L-form had an LD_{50} of 87.5 mg/kg, the D-form 442.2 mg/kg, and the *racemic* mixture 134.2 mg/kg. The reason thujone is so toxic may possibly be the presence of the cyclopropane ring moiety. Corresponding LD_{50} (i.p. mice) values for the closely related camphor was 3.0 g/kg and for fenchone was 6.16 g/kg (111). Other studies on the toxicity of thujone (141) have been published and its role in possible causes of Vincent Van Gogh's visions were hypothesized (2, 117).

Early work had demonstrated the toxicity of extracts from western red cedar (173). This research was continued in Sweden with the identification of the thujaplicins as the agents responsible for this toxicity and the chemical structure determinations (54, 145, 172). The toxicity, however, was against fungi rather than humans. Meanwhile, Nozoe and co-workers had isolated β-thujaplicin and its iron complex from the oil of *Chamaecyparis taiwanensis* (135, 137). These compounds, which were used medically, were called hinokitiol and hinokitin, respectively, in their literature (136). The durability of *Thuja plicata* wood and its chemistry were reviewed by Gardner (63).

Rodrigues et al. (149) reviewed the biological activities, as well as the structure and occurrence, of the sesquiterpene lactones. The biological activities discussed were anti-cancer, cytotoxic, antibiotic, chemophylaxis (against schistosomiasis), allergic contact dermatitis, antifeedants for insects, vertebrate poisons, and phytotoxins. The structural requirements for biological activity in these compounds were elucidated. Kuksis et al. (96) used GLC to study phytosterolemia.

9.5.3 Allergenic Extractives

The effects of allergenic extracts were, at first, thought to arise exclusively either through skin contact or by breathing wood dust. Mitchell (113) showed that this was not so, and enumerated 12 mechanisms by which extractives could effect allergic reactions: mechanical injury, toxic effects, pharmacological effects, contact dermatitis by irritancy, immunological effects, pseudo- and phytophotodermatitis, parasitophytodermatitis, heterophytodermatitis, and plant/animal dermatitis. Nevertheless, the most common cutaneous routes will be covered in the following three sections. The asthma and breathing disorder effects were present also in some of these compounds (the quinones, alkyl phenols, and terpenes), but most research has been on plicatic acid.

Mitchell and Rook (116) reviewed contact allergy from plants. Mitchell et al. (115) have investigated the role of many sesquiterpene lactones in allergic contact dermatitis.

Hausen (75) reviewed wood properties causing contact allergies among woodworkers, examining the extractives present in over 100 species. Sensitizing experiments with guinea pigs have shown that the contact allergens were always low-molecular weight extractives that were chemically reactive (74).

9.5.3.1 Quinones

Sandermann and Barghoorn (155) contributed an early review (61 references) of problem woods. In several cases they were able to show that specific compounds were responsible for the symptoms of illness. Also, they indicated that there were 42 problem woods for which further research was required. They subsequently isolated the γ,γ-dimethallyl-1,4-naphthoquinone (158) from *Tectona grandis* (156), the quinone from *Mansonia altissima* (157), and lapachonone from *Paratecoma peroba* (154).

Sandermann and Dietrichs (157) examined the heartwood extractives of *Mansonia* sp. Workers using this wood suffered from violent sneezing, nose bleeds, vertigo, fainting, and contact eczema. They showed that the dermatological allergy symptoms were caused by the presence of a quinone related to perezone. Several strophanthidine glycosides were also present in the wood and bark of this species. It was these compounds that had caused the symptoms of vertigo and sneezing.

The bicyclic quinone, juglone, from hickory and walnut (*Carya* and *Juglans* spp.), possesses several activities. It causes blackening, blistering, and skin peeling (152); it is a tranquilizer and sedative (186); it possesses antitumor activity (14); and it is fungitoxic, antibiotic, and allelopathic (14). Several *Diospyros* spp. cause severe allergies because of the presence of quinones (190). The derivatives, 7-methyl juglone and isodiospyrin have germicidal activities (165). Morgan et al. (128) continued their investigation of dermatitis from wood dusts with an investigation in three furniture factories. Several cases were caused by *Khaya anthotheca*, but they discovered, in one factory, several due to *Machaerium scleroxylon*. They claim that the compound responsible for this allergy is 3,4-dimethoxydalbergione and its quinol.

Schmalle et al. (164) determined the structure of 2,6-dimethoxy-1,4-benzoquinone from crystallographic data. This substance was a contact allergen that Hausen and Schulz (77) isolated from 25 different plants and woods, among which were *Tieghemella heckelii*, *Milettia laurentii*, and *Swietenia macrophylla*. Schmalle and Hausen (163) isolated the above benzoquinone and also acamelin, a furano-benzoquinone, from *Acacia melanoxylon*. These two benzoquinones are responsible for the allergy-inducing property of this wood. Schulz et al. (165) studied the sensitizing capacity of quinones.

9.5.3.2 Alkyl Phenols

The simple diphenol, pentadecylcatechol, is the major constituent of the irritant oils from *Rhus* and *Toxicodendron* spp. (see Chap. 7.1). It appears to be ubiquitously troublesome to the general public in poison ivy, oak, sumac, and the lacquer tree (57). It is only moderately troublesome to woodworkers because only a very few commercial woods contain these compounds (75). Hausen states that their mechanism of action involves subcutaneous oxidation of the catechol moiety to an *o*-quinone, which is the active substance (75). The chemical structure (the position and length of the side chain) and potency are discussed by Hausen (75), who states they are the strongest sensitizers known. The comparison with the mechanism of action of deoxylapachol is very interesting.

El Sohly and co-workers (52, 53) have described the separation (1, 104), characterization, synthesis, and antiallergenic properties of pentadecylcatechol and related compounds, which are also known as urushiols. El Sohly also isolated a new component from *Toxicodendron vernix* (1), and analyzed various plant parts in *T. radicans* (45). Various studies by other workers (85) have investigated the immune response in humans (25, 26, 101), mice (50), and guinea pigs (183).

9.5.3.3 Terpenes

Lin et al. (101) isolated five ingenane (diterpene) derivatives from the latex of *Euphorbia hermentiana*. Small amounts of this latex produced an irritant follicular dermatitis in several subjects. The irritant principles were isolated, shown to be the above ingenane-derivatives, and tested on the subjects. *Parthenium* spp. contain toxins, which cause a severe skin rash in humans (167). Rodriguez et al. (148) performed a dermatoxicological study on the extractives from *Parthenium argentatum*. The guayulin A and B present in the extractives are sesquiterpene esters. Only the former elicits contact dermatitis in tests with guinea pigs; its potency is comparable to the pentadecylcatechols from *Toxicodendron* sp. and to parthein. Parthenin, another sesquiterpene derivative, has been studied by Towers et al. (176).

Morton (123) has shown that the pepper tree, *Schinus terebinthifolius*, was a cause of breathing difficulties, especially when it was in bloom. She ascribes these properties to air-volatile agents, possibly the monoterpenes. The fruits, which were poisonous to birds, contained the triterpenes, terebinthone, and schinol.

Morgan and Wilkinson (130) described seven compounds that are known to cause sensitization. They noted a further nine woods that give various symptoms of sickness. Among *Khaya* spp., in particular, nine cases of dermatitis resulted from contact with *K. anthotheca*. They were able to show (131) in five cases that sensitization was due to the presence of anthothecol, a triterpene derivative. The subjects were sensitive to this compound, and not to the extractive-free wood. They believed that the sensitization to this timber was non-typical and possibly caused by variations in extractive contents of particular *K. anthotheca* timber.

The tricyclic diterpene carboxylic acid, abietic acid, is widely distributed (see Chap. 8.1) and commonly available commercially. Fisher implicated it in a case of allergic contact dermatitis because it is the major constituent of violinist's resin (58). It is also found in colophony and pine resins, on which Karlberg et al. (92) performed a dermatological study. Wahlberg (182) studied dermatitis caused by abietic acid and colophony. Abietic acid derivatives are also common, so some derivatives (161) of it have also been studied, notably dihydroabietyl alcohol (49). A white powder from the bark of *Betula* sp. causes skin eruptions in mill workers

(190). The major components of this powder are pentacyclic triterpene derivatives, the lupanes (152).

9.5.3.4 Phenolics

The furanocoumarin derivative, 5-methoxypsoralen, and other related compounds, are responsible for the phototoxicity of various essential oils on skin (9, 65, 169). Other studies with lapachol concern its toxicity to viruses (98) and parasites (102). The sensitizing capacity of naphthoquinones was studied on guinea pigs (166). The problem of photosensitization was present in livestock (93). The effects were caused by 5-methoxypsoralen, and also the bis-anthraquinone hypericin. The flavanoids are thought to be responsible for a contact eczema from *Robinia* sp. (190).

Morgan and Thomson (129) examined a case of dermatitis from *Distemonanthus benthamianus*, showing that it is caused by some extractives present in the wood. In particular, the dermatitis is caused by oxyanin-A and -B, and not by ayanin and distemonanthin, which are also present. All four compounds are flavanoids. Of a variety of *Khaya* spp. imported into the U.K., some are more potent than others. Morgan and Orsler (127) therefore developed a simple color test with a chromatography spray reagent to distinguish *K. anthotheca* from its relatives.

The ability of *Thuja plicata* (western red cedar) dust to cause asthmatic attacks in humans has been studied by Chan-Yeung and co-workers (36, 37). The causative agent in the dust is plicatic acid, the major non-volatile phenolic extractive from *Thuja plicata* heartwood. Mitchell and Chan-Yeung (114) studied contact allergies from *Frullania* sp. associated with this wood. Chan-Yeung followed up on a study of her patients with occupational asthma (30, 31) and she reviewed her research in this field (32, 33, 40, 41). Ashley et al. (8) made a respiratory survey of cedar mill workers. Chan-Yeung et al. (38) investigated the activation of complement by plicatic acid. David and Chan-Yeung studied histamine release in asthma (46). This research was pursued by Giclas (64). She showed that this reaction was due to a specific component of the human serum complement. Lam et al. (100) also studied asthmatic and bronchial reaction and specific antibodies in patients. Also, Tse et al. (177) showed the presence of a specific immunoglobin in the blood of affected workers.

Chan-Yeung et al. (42) have compared the respiratory effects in mill workers versus office workers, and Vedal et al. (178, 179) have related this to duration of exposure to the wood. Lam et al. (99) have studied cellular and protein changes in bronchial lavage fluid following late asthmatic reaction in patients with red cedar asthma.

Large amounts of *Thuja plicata* wood are exported to other countries either as lumber or as logs that are then sawn into lumber. For this reason, several researchers in other countries have been concerned with the health aspects of this wood. In Japan, Mue et al. (132, 133) studied 154 woodworkers for allergic symptoms to *Thuja plicata*. They found that a high incidence of these workers (24.7%) had bronchial asthma. Of these, 89% had a positive intradermal test, with an aqueous extract, as did 56% of the total. They suggested that the exposure to this wood may cause other symptoms as well as the asthma. In research on western red cedar elsewhere, Evans and Nicholls studied histamine release from lung tissue in vitro (56); Cockcroft et al. (43) studied non-specific bronchial hypersensitivity; Hamilton et al. (71) studied bronchial activity in *Thuja plicata*-induced asthma; Blainey et al. (17) studied respiratory tract reactions to this wood; Brooks et al. (22) performed an epidemiological study; and Gozalo Reques (66) studied two causes of asthma.

Malo (107) had shown that several mill workers in Quebec were affected by an occupational asthma from exposure to *Thuja occidentalis*. Subsequently, Cartier et al. (29) showed that this occupational asthma was caused by the presence of plicatic acid in this wood.

9.5.3.5 Other Allergenic Extractives

The disease known as hot-tub dermatitis can be caused by bathing in redwood (*Sequoia* sp.) tubs. The disease is caused by a bacterium (*Pseudomonas* sp.) that enters via hair follicles (180), and it is probably not related in any causal way to the use of redwood.

Maple-bark disease was described as a pulmonary hypersensitivity to *Cryptostroma corticale* (152). Yellow poplar (*Liriodendron tulipifera*) wood caused a severe contact dermatitis (168, 190).

9.5.4 Carcinogenic Extractives

9.5.4.1 Early Work

The carcinogenicity of safrole was the reason for the banning of several beverages from *Sassafras* sp. in the United States (152). The safrole was extracted from the root bark. The carcinogenicity was actually caused by conversion of it into 1-hydroxysafrole, which was a proximate carcinogen (19).

Sabine et al. (153) showed that the high incidence of spontaneous mammary tumors in certain inbred mice could possibly have been caused by bedding of shavings from *Juniperus virginiana*. Early observations were made of cancer in woodworkers (109), and of increased incidence of esophageal cancer (124).

9.5.4.2 Hausen's Contributions

Hausen (75) has written an excellent review of the adenocarcinoma of the nasopharynx in woodworkers. He introduces the subject by the original work of MacBeth (105) showing increased incidence of it in furniture workers in England. He quotes 32 other surveys on the excess occurrence of this carcinoma (mostly in European countries) and six or seven other surveys that did not find such an association. Some data on fine dust in other industries also suggests an increased incidence of various types of tumors. The studies on incidence rates in various countries will undoubtedly continue. The woods responsible are mainly hardwoods (30 authors in various countries had examined the species question). He shows diagrams of the nasal cavities, and possible ways that fine dust could be deposited in them. Hausen's section on theories on the etiological factors in woodworkers' adenocarcinoma is very thorough. He remarks on the very long exposure period to the dust for most workers.

MacBeth (105) believed that the primary causal factor must be some constituent of the wood itself. On the other hand, Michaels (112) examined the lungs from two deceased woodworkers. He believed that pathological changes in lung tissue were due to the presence of wood dust in the worker's environment. The former viewpoint is supported by Hausen who first examined other causative agents. These are molds and fungi, wood tar, tannins (both hydrolyzable and condensed), unsaturated aldehydes, and oxidation products. In particular, he believes that various concentrations of sinapaldehyde in different species and situations (perhaps even from burning the woods with poor ventilation) may be one of the chief causative agents of this cancer. Finally, he reviews the incidence of a different cancer, Hodgkin's disease, in woodworkers. He shows that a great deal of research remains to be performed on several interesting questions, although Greene et al. (67) have investigated this matter. The book by Hausen (75), and this particular section, must be studied by all interested researchers for an understanding of this complex field.

Later reviews have been equally instructive. Recently, Hausen (76) has again reviewed the above work. At least six theories have been put forward regarding etiological factors. In his opinion, the most interesting hypothesis is that the nasal cancer is caused by oxidation of degradation products from lignin, formation of quinones, absorption on various tissue sites, and eventual formation of cancer at the sites.

9.5.4.3 Tannin

The word tannin is used historically, for chemically it is incorrect. Hergert believes that the word tannin should be reserved for substances used in the production of leather; however, the word has been used extensively in a context that means "any brown solution of astringent taste." The mixture of correct and incorrect usages must be followed in this section for two reasons: First, the tannins (correct usage) are found ubiquitously in softwoods, hardwoods, barks, and roots; second, there is a considerable body of research on the effects of tannins (both usages) on humans.

Morton (122) has studied the epidemiology of esophageal cancer, particularly in those areas of the world that have remarkably high rates of this disease: Hunan province in China, Turkmen and Uzbek regions of Russia and Iran, Transkei region of South Africa, Djibouti, the Normandy peninsula of France, Curacao (Netherlands Antilles), northwestern Venezuela, coastal South Carolina, Bombay, western Kenya, and northern Chile. In her opinion, the increased incidence of esophageal cancer in these areas is caused by heavy ingestion of tannin-containing foods, beverages, and folk remedies; also, the disease may be caused by chewing betel nuts and khat (*Catha edulis*) leaves. She found these were high in condensed catechin tannins (now called proanthocyanidins). In South Carolina, an extra risk was found in workers exposed to airborne sawdust in mills and carpentry shops (118). In the Transkei and in Kenya, the cancer can be linked to a native preference for dark-colored strains of sorghum in beer (121); elsewhere to excessive intake of tea (without milk) and dry red wine. She postulates a direct correlation between tannin ingestion and esophageal cancer in a wide variety of dietary habits and sawdust-swallowing in many areas of the world (118, 121, 126). She implicates, too, the tannin in liquor aged in oak barrels or heated with oak chips.

In particular, Morton correlated adverse effects with the flowering of several species of trees such as the mango, the Brazilian pepper tree, the cajeput tree, and others. She has collected all these data into a book (120, 125, 126).

Kapadia et al. (91) have discussed the relationship between herbal tea consumption and cancer. They began with a review of the most common tea, that from the leaves of *Camellia sinensis*. Tea from other plants such as *Ilex paraguaiensis* (maté), and other lesser known teas (such as those from *I. varitoria*, *Eupatorium triplinerve*, and *Sassafras albidium*) were discussed. The resurgence of herbal teas used for medicinal purposes, which are available in health food stores and are used in areas high in risk for esophageal cancer, may be implicated. They concluded that consumption, in moderation, of ordinary tea (especially with milk) is relatively safe. Other teas that are tannin-rich and other products may be causes of increased esophageal cancer.

Morton (119) also studied among various plant extracts the red oak (*Quercus rubra*) extract that has been used as a folk medicine and beverage. She was interested in the esophageal cancer connection between this extract and its victims. These results were described in earlier research that had arisen as a consequence of Morton's work. Thus, Pradhan et al. (142) studied the aqueous extracts of *Krameria ixinia*, *K. triandra*, *Acacia villosa* and *Sorghum vulgare*. These extracts produced malignant fibrous histocytomas at the injection sites in black rats. However, the tannin-free extracts from these plants had little carcinogenicity. Kapadia et al. (90) showed that tannin-containing extracts produced tumors at some of the injection sites. The tannins from *Quercus falcata*, *Diospyros virginiana*, and *Camellia sinensis* were particularly potent. These data showed that both the uncondensed (gallate-based) and condensed (catechin-based) tannins were active.

Some extractives, however, have well known anti-cancer effects. Perdue and Hartwell (140) performed an extensive search for anti-cancer agents in plants. They found that 240 species contained such activity. Of these species, the activity of 84 was due to the tannins present. Fong et al. (59) fractionated extracts from four plants to test their anti-tumor activities. The fractions with the tannins removed were inactive, which showed that the tannins were the compounds responsible for this activity. Subsequently, Loub et al. (103) partially characterized the tannins from one of these plants, *Calygonium squamulosum*, one constituent (ellagic acid) of which possessed anti-tumor activity (14). Wood et al. (189) showed that ellagic acid neutralized the active form of the carcinogen benzo[a]-pyrene. Ellagic acid was the most reactive of the uncondensed tannins.

Obviously, in the past, the difficulty has been that the tannins studied have been difficult to purify and to characterize chemically. This was not the case with the most recent work (59), and precise characterization of structure will be possible in the future. In addition, these studies must include tests in animal models that are recognized as being useful models in place of humans. The differention of a true toxic effect from an irritant one will not be easy. Also, the tumors produced must be examined by an expert pathologist to prove carcinogenicity.

9.5.4.4 Other Carcinogenic Extractives

The natural carcinogens from plants have been reviewed (10, 82, 121). The mutagenic potential of other chemicals from wood (in the pulp and paper industry) have been evaluated (21). The anthrone derivative, chrysarobin, has been investigated for its tumor-promoting activity (48). Several studies have been published on the mutagenicity, and structure-activity relationships, in quercetin (13, 16) and related flavonoids (16, 23, 24, 51, 134, 174). Bhattacharjee et al. (15) tested the wood dusts from *Dalberghia sissoo* and *Mangifera indica* for in vitro hemolytic and macrophage cytotoxicity. Hecker studied the co-carcinogens and cryptic ones from the latex of *Hippomane mancinella* (80). Carcinogens found elsewhere in the environment, such as in foods, have been reviewed by Austwick and Mattocks (10). The herbal remedies of the Maritimes in Canada and their triterpene contents have been surveyed (83).

9.5.5 Hygiene and Safety

The seminal research and review by Senear (168) has been noted above. Subsequently, Sandermann and Barghoorn (155) published an early review that contained a table of 42 problem woods thought to cause illness. Woods and Colman (190) surveyed toxic woods. Among those causing woodcutter's eczema were *Maclura pomifera*, *Platanus americana*, and *Ulmus* sp. (152). Hanslian and Kadec (72) reviewed the toxicity of wood as problems in hygiene in 1963 and again in 1966 (73). Zafiropoulo et al. (192) surveyed the occurrence and frequency of wood allergies in France, showing that about 25 of the woods imported to France (exotic woods) could give various allergies. They reviewed the extractives work and skin allergy tests up to 1968. Interestingly, although *Thuja plicata* was correctly referred to as an exotic timber (192), the author can see them growing outside his window. One man's exotic timber is another's forest. This is an example of the complex trade in timbers with many and various trees, logs, and timbers exported from the producers to the consumers. The health problems are exported together with the wood. When an article is manufactured from the exotic timber, the safety and hygiene considerations are painfully learned over time and in many locations.

Schreiber reviewed the wood species that endangered health in East Germany. These were mainly tropical woods that give rise to dermatitis, hay fever, and allergies in sawmill workers. He made recommendations for hygiene and work safety precautions (162).

Thörnqvist and Lundström (175) studied a respiratory allergy caused by airborne fungal particles from piles of wood chips used as fuel. They found that hardwoods gave more particles than softwoods, and that storage times longer than three months were to be avoided. Antonsson and Lundberg (5) have studied health hazards in the wood-preserving industry in Sweden.

Howie et al. (84) studied the pulmonary sensitivity of a patient to the dust from *Gonystylus bancanus*. They showed that this is an immunological reaction rather than simple mechanical irritation of the bronchial mucosa. Mathias (108) gave a toxicological review of pathological symptoms caused by exposure to fresh wood. The symptoms were described and ascribed to resins and tannins. Schweisheimer (166) reported on a single case of breathing difficulties associated with working with *Pterocarpus angolensis*, *Thuja plicata*, and *Lovoia klaineana* woods. Booth et al. (18) studied two patients with severe respiratory difficulties that resulted from exposure to *Pouteria* spp. Carroll et al. (28) investigated the toxic extractives of the manchineel tree, *Hippomane mancinella*, which Bowder (20) had described rather vividly.

Wellborn (184) surveyed the health hazards for wood workers and listed 35 woods that have been implicated in respiratory ailments and skin or eye allergies. He believed that proper preventative and safety equipment, together with shop cleanliness and good hygiene, were absolutely essential in any woodworking area. His list included many of the above problem woods such as cocobolo, ironwood, and teak. However, there are some species for which the extractives responsible for the problems have not been isolated. On the other hand, some woods have been well studied. Chan-Yeung et al. (35) studied two cases of occupational asthma that were due to exposure to the dust from *Sequoia sempervirens*. They (35) believed that the extractives from the wood were responsible for the symptoms, and they specifically mentioned isosequiric acid, sugiresinol, and hydroxysugiresinol (see Chap. 7.3). However, a disease called sequoiosis is not due to extractives (44). The Japanese work (132, 133) on sensitivity of 154 workers to western red cedar is discussed above (Sect. 9.5.3.4). Mue et al. observed these workers for two years. Afterward, workers who avoided contact with western red cedar, nine of 14, were free from asthmatic attacks. Earlier Japanese work was on exposure to *Thuja standishii* dust (86).

Orsler (138) has reviewed the health problem associated with wood processing in England. The extractives are given as the cause of irritation in dermatitis and respiration. The increased incidence of nasal cancer is noted and, although the causes remain obscure, precautionary measures are given. Also, a table or irritant timbers is given containing eight well established irritants, eight occasional irritants, and 20 possible irritants, by botanical name. Many of the well established and occasional irritants are from species that have been known to cause problems. Baxter et al. (12) have shown that occupational exposure to wood dust was the cause of increased incidence of nasal cancer.

Hausen (75) recommends the use of protective clothing, but he believes that barrier creams are of doubtful efficacy. He advocates worker education, selection, and cleanliness. He discusses outbreaks of allergic contact dermatitis in the factory. Obviously, early and competent assistance from health professionals is necessary in any situation. The identification of the timber species involved and its replacement with less toxic varieties are suggested. Affected workers should be removed from exposure, treated professionally by allergists, and assigned elsewhere or retrained for other safe occupations.

McCann (109) notes the increased incidence of adenocarcinoma among woodworkers. Not only furniture workers, but also automobile (new model and die-builders) workers and others are involved. Precautionary measures are suggested to the hobbyist and fine woodworkers whom he addresses. Similarly, Melino et al. (110) have studied workplace risks and safety aspects in a tropical wood industry.

Hayes et al. (79) have studied the relationship between sinonasal cancer and wood-related occupations, as have Bhattacharjee et al. (15) and Anderson (3). Allergies of various types have been found to have been caused by oil of turpentine in Portugal (27). Chan-Yeung and Lam (39) and Chan-Yeung herself (34) have recently reviewed the status of occupational asthma.

9.5.6 Discussion and Further Research

9.5.6.1 Discussion

Obviously, this field has been developing rapidly in the last few decades. Much early work suffered because of insufficient techniques or isolation difficulties in chemical characterization of the extractive components. Now this is no longer the case; modern advances in techniques, instrumentation, and analytical methods have resulted in some excellent and solid research in this field. A great deal of research remains to be done and this is discussed in the section below.

From a survey of the above sections, we can see that the effects of many important compounds have already been investigated. However, these important compounds that were previously studied were those that were easily isolated and purified. For example, abietic acid, lapachol, and plicatic acid are all compounds that were readily available. Many other compounds in the same chemical class remain to be examined. For example, many resin acids have been isolated besides abietic. The flavanoids are another class of compounds that have received insufficient attention. There are a few examples of flavanoids (or stilbenes) in the above review but it is not clear what their effects or functions are, especially considering their widespread distribution. The possibility of the isolation of artifacts must also be considered. For example, Rowe (private communication) believes that 2,6-dimethoxybenzoquinone is relatively common among hardwoods (he found it in oak) and is a known allergen. Diphenylmethane types are very common among tree phenolics both in lignin, lignans, and other polyphenolics. The methane moiety is readily attacked by oxidative enzymes with cleavage to 2,6-dimethoxybenzoquinone in the case of pendant syringyl groups. Thus, this allergen is presumably an artifact and would be expected to be quite widespread in hardwoods and highly variable in amount. This might contribute to the rather erratic occurrence of sensitivity to various hardwoods.

Another point of development in this field has been the organisms involved. The main interest in this research has been the effects on man, but animals (93) have also suffered from exposure to toxic extractives. We have seen these effects studied on fish, microorganisms, and especially fungi. Rogers et al. (150) examined techniques for identifying extractives that were toxic to aquatic life; Heitmuller et al. (81) investigated the toxicity of 54 chemicals to minnows; Furtado et al. (62) studied the effects of wood extractives on marine organisms; the effects of extractives on wood decay (55) and antifungal components in essential oils (97) have been reported. The role of extractives as antifeedants was studied (61). The original triad of toxic, allergenic, and carcinogenic effects have been joined by others such as antifeedant, phytotoxic, mutagenic, teratogenic, cytotoxic, and antibiotic (-microbial or -fungal) ones. Studying the combination of remaining extractive classes with respect to the above effects presents a staggering picture of the complexity that this field will assume.

This complexity has been reflected in the approach taken by two schools of research. In the first, Morton started with a great deal of data on the epidemiology of cancer. She took these results and showed the importance of tannin in this effect. Much research resulted from her studies, and the effects (both pro and con) of specific polyphenols on these cancers was undeniable. The work of the second researcher, Hausen, has proven the effects of a wide variety of quinones on the above triad of effects. The quinones themselves and their mechanisms of action have been clearly delineated. The health and safety aspects of his book make it indispensable.

9.5.6.2 Future Research

There have been several lists of problem woods (74, 130, 138, 155, 162, 190, 192). A survey of these lists shows that some easily obtainable woods – e.g., *Thuja plicata* – have been well studied. Many others remain unstudied, and these together with the following collection makes them candidates for such research.

Such research, however, was difficult to perform and may contain pitfalls. For example, Sosman et al. (171) studied four cases of hypersensitivity to wood dust that they believed came from the dust per se. The woods involved were pine, oak, mahogany, and cedar. However, these woods were identified only by their common names, and they were not examined by expert wood anatomists. Wood identification cannot always be performed, for in many cases the experts require bud, flower, and cone samples for this work.

In the future, such research will require a multidisciplinary effort. If a wood gives health problem effects, the necessary research may be complex and difficult. Suppose that a symptom or various toxic effects exist in workers who are handling wood. Preferably, these symptoms will arise in more than one worker or location, and will be described in the appropriate literature, then:

1. The disease must be reported by expert medical practitioners as being caused by contact, inhalation, or other association with the wood.
2. The wood responsible must be examined anatomically and correctly described (by scientific name) by an expert botanist, who may need more than wood samples to do this identification.

3. The extractives must be removed, separated into classes, then into individual compounds, by a natural products chemist. The structure, including stereochemistry, of each component of interest must be assigned.
4. The individual extractives must be tested for appropriate activity, first in animals then, perhaps, in humans, but not, for instance, for cancer, by the expert medical practitioner. The extractive-free wood must be shown to be inactive.

The first researchers to perform these steps were Morgan and Thompson (129). They also showed that some compounds were present but not active. This is an interesting phenomenon and it remains for further research to elucidate the reasons why such closely related compounds are present in the wood. The possibilities for synergistic effects, and their measurements are too complex for present research, and future work will be necessary for this. When all these criteria are met and the individual species are extended to include an entire family, such as in the work of Mitchell (113) on the Compositae, then truly elegant research is obtained. His data on a single family may be extended by other workers to other families.

The above analytical techniques, facilitated by advances in instrumentation, will continue to be a feature of this field. The most sophisticated example must be the work of Schmalle et al. (164), in which they determined the structure of extractives by means of X-ray crystallography. Similar sophisticated methods will be two-dimensional nuclear magnetic resonance (NMR), and high resolution mass spectrography.

The immuno-assay methods of analysis for individual compounds will become very important. The sensitivity and selectivity of this method (78) have been discussed and its application to the analyses for leucotriene C_4 (191) and plicatic acid (29) have been reported.

We can see the outlines of the jigsaw-puzzle in this field. The surveys of actions of classes of compounds, such as the lignans (106), or families of plant extracts have been ably done. The majority of the review by Mitchell et al. (113) focuses on a particular class of chemicals, the sesquiterpene lactones, and their occurrence among Compositae species. They present some very interesting data on clinical aspects of allergic contact dermatitis from these species according to occupation, which leads to a discussion of taxonomy and chemotaxonomy in the investigation of dermatitis from plants. This review shows the usefulness of biochemical systematics in chemical dermatology investigations and reveals vast areas of further research that could be performed in this field. There are many other classes of compounds, such as the procyanidins, or other species and families, such as the Cupressaceae, which have just begun yielding interesting data. In addition, there are many families whose extractives have not been investigated. For example, in North America, Kingsbury (94) has reviewed the poisonous plants, and Rowe and Conner (152) have surveyed the extractives of hardwoods grown in the eastern United States. The extractives of some — for example, the most toxic, *Liriodendron tulipifera* — are well characterized. On the other hand, Rowe and Conner note several examples of families that are poorly studied and in need of further research. Notable examples of these families are the *Aquifoliaceae, Betulaceae, Hippocastanaceae, Nyssaceae, Rhizophoraceae, Tiliaceae, Ulmaceae,* and

Diospyros spp. There are a great many species from other continents (181) that are candidates for study, and future research on them should be equally rewarding.

References

1. Adawakar P D, El Sohly M A 1982 Isolation and characterization of a new urishiol component from poison sumac *Toxicodendron verni*. Phytochemistry 22:1280–1281
2. Albert-Puleo M 1981 van Gogh's vision: Thujone intoxication. J Am Med Assoc 246:42
3. Anderson R S 1986 Cancer and wood dust. Chem Eng News 64(51):2–3
4. Angarskaya M A 1974 [Relation between the biological activity and sugar component of cardiac glycosides.] Farmakol Toksikol (Kiev) 9:24–28
5. Antonsson A B, Lundberg B 1985 Investigations of health hazards in the wood preserving industry. Swedish Wood Pres Inst Rep 152. Stockholm
6. Arctander S 1961 Perfume and flavor chemistry. S Arctander Elizabeth N J
7. Aronow R 1976 Camphor poisoning. J Am Med Ass 235:1260–1261
8. Ashley M J, Corey P, Chan-Yeung M, MacLean L, Maledy H, Grzybowski S 1978 A respiratory survey of cedar mill workers. II. Influence of work-related and host factors on the prevalence of symptoms and pulmonary function abnormalities. J Occup Med 12:328–332
9. Ashwood-Smith M J, Poulton G A, Ceska O, Liu M, Furniss E 1983 An ultrasensitive bioassay for the detection of furocoumarins and other photosensitizing molecules. Photochem Photobiol 38:113–118
10. Austwick P, Mattocks R 1979 Naturally occurring carcinogens in food. Chem Ind (Feb 3):76–83
11. Baslas R K, Agha R 1985 Isolation of a hypoglycemic principle from the bark of *Ficus glomerata* Roxb. Himal Chem Pharm Bull (2):1–13
12. Baxter P J, Jones R D 1986 Occupation and cancer in London. An investigation into nasal and bladder cancer using the Cancer Atlas. Br J Ind Med 43:44–49
13. Beretz A, Cazenave J P, Anton R 1982 Inhibition of aggregation and secretion of human platelets by quercetin and other flavonoids structure activity relationships. Agents Actions 12:382–387
14. Bhargava U C, Westfall B A 1968 Antitumor activity of Juglans nigra black walnut extractives. J Pharm Sci 57:1674–1677
15. Bhattacharjee J W, Dogra R K S, Lal M M, Zaidi S H 1979 Wood dust toxicity in vivo and in vitro studies. Environ Res 20:455–464
16. Bjeldanes L F, Chang G W 1977 Mutagenic activity of quercetin and related compounds. Science 197:577–578
17. Blainey A D, Graham V A, Phillips M J, Davies R J 1981 Respiratory tract reactions to western red cedar. Human Toxicol 1:41–51
18. Booth B H, Le Foldt R H, Moffitt E M 1976 Wood dust hypersensitivity. J Aller Clin Immunol 57:352–357
19. Borchert P, Miller J A, Miller E C, Shires T K 1973 1'-Hydroxysafrole, a proximate carcinogenic metabolite of safrole in the rat and mouse. Cancer Res 33:590–600
20. Bowder J 1974 Poison apple of the everglades. US Nat Parks Conserv Mag Environ J (Feb):22–24
21. Boyle V J, Lardner C A, Frankle W E 1980 Evaluation of pulping and bleaching by products for their mutagenic potential. Tappi 62(12):59–62
22. Brooks S M, Edwards J J Jr, Apol A, Edwards F H 1981 An epidemiologic study of workers exposed to western red cedar and other wood dusts. Chest 80:30–32
23. Brown J P, Dietrich P S 1979 Mutagenicity of plant flavonols in the salmonella mammalian microsome test. Activation of flavonol glycoside by mixed glycosidases from rat fecal bacteria and other sources. Mutat Res 66:223–240
24. Brown S, Griffiths L A 1983 New metabolites of the naturally occurring mutagen quercetin, the pro-mutagen rutin and of taxifolin. Experientia (Basel) 39:198–200
25. Byers V S, Epstein W L 1979 In vitro studies of cellular immunity to poison oak in humans. J Invest Dermatol 72:269–270

26 Byers V S, Epstein W L, Castagnoli N, Baer H 1979 In vitro studies of poison oak immunity. I. In vitro reaction of human lymphocytes to urushiol. J Clin Invest 64:1437–1448
27 Cachai P, Brandao F, Menezes C M, Frazao S, Silva M 1986 Allergy to oil of turpentine in Portugal. Contact Derm 14:205–208
28 Carroll M N, Fox L E, Ariail W T 1957 Investigation of the toxic principles of *Hippomane mancinella* L. Toxic actions of extracts of *Hippomane manchineel* L. J Am Pharm Assoc (Sci Ed) 66(2):93–97
29 Cartier A, Chan H, Malo J L, Pineau L, Tse K S, Chan-Yeung M 1986 Occupational asthma due to eastern white cedar (*Thuja occidentalis*) with demonstration that plicatic acid is present in this wood dust and is the causal agent. J Allerg Clin Immunol 77:639–645
30 Chan-Yeung M 1973 Maximal expiratory flow and airway resistance during induced bronchoconstriction in patients with asthma due to western red cedar *Thuja plicata*. Am Rev Resp Dis 108:1103–1110
31 Chan-Yeung M 1977 Fate of occupational asthma. A follow-up study of patients with occupational asthma due to western red cedar *Thuja plicata*. Am Rev Resp Dis 116:1023–1030
32 Chan-Yeung M 1980 Wood dust hypersensitivity. In: Oehling A, Glazer I, Mathov E, Arbesmen C (eds) Advances in allergology and clinical immunology. Pergamon Press Oxford, 345–353
33 Chan-Yeung M 1982 Immunologic and nonimmunologic mechanisms in asthma due to western red cedar *Thuja plicata*. J Allerg Clin Immunol 70:32–37
34 Chan-Yeung M 1986 Occupational asthma. Clin Rev Allerg 4:251–266
35 Chan-Yeung M, Abboud R 1976 Occupational asthma due to California redwood *Sequoia sempervirens* dust. Am Rev Resp Dis 114:1027–1031
36 Chan-Yeung M, Barton G M, MacLean L, Grzybowski S 1971 Bronchial reactions to western red cedar *Thuja plicata*. Can Med Ass J 105:56–61
37 Chan-Yeung M, Barton G M, MacLean L, Grzybowski S 1973 Occupational asthma and rhinitis due to western red cedar *Thuja plicata* Donn. Am Rev Resp Dis 108:1094–1103
38 Chan-Yeung M, Giclas P C, Henson P M 1980 Activation of complement by plicatic acid, the chemical compound responsible for asthma due to western red cedar *Thuja plicata*. J Allerg Clin Immunol 65:333–337
39 Chan-Yeung M, Lam S 1986 State of the art – occupational asthma. Am Rev Resp Dis 133:686–703
40 Chan-Yeung M, Lam S, Koener S 1982 Clinical feature and natural history of occupational asthma due to western red cedar *Thuja plicata*. Am J Med 72:411–415
41 Chan-Yeung M, MacLean L, Paggiaro P L 1987 Follow up study of 232 patients with occupational asthma due to western red cedar (*Thuja plicata*). J Allerg Clin Immunol (in press)
42 Chan-Yeung M, Vedal S, Kus J, McCormack G, MacLean L, Dorken E, Enarson D, Tse K S 1984 Symptoms, pulmonary function, atopy and bronchial reactivity in western red cedar workers compared to office workers. Am Rev Resp Dis 130:1038–1041
43 Cockcroft D W, Cotton D J, Mink J T 1979 Nonspecific bronchial hyperreactivity after exposure to western red cedar. Am Rev Resp Dis 119:505–510
44 Cohen H I, Kosek J C, Eldridge F 1967 Sequoiosis, a granulomatous pneumonitis associated with redwood sawdust inhalation. Am J Med 43:785–794
45 Craig J C, Waller C W, Billets S, El Sohly M A 1978 New GLC analysis of urushiol congeners in different plant parts of poison ivy *Toxicodendron radicans*. J Pharm Sci 67:483–485
46 David L A, Chan-Yeung M 1979 Plicatic acid mediated histamine release in red cedar *Thuja plicata* asthma. Am Rev Resp Dis 119:209
47 Damas J, Remacle-Volon G 1982 Kinins and oedema induced by different carrageenans. J Pharmacol 13:225–239
48 Digiovanni J, Boutwell R K 1983 Tumor promoting activity of 1,8-dihydroxy-3-methyl-9-anthrone (chrysarobin) in female Sencar mice. Carcinogenesis 4:281–284
49 Dooms-Goossens A, Degreef H, Luytens E 1979 Dihydroabietyl alcohol abitol a sensitizer in mascara. Contact Derm 5:350–353
50 Dunn I S, Liberato D J, Dennick R G, Castagnoli N, Byers V S 1982 A murine model system for contact sensitization to poison oak or ivy urushiol components. Cell Immunol 68:377–388
51 Edwards J M, Raffauf R F, LeQuesne P W 1979 Antitumor plants. Part VII. Antineoplastic activity and cytotoxicity of flavones, isoflavones and flavanones. J Nat Prod 42:85–91

52 El Sohly M A, Adawadkar P D, Ma C Y, Turner C E 1982 Separation and characterization of poison ivy and poison oak urushiol components. J Nat Prod 45:532–538
53 El Sohly M A, Benigni D A, Torres L, Watson E S 1983 Synthesis and antiallergenic properties of 3-n-pentadecyl and 3-n-heptadecylcatechol esters. J Pharm Sci 72:792–795
54 Erdtman H, Gripenberg J 1948 Antibiotic substances from the heartwood of *Thuja plicata* Donn. Nature 161–719
55 Eslyn W E, Bultman J D, Jurd L 1981 Wood decay inhibition by tropical hardwood extractives and related compounds. Phytopathology 71:521–524
56 Evans E, Nicholls P J 1974 Histamine release by western red cedar *Thuja plicata* from lung tissue in vitro. Br J Ind Med 31:28–30
57 Fisher A A 1977 The notorious poison ivy family of Anacardiaceae plants. Cutis 20:570–582
58 Fisher A A 1981 Allergic contact dermatitis in a violinist. The role of abietic acid, a sensitizer in rosin colophony as the causative agent. Cutis 27:466, 468, 473
59 Fong H H S, Bhatti W, Farnsworth N R 1972 Antitumour activities of certain plants due to tannins. J Pharm Sci 61:1818
60 Food and Cosmetics Toxicology 1974 Food and cosmetics' toxicology. Food Cosmet Toxicol 12:843
61 Freeland W J, Janzen D H 1974 Strategies in herbivory by mammals. The role of plant secondary compounds. Am Nat 108:269–289
62 Furtado S E J, Jones E B G, Bultman J D 1977 The effect of certain wood extractives on the growth of marine microorganisms. C R Congr Int Corros Mar Salissures (4th):195–201
63 Gardner J A F 1963 The chemistry and utilization of western red cedar. Can Dept For Pub 1023: 26 pp
64 Giclas P C 1982 Effect of plicatic acid on human serum complement includes interference with CT inhibitor function. J Immun 129:168–172
65 Girard J, Unkovic J, Delahayes J, Lafille C 1979 Phototoxicity of Bergamot oil. Comparison between humans and guinea pigs. Dermatologica 158:229–243
66 Gozalo Reques F 1982 Asthma caused by red cedar *Thuja plicata*, 2 cases. Allergol Immunopathol (Madr) 10:441–447
67 Greene M H, Brinton L A, Fraume J F, Damico R 1978 Familial and sporadic Hodgkins disease associated with occupation wood exposure. Lancet (8090):626–627
68 Grmek M D 1966 Unpublished notes of Claude Bernard on the physiologic properties of arrow poisons (curare, upas, strychnine and others). Biol Med (Paris) Supp 1: 159 pp
69 Gossweiler B 1982 Poisoning by camphor today. Praxis 71:1475–1478
70 Gulati D K, Chambers C L, Rosenthal G A, Sabharwal P S 1981 Comparative toxicity of some naturally occurring and synthetic nonprotein amino acids. Environ Exp Bot 21:225–230
71 Hamilton R D, Crockett A J, Ruffin R E, Alpers J H 1979 Bronchial activity in western red cedar induced asthma. Aust NZ J Med 9:480–481
72 Hanslian L, Kadlec K 1963 Die Toxizität des Holzes als hygienisches Problem. Proc Conf Perspectives Basic Res Wood
73 Hanslian L, Kadlec K 1966 [Drevo z hlediska hygienickeho. (VII). Biologicky silne agresivni dreving.] Drevo 21:157–160
74 Hausen B M 1981 Wood constituents with sensitizing properties. Proc XVII IUFRO World Congr (Div 5). IUFRO Rome, 343–355
75 Hausen B 1981 Wood injurious to human health. A manual. de Gruyter Berlin, 189 pp
76 Hausen B M 1983 The risk of cancer in the nasopharynx of woodworkers. Proc Int Symp Wood Pulp Chem, Chem Struct Wood 129–131
77 Hausen B M, Schulz K H 1979 The sensitizing capacity of naturally occurring quinones. Occ Env Derm 27:18–30
78 Hawke H 1984 Antibodies to the rescue. Nature 310:272–273
79 Hayes R B, Olsen J H 1986 Wood-related occupations, wood dust exposure and sinonasal cancer. Am J Epiderm 124:569–577
80 Hecker W A E 1984 On the active principles from the spurge family. 10 skin irritant cocarcinogens and cryptic cocarcinogens from the latex of the manchineel tree. J Nat Prod 47:482–496
81 Heitmuller P T, Hollister T A, Parrish P R 1981 Acute toxicity of 54 industrial chemicals to sheephead minnows *cyprinodon variegatus*. Bull Environ Contam Toxicol 27:596–604
82 Hirono I 1980 Natural carcinogenic products of plant origin. CRC Crit Rev Toxicol 8:235–277

83 Hooper S N, Chandler R F 1984 Herbal remedies of the maritime (Canada) Indians. Phytosterols and triterpenes of 67 plants. J Ethnopharmacol 16:181–194
84 Howie A D, Boyd G, Moran F 1976 Pulmonary hypersensitivity to ramin, *Gonystylus bancanus*. Thorax 31:585–587
85 Ippen H 1983 Contact allergy to Anacardiaceae. A review and case reports of poison ivy allergy in central Europe. Derm Beruf Umwelt 31:140–148
86 Ito K 1963 Allergic disorders of upper respiratory tract by sawdust of *Thuja standishii*. 3. Allergic disorders due to the sawdust of *Thuja standishii* and other woods. J Sci Labour (Tokyo) 39:501–510
87 Jimenez J F, Brown A L, Arnold W C, Byrne W J 1983 Chronic camphor ingestion mimicking Reye's syndrome. Gastroenterology 82:394–398
88 Joly C, Bouillie C, Hummel M 1980 Acute poisoning by camphor administered externally in an infant. Ann Pediatr (Paris) 27:395–396
89 Jones E, Ekundayo O, Kingston D G I 1981 Plant anticancer agents. XI. 2,6-dimethoxybenzoquinone as a cytotoxic constituent of *Tibouchina pulchra*. J Nat Prod 44:493–494
90 Kapadia G J, Paul B D, Chung E B, Ghosh B, Pradhan S N 1976 Carcinogenicity of *Camellia sinesis* (tea) and some tannin-containing folk medicinal herbs administered sub-cutaneously in rats. J Nat Canc Inst 57:207–209
91 Kapadia G J, Rao G S, Morton J F 1983 Carcinogens and mutagens in the environment. III. Naturally occurring compounds. In: Stich H F (ed) Epidemiology and distribution. CRC Press Boca Raton, 3–12
92 Karlberg A T, Boman A, Wahlberg J E 1980 Allergenic potential of abietic acid, colophony and pine resin. Clinical and experimental studies. Cont Derm 6:481–487
93 Keeler R F 1975 Toxins and teratogens of higher plants. Lloydia 38:56–86
94 Kingsbury J M 1964 Poisonous plants of the United States and Canada. Prentice-Hall Englewood Cliffs, 626 pp
95 Koeppel C, Tenczer J, Schirop T, Ibe K 1982 Camphor poisoning. Abuse of camphor as a stimulant. Arch Toxicol 51:101–106
96 Kuksis A, Myher J J, Marai L, Little J A, McArthur R G, Roncari D A K 1986 Usefulness of gas chromatographic profiles of plasma total lipids in diagnosis of phytosterolemia. J Chrom 381:1–12
97 Kurita N, Miyaji M, Kurane R, Takahara Y 1981 Antifungal activity of components of essential oils. Agric Biol Chem 45:945–952
98 Lagrota M H, Wigg M D, Pereira L O, Fonseca M E, Pereira N A, Guimareas J C 1983 Antiviral activity of lapachol. Rev Microbiol 14:21–26
99 Lam S, Salari H, Chan H, LeRiche J, Chan-Yeung M 1987 Cellular and protein change in bronchial lavage fluid following late asthmatic reaction in patients with red cedar asthma. J Allerg Clin Immunol (in press)
100 Lam S, Tan F, Chan H, Chan-Yeung M 1983 Relationship between types of asthmatic reaction nonspecific bronchial reactivity and specific immunoglobin E antibodies in patients with red cedar *Thuja plicata* asthma. J Allerg Clin Immunol 72:134–139
101 Lin L J, Marshall G T, Kinghorn A D 1983 The dermatitis-producing constituents of *Euphorbia hermentiana* latex. J Nat Prod 46:723–731
102 Lopes J N, Cruz F S, Docampo R, Vasconcellos M E, Sampaio M C, Pinto A V, Gilbert B 1978 In vitro and in vivo evaluation of the toxicity of the 1,4-naphthoquinone and 1,2-naphthoquinone derivatives against *Trypanosoma cruzi*. Ann Trop Med Parasitol 72:523–531
103 Loub W D, Fong H H S, Theiner M, Farnsworth N R 1973 Partial characterization of antitumour tannins isolated from *Calygonium squamulosum*. J Pharm Sci 62:149–160
104 Ma C, El Sohly M A, Baker J K 1980 High performance liquid chromatographic separation of urushiol congeners in poison ivy and poison oak. J Chromatogr 200:163–169
105 MacBeth R 1965 Malignant disease of the paranasal sinuses. J Laryngol Otol 97:592–612
106 MacRae W D, Towers G H N 1984 Biological activities of lignans. Phytochemistry 23:1207–1220
107 Malo J L 1984 Occupational asthma in sawmills in eastern Canada. J Allerg Clim Immunol 74:261–266
108 Mathias A 1975 On pathology caused by wood. Arch Mal Prof 29:452–455
109 McCann M 1982 Wood dust poses high cancer risk. Art Hazards Newsletter (Oct/Nov):36

110 Melino C, Messineo A, Marracino F 1979 Work related risks and preventive measures in the tropical wood industry. Riv Med Lav Ig Ind 3:143–165
111 Merck Index 1983 10th ed. Windholz M, Budvari S, Blumetti R F, Otterbein E S (eds) Merck Company Rahway NJ
112 Michaels L 1967 Lung changes in wood workers. Can Med Assoc J 96:1150–1155
113 Mitchell J C 1975 Contact allergy from plants. In: Runeckles V C (ed) Recent advances in phytochemistry, vol 9. Phytochemistry in disease and medicine. Plenum Press New York, 119–138
114 Mitchell J C, Chan-Yeung M 1974 Contact allergy from *Frullania* and respiratory allergy from *Thuja plicata*. Can Med Assoc J 110:653–655
115 Mitchell J C, Fritig B, Singh B, Towers G H N 1970 Allergic contact dermatitis from *Frullania* and *Compositae*. The role of sesquiterpene lactones. J Invest Derm 54:233–239
116 Mitchell J C, Rook A 1979 Botanical dermatology. Plants and plant products injurious to the skin. Greenglass Ltd Vancouver, 787 pp
117 Monroe R R 1978 The episodic psychoses of Vincent van Gogh. J Nerv Ment Dis 166:480–488
118 Morton J F 1971 Plants poisonous to people in Florida and other warm areas. Hurricane House Miami FL, 113 pp
119 Morton J F 1973 Plant products and occupational materials ingested by esophageal cancer victims in South Carolina. Q J Crude Drug Res 13:2005–2022
120 Morton J F 1974 Folk remedy plants and esophageal cancer in Coro, Venezuela. Morris Arbor Bull 25:24–34
121 Morton J F 1977 Poisonous and injurious plants and fungi. In: Tedeshi C G, Eckert W G (eds) Environmental hazards in forensic medicine, vol III. WB Saunders Co Philadelphia London Toronto, 112 pp
122 Morton J F 1978 Economic botany in epidemiology. Econ Bot 32:111–116
123 Morton J F 1978 Brazilian pepper: Its impacts on peoples, animals and the environment. Econ Bot 32:353–359
124 Morton J F 1980 Search for carcinogenic principles. In: Swain T, Kleiman R (eds) Recent advances in phytochemistry, vol 14. The resource potential in phytochemistry. Plenum Press New York, 53–73
125 Ref. 124
126 Morton J F 1986 The potential carcinogenicity of herbal tea. Environ Carcin Rev (J Environ Sci Hlth) C4:203–223
127 Morgan J W W, Orsler R J 1967 A simple test to distinguish *Khaya anthotheca* from *K. ivoriensis* and *K. grandifoliola*. J Inst Wood Sci 18:61–64
128 Morgan J W W, Orsler R J, Wilkinson D S 1968 Dermatitis due to the wood dusts of *Khaya anthotheca* and *Machaerium scleroxylon*. Br J Ind Med 25:119–125
129 Morgan J W W, Thompson J 1967 Ayan dermatitis. Br J Ind Med 24:156–158
130 Morgan J W W, Wilkinson D S 1965 Dermatitis produced by wood. Proc IUFRO Congr Sec 41 vol 3. IUFRO Rome
131 Morgan J W W, Wilkinson D S 1965 Sensitization to *Khaya anthotheca*. Nature 207:1101
132 Mue S, Ise T, Ono Y, Akasaka K 1975 A study of western red cedar induced asthma. Ann Allerg 34:296–304
133 Mue S, Ise T, Ono Y, Akasaka K 1975 A study of western red cedar sensitivity workers allergy reactions and symptoms. Ann Allerg 35:148–152
134 Nagao M, Morita N, Yahagi T, Shimizu M, Kuoyanagi M, Fukuoka M, Yoshihira K, Matori S, Fujino T, Sugimura T 1981 Mutagenicities of 61 flavonoids and 11 related compounds. Environ Mutagen 3:401–420
135 Nozoe T 1952 Studies on hinokitiol. Part V. On the revision of the hinokitiol structure. Sci Rept Tohoku V 36:82–87
136 Nozoe T 1956 Natural tropolones and some related tropinoids. Fortschr Chem Org Naturst 13:232–301
137 Nozoe T, Katsura S 1944 On the structure of hinokitiol. J Pharm Soc Jpn 64:181–186
138 Orsler R J 1979 Health problems associated with wood processing. Build Res Estab Princes Risborough England Inf Paper IP13/79
139 Palant L G, Cherkasova G I, Talapin V I 1976 [Toxicological assessment of some borneol isomers.] Aktual Vopr Okhr Tr Khim Promst 143–146

140 Perdue R, Hartwell J 1969 The search for plant sources of anti cancer drugs. Morris Arbor Bull 20:33−35
141 Pinto-Scognamiglio W 1967 Current knowledge on the pharmacodynamic activity of the prolonged administration of thujone, a natural flavoring agent. Boll Chim Farm 106:292−300
142 Pradhan S N, Chung E B, Ghosh B, Paul B D, Kapadia G J 1974 Potential carcinogens. I. Carcinogenicity of some plant extracts and their tannin containing fractions in rats. J Nat Cancer Inst 52:1579−1582
143 Raffauf R F 1970 A handbook of alkaloids and alkaloid-containing plants. Wiley-Interscience New York
144 Rate R G, Leche J, Chervenak C 1979 Podophyllin toxicity. Ann Intern Med 90:723
145 Rennerfelt E 1948 Thujaplicin, a fungicidal substance in the heartwood of *Thuja plicata*. Physiol Plant 1:245−254
146 Reut N A, Danusevich I K, Zakharevskii A S, Kuz'mitskii B B, Kevra M K, Miklevich A V, Manukov E I, Reutsakaya G I, Sidorenko E R 1975 Toxicological properties of synthetic camphor prepared from pine tree oleoresins. Farmakol Toksikol Nov Prod Khim Sint Mater Resp Konf 3rd, 187
147 Rice K C, Wilson R S 1976 3-Isothujone, a small non-nitrogenous molecule with antinocioceptive activity in mice. J Med Chem 19:1054−1057
148 Rodriguez E, Reynold G N, Thompson J A 1981 Potent contact allergen in the rubber plant guayule *Parthenium argentatum*. Science 211:1444−1445
149 Rodriguez E, Towers G H N, Mitchell J C 1976 Biological activities of sesquiterpene lactones. Phytochemistry 15:1573−1580
150 Rogers I, Mahood H, Servizi J, Gordon R 1979 Identifying extractives toxic to aquatic life. Pulp Pap Mag Can 80(9):94−96, 98−99
151 Rosenthal G A, Bell E A 1979 Naturally occurring toxic nonprotein amino-acids. In: Rosenthal G A, Janzen D H (eds) Herbivores: their interaction with secondary plant metabolites. Academic Press New York, 353−386
152 Rowe J W, Conner A H 1979 Extractives in eastern hardwoods: a review. USDA For Serv Gen Tech Rep FPL-18, 69 pp
153 Sabine J R, Horton B J, Wicks M B 1973 Spontaneous tumours in C3H-A^{vy} and C3H-A^{ty}fB mice: high incidence in the United States and low incidence in Australia. J Nat Cancer Inst 50:1237−1242
154 Sandermann W, Barghoorn A W 1955 Über die Inhaltsstoffe von Makoré- und Peroba-Holz sowie ihre gesundheitsschädigende Wirkung. Holzforschung 9:112−117
155 Sandermann W, Barghoorn A W 1956 Gesundheitsschädigende Nutzhölzer. Ein Übersichtsbericht. Holz Roh- Werkst 14:37−40, 87−94
156 Sandermann W, Barghoorn A W 1963 Über Inhaltsstoffe aus Teak (*Tectona grandis* L.) I. Isolierung und Konstitution eines toxischen Teakchinons. Chem Ber 96:2182−2185
157 Sandermann W, Dietrichs H H 1959 Chemische Studien an Tropenhölzern. 3. Über die Inhaltsstoffe von *Mansonia altissima* und ihre gesundheitsschädigende Wirkung. Holz Roh- Werkst 17:88−97
158 Sandermann W, Simatupang M H 1962 A toxic quinone from teakwood. Angew Chem (Int Ed) 1:599
159 Sanduja R, Lo M Y R, Euler K L, Alam M, Morton J F 1984 Cardenolides of *Crytostegia madagascariensis*. J Nat Prod 47:260−265
160 Schafer R, Schafer W 1982 Percutaneous absorption of various terpenes − menthol, camphene, limonene, isobornyl acetate, alpha pinene − from foam baths. Arzneim Forsch 32:56−58
161 Schlewer G, Chabeau G, Reimeringer A, Foussereau J 1979 Study on the allergens of colophony type and its derivatives used in adhesive plasters. Dermatosen Beruf Umwelt 27:170−172
162 Schreiber C 1975 Gesundheitgefährdende Holzarten. I. Schadstoffe, Krankheiten und Empfehlungen für den Gesundheits- und Arbeitschutz. Holzindustrie 28:245−247, 279−283
163 Schmalle H W, Hausen B M 1980 Acamelin, a new sensitizing furano quinone from *Acacia melanoxylon*. Tetrahedron Lett 21:149−152
164 Schmalle H, Jarchow O, Hausen B M 1977 2,6-dimethoxy-1,4-benzoquinone, a new contact allergen in commercial woods. Naturwissenschaften 64:534

165 Schulz K H, Garbe I, Hausen B M, Simatupang M H 1977 The sensitizing capacity of naturally occurring quinones. Experimental studies in guinea pigs. I. Naphthaquinones and related compounds. Arch Dermatol Res 258:41−52
166 Schweisheimer W 1952 Wood dust and respiratory diseases. Wood (London) 17:181−200
167 Science 1979 US rubber shrub may have hidden thorn. Science 205:564
168 Senear F E 1933 Dermatitis due to woods. J Am Med Assoc 101:1527−1536
169 Shibamoto T, Mihara S 1983 Photochemistry of fragrance materials. II. Aromatic compounds and phototoxicity. J Toxicol Cutaneous Ocul Toxicol 2:263−371
170 Soderquist C J 1973 Juglone and allelopathy. J Chem Ed 50:782−783
171 Sosman A J, Schlueter D P, Fink J N, Barboriak J J 1969 Hypersensitivity to wood dust. New Eng J Med 281:977−980
172 Southam C M, Ehrlich J 1943 Effect of extract of western red cedar heartwood on certain wood decaying fungi in culture. Phytopathology 33:517
173 Sowder A M 1929 Toxicity of water-soluble extractives and relative durability of water-treated wood flour of western red cedar. Ind Eng Chem 21:981−984
174 Sugimura T, Minako N, Matsushima T, Yahagi T, Seino Y, Shirai A, Sawamura M, Natori S, Yoshihira K 1977 Mutagenicity of flavone derivatives. Proc Jpn Acad Ser B 53:194−197
175 Thörnqvist T, Lundström H 1982 Health hazards caused by fungi in stored wood chips. For Prod J 32(11):29−32
176 Towers G H N, Mitchell J C, Rodriguez E, Subba Rao N, Bennett F D 1977 Biology and chemistry of *Parthenium hysterohorus* L. A problem wood in India. J Sci Ind Res (Sect C) 36:672−684
177 Tse K S, Chan H, Chan-Yeung M 1982 Specific immunoglobin in E antibodies in workers with occupational asthma due to western red cedar. Clin Allergy 12:249−258
178 Vedal S, Enarson D, Kus J, McCormack G, MacLean L, Chan-Yeung M 1986 Symptoms and pulmonary function in western red cedar workers related to duration of employment and dust exposure. Arch Environ Health 41:179−184
179 Vedal S, Enarson D, Kus J, McCormack G, MacLean L, Chan-Yeung M, Tse K S 1986 Specific IgE, bronchial reactivity and atropy in western red cedar workers. J Allerg Clin Immunol 78:1103−1109
180 Vicker R 1981 An innocent soaking in a hot tub could be an irritating experience. Wall Street J Nov 12
181 Vietmeyer N D 1986 Lesser known plants of potential use in agriculture and forestry. Science 232:1379−1384
182 Wahlberg J E 1978 Abietic acid and colophony. Contact Derm 4:55
183 Watson E S, Murphy J C, El Sohly M A 1983 Immunologic studies of poisonous Anacardiaceae: oral desensitization to poison ivy and oak urushiols in guinea pigs. J Invest Dermatol 80:149−155
184 Wellborn S N 1977 Health hazards in woodworking: simple precautions minimize risks. Fine Woodwork (Winter):54−57
185 West L G, McLaughlin J L, Eisenbeiss G K 1977 Saponins and triterpenes from *Ilex opaca*. Phytochemistry 16:1846−1847
186 Westfall B A, Russell R L, Auyoung T K 1961 Depressant agent from walnut hulls. Science 134:1617−1618
187 Willaman J J, Li H L 1970 Alkaloid-bearing plants and their contained alkaloids 1957−1968. (Suppl) J Nat Prod 33:1−286
188 Willaman J J, Schubert B G 1961 Alkaloid-bearing plants and their contained alkaloids. USDA Tech Bull 1234, 287 pp
189 Wood A W, Huang M T, Chan R L, Newmark H L, Lehr R E, Yagi Y, Sayer J M, Jerina D M, Conney A H 1982 Inhibition of the mutagenicity of bay region diolepoxides of polycyclic aromatic hydrocarbons by naturally occurring plant phenols. Exceptional activity of ellagic acid. Proc Nat Acad Sci USA 79:5513−5517
190 Woods B, Colman C D 1976 Toxic woods. Br J Derm 94 (Suppl 13):1−97
191 Wynalda M A, Brashler J R, Bach M K, Morton D R, Fitzpatrick F A 1984 Determination of leucotriene C_4 by radio immunoassay with a specific antiserum generated from a synthetic hapten mimic. Anal Chem 56:1862−1865
192 Zafiropoulo A, Audibert A, Charpin J 1968 A propos des accidents dus à la manipulation des bois exotiques. Rev France Allerg 8:155−171

Chapter 10

The Utilization of Wood Extractives

Introduction

Humans, even back in the most primitive of times, found that the chemicals in woody plants could be converted to their use. In some cases these chemicals, such as some saponins and alkaloids, were highly toxic and could be used in arrow poisons and piscicides. The witch doctor's bag of tricks included many with healing or curative powers. Indeed, 20th-century medicine has benefited immensely by searching out native remedies and isolating the active principles. Even the discovery of the positive effects of salicylates (e.g., aspirin) derived from their occurrence in plants. Early humans also found that leather could be preserved by tanning with natural extracts of polyphenols. In recent times the same compounds have found many other varied uses, including adhesives. The oleoresins from pines and other species were critical for waterproofing and made possible the era of the giant wooden sailing vessels — hence the sobriquet, 'naval stores.' The resins, varnishes, lacs (including shellac), and so forth were important to waterproofing and became critical components of the great paintings of the past. Indeed they are still highly valued today in spite of the in-roads of synthetics. Turpentine was one of the essential solvents prior to the petrochemical era. Balsams, gums, Storax, myrrh, frankincense, and a host of other exudates found a wide asortment of uses. Rubber has had a significant impact on our 20th-century civilization. The derivation of chemical intermediates from natural sources will undoubtedly increase in areas in which they can be derived more economically than from increasingly expensive petrochemical sources. This chapter discusses many of these more significant aspects of the utilization of wood extractives.

10.1 Naval Stores

D. F. ZINKEL

10.1.1 Introduction

Natural products derived from conifer oleoresins, particularly those from pine, have been articles of commerce since before recorded history. These oleoresins, which consist of an essential oil and a resin, are the source of turpentine (the essential oil), rosin (the resin), and a host of degraded products of rosin such as tars, pitches, oils, and fossil resins.

One of the most ancient commercial derivatives is amber, a fossilized resin originating from a group of conifers loosely named *Pinus succinifera* that grew some 50 million years ago. Ancient sun worshippers valued amber for its sun-like radiance. Baltic amber was traded throughout the old world and became revered by early Greek, Roman, and other European civilizations as having powers to ward off evil spirits and give protection from illnesses and enemies (13).

The more traditional naval stores uses of pine oleoresin products date from biblical times when Noah was told to "Make yourself an ark of gopherwood;.... and cover it inside and out with pitch."(Genesis 6:14). The term "naval stores" first appeared in 17th-century English records, being used for the pine-derived, naval-oriented commodities such as pitch, tar, timber, planking, etc. The tar was used to waterproof lines and rigging, and pitch was used to caulk the seams of the ships to keep them seaworthy – uses that continued into the 19th century. Since the demise of commercial wooden ships, the term "naval stores" remains as the generic name for the commodities of turpentine and rosin. Rosin has also been called colophony, but that term is no longer in use.

Among other early uses, naval stores were used as a source of light (torches), as a liquid fire in early warfare, in medicinal preparations, in the preparation of the characteristic Greek wine retsina, in axle greases (as mixtures of pine tar with tallow), and as adhesives for the decorative application of marble and other stones to buildings.

Because of the emphasis in this chapter on a review of naval stores utilization, an extensive attribution to primary publications is not appropriate. For in-depth coverage of the facets of utilization, the reader is referred to a new book on the subject (20) and to the somewhat dated 1961 tome of Sandermann (14). Other important publications provide further information on the early history of naval stores (6), the utilization (11) and recovery of tall oil (3), the recovery of sulfate turpentine (4), comparative values of naval stores commodities and chemicals (19), and naval stores statistics (12).

10.1.2 Naval Stores Sources

Records from the 4th to the 2nd centuries B.C. described methods for preparing naval stores materials from conifers. One method chronicled the Macedonian method of making pitch by fire, a method used in this country in the 17th to 19th centuries. (A more modern version of this method is illustrated on the medicine tax stamp depicted in Fig. 10.1.1.) The process involved the controlled, air-deficient burning of a pile of resinous wood stacked in the shape of a hive and covered with earth. Pine tar was collected at the base of the hive; the higher-viscosity pitch was obtained by burning the tar in open kettles. Another method detailed the gathering of oleoresin and processing into tars by boiling in an open pot; turpentine could be wrung from a sheepskin stretched across the top of the pot.

Today's naval stores are classified into three types – gum, wood and sulfate – based on the methods for obtaining them. Although gum naval stores accounted for over 80% of total U.S. naval stores production in the 1930s, sulfate byproducts now account for some 85% of U.S. production with only 2% from gum naval stores operations; wood naval stores is the source of the remaining 13%. However, gum oleoresin accounts for over half of the world production of naval stores.

Fig. 10.1.1. Ancient process of making pine tar by fire is depicted in this medicine revenue stamp of the 1870s

10.1.2.1 Gum Naval Stores

Wounding pines and collecting the exudate is an operation referred to as turpentining. (The paper currency of Fig. 10.1.2 depicts the older, destructive method called boxing, and the stamp shows the more modern cup and gutter collection method.) The collected oleoresin – more commonly, yet technically incorrectly, called pine gum (see Sect. 10.2 for a discussion on true gums) – is then processed to rosin and turpentine. Early wounding techniques resulted in high tree mortality but improvements in turpentining methodology now leave the trees salable for wood products.

The species used for commercial turpentining in the United States are longleaf pine (*Pinus palustris*), slash pine (*P. elliottii*), and some shortleaf pine (*P. echinata*). Slash pine has been introduced into many subtropical parts of the world and is the primary species involved in the growing production of gum naval stores in Brazil and Argentina. Worldwide, about one-third of the species of pine have been tapped at some time in history. Other principal species are Masson (*P. massoniana*), Yunnan (*P. yunnanensis*), and khasya (*P. insularis*) pines in the People's Republic of China; maritime pine (*P. pinaster*) in Portugal, France, and Spain; Scots (*P. sylvestris*) and Siberian stone (*P. sibirica*) pines in northern Europe and the U.S.S.R.; aleppo (*P. halepensis*) and Calabrian (*P. brutia*) pines in Greece; khasya (*P. kesiya*) and Merkus (*P. merkusii*) pines in Indonesia, the Phillipines, and Southeast Asia; chir pine (*P. roxburghii*) in India and Pakistan; and a host of pines in Mexico including oocarpa (*P. oocarpa*), Mexican white pine (*P. ayacahuite*), Aztec (*P. teocote*), Mexican weeping pine (*P. patula*), Michoacan (*P. michoacana*), and Montezuma (*P. montezumae*) pines.

Following improvements in turpentining methods in the 20th century came improvements in processing, the most important being the development and widespread implementation of the Olustee process. This process, which results in

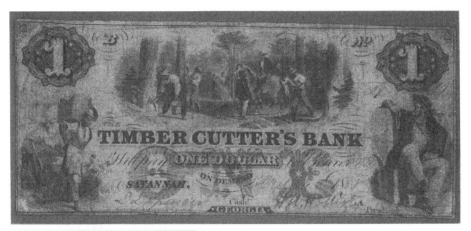

Fig. 10.1.2. Old and modern methods of turpentining as shown on currency and stamps

increased yields and higher quality turpentine and rosin, consists of the following steps:
1) Dissolving the crude oleoresin in turpentine to lower the density to less than that of water and to reduce the viscosity;
2) Filtering to remove trash such as bark, needles, and insects;
3) Washing with dilute oxalic acid to remove iron contaminants that promote oxidation and color formation; and
4) Distilling the oleoresin solution with steam sparge to separate turpentine and rosin.

10.1.2.2 Wood Naval Stores

This type of naval stores was an outgrowth of the tar-burning methods in that resinous wood is the raw material. The technology for the production of wood naval stores is an adaptation of the Yaryan process, a process originally designed to extract flaxseed with petroleum hydrocarbons. The raw material for the process is virgin pine stumps from which the sapwood has rotted away. The remaining resinous heartwood, containing about 25% extractives, is chipped, shredded, and extracted at elevated temperature and pressure (7 to 10 atmospheres) with petroleum solvents; methyl isobutyl ketone is also used. The extract is distilled to recover the solvent, turpentine, a pine oil, and a crude resin. This dark red resin is further purified by selective absorption on fuller's earth or by countercurrent extraction with furfural. Decolorized wood rosin and a dark pitchlike fraction are obtained as final products.

10.1.2.3 Sulfate Naval Stores

These materials are by-products from the kraft pulping process (Fig. 10.1.3). As the pine chips are cooked in the alkaline liquor to produce pulp, the volatilized gases are vented and condensed to yield sulfate turpentine. As pulping proceeds, the alkaline liquor saponifies the fats and converts fatty and resin acids to the sodium salts. During the recovery of the pulping chemicals, the aptly named black liquor is concentrated in multiple-effect evaporators. The insoluble soaps can be skimmed from the surface of the black liquor, either from the black liquor as separated from the pulp and/or during/after concentration in the evaporators. The skimmed soap is acidified to yield a material known as crude tall oil (CTO). The term "tall oil" is derived from the Swedish word "tallolja," which translates as "pine oil." However, such literal translation would have caused confusion with the essential oil known as pine oil − thus, the simple transliteration to tall oil.

In the early years, most of the crude tall oil was burned for the fuel value and as a method of disposal. But because it was recognized early that CTO was a potential source of fatty acids and rosin, numerous processes including acid refining, solvent extraction, adsorption, selective chemical transformations, and, particularly, distillation were developed to upgrade CTO. Although the distillation processes provided useful products, it was not until 1949 that production of high quality tall oil fatty acids and tall oil rosins by fractional distillation was commercially realized. Virtually all tall oil is now being fractionated.

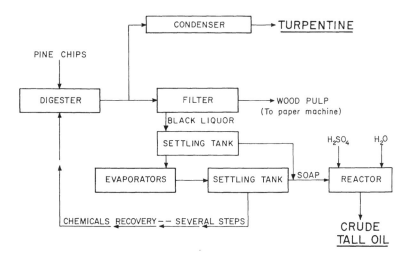

Fig. 10.1.3. Recovery of naval stores by-products from the kraft pulping process

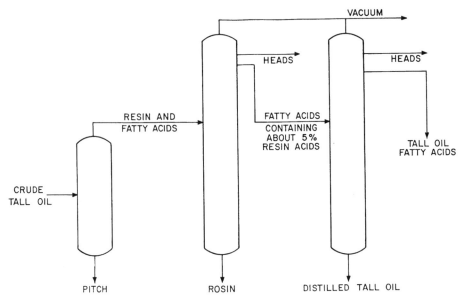

Fig. 10.1.4. Fractionation of crude tall oil

Crude southern U.S. tall oil contains 40%–55% resin acids, 40%–55% fatty acids, and 5%–10% neutral constituents. Distillation of 1 ton of southern CTO yields about 600 lb of fatty acids, 700 lb of rosin, and 700 lb of intermedite (i.e. distilled tall oil – fractions generally containing >5% to <90% fatty acids), head, and pitch fractions (Fig. 10.1.4). CTO from the northern United States and from Canada and other cold climate world sources frequently contains 20%–30% neutrals, resulting in yield and quality problems on distillation. United States fractionating capacity is nearly 1 million tons per year, about twice that of the rest of the world.

10.1.2.4 Potential New Sources

Several additional sources or methods can be considered as possibilities for naval stores production. These include by-products from solvent drying of lumber, conifer foliage extractives from logging slash, and induced lightwood.

Solvent drying of conifer woods with water-miscible solvents such as acetone, however, remains economically unfavorable because of solvent losses and energy requirements. A limited quantity of perfumery-grade pine needle oil can be obtained from a few *Pinus* species, but the needle oils from most species too closely resemble turpentines to command the premium prices that would justify the production costs. Although the rosins that could be derived from needles of many pines are essentially similar to the xylem rosins, the needle rosins from other species are of unusual composition. Such rosins could be the source of specialty fine chemicals but would not have any significant impact on rosin as a commodity material. The composition of the resin acids in pine needles may be an important characteristic in chemotaxonomic and genetic studies (18).

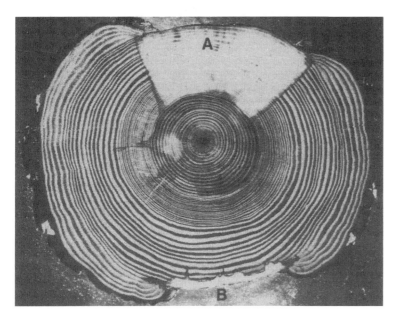

Fig. 10.1.5. Induced lightwood formation on treatment of longleaf pine: (**A**) with the herbicide paraquat. (**B**) compared with limited resin-soaking during normal turpentining

The technique of inducing lightwood formation in pines is the most significant development in naval stores technology since the fractional distillation of tall oil; lightwood formation can be induced in some other conifers but only to a limited extent. Lightwood (oleoresin-soaked wood) can be induced artificially (as opposed to natural heartwood formation) by treatment of the tree xylem with certain dipyridyl herbicides, particularly paraquat. Extractives contents as high as 40% have been found in some wood samples. Such oleoresin-soaked wood is known as lightwood, because it kindles and burns readily and, at one time, served as a source of light. Not only does the oleoresin soaking occur in the immediate area of treatment, as in turpentining, but the zone of lightwood formation extends to the pitch or the heartwood and many feet along the trunk above the point of treatment (Fig. 10.1.5). Commercialization of this technology at this time has been limited to inducing lightwood formation in the tree stem below ground level. This portion of the stem is recovered in the process of whole-tree harvesting. Recovery of the oleoresin is by extraction in wood naval stores extraction plants. Full implementation of the induced lightwood technology could readily double world rosin and turpentine production.

10.1.3 Turpentine

The historic use for turpentine as a solvent in varnishes and paints continues today in some countries − e.g., about 20% of Soviet production is still so used (15). However, most turpentine is now being used as a chemical raw material. Thus,

	gum	wood	sulfate
		%	
α-pinene	60 — 65	75 — 80	60 — 70
β-pinene	25 — 35	0 — 2	20 — 25
camphene	—	4 — 8	—
other terpenes	5 — 8	15 — 20	6 — 12

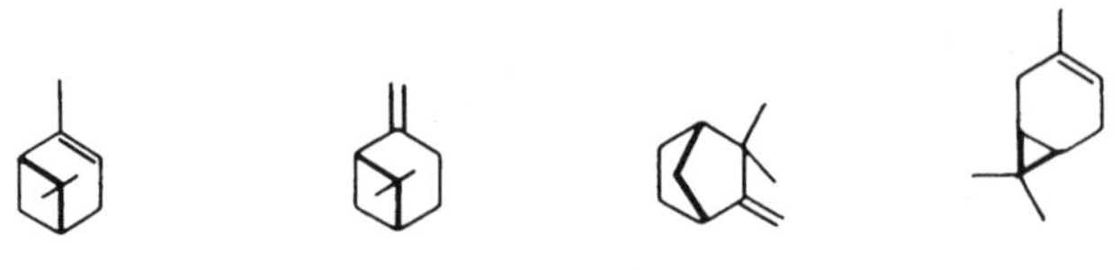

α-pinene   β-pinene   camphene   3-carene

Fig. 10.1.6. Typical composition of some commericial turpentines

the chemical composition is of particular importance. This composition depends upon a number of factors, the most important of which are the pine species of origin and the method of recovery.

In general, refined turpentines consist primarily of unsaturated bicyclic monoterpenes with smaller amounts of monocyclic monoterpene hydrocarbons, alcohols, and some sesquiterpenes and diterpenes (see Chap. 8.1 for a discussion of terpenes). Structures of the major bicyclic monoterpene components and typical compositions of various turpentines from the southeastern United States are given in Fig. 10.1.6. The composition of sulfate turpentine can be considerably different from that indicated in the figure because of the number of species (pines and other conifers) pulped, and the large geographical range from which they are taken. Western sulfate turpentine has appreciable amounts of 3-carene, as do turpentines from Asia and Europe. [At the Forest Products Laboratory, we have used a selected western sulfate turpentine from Douglas-fir (*Pseudotsuga menziesii*) wood furnish as a source of the diterpene, cembrene.] Wood turpentines are relatively consistent in composition because the stump raw material is mostly from a single species, longleaf pine (*Pinus palustris*). α-Pinene is the major component of these turpentines with significant amounts of β-pinene being present in gum and sulfate turpentine. β-Pinene is the major component in *P. radiata*, a minor U.S. species that is extensively planted in parts of the southern hemisphere, and in the Chinese pine, *P. yunnanensis* (9). The β-pinene of commercial turpentines is always the *l*-isomer, but the α-pinene varies. Among the U.S. southern pines, only slash pine (*P. elliottii*) has *l*-α-pinene; the α-pinene of other important naval stores species is either *d* or *dl*.

For use as a chemical raw material, turpentine is fractionated into its components by distillation. For example, southern U.S. sulfate turpentine is frac-

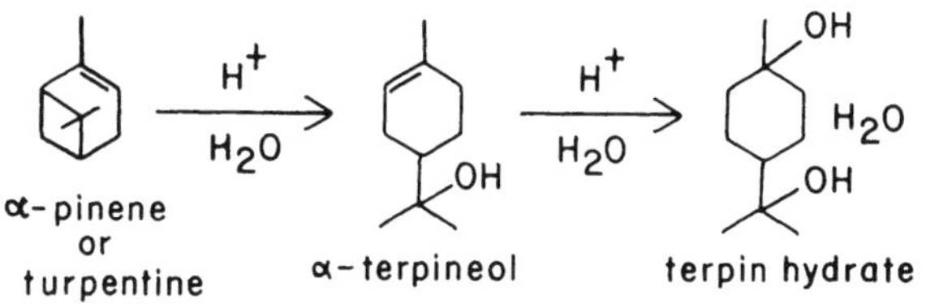

Fig. 10.1.7. Synthesis of pine oil (α-terpineol) and terpin hydrate

tionated into α-pinene, β-pinene, "dipentene," pine oil (mostly monoterpene alcohols), and anethole-methyl chavicol-caryophyllene fractions. The monoterpene hydrocarbon fractions are used to prepare synthetic pine oil, polyterpene resins, perfume and fragrance materials, and insecticides.

10.1.3.1 Pine Oil

The primary use of α-pinene, and therefore of turpentine, involves the hydration by aqueous mineral acids to synthetic pine oil, the primary constituent of which is α-terpineol (Fig. 10.1.7). Between 40% and 50% of the world turpentine production is so used. Good emulsifiers are needed in the hydration reaction because the hydration takes place at the α-pinene-aqueous acid interface. Variations in reaction conditions, fractionation, and blending give pine oils with different compositions and properties. Over 90% of the U.S. pine oil is used in cleaner and disinfectant formulations because the piney odor is pleasant and because pine oil is a good emulsifier, an excellent solvent, and an effective germicide of low toxicity.

Pine oils are effective frothing agents for flotation. They are widely used for metallic sulfide ores such as copper, zinc, nickel, iron, and lead. Other applications are in nonmetallic recovery for feldspar ores, mica, talc, glass sand, and phosphate rock.

Pine oil is also used as a penetrant and dispersing agent in processing various textile fibers. The process of α-pinene hydration can be controlled to favor further hydration, yielding the well-known expectorant, terpin hydrate.

10.1.3.2 Polyterpene Resins

The second largest use of turpentine is in the production of polyterpene resins. A variety of low-molecular weight (600–2000) polymers is produced from α-pinene, β-pinene, dipentene, and mixtures of monomers by cationic polymerization using Lewis acids such as aluminum chloride. The polymerization of the different monoterpenes yields polymers having differing repeating units; the final molecular weight is controlled by chain transfer. Chain termination has been postulated as a nonpropagating reaction involving camphenic carbenium ion rearrangement. The somewhat-higher molecular weight polyterpenes are prepared from β-pinene. About 40% of polyterpene production is from β-pinene with dipentene a close second. The various polyterpene resins are used primarily to impart tack in the compounding of adhesives. Polyterpene resins from β-pinene are used mostly in the preparation of solvent-based pressure-sensitive adhesives, and dipentene resins are used predominately in the manufacture of hot-melt adhesives. Other uses for the unique tackifying and thermoplastic properties of these resins are in coatings, sealants, and investment castings.

10.1.3.3 Flavors and Fragrances

Perhaps the most technically sophisticated area of turpentine utilization is in the small (10% – 15% of turpentine consumption) but growing production of flavor and fragrance chemicals. Although small amounts of turpentine components are directly used in flavors and fragrances, most of the turpentine components are converted to other chemicals. Of the turpentine components, β-pinene is the most important starting material. Its importance is indicated by the development of a β-pinene production method based on isomerization of α-pinene; the β-pinene is recovered from a continuously recycled process steam that has an equilibrium concentration of 95% α-pinene/5% β-pinene.

Tonnage quantities of *l*-menthol are produced from *l*-α-pinene by synthesis through a sequence of reactions involving *d*-citronellol (Fig. 10.1.8). Menthol is

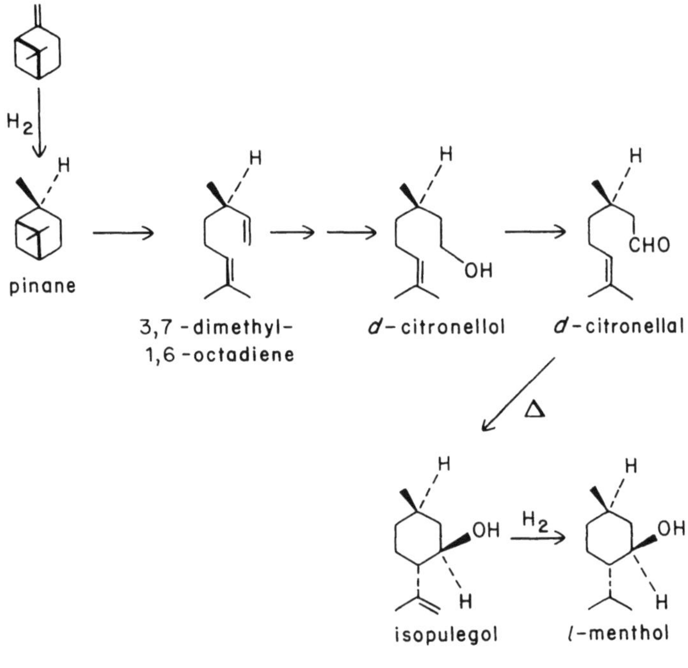

Fig. 10.1.8. Synthesis of menthol

used for its cooling effect (the *l*-isomer is most active) in cigarettes and cosmetics and for flavoring, particularly peppermint. There are several other commercial routes to menthol, some of which result in racemic menthol, which is then resolved, such as by fractional crystallization of acetates or formates.

The preparation of myrcene by the pyrolysis of β-pinene begins an important sequence in the synthesis of several flavor and fragrance chemicals, as depicted in the Fig. 10.1.9. A more recent alternate route to linalool and geraniol involves

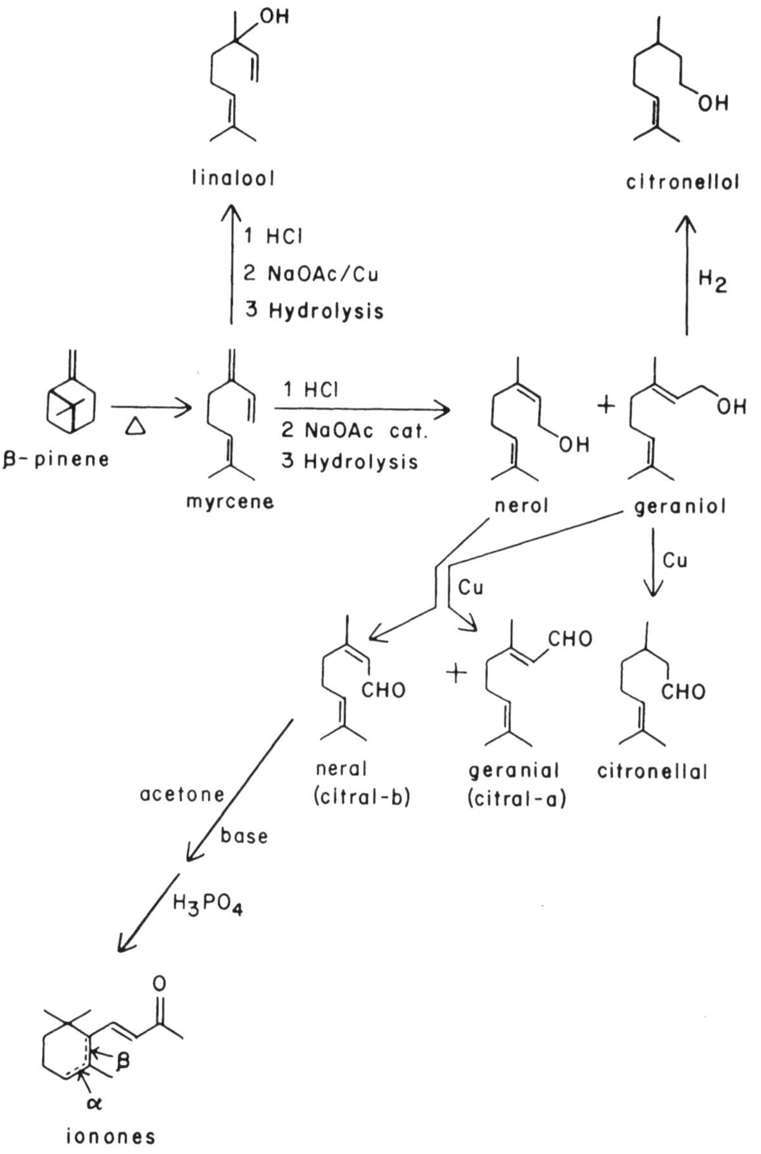

Fig. 10.1.9. Synthesis of some flavor and fragrance chemicals

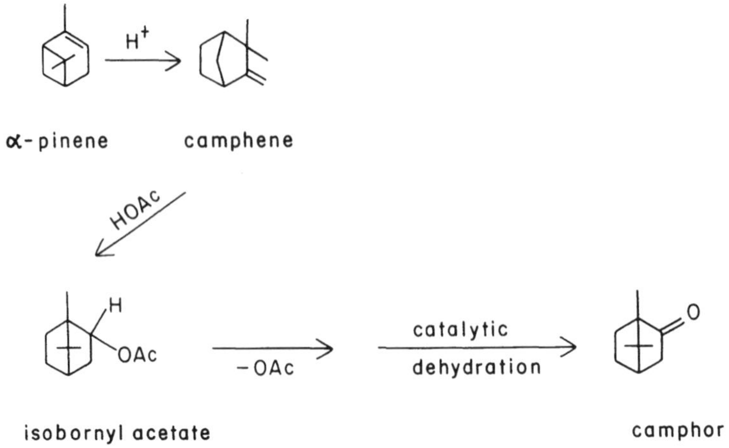

α-pinene   camphene

isobornyl acetate   camphor

Fig. 10.1.10. Synthesis of isobornyl acetate and camphor

the hydrogenation of α/β-pinene to pinane, oxidation of pinane to the hydroperoxide, followed by hydrogenation and thermal isomerization.

Linalool and its esters have lilac-like fragrances. Geraniol and nerol and their esters are rose-like, as is their hydrogenation product citronellol. Citronellol, in turn, is the precursor of rose oxide, another important chemical for perfumery. Rearrangement of geraniol-nerol over copper catalyst produces either citronellal or a mixture of citrals (geranial and neral) depending on the conditions used. Citronellal finds use as an odorant and can be reduced to ctronellol or hydrated (via the bisulfite adduct to prevent crystallization) to hydroxycitronellal, which has a lily-of-the-valley fragrance. Citral is used for its intense lemon character in the preparation of citrus flavors and fragrances.

Aldol condensation of citral with acetone followed by acid cyclization gives a mixture of ionones, of which the α-isomer is widely used for its violet odor (Fig. 10.1.9). However, the β-ionone is more important because it is used in the synthesis of vitamin A. In a similar fashion, pseudoionone can be used in the synthesis of vitamin E. Condensation of citral with 2-butanone gives methylionones, also valued as perfume ingredients. The dihydropseudoionone geranyl acetone, a perfumery chemical having a magnolia-like odor, can be prepared from either geranyl chloride or linalool.

Pyrolysis of α-pinene is the starting point for a number of chemicals. The *allo*-ocimene product can be hydrochlorinated and then hydrolyzed to give a mixture of tertiary alcohols that has a sweet floral odor for use in soaps and cosmetics. Tetrahydro-*allo*-ocimenyl acetate is used in citrus-type colognes for men. Similarly, dihydromyrcenol, prepared in several steps from 3,7-dimethyl-1,6-octadiene (the pyrolysis product of *cis*-pinane), is used in formulations needing a lime note.

Camphene is the starting material for the synthesis of camphor and of isobornyl acetate (Fig. 10.1.10). Condensation of camphene with phenolic compounds followed by hydrogenation gives a mixture of isobornyl cyclohexanol isomers, which have the odor of (costly) sandalwood. Methoxyeugenol, another com-

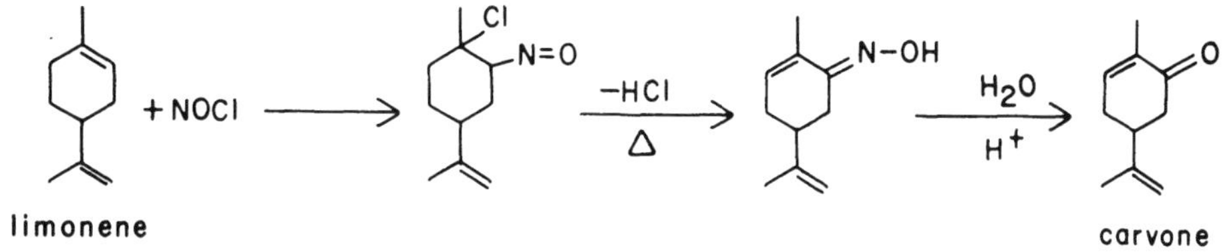

Fig. 10.1.11. Preparation of carvone from limonene

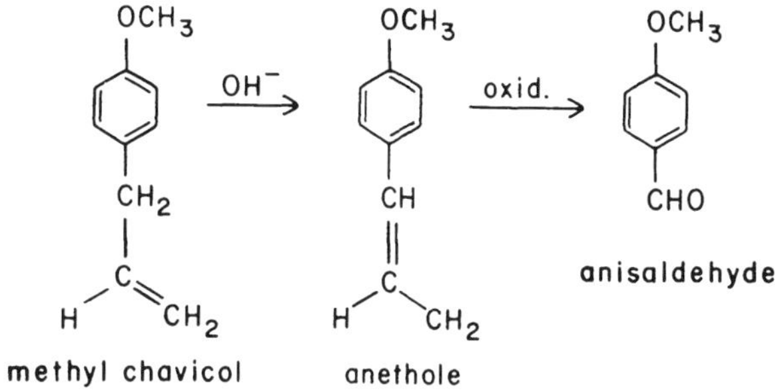

Fig. 10.1.12. Phenolic flavor and fragrance components from turpentine

pound having sandalwood-type odor can be made in a series of reactions from dihydromyrcene.

Limonene (*dl*-limonene is generally known as dipentene) is the starting material for synthesizing carvone (Fig. 10.1.11), but the *d*- and *l*-limonenes give products having entirely different flavor and fragrance characteristics. *d*-Limonene, a citrus industry by-product, gives *l*-carvone, the main component of spearmint oil; *l*-limonene gives *d*-carvone, a compound having the odor and flavor of dill.

Many other flavor and fragrance chemicals are being made from turpentine components including a number from 3-carene. In addition to terpene-derived flavors and fragrances, turpentine is the source of phenolic-type materials. Estragole (methyl chavicol) and anethole are both isolated during fractionation. The estragole can be isomerized to anethole, well-known for its licorice-anise flavor (Fig. 10.1.12). Oxidation of anethole produces anisaldehyde, which is used in perfumery because of its hawthorn or coumarin-type fragrance. Anisaldehyde is also used in the electroplating industry.

10.1.3.4 Insecticides

Certain natural products have long been known to have insecticidal and repellent properties. Turpentine was recommended at one time for repelling and killing insects. Pine oil was found to effectively enhance the activity of costly pyrethrum (introduced to the United States in the 1880s). A currently available repellent formulation consists of rectified oil of pine tar, isobornyl acetate, and citronella. An unusual selective and insecticidal property of menthol is being investigated for control of mites in bee hives (8).

The first important synthetic insectide from turpentine – isobornyl thiocyanoacetate (prepared from camphene) – went into production in 1940. Its use in the control of household and livestock pests peaked in the 1940s when imported pyrethrum was in short supply because of World War II. By 1979 it was no longer being produced.

Because of the early successes of DDT, the area of synthetic terpene insecticides turned to chlorinated products during the early 1940s. In 1944, chlorinated camphene (toxaphene) was demonstrated to be effective on houseflies and on cotton insects. Commercial production began in 1947. Toxaphene is a product of the average formula $C_{10}H_{10}Cl_8$ and consists of some 200 chlorinated components. Strobane, a terpene insecticide produced by chlorination of a mixture of camphene and α-pinene, was introduced at a later date to circumvent the toxaphene patents. Toxaphene had wide use for field crops and for animal parasites. However, all labels and uses for toxaphene in the United States were cancelled at the end of 1983 with the exception of scabies control for beef cattle and sheep.

chrysanthemic acid

The pyrethenoids, the active components of pyrethrum, consist of esters of chrysanthemic acid. Chrysanthemic acid can be synthesized from 3-carene or β-pinene, but the present commercial method is petroleum based.

10.1.3.5 Miscellaneous Uses

As noted previously, the major use of turpentine at one time was as a solvent, particularly in varnish formulations, and as a thinner for linseed-oil-base paints. Petroleum-based solvents have now replaced turpentine for these purposes. However, turpentine or turpentine fractions are packaged in small containers (up to 1 gallon) for the retail market for home use as paint thinners (including use with artist's oil paints) and as a general-purpose solvent. Turpentine fractionation products may recover some of the historical solvent use by replacing chlorofluorocarbons in the specialized circuit board cleaning market (1).

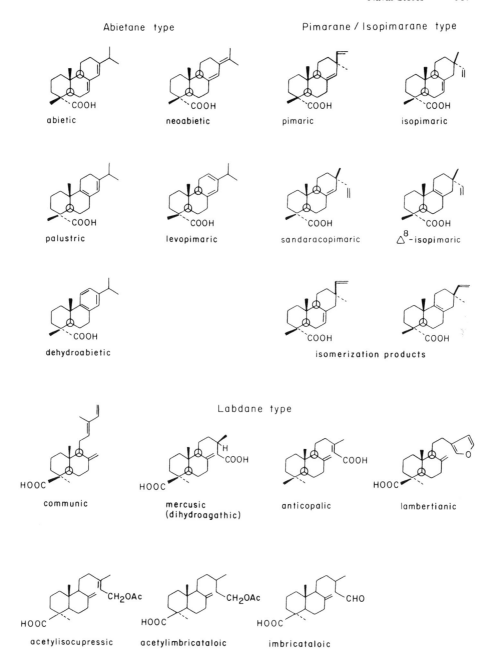

Fig. 10.1.13. Resin acids of pine oleoresins and rosins

10.1.4 Rosin

Rosins are the nonvolatile resin of pine oleoresins and consist primarily of diterpene resin acids with 5%–20% neutrals. Resin acids of the abietane and pimarane/isopimarane types predominate in commercial rosins, with lesser amounts of the labdane type. The most common abietane, pimarane/isopimarane, and labdane-type xylem-derived resin acids are given in Fig. 10.1.13. Abietic, neoabietic, palustric, levopimaric, dehydroabietic, pimaric, sandaracopimaric, and isopimaric are the principal tricyclic resin acids found in rosins (Δ^8-isopimaric acid is the primary resin acid in the oleoresins of pinyon pines but is a minor component in commercial rosins). The labdane acids of commercial rosins are communic (from slash pine, *Pinus elliottii*), dihydroagathic (from *P. merkusii*), and, in even much smaller amounts, imbricataloic and acetylisocupressic acids (slash and longleaf, *P. palustris*, pines). Labdane resin acids are the principal resin acids in some noncommercial xylem oleoresins such as lambertianic acid in sugar pine, *P. lambertiana*, and anticopalic acid in western white pine, *P. monticola*. Implementation of the lightwood technology with these species could provide a commercial source of labdane resin acids for fine chemical uses (17). The labdane resin acids are also particularly prevalent, both in variety of specific compounds and in proportion of total resin acids, in pine needle oleoresins.

The source (species), history, recovery, and processing affect the final composition of the resin acids in a rosin. All these factors are brought together in a comparison of the resin acid composition in different types of rosin (Table 10.1.1). In all the rosins, most of the levopimaric acid has been isomerized to the other abietadienoic acids, palustric, abietic, and neoabietic. At equilibrium, the thermal isomerization product of any of the abietadienoic acids is comprised of approximately 80% abietic acid, 14% neoabietic acid, and 5% palustric acid (mineral acid isomerization results in 95% abietic acid) (16). Thus, the thermal isomerization of the tall oil and wood rosins of Table 10.1.1 is near equilibrium. This is not the case for gum rosin. Indeed, some gum rosins from carefully controlled, continuous processing have levopimaric acid contents of some 10%. The presence of communic acid is indicative of the use of slash pine, which is the primary U.S. species for gum rosin and an important component of southeastern U.S. pulp mills; the wood rosin, being derived primarily from longleaf pine, does

Table 10.1.1. Composition of the common resin acids in some typical U.S. rosins[a]

Type	Resin acids (%)								
	pimaric	sandaraco-pimaric	communic	levopi-maric	palustric	isopi-maric	abietic	dehydro-abietic	neoabietic
Tall oil[b]	4.4	3.9	1.0	–	8.2	11.4	37.8	18.2	3.3
Wood	7.1	2.0	–	–	8.2	15.5	50.8	7.9	4.7
Gum	4.5	1.3	3.1	1.8	21.2	17.4	23.7	5.3	19.1

[a] % of acid fraction
[b] Also contains fatty acids and other minor resin acids such as the secodehydroabietic acids

not contain communic acid. The lower content of isopimaric acid in the tall oil rosin reflects the large portion of loblolly pine used for pulping. The elevated content of dehydroabietic acid is a result of oxidation of levopimaric and palustric acids that occurs in the chips before pulping and, to some extent, disproportionation that occurs during distillation. In addition, the alkaline oxidation of levopimaric acid gives rise to small amounts of the secodehydroabietic acids. Acidification of tall oil soaps and distillation of crude tall oil results in some isomerization of the ring double bonds in the pimaric- and isopimaric-type acids to the 8(9) position. The secodehydroabietic and the Δ^8-pimaric and Δ^8-isopimaric acids are concentrated in the intermediate fractions of tall oil distillation. In addition, small amounts of 7,15-pimaradienoic acid, an acid not found in the oleoresin, has been isolated recently from the intermediate fractions (10).

secodehydroabietic acid

The neutral fraction in rosin is usually a detriment in rosin utilization as most uses are based on the carboxyl group. Thus, rosins of high neutrals content are of lower value. Gum rosin neutrals consist predominately of diterpene hydrocarbons, alcohols, and aldehydes plus smaller amounts of the higher-boiling monoterpene and sesquiterpene components of turpentine. The same types of diterpene compounds are also present in tall oil rosin neutrals. However, sitosterol is the predominant neutral in crude tall oil and tall oil pitch.

Rosins are used little as such but are widely used in chemically modified forms. Most of the modifications involve the carboxyl group or the double bonds and are effected by transformations such as hydrogenation, disproportionation (the process is, to a significant part, a dehydrogenation), esterification, polymerization, salt formation, or reaction with maleic anhydride or formaldehyde.

10.1.4.1 Paper Sizes

The discovery in 1807 that water resistance in paper could be achieved by precipitating partially saponified rosin with alum on pulp led to a market that has been the major use of rosin for many years. In addition to the simple aqueous dispersion of sodium resinate, called paste size, rosin sizes in emulsion and dry form are also used. The composition of the rosin, the amount of neutrals and oxidized components as well as the resin acid composition, affects the efficacy and physical properties of a size. The efficiency of rosin sizes is improved by fortification – that is, the modification of the rosin by forming these maleic anhydride derivative of the abietic acids. Further modification of fortified size such as by reaction with formaldehyde can improve the product by decreasing its unwanted tendency to crystallize. Although dramatic improvements have been made in the technology of paper sizing with the rosin because of advancements in the knowledge of the physical and chemical nature of sizing, the basic application remains.

The market for rosin sizes consumes over 30% of rosin production. However, competition from synthetic sizes has increased, with particular pressure being seen in alkaline papermaking applications.

10.1.4.2 Polymerization Emulsifiers

At one time rosin found considerable use in the old yellow bar laundry soaps (38% of rosin production was used for this purpose in 1938); this use is almost negligible now. Rosin soaps, however, find extensive use as polymerization emulsifiers in the preparation of SBR (styrene-butadiene rubber; the majority of SBR production is used in tire stock) and are used in the manufacture of ABS (acrylonitrile-butadiene-styrene) resins as well as other specialty polymers. In addition to their role in the polymerization process, rosin soaps contribute important tack properties to the polymeric products. Because of the detrimental effect of the conjugated double-bond system of abietic-type resin acids on the rate of peroxy-catalyzed polymerization, the conjugated double bonds must be removed. This is accomplished by hydrogenation or, more often, disproportionation. Commercial hydrogenated rosins contain primarily dihydro- or tetrahydro-acids, the proportion depending on the catalysts and hydrogenation conditions used. In the dehydrogenation of rosin, hydrogen is removed from the abietic-type acids and transferred to the exocyclic vinyl group of the pimaric/isopimaric acids. A small proportion of the hydrogen is also transferred to abietic acids to form dihydro derivatives. The final dehydrogenated rosin product contains dehydroabietic acid as the major component with up to 25% dihydro acids. A somewhat similar disproportionated product is also made using catalysts such as iodine, iodides, or certain organic sulfur compounds.

10.1.4.3 Adhesives

Adhesives are composed primarily of a polymeric backbone and a tackifying resin with smaller amounts of other components as antioxidants, curing agents, and sequestering agents. About 20% (80–100 million pounds) of the resins used in adhesives are rosin-derived (contrasted to 50 million pounds of terpene polymers). The most important of the rosin derivatives used in adhesives are disproportionated or hydrogenated rosins, polymerized (dimer) rosins (Fig. 10.1.14), and the corresponding esters (especially the esters with glycerol or pentaerythritol).

Rosins have long been used in pressure-sensitive adhesives, the most familiar of which are the solvent-based rubber cements. The compounding of these

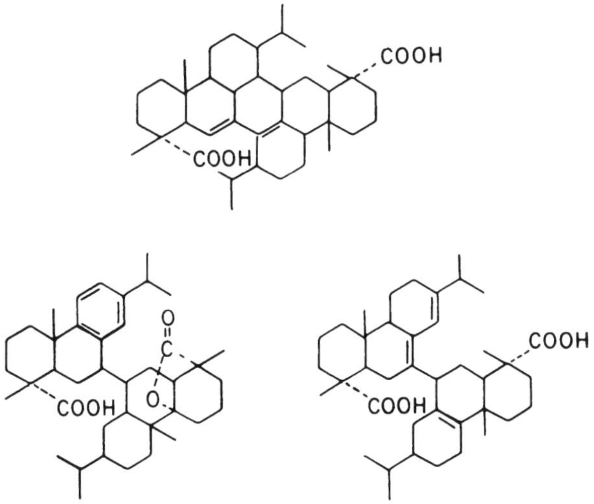

Fig. 10.1.14. Three known types of resin acid dimers (5, 7)

pressure-sensitive adhesives is a combination of art and application of the theory of two-phase resin-elastomer systems. Other uses for pressure-sensitive adhesives and mastics (very viscous adhesives) are in contact adhesives used in construction (for tiles, panel, drywall, and subfloor applications), automobiles, and laminating.

Hot-melt adhesives, as the name implies, are applied in the molten state (about 150°C) with the adhesion being completed on cooling. In hot-melt adhesives containing rosin derivatives, the primary polymers used are ethylene-vinyl acetate or ethylene-ethylacrylate copolymers, low molecular weight polyethylene, and amorphous polypropylene. The hot-melt formulations are used in shoe manufacture, the assembly of various products, carpet sealing tape, and, particularly, bonding paper, which includes corrugated paperboards, heat-sealed containers, book binding, and laminates.

10.1.4.4 Inks

One of the most important markets for rosin products is the ink industry. The value of ink products approaches $2 billion in the United States alone. Most printing inks are used in publishing or in packaging and are classified by their use in one of the four basic printing methods: letterpress, lithography (offset), flexography, and gravure. Growth in printing markets and in attendant opportunities for rosin products are expected in offset, flexographic, and gravure applications.

The diversity in specific applications requires a large variety of ink formulations. Rosin derivatives, often in conjunction with other resins and polymers, are components of the resin portion of the ink formulations, contributing binding, film-forming, and solvent characteristics to the final ink products. Modified rosins having high melting points are the most usual rosin-based components of printing inks. Although rosin derivatives, such as esters of dimerized rosin, find specialized uses in ink formulations, the most important derivatives are resin acid salts (particularly calcium and zinc resinates) or are based on maleic anhydride or fumaric acid adducts of rosin. Of particular importance are the various esters of these adducts with polyhydric alcohols. The specifics of the chemical modifications depend on the requisite solubility characteristics of the ink resin.

10.1.4.5 Other Market Areas

Rosin has a large variety of minor uses that can be attributed to its low cost, acidity, and ready availability. These uses can be conveniently divided into unmodified and modified rosin categories.

At one time, large amounts of unmodified rosin were used in linoleum, varnishes, electrical insulation, leather processing, laundry soaps, etc. Many of these markets still exist, albeit to a much smaller degree. Some of the many other uses that touch many aspects of our lives include:

In soldering fluxes, both as a paste or in hollow cores, to remove oxide films and to etch slightly the metal surface.
As a drossing compound to remove the scum that forms on top of molten metal, particularly in type-metal foundries.
As a depilatory to remove hair from hog carcasses by dipping the carcass in molten rosin, cooling, and stripping off the rosin coating.
As a gripping material for sports, such as mixing ground rosin with a noncaking agent (e.g., magnesium carbonate) for the familiar rosin bag used by baseball pitchers.
In tree-banding/wrapping compounds (and at one time in fly paper) as an adhesive (plasticized with nondrying oils), and
In the grinding of lenses to hold the glass blank.

Modified rosins, however, are of greater importance now than the nonmodified material. Typical modifications are hydrogenation-disproportionation, esterification, dimerization, and maleic/fumaric adducts.

Esters are the most important derivatives in the miscellaneous class of uses. One of the largest of these in the compounding of stick, bubble, and sugarless chewing gums. The food-grade rosin esters function as modifiers of the elastomeric component of the gum base, which may be a natural exudate like chicle (from *Manilkara zapota*) or synthetic polymers such as polyvinyl acetate. In another food-related use, rosin esters are compounded with citrus flavoring for soft drinks to raise the specific gravity of the flavoring to that of water. The resultant oil is dispersed and the dispersion stabilized with a colloid.

A number of applications of modified rosin esters depend on a surface-gloss effect. Although such use in structural surface coatings has decreased over the years, numerous formulations still incorporate rosin esters for gloss, leveling flow, brushability, and other characteristics. The esters also contribute gloss characteristics in emulsion floor polishes and in shoe polishes. Other food-related uses of rosin esters are as a surface coating of fresh fruit for gloss to enhance sales and in the interior-coating and sealing of cans used in food packaging. Other diverse markets include fragrance fixatives, fertilizer encapsulation, cellulosic-based thermoplastic molding compounds for doll heads, hot-melt road markings, and "lost wax" casting.

Other rosin derivatives expand the range of rosin markets. For example, the alcohol derivatives are used in plastic heat stabilizers, ethoxylated amines are used in corrosion prevention and specialty cleaning, and polyols are used for polyurethane foams. Amines (dehydroabietylamine) are used as antimicrobials (fungicides, bactericides, and algicides), as cationic ore flotation agents, and in the laboratory as a resolving agent for racemic mixtures.

10.1.5 Fatty Acids

Whereas oleoresin-derived products of turpentine and rosin have been naval stores materials from antiquity, tall oil fatty acids are a 20th-century newcomer. Their inclusion in the naval stores arena is a result of the co-occurrence with rosin in crude tall oil and recovery by distillation as a product along with tall oil rosin. Successful commercial fractional distillation of crude tall oil was first achieved in 1949, although distilled tall oil products had been available for several decades.

Tall oil fatty acids (TOFA) are available as commodity products having resin acid contents of 0.5% – 10% (fatty acid fractions having greater than 10% resin acids are clasified as distilled tall oil). American TOFA consists primarily of C_{18} acids with the monoenoic (oleic) and dienoic (linoleic and isomers) acids predominating. Only small amounts of trienoic and saturated acids are present. The linoleic portion is mostly the *cis,cis*-9,12-isomer accompanied by some isomerization products from processing. A 5,9-linoleic acid isomer, a 5,9,12-linolenic acid isomer, and mono-, di-, and tri-unsaturated eicosanoic acid derivatives (Chap. 6.3.3.2) are present in small amounts in tall oil and pine extractives. Scandinavian tall oil fatty acids, however, contain over 10% of the 5,9,12-linolenic acid.

Specialty tall oil fatty acid products are also available. One such product, obtained from tall oil heads, contains predominately saturated fatty acids, most of which is palmitic acid. High-quality oleic and linoleic acids are produced by crystallization of tall oil fatty acids. Another product consisting of about equal amounts of oleic acid and its *trans*-isomer, elaidic acid, is prepared by disproportionation of tall oil fatty acids.

Of the 1 billion pounds of saturated and unsaturated fatty acids produced in this country, about 35% of the total and 50% of the unsaturated fatty acids portion come from tall oil. Although originally having served as a low-cost substitute for vegetable oil-derived fatty acids, tall oil fatty acids have come into their own as chemical raw materials. Major uses are in intermediate chemicals and protective coatings.

10.1.5.1 Intermediate Chemicals

Some 40% of tall oil fatty acids are used as chemical intermediates, a catch-all category including a variety of derivatives. Polymerized fatty acids are one of the major type of intermediate chemicals from tall oil fatty acids, and one for which tall oil is the principle feedstock. The product, which consists primarily of dimers with smaller amounts of trimers and higher polymers as well as unreacted monomers, is further processed to concentrate the dimer fraction. This fraction is then used to prepare a variety of soaps, polyesters, polyamides, amines, and other nitrogen derivatives for a multitude of uses, including coatings, printing inks, adhesives, corrosion inhibitors, gelling agents (soaps), and lubricants (esters). Polyglycol esters from the reaction of tall oil fatty acids with oxiranes such as ethylene or propylene oxides have extensive use as emulsifiers and nonionic surfactants. Epoxy esters prepared by the reaction of hydrogen peroxide with tall oil fatty acid esters are used as plasticizers and stabilizers for vinyl resins.

Diels-Alder derivatives can be made from TOFA. For example, an interesting group of chemicals is based on the product of the reaction of TOFA-derived conjugated linoleic acids with acrylic acid. The C_{21} dibasic acid product and its derivatives have interesting properties that lead to diverse uses such as in detergents, textile lubricants, fabric softeners, corrosion inhibitors, inks, and floor polishes.

10.1.5.2 Protective Coatings

A protective or surface coating is a formulation that on application to a substrate forms a continuous, tack-free film. Essentially, the coating consists of a resinous medium in which pigments are dispersed; a solvent is often a part of the formulation to aid in application. Although plant oils were used in coatings with limited success some 2000 years B.C., and Theophilis described a cooking method for making varnish (a coating without pigments) in the 11th century, it was not until the mid-1600s that the value of certain metal salts (lead, cobalt, manganese, and iron) as driers was recognized. Ready-mixed paints, however, were not a commercial reality until the 1860s. Driers in this context are not concerned with evaporation of solvent but rather to their participation as catalysts in the autooxidative process that leads to polymerization of an oil or resin to form a hard film. Whereas many of the above chemical intermediate uses for tall oil fatty acids are based only on the carboxyl group, the autooxidative drying process involves the olefinic functionalities of the unsaturated fatty acids – particularly linoleic acid.

While the early paints used heat-bodied (thermally polymerized) vegetable oils – typically linseed oil – later formulations were based on alkyd resins. These alkyd resins used polyhydric alcohols – primarily glycerol and pentaerythritol – and polybasic acids such as phthalic anhydride with fatty oils as modifiers. The availability of inexpensive tall oil fatty acids in the 1950s made it possible to further manipulate formulations for a greater diversity of products. This is accomplished using a three-part resin formulation incorporating various polyhydric alcohols, polybasic acids, and tall oil fatty acids. This approach is successful even though tall oil fatty acids are only semidrying materials (a result of the large content of oleic acid), because of possible compensation in polybasic alcohol-acid portions of the formulations. The quick acceptance of tall oil fatty acids into paint products was also due to the low content of (triene) linolenic-type acids, which are detrimental because they result in undesirable yellowing of the coating. The versatility of the alkyd resins leads to end uses in industrial finishes (automotive and appliance) and architectural coatings. Further modifications such as reaction with silicone resins, replacement of polybasic acids with tolylene diisocyanate to give urethane alkyds, or incorporation of epoxy resins result in other special-purpose coatings. Tall oil-derived dimer acids are also used in a variety of coatings formulations.

10.1.5.3 Other Uses

Fatty acid soaps have been used as surfactants for thousands of years, beginning with the primitive wood ash-animal fat products and including the esoteric oil (Castile) soaps. Although synthetic detergents have replaced soaps for many cleaning uses, fatty acid soaps and derivatives are still important ingredients of many cleaning products. About 10% of tall oil fatty acids enter this market. Tall oil fatty acids are used extensively in liquid soaps and waterless hand cleaners. Whereas detergents based on TOFA derivatives such as ethoxylates, sulfated monoglycerides, or alkanolamides can be produced, there is little use in such detergents. Ethoxylated tall oil fatty acids are used, however, in textile processing.

Tall oil soaps are also used as lubricants in metal working. Lithium soaps are important components (gelling agents) of automotive greases. TOFA and sulfonated/sulfated derivatives find use in ore flotation applications such as with iron ore, phosphate rock, fluorspar, and barite. A variety of nitrogen derivatives are used as foaming agents, fabric softeners, and antibacterial agents. The use of dimethylamide derivatives as pulping aids brings TOFA full circle back to the pulping process. Other uses of tall oil fatty acids span the gamut from agricultural emulsions and oil-well fracturing compounds to tar removers and waterproofing agents.

10.1.6 Miscellaneous Products

The previous sections of this chapter have considered the utilization of turpentine, rosin, and tall oil fatty acids, the primary commodity products of the naval stores industry. In the course of processing the raw materials (crude tall oil and crude sulfate turpentine, and stumpwood extractives), by-products are obtained in addition to turpentine, rosin, and tall oil fatty acids, the primary commodity products.

The distillation of crude tall gives three by-product streams: heads, distilled tall oil, and pitch. Although palmitic acid can be recovered from the heads fraction, usually this fraction is burned for its fuel value. Distilled tall oils (Sect. 10.1.2.3) most often have the same types of uses as for rosin and for tall oil fatty acids. These include such markets as surface coatings, inks, polymerization emulsifiers, and detergents. Tall oil pitch, the nondistilled residue from crude tall betulin (R=CH₂OH)

oil distillation, is a darkly-colored resin of low acid number that has been described as being "too thin to walk on and too thick to pour." Because of these characteristics and its low cost, however, tall oil pitch is used in rubber reclaiming, bonding (making coal-dust briquets and extruded fire logs), and petroleum recovery, and in roofing cements and papers, construction adhesives, core oils, floor coverings (tiles and linoleum) and caulking compounds, soil stabilizers, and a host of other products. Surplus tall oil pitch is burned as a fuel.

The neutral components in crude tall oil are usually detrimental to product quality, often disproportionate to the initial content. Neutrals result in reduced acid number and, thus, in lower quality of distilled products simply by dilution. But more importantly, reaction during distillation of the alcohol components with fatty acids results in lower yields through formation of higher-boiling esters, which end up in the pitch fraction. To minimize these problems, tall oil soaps that are high in nonsaponifiable components (e.g. Scandinavian soaps, which contain as much as 30% nonsaponifiables due to the high content of birchwood in the pulp mill wood furnish) are extracted using a liquid-liquid system of water-acetone and hexane. Betulin (betulinol), sitosterol, and related neutral products from soap washing are being actively promoted for a host of potential uses including cosmetics and shampoos, lacquers and other protective coatings, emulsifiers, and polyurethanes. A number of commercially viable methods have been reported for the extraction of the phytosterols, principally sitosterol, from tall oil pitch and from soap washing neutrals. Fermentation of the sitosterol can give androst-4-ene-3,17-dione (Fig. 10.1.15), an intermediate for synthesizing steroid drugs such as spironolactone (a diuretic used to treat hypertension).

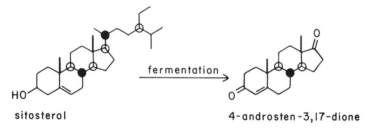

Fig. 10.1.15. Fermentation of sitosterol to a sterol intermediate (2)

As turpentine is essentially a chemical raw material, it is extensively fractionated. Unneeded fractions from this refining are used as solvents or burned as fuels.

The colored fraction from crude (FF grade) wood rosin decolorization does not have a generic name. The only material of this nature commercially available at the present time in the United States is the Hercules product, Vinsol. It consists of more than 50% phenolic components and about 25% resin acids and derivatives. Vinsol has a wide diversity of uses, which include emulsifiers for asphalts, tackifiers and bonding agents in adhesives, air-entrainment agents in cements, components of electric insulation, resin extenders in laminates, and foundry cores.

10.1.7 The Future for Naval Stores

The naval stores industry is in the midst of transition from the traditional, both in source of supply and in changes of use, to more sophisticated chemical products. One of the major problems inherent in the naval stores industry has been that of chronic cycles from excess production to shortages, the latter resulting in loss of markets because of irreversible shifts by users to substitutes. The availability of the lightwood technology, however, can help ameliorate the shortage problem, and in doing so, provide sufficient and dependable supplies of naval stores for new, large-volume markets.

References

1 Chemical Marketing Reporter 1988 Fluorocarbon replacement could be big terpene user. Chem Market Rep 233(5):23
2 Conner A H, Nagaoka M, Rowe J W, Perlman D 1976 Microbial conversion of tall oil sterols to C_{19} steroids. Appl Environ Microbiol 32:310–311
3 Drew J, Propst M 1981 Tall oil. Pulp Chem Assoc New York, 199 pp
4 Drew J, Russell J, Bajak H W 1971 Sulfate turpentine recovery. Pulp Chem Assoc New York, 147 pp
5 Fujii R, Arimoto K, Zinkel D F 1987 Dimeric components from the dimerization of abietic acid. J Am Oil Chem Soc 64:1144–1149
6 Gamble T (ed) 1921 Naval stores: history, production, distribution and consumption. Review Publishing Savannah, 286 pp
7 Gigante B, Jones R, Lobo A M, Marcelo-Curto M J, Prabhakar S, Rzepa H S, Williams D J, Zinkel D F 1986 The structure of an abietic acid dimer. J Chem Soc Chem Commun (13):1038–1039
8 Herbert E W Jr, Shimanuki A, Matthenius J C Jr 1988 An evaluation of menthol placed in hives for honeybees for the control of *Acarapis woodi*. Am Bee J 128:185–187
9 Luo J L, Zhu J, Li Z X, He T, Yin Y, Yin X B 1985 Investigation on high α-pinene *Pinus yunnanensis* trees. Chem Ind For Prod 5(4):1–11
10 Magee T V, Zinkel D F 1988 Composition of American distilled tall oil. J Am Oil Chem Soc (in press)
11 McSweeney E E 1987 Tall oil and its uses. Pulp Chem Assoc New York, 132 pp
12 Naval Stores Review 1987, 1986 International yearbook. Naval Stores Review New Orleans, 24 pp
13 Rice P 1980 Amber, the golden gem of the ages. Van Nostrand-Reinhold New York, 289 pp

14 Sandermann W 1960 Naturharze – Terpentinol – Tallol, Chemie und Technologie. Springer Heidelberg, 483 pp
15 Starostina E B, Nesterova E T, Sedel'nikova A I 1987 [Trends in the use of turpentine.] Gidroliz Lesokhim Promst (1):26–28
16 Takeda H, Schuller W H, Lawrence R V 1968 The termal isomerization of abietic acid. J Org Chem 33:1683–1684
17 Zavarin E, Wong Y, Zinkel D F 1978 Lightwood as a source of unusual naval stores chemicals. Proc Lightwood Res Coord Counc, 5th Ann Meeting, US For Serv East For Exp Sta Asheville NC, 19–30
18 Zinkel D F 1977 Pine resin acids as chemotaxonomic and genetic indicators. TAPPI Conf Papers, For Biol Wood Chem Conf Madison, WI, TAPPI Atlanta, 53–56
19 Zinkel D F 1981 Turpentine, rosin, and fatty acids from conifers. In: Goldstein I S (ed) Organic chemicals from biomass. CRC Press Boca Raton, 163–187
20 Zinkel D F, Russell J (eds) 1988 Naval stores: production, chemistry, utilization. Pulp Chem Assoc New York.

10.2 Gums

J. N. BeMiller

10.2.1 Introduction

While non-woody plant materials (seeds, fruits, stems, roots, tubers, fronds) have been examined, and are used, as commercial sources of carbohydrates such as sucrose, starch, and industrial gums, woody tissues in general have not been looked to as sources of carbohydrates other than cellulose (for the pulp and paper industry). Exceptions are the arabinogalactan of *Larix* (larch), the exudate gums (gums arabic, karaya, tragacanth, and ghatti), sucrose of maple sap (in maple syrup and sugar), and, to a limited extent, sago starch (Chap. 4).

The commercial potential of other extractable carbohydrates has been neither proposed nor promoted for one or more of the following reasons: a) they are present in amounts too small to make their extraction economical; b) their extraction is difficult; c) they are more conveniently and inexpensively obtained from other sources; d) the wood is more valuable for purposes other than being chipped and extracted; and/or e) they remain undetected as a result of not being sought (see Chap. 4). Larch arabinogalactan is not being utilized currently because no application has been found for which it is uniquely suited – that is, where its cost: functionality ratio makes it the gum of choice.

Exudate gums are not extracted from woody tissues but are gummy secretions from natural cracks, wounds made by insects, or man-made incisions in the bark of trunks and branches. The exudate gums of commerce are produced by various small trees and shrubs that grow on dry land that generally is not cultivated. There is no competition either for the wood of the trees from which the gums are obtained, except for that of the trees that produce gum ghatti, or for the land on which the trees and shrubs grow.

The drawbacks that restrict their use are the labor intensity of their collection, the resulting relatively high price, their inconsistent quality due to seasonal effects and variable storage conditions, the presence of foreign matter (impurities), adulteration, and the relatively limited and undependable amounts available, even though some *Acacia* trees, which produce gum arabic, and *Sterculia urens*, which produces gum karaya, are now grown on plantations.

The exudate gums have the advantage of having been tested by use over a long period in certain formulations and of having been approved for use in many food products, which is where they find their principal application. They do have desirable properties that make them the gums of choice in certain other products, but their use is limited by availability and cost. Substitutes for them are available.

A survey of the utilization potential of larch arabinogalactan and the current uses of the exudate gums follows. Brief discussions of the utilization of maple sap and sago starch are included with discussions of their occurrence in Chap. 4.

10.2.2 Larch Arabinogalactan

Larch arabinogalactan is the only carbohydrate known that is a true wood extractive with commercial potential. The polysaccharide, first described in 1898 (7, 34) was originally called epsilon-galactan (29), then larch arabogalactan, and now, more properly, larch arabinogalactan (6). As explained in the preceding section, it is not being utilized at this time because no application for which it would be the gum of choice because of properties or cost has been developed. Its structure (Chap. 4; 6, 7, 30, 31, 33 – 35), production (6, 34), properties (6, 34), and proposed uses (6, 22, 34) have been described and discussed in detail.

10.2.2.1 Source

Arabinogalactan occurs in high concentration (5% to 35%, dry weight basis) in the heartwood of many species of *Larix*, a deciduous conifer native to the temperate and subarctic zones of the Northern Hemisphere; in fact, the presence of this water-soluble, easily extractable polysaccharide is characteristic of *Larix* spp. Larch arabinogalactan will accumulate in masses under the bark as a result of injuries; however, this supply is limited and its collection is difficult.

The gum is contained in the lumen of the tracheids and also in ray cells. The yield of extractable gum depends upon the part of the tree from which it is extracted. In western larch (*Larix occidentalis*), the polysaccharide is unevenly distributed throughout the tree. The lower portion of the tree contains so much ($\leq 15\%$ to 25%, dry weight basis) that the but cut is good for neither lumber nor pulping. In *Larix laricina* (tamarack), however, the concentration of the arabinogalactan in the heartwood increases with increased height in the tree; the highest concentrations ($\approx 21\%$) are found in the branches (6). The concentration in all industrially important larch species is lowest in the pith and increases radially, reaching a maximum at the heartwood-sapwood boundary. The gum varies in chemical structure and molecular weight with location in the tree (6).

Arabinogalactan is found in small amounts in the wood of other trees. It occurs in the cambium layer and sapwood of hemlock. Minor amounts are also present in black spruce (*Picea mariana*), parana pine (*Araucaria angustifolia*), mugo pine (*Pinus mugo*), Douglas-fir (*Pseudotsuga menziesii*), incense cedar (*Libocedrus decurrens*), and juniper (*Juniperus* spp.) (6).

10.2.2.2 Production

Several extraction processes for production of larch arabinogalactan have been patented. All involve chipping or grinding larch heartwood – generally waste wood and the butt cut – and extracting the finely divided wood with water. The rate of extraction is dependent upon the size of the chips and the extraction temperature; the yield is dependent upon the amount of gum present in the raw material (6).

The only actual commercial production facility, when operating, used a countercurrent extraction process (4) and had a production capacity of more than 200 tons (180000 kg) per year. The solution obtained from the final tank in the battery, which contained 8% to 10% of arabinogalactan, was drum-dried to produce a technical-grade product.

In an alternative procedure described in the same patent (4), water is added to finely divided larch wood to give a moisture content of 200%. The water-saturated particles are then pressed (recovery of water extract $\approx 70\%$, time required for one cycle ≈ 10 min). A second cycle recovers $\approx 70\%$ of the remaining gum (recovery of gum from two cycles $\approx 90\%$).

Both procedures can be applied at elevated temperatures. As already mentioned, the rate of extraction is dependent upon the extraction temperature; however, the rate of extraction of impurities also increases with increasing temperature and to a greater extent than the rate of extraction of the gum. Higher temperatures also produce some hydrolysis. A hydrolyzed product can be produced by heating finely divided larch wood with dilute sulfuric acid in an autoclave at temperatures up to 160°C (3).

The commercial, technical-grade gum produced by the countercurrent extraction process (4) had a light-tan color, a slightly woody odor, and a mild, woody taste; it contained both low-molecular-weight and polymeric phenolic compounds, mono- and oligosaccharides, and trace amounts of volatile terpenes (6). The product was water soluble, producing low-viscosity, tan to brown solutions that were moderately turbid at concentrations above 10% (4).

10.2.2.3 Purification

At least five procedures for removal of impurities have been described (6); none have been completely successful in removing impurities in an economical way (6). In the best method, active magnesium oxide is mixed rapidly with a solution of the gum heated to almost the boiling temperature (21). The flocculent, brownish precipitate (phenolic compounds and iron salts), which forms almost immediately, is removed quickly by filtration. The almost colorless solution is then drum-dried to produce a white powder that is odorless and has only a slight taste. The remaining taste can be removed by treatment with charcoal (6). The final, purified arabinogalactan has the same solubility and solution viscosity as the technical-grade product, but without any of the surface-active properties of the latter.

Another patent (8) describes the removal of aromatic impurities (lignin, flavones, flavanones, and related compounds) by oxidation, which renders them soluble in the water-miscible organic solvent used to precipitate the polysaccharide. Impurities that give color to the product are the most difficult to remove and are best eliminated by careful removal of all bark prior to chipping.

10.2.2.4 Properties

Larch arabinogalactan has several unique properties; among them are solubility and solution viscosity.

Aqueous solutions of larch arabinogalactan of 60% concentration can be prepared easily and are fluid. At higher concentrations, pastes are formed. At even higher concentrations, a glass that is friable when the moisture content is lowered to less than 10% forms.

Larch arabinogalactan imparts lower solution viscosities than do equivalent concentrations of most other gums (10%). A 10% solution of either purified or technical-grade gum has a viscosity of only 1.74 cp at 20°C. Solution viscosities are unaffected by the presence of electrolytes or changes in pH over a wide range (pH 1.5 to 10.5), although hydrolysis will occur in acidic solutions. A 50% solution of the gum exhibits Newtonian flow up to Brookfield Viscometer spindle speeds of 300 rpm (13, 25).

The surface tension of a solution depends to a large extent on the purity of the gum used, as indicated in the following tabulation (9, 25):

Gum quality	Surface tension (dynes/cm at 20°C)	
	10% conc.	30% conc.
Technical-grade	65.1	59.9
Purified	72.3	71.7

Impurities in the technical-grade gum lower both the surface tension and the interfacial tension. Presumably, the same impurities are responsible for the dispersant properties of the technical-grade gum. The surface tension of solutions of the technical-grade gum decreases as the pH is lowered; addition of electrolytes also lowers the surface tension. The solution pH of technical-grade products is 4.0 to 4.5 (9, 25). The pH values of solutions of commercially purified preparations will vary with the method of purification (21).

The glycosidic bonds of arabinofuranose units are more acid-labile than are those of pyranose units; therefore, portions of the side chains can be removed by a mild acid treatment (36). Treatment of the gum with dilute acid at high temperatures produces furfural. Galactitol (dulcitol) is obtained by hydrogenolysis of the polysaccharide (40). Galactaric (mucic) acid is obtained by its treatment with nitric acid (1, 2, 28).

A more detailed summary of the physical and chemical properties of larch arabinogalactan is available (6).

10.2.2.5 Uses

Proposed uses of larch arabinogalactan (6) are few, presumably because, at the time its commercialization was being actively pursued (roughly over a two-decade period beginning just before 1960), emphasis in both the gum-producing and the gum-utilizing industries was on gums producing high-viscosity solutions.

Because of its high solubility and the low viscosity of its solutions, larch arabinogalactan was considered as a replacement for gum arabic. Larch arabinogalactan is an effective flavor fixative and is reported to be superior to gum arabic in certain food applications, for example, in gelatin-type desserts where conventional gum arabic-fixed flavors have a tendency to react with gelatin to give a cloudy product (19). It is also an effective low-caloric-value bulking agent that can be used in admixture with artificial sweeteners (32). It has much the same bodying, bulking, and other physical and organoleptic properties as sucrose and can be used to make low-calorie products such as maple syrup, chocolate candy, butter icing, orange jelly, and fudge sherbet (32). By mixing it with D-glucitol (sorbitol) or another polyhydroxy compound, it can be used as a sugar replacement in products containing a large amount of fat (11). The tough, gummy, sticky characteristics of cooked candies containing larch arabinogalactan can be controlled and regulated by the addition of mannitol and/or lactose prior to or during the cooking of a solution; by varying the ratio of gum to mannitol and/or lactose, many different types of textures of cooked candy can be prepared (12).

Larch arabinogalactan has been cleared by the U.S. Food and Drug Administration (10) for use in food in accordance with the following conditions: "It is used in the following foods in the minimum quantity required to produce its intended effect as an emulsifier, stabilizer, binder, or bodying agent: Essential oils, non-nutritive sweeteners, flavor bases, non-standardized dressings, and pudding mixes."

The same properties that make larch arabinogalactan useful in foods make it useful in cosmetics, pharmaceutical dispersions, and pharmaceutical oil-in-water and water-in-oil emulsions (25, 27). It is also useful as a tablet binder (26).

Salts formed between larch arabinogalactan sulfate and pharmaceutically active organic bases (for example, alkaloids, antihistamines, antibiotics, tranquilizers) have longer duration of, and more uniform, action than the bases themselves (37, 38).

Surveys of other proposed uses can be found in references (5, 6). Not included there is its use as an inhibitor for the setting and hardening of portland cement (39).

10.2.3 Gum Arabic (Gum Acacia, Acacia Gum)

Of the ancient, water-soluble commercial gums that are dried, gummy exudations collected by hand from the trunks and branches of various trees and shrubs, only gum arabic (Chap. 4) is yet in significant use.

10.2.3.1 Source, Production, and Purification

Gum arabic (14, 18) is obtained from various species of *Acacia* growing in the region just south of the Sahara desert, primarily in the Sudan. The gum is produced during periods of unfavorable conditions (lack of moisture and nutrition or extremely hot weather) and exudes through cracks, from wounds made by insects, or from man-made incisions in the bark. It is picked from the trees by hand and then taken to central collection stations, where it is sorted into grades for export. Grading is also done by hand and is based upon color and degree of foreign matter (tannin, other pigments, dust, insects, pieces of bark). Because gum collecting is a part-time occupation, the amount collected depends on economic conditions in the region in which the trees grow. Plantations of trees have now been established, and trees are being propagated using modern cloning techniques.

When gum arabic is received by importers, it is examined for clarity, color, and gross impurities. Representative samples are then ground, and the resultant powders are analyzed for ash content, water-insoluble impurities, viscosity, and flow characteristics. Approved lots are cracked, then sifted and blown to remove bark and other extraneous matter. The resulting cleaned "crystals" are further processed to grains or powders of different mesh sizes and blended to various viscosity grades. Impurities, including water-insoluble specks, bacteria and enzymes (oxidases, peroxidases, pectinases), remain. Therefore, products in which it is used must be sterilized, pasteurized, kept refrigerated, or protected with a preservative. The most highly purified and expensive grades are spray-dried products prepared from sterilized, filtered, or centrifuged solutions.

10.2.3.2 Properties

Like larch arabinogalactan, gum arabic can be used to make solutions of unusually high concentration and unusually low viscosity, with almost Newtonian flow, which make them quite different from most other gums (14, 18). Gum arabic can yield solutions of up to 50% concentration. A 20% solution resembles a thin sugar syrup in body and flow properties. The flow properties of its solutions are typically Newtonian at concentrations up to 40%.

Addition of salts to a solution of gum arabic results in a lowering of both viscosity and interfacial tension. The marked decrease in viscosity, which is proportional to the valence of the cation and its concentration, is usually of no consequence because gum arabic is almost never used as a thickening agent. Surface tension is lowered more by monovalent than by divalent cations. Solution viscosity and surface tension are also lowered when the pH is lowered below 4.6.

Solutions of gum arabic are clear and slightly acidic (pH 4.5–5.0). Maximum viscosity occurs at pH 4.6–6.3 (a rather broad range), with only a slight decrease in viscosity as the pH is raised from 6.3 to 9.0, although again gum arabic is seldom used as a thickener.

10.2.3.3 Uses

The major uses of gum arabic are in foods, most often in the preparation of emulsions. It is an effective emulsion stabilizer because of its protective colloid action and is widely used in the preparation of flavor oil-in-water emulsions. It produces stable emulsions with citrus and other essential oils over a wide pH range and in the presence of electrolytes, without the need for a secondary stabilizing agent.

Gum arabic is also widely used to make dry flavor powders by spray-drying a flavor oil-in-water emulsion. Gum arabic is especially suitable for this purpose because it completely surrounds the flavor oil droplets when the water is removed and, thereby, prevents evaporation, oxidation, and absorption of moisture from the air; the gum coating is very soluble in water, giving uniform flavor distribution. Use of gum arabic permits spray-drying at higher temperatures than is possible with other gums, and no special conditioning of the air is needed as the powder is cooled and conveyed from the spray dryer. Some gum arabic is used in the preparation of confectionaries (for example, gum drops) and lozenges.

10.2.4 Gum Karaya

Gum karaya is the dried exudation of *Sterculia* spp., especially *Sterculia urens*, and is sometimes known as Sterculia gum. Of the exudate gums, it ranks second to gum arabic in commercial utilization.

All gum karaya comes from India, where the trees are cultivated and production is closely supervised. Holes about 10 cm deep are drilled into the trunks of the trees. After the slowly exuding gum has hardened, it is collected by hand, and pieces are sorted and graded. The best grades are white and contain a minimum of extraneous matter. Lower grades are light yellow to brown in color and may contain up to 3% of impurities, such as bark and sand (16, 20).

Gum karaya does not dissolve in water to give a clear solution, but individual gum particles swell to form a viscous, discontinuous, mucilaginous, colloidal dispersion. Of the four exudate gums discussed in this section (gum arabic, gum karaya, gum tragacanth, and gum ghatti), gum karaya is the least soluble. Swelling gums like gum karaya and gum tragacanth (Sect. 10.2.5) give high viscosities at high concentrations, but the viscosity drops off rapidly at lower concentrations. Gum karaya, unlike gum tragacanth, will form viscous solutions at concentrations of up to 60% in 35% alcohol.

Particle size influences the type of dispersion obtained. When a coarse granulation is used, a discontinuous, grainy dispersion results. A finely powdered gum (<0.105 mm) can give an apparently homogeneous dispersion. In every case, individual gum particles swell to such an extent that a 3% to 4% concentration is the maximum that can be obtained by cold-water hydration, because a 3% to 4% concentration usually produces a heavy, gel-like paste. Higher maximum viscosity is obtained by cold-water than by hot-water hydration. Dispersions of finely powdered gum preparations exhibit a greater rate of viscosity increase and a higher maximum viscosity than do dispersions of coarser gum preparations. This behavior is generally typical of gum karaya, gum tragacanth, and gum ghatti, but is perhaps most pronounced with gum karaya.

The normal pH of a gum karaya dispersion is 4.3 to 4.8. The viscosity is maximum at about pH 8.5 (pH 7.0 to 11.0); but at alkaline pH values, the acetyl groups (content $\approx 8\%$) are removed. As acetyl groups are lost, the characteristic short-bodied (pseudoplastic) solutions become irreversibly transformed into ropy, stringy mucilages. The viscosity of gum karaya dispersions also decreases when salts are added. As with gum arabic, this lowering of viscosity and the concomitant lowering of surface tension improves its ability to stabilize emulsions.

About half of the gum karaya imported into the United States is used in the pharmaceutical industry; most of the other half is used in the food industry. The primary pharmaceutical use is as a bulk laxative. Coarse (0.55 to 2.4 mm) particles absorb water and swell to 60 to 100 times their original volume, forming a discontinuous, mucilagenous dispersion that is effective as a laxative. In ice pops and sherbets, gum karaya prevents the bleeding of free water, flavor, and sugar and the formation of large ice crystals. It may be added to cheese spreads to prevent water separation.

10.2.5 Gum Tragacanth

Gum tragacanth is the dried, gummy exudation from the stems of various *Astragalus* spp., which grow in Asia Minor (Iran, Turkey, Syria).

Like the other exudate gums, gum tragacanth exudes from natural breaks and wounds in the bark of the shrub and from man-made incisions. As with the other exudate gums, collecting and grading is done by hand. Initial grading is based upon color, texture, opacity, and the presence of foreign matter. When received by importers/processors, the gum is cleaned and then processed into viscosity-controlled grades of crystals, broken ribbons, grains, and powders of various mesh sizes (17, 24).

Gum tragacanth swells rapidly in water to yield highly viscous, colloidal sols or soft, pourable gels. Solution viscosity varies with the grade. At a concentration of 1%, the viscosity can vary from 100 to 5000 centpoises because gum tragacanth is made up of at least two components: a water-soluble portion constituting 60–70% of the gum and a water-insoluble, but swellable, minor component. Selection and blending is used to prepare specific viscosity grades. Higher viscosity grades form milky-white dispersions; clear solutions are obtained only from low-viscosity grades.

The most important property of gum tragacanth is its ability to produce dispersions of high viscosity that are relatively stable at acid pH values. At 25°C, viscosity reaches a maximum in about 24 hours; the same maximum may be obtained in about 2 h by warming to 50°C. A thick gel is produced at 2–4% concentration.

Gum tragacanth is compatible with relatively high salt concentrations. Many of its uses depend upon its ability to swell in water to give thick, viscous dispersions or pastes. A drawback, however, is the difficulty in preparing homogeneous dispersions, without lumps.

Gum tragacanth solutions are acidic, with a pH that is usually 5 to 6. Compared to other gums, gum tragacanth is fairly stable over a wide pH range (3.0–8.5). The maximum initial viscosity occurs at about pH 8, but the maximum stable viscosity occurs at about pH 5.

Gum tragacanth is an effective emulsion stabilizer because it delays coalescence of oil globules by increasing the viscosity of the external phase. As a result of this property and because it gives stable viscosities at acid pH values and has a long shelf life and good refrigerator stability, it was widely used in the past in such products as pourable salad dressings, relishes, sauces, condiment bases, and sweet pickle liquors. It has also been used as a stabilizer for water ices, ice pops, and sherbets, and as a thickener and emulsion stabilizer in barbecue and steak sauces. However, owing to its scarcity and cost, only small amounts are being imported and used today.

10.2.6 Gum Ghatti

Gum ghatti (Indian gum) is the dried exudation of *Anogeissus latifolia*, a large, deciduous tree found abundantly in the dry forests of India and, to a lesser extent, in Sri Lanka. As with the other exudate gums, harvesting and grading are done by hand (15, 23).

Only about 90% of the gum disperses in water, and this portion forms a colloidal dispersion rather than a clear solution. Its dispersions are intermediate in viscosity between those of gum arabic and gum karaya at equivalent concentrations. The viscosity of its dispersions increases as they are aged. Normally, a 5% concentration is the maximum that can be obtained by cold-water hydration. As with gum arabic, spray-dried products that contain no insolubles are available; their viscosity-imparting ability is somewhat lower than that of untreated preparations. Gum ghatti will form viscous solutions in 25% alcohol (15, 23).

Gum ghatti is little used in the United States. Its quality varies and only small amounts are available. It can stabilize more difficult emulsion systems than can gum arabic, which serves as the basis for most of its applications.

10.2.7 Predictions

The outlook for gums extraneous to the lignocellulosic cell wall of woody tissues is not bright. Even though efforts have been underway since early in the 1970s to improve the quality and increase the available quantity of gum arabic, its use is likely to decline because of a small and fluctuating supply and an increasing cost. New geographic areas have been opened to gum arabic production, and modern biological techniques have been applied to improvement of *Acacia* species and strains. Nevertheless, the supply of gum arabic has been variable and uncertain because of changing climatic, economic, and political conditions in the producing regions and because, with each shortage, some gum arabic is replaced with other gums and customers are lost permanently. The use of gums ghatti, karaya, and tragacanth have already declined to insignificant amounts in the United States for the same reasons.

A re-examination of the production, properties, and potential uses of larch arabinogalactan in light of today's needs is needed and could bring it to the market.

References

1. Acree S F 1921 Mucic acid, etc. Brit Patent 160777
2. Acree S F 1931 Mucic acid from western larch wood. US Patent 1816137
3. Acree S F 1937 Galactan product. US Patent 2073616
4. Adams M F 1967 Extracting arabinogalactan from larchwood. US Patent 3337526
5. Adams M F, Douglas C 1963 Arabinogalactan − a review of the literature. Tappi 46:544−548
6. Adams M F, Ettling B V 1973 Larch arabinogalactan. In: Whistler R L, BeMiller J N (eds) Industrial gums. Academic Press New York, 415−427
7. Bouveng H O 1961 Studies on some wood polysaccharides. Svensk Kem Tidskr 73:115−131
8. Dahl K 1970 Recovery of high-purity arabinogalactan from larch. US Patent 3509126
9. Ekman K H, Douglas C 1962 Some physicochemical properties of arabinogalactan from Western larch (Larix occidentalis Nutt.). Tappi 45:477−481
10. Federal Register 1965 Arabinogalactan. Federal Register February 25:2430
11. Frey R R 1973 Treatment of arabinogalactan. US Patent 3737322
12. Frey R R 1973 Additives for use in the preparation of cooked candies. US Patent 3738843
13. Glicksman M 1969 Arabinogalactan (larch gum, Stractan). In: Glicksman M (ed) Gum technology in the food industry. Academic Press New York, 191−198
14. Glicksman M 1983 Gum arabic (gum acacia). In: Glicksman M (ed) Food hydrocolloids, vol II. CRC Press Boca Raton FL, 7−29
15. Glicksman M 1983 Gum ghatti (Indian gum). In: Glicksman M (ed) Food hydrocolloids, vol II. CRC Press Boca Raton FL, 31−37
16. Glicksman M 1983 Gum karaya (*Sterculia* gum). In: Glicksman M (ed) Food hydrocolloids, vol II. CRC Press Boca Raton FL, 39−47
17. Glicksman M 1983 Gum tragacanth. In: Glicksman M (ed) Food hydrocolloids, vol II. CRC Press Boca Raton FL, 49−60
18. Glicksman M, Sand R E 1973 Gum arabic. In: Whistler R L, BeMiller J N (eds) Industrial gums. Academic Press New York, 197−263
19. Glicksman M, Schachat R E 1966 Gelatin-type jelly dessert mix. US Patent 3264114
20. Goldstein A M, Alter E N 1973 Gum karaya. In: Whistler R L, BeMiller J N (eds) Industrial gums. Academic Press New York, 273−287
21. Herrick I, Adams M F, Huffaker E M 1967 Refining arabogalactan. US Patent 3325473
22. Lawrence A A 1976 Natural gums for edible purposes. Noyes Data Corp Park Ridge NJ, 3−6
23. Meer G, Meer W A, Gerard T 1973 Gum ghatti. In: Whistler R L, BeMiller J N (eds) Industrial gums. Academic Press New York, 265−271
24. Meer G, Meer W A, Gerard T 1973 Gum tragacanth. In: Whistler R L, BeMiller J N (eds) Industrial gums. Academic Press New York, 289−299
25. Nazareth M R, Kennedy C E, Bhatia V N 1961 Studies on larch arabogalactan. I. J Pharm Sci 50:560−564
26. Nazareth M R, Kennedy C E, Bhatia V N 1961 Studies on larch arabogalactan. II. J Pharm Sci 50:564−567
27. Patel B N 1964 Hydrocolloids in cosmetic drugs. Drug Cosmetic Ind 95:337−341, 451
28. Schorger A W 1929 Mucic acid. US Patent 1718837

29 Schorger A W, Smith D F 1916 The galactan of Larix occidentalis. J Ind Eng Chem 8:494–499
30 Simson B W, Côté W A Jr, Timell T E 1968 Larch arabinogalactan. IV. Molecular properties. Svensk Papperstidn 71:699–710
31 Smith F, Montgomery R 1959 The chemistry of plant gums and mucilages. Reinhold New York, 627 pp
32 Stanko G L 1966 Artificial sweetener-arabinogalactan composition and edible foodstuff utilizing same. US Patent 3294544
33 Stephen A M 1983 Other plant polysaccharides. In: Aspinall G O (ed) The polysaccharides, vol 2. Academic Press New York, 130–134
34 Stout A W 1959 Larch arabogalactan. In: Whistler R L, BeMiller J N (eds) Industrial gums. Academic Press New York, 307–310
35 Timell T E 1965 Wood hemicelluloses: Part II. Adv Carbohyd Chem 20:409–483
36 White E V 1941 The constitution of arabo-galactan. I. The components and position of linkage. J Am Chem Soc 63:2871–2875
37 Wirth P C 1971 Arabinogalactan sulfonates of nitrogen-containing organic bases. Ger Offen 2117902
38 Wirth P C 1974 Salts of nitrogen bases and polysaccharide sulfates. US Patent 3832340
39 Xu L 1984 [Arabinogalactan from two larch timbers – contents, molecular properties, and effects on the setting of cement.] Linye Kexue 20:57–65
40 Zaitseva A F, Karpov A Z, Levin S V, Antonovskii S D 1959 [Dulcitol from larch arabogalactan.] Zh Prikl Khim 32:690–693

10.3 Significance of the Condensed Tannins

L. J. PORTER and R. W. HEMINGWAY

10.3.1 Introduction

The condensed tannins are biologically significant natural products primarily because of their ready complexation with proteins. These compounds are particularly important flavor components, being responsible for the astringency of many fruits and vegetable products. However, the greatest economic significance of the condensed tannins lies in human or animal nutrition, because condensed tannins complex with vegetable proteins and limit their digestibility. The complexation of condensed tannins with proteins is also the basis for their properties as insect, mollusc, bacteria, or fungal control factors as well as for their principal industrial use in leather manufacture. Because of their rapid condensation with formaldehyde, substantial efforts have been made over the past 30 years to use condensed tannins in adhesive polymers. Following the petroleum shortage of the early 1970s, substantial research efforts were directed to use of tannins in wood adhesives, but only the use of wattle tannins has been commercially successful to date. With lower phenol prices through the early 1980s, research has taken a healthy move to consideration of the use of these polymers in a much wider range of specialty chemicals.

10.3.2 Tannins as Human and Animal Nutritional Factors

Bate-Smith and Swain (19) were the first to recognize the widespread occurrence of proanthocyanidins and related phenolics in foods and to appreciate their possible role as nutritional factors. Price and Butler (198) reviewed the literature up to 1978 involving tannins and nutrition and compiled a useful list of fruits, cereals, beverages, forages, and other foodstuffs containing tannins. Fruits included: *Malus pumila, Musa sapientum, Rubus monogyna, Rubus idaeus, Vaccinium oxycoccus, Vaccinium membranaceum, Phoenix dactylifera, Vitis* spp., *Crataegus monogyna, Prunus* spp., *Pyrus communis, Diospyros kaki*, and *Fragaria ananassa*. Cereals (seeds and grains) included: *Hordeum* spp., *Vicia* spp., *Vigna unguiculata, Phaseolus* spp., *Pisum* spp., *Eleusine coracana, Brassica napus*, and *Sorghum* spp. Beverages included coffee (*Coffea* spp.), tea (*Camellia sinensis*), and wine (*Vitis* spp.). Forages included *Coronilla varia, Lespedeza sericea, Lotus* spp., *Orobrychis viciifolia*, and *Trifolium* spp. Other miscellaneous foodstuffs included *Pteridium equilinum, Ceratonia siliqua, Theobroma cacao*, and *Quercus* spp.

Since that time, several detailed studies have appeared which have identified and characterized the tannins in most of the above species and many more food crops. In particular, Foo and Porter (64) gave the tannin constitutions of 20 fruits, including most of those listed, and showed that procyanidins or mixed procyanidins and prodelphinidins occurred in all the fruits except *Fragaria ananassa* and the various *Rubus* cultivars. The latter fruit contained ellagitannins, whereas *Fragaria* contained mixed procyanidins and hydrolyzable tannins. In addition, *Cydonia oblonga* was shown to contain a epicatechin procyanidin homopolymer. Subsequently, it has been shown that the fruits of *Cocos nucifera* contain a procyanidin polymer with both epicatechin-4 and *ent*-epicatechin-4 units (56) and that fruits of *Diospyros kaki* contain a mixed epicatechin-4 and epigallocatechin-4 proanthocyanidin polymer in which the units are partly 3-*O*-galloylated (138, 172).

The proanthocyanidins of *Sorghum* spp. and *Hordeum* spp. have been studied in some detail. In particular, *Sorghum* species have been shown to contain a polymer based largely on epicatechin-4 units with a catechin terminal unit (26, 74), but some cultivars also contain epigallocatechin-4 units (26). Certain other cultivars contain significant concentrations of a novel proluteolinidin based on 5-*O*-glucosylated flavan units (73). The proanthocyanidins of *Hordeum* spp. will be considered later.

Reddy et al. (205) have published an augmented list of food legumes which contain tannins, listing some 21 species and cultivars. They have developed the thesis that high tannin bean varieties are of lower nutritional quality than lower tannin varieties due to inactivation of digestive enzymes and protein insolubilization through complexation with the tannins. They also demonstrated that tannins decrease protein digestibility and cause lowered feed efficiency and growth depression in experimental animals. In the case of legume seeds (peas and beans) these effects may be largely eliminated by removal of the seed coat where most of the tannins reside, or by a variety of other strategies to lower tannin content, such as soaking or chemical treatment (199). It should be noted that these findings are typical of many other studies in the literature that assess the nutritional effects of a high tannin content in other foodstuffs – particularly cereals.

They also point out that tannin content may be lowered by selective breeding. This trend may be observed for most domesticated food crops where evidence that man has selected low tannin varieties may be seen when cultivated are compared with wild strains.

The tannins of fodder legumes have also been studied in some detail by Foo et al. (63), who determined the structures of the proanthocyanidins from 13 species in the genera *Coronilla, Lotus, Onobrychis, Robinia, Trifolium,* and *Vicia.* All were mixed procyanidin-prodelphinidin polymers except *Trifolium repens* (a prodelphinidin) and *Vicia* (a procyanidin).

In contrast to the findings reported earlier, there is now clear evidence that tannins, particularly proanthocyanidins, are a beneficial and perhaps essential feature of the bovine diet. They have been shown to be a powerful anti-bloat factor and to improve the efficiency of leaf-to-meat protein conversion in cows; both effects operate through pre-digestion complex formation in the first gut (132).

Other foodstuffs whose tannins have been studied in some detail, but not already considered include: the cereal *Lens culinaris* (159) and sago starch (163) (*Metroxylon sagu*) in which proanthocyanidins are implicated in the post-harvest browning of both crops; *Manihot esculenta* (*M. utilissima*) where proanthocyanidins limit its use as a forage crop (206).

Many food additives contain proanthocyanidins. Schultz and Herrmann (221) surveyed the common spices and found proanthocyanidins in *Laurus nobilis, Cinnamomum zeylanicum, Illicium verum, Myristica fragrans, Pimenta dioica,* and *Juniperus communis,* but they were absent from all the others tested. Subsequently, the proanthocyanidins of cinnamon (*Cinnamomum zeylanicum*) and cassia (*Cinnamomum cassia*) were studied in detail by Nishioka and co-workers (148, 158). Both contained epicatechin procyanidin homo-oligomers; but cinnamon additionally contained high concentrations of doubly-linked (A-type) procyanidins.

Nishioka and co-workers also showed that the masticatory betel nut (*Areca catechu*) was rich in epicatechin procyanidin homo-oligomers (155). Karchesy and Hemingway (111) have shown that mature peanuts (*Arachis hypogea*) contain similar procyanidin oligomers together with A-type dimers, which together constitute 17% by weight of the skins. These have been shown to interfere with protein digestibility when the skins are used as a protein supplement for cattle and pigs, and have been implicated in causing leg deformities in poultry (111).

An enormous research effort has been expended on investigating the role of condensed tannins in the quality and characteristics of five beverages:

1. Their role in the taste and haze formation in beer.
2. Their contribution to the taste of red and white wines and their role in the physicochemical processes involved in the maturing of red wines.
3. The chemical transformations that take place in the flavanoids and tannins of green tea during its enzymatic conversion to black tea.
4. The contribution of oligomeric proanthocyanidins to the taste and quality of apple cider.
5. The concentration and type of procyanidins in fresh cacao beans and their fate in the natural fermentation process and subsequent chemical treatment that produces cocoa.

While each of these topics has an extensive and mature literature, it has only been in the last few years that the structure of the proanthocyanidins has been determined in these products. Thus, it has been established (26, 168) that beer contains low molecular weight (dimers and trimers) mixtures of procyanidins and prodelphinidins based on catechin-4 and gallocatechin-4 units originating from barley (*Hordeum* cv.). Grapes (*Vitis* cv.) contain predominantly procyanidins containing epicatechin-4 units, some of which are 3-*O*-galloylated (49, 126). The flavan-3-ols and proanthocyanidins of green tea (*Camellia sinensis*) are exceedingly complex and include a host of mono- and di-galloylated compounds based on epicatechin or epigallocatechin (48, 157). The characteristic red-brown color of infusions of black tea are largely due to oxidative coupling and benztropolone ring formation by the flavan-3-ol B-rings (47, 156). Apple (*Malus pumila*) cider contains epicatechin procyanidin homo-oligomers and -polymers (124). Lea and Arnold (125) have shown that the bitterness of cider is largely associated with a procyanidin tetramer, whereas astringency is characteristic of higher molecular oligomers. Fresh cacao (*Theobroma cacao*) beans contain a relatively low molecular weight (average degree of polymerization 6 units) epicatechin procyanidin homopolymer (152). The natural fermentation process rapidly (2–3 days) converts over 90% of the cacao procyanidins (L. J. Porter, unpublished results) to the red-brown amorphous products characteristic of cocoa (this process is also promoted by alkali treatment).

The best understood of the processes involving proanthocyanidins in beverages is probably that of beer. As stated earlier, proanthocyanidins are present in low concentrations in barley (*Hordeum* cv.) (26, 168) and significant amounts of them survive the malting process (143). Two haze phenomena may be recognized: chill haze, a reversible complex formed between the proanthocyanidins and the yeast proteins and polysaccharides; and permanent haze, which is due to reaction between cysteine residues and proanthocyanidins (21). It has been shown that at least four proteose fractions are present in beer and those of molecular weight higher than 10 000 are largely responsible for chill haze (142). Further studies have shown that molar tanning capacities of flavanoids are linearly related to procyanidin concentration, directly proportional to their degree of polymerization, and that catechin-(4→6)-catechin is more effective than catechin-(4→8)-catechin at protein precipitation (53). Haze formation may be overcome by the introduction of proanthocyanidin-free barley cultivars, but this is at the expense of taste quality (250).

10.3.3 Pharmacological and Physiological Properties

Proanthocyanidins and flavan-3-ols have been accorded both pharmacological and physiological properties. Morton (149) has reviewed the literature connecting both condensed and hydrolyzable tannins with liver and esophageal cancer and documents several instances where there has been established a correlation between the latter form of cancer and herbal teas containing high concentrations of proanthocyanidins (see also Chap. 9.5.4.3.). The same danger is maintained to exist from the consumption of tea without milk (which precipitates the tannin fraction) and the mutagenicity of the infusion has been demonstrated. Nevertheless, tannins, like radiation, can also function as a carcinostat, several examples of flavan-3-ols and proanthocyanidins acting as antitumor agents having been demonstrated.

Nishioka (154) has summarized the chemistry of a wide variety of lower molecular weight proanthocyanidins isolated in his laboratory and briefly summarized their biological activities. In particular, "Rhatannin," a 3-O-galloylated procyanidin polymer, has been shown to affect the urea nitrogen concentrations in the serum and liver of the rat, and also the activities of a range of liver enzymes.

Other physiological properties accorded flavan-3-ols or tannins are: epigallocatechin is considered by some workers to be a vitamin C co-factor (70); oligomeric procyanidins isolated from *Crataegus oxyacantha* berries have been found to lower blood pressure by capillary dilation, and to have hypnotic and sedative effects on experimental animals (207); *Diospyros kaki* proanthocyanidins have been shown to have a strong detoxifying effect against snake venom and bacterial toxins (67); 3-O-esters of catechin have been shown to have promising pharmacological properties involving stimulating cell-mediated immunity to the hepatitis B virus (153) and hypertensive activity (151).

Both hydrolyzable and condensed tannins, often with novel structures, have been found to be associated with plant extracts used by traditional Chinese medications. This has led to much work, especially by the Japanese, to explore the pharmacological properties of tannins in such extracts. Such studies have established that various tannins have hypotensive (101, 106, 249), anti-allergic (107), anti-bacterial, against carcinogenic bacteria (108), anti-coagulant (115), anti-inflammatory (107, 150, 167), anti-peptic (62), and anti-viral (109, 236) activity. These studies support the view that specific tannin molecules or preparations have potential for new pharmaceutical products in the future.

It can be categorically stated that hydrolyzable and condensed tannins together form a significant part of man's diet. They will also form a component of the diet of any herbivore. While they have a demonstrable anti-growth effect on animals, it can also be demonstrated that they have a beneficial role in the bovine diet.

Demonstration of diet related metabolic or physiological effects by purified tannins are probably of little value because, in practice, tannins or any other phenolics in the diet are ingested together with comparatively large amounts of proteins and polysaccharides to which they are strongly bound, especially at acidic pH values (21). In this context, ingestion of strong concoctions of a high tannin vegetable matter (149) is probably their least desirable form in the diet.

10.3.4 Tannins as Insect, Mollusc, Bacterial, and Fungal Control Factors

This section overlaps with Sects. 7.7.4.1 and 7.7.4.2, as much of the work relating to insect and fungal control is common with the work reported there.

A recent study by Pospisil (197) showed that tannin may be used as an exogenous antifeedant factor. In this study, Colorado potato beetle larvae refused to eat potato (*Solanum tuberosum*) leaves treated topically with tannic acid solution. In contrast, the larvae of *Heliothis* spp. exhibited attenuated growth on, but not antifeedant behavior toward a diet containing condensed tannin from *Gossypium hirsutum* leaves (37). The larvae of the chrysomelid beetle *Paropsis atomana* were unaffected by a diet containing high levels of tannins from *Eucalyptus* spp. leaves (66). Bernays and Woodhead (23) presented evidence that dietary tannins may be utilized as nutrients by certain phytophagous insects. The current evidence is that while tannins may have an antibiotic effect on some insect larvae, others have completely adapted to high tannin diet.

There has been much interest in investigating the possibility that the water-inhabiting snails (e.g., *Bulinus* spp., *Phyopsis* spp., *Planorbis* spp.), which act as intermediate hosts for the human parasitic blood flukes of the *Schistoma* genus, and which are responsible for the debilitating tropical disease schistomiasis (billarziasis), may be controlled by plant natural products. Marston and Hostettmann (133) have reviewed the literature on natural molluscicides and among the chemicals noted to be toxic to snails were tannins from *Krameria*, *Hamamelis*, and *Quercus*. Potent molluscicides have been isolated from the fruits and stem bark of *Acacia nilotica*, and this activity has been found to be largely due to epigallocatechin-7-*O*-gallate and 5,7-di-*O*-gallate (17). Many studies have shown that tannins have an inhibitory effect on both bacteria (128, 251) and fungi (137, 227). Both classes of organisms thrive by secretion of extracellular enzymes. It is almost certain that the antibiotic activity of the tannin results from inhibition of such enzymes. For instance, deactivation has been demonstrated for both hydrolyase (102, 105, 197) and peroxidase (25) enzymes. Presumably in response to such inhibitory effects, hydrolyzable tannins stimulate the production of tannase in bacteria feeding on *Castanea sativa* leaves (54).

In contrast, Mole and Waterman (147), using in vitro model studies, have demonstrated that both hydrolyzable and condensed tannins may stimulate tryptic hydrolysis due to tannin-induced structural changes in the substrate protein. Condensed tannins have also been shown to possess algicidal activity (18). Much work remains to be done on the ecological role of tannins, which should be a fruitful field of research now that tannins of defined structure are readily available.

10.3.5 Leather Tannage

Condensed tannins have been used to make leather goods since before recorded history – at least, earlier than 10 000 B.C. Despite falling markets for leather products in general and a declining proportion of leather production using natural vegetable tannins, the condensed and hydrolyzable tannins remain important products of world commerce. The use of tannins in leather manufacture remains one of the most important commercial applications of all extractives of woody plants.

Approximately 200 000 tons of quebracho (*Schinopsis* sp.) and 100 000 tons of wattle (*Acacia mearnsii*) tannins are produced world-wide annually. Although statistics are not so accurate, about 200 000 tons of vegetable tannins derived from other plants such as hemlock (*Tsuga* sp.), chestnut (*Castanea* sp.), oak (*Quercus* sp.) and myrobylan (*Terminalia* sp.) extracts add to this for a total world production thought to be about 500 000 tons per year. (see also Chap. 1.1).

The United States is not a major producer of leather goods; about 21 000 tons of tannins are imported annually. The major sources of tannins imported to the United States are wattle (*Acacia mearnsii*) tannins from South Africa (about 8000 tons/year) and quebracho (*Schinopsis* spp.) tannins from South America (about 5000 tons/year) (D. F. Galloway, personal communication 1984). Because of specific requirements for vegetable-tanned leathers in certain applications, the United States government stockpiles condensed tannins as strategic materials. In 1982, approximately 138000 tons of quebracho, 16000 tons of wattle, and 16000 tons of chestnut (*Castanea* spp.) tannins were held in reserve with an estimated total value of about US $115 million (U.S. Office of Management and Budget, personal communication 1984).

The leather-tanning industry of India used about 120000 tons of assorted vegetable tannins in 1970 (203). India imported about 25000 tons of wattle extracts to supplement domestically available tannins such as myrobalan nut tannins (3500 tons/year) and wattle tannins (2000 tons per year). Vegetable tannins (primarily mixtures of chestnut, wattle, and quebracho tannins) are also used widely in Italy (173). In the Peoples Republic of China, much of the leather manufactured is made using vegetable tannins. Research efforts are being accelerated on the use of tannins from indigenous plants such as *Larix* and *Pinus* spp. (Sun and Foo, personal communication 1986).

For comparison, about 400000 tons of phenol used in phenol-formaldehyde resins were produced in the United States annually in the period 1983 – 1985 (30). While demand for condensed tannins does not compare with basic commodity chemicals such as phenolic resins, they can be considered to be specialty polymers with sizable markets.

Research on the use of condensed tannins in leather manufacture has come to a near standstill in western industrialized countries. Sparrow's fundamental work on condensed tannin-collagen interactions (228) and studies on the combination vegetable-aluminum (110, 226, 234, 241, 242) tannage are important thrusts of current efforts on the use of condensed tannins in leather manufacture.

10.3.5.1 Mechanisms of Vegetable Tanning

Despite a long history of investigation, there remain substantial debates (with little definitive evidence) about the mechanisms of interactions between vegetable tannins and proteins in general. A comparison of current knowledge about the chemistry of vegetable tannins (Chaps. 7.2, 7.6, and 7.7) with our understanding in the middle 1950s, thoroughly reviewed by White (254, 255), shows the phenomenal advance that has been made over the past 30 years. We are now able to isolate and accurately define the structures of hydrolyzable and condensed tannin molecules capable of tanning leather. Haslam (Chap. 7.2) presents interesting examples of interactions of tannins of defined structure with proteins that demonstrate the importance of molecular shape and conformational flexibility in addition to molecular size in determining the relative astringency of different hydrolyzable and condensed tannins (21). Asquith and Butler (11) have recently demonstrated that tannin/protein interactions are highly specific with binding efficiency dependent both on the structure of the tannin and on differences in proteins. Porter (Chap. 7.7) has also outlined some of the more general principles that can be drawn from the present body of experimental results. The rapid expansion of our ability to isolate and define the structures of complex hydrolyzable (Chap. 7.2) and condensed (Chap. 7.6 and 7.7) tannins, together with modern tools for study of the structure and conformation of proteins, suggest that important advances in our understanding of tannin-protein interactions are imminent. Because of the importance of these interactions to biological activity, animal nutrition, and industrial significance of vegetable tannins, this aspect of tannin chemistry is presently a most exciting area of study.

Early studies on the binding of vegetable tannins to animal skins were summarized by White (254, 255), Lollar (129), and, more recently, Santappa and Rao (219). The most recent book on this subject was written in 1956 (160). Russell, Shuttleworth, and Williams-Wynn (209–212, 222–224) of the Leather Industries Research Institute in Grahamstown, South Africa, were particularly active in the study of the mechanisms of leather tannage during the middle 1960s. They concluded that vegetable tannins were bound to proteins by a reversible crosslinking mechanism in which hydrogen bonding provided the dominant forces, and that hydroxyl associations with the peptide linkages predominated. Their conclusions were supported by Sykes and Roux's studies on the affinity of wattle tannins for collagen, collagen that had been derivatized, and cellulose (235). Endres (57) concluded, however, that covalent bonding of quinones to amine functions was an important mechanism. Similar conclusions were reached by Gustavson (75, 76), and this stimulated an interesting debate between Gustavson and Shuttleworth. However, Kedlaya et al. (113), studying the interaction of catechin and its oxidation products from reaction with potassium ferricyanide, concluded that covalent bonds resulting from oxidative coupling to collagen were not the predominant mechanisms involved in vegetable tanning processes. More recent work by Sparrow et al. (228) studying the interaction of low molecular weight phenolics with soluble collagen by a continuous-flow dynamic dialysis technique showed that low molecular weight phenolics are bound to collagen with low interaction affinity but with high capacity and without any evidence for site saturation. The weak binding energy supported the conclusion that the interaction was due to hydrogen bonding as had been proposed by Shuttleworth (223). Similar conclusions were reached by Heidemann and Srinivasan (86) in an infrared study of deuterium-exchanged collagen and synthetic polypeptides treated with wattle (*Acacia mearnsii*) or myrobalan (*Terminalia* spp.) tannins as well as model phenols.

Curiously, the above discussions (129, 219) on the interactions of vegetable tannins with collagen have paid scant attention to the nature of collagen polymers. A great deal is now known about the structures of collagen isolated from different tissues (165, 231). Highberger (100) has prepared an especially interesting account of early efforts by Pauling and other eminent chemists to define the structure of collagen and other skin proteins. Current knowledge of the structure of these proteins can only be briefly reviewed here with particular reference to how collagen could serve as a binding matrix for vegetable tannins. The regular occurrence of glycine at every third amino acid residue results in a tightly wound, three-stranded helix in which the amide linkages are deeply buried within the core

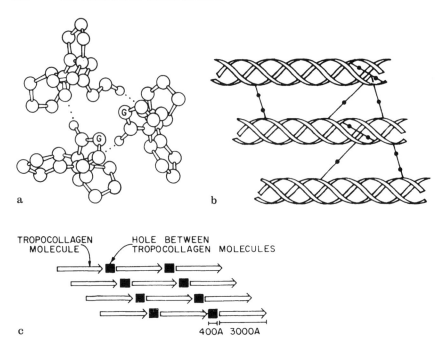

Fig. 10.3.1. Structural features of collagen (adapted from Orten and Newhaus (165) and Stryer (231))

of the helix (Fig. 10.3.1). The strands of the triple helix are held together by hydrogen bonding (the peptide N−H of glycine to the C=O of other amide linkages) and, in addition, occasional inter- and intra-molecular covalent bonds. Most of the other units in the polymer are made up of proline and hydroxyproline; about 20% and 5%, respectively, in the case of calf skin. The melting point and shrinkage temperature of collagen isolated from different tissues (indices of the stability of the binding among triple helices) are highly correlated with the hydroxyproline contents of the polymers. These hydroxyproline units afford the primary means of hydrogen bonding among triple helices. Collagen also contains a few 5-hydroxylysine residues to which gluco-(1β-O-2)-galactosyl units are bound through ether linkages. Individual triple-helices are bound together in a quarter staggered helical array which results in arranged 400 Å gaps about the aligned molecules of about 3000 Å in length (Fig. 10.3.1).

These features of the structure of collagen suggest a number of important factors relevant to vegetable tanning. The size of the 400 Å wide helical gap around fiber bundles of collagen must be very important to the penetration of tannin molecules into the fiber structure. The amide linkage is deeply buried in the core of the triple helix so the N−H and C=O functions must be quite inaccessible to hydrogen bonding with the phenolic hydroxyls of tannins. The most important sites for hydrogen bonding must be with the hydroxyproline residues that occur on the surfaces of the triple helices. However, the surface of the fiber is dominated by the aliphatic part of the prolyl residues so that hydrophobic binding, as recently elucidated by Oh (161), also seems very important in the mechanism of leather tannage.

In addition to reinforcing the stability of the triple helix (as evidenced by increases in shrinkage temperature after tanning), the most important aspect of leather tannage involves the resistance to collagenase enzymes imparted by the presence of vegetable tannins. There seems to be no reference to any study of how vegetable tannins inhibit the activity of these enzymes. Further study on the binding of vegetable tannins of defined structure to collagen similar to the electron microscopic studies of Mohanaradhakrishnan and Ramanathan (146), the studies of spectral properties of collagen-tannin complexes similar to those of Heidemann and Srinivasan (86), and studies on the interactions of vegetable tannins with collagenase enzymes would seem to be rewarding.

10.3.5.2 Vegetable Tanning Processes

Kay (112) has prepared a summary description of the properties of vegetable tannins from many plants: factors influencing their penetration into hides; interrelationships between penetration into and fixation onto the hide; the importance of such factors as pH, temperature, and tannin solution concentration on tannin fixation; and some processes typical of American practice. Optimizing the vegetable tanning processes clearly remains a complicated art of balancing opposing responses to process variables. For example, the amounts of tannin fixed increases but penetration rate decreases significantly as the tannin solution pH is decreased from 5.0 to 3.0. The presence of non-tans and salts such as sodium chloride or particularly sodium sulfate can increase the penetration rate but decrease the extent of fixation. Because of the extreme complexity of these interactions, there are no "standard" vegetable tanning processes but rather each tannery seems to perfect a process that optimizes each local situation. Kay (112) has prepared rather detailed descriptions of the manufacture of sole leather in multiple vats and drum tanning processes that can be considered to represent basic processes used in the United States. Readers are referred to his chapter (112) as well as others in the book edited by O'Flaherty, Roddy, and Lollar (160) for details.

The researchers of the Central Leather Research Institute in Madras, India, continue to be very active in development of new formulations for tanning of various hides with vegetable tannins (201, 219). The importance of carbohydrate gums (that generally co-occur with the tannins in most plant extracts) on penetration rate, leather color, shrinkage temperature, and tannin fixation in the leather have been described in an interesting paper by Gosh (71). Wattle-myrobalan (*Acacia mearnsii-Terminalia* spp.) tannin mixtures are widely used in India (52, 157). They are combined with a variety of pre- and post-treatments designed to improve selected properties. A urea-formaldehyde and syntan pretreatment followed by tanning with a wattle-myrobalan mixture gave a rapid and excellent tannage of sheep skins (69). Wattle-myrobalan mixtures as well as chestnut (*Castanea* spp.), cutch (*Acacia catechu*), and quebracho (*Schinopsis* spp.) tannin mixtures were used in conjunction with an aluminum sulfate secondary treatment to produce shoe sole leathers with particularly low water absorbency (164). A chrome tannage is followed by treatment with wattle tannins and a final aluminum sulfate treatment for making water-resistant shoe uppers (204).

Pirich (173) provides a very interesting account of the use of vegetable tannins in Italy and compares processes used there with those used in the United States. The Italian industry uses blends of chestnut, quebracho, and wattle tannins in the manufacture of shoe sole leathers. The quebracho tannin is particularly useful because of the fast diffusion into the skin and good capacity to fix to the fiber. The wattle tannin also has a fast diffusion rate but its main advantage is the light color of leather that can be obtained. The chestnut tannin is the major component used in Italian practice. Chestnut tannin has a particularly high fixation level and good color characteristics that are relatively stable to light. The chestnut tannin is also considered to improve substantially on the tanning properties of the quebracho and wattle tannins. In Italy, skins are first treated with a light syntan pretannage followed by a pit tannage in which the tannin blend is composed of about 50% chestnut tannin, 15% of a sweetened (treated with sulfite) chestnut tannin, 18% cold water-soluble quebracho extract, and 17% of wattle tannin. The pit tannage is followed by a drum tannage with a blend of 60% chestnut extract, 30% sweetened chestnut extract, and 10% cold water-soluble quebracho extract. Pirich (173) also describes a tanning blend used in an ultra-rapid, drum tanning process which is made up of 45% chestnut extract, 25% cold water-soluble quebracho extract, 15% sweetened chestnut extract, and 15% wattle tannin.

Wattle and quebracho tannins (both imported) are the predominant vegetable tannins used in the United States. Only a few cottage-industry-scale tanneries use vegetable tannins obtained from domestic sources such as oak (*Quercus* spp.) or hemlock (*Tsuga* spp.) extracts. Because the United States leather tanning industry uses significant proportions of the chromium imported to the United States and because environmental concerns with its use are present as well, there is a glimmer of renewed interest in the use of vegetable tannins from indigenous plants of North America. ITT-Rayonier marketed an acetone-water-bisulfite extract from western hemlock (*Tsuga heterophylla*) bark for a short time but was not successful despite the fact that excellent leather could be obtained (93). Another recent example is the use of condensed tannins from peanut (*Arachis hypogea*) or pecan (*Carya illinoensis*) nut skins or pith. Surprisingly large amounts of these high-tannin residues are produced in the United States and these tannins have a structure and molecular weight properties that are well suited to leather tanning (157). Bliss and Galloway (personal communication) have shown that these tannins can be used to replace quebracho tannins in wattle tannin mixtures to obtain leathers of exceptional quality.

Vegetable tannins are often used as a retannage of chromium tanned leathers for the production of shoe uppers. Williams-Wynn (256) has summarized the results of a series of studies made on this type of tanning process at the Leather Industries Research Institute in South Africa. Chromium pretannage increases the reactivity of vegetable tannins, apparently by increasing the availability of amino groups as well as by formation of coordination complexes of both the tannin and non-tans with the chromium. When these leathers are dried and aged, further reaction takes place resulting in an increase in free sulfate and high acidity. The low pH causes hydrolytic degradation of the collagen with loss of strength of the leather. Williams-Wynn (256) offers suggestions of ways to modify this tanning process to improve the properties of chromium-vegetable tanned leathers. They include reducing the degree of chromium tannage and/or use of anionic syntans prior to the vegetable retannage. Formate masking was very effective in reducing the loss of chromium in the retannage, in reducing the amount of vegetable tannin fixed in the retannage and in reducing the acidity of the resulting leathers.

Retannage with condensed tannins (wattle or quebracho) was much better than with the hydrolyzable tannins (myrobalans) or chestnut tannins. Wattle tannins reacted more slowly than quebracho tannins. Use of a wattle retannage with neutralization gave the most stable leathers. The wattle tannin were also superior because of the lower level of non-tans as they also reacted with the chromium and increased the amounts of chromium stripped off in the vegetable retannage (13).

The LIRITAN system developed by Shuttleworth and coworkers has been widely adopted by industry because of environmental pressures on disposal of tanning liquors (224). This "no-effluent" system requires substantial changes from traditional pit tanning processes but is readily employed in traditionally designed plants. The skins are pickled with a 2.5% Calgon solution that needs to be discarded only about once a year. The pickled skins are washed in two color vats, one day each, and then tanned in a 6- to 8-hot-pit system. For manufacture of sole leather, the strength of the tanning liquor is kept comparatively high (117 °Bk) at 95 °C. (°Bk is the industry measure of the strength of the tannage.) Either pure wattle tannin or blends can be used, but it is important that other tannin extracts used have a low proportion of non-tans. The tanning liquor is circulated through a series of vats for 8 days. The liquor does not need to be discarded as an equilibrium in the proportion of tannins to non-tans is established. For the production of harness or saddle leathers, the process is the same except that the tanning liquor concentrations are slightly lower (100 °Bk), and the pH is maintained at 3.8. A rapid tannage, combination pit-drum process has also been developed using the same basic process. Atkinson (12) has described other drum tanning processes. The WEDDEV process that uses wattle tannins in a spray-dried powder form in drum tannage also results in minimal effluent.

The LIRITAN process stimulated interest in vegetable tannin with high concentration solutions and dry tannage as proposed in the pit-drum process (224). Toth (243) made a series of studies of the kinetics of wattle (*Acacia mearnsii*), sulfited quebracho (*Schinopsis* spp.), and chestnut (*Castanea* spp.) penetration into hides. Penetration rate was not highly correlated with solution concentration but was strongly correlated with depth from the surface of the pelt. To increase the rate of tannin penetration into full thickness cattle hides, Luvisi and Filachione (130) suggested a 24-hour pretannage with a 15% glutaraldehyde solution. This material could then be retanned with wattle, sulfited quebracho, or chestnut extracts at solution concentrations as high as 120 °Bk within 8 hours. The properties of these leathers were especially good, with shrinkage temperatures between 86 ° and 93 °C. A triple glutaraldehyde vegetable-glutaraldehyde tanning process increased the shrinkage temperature a further 5 °C and gave leathers having very low water-solubility. Industrial-plant-scale experiences with rapid vegetable tannage processes are reviewed by Exner and Kupka (58).

The most recent major advance in vegetable tanning processes is the development of a combination vegetable-aluminum tannage (85, 225, 233). Leathers produced with this type of tannage have heat stability even greater than commercial chromium tanned leathers (226). Sykes and coworkers (234) have made important contributions to our understanding of the mechanisms involved in vegetable-aluminum tannage by studying the relationships of shrinkage temperature, hydrogen bonding in and to collagen, and the complexation between aluminum and model phenolic compounds as well as to collagen. They showed that aluminum forms complexes with phenolics with two adjacent hydroxyls (catechols) and that the presence of a third vicinyl hydroxyl (pyrogallol) substantially increases the stability of the complex. The aluminum coordinates with more than one pair of adjacent hydroxyl groups, permitting extended complexes that account for the substantial increases in shrinkage temperature. Aluminum also can complex with carbonyl and amino functions, so effective crosslinking can be initiated either through the aluminum-collagen or the vegetable tannin-collagen complex. The high degree of thermal stability achieved with vegetable-aluminum tannage is due to the complexation between the aluminum and the vegetable tannin. This concept was highlighted by further work by Kallenberger and Hernandez (110) who showed that many multivalent metals enhance the shrinkage temperature of vegetable tanned hides. They concluded that the metal retannage involves the formation of extended tannin-metal complexes that effectively crosslink the system rather than the induction of the formation of covalent bonds between the tannin and collagen or within the collagen polymer, as occurs in chromium tannage. Of all processes developed for use of vegetable tanning, the vegetable-aluminum tannage has the most promise of displacing chromium tannin processes (226, 233, 234, 241, 242). For these reasons, further study of the complexation of various proanthocyanidins of known structure with aluminum would seem to be an interesting and important area of research.

10.3.5.3 Treatment of Vegetable Tanning Spent Liquors

Because of the extreme problem presented by treatment and disposal of tannery wastes, reduction of the environmental impacts of these effluents has been the subject of research and development since at least the early 1900s (59). In a vegetable tanning process, about 90% of the volume of effluent comes from the beam house and various washing processes. This effluent contains flesh, hair, blood, and high concentrations of salts but it is comparatively low in biological oxygen demand (500–2000 ppm). The spent tan waste, which accounts for less that 10% of the effluent volume, contains materials with extremely high biological oxygen demand (10000 to 45000 ppm) and especially stable colored species. For the past 10 years, research on vegetable tannins in the United States has been largely focused on development of environmentally sound ways of treating spent liquors. Problems in treating vegetable tanning liquors have been a significant factor in displacement of vegetable tannage by other processes such as use of syntans and chromium tannage. The least expensive, and usually an effective, way of handling spent tan liquors has been through the use of lagoons. Where sufficient land was available, spraying the effluent from lagoons onto fields was a satisfactory treatment in most instances. However, as environmental requirements have become more stringent, other approaches have become necessary.

A very effective approach is, of course, to employ a closed circuit vegetable tanning system as has been done with the LIRITAN process described above (224). In addition, processes have been developed permitting rapid drum tanning that substantially reduce the volume of liquor that must be treated (9, 12). Tomlinson and coworkers (240) have reviewed a number of early approaches that have been suggested to remove chromophoric materials from these liquors. In addition, they studied several new approaches to the reduction of color from tanning liquors typically used in the United States. Oxidations with sodium hypochlorite were not feasible, as about 80 g of chlorine equivalent were required to decolorize 1 l of typical spent liquor. Precipitation with lime did not produce a sludge which settled, and treatment with alum was also not as effective, as was indicated by combining spent alum tannage with vegetable tanning liquors to precipitate the tannins (238). However, color could be removed effectively by precipitation with a cationic polymer (Nalcolyte 605) added at a rate of 30–40 ml/l over a pH range of 2.0–3.0. This process removed about 90% of the color, about 50% of the chemical oxygen demand (COD) and about 90% of the suspended solids. This process requires the handling of large amounts of sludge (about 30% of the volume of the liquor treated). The sludge precipitates at a solids content of only about 8% and this must be dewatered further prior to disposal.

Tomlinson's approach (240) was studied further by Suddarth and Thackston (232) who examined the use of several different cationic polymers for coagulation of spent tanning liquors. Several cationic polymers removed 90% of the color but dosage rates of about 500 ml/l were required. The optimum pH was about 4.0 and, with these higher dosage rates, the floc was precipitated to a higher solids content than was reported by Tomlinson (240). Eye (61) and Hagen (77) studied the removal of color from vegetable tannery wastes through use of anionic polymers in conjunction with lime precipitation. Only small amounts of polymer (2 to 5 mg/l) are required, but about 2 g/l of lime are needed to maintain the pH at about 12.

Briggs and coworkers (28) were able to remove about 90% of the tannin by absorption on Amberlite XAD-2 resin. The resin could be regenerated and the absorbed materials collected by washing with acetone, which could be recovered by distillation. The recovered tannin had a purity of 80% and preliminary economic feasibility analyses indicated that its recovery and reuse could result in a significant net operating profit for the treatment process.

Anaerobic and aerobic biological oxidations have also received attention. Parker (169) studied a deep lagoon system that was capable of removing 81% of BOD and 58% of the COD from combined batepool, soak-waste, and spent tanning liquors. The system did not remove the color from the spent liquors. Similar results were obtained by Eye and Liu (61). Eye and Ficker (60) examined the possibility of an anaerobic process with the idea of generating methane as a by-product. Laboratory tests showed that anaerobic degradation was very effective in reducing total COD, degrading the condensed tannins added (90+% conversion), and removing the color of the liquor when the addition of tannins was stopped and the system allowed to recover for 13 days. So long as the total tannin addition did not exceed 3.3 g/l, there was no buildup of COD or volatile organic acids. There was evidence of inhibition of the bacteria at higher tannin concentrations.

When a total of 10 g of tannin had been added to the 3 l test vessels, the total gas production was about 1.3 l/day with a methane content of about 60%. At these levels of tannin addition, there was little difference between quebracho and wattle tannin treatments. The quebracho tannin was more inhibitory to the system than the wattle tannin at higher concentrations. With proper control, this approach would seem to be an effective process for reduction of both COD and color removal.

Arora and Chattopadhya (10) also studied anaerobic degradation of waste streams including the liming liquor, deliming wash, and the exhausted tannin liquor in a ratio of 1:3:2. Batch treatment of spent liquors containing a COD of 16 g/l and a BOD of 11 mg/l for 0.5 days on an anaerobic contact filter removed 80% to 90% of the COD. It was necessary to remove chlorides by pretreatment as they were toxic to the system. The system produced no sludge and allowed for production of high-methane-content gases.

In summary, use of vegetable tannins for leather manufacture remains an important commercial application for extracts of woody plants and, because of special properties provided (particularly in tanning heavy leathers such as those used in sole leathers), their use for this application can be expected to continue. Uncertainties about the price and availability of chromium (38) and concerns about its biological properties suggest that the leather tanning research community in countries such as the United States should find it prudent to continue to explore new approaches to use of vegetable tannins. Because of the heavy use of tannins imported from South Africa and South America, study of the possible use of tannins that can be obtained from indigenous plants and particularly waste products from processing nuts or forest products might be rewarding.

10.3.6 Condensed Tannins in Wood Adhesives

Most of the recent efforts to develop uses for the condensed tannins have centered on their application in wood adhesives. Reviews by Pizzi (182, 186) and others (15, 78, 87, 93, 208) provide references to several hundred papers and patents on this subject. Despite world-wide research efforts on other sources of tannins, particularly since the 1972–1973 petroleum shortage, the "mimosa" or "wattle" tannins extracted from the bark of black wattle (*Acacia mearnsii*) remain the major source of condensed tannins exploited commercially for adhesive manufacture. Of the approximately 100000 tons of wattle tannin produced annually, only about 10000 tons are used in wood adhesives, predominantly in South Africa but also in Australia and New Zealand (186). The extensive use of wattle tannins by the wood products industry of South Africa is impressive indeed, as these tannins have partly replaced phenol and resorcinol usage in adhesives for bonding of particleboard, plywood, and laminated timbers (182, 186, 213). Three factors have contributed to the success in use of wattle tannin-based adhesives, namely the comparatively high costs of phenol and resorcinol in the Southern Hemisphere, their resorcinolic functionality and low molecular weight; and, perhaps most importantly, the commitment by the research and industrial communities of these countries to reduce the reliance of the forest products industry on petroleum-based adhesives.

A similar climate exists for development of adhesives from *Pinus radiata* bark in Australia and New Zealand. A 22 ton/day bark extraction plant was built by New Zealand Forest Products Ltd. at Kinleith, New Zealand. These extracts have proved to be more difficult to use than wattle tannin due to their comparatively high molecular weight, the high viscosity of most extract preparations, their rapid rate of reaction with formaldehyde, and the often higher proportion of carbohydrate impurities. Current information (L. J. Porter, 1987) is that production of tannin by New Zealand Forest Products Ltd. has now ceased. In an attempt to make more uniform extracts with lower proportions of carbohydrates, ultrafiltration (257, 258) fractionation on Amberlite XAD-B gel (239), and fermentation (220) purifications have been investigated. Various reactions such as sulfonation and either acid- or base-catalyzed cleavage have been employed to reduce the viscosity of these extracts. A number of adhesive formulations based on *P. radiata* bark extracts have been developed. However, technical difficulties continue to inhibit the commercial use of *Pinus radiata* bark extracts in wood adhesives.

Commercial success has also not yet been enjoyed in North America with the use of tannins extracted from conifer tree barks, in part because of the comparatively low cost of phenol (39) and in part because of their phloroglucinolic functionality and comparatively high molecular weight. Several forest products companies have made major commitments to use of tannins extractable from Douglas-fir (*Pseudotsuga menziesii*) (78), western hemlock (*Tsuga heterophylla*) (93), spruce (*Picea* spp.) (127), fir (*Abies* spp.) (6), and a number of species of pine (*Pinus* spp.) (8, 14, 104) tree barks without commercial success. Weyerhaeuser Co. marketed a series of ground bark fractions from Douglas-fir named Silvacons in the 1950s–1960s (78). Bohemia Lumber Co. also marketed a bark powder fraction from Douglas-fir as a plywood adhesive extender in addition to granulated cork and wax during the 1970s (F. Trocino, 1980–1984, personal communication). ITT-Rayonier marketed a wide range of specialty polymers made from extracts from western hemlock bark in the 1960s, but production ceased in 1972 (93). None of these products are manufactured now. In North America, low prices for phenol limit prospects for success in the development of plywood or particleboard adhesives containing tannins over the short term (39). However, economics are favorable for use of condensed tannins as substitutes for resorcinol used in cold-setting wood adhesives. Because the chemistry and thus approaches to use of wattle and conifer bark tannins differ so dramatically (Chaps. 7.6 and 7.7), discussions of their applications in wood adhesive formulations are best presented separately.

10.3.6.1 Wattle Tannin-Based Particleboard Adhesives

The first attempts to use wattle tannins in particleboard adhesives (114, 170) followed reports by Dalton (50, 51) on the use of *Pinus radiata* bark extracts as substitutes for phenol-formaldehyde resins in plywood adhesives. Even though the molecular weight of wattle tannin is comparatively low (208), solutions of bark extracts at solids contents required for adhesives (40%–58%) exhibit excessively high viscosities. High-temperature alkaline treatments reduced their viscosity, and these products could then be used in wattle tannin-based particleboard adhesives that gave panels with excellent water resistance (215, 216). The alkaline treatments were thought to hydrolyze co-occurring high-molecular-weight carbohydrates considered to be responsible for the high viscosity. Reaction conditions must be optimized because extended heating at high temperature results in subsequent increases in viscosity thought to be the result of base-catalyzed condensations of the tannins (183). Further development of wattle tannin-based particleboard adhesives by Plomley and Stashevski (196) and Stashevski and Deppe (229) led to wattle tannin-bonded panels that resisted degradation in boiling water and clearly had true exterior performance characteristics. Similar development work on these types of adhesives was done by Saayman at the Leather Industries Research Institute in South Africa (218). Wattle tannin-bonded, exterior quality panels were produced commercially both in South Africa and New Zealand in 1965.

Bond quality could be improved considerably through use of either phenol-resorcinol-formaldehyde or phenol-formaldehyde polymers as crosslinking agents (175). However, comparatively high resin solids (about 16% on the face and 11% in the core, based on wood dry weight) are required to produce boards with exterior use properties. One of the more interesting aspects of these results is that wattle tannin/urea-formaldehyde copolymers with urea-formaldehyde contents of 15% to 20% of the resin solids gave bonds that resisted degradation even after boiling the panels for 72 hours. Some typical board properties obtained with different adhesive formulations are given in Table 10.3.1.

Further research (187, 195) centered on formulations that would permit shorter press times and extend the comparatively short pot life by adding paraformaldehyde or formalin solutions emulsified in oil to wattle tannin and wood chips. Additionally, use of zinc acetate (2% of tannin solids) as a catalyst permitted marked reductions in the press temperature (from 170° to 130°C) required to produce exterior quality boards (188). Pizzi and Merlin (191) also demonstrated that reaction of wattle tannins with 3% to 5% sodium sulfite would provide

Table 10.3.1 Properties of exterior particleboards bonded with wattle tannin-based adhesives (136)

Property	Formulation[1]				
	A	B	C	D	E
Panel density (g/cm^3)	0.708	0.713	0.743	0.732	0.748
Internal bond – dry (MPa)	1.28	1.07	0.90	0.72	1.28
– 2-h boil (MPa)	0.73	0.55	0.60	0.21	0.94
Thickness swelling – 24-h soak (%)	4.6	7.7	4.9	14.5	5.0
– 2-h boil (%)	12.6	13.0	17.3	34.6	12.9

[1] a. A tannin solution modified by reaction with acetic anhydride, phenyl acetate, and NaOH crosslinked with 8% paraformaldehyde; 15% wax emulsion, 2.75% insecticide based on tannin solids. b. As in a.), but panels tested after 5 years of outside weathering. c. An extract crosslinked with 8.6% of a 63.8% solids urea-formaldehyde resin and 6% paraformaldehyde; 13.5% wax emulsion, 2.5% insecticide based on tannin solids. d. An extract crosslinked with 8% paraformaldehyde and 5.5% zinc acetate as a catalyst; 15% wax emulsion, 2.75% insecticide based on tannin solids. e. An extract crosslinked with 8% of a 70% solids content phenol-resorcinol-formaldehyde resin and 8.5% paraformaldehyde; 13.5% wax emulsion, 2.5% insecticide based on tannin solids

reductions in solution viscosity comparable to those obtained by hot alkaline treatment and that the sulfonated tannin could be used in adhesives for exterior quality particleboard. Pizzi (184) showed that condensed tannins from either wattle or conifer barks could be used to manufacture exterior particleboard adhesives by crosslinking with 4,4'-diphenylmethane diisocyanate (MDI).

Comparatively little attention was given to the use of tannin-bonded particleboards for interior use until restrictions on formaldehyde emissions from urea-formaldehyde bonded products became a critical concern (178). Wattle tannin/urea-formaldehyde adhesive formulations for interior particleboards do reduce the large, short-term formaldehyde release normally observed soon after pressing, but not the slow emission that is common after extended storage (33). The scavenging effects of wattle tannins are limited to comparatively small amounts of formaldehyde because of the highly condensed nature of the tannin/urea-formaldehyde polymer after curing, as was also observed by Marutzky and Dix (136).

10.3.6.2 Wattle Tannin-Based Plywood Adhesives

The development of wattle tannin-based plywood adhesives occurred concurrently with the particleboard adhesives and, although quite different adhesive properties are required to bond these two types of furnish, many of the basic premises on which their development were based are the same. A series of wattle tannin-based plywood adhesives that provide exterior bond qualities have been described in detail by Pizzi (182, 186). All recent formulations use low-molecular-weight polymers, as first suggested by MacLean and Gardner (131), rather than formaldehyde for crosslinking agents.

Exterior plywood adhesives based on condensation of wattle tannins with urea-formaldehyde crosslinking agents are perhaps of greatest interest because of their low cost (174, 177). An example formulation is 100 parts of a 50% aqueous solution of wattle tannin were mixed with 0.25 parts of a defoamer and 8.6 parts of a 63.8% solids content urea-formaldehyde resin. A gummy precipitate formed, which was redissolved by addition of 5 to 10 parts of water and stirring. This resin could be spray-dried for extended storage. The adhesive was formulated by combining 100 parts of the above resin with 4.8 parts of 96% paraformaldehyde, 7.9 parts of 200-mesh coconut shell powder and adjusting the pH to 4.9. Plywood panels were pressed at 125 °C for 6 minutes at 1.7 Mpa pressure. On the basis of British Standard Knife Test, panels had 9 or 10 ratings (90% to 100% wood failure) when tested dry, after 24 hours of soaking in cold water, and after 72 hours of boiling in water. These test results met the requirements for marine-grade plywood (178). The amounts of urea-formaldehyde resin used in crosslinking were important to bond quality; amounts greater than 15% to 20% resulted in low water resistance and amounts less than about 5% resulted in poor strength.

Wattle tannin-based plywood adhesives were also crosslinked with 10% to 20% (by weight of solids) of either a phenol-formaldehyde resol or a phenol-resorcinol-formaldehyde polymer with approximately 20% resorcinol, to which paraformaldehyde was added to provide crosslinking between the resorcinol and

A-ring of the tannin (194). Resin synthesis conditions varied in the approach to addition of formaldehyde. For example, the tannins could be crosslinked by reaction of a phenol-resorcinol-formaldehyde resin carrying little or no methylol functionality by addition of paraformaldehyde or by reaction of a phenol-formaldehyde prepolymer carrying a comparatively high methylol functionality. Adhesive formulations were similar to those described above. Typically, 100 parts of a 55% solids content water solution of wattle tannin were combined with 0.25 parts of defoamer, 7 parts of paraformaldehyde and 9 to 10 parts of coconut shell powder; the pH was adjusted to 6.5 to 7.4 by addition of 40% NaOH. These adhesives also provided exterior quality glue lines. They are exceptional in their tolerance to high moisture content veneer and permit fast curing rates, subjects of particular interest in the plywood industry today.

Another approach involved the use of metal catalysts, such as zinc acetate, to speed the curing of these polymers (103, 179). Addition of 0.01 mole of zinc acetate/flavanoid unit approximately doubled the gelation rate of wattle tannin-based adhesives. Plywood adhesives crosslinked with paraformaldehyde in the presence of zinc acetate catalysts and with no fortifying resins were used in bonding of low-density, easy-to-glue species. Although less severe than in the case of particleboard resins, the high viscosity of the natural tannin extracts can be troublesome. Rather than treating wattle extracts at high temperature in base, it was possible to obtain dramatic reductions in solution viscosities by reaction with sodium sulfite. While commonly thought to introduce too much water solubility into the tannin for use in water-resistant adhesives, sulfonated wattle tannins crosslinked with phenol-resorcinol-formaldehyde resins can be used in the manufacture of marine-grade plywood (180).

10.3.6.3 Wattle Tannin-Based Laminating Adhesives

Perhaps the greatest success in use of condensed tannins from wattle bark for adhesives in recent years involves exploitation of their resorcinol functionality in formulations for cold-setting, wood-laminating adhesives. Early attempts to formulate these adhesives from wattle tannins involved their reaction with resorcinol, resorcinol-formaldehyde oligomers, or phenol-resorcinol-formaldehyde prepolymers (193) much the same approach as that described by Herrick and Conca (95) for formulation of cold-setting, phenolic resins from western hemlock bark tannins. In later research, less expensive formulations were developed. Urea-formaldehyde-resorcinol condensates reacted with the tannin, tannin-urea formaldehyde condensates reacted with resorcinol or mixtures of tannins, and resorcinol reacted with urea-formaldehyde resins all provided adhesives that would cure rapidly at ambient temperature after addition of paraformaldehyde at pH 7.9 to 8.3. Adhesives formulated with these resins provided shear strength and wood failure on beechwood strips passing British Product Standard BS 1088-1957 (180, 190, 244). Pizzi and Daling (190) investigated the use of sulfonated wattle tannins in an attempt to find a simplified solution to problems with high extract viscosity and dry-out of glue lines during assembly. Apparently, cleavage of the pyran ring with sulfonation at C-2 caused the tannin to react more like resorcinol so the amounts of resorcinol used as crosslinking agent could be reduced. Reactions of wattle tannins with 7% to 10% sodium sulfite based on tannin solids at reflux for 6 hours gave the best results. These resins also proved to be more resistant to dry-out than those described previously. Crosslinking sulfonated wattle tannins with a urea-formaldehyde-resorcinol prepolymer also had promise.

An important breakthrough occurred when Pizzi's adhesives group (192) at the National Timber Research Institute in Pretoria, South Africa, applied Kreibich's "honeymoon" concept (20, 116) to the use of wattle tannin-based, cold-setting adhesives for fingerjointing of lumber. Not only was it possible to replace over 50% of phenol-resorcinol-formaldehyde resins, but the tannin-based adhesives cured more rapidly than fast-cure resins formulated with expensive *m*-aminophenol ($4/lb). A number of different approaches to use of wattle tannins in both the *A*-side (hardener) and *B*-side (catalyst) resin components have proved to work well. In early trials, good results were obtained with use of a wattle tannin/resorcinol-formaldehyde resin and hardener on the *A*-side and a wattle tannin adhesive with *m*-aminophenol or simply a wattle tannin extract at pH 12.6 on the *B*-side. Use of a wattle tannin/resorcinol-formaldehyde resin at a pH of 11.6 with no hardener present on the *B*-side also gave promising results. Further development (189) resulted in approval by the South African Bureau of Standards for three systems: 1) *A*-side, PRF, pH 7.9 to 8.2, extra paraformaldehyde; *B*-side, PRF, pH 11.5, no hardener. 2) *A*-side, PRF, pH 7.9 to 8.2, extra paraformaldehyde; *B*-side wattle tannin extract 50%–55% solids, pH 12.0 to 12.5. 3) *A*-side, tannin/resorcinol-formaldehyde resin pH 7.5 to 7.9 with paraformaldehyde; *B*-side tannin/resorcinol-formaldehyde at pH 12.0 to 12.5, no hardener. These resins have been used to bond *Eucalyptus saligna* as well as pine (*Pinus* spp.) and are currently used commercially in the majority of the timber laminating plants of South Africa.

Two wattle tannin-based adhesive formulations for preparation of finger-jointed lumber by radio-frequency curing have also been described (176). Although the resin crosslinked with a resorcinol-formaldehyde co-polymer gave the best results, crosslinking with a phenol-resorcinol-formaldehyde prepolymer also gave bond strengths and wood failures exceeding the requirements of South African Bureau of Standards 096-1976 specifications for finger-joint bond quality. The latter adhesives were lower in cost than a conventional phenol-formaldehyde resin in South Africa.

10.3.6.4 Other Wattle Tannin-Based Adhesives

Wattle tannins have been used as fortifiers in starch-based adhesives for corrugated carton manufacture in order to improve the moisture resistance of boxes used in shipment of fruit, for example (144, 217). Addition of 4% wattle tannin based on resin solids substantially improved the durability of these cartons. Wattle tannin urethane-based varnishes with excellent durability have been developed by Saayman (214). Other adhesive applications based on reactions of wattle tannins with isocyanates have been developed for specialty applications such as the bonding of aluminum (181).

10.3.6.5 Conifer Bark and Related Tannins as Particleboard Adhesives

During the period of 1950 to 1965 most of the published work on use of conifer bark tannins as particleboard adhesives centered on the use of *Pinus radiata* extracts (29–31, 50, 51, 79). Extracts made with hot water, sodium carbonate, and/or sodium sulfite solutions were concentrated to solids contents of 40% to 45%, the viscosity of the solutions was adjusted by various means, and paraformaldehyde was used for crosslinking in order to obtain an acceptable pot life. Particleboards made using 6% resin solids on the dry wood particles and pressed for 10 to 12 minutes at 160°C gave panels with properties superior to those of urea-formaldehyde-bonded panels, but their water resistance was not high enough to be used under conditions of "exterior" exposure. Anderson and his coworkers followed work on use of *Pinus radiata* tannins (5) with a series of studies on the use of a variety of North American species including white-fir (*Abies concolor*) (6), ponderosa pine (*Pinus ponderosa*) (8), Douglas-fir (*Pseudotsuga menziesii*), and western hemlock (*Tsuga heterophylla*) tree barks (4, 7). Three-layer boards (wood particle face and back with a bark particle core) using 8% resin solids and pressed for 3 minutes at 180°C passed Canadian Product Standards CS 236-66 for phenolic-bonded exterior panels of Class B. Typical strength and water resistance properties of 0.95-cm-thick, 0.67 to 0.70-g/cm^3 density boards are given in Table 10.3.2. Similar efforts to use extracts of Englemann spruce (*Picea engelmannii*) and lodgepole pine (*Pinus contorta*) bark extracts by crosslinking with either paraformaldehyde or oxazolidine did not give water-resistant panels (36). These panels were pressed at 149°C for 15 minutes, which may have been too low a curing temperature. Steiner and Chow (230) showed that water-resistant bonds could not be obtained with western hemlock bark tannins unless press temperatures were above 180°C.

Chen studied the use of sodium hydroxide solution extracts from southern pine (*Pinus spp.*) bark, peanut (*Arachis hypogea*) hulls, and pecan (*Carya illinoensis*) nut pith for bonding of particleboards, oriented strandboards, and composites made with a flakeboard core and southern pine veneer face and backs (40, 45). All the natural product-based adhesives, but particularly the peanut hull extract, cured more rapidly than a commercial flakeboard phenolic resin. Use of tannins permitted reductions in pressing time of about 1 minute. The natural product-based adhesives gave boards with higher internal bond and modulus of rupture and retained these strength properties better after 2 hours of boiling in

Table 10.3.2. Properties of wood-face, bark-core, three-layer particleboards bonded with conifer bark tannins (173)

Property	Value
Modulus of rupture (MPa)	17.7–19.8
Internal bond (MPa)	0.95–1.23
Water absorbtion (cold %)	20.7–27.0
Thickness swelling (%)	7.6–9.8
Linear expansion (%)	0.19–0.56

Table 10.3.3. Properties of particleboards bonded with adhesives in which 40% of the phenol-formaldehyde resin is replaced by tannins (176, 177)

Property	Value by tannin source[a]		
	Peanut	Pecan	Pine bark
Modulus of rupture – dry (MPa)	17.1	16.8	14.0
– after boiling (MPa)	5.9	5.3	5.0
Internal bond (MPa)	1.78	1.82	1.48
Thickness swelling – 24-h soak (%)	18.4	19.3	18.0

[a] Peanut, *Arachis hypogea*; pecan, *Carya illinoensis*; pine, *Pinus taeda*

water than did the panels bonded with two commercial phenolic resins. Typical properties are given in Table 10.3.3 for southern pine particleboards of 1.6 cm thickness and 0.80 g/cm³ density that were bonded with resins in which 40% of the phenolic resin had been replaced by sodium hydroxide extracts of peanut hulls, pecan pith, and southern pine bark.

These adhesives performed equally well when used for bonding of oriented strandboards made from Douglas-fir flakes. In addition, composites made from oak or southern pine flakeboards to which southern pine veener was bonded on the face and back passed the 6-cycle durability test of the American Plywood Association when the core was bonded with the peanut hull-based adhesives, even when the 0.64-cm cores were pressed for only 75 seconds. Chen's work on these adhesives is described further in a series of four U.S. Patents (41–44).

New Zealand Forest Products Ltd. continued their research on the use of radiata pine tannins and had a 1 ton/day pilot plant operating by 1974 and a full-scale plant in operation from mid-1981 to 1987. These tannin extracts were marketed under the trade name Tannaphen (Woo, 1982, personal communication). The extracts were used in bonding of both plywood (described below) and particle board. The Tannaphen extract (100 parts) is combined with a defoamer (0.5 parts), paraformaldehyde (6–10 parts), and parafin wax (10 parts); it is sprayed on wood chips at a loading of 10% to 12% by weight of solids on the face and 7% to 9% by weight of solids on the core particles. Panels 20 mm thick are pressed for 7 minutes at 170°C to obtain panels with the properties shown in Table 10.3.4. These panels were marketed primarily as a flooring material in New Zealand. While not classified as exterior panels, these Tannaphen-bonded boards re-

Table 10.3.4. Properties of particleboards bonded with Tannaphen and paraformaldehyde (182)

Property	Value
Modulus of rupture – dry (MPa)	26.7
– 72-h boiling (MPa)	10.2
Modulus of elasticity – dry (GPa)	3.9
Internal bond – dry (MPa)	0.85
Thickness swelling – 24-h cold soak (%)	12.8
– 2-h boiling (%)	23.5

Table 10.3.5. Properties of Tannaphen-isocyanate resin bonded particleboards (183)

Property	Value
Modulus of rupture – dry (MPa)	19.1
Internal bond – dry (MPa)	1.11
– 2-h boil (MPa)	0.17
Thickness swelling – 24-h soaking (%)	12.2

tained their strength properties far better than urea-formaldehyde bonded boards after 6 months of outdoor exposure.

Dix and Marutzky (55) made a thorough study of the use of Tannaphen in particleboard adhesives. Homogeneous boards at a density of 0.69 g/cm^3 were bonded with 9% Tannaphen on dry wood weight by crosslinking with paraformaldehyde. These panels had a modulus of rupture of 14.6 MPa and an internal bond strength of 0.719 MPa with 16.5% thickness swelling after 24 hours of soaking in cold water. However, the panels failed the moisture durability requirements of the 2-hour boil test for V100 class panels. Fortification of the Tannaphen with phenol-formaldehyde resins of different types improved these board properties only marginally. The Tannaphen was also crosslinked with urea-formaldehyde resins at a wide range of dilution ratios. Blending of the tannin with UF resins resulted in inferior board properties at all dilution ratios. Board properties were best when bonded with either a straight UF or a Tannaphen-paraformaldehyde formulation. However, particleboards with water resistance superior to that required by V100 class boards could be made by blending Tannaphen with MDI resins. Panels at a density of 0.70 g/cm^3 bonded with 6% Tannaphen and 3% MDI on dry wood weight had the properties given in Table 10.3.5. The ratio of tannin to isocyanate resin could be reduced to 3:1 w/w (i.e., 2.2% isocyanate) using a resin content of 9% total solids on dry wood weight to obtain approximately the same properties as were obtained with boards bonded with 4.5% of a pure diisocyanate resin. Similar results were obtained with three-layer particleboards.

Pizzi (185) had shown earlier that extracts obtained from *Pinus patula* bark could be used in adhesive formulations for exterior quality wood particleboards when combined with isocyanate resins. Pine tannin-MDI bonded panels using 15% resin solids at mass ratio of 70:30 tannin to MDI in three-layer boards, where the two resin components were applied separately, had an internal bond strength when measured dry of 0.84 MPa and 0.42 MPa after 2 hours of boiling. After 2 hours of boiling, the panels had a thickness swelling of 15% were measured wet and only 4.3% when measured dry, properties far in excess of the minimum requirements for external particleboard. A comparison of the properties of these panels and those of Dix and Marutzky (55) (Table 10.3.5) with those of panels bonded only with 3% MDI adhesives indicated that the tannin component of these adhesives added substantially to the board properties, particularly with regard to thickness swelling resistance after 2 hours of boiling. Pizzi (185) found several other adhesive formulations that passed requirements for exterior particle boards in which substantial amounts of pine tannins could be used (Table 10.3.6). While the combinations of pine tannins with melamine and wattle-formaldehyde

Table 10.3.6. Properties of particleboards bonded with pine tannins as 50% replacements for synthetic resins (184)

Property	Formulation[a]		
	MF/PT	PF/PT	WF/PT
Panel density (lb/ft³)	42.2	41.5	43.5
Internal bond – dry (MPa)	1.31	0.94	0.83
– 2-h boil (MPa)	0.43	0.29	0.157
Thickness swelling – 2-h boil wet (%)	26.4	34.1	21.2
– 2-h boil dry (%)	12.6	20.1	10.1

[a] MF/PT: melamine-formaldehyde/pine tannin, 50/50 by mass, applied separately; PF/PT: phenol-formaldehyde/pine tannin, 50/50 by mass, applied separately; WF/PT: wattle-formaldehyde/pine tannin, 50/50 by mass, applied separately

resins gave good properties, these adhesives were not as effective as the combination of pine tannins with isocyanates.

It would seem that condensed tannins would have good potential for use in adhesives for oriented strand- or flakeboard particularly because of their known tolerance to comparatively high moisture content furnish and because of the potential for more rapid cure rates. At current prices for phenol-formaldehyde resins in North America, interest in these types of adhesives is not particularly strong. However, considering the rapid expansion of this industry, good potential exists if prices for phenol-formaldehyde resins should increase substantially. It should be recognized that amounts of resins applied to oriented strand- or flakeboards in North America are much lower that those used in the above-described exterior particleboard adhesive applications.

10.3.6.6 Conifer Bark and Related Tannins as Plywood Adhesives

Dalton's early work (50, 51) on the use of *Pinus radiata* bark extracts was a major catalyst in the development of interest in using condensed tannins as wood adhesives. Booth et al. (24) and Herzberg (99) describe plywood adhesive formulations in which *Pinus radiata* bark extracts are crosslinked with paraformaldehyde or combinations of paraformaldehyde and formalin solutions. Steiner and Chow (230) also describe similar formulations based on extracts from western hemlock (*Tsuga heterophylla*) bark. They found that press temperatures of 180 °C were required to obtain bonds that would pass exterior grade plywood wood failure requirements when these tannins were crosslinked with paraformaldehyde. These adhesive formulations were never successful on an industrial scale, probably because of difficulties in controlling viscosity, pot-life, and curing rates. MacLean and Gardner (131), studying the use of western hemlock bark tannins, focused on the principal problems that remain with most formulations even today. To obtain more control of their rapid polymerization, they used hexamine as the aldehyde donor. However, water-resistant bonds were not obtained. They reasoned that this was because of inadequate crosslinking due to the requirement of bridging large, sterically hindered polymers with a small molecule such as formaldehyde and, therefore, recommended the use of phenol-formaldehyde resols as crosslinking agents. Herrick and Bock (94) made extensive studies of this approach. They were able to replace 60% of the phenol in adhesives for bonding exterior grade Douglas-fir plywood with western hemlock bark extracts. ITT-Rayonier marketed their adhesives for a time but low prices for phenol in North America prevented success (93). McGraw and Hemingway (140, 141) have made detailed studies of the reactions of model compounds for tannins with hydroxylbenzyl alcohols in an attempt to improve the properties of tannin-resol copolymer adhesives.

Brandts and Lichtenberger (27) found that adhesive properties and bond quality were markedly improved if alkaline extracts were heated to high temperature, the extract precipitated by acidification, and the precipitate washed to remove impurities. When these purified extracts were used to replace 60% of the phenol in plywood adhesives for bonding of birch plywood, wood failure ranged from 27% to 75% at wet shear strengths of 2.27 to 2.31 MPa after specimens were boiled in water. Hemingway and McGraw (91) found that southern pine (*Pinus taeda*) bark alkaline extracts that had been treated in this way reacted with formaldehyde at much the same rate as phenol, so the extract and phenol were combined and reacted with formaldehyde to provide resins for bonding southern pine plywood. Here again, wood failures in a range of 50% to 70% were obtained after vacuum-pressure water soak, well below the requirements of U.S. Product Standard PS-1-66.

A number of investigators have examined the use of bark ground to a fine powder as a reactive extender to partially replace phenol requirements in plywood adhesives. Herritage (97, 98) heated finely ground bark in sodium hydroxide solutions and used these products to replace phenol-formaldehyde resins in amounts of about 20% of that normally used in adhesives for gluing Douglas-fir plywood. Similarly, Hamada (80) used finely ground wattle (*Acacia mearnsii*) or hemlock (*Tsuga heterophylla*) bark powders as 15% – 20% phenol replacements in adhesives for bonding of luan (*Shorea* spp., *Parashorea* spp., or *Pentacme* spp.) plywood. While excellent bonds were obtained using wattle bark, no bonds of exterior quality could be obtained using the hemlock bark powder as a reactive extender. Hartman (82) was successful in developing a 100% natural-product-based plywood adhesive by combining a slurry of finely-ground bark in sodium hydroxide with an equal weight of a 50% wattle tannin solution. Southern pine plywood glued with his formulation passed requirements for exterior sheathing-grade plywood (PS-1-66). Hemingway and McGraw (88) attempted a 40% phenol replacement by combining finely ground Douglas-fir bark powder with phenol in an alkaline solution, heating the mixture to reduce the solution viscosity and aldehyde reactivity, and then crosslinking the mixture with paraformaldehyde. Southern pine plywood had wood failure of 60% to 70% at wet shear strengths of about 2.1 MPa after vacuum pressure soaking (PS-1-66). Pauley (171) obtained similar results in attempts to use finely ground cork from Douglas-fir as the sole source of aldehyde reactive phenols. Exterior-quality bonds were obtained when 3,5-xylenol was added to the formulation. Liiri et al. (127) replaced about 20% of the phenolic resin and obtained exterior quality bonds in birch plywood using either sodium hydroxide extracts from Norway spruce (*Picea abies*) bark or dispersions of finely ground bark in alkaline solutions.

Research directed to the use of bark extracts from various species of pines has continued despite the marked reduction in prices for petroleum in the late 1970s and early 1980s. A number of new plywood adhesive formulations based on extracts of *Pinus radiata* have been described recently. Weissmann and Ayla (253) used sulfonated tannin extracts at 40% solids and fortified these extracts with a phenol-formaldehyde resin (Kauresin 260 produced by BASF) at levels to 10% to 50% by weight of solids. Both paraformaldehyde and hexamethylenetetramine were examined as aldehyde sources. Exterior grade plywood bonds were obtained at 10% and 30% fortifying levels.

Woo (1982, personal communication) has described the exterior plywood adhesive formulation based on Tannaphen and used commercially by New Zealand Forest Products Ltd. since 1981 as: Tannaphen – 56 parts, a phenol-formaldehyde fortifying resin – 21 parts, olivestone flour – 18 parts, and paraformaldehyde – 5 parts on weight of solids. These adhesives are tolerant of open assembly times of 20 minutes, closed assembly times of up to 4 hours, and veneer moisture contents of up to 15%. Both shear strengths and wood failure were comparatively high even after 5 weeks in water at 25°C.

Jenkin (1982, personal communication) reduced the viscosity of sodium carbonate/sodium sulfite extracts from radiata pine bark by adjusting the pH to 4.0 to 4.5, adding 1% to 2% of phenol, and heating this mixture to 90° to 100°C for 15 to 30 min. The molecular weight of the tannin increased during this reaction, but, at the same time, the viscosity of the solutions was decreased substantially. Jenkin, like others, attributed the reduction of viscosity to hydrolysis of co-extracted carbohydrate gums. Crosslinking this product with paraformaldehyde gave bonds with high shear strength after Vacuum-Pressure-Soak treatment, but wood failures were low. Two approaches to fortification of the tannin adhesive gave radiata pine plywood with bonds that exceeded the minimum 85% wood failure requirements of U.S. Product Standard 1-74 (4 h boiling, 20 h drying at 62°C, and 4 h of boiling). One approach involved addition of a low-molecular-weight phenol-formaldehyde novolac that apparently co-condensed with the tannin in reaction with paraformaldehyde. The other approach involved grafting of about 7.5% phenol on the tannin by an undisclosed reaction. These adhesives did not give bonds on southern pine veneer that would pass requirements of U.S. Product Standard PS-1-66 for exterior grade sheathing plywood, as wood failure averaged only between 60% and 75% (Jenkin and Hemingway, 1983, unpublished results).

Chen (46) used sodium hydroxide extracts from southern pine (*Pinus* spp.) bark, peanut (*Arachis hypogea*) hulls, and pecan (*Carya illinoensis*) pith as 20% replacements for phenol-formaldehyde resins in plywood adhesives. Bond quality was highly dependent on veneer moisture content and assembly time. Exterior quality bonds (U.S. Product Standard PS-1-74) in southern pine plywood were obtained if the veneer moisture content was 7.0% with 60-minute assembly times. Phenol-formaldehyde resin replacements could be increased to 40% by peanut hull and pecan pith extracts when the extraction solutions containing 5% of sodium hydroxide were used (45). Southern pine veneer (0.32 cm) was spread at a rate of 41.5 to 42.5 g/1000 cm^2 of double glue line and the layups were stored in a

1014   The Utilization of Wood Extractives

Table 10.3.7. Wood failure of southern pine plywood after PS-1-74 tests (198)

Assembly time (min.)	Press time (min.)	Wood failure by adhesive type (%)		
		Phenolic	Peanut[a] hull	Pecan[a] pith
20	1.5	0	25	0
	2.0	32	82	62
	3.0	75	87	78
60	1.5	0	70	30
	2.0	70	84	80
	3.0	75	89	75

[a] Peanut, *Arachis hypogea*; pecan, *Carya illinoensis*

100 °C convection oven for 20 or 60 minutes prior to hot pressing. The panels were pressed at 149 °C for 1.5, 2.0 and 3.0 minutes at 1.38 MPa. The rapid curing rates of both natural product-based adhesives, especially of the peanut hull containing adhesives, are evident from results shown in Table 10.3.7.

Extracts from the bark of *Pinus brutia* are among the most promising of those from conifer barks because of the comparatively high yield of extract obtainable by mild extraction conditions (i.e. hot water solubility of 25%) and the comparatively low molecular weight of the tannin extracted (i.e. Mn of 1885 and Mw 3787 as peracetates or approximately half the molecular weight of most conifer bark tannins) (14, 16). Aqueous solutions of these extracts at 50% solids have suitably low viscosity to permit their reaction with paraformaldehyde without prior treatment. When combined with paraformaldehyde in amounts of 4% to 10% of extract solids, water-resistant bonds meeting German Product Standards (DIN 68-602) are obtained. The amount of wood failure on beech wood strips was low even after pressing the samples at 150° to 160 °C for 15 minutes. However, fortification of the tannin with only 10% of a phenol-formaldehyde resin and crosslinking with hexamethylenetetramine gave shear strengths of 12.4 MPa and wood failure of 100% when tested dry and 5.9 MPa and 97% wood failure after the specimens had been boiled for 6 hours in water and fractured when wet. *Pinus brutia* tannin-based adhesives named "Catex" adhesives have been developed in Turkey at least to pilot-plant scale. The bark of *Pinus canariensis* is another exceptionally good source of tannin and preliminary studies of adhesive formulations with this extract indicate considerable potential (253).

10.3.6.7 Conifer Bark and Related Tannins in Cold-Setting Phenolic Resins

With the exception of the early work by Herrick and Conca (95) there has, until recently, been surprisingly little research directed to the use of conifer bark tannins in cold-setting wood-laminating adhesives. Current prices for the resorcinol that generally makes up about 25% of the weight of cold-setting phenolic resins are about $1.80 to 1.85/lb. Therefore, comparatively expensive reactions and processing costs can be economically feasible in the development of these types of adhesives, while only extraction and drying costs make replacement of phenol in plywood and particleboard adhesives marginal in North America. Ayla (14) made some preliminary studies of the use of *Pinus brutia* tannins that showed considerable promise. Tannin extracts were reacted with 1% to 5% sodium bisulfite based on tannin solids and concentrated to 50% solids. When combined with an equal weight of a phenol-resorcinol-formaldehyde resin at 58% solids (Kauresin 440 produced by BASF) and applied to beechwood panels, shear strengths of bonds exceeded the minimum requirements of German Product Standard DIN 68-602 when tested dry or wet after boiling in water.

Kreibich and Hemingway have developed three approaches to use of extracts from loblolly pine (*Pinus taeda*) bark in cold-setting, wood-laminating adhesives (117–120). The first approach examined exploited the facile acid-catalyzed cleavage of the interflavanoid bond in pine bark tannins (90) to reduce the molecular weight, using resorcinol as a nucleophile (89). Loblolly pine bark extracts (2 parts) were combined with resorcinol (1 part) and heated at 120°C in the presence of acetic acid catalyst to produce flavan-4- and oligomeric procyanidin-4-resorcinol adducts. Extraction of the reaction product with ethyl acetate separated the tannin and unreacted resorcinol from the carbohydrates that had a deleterious effect on bond quality. The ethyl acetate soluble extract was used to completely replace all resorcinol in a conventional phenol-resorcinol-formaldehyde timber laminating adhesive. Bond quality (shear strength and wood failure) exceeded the requirements of the American Institute for Timber Construction (117).

In the second approach (120), condensed tannins were purified of co-occurring carbohydrates and used as resorcinol replacements in a "honeymoon" system (116) as described previously in use of wattle tannin adhesives. One surface was spread with a commercial phenol-resorcinol-formaldehyde laminating adhesive, to which additional formaldehyde was added. The other surface was spread with pine bark tannin extract in sodium hydroxide solutions. Bonds meeting the requirements of the American Institute of Timber Construction were also obtained using this approach. It was necessary to remove the carbohydrates, however. Addition of the separated carbohydrates at comparatively low levels (about 10%) resulted in bonds with low wood failure.

The most economical approach to utilization of tannin extracts from southern pine bark as resorcinol replacements in cold-setting phenolic adhesive was through use of products obtained by extraction of bark with sodium sulfite (118). The interflavanoid bond is cleaved during the course of this extraction and oligomeric procyanidin and flavan-4-sulfonates are produced with a marked reduction in the molecular weight and viscosity of the solutions (65). The sulfonate function is a good leaving group under mild conditions of temperature and alkaline pH (139). Therefore, water-resistant bonds are obtained when these sulfite extracts are combined with conventional phenol-resorcinol-formaldehyde resins, either as mixed system adhesives or when applied in "honeymoon" systems. The adverse effects of reducing sugars on wood failure after vacuum-pressure soaking of test specimens may be countered by the formation of sulfite adducts with the aldehyde. Bonds passing the requirements of the American Institute of Timber Construction Product Standards are obtained with southern pine bark sulfite extracts are used to replace 50% of the phenol-resorcinol-formaldehyde-based adhesives (118).

The same type of extract has recently been used in a cold-setting, wood finger-jointing adhesive formulated in a "honeymoon" system (119). End-joints cured at ambient temperature develop sufficient strength (3.5 to 7.0 MPa) within 15 to 20 minutes to permit handling and ultimate strengths of well over 35 MPa with exceptionally good wood failure. With some minor alterations of plant design, these adhesives have potential for replacing the costly radio-frequency curing systems used in North American plants with an end-jointing system much the same as is practiced in South Africa (186) using wattle tannin-based adhesives (Sect. 10.3.6.3).

10.3.7 Specialty Polymer Applications

Comparatively little attention has been directed to other specialty polymer uses for condensed tannins over the past 10 years, possibly due to the strong focus on their application as wood adhesives (88, 186). This seems unfortunate because condensed tannins offer excellent opportunities for use as specialty polymers due to the high degree of nucleophilicity of the phloroglucinol or resorcinol A-rings, the excellent potential for complexation to the catechol or pyrogallol B-rings and, in the procyanidins or prodelphinidins, the comparative ease with which the interflavanoid bond can be cleaved to permit the formation of derivatives substituted at the C-4 carbon.

Aldehyde condensation products have been used as a grouting medium to reduce water flow in earth dams and to stabilize soils for building foundations (245). They have also recently been shown to be effective as resorcinol replacements in rubber compounding both from the standpoint of improving properties of rubbers as well as improving bonding to a variety of cord materials when used as replacements for resorcinolic resins in cord adhesives (81). Condensed tannin-formaldehyde condensation products have also been used as cation exchange resins (135). Stronger binding capacity was obtained by further reaction with monochloroacetic acid, sulfuric acid, or chlorosulfonic acid. The tannins in granulated barks are reacted with formaldehyde to fix the tannin in a porous medium, and the catechol B-rings are then used to complex heavy metal ions to scavenge metals in mine tailings (202). Similar approaches have been taken by the Japanese to develop an ion exchange resin to remove iron from *sake* (252). Condensed tannins (1) or quaternary ammonium derivatives produced through reaction with ethanolamine (1, 2, 208), dimethylamine and formaldehyde, or diethylaminoethyl chloride and methyl iodide have been used as a basis for flocculants or coagulation aids for water treatments (200).

Taking advantage of the many aromatic hydroxyl functions in conifer bark tannins, Hartmann (83) used ground bark as a polyol for reaction with isocyanates to prepare urethane foams with particularly good flammability resistance. Most uses for conifer bark tannins that involve reactions with the hydroxyl functions center on their complexation with cations. When sulfonated, condensed tannins can also be used as water-soluble heavy metal complexes. One of the more interesting of these applications is the development of water-soluble heavy-metal micronutrient complexes that have been used to correct iron deficiency in citrus plantations (96, 247). They can be used to prevent scale formation in boiler waters and to reduce corrosion in these systems (208). Their strong binding capacity for lead has permitted monitoring of lead pollution along roadways by analysis of the bark of nearby trees (166, 237). Copper complexes with condensed tannins are effective biocides that have been used in the preservation of fishing nets and cotton fabric (35, 68). Current work on the use of copper complexes of condensed tannin derivatives from southern pine (*Pinus taeda*) bark as biocides has demonstrated that these materials have considerable potential as wood preservatives (121).

Condensed tannins are excellent clay dispersants. Sulfite extracts from conifer tree barks are very effective in reducing the viscosity and increasing the gel strength of muds used in well drilling (72, 92, 145, 246, 248). Condensed tannins still face strong competition from lignosulphonates for this application, particularly because of the comparatively low thermal stability and salt tolerance of the tannin (96). However, reaction of tannins with chromium increases their thermal stability considerably to permit their use in muds for wells drilled as deep as 6000 feet (208). Sulfonated condensed tannin derivatives have found use as dispersants in other specialty applications, such as ceramic clays, pigments, carbon black, and pesticides (93, 96).

So far there have been few applications that directly center on taking advantage of the comparative ease of formation of reactive carbocations or quinone methides through cleavage of the interflavanoid bond in polymeric procyanidins or prodelphinidins (Chap. 7.6). Although the mechanism of the reaction was not defined, it seems likely that the strong adsorption capacity of conifer barks for methylmercaptan (the major odoriferous compound produced in the manufacture of paper pulp by the kraft pulping process) involves reaction with the condensed tannins to produce sulfide derivatives at C-4 of the flavanoid units (134). One specific example of the use of this approach is in the production of biocides by reaction of southern pine bark tannins with fatty thiols (122). Epicatechin fatty sulfides are very effective against gram-positive bacteria with minimum inhibitory concentrations ranging from 5 to 50 ppm. These compounds show much less toxicity to gram-negative bacteria. Their toxicity to wood-decay organisms is quite species specific. Among the white rotters, the epicatechin decylsulfide derivative was toxic to *Phanerochaete chrysosporium* and *Coriolus versicolor* with LD_{50} of 8 and 50 ppm, respectively. Among the brown rotters, it was quite effective against *Poria placenta* with an LD_{50} of 20 ppm; and among the soft rotters, it was effective against *Chaetomium globosum* and *Scytalidium lignicola* with LD_{50} of 10 and 2 ppm, respectively. A combination of this approach with complexing by copper (II) should broaden the spectrum of the toxicity of these compounds to microorganisms.

The higher-molecular-weight condensed tannins are well known to complex with proteins and to inhibit enzymes (Sect. 10.3.2). The growth of two wood-destroying fungi, interestingly *Gloeophyllum trabeum* (84) [which was fairly resistant (LD_{50} 100 ppm) to epicatechin decylsulfide] and *Aspergillis parasiticus* (123), is inhibited by the presence of condensed tannins. It might also be expected that use of partial cleavage products with fatty thiols, retaining a portion of the polymeric procyanidin species, will also broaden the spectrum of toxicity to wood-destroying fungi because of complexation of the higher-molecular-weight oligomers with extracellular enzymes on the basis of tannin protein interaction. The effect of condensed tannins on animals was reviewed in Sect. 10.3.1, but it should be mentioned here that bark extract solutions that contain high proportions of condensed tannins from a number of tree species have been shown to deter termites feeding on cellulose (84).

Flavans, such as catechin (32, 259) and higher-molecular-weight procyanidins (162, 259), are also known to have moderate phytotoxic properties. This, together with the pleasing appearance of bark particles, explains their use as ground cover in gardens. Allan and coworkers (2) have carried this application further by reacting bark with herbicides to produce slow-release pesticides. When applied to Douglas-fir (*Pseudotsuga menziesii*) plantations, growth of the trees was remarkably increased because of reduced competition. Significant amounts of bark are used as soil mulch (3). The bark is generally composted to reduce the phytotoxicity of the flavanoids (22) and it is also often reacted with ammonia or other nitrogen-containing compounds to ensure that there is no significant drain on nitrogen availability (34). With increasing worldwide interest in the role of acid deposition on the growth of plants, the interrelationships among condensed tannins in litter, inorganic nutrient availability, and soil microflora seem worthy of further study, particularly as these phenomena relate to pH.

In summary, the current low cost of phenol has forced a significant reduction in the worldwide research effort on utilization of condensed tannins in adhesives. Those research programs that are being continued are not focused on use of these polymers for replacement of phenol in applications such as particleboard or plywood adhesives. Rather, efforts are now being directed to much higher-value uses, such as resorcinol replacements in a variety of products, biocides, and pharmaceuticals. While this broadening of endeavor is certainly a healthy change, it is also important that research on the chemistry and utilization of these abundant renewable phenolic polymers be maintained in preparation for the certain eventual return of high costs and scarcity of basic petroleum-derived phenolics.

Acknowledgements. RWH is indebted to Ginger Rutherford who assisted in gathering reference materials, Julia Wilson, Nancy Greene, and Sue Moore who typed a number of drafts, and Drs. L. Y. Foo, G. W. McGraw, J. J. Karchesy, and P. E. Laks who made helpful suggestions.

References

1. Allan G G 1969 Reaction products of lignin and bark extracts and process for same. U S Pat No 3470148
2. Allan G G, Beer J W, Cousin M J, Powell J C 1978 Wood chemistry and forest biology, partners in more effective reforestation. Tappi 61(1):33–35
3. Allison R C 1980 Organic and fuel uses for bark and wood residues. For Prod Res Soc Proc P-80-27 Madison WI, 45 p
4. Anderson A B 1976 Bark extracts as bonding agents of particleboards. In: Goldstein I S (ed) Wood technology, chemical aspects. Am Chem Soc Washington DC, 235–242
5. Anderson A B, Breuer R J, Nicholls G A 1961 Bonding particleboards with bark extracts. For Prod J 11:226–227
6. Anderson A B, Wong A, Wu K-t 1974 Utilization of white fir bark and its extract in particleboard. For Prod J 24(7):40–45

7 Anderson A B, Wong A, Wu K-t 1975 Douglas-fir and western hemlock bark extracts as bonding agents for particleboards. For Prod J 25(3):45–48
8 Anderson A B, Wu K-t, Wong A 1974 Utilization of ponderosa pine bark and its extract in particleboard. For Prod J 24(8):48–53
9 Angelinetti A R, Cantera C S, Sofia A, Martinez J 1978 Rapid tanning of shoe leather using a quebracho extract. Rev Tech Ind Cuir 70(8):237–240
10 Arora H C, Chattopadhyda S N 1980 Anaerobic contact filter process: a suitable method for the treatment of vegetable tanning effluents. Water Pollut Control (Maidstone Engl) 79:501–506
11 Asquith T N, Butler L G 1986 Interactions of condensed tannins with selected proteins. Phytochemistry 25:1591–1593
12 Atkinson J H, Scowcroft F 1974 Fast and economical vegetable tanning method applicable to the present. Rev Tech Ind Cuir 66(5):161–164
13 Atkinson J H 1978 Modern methods for producing vegetable tanned leather. In: Radnotti L (ed) Proc 6th Cong Leather Ind OMIKK-Technoinform, Budapest 1:1–13
14 Ayla C 1984 *Pinus brutia* tannin adhesives. J Appl Polym Sci Appl Polym Symp 40:69–78
15 Ayla C, Weissmann G 1984 New developments in the use of tannin formaldehyde resins for the manufacture of particleboard. Plant Res Devel 19:107–121
16 Ayla C, Weissmann G 1982 Gluing tests with tannin formaldehyde resins from bark extracts of *Pinus brutia*. Holz Roh-Werkst 40:13–18
17 Ayoub S M H 1985 Flavanol molluscicides from the Sudan acacias. Int J Crude Drug Res 23:87–90
18 Ayoub S M H, Yankov L K 1985 Algicidal properties of tannins. Fitoterapia 56:227–229
19 Bate-Smith E C, Swain T 1953 Identification of leucoanthocyanins as "tannins" in foods. Chem Ind 377–378
20 Baxter G F, Kreibich R E 1973 A fast-curing phenolic adhesive system. For Prod J 23(1):17–22
21 Beart J E, Lilley T H, Haslam E 1985 Polyphenol interactions. Part 2. Covalent binding of procyanidins to proteins during acid-catalyzed decomposition; observations on some polymeric proanthocyanidins. J Chem Soc Perkin Trans II 1439–1443
22 Bennett P G, Jones D L, Nichols D G 1978 Utilization of *Pinus radiata* bark. Appita 31:275–279
23 Bernays E A, Woodhead S 1982 Plant phenols utilized as nutrients by a phytophagous insect. Science 216:201
24 Booth H E, Herzberg W J, Humphreys F R 1958 *Pinus radiata* (Don) bark tannin. Aust J Sci 21:19–20
25 Borisova V N, Dvoinos L M 1972 Effect of tannin on peroxidase activity of fungi. Mikol Fitopatol 6:152–154
26 Brandon M J, Foo L Y, Porter L J, Meredith P 1982 Proanthocyanidins of barley and sorghum: composition as a function of maturity of barley ears. Phytochemistry 21:2953–2957
27 Brandts F G, Lichtenberger J A 1965 Resins from bark. Canadian Pat No 708936
28 Briggs T M, Hauck R A, Eye J D 1973 Tannin removal and recovery from vegetable tanning waste liquor. J Am Leather Chem Assoc 68:176–188
29 Bryant L H, Humphreys F R, Martin P J 1953 Improvements in wood products and methods of making the same. Aust Pat No 150592
30 Bryant L H, Humphreys F R 1958 Building boards from sawmill waste. Composite Wood 5:41–46
31 Bryant L H, Humphreys F R, Martin P J 1950 Structural and insulating boards from sawmill waste. Nature 165:694
32 Buta J G, Lusby W R 1986 Catechins as germination and growth inhibitors in *Lespedeza* seeds. Phytochemistry 25:93–95
33 Cameron F A, Pizzi A 1985 Tannin-induced formaldehyde release depression in UF particleboard. Nat Timber Res Inst CSIR Report 247:1–5
34 Casebier R L, Sears K D 1976 Aminated sulfites of polyphenolic bark. U S Pat No 3966706
35 Cecily P J, Kunjappan M K 1973 Preservation of cotton fish net twines by tanning. II. Fixation of tannin. Fish Tech 10:24–32
36 Chaing K K 1973 Engelmann spruce and lodgepole pine bark as adhesives in particleboard. Ph D Thesis. Colorado State Univ Fort Collins CO
37 Chan B G, Waiss A C, Lukefar M 1978 Condensed tannins, an antibiotic chemical from *Gossypium hirsutum*. J Insect Physiol 24:113–118

38 Chemical and Engineering News 1985 Protection of critical metals supply probed. Chem Eng News 63(2):29
39 Chemical and Engineering News 1985 Key chemicals: phenol. Chem Eng News 63(44):12
40 Chen C M 1982 Copolymer resins of bark and agricultural residue extracts with phenol and formaldehyde: 20 percent weight of phenol replacement. For Prod J 32(2):35–40
41 Chen C M 1980 Organic phenol extract compositions of peanut hull agricultural residues and method. U S Pat No 4 200 723
42 Chen C M 1980 Phenol-aldehyde resin composition containing pecan pith extract and an aldehyde. U S Pat No 4 201 699
43 Chen C M 1980 Phenol-aldehyde resin composition containing peanut hull extract and an aldehyde. U S Pat No 4 201 700
44 Chen C M 1980 Organic phenol extract compositions of pecan pith agricultural residues and method. U S Pat No 4 201 851
45 Chen C M 1981 Gluability of copolymer resins having higher replacement of phenol by agricultural residue extracts. I EC Prod Res Dev 20:704–708
46 Chen C M 1982 Bonding particle boards with fast curing phenolic agricultural residue extract copolymer resins. Holzforschung 36:109–116
47 Collier P D, Bryce T, Mallows R, Thomas P E, Frost D J, Korver O, Wilkins C K 1973 The theaflavins of black tea. Tetrahedron 29:129–142
48 Coxon D T, Holmes A, Ollis W D, Vora V C, Grant M S, Tee J L 1972 Flavanol digallates in the green tea leaf. Tetrahedron 28:2819–2826
49 Czochanska Z, Foo L Y, Porter L J 1979 Compositional changes in lower molecular weight flavans during grape maturation. Phytochemistry 18:1819–1822
50 Dalton L K 1953 Resins from sulfited tannins as adhesives for wood. Aust J Sci 4:136–145
51 Dalton L K 1950 Tannin-formaldehyde resins as adhesives for wood. Aust J Sci 1:54–70
52 Deb J C, Prasad L M, Hore G N, De J M 1973 Manufacture of rapid vegetable tanned sheep and goat skin for leather goods. Leather Sci (Madras) 20:284–287
53 Delcour J, Schoeters M M, Meysman E W, Dondeyne P, Moerman E 1984 The intrinsic influence of catechins and procyanidins on beer haze formation. J Inst Brew 90:381–384
54 Deschamps A M, Otuk G, Lebeault J 1983 Production of tannase and degradation of chestnut tannin by bacteria. J Ferment Tech 61:55–59
55 Dix B, Marutzky R 1984 Gluing of particleboards with tannin formaldehyde resins from bark extract of *Pinus radiata*. Holz Roh-Werkst 42:209–217
56 Ellis C J, Foo L Y, Porter L J 1983 Enantiomerism: a characteristic of the proanthocyanidin chemistry of the Monocotyledonae. Phytochemistry 22:483–487
57 Endres H 1961 Die Gerbwirkung niedermolekularer Polyhydroxyphenole. Ds Leder 12:294–297
58 Exner R, Kupka L 1982 Rapid tannage technology for vegetable tanned leather. In: Radnotti L (ed) Proc 7th Cong Leather Ind OMIKK-Technoinform, Budapest I(a):119–130
59 Eye J D 1958 The treatment and disposal of tannery wastes. In: O'Flaherty F, Roddy W T, Lollar R M (eds) The chemistry and technology of leather, vol 3. Reinhold New York, 461–485
60 Eye J D, Ficker F C 1982 A pilot plant study of the effects of quebracho and wattle on anaerobic digestion. J Am Leather Chem Assoc 77:137–148
61 Eye J D, Liu L 1971 Treatment of wastes from a sole leather tannery. J Water Pollut Control Fed 43(11):2291–2303
62 Ezaki N, Kato M, Takizawa N, Morimoto S, Nonaka G, Nishioka I 1985 Pharmacological studies on *Linderae umbellatae* Ramus. IV. Effects of condensed tannin related compounds on peptic activity and stress-induced gastric lesions in mice. Planta Med 34–38
63 Foo L Y, Jones W T, Porter L J, Williams V M 1982 Proanthocyanidin polymers of fodder legumes. Phytochemistry 21:933–935
64 Foo L Y, Porter L J 1981 The structure of tannins of some edible fruits. J Sci Food Agric 32:711–716
65 Foo L Y, McGraw G W, Hemingway R W 1983 Condensed tannins: preferential substitution at the interflavanoid bond by sulfite ion. J Chem Soc Chem Commun 12:672–673
66 Fox L R, Macauley B J 1977 Insect grazing on eucalyptus in response to leaf tannins and nitrogen. Oecologia 29:145–162
67 Fukami M, Hattori Z, Okonoyi T 1978 Detoxifying effects of persimmon fruit tannin on snake venom and bacterial toxins. Sankyo Kenkyusho Nempo 30:104–111

68 Furry M S, Humfeld H 1941 Mildew resistant treatment on fabrics. Ind Eng Chem 33:538–545
69 Ganesan A, Rao M V 1977 Vegetable tanning of Pujab wool sheep skins. Leather Sci (Madras) 24:144–146
70 Gazave J M, Parrot J L, Roger C, Achard M 1975 Investigations on vitamin C2 (sparing factor of ascorbic acid). A study of its metabolism. In: Farkas L, Gabor M, Kallay F (eds) Topics in flavonoid chemistry and biochemistry. Elsevier Amsterdam, 197–200
71 Gosh D, Vijayalakshimi K, Nayudamma Y 1967 Studies on the effect of nontans in vegetable tanning. Leather Sci (Madras) 14:154–160
72 Gray K R, Van Blaircom L E 1961 Bark treatment process and products. U S Pat No 2999108
73 Guger R, Magnolato D, Self R 1986 Glucosylated flavanoids and other phenolic compounds from sorghum. Phytochemistry 25:1431–1436
74 Gupta R K, Haslam E 1978 Plant proanthocyanidins. Part 5. Sorghum polyphenols. J Chem Soc Perkin Trans I:892–896
75 Gustavson K H 1966 The function of the basic groups of collagen in its reaction with vegetable tannins. J Soc Leather Trades Chem 50:144–160
76 Gustavson K H 1966 Differences in the reaction of condensed and hydrolysable tannins with collagen. J Soc Leather Trades Chem 50:445–454
77 Hagen J R 1972 Investigation into the removal of color from biologically treated vegetable tannery wastes. MS Thesis, Univ Cincinnati Cincinnati OH
78 Hall J A 1971 Utilization of Douglas-fir bark. U S For Serv Misc Rep PNW, 138 p
79 Hall R B, Leonard J H, Nicholls G A 1960 Bonding particleboards with bark extracts. For Prod J 10:263–272
80 Hamada R, Ikeda S, Satake Y 1969 Utilization of wood bark of plywood adhesives. I. Effect of addition of bark of *Acacia mollissima* on the properties of phenol formaldehyde resin. Mokuzai Gakkaishi 15:165–170
81 Hamed G R, Chung K H, Hemingway R W (1989) Condensed tannins as substitutes for resorcinol in bonding polyester and nylon cord to rubber. In: Hemingway R, Conner A H (eds) Adhesives from renewable resources. ACS Symp Ser Washington DC 242–253
82 Hartman S 1977 Alkali treated bark extended tannin aldehyde resinous adhesives. U S Pat No 4045386
83 Hartman S 1977 Polyurethane foams from the reaction of bark and diisocyanate. In: Goldstein I S (ed) Wood technology, chemical aspects. Am Chem Soc Washington DC, 257–269
84 Harun J, Labosky P 1985 Antitermitic and antifungal properties of selected bark extractives. Wood Fiber Sci 17:327–335
85 He X-q, Jiang W-q, Li J-z, Lan Z-c 1983 The properties of several commercial vegetable tannin extracts in vegetable-aluminum combination tannage. Chem Ind For Prod 13(2):1–13
86 Heidemann E, Srinivasan S R 1976 Vegetable tanning. IV. Deuterium exchange between collagen and synthetic polypeptides treated with phenolic compounds. Ds Leder 18:145–151
87 Hemingway R W 1981 Bark: its chemistry and prospects for chemical utilization. In: Goldstein I S (ed) Organic chemicals from biomass. CRC Press Boca Raton FL, 190–248
88 Hemingway R W 1978 Adhesives from southern pine bark. A review of past and current approaches to resin formulation problems. In: Proc Complete Tree Util Southern Pines. For Prod Res Soc Madison WI, 443–457
89 Hemingway R W, Kreibich R E 1984 Condensed tannin-resorcinol adducts and their use in wood-laminating adhesives: an exploratory study. J Appl Polym Sci Polym Symp 40:79–90
90 Hemingway R W, McGraw G W 1983 Condensed tannins: kinetics of acid-catalyzed cleavage of procyanidins. J Wood Chem Technol 3:421–435
91 Hemingway R W, McGraw G W 1977 Southern pine bark polyflavonoids: structure, reactivity, use in wood adhesives. In: Proc 1977 TAPPI Forest Biol/Wood Chem Symp. TAPPI Atlanta, 261–269
92 Hergert H L, Van Blaircom L E, Steinberg J C, Gray K R 1968 Isolation and properties of dispersants from western hemlock bark. For Prod J 15:485–491
93 Herrick F W 1980 Chemistry and utilization of western hemlock bark extractives. J Agric Food Chem 28:228–237
94 Herrick F W, Bock L H 1958 Thermosetting exterior plywood type adhesives from bark extracts. For Prod J 8:269–274
95 Herrick F W, Conca R J 1960 The use of bark extracts in cold setting waterproof adhesives. For Prod J 10:361–368

96 Herrick F W, Hergert H L 1977 Utilization of chemicals from wood: retrospect and prospect. In: Loewus F F, Runeckles V C (eds) Recent advances in phytochemistry, vol II. Structure, biosynthesis, and degradation of wood. Plenum Press New York, 443–515
97 Heritage C C 1951 Phenolic adhesive and method of bonding wood plies. US Pat No 2574284
98 Heritage C C 1951 Ingredient of adhesives and a process of making it. US Pat No 2574785
99 Herzberg W J 1960 *Pinus radiata* tannin-formaldehyde resin as an adhesive for plywood. Aust J Appl Sci 11:462–472
100 Highberger J H 1956 The chemical structure and macromolecular organization of the skin proteins. In: O'Flaherty F, Roddy W T, Lollar R M (eds) The chemistry and technology of leather, vol 1. Reinhold New York, 65–193
101 Hikino H, Shimoyama N, Kasahara Y, Takahashi M, Konno C 1982 Structures of mahuannin A and B, hypotensive principles of *Ephedra* roots. Heterocycles 19:1381–1384
102 Hill G K 1982 Studies of the enzymic activity of *Phomopsis viticola* Sace, the cause of black spot disease. Wein-Wiss 37:333–339
103 Hillis W E, Urbach G 1959 Reaction of polyphenols with formaldehyde. J Appl Chem 9:665–673
104 Hillis W E, Yazaki Y 1985 The potential of pine bark for adhesives. In: Proc IUFRO-NTRI Symp For Prod Int Achievements and the Future, vol 6. IUFRO Rome, 1–13
105 Huth G, Schloesser E 1982 Role of extracellular microbial hydrolases in pathogenesis. Meded Fac Landbouwwet Rijksunn Gent 47:855–861
106 Inokuchi J, Okabe H, Yamauchi T, Nagamatsu A, Nonaka G, Nishioka I 1985 Inhibitors of angiotensive-coverting enzyme in crude drugs. II. Chem Pharm Bull 33:264–269
107 Kakegawa H, Matsumoto H, Endo K, Satoh T, Nonaka G, Nishioka I 1985 Inhibitory effects of tannins of hyaluronidase activation and on the degranulation from rat mesentery mast cells. Chem Pharm Bull 33:5079–5082
108 Kakiuchi N, Hattori M, Namba T, Nisizawa M, Yamagishi T, Okuda T 1985 Inhibitory effect of tannins on reverse transcriptase from RNA tumor virus. J Nat Prod 48:614–621
109 Kakiuchi N, Hattori M, Nishizawa M, Yamagashi T, Okuda T, Namba T 1986 Studies on dental caries prevention by traditional medicines. VIII. Inhibitory effect of various tannins on glucan synthesis by glucosyltransferase from *Streptococcus mutans*. Chem Pharm Bull 34:720–725
110 Kallenberger W E, Hernandez J F 1983 Preliminary experiments in the tanning action of vegetable tannins combined with metal complexes. J Am Leather Chem Assoc 78:217–222
111 Karchesy J J, Hemingway R W 1986 Condensed tannins: (4β→8;2β→0→7)-linked procyanidins in *Arachis hypogea* L. J Agric Food Chem 34:966–970
112 Kay A N 1958 The process of vegetable tannage. In: O'Flaherty F, Roddy W T, Lollar R M (eds) The chemistry and technology of leather, vol 2. Reinhold New York, 161–200
113 Kedlaya K J, Ramanathan H, Nayudamma V 1971 Mechanism of vegetable tannage. J Am Leather Chem Assoc 66:256–262
114 Knowles E, White T 1954 Tannin extracts as raw materials for the adhesives and resin industries. Adhes Resins 10:226–228
115 Kosuge T, Ishida H 1985 Studies on active substances in herbs used for Oketsu ("Stagnant blood") in Chinese medicine. IV. On the anticoagulative principle in *Rhei rhizoma*. Chem Pharm Bull 33:1503–1506
116 Kreibich R E 1974 High-speed adhesives for the wood gluing industry. Adhes Age 17:26–33
117 Kreibich R E, Hemingway R W 1985 Condensed tannins: resorcinol adducts in laminating adhesives. For Prod J 35(3):23–25
118 Kreibich R E, Hemingway R W 1987 Condensed tannin sulfonates in cold-setting wood-laminating adhesives. For Prod J 37(2):43–46
119 Kreibich R E, Hemingway R W (1989) Tannin-based, wood finger-jointing adhesives. In: Hemingway R W, Conner A H (eds) Adhesives from renewable resources. ACS Symp Ser Washington DC 203–216
120 Kreibich R E, Hemingway R W 1985 The use of tannin in structural laminating adhesives. In: Proc IUFRO-NTRI Symp For Prod Res International Achievements and the Future, vol 7. IUFRO Rome, 1–12
121 Laks P E, Hemingway R W (1989) Flavonoid biocides: Wood preservatives based on condensed tannins. Holzforschung 42(5):299–306
122 Laks P E 1987 Flavonoid biocides: phytoalexin analogues from condensed tannins. Phytochemistry 26:1617–1622

123 Lansden J A 1982 Aflatoxin inhibition and fungistasis by peanut tannins. Peanut Sci 9:17–20
124 Lea A G H 1978 The phenolics of ciders: oligomeric and polymeric procyanidins. J Sci Food Agric 29:471–477
125 Lea A G H 1978 The phenolics of ciders: bitterness and astringency. J Sci Food Agric 29:478–483
126 Lea A G H, Bridle P, Timberlake C F, Singleton V L 1979 The procyanidins of white grapes and wines. Am J Enol Vitic 30:289–300
127 Liiri O, Sairanen H, Kilpelainen H, Kivisto A 1982 Bark extractives from spruce as constituents of plywood bonding agents. Holz Roh-Werkst 40:51–60
128 Lodhi M A K, Killingbeck K T 1980 Allelopathic inhibition of nitrification and nitrifying bacteria in a ponderosa pine (*Pinus ponderosa* Dougl.) community. Am J Bot 67:1423–1429
129 Lollar R M 1958 The mechanism of vegetable tannage. In: O'Flaherty F, Roddy W, Lollar R M (eds) The chemistry and technology of leather, vol 2. Reinhold New York, 201–220
130 Luvisi F P, Filachione E M 1968 The effect of glutaraldehyde on vegetable tanning. J Am Leather Chem Assoc 63:584–600
131 MacLean H, Gardner J A F 1952 Bark extracts in adhesives. Pulp Pap Mag Can 53:111–114
132 Mangan J L, Vetter R L, Jordan D J, Wright P C 1976 The effect of the condensed tannins of sainfoin (*Onobrychis viciaefolia*) on the release of soluble leaf protein into the food bolus of cattle. Proc Nutr Soc 35:95A–97A
133 Marston A, Hostettmann K 1985 Plant molluscicides. Phytochemistry 24:639–652
134 Martin R E, Crawford J L 1973 Sorption of sulfate mill odors by bark. In: Mater J (ed) Techniques for processing bark and utilization of bark products. For Prod Res Soc Madison WI, 42–47
135 Marutzky R, Dix B 1982 Kationenaustauscher auf der Basis von Tannin-formaldehyde Harzen. Holz Roh-Werkst 40:433–436
136 Marutzky R, Dix B 1984 Formaldehyde release of tannin-bonded particleboards. J Appl Polym Sci Polym Symp 40:49–57
137 Marwan A G, Nagel C W 1986 Microbial inhibitors of cranberries. J Food Sci 51:1009–1013
138 Matsuo T, Ito S 1978 The chemical structure of kaki-tannin from immature fruit of the persimmon (*Diospyros kaki* L.). Agric Biol Chem 42:1637–1643
139 McGraw G W, Laks P E, Hemingway R W (1988) Condensed tannins: Desulfonation of hydroxybenzyl sulfonic acids related to proanthocyanidin derivatives. J Wood Chem Technol 8(1):91–109
140 McGraw G W, Hemingway R W 1982 Electrophilic aromatic substitution of catechins: bromination and benzylation. J Chem Soc Perkin Trans I:973–978
141 McGraw G W, Ohara S, Hemingway R W (1989) In: Hemingway R W, Conner A H (eds) Adhesives from renewable resources. ACS Symp Ser Washington DC 185–202
142 McMurrough I, Hemigan G P, Cleary K 1985 Interactions of proteases and polyphenols in worts, beers and model systems. J Inst Brew 91:93–100
143 McMurrough I, Loughrey M J, Hennigan G P 1983 Content of (+)-catechin and proanthocyanidins in barley and malt grain. J Sci Food Agric 34:62–71
144 McKenzie A E, Yuritta J P 1972 Starch tannin corrugating adhesives. Appita 26:30–34
145 Miller R W, VanBeckum G W 1960 Bark and fiber products for oil well drilling. For Prod J 10:193–195
146 Mohanaradhakrishanan V, Ramanathan N 1967 Effect of certain constituents of vegetable tanning liquors on the structure of collagen. Leather Sci 14:285–294
147 Mole S, Waterman P G 1985 Stimulatory effects of tannins and cholic acid on tryptic hydrolysis of proteins: ecological implications. J Chem Ecol 11:1325–1331
148 Morimoto S, Nonaka G, Nishioka I 1986 Tannins and related compounds. 37. Isolation and characterization of flavan-3-ol glucosides and procyanidin oligomers from cassia bark (*Cinnamomum cassia* Blume). Chem Pharm Bull 34:633–642
149 Morton J F 1986 The potential carcinogenicity of herbal tea. Environ Carcino Revs C4:203–223
150 Mota M L R, Thomas G, Barbosa Filho J M 1985 Anti-inflammatory actions of tannins isolated from the bark of *Anacardium occidentale* L. J Ethnopharm 13:289–300
151 Muthullingam P, Daraiswany G, Muralimanohar Y, Selvarangan R 1983 Alternative technology for goat upper leather. Leather Sci (Madras) 30(5):155–157
152 Newman R H, Porter L J, Foo L Y, Johns S R, Willing R I 1987 High resolution ^{13}CNMR studies of proanthocyanidin polymers (condensed tannins). Mag Res Chem 25:118–124

153 Nicole A, Fasel-Felley J, Perrisond D, Frei P C 1985 Influence of palmitoyl-3-catechin and heptyl-3-catechin on leukocyte migration inhibition test carried out in the presence of PPD and hepatitis B surface antigen (HBsAg). Int J Immunopharmacol 7:87–92
154 Nishioka I 1983 Chemistry and biological activities of tannins. Yakugaku Zasshi 103:125–141
155 Nonaka G, Hsu F, Nishioka I 1981 Structures of dimeric, trimeric, and tetrameric procyanidins from *Areca catechui* L. J Chem Soc Chem Commun 781–783
156 Nonaka G, Hashimoto F, Nishioka I 1986 Tannins and related compounds. 36. Isolation and structures of the aflagallins, new red pigments from black tea. Chem Pharm Bull 34:61–65
157 Nonaka G, Kawahara O, Nishioka I 1983 Tannins and related compounds. XV. A new class of dimeric flavan-3-ol gallates, theasinensins A and B, and proanthocyanidin gallates from green tea leaf. Chem Pharm Bull 31:3906–3914
158 Nonaka G, Morimota S, Nishioka I 1983 Tannins and related compounds. 13. Isolation and structures of trimeric, tetrameric, and pentameric proanthocyanidins from cinnamon. J Chem Soc Perkin Trans I:2139–2145
159 Nozilillo C, de Bezada M 1984 Browning of lentil seeds, concomitant loss of viability, and the possible role of soluble tannins in both phenomena. Can J Plant Sci 64:815–824
160 O'Flaherty F, Roddy W T, Lolla R M 1978 The chemistry and technology of leather, vol 1–4. Reinhold New York
161 Oh H I, Hoff J E, Armstrong G S, Hoff L A 1980 Hydrophobic interaction in tannin-protein complexes. J Agric Food Chem 28:394–398
162 Ohigashi H, Minami S, Fukui H, Koshimizu K, Mizutani F, Sugiura A, Tomana T 1982 Flavanols as plant growth inhibitors from the roots of peach *Prunus persica* Batch cv. 'Hakuto'. Agric Biol Chem 46:2555–2561
163 Okamoto A, Ozawa T, Imagawa H, Arai Y 1985 Polyphenolic compounds related to the browning of sago starch. Nippon Nogei Kagaku Kaishi 59:1257–1261
164 Olivannan M S, Ramaswamy T 1971 Waterproofing of sole leather with basic aluminum sulphate dipping. Effect of pretanning with various vegetable tanning materials. Leather Sci (Madras) 18:186–195
165 Orten J M, Neuhaus O W 1982 Connective, nerve muscle and other tissues. In: Human biochemistry. C V Mosby St Louis, 398–412
166 Osibanjo O, Ajayi S O 1980 Trace metal levels in tree barks as indicators of atmospheric pollution. Environ Int 4:239–244
167 Otsuka H, Fujioka S, Komiya T, Mizuta E, Takamoto M 1982 Studies on anti-inflammatory agents. VI. Anti-inflammatory constituents of *Cinnamomum sieboldii* Meissn. Yakugaku Zasshi 102:162–172
168 Outtrup H, Schaumberg K 1981 Structure elucidation of some proanthocyanidins in barley by 1H 270 MHz NMR spectroscopy. Carlsberg Res Commun 46:43–52
169 Parker C E 1970 Anaerobic-aerobic lagoon treatment for vegetable tanning wastes. Water Pollut Control Research US-EPA Rept WPD-199-01-67. Environ Protect Agency Washington DC, 85 p
170 Parrish J R 1958 Particleboard from wattle wood and wattle tannins. J S Afr For Assoc 32:26–31
171 Pauley R D 1956 Bark components as resin ingredients. US Pat No 2773847
172 Piretti M M, Pistore R, Razzoboni C 1958 On the chemical constitution of kaki tannin. Ann Chim 75:137–144
173 Pirich E P 1984 Experiences with vegetable tanning systems and the extract blends used. J Am Leather Chem Assoc 68:209–219
174 Pizzi A 1977 Hot-setting tannin-urea-formaldehyde exterior wood adhesives. Adhes Age 20(12):27–29
175 Pizzi A 1978 Wattle-base adhesives for exterior grade particleboards. For Prod J 28(12):42–47
176 Pizzi A 1978 Wattle tannin adhesives for radio-frequency curing. J Appl Polym Sci 22:3603–3606
177 Pizzi A 1979 The chemistry and development of tannin/urea-formaldehyde condensates for exterior wood adhesives. J Appl Polym Sci 23:2777–2792
178 Pizzi A 1979 "Hybrid" interior particleboard using wattle tannin adhesives. Holzforsch Holzverwert 31:86–87
179 Pizzi A 1979 Phenolic and tannin-based adhesive resins by reaction of coordinated metal ligands. II. Tannin adhesive preparation, characteristics and application. J Appl Polym Sci 24:1257–1268
180 Pizzi A 1979 Sulphited tannins for exterior wood adhesives. Coll Polym Sci 257:37–40

181 Pizzi A 1979 Tannin based polyurethane adhesives for aluminum-aluminum jointing. J Appl Polym Sci 23:1889–1891
182 Pizzi A 1980 Tannin-based adhesives. J Macromol Sci Rev Macromol Chem C 18(2):247–315
183 Pizzi A 1981 Mechanism of viscosity variations during treatment of wattle tannins with hot NaOH. Int Adhesion Adhesives 1:213–214
184 Pizzi A 1981 A universal formulation for tannin adhesives for exterior particleboard. J Macromol Sci Chem Ed A16(7):1243–1250
185 Pizzi A 1982 Pine tannin adhesives for particleboard. Holz Roh-Werkst 40:293–301
186 Pizzi A 1983 Tannin-based wood adhesives. In: Pizzi A (ed) Wood adhesives: chemistry and technology. Marcel Dekker New York, 178–246
187 Pizzi A, Cameron F A 1981 A new hardener for tannin adhesives for exterior particleboard. Holz Roh-Werkst 39:255–259
188 Pizzi A, Cameron F A 1981 Decrease of pressing temperature and adhesive content by metallic ion catalysis in tannin-bonded particleboard. Holz Roh-Werkst 39:463–467
189 Pizzi A, Cameron F A 1984 Fast-set adhesives for glulam. For Prod J 34(9):61–68
190 Pizzi A, Daling G M E 1980 Laminating wood adhesives by generation of resorcinol for tannin extracts. J Appl Polym Sci 25:1039–1048
191 Pizzi A, Merlin M 1981 A new class of tannin adhesives for exterior particleboard. Int Adhesion Adhesives 1:261–264
192 Pizzi A, Rossouw D duT, Knuffel W E, Singmin M 1980 "Honeymoon" phenolic and tannin-based fast-setting adhesive systems for exterior grade fingerjoints. Holzforsch Holzverwert 32:140–150
193 Pizzi A, Roux D G 1978 The chemistry and development of tannin-based weather and boil-proof cold-setting and fast-setting adhesives for wood. J Appl Polym Sci 22:1945–1954
194 Pizzi A, Scharfetter H O 1978 The chemistry and development of tannin-based adhesives for exterior plywood. J Appl Polym Sci 22:1745–1761
195 Pizzi A, Scharfetter H, Kes E W 1981 Adhesives and techniques open new possibilities for the wood processing industry. Holz Roh-Werkst 39:85–89
196 Plomley K F, Stashevski A M 1969 Waterproof particleboard. CSIRO For Prod Div NewsLetter 363. Melbourne Australia
197 Pospisil J 1982 The response of *Leptinotarsa decemlineata* (Coleoptera) to tannin as antifeedant. Acta Entomol Bohemoslov 79:429–434
198 Price M L, Butler L G 1980 Tannins and nutrition. Purdue Univ Exp Sta Bull 272 W Lafayette IN, 36 pp
199 Price M L, Butler L G, Rogler J C, Featherston W R 1979 Overcoming the nutritionally harmful effects of tannin in sorghum grain by treatment with inexpensive chemicals. Agric Food Chem 27:441–445
200 Pulkkinen E, Peltonen S 1978 Cationic flocculant from a phenolic acid fraction of conifer tree bark. Tappi 61(5):97–100
201 Rageeb H A 1978 Towards pragmatism in tannin systems. Tanner 32:467–469
202 Randall J M 1977 Variations in effectiveness of barks as scavengers for heavy metal ions. For Prod J 27(11):51–56
203 Rao C K 1971 Use of indigenous vegetable tannin extracts in leather industry. Tanner 25:589–591
204 Rao M S, Kamat D H, Selvarangan R 1980 Upper leather from raw sheep skins (red hair). Leather Sci (Madras) 27(8):263–265
205 Reddy N R, Pierson M D, Sathe S K, Salunkhe D K 1985 Dry bean tannins: a review of nutritional implications. J Am Oil Chem Soc 62:541–549
206 Reed J D, McDowell R E, Van Soest P J, Horvath P J 1982 Condensed tannins. A factor limiting the use of cassava forage. J Sci Food Agric 33:213–220
207 Rewerski W, Piechocki T, Rylski M, Lewak S 1971 Pharmacological properties of oligomeric procyanidin ex *Crataegus oxyacantha* (hawthorn). Arzneim-Forsch 21:886
208 Roux D G, Ferreira D, Hundt H K L, Malan E 1975 Structure, stereochemistry and reactivity of natural condensed tannins as basis for their extended industrial application. J Appl Polym Sci Symp 28:335–353
209 Russell A E, Shuttleworth S G, Williams-Wynn D A 1967 Further studies on the mechanism of vegetable tannage. Part II. Effect of urea extraction on hydro-thermal stability of leather tanned with a range of organic tanning agents. J Soc Leather Trades Chem 51(6):222–230

210 Russell A E, Shuttleworth S G, Williams-Wynn D A 1967 Further studies on the mechanism of vegetable tannage. Part III. Solvent reversibility of wattle tannage. J Soc Leather Trades Chem 51(10):349–361

211 Russell A E, Shuttleworth S G, Williams-Wynn D A 1968 Further studies on the mechanism of vegetable tannage. Part IV. Residual affinity phenomena on solvent extracts of collagen tanned with vegetable extracts and syntans. J Soc Leather Trades Chem 52(6):220–238

212 Russell A E, Shuttleworth S G, Williams-Wynn D A 1968 Further studies on the mechanism of vegetable tannage. Part V. Chromatography of vegetable tannins on collagen and cellulose. J Soc Leather Trades Chem 52:459–486

213 Ryder A G D, Barnes J L 1985 Wattle tannin bonded composite panel products – current developments and future trends. In: Proc IUFRO-NTRI Symp For Prod Int Achievements and the Future, vol 6. IUFRO Rome, 1015

214 Saayman H M 1974 Wood varnishes from wattle bark extract. J Oil Colour Chem Assoc 57:114

215 Saayman H M 1967 Manufacture of experimental particleboards using black wattle tannin adhesives. Leather Ind Res Inst Bull 446:1–8

216 Saayman H M 1968 Black wattle tannin adhesives. Leather Ind Res Inst Bull 489:1–7

217 Saayman H M, Brown N N 1977 Wattle base-starch adhesives for corrugated containers. For Prod J 27(4):21–25

218 Saayman H M, Oatley J A 1976 Wood adhesives from wattle bark extract. For Prod J 26(12):27–33

219 Santappa M, Rao V S S 1982 Vegetable tannins – a review. J Sci Ind Res 41:705–718

220 Schmidt O, Ayla C, Weissmann G 1984 Microbiological treatment of hot water extracts from Norway spruce bark for production of adhesives. Holz Roh-Werkst 42:287–292

221 Schulz J M, Herrmann K 1980 Occurrence of catechins and proanthocyanidins in spices. Z Lebensm-Unters Forsch 171:278–280

222 Shuttleworth S G 1967 Further studies on the mechanism of vegetable tannage. Part I. Experiments on wattle tanned leather. J Soc Leather Trades Chem 51(4):134–143

223 Shuttleworth S G, Russell A E, Williams-Wynn D A 1968 Further studies on the mechanism of vegetable tannage. Part VI. General conclusions. J Soc Leather Trades Chem 52(12):486–491

224 Shuttleworth S G, Ward G J 1976 The Liritan minimum effluent vegetable tanning system. J Soc Leather Chem Assoc 71:336–343

225 Slabbert N P 1980 Metal complexes and mineral tanning. Ds Leder 31:79–82

226 Slabbert N P 1981 Mimosa-Al tannages: An alternative to chrome tanning. J Am Leather Chem Assoc 76:231–244

227 Somers T C, Harrison A F 1967 Wood tannins: isolation and significance in host resistance to *Verticillium* wilt disease. Aust J Bio Sci 20:475–479

228 Sparrow N A, Russell A E, Glasser L 1982 The measurement of the binding of tannin subunits to solublecollagen by continuous-flow dynamic dialysis. J Soc Leather Tech Chem 66:97–106

229 Stashevski A M, Deppe H J 1973 The application of tannin resin as adhesives for wood particleboard. Holz Roh-Werkst 31:417–419

230 Steiner P R, Chow S 1975 Factors influencing western hemlock bark extracts use as adhesives. West For Products Lab Rept VP-X-153 Vancouver B C, 25 p

231 Stryer L 1981 Connective tissue proteins, collagen, elatin and proteoglycans. In: Biochemistry. W H Freeman San Francisco, 185–203

232 Suddath J C, Thackston E L 1982 Improved color removal in spent vegetable tanning liquors. Proc Ind Waste Conf 36(16):801–813

233 Sykes R L, Cater C W 1980 Tannage with aluminum salts. Part 1. Reactions involving simple polyphenolic compounds. J Soc Leather Tech Chem 64:29–31

234 Sykes R L, Hancock R A, Orszulick S T 1980 Tannage with aluminum salts. Part II. Chemical basis of the reactions with polyphenols. J Soc Leather Tech Chem 64:32–37

235 Sykes R L, Roux D G 1957 Study of the affinity of black wattle extract constituents. Part IV. Relative affinity of polyphenols for swollen chemically modified collagen. J Soc Leather Trades Chem 41:14–23

236 Takechi M, Tanaka Y, Takehara M, Nonaka G, Nishioka I 1985 Structure and antiherpetic activity among the tannins. Phytochemistry 24:2245–2250

237 Tanaka J, Ichikuni M 1982 Monitoring of heavy metals in airbourne particles by using bark samples of Japanese cedar collected from the metropolitan region of Japan. Atmos Environ 16:2015–2018
238 Thackston E L 1973 Secondary waste treatment for a small diversified tannery. US-EPA Rept EPA-R2-73-209 Environ Protect Agency Washington DC, 75 p
239 Tisler V, Galla E, Pulkkinen E 1986 Fractionation of hot water extract from *Picea abies* Karst bark. Holz Roh-Werkst 44:2311–2316
240 Tomlinson H D, Thackston E L, Koon J H, Krenkel P A 1970 Removal of color from vegetable tanning solution. J Water Pollut Control Fed 31(3):562–576
241 Tonigold L, Heidemann E 1982 Tannage with a combination of vegetable tanning agents, syntans and aluminum salts. In: Radnotti L (ed) Proc 7th Cong Leather Ind OMIKK-Technoinform, Budapest I(b):428–434
242 Tonigold L, Heidemann E 1982 Limitations and possibilities of chromium-free tanning. Ds Leder 33(8):131–136
243 Toth G 1982 Penetration rate of mimosa tannin in pelt. Contributions to dry tanning. Ds Leder 33(2):17–28
244 Van der Westhuizen P K, Pizzi A, Scharfetter H 1977 A fast-setting wattle adhesive for fingerjointing. In: Proc IUFRO Symp Wood Gluing. IUFRO Rome, 1–8
245 VanBlaircom L E, Dewegert H R, Smith N H 1970 Grouting composition. US Pat No 3,490, 933
246 VanBlaircom L E, Gray K R 1960 Drilling mud composition. US Pat No 2964469
247 VanBlaircom L E, Johnston F A 1966 Iron complexed sulfonated polyflavonoids and their preparation. US Pat No 3270003
248 VanBlaircom L E, Tokos G M 1961 Process for forming a chemical product from bark. US Pat No 2975126
249 Villain C, Damas J, LeComte J, Foo L Y 1985 Pharmacological properties of a novel flavan derivative isolated from the New Zealand conifer *Phyllocladus alpinus*: cardiovascular activities of phylloflavan in rats. In: Farkas L, Gabor M, Kallay F (eds) Flavonoids and bioflavonoids. Elsevier Amsterdam, 287–292
250 vonWettstein D, Jende-Strid B, Ahrenst-Larson B, Erdal K 1980 Proanthocyanidin-free barley prevents the formation of beer haze. MBAA Tech Quart 17:16–23
251 Waage S K, Hedin P A, Grimley E 1984 A biologically active procyanidin from *Machaerum floribundum*. Phytochemistry 23:2785–2787
252 Watanabe T, Tosa T, Sakata N, Nunokawa Y, Shiinki S, Kikami S 1984 Removal of iron from sake by immobilized tannin and analysis of components of treated sake. J Brew Soc J 79:193–197
253 Weissmann G, Ayla C 1984 Investigation of the bark of *Pinus canariensis* Smith. Holz Roh-Werkst 42:457–459
254 White T 1956 The chemical principles of vegetable tannage. J Soc Leather Trades Chem 40:78–88
255 White T 1958 The chemistry of vegetable tannins. In: O'Flaherty F, Roddy W R, Lollar R M (eds) The chemistry and technology of leather, vol II. Reinhold New York, 98–160
256 Williams-Wynn D A 1970 Consideration of the vegetable tannin/chromium-collagen system. J Soc Leather Trades Chem 54:27–35
257 Yazaki Y 1985 Improved ultrafiltration of extracts from *Pinus radiata* bark. Holzforschung 39:79–83
258 Yazaki Y, Hillis W E 1980 Molecular size distribution of radiata pine bark extracts and its effects on properties. Holzforschung 34:125–130
259 Yazaki Y, Nichols D 1978 Phytotoxic components of *Pinus radiata* bark. Aust For Res 8:185–198

10.4 Rubber, Gutta, and Chicle

F. W. Barlow

10.4.1 Introduction

Natural rubbers can be extracted from more than 200 different species of plants, including such common ones as *Taraxacum koksaghyz* (dandelion) and *Parthenium argentatum* (guayule). However, only one source is commercially significant today, *Hevea brasiliensis*. It is outstanding in yield, longevity, and low resin content. Guayule rubber, although not now in commercial production, had a significant portion of the market in the early part of the century and was promoted during World War II as a substitute for common natural rubber.

The natural rubber industry includes the planting of the trees, tapping them for latex, coagulating the rubber from the latex, drying the coagulated rubber, and packaging it in an easy-to-use form. A small portion of the latex − about 10% − is concentrated for easier overseas shipping, and used to make products like rubber gloves. The industry is unique in many ways. Originally rubber came from plantations in Brazil. More than 50 years ago the South American leaf blight devastated those plantations, so most rubber now comes from Southeast Asia − half a world away. Rubber is an essential component of today's civilized living: Imagine cars fitted with leather or plastic tires. Because of this vital need for rubber, German scientists developed the leading synthetic rubber, styrene-butadiene rubber, in the 1920s and 1930s to make that country self-sufficient in time of war. From that beginning synthetic rubber plants were established throughout the world and today rubber usage is approximately 2/3 synthetic. The natural rubber industry, however, retains much of its vigor because for some product applications such as automotive mounts, rubber bands, and certain tire compounds it is the preferred polymer. The industry also benefits from the vast amount of research expended on it − for example, the Rubber Research Institute of Malaysia is the largest research organization devoted to a single tropical crop. Finally, there are some intriguing possibilities for the future. All common synthetic rubbers depend upon petroleum products − a nonrenewable resource. Natural rubber does not have this disadvantage and is relatively non-polluting in preparation. Natural rubber is essentially a hydrocarbon $(C_5H_8)_n$. With a continuing increase in yield it might yet be a source of energy.

With the multiplicity of elastomers now available, a definition for rubber is difficult to formulate. One that many would accept is that developed by the American Society for Testing and Materials (2).

> "Rubber − a material that is capable of recovering from large deformations quickly and forcibly, and can be, or already is, modified to a state in which it is essentially insoluble (but can swell) in boiling solvent, such as benzene, methyl ethyl ketone, and ethanol-toluene azeotrope.
> A rubber in its modified state, free of diluents, retracts within 1 min to less than 1.5 times its original length after being stretched at room temperature (18° to 29°C) to twice its length and held for 1 min before release."

Because they are rubber-like, have similar chemical compositions and are extracted from trees in a somewhat similar fashion as natural rubber, gutta percha, balata, and chicle are reviewed in this article. Their respective product applications are diverse; submarine cable covering, golf ball covers, and chewing gum base. It would appear though that production of these materials has reached, if not passed, its zenith. Gutta percha and balata are made synthetically and chicle now has a strong competitor in refined pine resins. With scattered and irregular production, reliable recent data on these products are difficult to obtain.

10.4.2 Historical Development of Natural Rubber Production

10.4.2.1 Hevea brasiliensis

Natural rubber (*Hevea brasiliensis*), like corn, was a gift of the New World to the Old. On Christopher Columbus's second voyage in 1496, he noticed the natives in Haiti playing with a ball that bounced (14). The use of rubber balls in games has now been traced back as far as the 11th century, to the times of the Mayans. The natives obtained latex, a white creamy fluid containing rubber globules in suspension, by cutting into the bark of the rubber tree and letting the latex exude. Balls could be built up by coating the end of a stick with latex and then drying it by an open fire. By repeated coatings and subsequent drying, a ball of the desired size could be made. They also applied latex to footwear or clothing and when dry the articles were waterproof.

Little was done over several centuries to exploit or develop natural rubber. The tree was identified by Fresneau, a French engineer living in Guiana, as *Hevea brasiliensis*. But European attempts to use the latex were thwarted. Raw latex coagulates rather quickly, so a difficulty in working with rubber in Europe was that the latex coagulated on the eastward journey from the Caribbean or shortly after its arrival. It was difficult to shape coagulated latex into useful articles. Things improved when two French chemists, Herissant and Macquer, found that turpentine and ether could dissolve rubber. For English-speaking people, the name rubber was coined in the 1770's by the English chemist, Joseph Priestly. He had noticed that the material would rub out pencil marks, hence the name rubber.

Although the discovery that rubber could be made into a rubber cement by solution was helpful, the toughness of the coagulated latex still hindered greater use. A step forward in processing was the invention of a masticator in England. Developed by Thomas Hancock in 1820, this machine consisted of a spiked cylinder that rotated within another hollow spiked cylinder. In the annular space between the two cylinders the rubber was torn and cleaved by the spikes. Although not discretely shredded, the rubber was much more plastic after this treatment and could be formed more easily; the freshly cut surfaces would adhere tightly to each other so articles could be built up (3).

The ability to put rubber into solution was followed up by a Scotsman, MacIntosh, who made a waterproof fabric suitable for raincoats (14). He dissolved rubber in benzene and naphtha, then placed the dried sheet between two layers of fabric.

The biggest break-through in natural rubber utilization was the discovery of vulcanization. This was the phenomenon that changed rubber from a thermoplastic material of little strength, soft and sticky in summer and stiff in winter, to a product that lost its stickiness, had good tensile strength, and remained firm through seasonal temperature variations. Charles Goodyear in America and Thomas Hancock in England were both working on this development at about the same time. In 1839, Goodyear discovered the essence of vulcanization when he heated rubber and sulfur together above the melting point of sulfur. Hancock was granted an English patent in 1844 covering much the same ground.

Natural rubber usage then increased dramatically, from about 30 tons per year in 1825 to 1500 tons 25 years later. The first reference to rubber being used in tires was a patent granted to Robert Thomas in England in 1845. It was for a rubber envelope containing an air bag and identified as a pneumatic tire. The use of rubber in tires got a boost when an Irish veterinarian, Dunlop, rediscovered the pneuematic tire for use on bicycles. About this time the bicycle industry was flourishing and within a decade the motor car industry started accelerating, and the use of rubber tires soared.

Developments in the natural rubber industry have been chronicled many times. Perhaps the most comprehensive review (although not devoted to natural rubber) is the "History of the Rubber Industry" (13). This was published in 1952, edited by Schridowitz and Dawson. Particularly interesting in this volume is R. W. Lunn's section on vulcanization and L. J. Lambourn and A. G. Perret's contribution on tires. A brief recapitulation, but covering all the significant steps to 1975, is in Colin Barlow's "The Natural Rubber Industry" (3). A capsule review of natural rubber and its applications is in the "Vanderbilt Rubber Handbook" (14). Certainly the largest reservoir of information is that held by two sister organizations:The Rubber Research Institute of Malaysia (RRIM) at Kuala Lumpur, Malaysia, and the Tun Abdul Razak Laboratory of the Malaysian Rubber Producers Research Association (MRPRA) at Brickendonbury, England.

"Wild rubber" from Africa and Central and South America was the main source of rubber until about 1900. The increasing demand for rubber and this dependence on wild sources caused shameful practices in the treatment of the natives who collected it. Some worked in virtual slavery. Although somewhat overdrawn, those conditions are described in a book by Vicki Baum, "The Weeping Wood" (4). Somewhat earlier, Howard and Ralph Wolf described early collection methods for rubber and the difficult conditions for rubber factory workers in "Rubber, A Story of Glory and Greed" (17). These conditions and problems were gradually eliminated by the development of the rubber plantation industry.

Three Englishmen played a large part in the establishment of the rubber plantation industry: Sir Clements Markham, Sir Henry Wickham, and Sir Henry Ridley. Sir Clements was a government official in India who had successfully introduced cinchona, the source for quinine used in treating malaria, into that country from South America. He thought rubber trees might thrive in Southeast Asia. Accordingly, specimens of *Hevea* were studied at Kew Gardens in England and *Hevea brasiliensis* selected as the most promising for test plantings in Malaya, Ceylon (now Sri Lanka), and Borneo. In 1876 Henry Wickham was requested to bring the rubber seeds from Brazil. Some 70000 seeds were collected and sent to

England (16). Contrary to legend, the seeds were not smuggled out of Brazil; there were no export restrictions on such seeds at that time (C.C. Davis, personal communication). Germination of the seeds was poor. About 2000 seedlings were shipped from Kew to Ceylon and Singapore. Mortality was extremely high – for example, of fifty seedlings shipped to Singapore only five arrived alive and these died shortly thereafter (3). Later a shipment of 22 plants from Ceylon to Singapore did arrive alive and thrived. Those plants and the remaining survivors at Ceylon were the stock from which the plantation industry was established.

Growth of the plantation industry was slow and it is doubtful whether plantations would be in place today if it had not been for a man called "Madman" Ridley during his lifetime. Henry N. Ridley was a dynamic, innovative, and industrious man who became Director of the Botanic Gardens at Singapore (16). In 1888 he invented the chevron system of tapping, which is still in use today. This involves excising a small amount of bark at regular intervals to yield latex without injuring the tree. The cut is made diagonally on the surface of the trunk; the latex exudes along the cut and flows down to the lower end where it is collected in a cup. Ridley discovered what is known as wound response, that condition in which latex flows most readily after initial tapping and then tapers off. By Ridley's system, the tree could be tapped every few days for the surface previously excised healed rapidly and allowed further tapping. Previous methods often injured, if they did not kill, the trees. He demonstrated that rubber trees could be grown as a regular crop in cleared areas, and the latex therefrom could be converted into clean, more uniform rubber. Ridley was knighted for his contribution to the rubber plantation industry.

By 1900, rubber acreage was estimated at 5000 acres in Southeast Asia, mostly on estates. World rubber consumption was about 54000 tons; by 1915 this had grown to an estimated 162000 tons and to 838000 tons in 1930 (3). About 66% of this came from plantations or estates, the remainder from smallholders. These are private individuals who farm 40 hectares or less (approximately 100 acres). In the same period, there were wide fluctuations in its price. The average spot price in the London market for a popular grade -RSS 1 was as low as 47.7 Malaysian cents per kilogram in 1930 from a high of 969.7 in 1910 (3).

Until the end of the last century rubber was produced as balls of different sizes. In 1899, John Parkins developed the acid coagulation method of making dry rubber. In this procedure the rubber was coagulated from the latex in a sheet form by adding acid and the sheet was then dried and smoked. By hanging the coagulated sheets over poles and supporting them above smokey fires, the water was driven off and smoke ingredients that condensed on the rubber surface inhibited mold. These rubbers were the first smoked sheets and were more easily handled than balls in the rubber goods factories.

10.4.2.2 Guayule

Guayule (*Parthenium argentatum*) is a shrub resembling sagebrush that grows wild in arid regions, especially in northwest Mexico and the southwest United States. It has also been grown in Israel, Spain, and Turkey. An early rubber goods company, the New York Belting and Packing Co., was already using guayule in 1888. They extracted the rubber from the bark of the shrub by boiling it. The rubber was of good quality but expensive. By 1910 half of the United States' supply of natural rubber came from guayule processing plants in Mexico (11). Some 8000 acres were planted at Salinas, California, in 1925 to take advantage of the relative high price of rubber at that time, about $1 per pound. When the price of natural rubber plummeted to $0.03/lb in 1932, the industry perished.

The flow of natural rubber from Southeast Asia was cut off in World War II. As an emergency measure, the U.S. Department of Agriculture planted tens of thousands of acres of guayule for the Emergency Rubber Project. In all, about 8000 tons of such rubber was produced during World War II, about 6000 tons from Mexico and 2000 tons form the United States (7). Since then little has been done to increase the production of this rubber. Mexico built a small plant in 1976 in Saltillo to produce the rubber from shrubs growing wild in the area. It had two objectives: to lessen dependence on imported natural rubber and to provide work in northern Mexico. In terms of production quantity, it must still be considered a pilot plant.

With a few minor changes, refined guayule rubber can be compounded to equal the quality of natural rubber vulcanizates. Guayule has about the same structure as *Hevea brasiliensis* and, since it is gel-free it should process more easily. It is the production process for guayule that hinders its acceptance. The whole bush is harvested and for each pound of refined rubber made, about 3.5 lb of wood fiber, resin, and leaves have to be handled. The resin does contain a hard wax that may have commercial possibilities. One development that might help guayule production is the use of metabolism regulators. Some triethylamines when sprayed on the plants about three weeks before harvest increase yields a remarkable 200% to 600%. It has been estimated that using such chemicals could increase overall yields 30% to 35% and cut the growing time by 1 to 2 years (9).

10.4.3 Natural Rubber Production Processes

10.4.3.1 Rubber Tree Growing

Today no wild rubber is collected; all rubber is produced on estates or smallholdings. At maturity the *Hevea* tree will average about 60 feet in height with a girth of 3 to 4 feet. The bark is smooth, light brown, and mottled with a gray-green coloration. The canopy is so dense that few plants can flourish underneath it. The tree has a deep tap root, anchoring it against the wind and helping in times of drought. It grows in a wide range of soils but flourishes best in loamy sandy clays with good drainage. *Hevea brasiliensis* grows in a band 10° to 15° on each side of the Equator, where there is heavy rainfall throughout the year and the temperature is subtropical. Countries like Malaysia, a foremost producer, get 80 to 100 inches of rain a year and the temperature will range from 70° to 90°F where rubber is grown. The tree grows best elevations less than 1000 feet above sea level.

Hevea trees are ready for tapping about 5 to 6 years after planting. This has been a focus of research at the RRIM, and certain clones can now be tapped in

3 to 4 years. The maximum latex flow occurs at 13 to 16 years. Rubber trees are considered to have a useful life of 30 to 35 years. By then, latex production is so small that economics indicate that harvesting and replanting are in order. Excessive tapping, called slaughter tapping, gets the last latex out of the tree while killing it. Rubber wood has been used in Malaysia to make furniture. The wood has also been considered for papermaking but the lactiferous vessels cause some problems.

Plantings can be established by using seedlings or bud-grafting. In the first method, seeds are germinated in a bed, then moved to a nursery, and finally to their permanent location. The choicest seedlings are those from seeds that result from the artificial pollination of high-yielding trees. In bud-grafting, a bud from a high-yielding tree is inserted under the bark of the lower stem of a young tree. Skillfully done, these grafts take and the rest of the tree above the graft can be removed so that the bud develops as the main stem. One of the greatest advantages to bud-grafting is that strains with good root systems might be used as a base stock with the high-yielding bud to give a better all-round tree.

A major research focus in the agricultural research institutes in Southeast Asia has been the development of clones with high yields. Since 1946 (end of the Japanese occupation), practically all major rubber plantations have been replanted with the high-yielding clones developed by the RRIM. Over the decades, yield possibilities have increased dramatically. For example in 1930, clone Pil 384 could be expected to yield about 1000 lb/acre/year. In 1970 clone RRIM 703 had a yield of 3300 lb/acre/year. Second and third generations of the RRIM 600 series clones in pilot scale plantings have been producing up to 6000 lb/acre/year. It should be noted that yield figures can depend on the planting density, the age of the trees, the tapping system, and so on, so exact comparisons are difficult.

After a year in the nursery, the young plants are transplanted to their final fields, and as they mature they are thinned out to 150 to 200 trees per acre. Research on higher density plantings are continuing; 225 trees per acre seems possible in the near future. Fertilizer requirements vary depending upon the soil, but phosphate fertilizers are most commonly used.

A major concern in growing rubber is the control of diseases. The South American leaf blight devastated *Hevea brasiliensis* in Central and South America. The blight, caused by the fungus, *Dothidella olei*, is still difficult to control. Emergency measures are in place to eradicate it, probably by burning, if any outbreak occurs in major producing areas. Two species, *Hevea benthamiana* and *Hevea pauciflora*, are known to be resistant to leaf blight. Possibly these species can be cross-bred with *brasiliensis* to provide blight resistance and good yield. Most diseases of rubber trees are caused by parasitic fungi, and most frequently the root is attacked. White- and red-root disease are caused by *Rigidoporus lignosus* and *Ganoderma pseudoferreum*. New leaves, which appear in the spring, can be defoliated by powdery mildew, caused by *Oidium heveae*. Young rubber trees, 3 to 6 years old, can have the stem attacked by pink disease, caused by *Corticium salmonicolor*. In the attacks, the upper canopy wilts and dies. New methods of chemical control have largely eliminated earlier control practices, which involved uprooting and destroying affected trees.

10.4.3.2 Latex Collection

Latex is obtained from the tree by a process called tapping. Near the cambium layer in the bark are the latex vessels, a network of capillary tubes that occur in all parts of the tree but are concentrated in the bark. Here the plant food is changed into rubber hydrocarbon. A cut is made into the thin lay of lactiferous vessels, the cambium lay being carefully avoided. Latex exudes from this cut and is collected.

Practically all latex is collected by the method developed by Ridley. To start the process, a cut is made perhaps 4 to 5 feet above the ground, one quarter of the circumference of the tree at 25° to the horizontal. The cut is deep enough to reach the lactiferous vessels and 1 to 2 mm of wood is removed in the cut. At the lower end of the cut a vertical channel is made, at the end of which a metal spout directs the latex flow into a ceramic or glass cup held to the tree by a spring coiled around its girth. On the next tapping day (every other day is common), a new cut is made along the lower edge of the previous one. When the successive cuts have descended to an uncomfortable position at which to work, the process is restarted on the other side of the vertical channel. When this side, too, is completed, tapping is started on the other half of the tree. The tapping panel therefore has a herringbone or chevron pattern. By this tapping method the tree is kept in good health, the cuts are repaired easily by the tree, and the whole process is repeated again and again.

Tapping is done in the early morning hours when flow is at a maximum. After about 5 hours, the flow rate decreases, the cut latex vessels start to choke with coagulated latex, and the flow stops. At the time of tapping, a small amount of ammonia is placed in the cup to prevent premature coagulation.

After the flow has stopped, the latex – called 'whole field latex' – is collected in 5-gallon containers and trucked to bulking stations. The collected latex is further stabilized by adding small amounts of ammonia, sodium sulfite, or formaldehyde. Latex can and does coagulate at times in the cup. Called 'cup lump', it is removed and saved along with the coagulated material over the last cut, a raw rubber known as 'tree lace'. Earth scrap is latex that has spilled to the ground by the tree, has coagulated there, and is collected every few months or so. These materials from the field are the crude forms of rubber that are processed into the commercial commodities that serve as the raw material for rubber goods producers.

10.4.3.3 Production of Latex Concentrate

Whole field latex contains 30% to 40% total solids by weight. Roughly 10% of whole field latex is concentrated rather than made into dry rubber. Concentration avoids the cost of shipping water. The largest proportion of concentrate is exported to consumer industries for conversion into such products as adhesives, gloves, elastic thread, and prophylactics.

Latex is concentrated to 60% to 68% dry rubber content. This is done by either centrifuging, evaporating, or creaming. By far, the most popular method is centrifuging. Usually in a single centrifuging, the latex is concentrated to about 60% dry rubber content; the residual serum has perhaps 5% to 6% rubber, which can be recovered and is a fast-curing rubber known as 'skim rubber'. By repeating the process, dry rubber content can be brought up to 66% to 68%. Triple centrifuging is rarely carried out, as power consumption and handling difficulties increase rapidly. Evaporated latex is made by passing latex through film evaporators at elevated temperatures.

Creamed latex is made by the use of a creaming agent such as ammonium alginate. The creaming agent helps form a rubber-rich layer at the top of the holding tank. This is a slow process; several weeks may be required before the desired concentration is reached in the top layer, which is drawn off for shipment. Creamed latex has a minimum dry rubber content of 62% and is preferred by rubber thread manufacturers because of its good filterability.

Most latex concentrate produced today would fall into one of the four types specified by ASTM specification D 1076 (1): centrifuged latex with normal ammonia, centrifuged with low ammonia, creamed latex with normal ammonia, and creamed with low ammonia. The minimum content for latex solids varies from 61.5% to 64.0% and dry rubber content minima from 60.0% to 62.0%. Total alkalinity is calculated as NH_3 based on the water phase of the latex and is 1.6% minimum for normal ammonia latex, 1.0% for the minimum ammonia type. Low-ammonia latex (often made with a second stabilizer) has less odor and requires less neutralization in subsequent processing, but other factors hinder its use. A very small amount of latex is vulcanized in the liquid state and sold for special compounding.

10.4.3.4 Production of Dry Rubber

Latex not sold as latex concentrate and cup lump, tree lace, and earth scrap are used to produce various grades of dry rubber.

At the rubber-producing factory, the collected latex is strained to remove trash such as leaves and is blended to produce a more uniform product. To reduce the incidence of air bubbles in the dried rubber, the latex is diluted with water before acid coagulation. The dilution is down to about 15% solids for sheet rubber, around 22% for a light-colored rubber called pale crepe. The coagulant is usually formic or acetic acid, 5 parts of a 0.5% formic acid or 1.0% acetic acid solution added to 100 parts of the diluted latex. The coagulating process can take anywhere from 4 to 24 hours, but conditions are usually selected so the process is completed overnight. the coagulated rubber is then in a soft gelatinous slab form and ready for further processing.

Very light colored rubbers, called pale crepe or sole crepe, are sold at a premium price and handled differently from other rubbers. Preferred field latices for these products are those with a low yellow pigment content and with resistance to darkening by enzymatic action. Before coagulation the latex is treated with sodium bisulfite to prevent darkening by oxidation or bleached with xylyl mercaptan. After coagulation, the sheets are passed through sets of rollers that squeeze out the water and then are slowly dried at temperatures low enough to avoid darkening. A still lighter product can be made by fractional coagulation. In this method only a small amount of the total coagulant is added. The coagulum removed after this addition has most of the naturally occurring colorants such as carotene. When the remaining acid is added, the second coagulum is very light in color.

A major dry natural rubber type is "ribbed smoked sheet." For this the coagulum is passed between pairs of even-speed steel rollers that squeeze the water out. The final set of rollers has channels cut in them so that the sheet emerging from them has a ribbed surface. The ribbed surface facilitates drying. This rubber is dried in smokehouses. The ribbed sheets are hung over poles mounted on trolleys. Rubber tree wood fires produce the smoke, which dries the rubber and gives the rubber some age resistance from its components such as cresols. Drying takes 48 to 96 hours, with entrance temperatures at about 40 °C and exit temperatures at about 60 °C. Some ribbed sheet rubber is dried in hot air out of contact with smoke. This produces a lighter colored rubber, which commands a premium and is referred to as air-dried sheets.

The cup lump, tree lace, and earth scrap can be worked up into rubbers called 'brown crepes'. These raw materials vary a great deal in quality and are often dirty, Preliminary treatment frequently includes soaking overnight in a dilute solution of sodium bisulfite to reduce the surface dirt and lighten the color. The material is then macerated and conveyed to a series of roll mills driven at friction speed and fitted with water sprays that clean up the rubber and blend it. The rubber is then sheeted off and dried like smoked sheet. 'Remilled blanket crepes' refers to rubbers similar to brown crepes but which are made of smoked sheet cuttings, smallholders' partially dried sheets, and other pieces. The production of brown crepes and remilled blanket crepe is diminishing as demand for technically specified rubber increases and technically specified rubbers use the same raw rubbers that go into crepes and blanket crepes.

After World War II, rubber growers conducted surveys of customers to find out how their product could compete most effectively with the synthetic rubbers then available. As a result of these surveys, rubber producers developed technically specified rubbers. First introduced by Malaysia about 1965 as SMRs (Standard Malaysian Rubbers), they were accepted so well that other countries followed with similar products such as SIR (Standard Indonesian Rubber) and TTR (Thai Tested Rubber).

These products provide the user with a cleaner, more uniform, better packaged rubber while at the same time speeding up the production process from weeks to days. Large-scale commercial equipment is used, and the process is more closely monitored. The manufacture differs essentially from that used for other grades in the preparation for drying and in the drying operation itself. Such machines as hammermills and granulators are used to break up the coagulum, cup lump, etc., into small discrete particles, which are then dried in screen-bottomed trays in tunnel dryers. The granules are dried in approximately 5 hours at 100 °C under closely controlled conditions. The dried rubber in the trays is then compressed into 33-1/3 kg blocks, polyethylene wrapped, and crated when cooled.

10.4.4 Packing and Market Grades

10.4.4.1 Packing

Natural rubber is a durable material, but packing it does present some problems. With high ambient temperatures it can flow into crevices under pressure and foreign material can often adhere to it easily. Originally it was shipped in tea boxes from the Far East but wood splinters became embedded in the rubber. Some pale crepe rubbers, carefully wrapped, are still shipped in boxes. Most sheet and brown crepe grades are shipped in so-called bareback bales. The rubber sheets are piled up and compressed in a hydraulic press to obtain a 5-ft^3 bale weighing about 250 lb. The rubber wrapper sheet on the outside of the bale is coated with a dispersion of whiting, talc, or soapstone so that the bales will not stick together in transit.

Technically specified rubbers are packaged differently. The dried crumb is compressed into a 33-1/3 kg bale measuring 66×33×18 cm. These blocks are wrapped with thin low-melting-point polyethylene bags identifying the rubber inside. Low-melting-point film is used so it will melt in subsequent Banbury mixing. After cooling they are packed in 1-ton-capacity crates for shipment. Cooling is necessary to prevent the lower bags from splitting their envelopes under pressure and allowing the rubber to fuse.

10.4.4.2 Grading

In the International Grade classification scheme, rubbers are divided into eight types, depending upon the source of raw material (e.g., bark crepe), color (brown, pale), and preparation method (e.g., ribbed smoked). Color is further graded by a numerical rating in which IX is the lightest and $^\#$4 the most opaque. This system is based on visual aspects of the rubber so its value is limited. For example, the darker the rubber, the greater is the probability of objectionably high dirt levels, yet color itself is not a criterion of quality for most uses. Typical grades in this scheme would be Ribbed Smoked Sheet $^\#$2, Thin Brown Crepe $^\#$4, and Pale Crepe LX.

The main grading criterion for Technically Specified Rubber is dirt content, identified as the material retained on a 45-μm sieve after the rubber has been dissolved in a suitable solvent and passed through the screen; the residue has to be dried free of solvent before weighing. The dry content is measured in increments of 0.01%, the number after the national acronym specifying the maximum amount of dirt in that grade. For example, SIR 20 is a Standard Indonesian Rubber whose dirt content does not exceed 0.20% by weight of the rubber. Most countries produce grades with dirt limits of 0.05%, 0.10%, 0.20% and 0.50%. Other cirteria of quality are listed in the specifications for these rubbers. National specifications from producing nations are similar to the ASTM and ISO specifications. Malaysia has been the leader in establishing quality controls; the specifications put maximum limits on the ash, nitrogen, and volatile matter content of the rubber. Two processability measures are also included: a minimum Plasticity

1038 The Utilization of Wood Extractives

Retention Index (PRI), a measure of resistance to heat oxidation, and Wallace Plasticity (P_O), a measure of plasticity under controlled conditions.

One type of modified rubber that can be most useful to compounders requires an additional grade assignment. It is called 'controlled viscosity' or CV rubber. A disadvantage of natural rubber is its tendency to harden in storage due to cross-linking reactions that connect aldehyde groups on the rubber molecule with other groups, perhaps on the non-rubbers present. This hardening can be stopped or blocked by adding a chemical such as hydroxylamine hydrochloride to the latex or wet stage. Malaysia identifies its main brand of viscosity stabilized rubber as SMR 5CV with the viscosity stabilized at 60±5 Mooney units (Mooney Viscosity at 100°C on rubber taken directly from the bale). With this treatment, CV rubber would not harden by more than 5 units on storage, plain SMR 5 might harden by 30 to 50 units.

10.4.5 Properties

10.4.5.1 Natural Rubber

The prevalence of rubber in wood has been studied by Sandermann et al. (12). Although rubber is found in more than 2000 genera in the stalks, stems, roots, and leaves; of 150 wood species investigated, only eight had rubber in the xylem. Of these eight, only four had a rubber content of more than 1%. Rubber is present only in the parenchymatic tissue of the xylem.

The mechanics of latex formation in the rubber tree has been the subject of considerable study. It is now known that the monomer isopentyl pyrophosphate is the one used in the biosynthesis of natural rubber. Catalysts for these transformation are enzymes in the latex and tissues of *Hevea*. The structure for natural-rubber is *cis*-1,4-polyisoprene (1); and the *trans* form for the structure of gutta

Table 10.4.1. Proximate analysis of natural rubber

Component	Content (%)
Moisture	0.4
Acetone extract	2.7
Protein (calc. from N_2)	2.8
Ash	0.5
Rubber hydrocarbon	93.6
Total	100.0

percha and balata (2). A proximate analysis of a representative sample of natural rubber is given in Table 10.4.1.

Non-rubbers in the latex include fatty acid, protein, sterols, and esters; these remain in the dry rubber. Trace elements include copper, manganese, iron, potassium, and magnesium. The proteins and fatty acids help in the vulcanization reaction. Slight variations in these properties can be expected because the non-isoprenic content will vary.

Latex can be defined as a colloidal dispersion of rubber in a water solution or serum. The dispersed rubber globules are constantly in motion (Brownian motion) and have a diameter of 1 to 2 microns. Latex has moderate stability because the particles all have a negative charge and therefore repel each other.

Natural rubber is a mixture of 'sol' and 'gel' rubber, two rather imprecise terms. Sol rubber refers to that portion of the rubber that is soluble in carbon disulfide, chlorinated hydrocarbons, and aliphatic and aromatic hydrocarbons, but insoluble in lower alcohols and ketones. Gel rubber is insoluble in such solvents. The gel portion is highly branched and lightly cross-linked. With a high gel content the rubber is difficult to work because of the high shearing resistance in roll mills or internal mixers.

Rubber is a high-molecular weight polymer. Molecular weight can be estimated from solution viscosity measurements. Such determinations show a range of polymers with number-average molecular weights varying from 200000 to 400000. The weight average molecular weight of pale crepe has been measured as 1.85×10^6.

Under ordinary conditions natural rubber is an amorphous material. When frozen or stretched, it crystallizes in the *cis* form. It is crystallization that gives rubber its self-reinforcing effect. As the rubber is stretched, crystallization increases, and this raises the ultimate tensile strength. Shipments of rubber when held for extended periods of time at cool temperatures – say 5 °C – can become frozen. In this state, rubber is rock hard and impossible to plasticize mechanically. Luckily, the crystallization process is completely reversible. At plants, the rubber can be stored in a hot room with temperatures over 45 °C and its original softness brought back. The time required for this is largely dependent upon how well the bales are separated, for natural rubber is a poor conductor of heat.

A unique quality of rubber is the Gough-Joule effect. Unlike other materials, when rubber is stretched rapidly it heats up. If rubber is stretched under a load, it will retract as the temperature is raised and it is held at the other end.

Table 10.4.2. Some physical properties of natural rubber

Density	0.92
Refractive index (20 °C)	1.52
Coefficient of cubical expansion	0.00062/°C
Heat of combustion	10 700 cal/g
Thermal conductivity	0.00032 cal/sec/cm^2/°C
Dielectric constant	2.37
Volume resistivity	10^{15} ohms/cm^3
Dielectric strength	1000 volts/mil

Some of the physical properties of natural rubber are given in Table 10.4.2. Slight variations in these properties can be expected because the non-isoprenic content can vary.

10.4.5.2 Modified Natural Rubber

Over the years, natural rubber has been modified in many ways. One of these was constant viscosity rubber, described earlier. Some of these modifications, such as chlorinated rubber, have been successful for a while but were then superseded by other materials. Others are produced in small volumes and, at times, stocks may be non-existent. A number of modified natural rubbers have been produced in the last decade, and the presumption is they may still be obtained.

10.4.5.2.1 Deproteinized Rubber

As normally produced, natural rubber contains a small amount of protein – perhaps 2.5% by weight. Latex can be treated with a bioenzyme that hydrolyzes the protein into water-soluble compounds. The latex is then diluted and the rubber is coagulated with acid and processed into a block rubber called DPNR. Compounds made with this rubber have reduced water absorption. With a careful selection of the vulcanizing system, vulcanizates can be produced with low creep properties under stress, a quality sought after in such items as rubber building mounts.

10.4.5.2.2 Depolymerized Rubber

There is only one North American producer of depolymerized rubber, Hardman, Inc., of Belleville, New Jersey. Manufacturing information of this product, sold under the trade name DPR, is proprietary. It is a flowable polymer that looks like molasses. Rubber compounds using the material make excellent molds for they release waxes, gypsum, and ceramics easily.

10.4.5.2.3 Peptized Rubber

Some Far East rubber traders sell a softened or peptized rubber. A small quantity of a softening agent or peptizer is added to the latex used to make sheet or crepe rubber. This type of rubber breaks down easily, reducing mastication time and cost.

10.4.5.2.4 Oil Extended Natural Rubber (OENR)

Oil can be incorporated into natural rubber – called oil extended natural rubber – by several procedures: simple absorption of oil by rubber sheet, by absorption of crumb immersed in oil, by Banbury mixing or extruder technics, or by co-coagulating latex with an oil emulsion. The limit is about 65 parts of oil per 100 of rubber but usually the proportions are much less: 10 to 20 parts per 100 of rubber. Commercial compounds using high oil and high carbon black loadings are easier to mix if a portion of the oil is already in the rubber. Due to pricing difficulties, among other reasons, there is little market for these rubbers overseas.

10.4.5.2.5 MG Rubbers

MG rubbers are graft copolymers. A secondary polymer, polymethyl methacrylate, is grafted on to the primary polymer, natural rubber. The grafting is done by activating the natural rubber latex while simultaneously grafting and copolymerizing the methyl methacrylate monomer. Two grades of this rubber are produced in Malaysia: MG30 and MG49. The suffix number refers to the percentage of methyl methacrylate in the blend.

MG rubbers show the hybrid qualities of the blend. When compounded, MG rubbers are self-reinforcing polymers with high hardness, impact resistance, and good elongation. Compatible with natural rubber in all proportions they can be molded into such items as cutting blocks. They also find use in adhesives permitting bonds to be made between natural or synthetic rubbers and PVC, synthetic fibers, leather, etc.

10.4.5.2.6 SP Rubbers

Still further modifications of natural rubber are the superior processing or SP rubbers. Here a portion of the latex is vulcanized before coagulation. In a sense, the vulcanized globules serve as small particle reinforcement of the rubber. Such rubbers help keep close dimensional tolerances or smooth finish requirements. Extrusion and/or calender output is improved, and die swell is lessened in lightly loaded compounds. In the interest of economical overseas shipments from the Far East, masterbatches of these rubbers, called Processing Aids (PA), are used. For example, PA80 is a rubber that has 80 parts of vulcanized rubber and 20 parts of raw rubber. A softer material called PA57 has 57 parts of vulcanized rubber, 14 parts of raw rubber, and 29 parts of oil.

10.4.6 Natural Rubber Utilization

There are four forms in which natural rubber can be used: as unvulcanized latex, unvulcanized dry rubber compound, vulcanized latex, or vulcanized dry rubber compound.

In terms of volume there is little use at present of unvulcanized latex or unvulcanized dry rubber compounds. Unvulcanized latex is used to make carpet back-sizing; unvulcanized dry rubber is used in non-curing pressure-sensitive tape formulations. There are possibilities for developments that would increase consumption of these unvulcanized materials. For example, studies have shown that spraying a latex compound on soil may successfully prevent erosion. Elastomers added to road-paving mixtures (asphalt) have significantly increased the time before replacement or repairs were needed. In view of the high cost of new road-building, unvulcanized dry natural might be used in such applications.

10.4.6.1 Vulcanization

As noted above, natural rubber would have limited usefulness if it could not be cured or vulcanized. The two terms are used interchangeably; the word vulcanization was derived from Vulcan, the Roman god of fire, as vulcanization is rarely effected without heat. The ASTM definition of vulcanization is (2):

"Vulcanization – an irreversible process during which a rubber compound, through a change in its chemical structure (for example, crosslinking), becomes less plastic and more resistant to swelling by organic liquids while elastic properties are confirmed, improved or extended over a greater range of temperature."

Natural rubber is vulcanized commercially in three ways (all using heat to promote the reaction): 1) mixed with sulfur or sulfur donors, 2) mixed with peroxides, or 3) mixed with urethane type crosslinkers.

The rubber compound, which usually contains other ingredients as well as the vulcanizing agent, is molded or otherwise formed into the shape desired, and heat is applied until the desired properties are achieved.

By far the most vulcanizing is done with sulfur or sulfur donors as the curing agent. Sulfur donors are certain organic compounds that break down during the heat treatment, releasing their elemental sulfur to react chemically with the rubber. Due to the variations in non-rubber constituents in natural rubber and the interactions between it, sulfur, and various vulcanization helpers called activators and accelerators, the vulcanization process is very complex chemically. Essentially, in sulfur or sulfur donor curing, the sulfur atoms unite with the polymer chains at double bonds linking them together in a network as shown in Fig. 10.4.1 (A).

Sulfur in the cyclic mode – the S_y sulfur – does not contribute to the qualities desired by vulcanization. The S_x bridge or crosslink may have one to eight atoms of sulfur. Heat resistance of the compound is improved if the number of atoms in the S_x crosslink is kept low – say one or two atoms. This is more easily accomplished by using sulfur donors rather than elemental sulfur. The amount of sulfur bound to the rubber is relatively small, perhaps 2 to 3 parts per 100 of rubber. With this ratio, only about 1 in 200 monomer units in a chain is crosslinked. Natural rubber can react with much larger quantities of sulfur to form hard ebonite products, such as combs, which may contain 30 parts of sulfur per 100 of rubber.

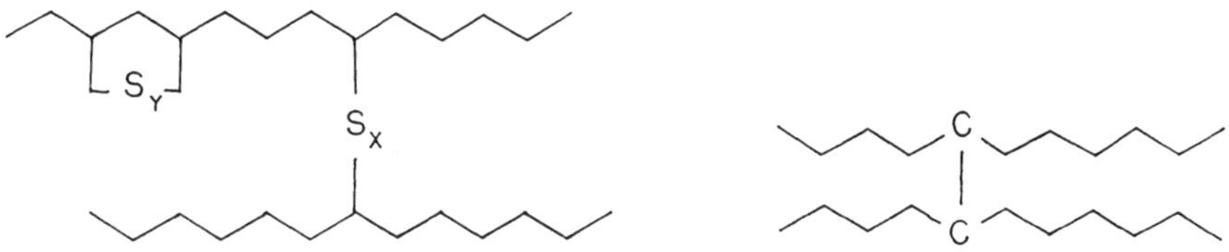

a

b

1. $ROOR \longrightarrow 2RO\cdot$

2. $RO\cdot + PH \longrightarrow P\cdot + ROH$

3. $2P\cdot \longrightarrow P-P$

c

Fig. 10.4.1. Vulcanization structures and reactions. (polyisoprene)

Peroxide curing of natural rubber might be used, for example, where excellent heat resistance is needed in the rubber product. This ruggedness occurs because there are direct carbon-to-carbon links from one chain to another, there is no intervening atom, and the bond energies between carbon atoms is greater than the carbon-sulfur bond or the bond between two sulfur atoms. The type of bond is illustrated in Fig. 10.4.1 (B) and the reactions involved in Fig. 10.4.1 (C).

Organic peroxides like dicumyl peroxide split under irradiation or heat to produce oxy radicals as in Reaction 1. These radicals are very reactive and will abstract hydrogen from the natural rubber molecule according to Reaction 2. The resultant polymer radical is also very reactive and joins with another such radical to form a stable crosslinked product (Reaction 3).

A more recent development has been the use of urethane crosslinkers. These are expensive but may be used as co-vulcanizing agents with sulfur for economy. They impart excellent fatigue resistance on aging. Urethane crosslinkers are formed from nitrosophenols and diisocyanates; the resultant diurethanes are stable at processing temperatures but spilt into their components at curing temperatures. Pendent amino-phenol groups are formed on the rubber molecules, which are then linked by the diisocyanate.

10.4.6.2 Transportation Items

Transportation items consume about 68% of the total production of natural rubber. The largest elastomer demand in this sector is for passenger and truck tire treads. Prior to World War II, all passenger and truck tire treads were made from natural rubber. Natural rubber producers lost this market — at least the passenger tread portion — to synthetic rubber. Perhaps the greatest reason for this was the greater year-round wear resistance that synthetic treads gave for driving conditions in the United States, the world's largest market. More recently, natural rubber has been used as one of the elastomers in the tread stock of winter passenger tires. Under winter conditions in the northern United States and Canada, natural outwears synthetic rubber and can match or outperform it in traction and skid resistance.

Although the use of natural rubber in passenger treads may be small, other applications in tires are numerous. This is especially true where radial tires are involved, for here it plays a vital role in bonding the wire belt to the rest of the tire. In radial passenger tires, natural rubber might constitute 50% of the polymer content of the black sidewalls and 40% of white sidewalls (where a pale crepe might well be chosen to keep the compound as white as possible). In passenger tires carcasses, the coating stock over the wire belt would probably be 100% natural rubber and the other body plies 60% or more for the excellent adhesion it gives and the low heat build-up on flexing. On a proportional basis by weight, truck and bus tires have a higher percentage of natural rubber in their polymer content than passenger tires. For this use, the lower heat build-up of natural rubber compared to synthetic is the main advantage.

Although the consumption of polymer is small, an important use of natural rubber is in off-the-road tires, such as earthmover tires and aircraft tires. Earthmover treads can be compounded with natural rubber to have excellent tear resistance and to reduce torn treads and chunkouts. In aircraft tires, the natural tread absorbs well the blow in landing. Natural rubber is little used in retreads although frequently used in the cushion gum, the thin rubber compound layer that joins the tread to the tread base in retreading.

10.4.6.3 Mechanical Rubber Goods

Mechanical goods items such as hose, belting, and molded products such as engine mounts use 13% to 14% of the total rubber produced. A prime user of natural rubber in the mechanical goods field is belting, especially conveyor belts. In Table 10.4.3 is given a typical formula for a conveyor belt cover designed especially for a belt to carry ore in an outdoor mining operation. Features that natural rubber gives to this application include good abrasion resistance to the cutting and tearing effect of the ore and flexibility under winter conditons.

An ubiquitous, if not large-volume, use of this polymer is in molded articles such as engine mounts in automobiles. Here natural rubber helps because of its low dynamic modulus and the low temperature stability of the dynamic modulus.

10.4.6.4 Footwear

One of the first uses of natural rubber was footwear and much is still used in this product. Probably about 5% of natural rubber production goes into this application. This is especially true in the Far East where rubber and labor costs are cheaper in a labor-intensive competitive industry. Besides its use in the well-known crepe rubber soles, natural rubber is used in sport and protective footwear

Table 10.4.3. Natural rubber conveyor belt cover

Component	Parts	Functional class
Natural rubber SMR CV	100.0	Elastomer
N220 Carbon black	44.0	Black reinforcer
Zinc oxide	3.5	Activator
Stearic acid	2.0	Activator
Rubber process oil	4.0	Softener
Agerite Resin D[a]	2.5	Age resister
Paraffin wax	1.0	Age resister
CBS[b]	0.5	Accelerator
Sulfur	2.5	Vulcanizing agent
Total	160.0	

[a] Vanderbilt Co. trade name
[b] *n*-cyclohexylbenzothiazole-2 sulfenamide

in the uppers and outsoles, often in blends with other rubbers. Shoes are built up piece by piece on a last, and tackiness of natural rubber helps keep the assembly together until it is bonded in the vulcanization reaction.

10.4.6.5 Miscellaneous Uses

The remaining natural rubber production – perhaps 10% to 15% – is channeled into miscellaneous products. These range from latex products such as condoms and surgeons' gloves to adhesives, tennis balls, and medical equipment made from dry rubber. An everyday product here is medical adhesive tape. Such adhesives contain large quantities of tackifiers and plasticizers and an antioxidant, their combined weight often outweighing the rubber 2 to 1. The tackifier might be ester gum, the plasticizer refined lanolin, and the filler zinc oxide. The non-toxic and non-allergenic nature of natural rubber makes it a preferred choice in this application.

10.4.6.6 Health and Safety Factors

In dealing with materials and the effect of skin contact with them or ingestion of them, a common term is GRAS, the acronym for "generally recognized as safe." Natural rubber can be generally recognized as safe. It has a long and successful history in this connection. It is frequently used in such personal care and medical products as hot-water bottles, ear plugs, adhesive bandages, surgical drainage tubing, and vaginal diaphragms. If a rubber product does give trouble, it is probably one of the components that is added to the rubber that is the cause. Undoubtedly, some softeners and tackifiers are not used in medical adhesive bandages because of their propensity to cause allergic reactions.

There are few safety requirements for this natural product. Care, of course, should be taken in handling and stacking bales and crates so that they are securely protected from slipping or falling at all times. The most important consideration is to recognize that rubber is a very combustible material and must be adequately protected from fire.

10.4.7 Natural Rubber Economy

By far the major portion of natural rubber is produced in three southeast Asia countries: Malaysia, Indonesia, and Thailand. Together they account for about 80% of world production. India, Sri Lanka, and Liberia also have significant production. Producing rubber is a labor-intensive operation and the industry supports millions of workers in these geographical areas.

Most of the world's rubber supply is grown by smallholders, not on estates. World rubber production has increased at only a modest rate over the last two decades. Barlow (3) gives natural rubber production in 1960 as 2 million metric tons and 3.1 million in 1970; the International Rubber Study Group (5) reports production in 1980 as 3.8 million. This is a growth rate of only about 3% a year. This rate does not match increasing total rubber consumption world-wide, and there has been a steady slippage of natural rubber's share of the market from 100% in 1940 to about 30% today.

A healthy natural rubber industry is desirable for the world economy. There are several reasons for this:

1. It provides millions of jobs. When countries like Malaysia, who are dependent on natural rubber exports for a large share of their income, face distressed prices for their product, thousands of people are out of work and social stresses increase intensely.
2. In overall technical utility, it matches synthetic rubber and is non-toxic.
3. Its production does not pollute the atmosphere the way that synthetic rubber does.
4. It does not use up irreplaceable world resources, such as petroleum, as synthetic rubber production does.

One of the obstacles to a healthy natural rubber industry is the price instability that has plagued it through the years. In the five years from 1979 through 1983 the average yearly price for a bellwether grade on the New York market varied from a low of $ 0.45 to a high of $ 0.74/lb. In 1975 the average price was as low as $ 0.30. These variations were not due solely to those factors commonly associated with supply and demand. For example, presumably as a hedge against inflation, one country's dealers were reported to have traded in one year three times that country's annual consumption. Wide price swings tend to discourage small farmers from cultivating rubber because of the fear of low returns. Consumers find it tempting to switch to synthetic rubber if the price of natural remains relatively high.

An attempt to improve this situation was the International Natural Rubber Agreement that went into effect in October 1980 for a five-year term. Developed with the help of the United Nations Committee for Trade and Development (UNCTAD) (6), it was a cooperative venture among producing and consuming nations to attain price stability and thus promote greater production. This was to be achieved by the establishment of a buffer stock of 5.5 million tons of rubber at the maximum with provisions to buy or sell rubber by a buffer stock manager. If the price average of certain popular grades at the main trading centers moved out of a certain initial band, the manager at a first set of trigger points was allowed to buy or sell rubber for or from the buffer stock to stabilize the price. If it moved still further to a second set of trigger points, he was obliged to buy or sell rubber. Provision was made to raise or lower the original band if a continued rise or fall in the price made that advisable.

With extensions, that agreement ended in October 1987. At that time negotiations were under way for a new price/supply agreement but the form of the agreement had not been worked out. The success of the expired agreement had been mixed. On the one hand, it probably prevented a drastic drop in the price of rubber. On the other hand, the buffer stock became relatively high (it is now being slowly sold off), and consuming nations representatives felt there had not been the sustained national endeavors they had expected to replant and expand rubber production (10).

Apart from difficulties due to price fluctuations, natural rubber producers face consistent heavy competition from synthetic rubber plants. Facilities built in the wartime emergency are still being used and are fully amortized. On technical grounds alone, there is little doubt natural rubber would command more of the world's rubber market, and synthetic rubber depends upon petroleum feedstocks. But availability of petroleum feedstocks world-wide gives any country the opportunity to establish their own synthetic plants. This might be done on the grounds of self-sufficiency, even if the market were not really large enough to make the plant economically viable by common standards. Of course, the drop in petroleum prices in the last few years has made synthetic rubber feedstocks more reasonable.

Forecasting the future trend of any commodity in terms of availability, production, and price is somewhat hazardous, and for natural rubber is probably more hazardous than most. Some conditions exist, however, that would indicate natural rubber will be around for some time at reasonable prices. Although tire production in the United States may be on a plateau, the increasing industrialization of the Third World countries keeps world demand for elastomers rising and at a faster rate than natural production. Such countries, at least in the short term, increase production of trucks rather than passenger automobiles and truck tires require more natural rubber than do passenger tires.

On the production side, yield of rubber can be increased dramatically as newer high-yielding clones are planted and come into production at shorter maturities. Existing trees can be stimulated to increase yield markedly by the use of yield stimulants. These are organic compounds that are applied to the bark of the tree and gradually release ethylene. For some reason the tree becomes, in a sense, a bleeder, with a much heavier flow before coagulation. The increase in yield depends upon the clones being tapped and the method of tapping, but increases of 10% to 50% are not uncommon. The effect of long term use of these materials — say 10 to 20 years — is not yet known. Besides improved yield from better trees and stimulation, labor costs for tapping and collection, now estimated at about

30% of rubber cost, may be cut. Polyethylene bags, for example, instead of cups allow more latex to be collected between tappings. Finally, the tapping process itself may be revolutionized. Tapping is not particularly attractive work and efforts have been made to mechanize it. Encouraging results have been obtained with a hand-operated device that makes a micro-cut quickly. Refinements of these experimental methods may well make tapping much quicker and less laborious.

10.4.8 Gutta Percha and Balata

Gutta percha and balata can be considered together, as they have the same structural formula and at times are used interchangeably. Gutta percha was known to the Western world in the early 1600s. A sample reached London in 1843 and shortly thereafter was found to be a good material for insulating submarine cables. Until the end of the 19th century, all gutta percha was obtained from wild forest trees. By 1925 most production came from forests converted to gutta; this was accomplished by simply felling other trees in the tract. In the 1950s, synthetic *cis*- and *trans*-polyisoprene appeared on the market. The *trans* variety became a substitute for both gutta percha and balata.

10.4.8.1 Production

Gutta percha is obtained from trees of the *Sapotaceae* family. These grow in a band about 6° each side of the Equator between 100° and 120° longitude. This area encompasses the states of Malaysia, Borneo, and Sumatra, where most of it is grown. Practically all gutta percha is now obtained from plantations where the best producer is *Palaquium gutta*. When it is favorably situated (it needs considerable light), it grows very rapidly. One 5-year test in Malaysia found an annual girth increase of 1.57 inches. In the tree, the lactiferous vessels are not in long tubes like those in *Hevea brasiliensis* but exist as scattered cavities throughout the leaves, bark, and pith, but not the heartwood.

In the tapping of wild trees, the gum was extracted from the stem by making cuts 45° to the axis of the stem, 5 to 10 inches long. At the lower end of this cut, a vertical cut would be made, with a small cup at its end, attached to the tree. Most of the latex coagulated in the cup and was removed therefrom, but some that ran into the cup coagulated later.

On plantations, the gutta is obtained from the leaves and the stems. This material is crushed in edge-runner rolls and then boiled in water. The woody material sinks to the bottom and the lighter gum rises to the surface, where it is skimmed off. It is worked on roll mills to give a product known as yellow gutta. This has about 10% resin; a more refined form, white gutta, has about 1% resin. Refining of gutta can be done by taking advantage of the fact that the resin is more soluble in petroleum spirits than the gum is. After the crude gum has been dissolved in warm solvent, the mix is cooled and the gutta crystallizes out and can be removed by centrifuges. Residual solvent in the solids can be reduced by using a distilling masticator. Gutta percha quality is inversely proportional to resin content.

Balata is obtained from *Mimusops balata* trees, more commonly known as the bulle, bully, or bullet tree. It grows wild in Brazil and in the Surinam and Guiana areas on the northeast coast of South America. The trees are not tapped until they are 20 years old and, as they take 50 years to mature, do not lend themselves well to plantation growing. Balata latex is obtained by tapping the trees, collecting the latex in canvas bags made leakproof by letting an earlier coating of latex dry inside. Coagulation is effected by the heat of the sun on latex held in dishes or trays. The coagulated layer is removed as a sheet and allowed to dry. Local practice sometimes includes pressing into a block. Raw balata is dirty and high in resin. It is purified by solvent extraction processes so that it contains less than 2% by weight of resins and non-rubber material.

10.4.8.2 Properties and Uses

Gutta percha and balata are *trans* forms of polyisoprene; the *cis* form is natural rubber. Both are tough and horny materials at room temperature, but are thermoplastic, soft at 60 °C, quite soft and sticky at 100 °C. Gutta percha is one of the best non-conductors of heat and electricity. Gutta percha has a higher specific gravity, 0.96 at 20 °C, than natural rubber, 0.92 at the same temperature. Gutta percha has a longer identity period, 9.6 Å, compared with natural rubber, 8.3 Å, in their X-ray diffraction patterns. Balata can be considered a natural substitute for gutta percha.

A first and most successful use of gutta percha was as an insulator for submarine cables. Although it oxidizes in air, it lasts indefinitely in water. Its use in this application has now been superseded by polyethylene. Pinpointing present day uses of gutta percha and balata is somewhat difficult as they have been used interchangeably and now have the competition of synthetic *trans* polyisoprene. Balata has been used to impregnate canvas to be used as belts. Both gutta percha and balata provide very tough golf ball covers. Both serve well as cutting blocks, especially in cutting glove leathers. They provide a hard base, do not blunt the edge of the knife, and can be reworked and remolded. In 1979, 3174 tons of balata and 536 tons of gutta percha were imported into the United States (15).

10.4.9 Chicle

Chicle is the coagulated latex from red and white sapodilla trees (*Achras sapota*). Its main use is as chewing gum base.

Over 1000 years ago, the Mayans used chicle as a chewing gum; their descendants continued the practice into the 19th century. It resurfaced as a chewing gum base in about 1870–1880. The well-known Mexican General Antonio Lopez de Santa Anna felt it could have commercial value as a rubber and had an American inventor, Thomas Adams, investigate this for him. Adams found it unsuitable as a rubber but saw its possibilities as an improvement over the then-current chewing gum base. Samples were made up, they were well liked, and the industry switched to chicle as a preferred gum base (8).

The sapodilla tree grows in Central and South America, particularly in Mexico, Guatemala, and Venezuela. Most of the trees are in the Yucatan Peninsula of Mexico. The trees frequently grow to 100 feet high, have a dense wood, and are not tapped until they are over 20 years old. Tapping is not unlike that practiced with natural rubber: Herringbone-like cross-cuts lead to a center channel. Yields however are vastly different. After about 6 hours, perhaps 2.5 lb of chicle gum might be produced; during the same interval there might be 0.5 lb of latex collected, 30% of which is natural rubber. Sapodilla trees are tapped only every 3 or 4 years. A first step in processing the latex is to boil it until the water content is reduced to 33% or less. Further refining includes washing it with strong alkali, then neutralizing with sodium acid phosphate. After washing and drying the resulting final product is an amorphous, pale pink powder.

Chewing gum is enjoying increased usage in the world and would be in short supply if chicle were the only base. Other bases now being used include refined pine tree resins from the southeastern U.S., jelutong from Indonesia and Malaysia, and synthetic rubbers such as polyisobutylene and styrene-butadiene rubber. Chicle consumption is estimated in the United States at 1000–2000 tons per year.

References

1. American Society for Testing and Materials 1987 Standard specification for rubber – concentrated, ammonia preserved, creamed and centrifuged natural latex. D 1076-80. ASTM Philadelphia
2. American Society for Testing and Materials 1987 Standard terminology relating to rubber. D 1566-87a. ASTM Philadelphia
3. Barlow C 1978 The natural rubber industry. Oxford University Press Kuala Lumpur Malaysia
4. Baum V 1943 The weeping wood. Doubleday Doran Garden City NY
5. International Rubber Study Group 1983 Rubber statistical bulletin 38(3):6
6. Malaysian Rubber Producers Research Association 1980 Rubber developments 33(3):62–63
7. National Academy of Sciences 1977 Guayule: An alternative source of natural rubber. Nat Acad Sci Washington DC
8. National Association of Chewing Gum Manufactureres (n d) The story of chewing gum. NACGM, New York
9. Nickell L G 1982 Plant growth substances. Kirk-Othmer Encyclopedia of Chemical Technology (3rd ed) 18:16
10. Rubber and Plastic News 1987 INRO begins selling some buffer stock NR to pay for expenses. Rubber Plastic News 17(8):7
11. Rubis DD, Cossens LJ 1987 Guayule. McGraw-Hill Encyclopedia of Science and Technology 8:248
12. Sandermann W, Dietrichs H H, Simatupang M H, Puth M 1963 Untersuchungen über Kautschukhaltige Hölzer. Holzforschung 17:161–168
13. Schridowitz O, Dawson T R (eds) 1952 History of the rubber industry. Heffer and Sons Cambridge
14. Semegen S T, Cheong S F 1978 Natural rubber. The Vanderbilt rubber handbook. R T Vanderbilt Co Norwalk CT, 18 pp
15. St Cyr D R 1982 Rubber, natural. Kirk-Othmer Encyclopedia of Chemical Technology (3rd ed) 20:489
16. Subramanian A 1987 Natural rubber. In: Morton M (ed) Rubber technology (3rd ed) Van Nostrand Reinhold New York, 180
17. Wolf H, Wolf R 1936 Rubber: A story of glory and greed. Covici Friede New York

10.5 Other Extractives and Chemical Intermediates

E. P. SWAN

10.5.1 Introduction

> "And when they were come into the house, they saw the young child with Mary his mother, and fell down, and worshipped him and when they had opened their treasures, they presented unto him gifts; gold, and frankincense, and myrrh."
>
> (Matthew 2:11)

The most famous extractives ever known in the West, frankincense and myrrh, were given by the three wise men to the Christ child nearly 2000 years ago. The text makes it clear that these extractives were treasures of value equal with gold. Unfortunately, this is no longer the case with most extractives: only the rarest ones command a price equal to gold's US $490 per troy ounce. Nevertheless, frankincense, the gum resin from five *Boswellia* spp., and myrrh, the exudate from *Commiophora* spp., are still produced today (17, 30), capping a long history of use in ancient times in worship, medicine, and embalming – proof of the durability of the demand for these extractive products (see Chap. 1.1). Other familiar wood products that have been produced throughout history are camphor oil (from *Cinnamomum camphora*) and sandalwood oils (from *Santalum album*) (2, 8).

Research in modern times has identified many thousands of compounds in extractives research. These extractives are produced in varying quantities depending on the demand and their natures. Many classes of wood extractives are covered in other sections of this book: natural exudates in Chap. 1.1; tall-oil constituents as extractives from softwoods by kraft pulping liquor in Sect. 10.1; and pharmaceuticals such as cascara extract in Sect. 10.6. In this section I will discuss successful softwood and hardwood extractives utilization and prospects for their continued and expanded uses.

10.5.2 Conifer Extractives Utilization

10.5.2.1 Balsams, Copals, Amber, and Other Products

Balsams are sticky, clear exudates from wood and resin pockets in the bark of certain conifers. The most useful one, Canada balsam, has been obtained from *Abies balsamea* (2). Canada balsam is mainly used in microscopy. It is produced on a small scale from the gum exuded from the blister resin pockets in the winter. A similar product, Oregon balsam, was obtained from *Pseudotsuga menziesii* (2). Production and value data for Canada balsams are available only up 1965 (6) because of infrequently collected statistics. The amount produced in 1965 was over 51 000 pounds. Sandermann (25) discussed the physical and chemical properties of 33 different balsams, their major components, and qualitative tests to distinguish between them. He described the sources by regions and provided some old production data with particular emphasis on the various copals and amber. The review by Weaver (30) is more recent, and it thoroughly describes the variety of natural resins. It gives sources, 1981 prices, and selected chemical reactions of resin components. The properties given were the acid, saponification and ester numbers, the wetting or softening points, ash content, if any, and solubility in various solvents.

Larch or venetian turpentine is another softwood extractive product used in the 1950s (25). At that time, usage, largely as a solvent, was about 20 tons per year. Modern data on production volumes and prices of this product are not available.

10.5.2.2 Cedar Wood Oils

The major recent development in the production of cedar wood oils has been the establishment of industries in China and Texas. The oil produced in China (10, 11) is probably from *Juniperus chinensis* (J W Rowe, personal communication, 1984). In 1986, 420000 kg were imported into the United States at a value of $847000 (29). At the same time, the United States' production of cedar wood oil allowed exports of 1 million kg, worth $ 5.6 million (16).

The United States' production of oil was derived from two species of wood. Virginia cedar wood oil is obtained by steam distillation of sawdust, waste wood, oil stumps, and chopped logs of eastern red cedar, *Juniperus virginiana* (19). Texas cedar wood oil is produced by steam distillation of chipped heartwood of *Juniperus mexicana* (15). The two oils are similar in chemical composition and they compete in the market place. The production of Virginia oil is about 140000 to 180000 kg/year and of the Texas oil about 450 to 900 million kg/year. The oils are used in the fragrance industries in products such as soaps, air fresheners, floor polishes, and sanitation supplies. The largest use for the oils is as a source of cedrol, their main component. Cedrol (**1**) is used to make cedryl acetate, which is used in perfumes. Cedar wood oils sell in the United States for $5 to $7/kg.

1 cedrol

Production of a third type of cedar wood oil distillate has been started. This is the oil from *Thuja plicata*, western red cedar. A single producer, Cedar-al Products, Clallam Bay, Washington, has been operating for a few years. Price and volume data on this product are unavailable. This oil is distinctly different from that made from the foliage. The wood oil is produced for its pleasant odor. Some of the production is used to impart this odor to cedar-chip-filled pet beds.

10.5.2.3 Non-Commercial Extractives

Non-commercial extractives are the so-called "Cinderella" products, for they are waiting in the kitchen, ready for a "Prince Charming" entrepreneur to utilize them. They are available because there is some novel feature of their isolation or occurrence that means that large amounts are potentially available. Some are not used at the moment because they are too expensive, untested, or unknown; others are of historical interest.

10.5.2.3.1 Conidendrin

This phenyl-tetralin-type lignan can be made from the sulfite pulping liquors from cooking of *Tsuga heterophylla* and *Abies amabilis* woods. A simple solvent extraction of the liquor causes separation of conidendrin crystals. More conidendrin (2) is found in the pulping liquors than in the wood because conidendrin is also formed by a dehydration reaction of a precursor, hydroxymatairesinol. Conidendrin and some derivatives were tested for use as antioxidants and for pharmaceutical applications without success.

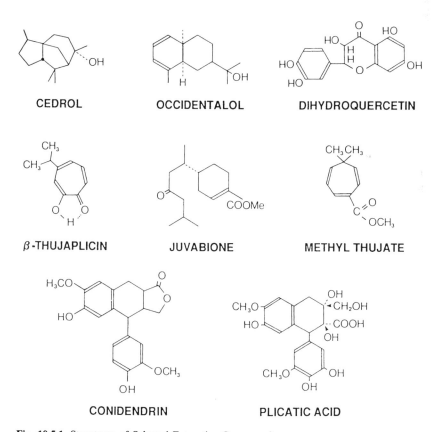

Fig. 10.5.1. Structures of Selected Extractive Compounds

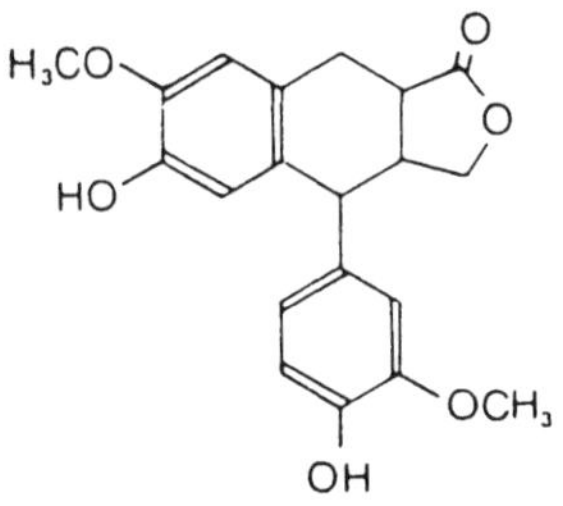

2 conidendrin

10.5.2.3.2 Dihydroquercetin

This pentahydroxy flavanolone (3) is of widespread occurrence in several species (13). It is most easily isolated by a simple extraction from *Pseudotsuga menziesii* bark. The yields of the crystalline product are about 5% from the bark. It may be readily oxidized to quercetin. These compounds have been tested as antioxidants, natural coloring materials, and drugs (13).

10.5.2.3.3 Juvabione and Related Insect Hormones

Since the classic work of Slama and Williams (27), who first demonstrated the presence of juvabione (4) in *Abies balsamea*, their work has been expanded to show that this keto-sesquiterpene methyl ester and its relatives are present in the heartwoods of other *Abies* species and in *Pseudotsuga taxifolia*. They may be recovered by a simple solvent extraction of groundwood pulp. They are used in forest-insect traps.

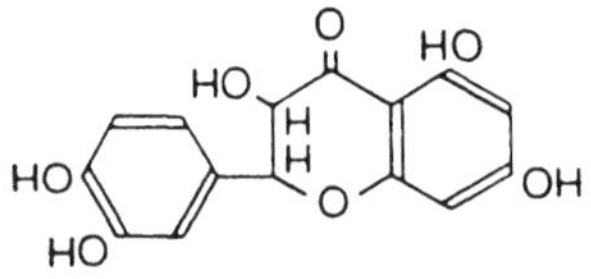

3 dihydroquercetin 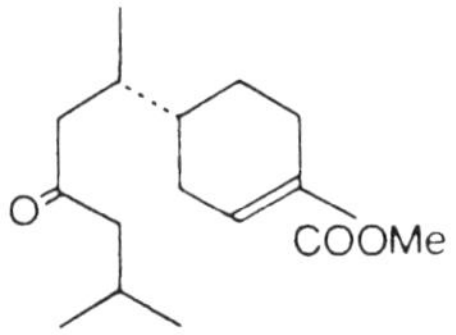   **4 juvabione**

10.5.2.3.4 Occidentalol

Roy et al. (24) have shown that this sesquiterpene alcohol was potentially available from kiln drying *Thuja occidentalis* in eastern Canada. Occidentalol (5) may find a use because of its pleasant odor.

10.5.2.3.5 Plicatic Acid

This highly hydroxylated phenyl-tetralin-type lignan acid is available from *Thuja plicata*. It is recovered in about 2% yield from the heartwood. Because it is one of the strongest known natural acids, plicatic acid (6) is readily separable from

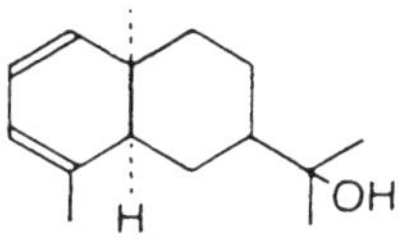

5 occidentalol

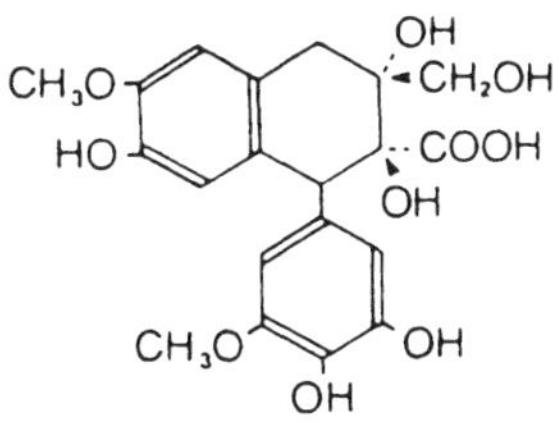

6 plicatic acid

the other water-soluble compounds by precipitation of its potassium or calcium salts (5). This strong acid can be isolated from the water extractives of the heartwood by various treatments of the crude water extract (5). It has been used in medical research (see Chap. 9.5).

10.5.2.3.6 Thujaplicins, Thujic Acid, and Methyl Ester

These monoterpenes (and derivatives) are all available from *Thuja plicata*. They can be prepared either via a dry-kiln condensate preparation route or by fractional distillation of the steam-distilled oil (5). Since this oil is now being produced commercially, the latter route is preferable. The structure of β-thujaplicin (7) is shown. The γ-isomer is also found in about the same amounts, together with much smaller amounts of the α-isomer. These compounds are natural fungitoxins with their activities responsible for the durability of *Thuja plicata* heartwood. Aqueous solutions of them corrode stainless steel. Digester corrosion thus caused is no longer a problem because of design changes (5). Methyl thujate (8) is responsible for the pleasant odor of the heartwood. Free thujic acid is also found in the wood. Some esters and amides of thujic acid were prepared for their interesting attractant and repellant qualities towards three insects: the German and American cockroaches and the yellow fever mosquito (14). Similarly, the oil from *Cedrus deodora* is effective against Indian mosquitoes (12).

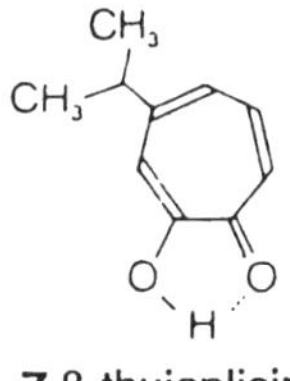

7 β-thujaplicin

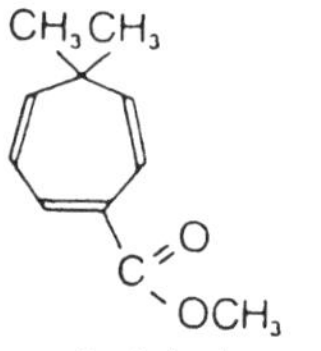

8 methyl thujate

10.5.2.3.7 Waxes

These fatty alcohol esters of fatty acids are only minor constituents of most woods. However, for a time, wax was prepared by solvent extraction of *Pseudotsuga menziesii* bark. The extraction plant that operated for six years by Bohemia, Inc., Eugene, Oregon, closed in 1982 for economic reasons. The plant produced a single grade of wax that competed with imported waxes. Besides the above ester, the crude wax contains an interesting mixture of estolides of hydroxy fatty acids, and fatty alcohol esters of caffeic and ferulic acids.

10.5.3 Utilization of Hardwood Extractives

Hardwood extractives are produced in minor amounts all over the world. Some, such as cinnamon and licorice, are familiar items of our diets. Others, such as crude medicinal extracts, are produced and consumed locally. Table 10.5.1 lists these extractives alphabetically. Production data on these items do not exist.

10.5.4 Prospects

This chapter is restricted to those intermediates not otherwise described in their own chapters. For a discussion of natural dyes, pharmaceuticals, tannins, and other products, the appropriate authors in this book should be consulted.

Bratt (7) points out that basic chemical products can be produced economically only from coal or oil. However, specialty chemical products can either be synthesized or produced by modification of silvichemicals. The modification process has been used with tall oils, with the odiferous essential oils, and with extracts produced on a cottage industry scale for pharmaceutical purposes. There have been two review articles written recently: Lawrence (19) examined the world production of essential oils in 1984; and Matsubara (20) discussed the utilization of terpenes in Japan.

The market for pleasant-smelling wood oils, such as those from sandalwood or from various members of the Cupressaceae, will continue to expand (17). The reasons for this are that they have been successful since historical times, and the population of the world is still increasing. These essential oils must generally be produced in "cottage industry"-type (1) factories since commercial extraction on a large scale is considered uneconomical, not only in the Philippines (1) but also elsewhere (19).

Successful products are obtained from woods with high contents of extractives of unique structure. The best example of this is the American cedar wood oil, whose main component is cedrol. The unique structure of this sesquiterpene alcohol means that it is available only by extraction and not by synthesis. Cedrol is available only from a few wood oil sources. Similarly, the alkaloid taxol is only available from the bark of yew trees. It may be necessary to harvest the product from 10 000 *Taxus brevifolia* plants in order to obtain sufficient for testing as an anticancer agent (26). This was not true for the waxes from Douglas-fir bark since many other waxes are commercially available. Similarly, the natural dye haemaxylin was thought to be supplanted by cheaper synthetics (see Chap. 1.1). However, McLaren (21) states that these compounds are still used in industry to color synthetic fibers. There is also a great deal of interest in using natural pigments for food coloring (28). These conditions of component uniqueness and non-synthetic competition will probably continue. At the same time, the flavor and fragrance industry is developing new products and lines. The industry has become more international in outlook and changes in it are expected to continue (9).

Development of commercially important extractive compounds from many other softwood and hardwood species throughout the world may be possible. How many other woods have a pleasant odor and yet are not exploited for their wood oil? Odor, of course, is merely one property that gives value to an oil. Toxic-

Table 10.5.1. Partial listing of hardwood extractives and references to them

Hardwood products	Reference
Acacia spp. extracts	Sect. 1.1.2.2
American storax	23
Balm of Gilead	23
Birdlime (Japanese *Ilex* sp.)	23
Birch oil	23
Cinnamon	Sect. 1.1.2.5
Copaiba balsam	Sect. 1.1.2.2
Frankincense	Sect. 1.1.2.1
Glaucine	23
Blackberry tannins	23
Jatropha curcas products	18
Kamala extracts	23
Licorice root	
Magnolia medicinal oil (Japan)	23
Mangrove tannin	23
Maple syrup	23
Myrrh	Sect. 1.1.2.1
Oak tannin	23
Osage orange dyes	23
Oxalic acid	3, 4, 22
Rauwolfia spp. extracts	23
Sassafras oil	23
Styrax	2
Tolu balsam	23
Vinca spp. alkaloids[a]	23
Wood oils – camphor	2
– bois de rose	2
– sandalwood	2

[a] Vincristine from this source sold at US $200000/kg (abt. 13 times the price of gold)

ity is another property of interest. The oil from *Abies* spp. could be produced for their contents of insect juvenile hormones. The fir oil might have applications as an aereal spray to kill, or lessen, infestations of forest insects. Perhaps oil from *Thuja plicata* may be used to repel or kill fleas and ticks in pets or livestock. Other wood oil may control infestations of ticks, nematodes, and worms in various agricultural applications. The potential usefulness of these oils is considerable. Undoubtedly, any natural product that replaces synthetic halogenated organics will find a ready market because of continued public obloquy about the synthetics. The testing of these compounds as agents to control human, animal, and plant diseases remains an awesome task for the future.

The production of natural dyes for cottage industries producing custom cloth will continue. The production of cinnamon and liquorice root should survive as long as mankind has a sweet tooth. Despite the availability of synthetic vanillin from sulfite pulping liquors (16), vanilla beans are still produced for those who demand the genuine article. It is these facts that lead to the optimistic conclusion that the natural dyes, pharmaceuticals, oleoresins, and gums will continue to fill an important niche.

References

1. Anzaldo F E 1981 Production and utilization of essential oils in the Philippines. NSDB Tech J 6(1):13–19
2. Arctander S 1960 Perfume and flavor materials in natural origin. Arctander Elizabeth NJ, 2000 pp
3. Bahtia K, Ayyar K S 1980 Barks of *Terminalia* species, a new source of oxalic acid. Indian For 106:363–367
4. Bahtia K, Lial J, Swaleh M 1981 Storage of *Terminalia* bark in the open: its effect on oxalic acid and tannin content. Indian For 107:519–523
5. Barton G M, MacDonald B F 1971 The chemistry and utilization of western red cedar. Can For Serv Publ 1023 Ottawa, 16 pp
6. Bender F 1967 Canada balsam, its preparation and use. Can For Rur Dev For Dept Publ, 7 pp
7. Bratt L C 1979 Wood-derived chemicals: trends in production in the US. Pulp Pap 53(6):102–108
8. Burley J, Lockhart L A 1985 Chemical extractives and exudates from trees. In: Fuller K W, Gallon J R (eds) Plant products and the new technology. Oxford University Press Oxford
9. Chemical and Engineering News 1987 Flavors and fragrances industry taking on a new look. Chem Eng News 65(July 20):35–38
10. Chemical Marketing Reporter 1982 China cedarwood oil output better than some have said. Chem Market Rep 22(Nov 22):18
11. Chemical Marketing Reporter 1983 Essential oil imports increase. Chem Market Rep 23(Dec 5):36–37
12. Divijendra S, Rao S M, Tripathi A K 1984 Cedar wood oil as a potential insecticidal agent against mosquitoes. Naturwissenschaften 71:265–266
13. Fengel D, Wegener G 1984 Wood chemistry ultra structure reactions. de Gruyter New York, 612 pp
14. Hach V, McDonald E C 1973 Terpenes and terpenoids. IV. Some esters and amides of thujic acid. Can J Chem 51:3230–3235
15. Halverson H N, Swan E P 1982 Market study for British Columbian cedar foliage oil. BC Sci Coun Rep Richmond BC, 40 pp
16. Hergert H C 1960 Chemical composition of tannins and polyphenols from conifer wood and bark. For Prod J 11:610–617
17. Hillis W E 1986 Forever amber. A story of the secondary wood components. Wood Sci Tech 20:203–227
18. Levingston S, Zamora R 1983 Medicine trees of the tropics. Unasylva 35(140):7–10
19. Lawrence B M 1985 A review of the world production of essential oils (1984). Perf Flav 10(5):2–16
20. Matsubara Y 1987 Recent advances of chemistry and utilization of terpenes. Chem Ind For Prod (China) 7(2):1–18
21. McLaren K 1983 The colour science of dyes and pigments. A Hilger Ltd Bristol, 280 pp
22. Prabhu V, Theagarajan K S 1977 Utilization of eucalyptus hybrid (Mysore gum) bark for production of oxalic acid. Indian For Bull 103:477–479
23. Rowe J W, Conner A H 1979 Extractives in eastern hardwoods: a review. US Dep Agr Gen Tech Rep FPL-18, 60 pp
24. Roy D N, Konar S K, Purdy J R 1984 Sublimation of occidentalol, a sesquiterpene alcohol from eastern white cedar (*Thuja occidentalis* L.) in a drying kiln. Can J For Res 14:401–411
25. Sandermann W 1960 Naturharze Terpentinöl – Tallöl. Chemie und Technologie. Springer Berlin, 20 pp
26. Science 1987 Combing the earth for cures to cancer, AIDS. Science 237:969–970
27. Slama K, Williams C M 1966 "Paper factor" as an inhibitor of the embryonic development of the european bug *Pyrrhocoris apterus*. Nature 210:329–330
28. Timberlake C F, Henry B S 1987 Plant pigments as natural food colours. Endeavour New S 10(1):31–35
29. US Department of Agriculture 1987 Foreign agriculture circular: essential oils. US Dep Agr FTEA 2–87, 20 pp
30. Weaver J C 1982 Resins, natural. In: Kirk-Othmer Encyclopedia of Chemical Technology (3rd ed) 20:197–206

10.6 Pharmacologically Active Secondary Metabolites from Wood

C. W. W. BEECHER, N. R. FARNSWORTH, and C. GYLLENHAAL

10.6.1 Introduction

Plant-derived drugs are of major importance to the health and well-being of mankind. Although many of these useful drugs have been synthesized, very few are produced commercially by synthesis. Thus, the plant source remains extremely important, not only in producing drugs, but in providing novel biologically active model compounds from which potentially more potent and less toxic drugs may be synthesized (171, 575).

In the United States alone, for example, 25% of all prescription drugs dispensed from community pharmacies from 1959–1980 (the latest data available) contained plant extracts or active principles prepared from higher plants. This figure did not vary by more than ±1.0% in any of the 22 years surveyed (310, 312), and in 1980, consumers in the United States paid more than $8 billion for prescriptions containing active principles extracted from higher plants (313).

On a global basis we have recently compiled data showing that at least 119 drugs are currently in use that are extracted from higher plants. These are obtained from only 91 species of plants (311). These 119 useful drugs are listed in Table 10.6.1, together with their most prominent uses. A large number of these drugs are not used in the United States for a variety of reasons. In some cases, the disease or malady for which they are useful does not exist to any degree in United States – e.g., emetine is useful to treat amoebic dysentery, which is not a condition of concern in most developed countries. Relatively few of these useful drugs accumulate to an appreciable degree in the wood of higher plants.

This review is intended to identify drugs currently in use that are found in woody parts of higher plants and to survey the current literature for additional pharmacologically active secondary metabolites extracted from wood that have interesting and/or potential future application as drugs for humans. The reader is directed to an interesting publication surveying potential useful secondary metabolites from Eastern hardwoods (884).

10.6.2 Sources of Information

Data for this review were obtained by searching the NAPRALERT computer database on natural products, which was housed on an Alpha-Micro minicomputer by the Program for Collaborative Research in the Pharmaceutical Sciences, College of Pharmacy, University of Illinois at Chicago. NAPRALERT contains data from over 75000 literature sources that treat the phytochemistry, biological activities, and medicinal folklore associated with 35000 organisms, both plants and animals. About 70% of the organisms in the database are higher plants. Approximately 195000 results of pharmacological tests and 190000 compounds identified from natural sources are contained in the database (628). Each chemical, pharmacological, or folkloric record is entered independently, and can be cross-correlated with any other record type, so that data from NAPRALERT searches are printed out in the form of a table of results, rather than as a series of abstracts from which information must be gleaned individually.

The search process used to generate the tables shown in this section took place in several stages. In the first stage, compounds isolated from plants in predominantly woody families, or woody genera in predominantly non-woody families of gymnosperms and angiosperms were requested. This search was conducted only for information published from 1980–1985. Determinations of which plant families and genera were likely to include woody species were made by referring to a number of standard botanical references, including systematic treatments, florae, and articles on wood anatomy in predominantly non-woody families. The print-out that resulted from this search was inspected to determine which compounds were likely to occur in the wood of the species listed, by noting the part of the plant from which the compound was isolated. Wood is seldom selected for individual analysis by phytochemists, since it is often perceived as being less likely to contain interesting compounds than other plant parts, such as leaves and bark. However, many plant organs that are analyzed do contain substantial amounts of wood, and it is possible that compounds isolated from woody stems, twigs, branches, or roots are actually contained in the woody, rather than non-woody, tissue. Therefore, compounds isolated from any plant part that contains a substantial proportion of wood were selected for further study in this phase of the search. These plant parts were the following: wood, sapwood, heartwood, trunkwood, rootwood, roots, stems, trunks, branches, and twigs. Roots of perennial herbs, which may contain woody secondary growth, and roots, stems, and twigs of woody shrubs were also included. Wood from plants with unusual wood anatomy, such as woody members of the Compositae, was also included.

The compounds selected in the first stage of the search were then used in the second stage, which consisted of a listing of all of the experimental results of pharmacological or biological activity testing of each compound. In this stage, the entire database was searched. The print-out that resulted from this search contains a brief summary of each experimental test to which the compound has been submitted. The summaries include information on the test system for in vitro tests, or the test animal in the case of in vivo tests, the dosage and dosage schedule, the test protocol, the results of the test, and other information such as comments concerning the validity of the testing procedures. For each compound, all types of pharmacological or biological tests in which the compound was active were noted. Only those biological tests, however, that are directly and obviously related to development of new drugs were included in the final table. Thus, tests for mutagenicity or toxicity, and tests on enzymatic systems that cannot be directly related to effects at the organism level were omitted from subsequent analysis.

Because compounds that show promise for some type of pharmacologically significant activity may have been submitted to numerous tests by different investigators, only one example of each type of pharmacological activity was selected, in order that the bibliography might remain as compact as possible. An effort was made to select the best possible experimental protocol for the reference listed in the table of pharmacological activities (Table 10.6.2). Thus, in vivo tests were selected rather than in vitro tests, when possible, and tests on human subjects were selected rather than those on other subjects. Superior experimental designs were selected over inadequate or unclear experimental designs.

Data on the compounds and their pharmacological activities were then put into a microcomputer database system for further manipulation. The general chemical class of each compound was determined from the NAPRALERT system of coding chemical names, and was included in the database.

In the third phase of the search, a list of all plants in which each compound occurs was requested from NAPRALERT. In this case, the entire database was searched. Printouts were inspected to select only those compounds that occurred in wood or in woody plant parts, and the plants in which the compounds occurred were noted. During this phase of the search, it was noted that compounds (e.g., terpenes) that are components of essential oils or resins were sometimes listed as having been isolated from the wood, and sometimes from the wood essential oil or the resin of a particular species. Thus, for the sake of consistency, plant species in which the compound occurred in the wood essential oil or the resin of conifers were included in the list of plants from which the compound had been isolated. Several compounds that were found to be very widely distributed among woody plants (e.g., β-sitosterol) were deleted from the analysis at this point for the sake of brevity. Data on the compounds and the plants from which they had been isolated were put into a microcomputer database system and used to construct Table 10.6.3.

10.6.3 Currently Used Drugs Produced in Wood

There are currently about 25 drugs that occur appreciably in the woody parts of higher plants. Most of these are alkaloids.

10.6.3.1 Alkaloids

10.6.3.1.1 Tropane

Atropine and hyoscyamine are widely employed as anticholinergic drugs, primarily for the treatment of stomach spasms. Atropine is also applied topically to the eye to produce dilation during ophthalmologic examinations. Scopolamine, although having an anticholinergic effect, has a more pronounced effect on the higher nervous centers and is employed as a sedative and to treat motion sickness.

10.6.3.1.2 Quinolizidine

Sparteine and its enantiomer pachycarpine are both employed as oxytocics in childbirth to initiate uterine contractions and to decrease post-partum hemorrhage. They are often employed as substitutes for the ergot alkaloids and oxytocin.

10.6.3.1.3 Phenylalanine Derivatives

10.6.3.1.3.1 Simple Tyramine Derivatives

Ephedrine, pseudoephedrine, and norpseudoephedrine are now produced primarily by synthesis. Ephedrine is used as a bronchodilator for treating the symptoms of asthma and for treating coughs. Pseudoephedrine and norpseudoephedrine are used for the same purposes and are included in formulae for over-the-counter remedies for treating symptoms of the common cold. All three of these alkaloids cause a rise in blood pressure and this is a deterrent to their use in patients with hypertension.

10.6.3.1.3.2 Protoberberine Type

Berberine is used widely throughout the world, but not to any great extent in the United States. It is an abundant alkaloid and is easily extracted from a number of plant sources. It is particularly useful in the treatment of bacillary dysentery, for it has good antibacterial activity and is poorly absorbed from the gastrointestinal tract when administered orally.

10.6.3.1.3.3 Phthalideisoquinoline Type

Hydrastine is a minor product, useful when applied topically as an astringent.

10.6.3.1.3.4 Benzo(c)phenanthridine Type

Sanguinarine is currently used in the United States in dental preparations (mouth washes, toothpastes), where there is some evidence that the products containing this alkaloid prevent dental plaque formation.

10.6.3.1.3.5 Ipecac Type

Emetine (and synthetic dehydroemetine) are employed primarily in developing countries from the treatment of amoebic dysentery. Ipecac root products (primarily ipecac syrup), which contain mainly emetine and cephaeline are widely used as emetics in cases of poisoning.

10.6.3.1.4 Tryptophane Derivatives

10.6.3.1.4.1 Indole

Reserpine, rescinnamine, and deserpidine have been employed since the early 1950s for the treatment of essential hypertension and as tranquilizers. Ajmalicine is used especially in Europe, in conjunction with other agents to treat hypertension and other cardiovascular disorders.

10.6.3.1.4.2 Quinoline

Quinine is employed primarily for the treatment of *Plasmodium falciparum*-induced cases of malaria, especially when the cases are resistant to synthetic chemotherapy. Quinidine is used for the treatment of cardiac arrhythmias.

10.6.3.2 Quinoids

Sennosides, primarily sennosides A and B, derived from *Cassia angustifolia* and *C. acutifolia*, are employed as a mixture for the alleviation of constipation.

10.6.3.3 Lignans

Podophyllotoxin is used externally to treat condylomata acuminata (soft warts). Etoposide is a semi-synthetic derivative of epipodophyllotoxin glycoside, currently being used for the treatment of cancer.

10.6.3.4 Triterpenes

Glaucarubolide is a quassinoid-type degraded triterpene that is used orally for the treatment of amoebic dysentery. Glycyrrhizin, a triterpene glycoside, is widely employed as a sweetening agent (50 times sweeter than sucrose). It also imparts a typical licorice-like taste. Extremely high doses of glycyrrhizin produce aldosteronism in humans.

10.6.3.5 Pyrones

Kawain occurs naturally in the rhizomes of *Piper methysticum*. Synthetic kawain is used in Europe in geriatric patients to improve well being and as a calmative agent.

10.6.3.6 Coumarins

Xanthotoxin (8-methoxypsoralen) is used as a repigmentation agent for the treatment of vitiligo and leukoderma. Subjects using xanthotoxin should avoid exposure to direct sunlight since severe blistering can result. This agent is also hepatotoxic in high doses.

10.6.4 Potential Drugs Derived from Secondary Metabolites of Wood

It is obvious that the many secondary metabolites from wood listed in Table 10.6.2 have a myriad of biological effects. It will not be possible to review all of them thoroughly. Thus, only a few of the most recently discovered compounds, which have unusual structures and which appear to have the most interesting potential as drugs or as models for the synthesis of related compounds as drugs, will be briefly discussed.

10.6.4.1 Alkaloids

10.6.4.1.1 Tropane

Anisodamine, **1**, and anisodine, **2**, have been studied extensively (195, 199, 420, 476, 723, 724, 930, 1059) in the People's Republic of China as anticholinergic agents, especially in the treatment of septic shock.

10.6.4.1.2 Isoquinoline

Of great interest as a potential antitumor agent for human neoplastic disease has been the alkaloid camptothecin, **3**, first isolated from the Chinese tree *Camptotheca acuminata*. Although this alkaloid has been subjected to some limited clinical trials, its effects have not paralleled those seen in animal studies. However, interest in camptothecin and certain of its analogues continues (441, 629, 630, 1060, 1120). Harringtonine, **4**, and homoharringtonine, **5**, obtained from several *Cephalotaxus* species, are both currently in clinical trials in the United States and in the People's Republic of China for the treatment of neoplastic diseases (155, 441–443, 612, 626, 787, 1084, 1094, 1095, 1097, 1109). Nitidine, **6**, a benzophenanthridine alkaloid of interest as an antitumor agent to the drug development program of the United States National Cancer Institute (NCI), has been clinically evaluated, but has been shown to be too toxic for general acceptance as an antitumor agent (447).

10.6.4.1.3 Indole Alkaloids

Ellipticine, 7, and a number of its derivatives, especially quaternized derivatives, show great promise as antitumor agents (832). These derivatives are of interest in the NCI program in the United States and even more so in France.

10.6.4.1.4 Ansamacrolides

A large number of ansamacrolides, exemplified by maytansine, 8, have potent antitumor activity. Maytansine has been subjected to clinical studies in the NCI program and has been disappointing as an anticancer agent. Furthermore, it has been shown to be quite toxic. Analogue development may produce a more useful and less toxic antitumor agent in the future (881).

10.6.4.1.5 Quinazoline

Vasicine, 9, from *Adhatoda vasica*, is currently being tested in Phase I clinical studies in India as an oxytocic agent – i.e., to initiate contractions in childbirth and to reduce post-partum bleeding (66).

10.6.4.1.6 Pyrazine Derivatives

Tetramethylpyrazine (TMP), **10**, has been isolated from a variety of plant species (Table 10.6.3) and has been shown experimentally to have hypotensive (767) activity, antiarrhythmic, and antibronchoconstrictor (763) activity, and antimicrobial activity (748), and to improve coronary blood flow (1137).

10.6.4.2 Quinoids

Lapachol, **11**, has interesting antitumor activity (264, 351) as well as in vitro antiviral activity against polio type 1, ECHO type 19, coxsackiae B-4, and several strains of influenza viruses (596). Three novel furonaphthoquinones, **12**, **13**, and **14**, are reported to have cytotoxic effects with ED50 values of 1.0–2.0 µg/ml against the P388 leukemia in vitro (861).

12 $R_1 = R_2 = H$; $R_3 = O$

13 $R_1 = R_2 = H$; $R_3 = OH$

14 $R_1 = H$; $R_2 = OH$; $R_3 = OH$

or $R_1 = OH$; $R_2 = H$; $R_3 = OH$

10.6.4.3 Lignans

Magnolol, **15**, honokiol, **16**, and 3,4'-diallyl-2'-hydroxy-4-methoxybiphenyl, **17**, are three lignans isolated from *Magnolia grandiflora* that have in vitro antibacterial and antifungal activity (216, 590). The bark of this plant is used in Japan to prevent dental caries (721). Magnolol and honokiol produce sedation, ataxia, muscle relaxation, and a loss of the righting reflex with an increase in dose from 50 to 100 mg/kg in mice. They also suppress spinal reflexes in young chicks in a manner similar to mephensin, but with a much longer duration of action. Pretreatment of mice with magnolol 100 mg/kg inhibited tonic extensor convulsions and death produced by intraventricular injection of penicillin G (1063).

Pharmacologically Active Metabolites 1067

	R	R₁	R₂
15	OH	OH	H
16	OH	H	H
17	OH	H	OCH₃

	R₁	R₂	R₃	R₄	R₅
18	Ac	H	Ac	H	CH₃
19	H	Ac	Ac	H	CH₃
20	Ac	H	Ac	H	CH₂-OH

Magnolol also prevents stress-induced ulcer formation and is antisecretory, probably due to its central depressant action (1064).

Phyllanthoside, **18**, and phyllanthostatins 1 (**19**), 2 (**20**), and 3 (**21**) have interesting antitumor activity in vivo against several murine tumor systems, including the B16 melanoma, at doses of 6–8 mg/kg. Phyllanthoside has been selected for preclinical development in the United States NCI program (818).

10.6.4.4 Diterpenes

Jatrophone, **22**, also has a broad antitumor spectrum and has been of interest in the NCI program (586). Tripdiolide, **23**, and related compounds such as triptolide, **24**, have also been of interest as antitumor agents in the NCI program (283). Four novel furanoditerpenes of the entclerodane type, plaunol B (**25**), C (**26**), D (**27**), and E (**28**) have been isolated from *Croton sublyratus* and show interesting activity against Shay-induced ulcers in rats at 3–10 mg/kg (552).

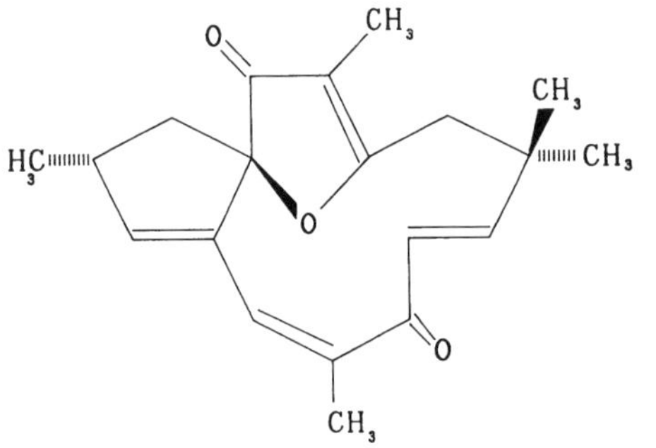

22

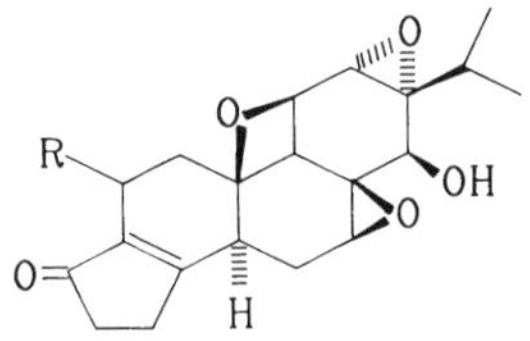

23 R=OH
24 R=H

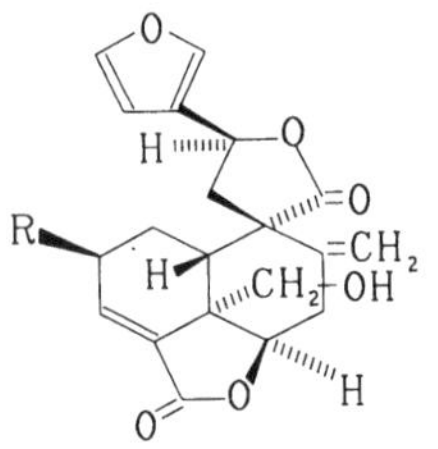

25 R=H
26 R=OH

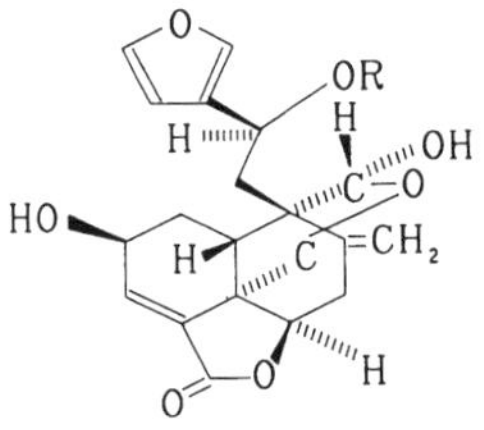

27 R=H
28 R=Ac

Of considerable interest is the New York Times report on March 1, 1988, of E. J. Corey's group's synthesis of the diterpene, ginkgolide B. Ginkgo extracts have a long history of use in Oriental medicine and have found therapeutic use in treating circulatory problems, asthma, toxic shock, and Alzheimer's disease, and as a possible replacement for cyclosporin used to prevent rejection of organ transplants.

10.6.4.5 Triterpenes

The most interesting of the varied types of triterpenes from the point of view of biological activity are the quassinoids. They exhibit antiamoebic (353), antimalarial (1020), antimitotic (287), antitumor (59, 403, 950), and cytotoxicity (59, 821, 823) effects. Bruceantin, **29**, is being developed as a potential cancer chemotherapeutic agent and is currently in Phase 1 clinical trials in the United States (950).

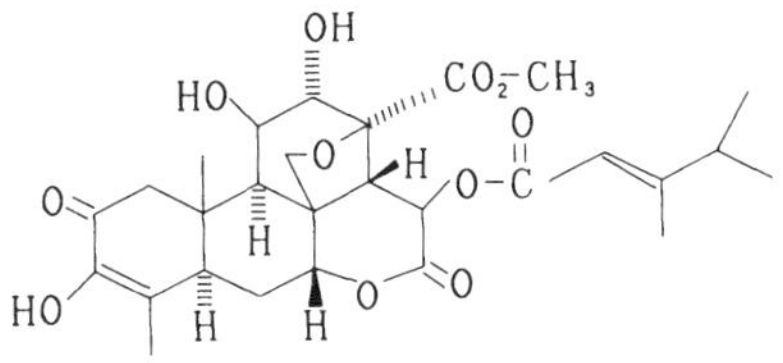

29

10.6.4.6 Flavonoids

The unusual flavonoid derivative kuwanon G, **30**, is reported to lower blood pressure in rabbits when administered intravenously at a dose of 1.0 mg/kg of body weight (743). (+)-Catechin [(+)-cyanidan-3β-ol], **31**, is remarkably bioactive. Among its recently reported biological effects are the following: antianaphylactic (517), anticoagulant (569), antihepatotoxic (963), platelet aggregation inhibition (118), and serotonin secretion inhibition (100) effects.

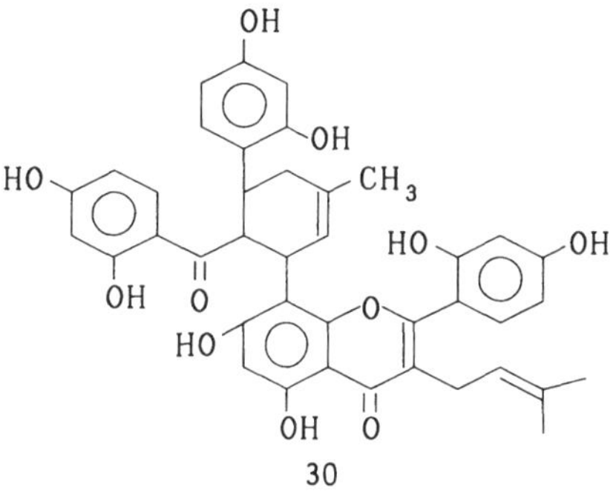

30 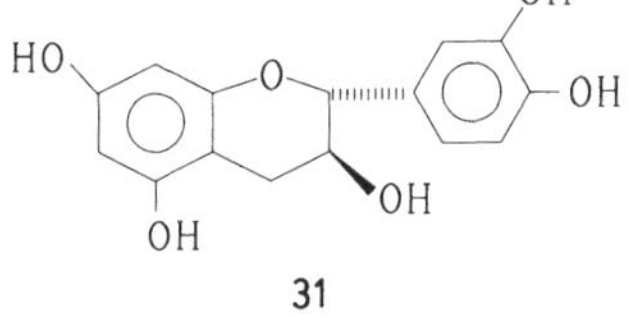   **31**

10.6.5 Summary and Conclusions

It is apparent from the data presented in Tables 10.6.2 and 10.6.3 that the woody parts of plants are a rich source of biologically active secondary metabolites with greater promise than one would have previously imagined. However, few of these metabolites have been isolated in studies by chemists interested in wood per se. Perhaps the data presented in this review will stimulate more chemists to look specifically at wood as a source of useful biodynamic agents.

Acknowledgements. The operation of the NAPRALERT database was supported by National Science Foundation grant number BSR A31 129 7. Mary Lou Quinn and her staff performed the computer searches used in constructing the tables and bibliography. Carol Lewandowski and her staff aided in editing the bibliography. Dr. Michael J. Huft assisted in microcomputer procedures.

Table 10.6.1. Drugs Used Globally that are Obtained from Plants

Drug	Use
Acetyldigoxin	Cardiotonic
Adoniside	Cardiotonic
Aescin	Anti-inflammatory
Aesculetin	Antidysentery
Agrimophol	Anthelmintic
Ajmalicine	Circulatory disorders
Allantoin[a]	Vulnerary
Allyl isothiocyanate[a]	Rubefacient
Anabasine	Skeletal muscle relaxant
Andrographolide	Bacillary dysentery
Anisodamine	Anticholinergic
Anisodine	Anticholinergic
Arecoline	Anthelmintic
Asiaticoside	Vulnerary
Atropine	Anticholinergic
Benzyl benzonate[a]	Scabicide
Berberine	Bacillary dysentery
Bergenin	Antitussive
Borneol[a]	Antipyretic; analgesic; anti-inflammatory
Bromelain	Anti-inflammatory; proteolytic agent
Caffeine	CNS stimulant
Camphor	Rubefacient
(+)-Catechin	Haemostatic
Chymopapain	Proteolytic; mucolytic
Cissampeline	Skeletal muscle relaxant
Cocaine	Local anaesthetic
Codeine	Analgesic; antitussive
Colchiceine amide	Antitumor agent
Colchicine	Antitumor agent; antigout
Convallatoxin	Cardiotonic
Cucurmin	Choleretic
Cynarin	Choleretic
Danthron (1,8-dihydroxy-anthraquinone)[a]	Laxative
Demecolcine	Antitumor agent
Deserpidine	Antihypertensive; tranquilizer
Deslanoside	Cardiotonic
L-Dopa[a]	Anti-Parkinsonism
Digitalin	Cardiotonic
Digitoxin	Cardiotonic
Digoxin	Cardiotonic
Emetine	Amoebicide; emetic
Ephedrine	Sympathomimetic
Etoposide[b]	Antitumor agent
Galanthamine	Cholinesterase inhibitor
Gitalin	Cardiotonic
Glaucarubin	Amoebicide
Glaucine	Antitussive

Table 10.6.1 (continued)

Drug	Use
Glaziovine	Antidepressant
Glycyrrhizin (glycyrrhetic acid)	Sweetener; Addison's disease
Gossypol	Male contraceptive
Hemsleyadin	Bacillary dysentery; antipyretic
Hesperidin	Capillary fragility
Hydrastine	Haemostatic; astringent
Hyoscyamine	Anticholinergic
Kainic acid	Ascaricide
Kawain	Tranquilizer
Khellin	Bronchodilator
Lantasides A, B, C	Cardiotonic
α-Lobeline	Smoking deterrent; respiratory stimulant
Menthol[a]	Rubefacient
Methyl salicylate[a]	Rubefacient
Monocrotaline	Antitumor agent (topical)
Morphine	Analgesic
Neoandrographolide	Bacillary dysentery
Nicotine	Insecticide
Nordihydroguaiaretic acid	Antioxidant (lard)
Noscapine (narcotine)	Antitussive
Oubain	Cardiotonic
Pachycarpine ((+)-sparteine)	Oxytocic
Palmatine (fibraurine)	Antipyretic; detoxicant
Papain	Proteolytic; mucolytic
Papaverine[a]	Smooth muscle relaxant
Phyllodulcin	Sweetener
Physostigmine (eserine)	Cholinesterase inhibitor
Picrotoxin	Analeptic
Pilocarpine	Parasympathomimetic
Pinitol[a]	Expectorant
Podophyllotoxin	Condylomata acuminata
Protoveratrines A & B	Antihypertensives
Pseudoephedrine, norpseudoephedrine	Sympathomimetic
Quinidine	Antiarrhythmic
Quinine	Antimalarial; antipyretic
Quisqualic acid	Anthelmintic
Rescinnamine	Antihypertensive; tranquilizer
Reserpine	Antihypertensive; tranquilizer
Rhomitoxin	Antihypertensive; tranquilizer
Rorifone	Antitussive
Rotenone	Piscicide
Rotundine ((+)-tetrahydropalmatine)	Analgesic; sedative; tranquilizer
Rutin	Capillary fragility
Salicin	Analgesic
Sanguinarine	Dental plaque inhibitor
Santonin	Ascaricide
Scillarin A	Cardiotonic
Scopolamine	Sedative
Sennosides A & B	Laxative
Silymarin	Antihepatotoxic
Sparteine	Oxytocic
Stevioside	Sweetener
Strychnine	CNS stimulant

Table 10.6.1 (continued)

Drug	Use
Teniposide[b]	Antitumor agent
9-Tetrahydrocannabinol	Antiemetic; decreases ocular tension
Tetrandrine	Antihypertensive
Theobromine	Diuretic; vasodilator
Theophylline	Diuretic; bronchodilator
Thymol	Antifungal (topical)
Trichosanthin	Abortifacient
Tubocurarine	Skeletal muscle relaxant
Valepotriates	Sedative
Vasicine (peganine)	Oxytocic
Vincamine	Cerebral stimulant
Vinblastine (vincaleukoblastine)	Antitumor agent
Vincristine (leurocristine)	Antitumor agent
Xanthotoxin (ammoidin; 8-methoxypsoralen)	Leukoderma; vitiligo
Yohimbine	Aphrodisiac
Yuanhuacine	Abortifacient
Yuanhuadine	Abortifacient

[a] Now also produced by synthesis
[b] Synthetic modification of a natural product

Table 10.6.2. Pharmacological activities of compounds isolated from woody tissue.

Activity	Chemical	Class	Ref.
Abortifacient	Vasicine	Alkaloid	66
	Protopine	Isoquinoline alkaloid	1142
	Quinine	Quinoline alkaloid	655
Adrenergic receptor blocker	Yohimbine, alpha	Indole alkaloid	815
Adrenolytic	Sanguinarine	Isoquinoline alkaloid	529
Algicidal activity	Plumieride coumarate glucoside	Monoterpene	224
Analeptic	Yohimbine	Indole alkaloid	554
Analgesic	Prosopinine	Alkaloid	207
	Ellagic acid	Coumarin	209
	Yohimbine	Indole alkaloid	781
	Ephedrine	Isoquinoline alkaloid	427
	Protopine	Isoquinoline alkaloid	1125
	Terpinen-4-ol	Monoterpene	117
	Skimmianine	Quinoline alkaloid	185
	Ginsenoside RC	Triterpene	551
	Ginsenoside RE	Triterpene	551
	Ginsenoside RG-1	Triterpene	551
	Anisodamine	Tropane alkaloid	195
Anesthetic	Sanguinarine	Isoquinoline alkaloid	393
Anthelmintic	Vasicine	Alkaloid	239
	Vasicinone	Alkaloid	239
	Vanillic acid	Benzenoid	1004
	Matrine	Quinolizidine alkaloid	1007

Table 10.6.2 (continued)

Activity	Chemical	Class	Ref.
Antiallergenic	Coclaurine, (DL)-N-demethyl	Isoquinoline alkaloid	1140
	Ephedrine	Isoquinoline alkaloid	293
	Terpinen-4-ol	Monoterpene	117
Antiamoebic	Sanguinarine	Isoquinoline alkaloid	122
	Bruceantin	Triterpene	353
	Glaucarubolone	Triterpene	353
Antianalgesic	Yohimbine	Indole alkaloid	555
Antianaphylactic	Gallic acid	Benzenoid	517
	Catechin, (+)	Flavonoid	517
	Epicatechin, (−)	Flavonoid	517
Antiarrhythmic	Pyrazine, tetramethyl-	2-N heterocyclic alkaloid	763
	Scopoletin	Coumarin	763
	Umbelliferone	Coumarin	8
	Allocryptopine, alpha	Isoquinoline alkaloid	34
	Coclaurine, (DL)-N-demethyl	Isoquinoline alkaloid	400
	Ferulic acid	Lignan	176
	Cyclovirobuxine D	Triterpene alkaloid	1054
Antiasthmatic	Coclaurine, (DL)-N-demethyl	Isoquinoline alkaloid	1111
	Carveol, dihydro	Monoterpene	976
	Menth-2-en-7-ol, para-	Monoterpene	976
	Myrcene	Monoterpene	976
	Terpinen-4-ol	Monoterpene	976
	Terpineol, alpha	Monoterpene	976
	Anisodamine	Tropane alkaloid	199
Antiatherosclerotic	Oleanolic acid	Triterpene	1041
Antibacterial	Pyrazine, tetramethyl-	2-N heterocyclic alkaloid	748
	Prosopinine	Alkaloid	207
	Benzaldehyde, para-hydroxy-	Benzenoid	992
	Gallic acid	Benzenoid	268
	Gentisic acid	Benzenoid	268
	Hydroquinone	Benzenoid	988
	Protocatechuic acid	Benzenoid	268
	Resveratrol	Benzenoid	578
	Aesculetin	Coumarin	419
	Pimaric acid, iso-	Diterpene	409
	Abyssinone I	Flavonoid	514
	Abyssinone II	Flavonoid	514
	Abyssinone IV	Flavonoid	514
	Abyssinone V	Flavonoid	514
	Apigenin	Flavonoid	772
	Erycristagallin	Flavonoid	693
	Genistein	Flavonoid	1042
	Aricine	Indole alkaloid	439
	Girinimbine	Indole alkaloid	177
	Pericalline	Indole alkaloid	879
	Acanthine, oxy-	Isoquinoline alkaloid	48
	Berberine	Isoquinoline alkaloid	437
	Chelerythrine	Isoquinoline alkaloid	120
	Liriodenine	Isoquinoline alkaloid	451

Table 10.6.2 (continued)

Activity	Chemical	Class	Ref.
	Lysicamine	Isoquinoline alkaloid	451
	Oxonantenine	Isoquinoline alkaloid	451
	Sanguinarine	Isoquinoline alkaloid	498
	Caffeic acid	Lignan	269
	Chlorogenic acid	Lignan	268
	Eugenol	Lignan	37
	Ferulic acid	Lignan	421
	Forsythoside B	Lignan	295
	Honokiol	Lignan	721
	Magnolol	Lignan	590
	Sinapic acid	Lignan	268
	Carvacrol	Monoterpene	351
	Gardenoside	Monoterpene	464
	Geranyl hydroquinone	Monoterpene	988
	Loganin	Monoterpene	464
	Terpinen-4-ol	Monoterpene	117
	Brazilin	Oxygen heterocycle	842
	Hematoxylin	Oxygen heterocycle	842
	Uvafzelin	Oxygen heterocycle	450
	Erythrabysin II	Pterocarpan	514
	Alkannin	Quinoid	988
	Emodin	Quinoid	522
	Germichrysone	Quinoid	296
	Lapachol	Quinoid	357
	Physcion	Quinoid	106
	Shikonin	Quinoid	988
	Singueanol I	Quinoid	296
	Singueanol II	Quinoid	296
	Dictamnine	Quinoline alkaloid	1077
	Rutacridone epoxide	Quinoline alkaloid	714
	Torosachrysone	Quinone	296
	Pristimerin	Triterpene	111
Anticoagulant	Allantoin	2-N heterocyclic alkaloid	660
	Berberine	Isoquinoline alkaloid	892
	Catechin, (+)	Isoquinoline alkaloid	569
Anticonvulsant	Scoparone	Coumarin	16
	Scopoletin	Coumarin	16
	Eugenol	Lignan	242
	Magnolol	Lignan	1063
	Safrole	Lignan	242
	Skimmianine	Quinoline alkaloid	303
Antidiarrheal	Berberine	Isoquinoline alkaloid	1115
Antifatigue	Ginsenoside RC	Triterpene	512
	Ginsenoside RF	Triterpene	512
Antifertility	Genistein	Flavonoid	435
	Hederagenin	Triterpene	485
Antifilarial	Ecdysone, beta	Steroid	1103
Antifungal	Benzoic acid	Benzenoid	775
	Gallic acid	Benzenoid	31
	Protocatechuic acid	Benzenoid	31

Table 10.6.2 (continued)

Activity	Chemical	Class	Ref.
Antifungal	Pyrogallol	Benzenoid	889
	Resveratrol	Benzenoid	578
	Vanillin	Benzenoid	494
	Aesculetin	Coumarin	103
	Marmesin	Coumarin	995
	Abyssinone I	Flavonoid	718
	Abyssinone II	Flavonoid	514
	Abyssinone III	Flavonoid	718
	Abyssinone V	Flavonoid	718
	Abyssinone VI	Flavonoid	718
	Abyssinone VII	Flavonoid	718
	Daidzein	Flavonoid	571
	Formononetin	Flavonoid	571
	Taxifolin	Flavonoid	654
	Berberine	Isoquinoline alkaloid	645
	Liriodenine	Isoquinoline alkaloid	451
	Lysicamine	Isoquinoline alkaloid	451
	Oxonantenine	Isoquinoline alkaloid	451
	Protopine	Isoquinoline alkaloid	331
	Sanguinarine	Isoquinoline alkaloid	1037
	Borbonol	Lactone	1129
	Caffeic acid	Lignan	31
	Coumaric acid, para-	Lignan	31
	Ferulic acid	Lignan	31
	Honokiol	Lignan	216
	Sinapic acid	Lignan	103
	Aloe emodin	Quinoid	342
	Aucuparin	Quinoid	531
	Emodin	Quinoid	342
	Rhein	Quinoid	342
	Dictamnine	Quinoline alkaloid	1077
	Costunolide	Sesquiterpene	996
	Diospyrin, iso	Steroid saponin	667
	Mangiferin	Xanthone	397
Antihemorrhagic	Ellagic acid	Coumarin	217
	Quercitrin	Flavonoid	570
	Daucosterol	Steroid	731
Antihemorrhagic shock effect	Anisodamine	Tropane alkaloid	969
	Scopolamine	Tropane alkaloid	1107
Antihepatotoxic	Protocatechuic acid	Benzenoid	549
	Naringenin	Flavonoid	426
	Quercitrin	Flavonoid	492
	Taxifolin	Flavonoid	426
	Berberine	Isoquinoline alkaloid	1026
	Catechin, (+)	Isoquinoline alkaloid	963
	Caffeic acid	Lignan	549
	Coumaric acid, para-	Lignan	549
	Ferulic acid	Lignan	580
	Sinapic acid	Lignan	549
	Ecdysone, beta	Steroid	986

Table 10.6.2 (continued)

Activity	Chemical	Class	Ref.
	Glycyrrhetinic acid	Triterpene	550
	Oleanolic acid	Triterpene	425
Antihypercholesterolemic	Cycloartenol	Triterpene	496
	Oleanolic acid	Triterpene	1041
	Ursolic acid	Triterpene	794
Antihyperglycemic	Epicatechin, (−)	Flavonoid	971
	Manniflavanone	Flavonoid	751
	Sanguinarine	Isoquinoline alkaloid	393
	Lupeol	Triterpene	953
	Lupeol acetate	Triterpene	953
	Ursolic acid	Triterpene	1101
Antihypertensive	Ellagic acid	Coumarin	209
	Sanggenone C	Flavonoid	493
	Sanggenone D	Flavonoid	491
	Reserpine	Indole alkaloid	1135
	Tabernulosine	Indole alkaloid	12
	Sanguinarine	Isoquinoline alkaloid	393
	Oleuropein	Monoterpene	219
	Cyclovirobuxine D	Triterpene alkaloid	1054
Antihypothermic	Ephedrine	Isoquinoline alkaloid	427
Antiimplantation	Reserpine	Indole alkaloid	1147
	Nicotine	Pyridine alkaloid	692
Antiinflammatory	Hentriacontane, N-	Alkane	875
	Triacontane, N-	Alkane	876
	Gallic acid	Benzenoid	621
	Aesculetin	Coumarin	559
	Epicatechin, (−)	Flavonoid	984
	Flavone, 5-hydroxy-4'-7-dimethoxy	Flavonoid	329
	Leucodelphinidin	Flavonoid	289
	Procyanidin B-2	Flavonoid	779
	Procyanidin B-4	Flavonoid	779
	Berberine	Isoquinoline alkaloid	780
	Coptisine	Isoquinoline alkaloid	778
	Jatrorrhizine	Isoquinoline alkaloid	778
	Obaberine	Isoquinoline alkaloid	894
	Protopine	Isoquinoline alkaloid	698
	Acetyl shikonin	Quinoid	617
	Daucosterol	Steroid	990
	Ecdysone, beta	Steroid	1003
	Hederagenin	Triterpene	991
	Oleanolic acid	Triterpene	991
	Anisodamine	Tropane alkaloid	476
	Euxanthone	Xanthone	365
	Jacareubin	Xanthone	365
	Jacareubin, 6-deoxy	Xanthone	365
	Mangiferin	Xanthone	597
	Xanthone, 1-3-5-trihydroxy-2-methoxy	Xanthone	726

Table 10.6.2 (continued)

Activity	Chemical	Class	Ref.
	Xanthone, 1-5-dihydroxy	Xanthone	726
	Xanthone, 3-8-dihydroxy-1-2-dimethoxy	Xanthone	726
Antiischemic	Anisodamine	Tropane alkaloid	930
Antilipase	Daidzein	Flavonoid	760
Antilipolytic	Ginsenoside RC	Triterpene	758
	Ginsenoside RE	Triterpene	758
Antimalarial	Aesculetin	Coumarin	1093
	Berberine	Isoquinoline alkaloid	1093
	Protopine	Isoquinoline alkaloid	1138
	Cinchonidine	Quinoline alkaloid	917
	Cinchonine	Quinoline alkaloid	917
	Quinidine	Quinoline alkaloid	917
	Quinine	Quinoline alkaloid	917
	Bruceantin	Triterpene	832
	Chaparrinone	Triterpene	1020
Antimitotic	Hydroquinone	Benzenoid	1022
	Xanthotoxin	Coumarin	5
	Tripdiolide	Diterpene	287
	Triptolide	Diterpene	287
	Ellipticine	Indole alkaloid	1019
	Harringtonine	Isoquinoline alkaloid	788
	Eugenol	Lignan	676
	Ferulic acid	Lignan	1082
	Maytansine	Macrocyclic lactam lactone	287
	Carvacrol	Monoterpene	676
	Bruceantin	Triterpene	287
	Ginsenoside RB-1	Triterpene	186
Antimutagenic	Benzoic acid. para-hydroxy-	Benzenoid	1033
	Gallic acid	Benzenoid	966
	Pyrogallol	Benzenoid	159
	Umbelliferone	Coumarin	762
	Delphinidin	Flavonoid	446
	Epicatechin, (−)	Flavonoid	773
	Quercitrin	Flavonoid	446
	Chlorogenic acid	Lignan	1082
Antioestrogenic	Ferulic acid	Lignan	258
Antioxidant	Carnosol	Diterpene	720
	Epicatechin, (−)	Flavonoid	433
	Okanin	Flavonoid	290
	Taxifolin	Flavonoid	841
	Ellipticine	Indole alkaloid	1019
	Caffeic acid	Lignan	1108
	Chlorogenic acid	Lignan	1108
	Eugenol	Lignan	923
	Oleanolic acid	Triterpene	977
Antiprotozoan	Maytansine	Macrocyclic lactam lactone	527

Table 10.6.2 (continued)

Activity	Chemical	Class	Ref.
Antipyretic	Coumarin	Coumarin	873
	Skimmianine	Quinoline alkaloid	202
Antiradiation	Leucodelphinidin	Flavonoid	70
	Berberine	Isoquinoline alkaloid	636
	Safrole, iso	Lignan	761
Antireserpine	Ephedrine	Isoquinoline alkaloid	427
Antischistosomal	Bergapten	Coumarin	2
Antischock effect	Anisodamine	Tropane alkaloid	1112
Antisecretory effect	Berberine	Isoquinoline alkaloid	1145
	Magnolol	Lignan	1062
Antispasmodic	Pyrazine, tetramethyl-	2-N heterocyclic alkaloid	768
	Umbelliferone	Coumarin	14
	Afzelin	Flavonoid	209
	Morin	Flavonoid	101
	Naringenin	Flavonoid	101
	Reserpine	Indole alkaloid	1134
	Yohimbine	Indole alkaloid	525
	Chelerythrine	Isoquinoline alkaloid	1028
	Coclaurine, (DL)-N-demethyl	Isoquinoline alkaloid	1140
	Protopine	Isoquinoline alkaloid	1125
	Sanguinarine	Isoquinoline alkaloid	393
	Synephrine	Isoquinoline alkaloid	906
	Honokiol	Lignan	1102
	Carvacrol	Monoterpene	1031
	Singueanol I	Quinoid	296
	Allohimachalol	Sesquiterpene	803
	Centdarol	Sesquiterpene	803
	Centdarol, iso	Sesquiterpene	803
	Himachalol	Sesquiterpene	803
	Ginsenoside RC	Triterpene	512
	Ginsenoside RF	Triterpene	512
	Himadarol	Triterpene	803
Antistress	Syringin	Lignan	898
	Ginsenoside RB-1	Triterpene	489
	Ginsenoside RC	Triterpene	489
	Ginsenoside RE	Triterpene	489
Antitachycardia	Harringtonine, homo	Isoquinoline alkaloid	626
Antithiamine	Caffeic acid	Lignan	1108
Antithyroid	Ellagic acid	Coumarin	68
Antitrichomonal	Sanguinarine	Isoquinoline alkaloid	121
	Eugenol	Lignan	901
Antitrypanosomal	Ellipticine	Indole alkaloid	102
	Ellipticine, 9-methoxy	Indole alkaloid	102
	Tingenone	Triterpene	355
Antituberculosis	Berberine	Isoquinoline alkaloid	1051
	Cassameridine	Isoquinoline alkaloid	451
	Lysicamine	Isoquinoline alkaloid	451

Table 10.6.2 (continued)

Activity	Chemical	Class	Ref.
	Oxonantenine	Isoquinoline alkaloid	451
	Honokiol	Lignan	216
	Magnolol	Lignan	216
	Uvafzelin	Oxygen heterocycle	450
	Rutacridone epoxide	Quinoline alkaloid	714
Antitumor	Gallic acid	Benzenoid	508
	Hydroquinone	Benzenoid	57
	Aesculetin	Coumarin	715
	Daphnoretin	Coumarin	396
	Ellagic acid	Coumarin	610
	Xanthotoxin	Coumarin	1036
	Jatrophone	Diterpene	586
	Tripdiolide	Diterpene	280
	Leucodelphinidin	Flavonoid	70
	Leucopelargonidin	Flavonoid	508
	Morin	Flavonoid	715
	Ellipticine	Indole alkaloid	627
	Ellipticine, 9-methoxy	Indole alkaloid	982
	Chelerythrine	Isoquinoline alkaloid	528
	Dopamine	Isoquinoline alkaloid	1070
	Harringtonine	Isoquinoline alkaloid	443
	Harringtonine, deoxy	Isoquinoline alkaloid	167
	Harringtonine, homo	Isoquinoline alkaloid	612
	Nitidine	Isoquinoline alkaloid	447
	Austrobailignan 1	Lignan	73
	Chlorogenic acid	Lignan	610
	Ferulic acid	Lignan	610
	Phyllanthoside	Lignan	818
	Maytancyprine, nor-	Macrocyclic lactam lactone	952
	Maytanprine	Macrocyclic lactam lactone	1143
	Maytansine	Macrocyclic lactam lactone	881
	Maytansine, nor-	Macrocyclic lactam lactone	951
	Loganin	Monoterpene	464
	Alizarin-1-methyl ether	Quinoid	184
	Anthraquinone, 2-hydroxy-methyl	Quinoid	184
	Lapachol	Quinoid	264
	Lapachone, beta	Quinoid	63
	Shikonin	Quinoid	907
	Camptothecin	Quinoline alkaloid	443
	Quinidine	Quinoline alkaloid	1024
	Phyllanthostatin 3	Sesquiterpene	818
	Bruceantin	Triterpene	950
	Bruceantinol	Triterpene	950
	Bruceantinoside A	Triterpene	770
	Brucein A, iso	Triterpene	403
	Brucein B	Triterpene	950
	Brucein C	Triterpene	830
	Chaparrinone	Triterpene	556

Table 10.6.2 (continued)

Activity	Chemical	Class	Ref.
	Chaparrinone, 6-alpha-senecioyl-oxy	Triterpene	59
	Ginsenoside RB-1	Triterpene	416
	Ginsenoside RE	Triterpene	474
	Ginsenoside RF	Triterpene	474
	Ginsenoside RG-1	Triterpene	55
	Glaucarubolone	Triterpene	403
	Squalene	Triterpene	757
	Anisodamine	Tropane alkaloid	476
Antitussive	Arbutin	Benzenoid	611
	Yohimbine	Indole alkaloid	76
	Ephedrine	Isoquinoline alkaloid	76
	Terpinen-4-ol	Monoterpene	117
	Humulene, alpha	Sesquiterpene	631
	Atropine	Tropane alkaloid	76
Antiulcer	Allantoin	2-N heterocyclic alkaloid	215
	Plaunol B	Diterpene	552
	Plaunol C	Diterpene	552
	Plaunol D	Diterpene	552
	Plaunol E	Diterpene	552
	Naringenin	Flavonoid	799
	Quercitrin	Flavonoid	89
	Caffeic acid	Lignan	473
	Magnolol	Lignan	1062
	Matrine	Quinolizidine alkaloid	1105
	Costunolide	Sesquiterpene	1098
	Lupeol acetate	Triterpene	388
	Taraxerol	Triterpene	388
	Ursolic acid	Triterpene	388
	Jacareubin	Xanthone	365
	Jacareubin, 6-deoxy	Xanthone	365
Antiviral	Zeatin	Alkaloid	456
	Zeatin riboside	Alkaloid	456
	Gallic acid	Benzenoid	268
	Gentisic acid	Benzenoid	268
	Protocatechuic acid	Benzenoid	268
	Morin	Flavonoid	97
	Quercitrin	Flavonoid	97
	Berberine	Isoquinoline alkaloid	72
	Chelerythrine	Isoquinoline alkaloid	120
	Sanguinarine	Isoquinoline alkaloid	120
	Caffeic acid	Lignan	268
	Chlorogenic acid	Lignan	268
	Lapachol	Quinoid	596
	Mangiferin	Xanthone	888
Antiyeast	Pyrogallol	Benzenoid	889
	Aesculetin	Coumarin	103
	Bergapten	Coumarin	526
	Cassameridine	Isoquinoline alkaloid	451
	Chelerythrine	Isoquinoline alkaloid	777
	Lysicamine	Isoquinoline alkaloid	451

Table 10.6.2 (continued)

Activity	Chemical	Class	Ref.
	Oxonantenine	Isoquinoline alkaloid	451
	Sanguinarine	Isoquinoline alkaloid	1037
	Eugenol	Lignan	134
	Ferulic acid	Lignan	86
	Honokiol	Lignan	216
	Magnolol	Lignan	216
	Sinapic acid	Lignan	103
	Carvacrol	Monoterpene	710
	Geranyl hydroquinone	Monoterpene	988
Anxiety induction	Yohimbine	Indole alkaloid	431
Aphrodisiac	Yohimbine	Indole alkaloid	487
	Syringin	Lignan	495
Arrhythmogenic effect	Allocryptopine, alpha	Isoquinoline alkaloid	39
	Protopine	Isoquinoline alkaloid	39
	Cyclovirobuxine D	Triterpene alkaloid	623
Automatotropic effect, negative	Coclaurine, (DL)-N-demethyl	Isoquinoline alkaloid	400
Barbiturate potentiation	Bergapten	Coumarin	1081
	Ellagic acid	Coumarin	209
	Reserpine	Indole alkaloid	243
	Ephedrine	Isoquinoline alkaloid	427
	Sanguinarine	Isoquinoline alkaloid	243
	Robustine	Quinoline alkaloid	305
	Skimmianine	Quinoline alkaloid	303
	Daucosterol	Steroid	862
	Ginsenoside RE	Triterpene	551
	Atropine	Tropane alkaloid	291
	Scopolamine	Tropane alkaloid	243
	Euxanthone	Xanthone	365
	Jacareubin	Xanthone	365
	Xanthone, 1-3-5-trihydroxy-2-methoxy	Xanthone	726
	Xanthone, 1-5-dihydroxy	Xanthone	726
	Xanthone, 3-8-dihydroxy-1-2-dimethoxy	Xanthone	726
Blood pressure effect biphasic	Sanguinarine	Isoquinoline alkaloid	393
Body weight loss	Ginsenoside RC	Triterpene	551
Body weight loss inhibitor	Ginsenoside RE	Triterpene	551
	Ginsenoside RG-1	Triterpene	551
Bradycardia	Puerarin	Flavonoid	306
	Harringtonine, homo	Isoquinoline alkaloid	626
	Magnoflorine	Isoquinoline alkaloid	174
	Tembetarine	Isquinoline alkaloid	174
Bronchoconstrictor	Sanguinarine	Isoquinoline alkaloid	393
Bronchodilator	Pyrazine, tetramethyl-	2-N heterocyclic alkaloid	763
	Vasicinone	Alkaloid	115
	Gallic acid	Benzenoid	517
	Scopoletin	Coumarin	763

Table 10.6.2 (continued)

Activity	Chemical	Class	Ref.
	Coclaurine, (DL)-N-demethyl	Isoquinoline alkaloid	1140
	Anisodamine	Tropane alkaloid	199
Calcium transport inhibitor	Terpineol, alpha	Monoterpene	1139
Capillary antihemmorrhagic	Leucodelphinidin	Flavonoid	1005
Capillary permeability decreased	Prosopinine	Alkaloid	207
	Leucodelphinidin	Flavonoid	289
	Coclaurine, (DL)-N-demethyl	Isoquinoline alkaloid	1140
	Cymene, para-	Monoterpene	1100
	Acetyl shikonin	Quinoid	617
Carcinogenesis inhibition	Umbelliferone	Coumarin	959
	Coumarin	Coumarin	959
	Caffeic acid	Lignan	1065
	Ferulic acid	Lignan	1065
	Glycyrrhetinic acid	Triterpene	738
Cardiac depressant	Sanguinarine	Isoquinoline alkaloid	393
Cardiac output decreased	Harringtonine, homo	Isoquinoline alkaloid	626
Cell attachment enhancement	Milliamine A	Diterpene	1104
Cell cycle cytotoxicity (G1 phase)	Harringtonine	Isoquinoline alkaloid	1097
	Harringtonine, homo	Isoquinoline alkaloid	1097
Cell differentiation induction	Genistein	Flavonoid	62
	Harringtonine, homo	Isoquinoline alkaloid	442
Cell growth enhancement effect	Ginsenoside RB-1	Triterpene	541
Cell proliferation inhibitor	Ellipticine	Indole alkaloid	1039
	Harringtonine	Isoquinoline alkaloid	445
	Harringtonine, homo	Isoquinoline alkaloid	442
	Eugenol	Lignan	453
	Camptothecin	Quinoline alkaloid	445
Cellular respiration inhibition	Berberine	Isoquinoline alkaloid	304
	Sanguinarine	Isoquinoline alkaloid	304
	Eugenol	Lignan	453
	Emodin	Quinoid	196
Chelating effect	Thujaplicin, alpha	Monoterpene	26
	Thujaplicin, beta	Monoterpene	26
	Thujaplicin, gamma	Monoterpene	26
Choleretic	Benzoic acid	Benzenoid	957
	Gallic acid	Benzenoid	957
	Berberine	Isoquinoline alkaloid	1026
	Protopine	Isoquinoline alkaloid	791
	Eugenol	Lignan	1099
	Loganin	Monoterpene	997
	Morroniside	Monoterpene	997
	Costunolide	Sesquiterpene	1098
Cholesterol inhibition	Resveratrol	Benzenoid	54

Table 10.6.2 (continued)

Activity	Chemical	Class	Ref.
Chronotropic effect negative	Scopoletin	Coumarin	766
	Cyclovirobuxine D	Triterpene alkaloid	623
Chronotropic effect, positive	Coclaurine, (DL)-N-demethyl	Isoquinoline alkaloid	400
Clastogenic	Caffeic acid	Lignan	967
	Chlorogenic acid	Lignan	967
	Picrasin B	Triterpene	769
CNS depressant	Ammodendrine	Alkaloid	412
	Scoparone	Coumarin	17
	Reserpine	Indole alkaloid	561
	Ephedrine	Isoquinoline alkaloid	427
	Protopine	Isoquinoline alkaloid	1142
	Eugenol	Lignan	242
	Eugenol methyl ether	Lignan	242
	Honokiol	Lignan	1063
	Magnolol	Lignan	1063
	Safrole	Lignan	242
	Skimmianine	Quinoline alkaloid	185
	Ecdysone, beta	Steroid	1003
	Ursolic acid	Triterpene	1122
	Cyclovirobuxine D	Triterpene alkaloid	1054
	Anisodamine	Tropane alkaloid	810
	Anosidine	Tropane alkaloid	809
	Atropine	Tropane alkaloid	810
	Scopolamine	Tropane alkaloid	810
	Euxanthone	Xanthone	365
	Jacareubin	Xanthone	365
	Jacareubin, 6-deoxy	Xanthone	365
Coagulant	Berberine	Isoquinoline alkaloid	892
Conditioned avoidance response decreased	Ginsenoside RF	Triterpene	512
	Scopolamine	Tropane alkaloid	740
Convulsant	Sanguinarine	Isoquinoline alkaloid	393
Coronary blood flow decreased	Magnoflorine	Isoquinoline alkaloid	174
Coronary blood flow increased	Pyrazine, tetramethyl-	2-N heterocyclic alkaloid	1137
	Puerarin	Flavonoid	306
	Allocryptopine, alpha	Isoquinoline alkaloid	39
	Protopine	Isoquinoline alkaloid	39
	Tembetarine	Isquinoline alkaloid	174
	Cyclovirobuxine D	Triterpene alkaloid	623
Coronary resistance increased	Puerarin	Flavonoid	306
Corticosterone induction	Ginsenoside RC	Triterpene	732
	Ginsenoside RD	Triterpene	732
Cyclic AMP stimulant	Reserpine	Indole alkaloid	44
Cytochrome P-450 binding	Brucine	Indole alkaloid	806
	Sanguinarine	Isoquinoline alkaloid	806

Table 10.6.2 (continued)

Activity	Chemical	Class	Ref.
	Nicotine	Pyridine alkaloid	806
	Scopolamine	Tropane alkaloid	806
Cytochrome P-450 induction	Ellipticine	Indole alkaloid	792
	Safrole	Lignan	254
	Safrole, iso	Lignan	890
Cytotoxic	Hordenine	Alkaloid	1136
	Hydroquinone	Benzenoid	57
	Phenol	Benzenoid	776
	Uzarigenin	Cardenolide	563
	Bergapten	Coumarin	1126
	Cleomiscosin A	Coumarin	601
	Scopoletin	Coumarin	73
	Xanthotoxin	Coumarin	1126
	Baccatin I, 1-beta-hydroxy	Diterpene	682
	Baccatin III	Diterpene	682
	Huratoxin	Diterpene	673
	Jatrophone	Diterpene	586
	Pimaric acid, iso	Diterpene	409
	Plaunol B	Diterpene	817
	Tripdiolide	Diterpene	849
	Alstonine	Indole alkaloid	99
	Canthin-6-one	Indole alkaloid	45
	Canthin-6-one, 11-hydroxy	Indole alkaloid	403
	Coronaridine, (−)	Indole alkaloid	380
	Eglandine-N-oxide, 10-methoxy	Indole alkaloid	380
	Ellipticine	Indole alkaloid	1019
	Ellipticine, 9-methoxy	Indole alkaloid	354
	Heyneatine, (−)	Indole alkaloid	380
	Pericalline	Indole alkaloid	380
	Anonaine	Isoquinoline alkaloid	135
	Cephalotaxine	Isoquinoline alkaloid	344
	Chelerythrine	Isoquinoline alkaloid	979
	Coptisine	Isoquinoline alkaloid	979
	Harringtonine	Isoquinoline alkaloid	999
	Harringtonine, iso	Isoquinoline alkaloid	344
	Liriodenine	Isoquinoline alkaloid	136
	Sanguinarine	Isoquinoline alkaloid	122
	Austrobailignan I	Lignan	73
	Justicidin B	Lignan	818
	Phyllanthoside	Lignan	818
	Syringaresinol, (+)	Lignan	387
	Maytancyprine, nor-	Macrocyclic lactam lactone	952
	Maytansine, nor-	Macrocyclic lactam lactone	951
	Alizarin-1-methyl ether	Quinoid	184
	Anthraquinone, 2-hydroxy-methyl	Quinoid	184
	Benzoquinone, 1-4,2-6-dimethoxy	Quinoid	402
	Benzoquinone, 2-6-dimethoxy	Quinoid	914
	Lapachol	Quinoid	861
	Lucidin omega-methyl ether	Quinoid	184
	Camptothecin	Quinoline alkaloid	1060

Table 10.6.2 (continued)

Activity	Chemical	Class	Ref.
	Camptothecin, 9-methoxy	Quinoline alkaloid	380
	Quinidine	Quinoline alkaloid	1024
	Costunolide	Sesquiterpene	1057
	Daucosterol	Steroid	620
	Brucein A	Triterpene	823
	Brucein A, iso	Triterpene	403
	Brucein B	Triterpene	823
	Celastrol	Triterpene	594
	Chaparrinone, 6-alpha-senecioyl-oxy	Triterpene	59
	Ginsenoside RB-1	Triterpene	747
	Glaucarubolone	Triterpene	403
	Prieurianin	Triterpene	638
	Tingenone	Triterpene	593
Diuretic	Hentriacontane, N-	Alkane	875
	Scopoletin	Coumarin	798
	Apigenin	Flavonoid	987
	Chrysophanic acid	Quinoid	1091
	Rhein	Quinoid	1091
	Friedelin	Triterpene	876
DNA binding effect	Ellagic acid	Coumarin	271
	Jatrophone	Diterpene	238
	Ellipticine	Indole alkaloid	792
	Harman	Indole alkaloid	675
	Berberine	Isoquinoline alkaloid	646
	Sanguinarine	Isoquinoline alkaloid	304
	Caffeic acid	Lignan	937
	Elemicin	Lignan	856
	Estragole	Lignan	856
	Eugenol methyl ether	Lignan	856
	Safrole	Lignan	856
	Chrysophanic acid	Quinoid	983
	Emodin	Quinoid	983
	Camptothecin	Quinoline alkaloid	339
	Dictamnine	Quinoline alkaloid	820
	Tingenone	Triterpene	164
DNA disruption effect	Ellipticine	Indole alkaloid	324
	Camptothecin	Quinoline alkaloid	630
DNA intercalating effect	Sanguinarine	Isoquinoline alkaloid	647
DNA polymerase inhibition	Daphnoretin	Coumarin	396
DNA scission effect	Ellipticine	Indole alkaloid	1151
	Camptothecin	Quinoline alkaloid	629
DNA stimulation	Hydroquinone	Benzenoid	687
	Phenol	Benzenoid	687
DNA synthesis inhibition	Phenol	Benzenoid	776
	Bergapten	Coumarin	1126
	Daphnoretin	Coumarin	396
	Xanthotoxin	Coumarin	1126
	Pimaric acid, iso	Diterpene	409

Table 10.6.2 (continued)

Activity	Chemical	Class	Ref.
	Tripdiolide	Diterpene	287
	Alstonine	Indole alkaloid	98
	Ellipticine	Indole alkaloid	562
	Ellipticine, 9-methoxy	Indole alkaloid	613
	Dopamine	Isoquinoline alkaloid	1070
	Harringtonine	Isoquinoline alkaloid	788
	Harringtonine, deoxy	Isoquinoline alkaloid	1094
	Harringtonine, iso	Isoquinoline alkaloid	1094
	Nitidine	Isoquinoline alkaloid	308
	Camptothecin	Quinoline alkaloid	1120
	Costunolide	Sesquiterpene	1083
	Bruceantin	Triterpene	395
	Ginsenoside RB-1	Triterpene	186
	Tingenone	Triterpene	355
Dopamine receptor blocking effect	Salsolinol	Isoquinolizidine alkaloid	146
Estrogen synthetase induction	Chrysin	Flavonoid	530
Estrogen synthetase inhibitor	Apigenin	Flavonoid	530
Estrogenic effect	Glycoperine	Quinoline alkaloid	27
Expectorant	Cymene, para-	Monoterpene	154
	Terpinen-4-ol	Monoterpene	117
	Cadinol, alpha	Sesquiterpene	634
	Anisodamine	Tropane alkaloid	1123
Fertilization inhibition	Reserpine	Indole alkaloid	179
Gall bladder effects	Berberine	Isoquinoline alkaloid	356
Gastric antisecretory	Taraxerol	Triterpene	388
Gene conversion effect	Caffeic acid	Lignan	967
	Chlorogenic acid	Lignan	967
Glycolysis inhibition	Sanguinarine	Isoquinoline alkaloid	138
	Rhein	Quinoid	196
Glycolysis stimulation	Emodin	Quinoid	196
Hemoglobin induction	Apigenin	Flavonoid	548
	Chrysin	Flavonoid	548
	Daidzein	Flavonoid	548
	Liquiritigenin	Flavonoid	548
	Morin	Flavonoid	548
	Naringenin	Flavonoid	548
Hemolytic	Ellipticine	Indole alkaloid	424
	Ginsenoside RC	Triterpene	512
Hemostatic	Quercitrin	Flavonoid	1096
	Syringin	Lignan	1090
Hypertensive	Choline	Acyclic nitrogen base	328
	Alstonine, tetrahydro	Alkaloid	880
	Hydroquinone	Benzenoid	149
	Yohimbine	Indole alkaloid	47
	Allocryptopine, alpha	Isoquinoline alkaloid	39

Table 10.6.2 (continued)

Activity	Chemical	Class	Ref.
	Ephedrine	Isoquinoline alkaloid	427
	Protopine	Isoquinoline alkaloid	39
	Sanguinarine	Isoquinoline alkaloid	273
Hypoglycemic	Quinine	Quinoline alkaloid	406
Hypolipemic	Daucosterol	Steroid	684
	Diosgenin	Steroid saponin	708
	Oleanolic acid	Triterpene	790
Hypotensive	Pyrazine, tetramethyl-	2-N heterocyclic alkaloid	767
	Choline	Acyclic nitrogen base	862
	Bergapten	Coumarin	340
	Ellagic acid	Coumarin	209
	Psoralen	Coumarin	340
	Scoparone	Coumarin	1149
	Scopoletin	Coumarin	52
	Apigenin	Flavonoid	987
	Kuwanon G	Flavonoid	743
	Kuwanon H	Flavonoid	744
	Puerarin	Flavonoid	306
	Coronaridine	Indole alkaloid	811
	Ellipticine	Indole alkaloid	423
	Harman	Indole alkaloid	340
	Reserpine	Indole alkaloid	340
	Allocryptopine, alpha	Isoquinoline alkaloid	39
	Berberine	Isoquinoline alkaloid	1027
	Coclaurine, (DL)-N-demethyl	Isoquinoline alkaloid	1140
	Harringtonine, homo	Isoquinoline alkaloid	626
	Magnoflorine	Isoquinoline alkaloid	911
	Menisperine	Isoquinoline alkaloid	340
	Carveol, dihydro	Monoterpene	976
	Menth-2-en-7-ol, para-	Monoterpene	976
	Myrcene	Monoterpene	976
	Terpinen-4-ol	Monoterpene	976
	Seneciphylline	Pyrrolizidine alkaloid	874
	Lupanine	Quinolizidine alkaloid	672
	Lupanine, 13-hydroxy	Quinolizidine alkaloid	672
	Matrine	Quinolizidine alkaloid	340
	Sparteine	Quinolizidine alkaloid	1146
	Lupeol	Triterpene	244
	Anisodamine	Tropane alkaloid	1112
Hypothermic	Ephedrine	Isoquinoline alkaloid	427
	Estragole	Lignan	242
	Eugenol	Lignan	242
	Eugenol methyl ether	Lignan	242
	Safrole	Lignan	242
	Terpinen-4-ol	Monoterpene	117
	Skimmianine	Quinoline alkaloid	827
	Matrine	Quinolizidine alkaloid	213
	Ursolic acid	Triterpene	1122
Immunostimulant	Nicotine	Pyridine alkaloid	565
	Squalene	Triterpene	757

Table 10.6.2 (continued)

Activity	Chemical	Class	Ref.
Immunosuppressant	Ellipticine	Indole alkaloid	418
	Ellipticine, 9-methoxy	Indole alkaloid	418
	Harringtonine	Isoquinoline alkaloid	787
	Harringtonine, homo	Isoquinoline alkaloid	443
	Quinine	Quinoline alkaloid	113
Inotropic effect negative	Isoflavone, 7-hydroxy-4'-8-dimethoxyiso	Flavonoid	192
Inotropic effect positive	Epicatechin, (−)	Flavonoid	924
	Sanguinarine	Isoquinoline alkaloid	15
	Synephrine	Isoquinoline alkaloid	198
	Tembetarine	Isquinoline alkaloid	174
	Cyclovirobuxine D	Triterpene alkaloid	623
Insecticide	Ryanodine, 9-21-didehydro	Alkaloid	1064
	Ryanodine	Diterpene alkaloid	1064
	Sesamin, (+)	Lignan	410
Insulin biosynthesis inhibition	Ginsenoside RB-1	Triterpene	1047
	Ginsenoside RG-1	Triterpene	1047
Insulin degradation inhibition	Arbutin	Benzenoid	1073
	Gallic acid	Benzenoid	1073
	Pyrogallol	Benzenoid	1073
Interferon secretory inhibition	Berberine	Isoquinoline alkaloid	893
Intestinal antisecretory	Berberine	Isoquinoline alkaloid	1115
Intraocular pressure increased	Sanguinarine	Isoquinoline alkaloid	273
Intraocular pressure reduction	Berberine	Isoquinoline alkaloid	614
	Protopine	Isoquinoline alkaloid	614
Labor induction effect	Vasicine	Alkaloid	66
	Sparteine	Quinolizidine alkaloid	236
Laxative	Choline	Acyclic nitrogen base	736
	Sennoside B	Quinoid	670
LH-release inhibition	Genistein	Flavonoid	516
	Reserpine	Indole alkaloid	926
Lipid peroxide formation inhibition	Resveratrol	Benzenoid	542
Lipolytic	Ginsenoside RC	Triterpene	758
	Ginsenoside RG-1	Triterpene	758
Liver regeneration stimulation	Estragole	Lignan	938
	Safrole	Lignan	347
	Safrole, iso	Lignan	347
Lysosomal membrane stabilization	Mangiferin	Xanthone	285
Lysozyme induction	Apigenin	Flavonoid	548
	Formononetin	Flavonoid	548
	Genistein	Flavonoid	62

Table 10.6.2 (continued)

Activity	Chemical	Class	Ref.
Macrophage cytotoxicity enhancement	Anisodamine	Tropane alkaloid	1124
	Berbamine	Isoquinoline alkaloid	284
Metastasis stimulant	Ellagic acid	Coumarin	20
Microsomal metabolizing system induction	Safrole	Lignan	925
	Safrole, iso	Lignan	925
Microsomal metabolizing system inhibition	Glycyrrhetinic acid	Triterpene	550
Microtubule assembly inhibition	Maytansine	Macrocyclic lactam lactone	316
Microtubule depolymerization	Maytansine	Macrocyclic lactam lactone	457
Mitogenic	Eugenol	Lignan	123
Mitotic activity	Sanguinarine	Isoquinoline alkaloid	393
Molluscicidal	Oleuropein	Monoterpene	577
Mutagenic	Ellagic acid	Coumarin	1082
Myocardial ischemic improvement	Protocatechuic acid	Benzenoid	863
	Puerarin	Flavonoid	306
Myocardial oxygen consumption decreased	Protocatechuic acid	Benzenoid	1137
Narcotic antagonist	Berberine	Isoquinoline alkaloid	145
Nematicidal	Sanguinarine	Isoquinoline alkaloid	777
Nerve growth factor effect	Reserpine	Indole alkaloid	201
	Ginsenoside RB-1	Triterpene	897
Neuromuscular blocking	Scopoletin	Coumarin	764
	Afzelin	Flavonoid	209
Neuromuscular stimulation	Matrine	Quinolizidine alkaloid	908
Nociceptive	Yohimbine	Indole alkaloid	781
Norepinephrine release inhibition	Reserpine	Indole alkaloid	201
Oestrogenic	Daidzein	Flavonoid	520
	Formononetin	Flavonoid	553
	Genistein	Flavonoid	553
Oestrogenic effect	Yohimbine	Indole alkaloid	218
Opiate receptor binding inhibition	Pericalline	Indole alkaloid	53
	Salsolinol	Isoquinolizidine alkaloid	146
Ovicidal	Coumarin	Coumarin	716

Table 10.6.2 (continued)

Activity	Chemical	Class	Ref.
Paradoxical sleep enhancement	Yohimbine	Indole alkaloid	509
Parasympatholytic	Protopine	Isoquinoline alkaloid	35
	Anisodamine	Tropane alkaloid	723
Phagocytosis stimulant	Ferulic acid	Lignan	1092
Phlogistic	Sanguinarine	Isoquinoline alkaloid	393
Plaque formation suppressant	Sanguinarine	Isoquinoline alkaloid	1069
Platelet aggregation inhibition	Abyssinone II	Flavonoid	514
	Abyssinone VI	Flavonoid	514
	Morin	Flavonoid	100
	Naringenin	Flavonoid	100
	Catechin, (+)	Isoquinoline alkaloid	100
	Caffeic acid	Lignan	566
	Ferulic acid	Lignan	1116
	Erythrabysin II	Pterocarpan	514
	Isoflavone, 7-hydroxy-4'-8-dimethoxyiso	Flavonoid	192
Platelet lipoxygenase inhibition	Umbelliferone	Coumarin	919
Prolactin stimulant	Yohimbine	Indole alkaloid	677
	Reserpine	Indole alkaloid	64
	Ferulic acid	Lignan	366
Prostaglandin induction	Caffeic acid	Lignan	566
	Rhein	Quinoid	330
	Sparteine	Quinolizidine alkaloid	7
Prostaglandin synthetase stimulation	Protocatechuic acid	Benzenoid	94
	Benzoic acid. para-hydroxy-	Benzenoid	94
	Ellagic acid	Coumarin	896
Prostaglandin synthetase inhibition	Hydroquinone	Benzenoid	94
	Morin	Flavonoid	94
	Taxifolin	Flavonoid	94
Protein binding	Sanguinarine	Isoquinoline alkaloid	273
Protein synthesis inhibition	Daphnoretin	Coumarin	396
	Tripdiolide	Diterpene	287
	Harringtonine	Isoquinoline alkaloid	1097
	Harringtonine, deoxy	Isoquinoline alkaloid	1095
	Harringtonine, homo	Isoquinoline alkaloid	1095
	Harringtonine, iso	Isoquinoline alkaloid	1084
	Gardenoside	Monoterpene	183
	Brucein A, iso	Triterpene	823
	Brucein B	Triterpene	823
	Chaparrinone	Triterpene	43
	Glaucarubolone	Triterpene	823
	Tingenone	Triterpene	355
Protein synthesis stimulation	Lapachol	Quinoid	643

Table 10.6.2 (continued)

Activity	Chemical	Class	Ref.
Radiomimetic	Leucodelphinidin	Flavonoid	70
Radioprotective	Scopoletin	Coumarin	912
Radioprotective effect	Umbelliferone	Coumarin	912
Respiratory depressant	Seneciphylline	Pyrrolizidine alkaloid	874
Respiratory stimulant	Scopolamine	Tropane alkaloid	847
RNA synthesis inhibition	Bergapten	Coumarin	1126
	Daphnoretin	Coumarin	396
	Xanthotoxin	Coumarin	1126
	Tripdiolide	Diterpene	287
	Myristic acid	Fatty acid	510
	Ellipticine, 9-methoxy	Indole alkaloid	613
	Harringtonine	Isoquinoline alkaloid	1109
	Harringtonine, homo	Isoquinoline alkaloid	441
	Nitidine	Isoquinoline alkaloid	308
	Gardenoside	Monoterpene	183
	Camptothecin	Quinoline alkaloid	1120
	Bruceantin	Triterpene	395
	Tingenone	Triterpene	355
Serotonin antagonist	Ferulic acid	Lignan	1116
Serotonin secretion inhibition	Morin	Flavonoid	100
	Naringenin	Flavonoid	100
	Catechin, (+)	Isoquinoline alkaloid	100
Sister chromatid exchange stimulation	Harringtonine	Isoquinoline alkaloid	155
	Camptothecin	Quinoline alkaloid	445
Skeletal muscle relaxant	Pyrazine, tetramethyl-	2-N heterocyclic alkaloid	767
	Eugenol	Lignan	242
	Eugenol methyl ether	Lignan	242
	Honokiol	Lignan	1063
	Magnolol	Lignan	1063
Skeletal muscle stimulant	Matrine	Quinolizidine alkaloid	908
Skin pigmentation effect	Psoralen	Coumarin	153
Smooth muscle relaxation	Gallic acid	Benzenoid	735
	Scopoletin	Coumarin	764
	Apigenin	Flavonoid	88
	Coronaridine	Indole alkaloid	812
	Allocryptopine, alpha	Isoquinoline alkaloid	39
	Coclaurine, (DL)-N-demethyl	Isoquinoline alkaloid	1140
	Protopine	Isoquinoline alkaloid	791
	Estragole	Lignan	942
Smooth muscle stimulant	Yohimbine	Indole alkaloid	669
	Choline	Acyclic nitrogen base	200
	Vasicine	Alkaloid	346
	Allocryptopine, alpha	Isoquinoline alkaloid	39
	Berberine	Isoquinoline alkaloid	678
	Ephedrine	Isoquinoline alkaloid	427

Table 10.6.2 (continued)

Activity	Chemical	Class	Ref.
	Lupanine	Quinolizidine alkaloid	672
	Matrine	Quinolizidine alkaloid	1007
	Sparteine	Quinolizidine alkaloid	672
Sodium pump inhibition	Sanguinarine	Isoquinoline alkaloid	156
Spasmolytic	Scopoletin	Coumarin	765
	Apigenin	Flavonoid	987
	Daucosterol	Steroid	532
Spermicidal	Reserpine	Indole alkaloid	1147
	Nicotine	Pyridine alkaloid	727
	Quinine	Quinoline alkaloid	211
	Atropine	Tropane alkaloid	866
Spontaneous activity reduction	Ellagic acid	Coumarin	209
	Eusiderin	Lignan	644
Spontaneous activity stimulation	Ephedrine	Isoquinoline alkaloid	427
Sympatholytic	Scopoletin	Coumarin	765
	Protopine	Isoquinoline alkaloid	35
Tachycardia	Ephedrine	Isoquinoline alkaloid	427
Taenifuge	Xanthotoxin	Coumarin	5
Thrombocytopoietic	Vasicine	Alkaloid	67
Tranquilizing effect	Scoparone	Coumarin	340
Tubulin binding effect	Maytansine	Macrocyclic lactam lactone	615
Tubulin polymerization inhibition	Maytansine	Macrocyclic lactam lactone	457
Tumor promotion inhibitor	Taxifolin	Flavonoid	739
Uterine relaxation effect	Protopine	Isoquinoline alkaloid	791
	Sanguinarine	Isoquinoline alkaloid	393
	Synephrine	Isoquinoline alkaloid	564
	Retamine	Quinolizidine alkaloid	904
	Retamine	Quinolizidine alkaloid	825
Uterine stimulant	Grandiflorenic acid	Terpenoid	96
Uterine stimulant effect	Vasicine	Alkaloid	1046
	Rutaecarpine	Indole alkaloid	543
	Berberine	Isoquinoline alkaloid	572
	Magnoflorine	Isoquinoline alkaloid	190
	Lupanine	Quinolizidine alkaloid	672
	Lupanine, 13-hydroxy	Quinolizidine alkaloid	672
	Sparteine	Quinolizidine alkaloid	399
Vasoconstriction	Leucodelphinidin	Flavonoid	289
Vasodilation	Scopoletin	Coumarin	52
	Yohimbine	Indole alkaloid	150
	Berberine	Isoquinoline alkaloid	1027
	Carveol, dihydro	Monoterpene	976

Table 10.6.2 (continued)

Activity	Chemical	Class	Ref.
Vasodilation	Menth-2-en-7-ol, para-	Monoterpene	976
	Myrcene	Monoterpene	976
	Terpinen-4-ol	Monoterpene	976
	Terpineol, alpha	Monoterpene	1139
Vasodilator	Sanguinarine	Isoquinoline alkaloid	393
	Anisodamine	Tropane alkaloid	724
	Anosidine	Tropane alkaloid	809
Water consumption increase	Ursolic acid	Triterpene	1101
Wound healing acceleration	Allantoin	2-N heterocyclic alkaloid	490

Table 10.6.3. Taxa in which compounds of potential pharmacological interest have been found to occur in woody tissue

Chemical	Occurrence	Ref.
Abyssinone I	Erythrina abyssinica	514
Abyssinone II	Erythrina abyssinica	514
Abyssinone III	Erythrina abyssinica	514
Abyssinone IV	Erythrina abyssinica	514
Abyssinone V	Erythrina abyssinica	514
Abyssinone VI	Erythrina abyssinica	514
Abyssinone VII	Erythrina abyssinica	718
Acanthine, oxy-	Berberis asiatica	1072
	Berberis coriaria	648
	Berberis floribunda	187
	Berberis lycium	606
	Berberis thunbergii	1072
	Berberis vulgaris	1072
	Mahonia repens	972
Acetyl shikonin	Lithospermum erythrorhizon	1002
Aesculetin	Ruta pinnata	870
	Xanthoceras sorbifolia	826
Alizarin-1-methyl ether	Morinda parvifolia	184
Alkannin	Alkanna tinctoria	793
Allantoin	Pueraria hirsuta	717
Allocryptopine, alpha	Zanthoxylum integrifoliolum	470
Allohimachalol	Cedrus deodara	803
Aloe emodin	Picramnia parvifolia	831
Alstonine	Catharanthus roseus	934
	Rauvolfia volkensii	230
	Rauvolfia vomitoria	1072

Table 10.6.3 (continued)

Chemical	Occurrence	Ref.
Alstonine, tetrahydro	Alstonia yunnanensis	197
	Catharanthus roseus	332
	Rauvolfia obscura	1012
	Rauvolfia volkensii	30
	Rauvolfia vomitoria	477
Ammodendrine	Cytisus scoparius	1074
Anisodamine	Anisodus tanguticus	420
	Physochlaina physaloides	624
Anisodine	Anisodus tanguticus	1141
	Przewalskia tanguticus	1089
Anonaine	Annona acuminata	135
	Annona squamosa	864
	Magnolia grandiflora	858
	Monodora tenuifolia	960
Anthraquinone, 2-hydroxy-methyl	Morinda parvifolia	184
Apigenin	Crudia amazonica	143
	Haplormosia monophylla	417
	Pueraria lobata	932
	Pueraria pseudo-hirsuta	932
	Pueraria thomsonii	932
Arbutin	Arctostaphylos uva-ursi	483
	Vaccinium vitis-idaea	851
Aricine	Aspidosperma marcgravianum	683
	Rauvolfia mombasiana	479
	Rauvolfia oreogiton	29
	Rauvolfia volkensii	30
Atropine	Scopolia species	737
Aucuparin	Malus sylvestris	531
	Sorbus americana	722
Austrobailignan I	Amyris pinnata	73
	Austrobaileya scandens	1006
Baccatin I, 1-beta-hydroxy	Taxus baccata	681
Baccatin III	Cephalotaxus mannii	834
	Taxus baccata	681
Benzaldehyde, para-hydroxy-	Plocama pendula	360
	Pterocarpus marsupium	286
Benzoic acid	Dalbergia cochinchinensis	279
	Dalbergia spruceana	223
	Paeonia albiflora	696
	Uvaria angolensis	753
Benzoic acid. para-hydroxy-	Fagara macrophylla	294
	Fagara zanthoxyloides	294
	Paratecoma peroba	152
	Pterocarpus santalinus	1035
	Tabebuia impetiginosa	151

Table 10.6.3 (continued)

Chemical	Occurrence	Ref.
	Vitis vinifera	962
	Zanthoxylum lemairei	294
	Zanthoxylum leprieurii	294
	Zanthoxylum rubescens	294
	Zanthoxylum viride	294
Benzoquinone, 1-4,2-6-dimethoxy	Acacia melanoxylon	414
	Peddiea fischeri	402
	Picrasma crenata	828
	Picrasma quassioides	461
	Vochysia thyrsoidea	253
Benzoquinone, 2-6-dimethoxy	Gyrinops walla	914
Berbamine	Berberis species	622
	Limaciopsis loangensis	175
Berberine	Berberis species	622
	Phellodendron wilsonii	191
	Tinospora glabra	118
	Zanthoxylum monophyllum	965
Bergapten	Esenbeckia litoralis	282
	Ruta oreojasme	362
	Ruta pinnata	359
	Thamnosma texana	750
	Toddalia asiatica	220
Borbonol	Persea borbonia	1129
	Persea schiedeana	1129
	Persea steyermarkii	1129
Brazilin	Caesalpinia sappan	429
	Haematoxylon brasiletto	842
Bruceantin	Brucea javanica	821
Bruceantinol	Brucea javanica	821
Brucein A	Brucea javanica	245
Brucein A, iso	Soulamea soulameoides	403
Brucein B	Brucea javanica	245
Brucein C	Brucea javanica	245
Brucine	Strychnos nux-vomica	90
Cadinol, alpha	Araucaria araucana	144
	Neocallitropsis araucarioides	886
Caffeic acid	Erica australis	169
	Pterocarpus santalinus	1035
	Vitis vinifera	962
Camptothecin	Camptotheca acuminata	1049
	Nothapodytes foetida	374
Camptothecin, 9-methoxy	Merrilliodendron megacarpum	58
	Nothapodytes foetida	374
Canthin-6-one	Picrasma excelsa	1044
	Zanthoxylum dipetalum	325
	Zanthoxylum elephantiasis	206

Table 10.6.3 (continued)

Chemical	Occurrence	Ref.
Canthin-6-one, 11-hydroxy	Soulamea soulameoides	403
Carnosol	Podocarpus koordersii	161
Carvacrol	Juniperus californica	816
	Juniperus procera	816
	Melia azedarach	110
	Paeonia albiflora	696
Carveol, dihydro	Ephedra sinica	976
Cassameridine	Litsea kawakamii	632
Catechin, (+)	Acacia giraffae	649
	Cassia fistula	802
	Prunus persica	755
	Salix caprea	654
Celastrol	Catha edulis	235
	Celastrus scandens	915
	Mortonia gregii	785
	Mortonia palmeri	785
	Tripterygium wilfordii	178
Centdarol	Cedrus deodara	803
Centdarol, iso	Cedrus deodara	803
Cephalotaxine	Cephalotaxus harringtonia	255
	Cephalotaxus wilsoniana	833
Chaparrinone	Simaba multiflora	59
Chaparrinone, 6-alpha-senecioyl-oxy	Simaba multiflora	59
Chelerythrine	Berberis chitria	394
	Berberis verruculosa	394
	Berberis vulgaris	394
	Fagara zanthoxyloides	749
	Zanthoxylum coriaceum	985
	Zanthoxylum dipetalum	325
	Zanthoxylum monophyllum	965
	Zanthoxylum williamsii	965
Chlorogenic acid	Panax ginseng	603
	Panax quinquefolius	603
	Vitis vinifera	962
Choline	Abroma augusta	112
	Abrus precatorius	349
	Atriplex halimus	902
	Atriplex hastata	902
	Atriplex littoralis	902
	Dictamnus albus	1011
	Lycium barbarum	283
	Lycium chilense	650
	Lycium tenuispinum	650
Chrysin	Prunus domestica	713

Table 10.6.3 (continued)

Chemical	Occurrence	Ref.
Chrysophanic acid	Cassia quinquangulata	754
	Dipterocarpus glandulosus	386
	Dipterocarpus hispidus	386
	Dipterocarpus insignis	386
	Dipterocarpus zeylanicus	386
	Hopea brevipetiolaris	386
	Hopea jucunda	386
	Rhamnus sectipetala	719
	Shorea affinis	386
	Shorea congestiflora	386
	Shorea dyerii	386
	Shorea lissophylla	386
	Shorea oblongifolia	386
	Shorea ovalifolia	386
	Shorea stipularis	386
	Shorea trapezifolia	386
	Shorea zeylanica	386
	Stemonoporus affinis	386
	Stemonoporus canaliculatus	386
	Stemonoporus cordifolius	386
	Stemonoporus oblongifolius	386
	Stemonoporus petiolaris	386
	Stemonoporus reticulatus	386
	Ventilago calyculata	405
	Ventilago maderaspatana	852
Cinchonidine	Cinchona ledgeriana	46
	Cinchona pubescens	961
Cinchonine	Cinchona ledgeriana	46
	Cinchona pubescens	961
Cleomiscosin A	Simaba multiflora	60
	Soulamea soulameoides	403
Coclaurine, (DL)-N-demethyl	Gnetum parvifolium	1111
Coptisine	Nandina domestica	458
Coronaridine	Ervatamia hainanensis	318
	Ervatamia heyneana	679
	Stemmadenia glabra	214
	Tabernaemontana amblyocarpa	813
	Tabernaemontana olivacea	10
	Tabernaemontana quadrangularis	9
Coronaridine, (-)	Ervatamia heyneana	380
	Muntfara sessilifolia	789
	Tabernaemontana ambylocarpa	814
Costunolide	Vanillosmopsis erythropappa	129
	Zanthoxylum wutaiense	222
Coumaric acid, para-	Dipterocarpus macrocarpus	573
	Picea koraiensis	378
	Pterocarpus santalinus	1035

Table 10.6.3 (continued)

Chemical	Occurrence	Ref.
Coumarin	Pterocarpus santalinus	361
	Ptychopetalum olacoides	1018
	Ruta pinnata	871
	Simaba multiflora	56
	Uvaria afzelii	449
Cycloartenol	Andira parviflora	143
	Pseudotsuga menziesii	221
	Swietenia mahagoni	49
	Ulmus glabra	618
Cyclovirobuxine D	Buxus microphylla	307
Cymene, para-	Boswellia frereana	968
	Larix cajanderi	132
	Larix sibirica	535
	Magnolia kobus	338
	Magnolia salicifolia	335
	Metasequoia glyptostroboides	337
	Ocotea petalanthera	370
	Picea glehnii	592
	Pinus caribaea	364
	Pinus pumila	404
	Pinus radiata	1127
	Pinus sylvestris	87
	Pinus taeda	978
Daidzein	Dalbergia ecastophyllum	671
	Lespedeza cyrtobotrya	695
	Machaerium villosum	589
	Pterodon apparicioi	608
	Pueraria lobata	932
	Pueraria pseudo-hirsuta	932
Daphnoretin	Wikstroemia indica	602
Daucosterol	Betula alleghaniensis	921
	Betula lenta	921
	Camptotheca acuminata	1087
	Clerodendrum fragrans	941
	Combretum quadrangulare	955
	Datisca glomerata	909
	Maytenus confertiflora	1056
	Maytenus diversifolia	745
	Melia azedarach	913
	Melia birmanica	85
	Pergularia daemia	922
	Simaba multiflora	59
	Tripterygium wilfordii	848
Delphinidin	Ephedra andina	392
	Ephedra frustillata	392
	Quillaja saponaria	877
Dictamnine	Adiscanthus fusciflorus	1038
	Aegle marmelos	1072
	Afraegle paniculata	868

Table 10.6.3 (continued)

Chemical	Occurrence	Ref.
	Dictamnus albus	348
	Dictamnus angustifolius	28
	Dictamnus caucasicus	537
	Esenbeckia flava	282
	Esenbeckia litoralis	282
	Flindersia dissosperma	1072
	Flindersia maculosa	1072
	Geijera balansae	691
	Glycosmis bilocularis	140
	Haplophyllum suaveolens	463
	Ruta chalepensis	1034
	Zanthoxylum ailanthoides	1072
	Zanthoxylum arnottianum	466
	Zanthoxylum cuspidatum	469
	Zanthoxylum inerme	469
	Zanthoxylum integrifoliolum	470
	Zanthoxylum wutaiense	468
Diosgenin	Balanites aegyptiaca	251
	Balanites roxburghii	1032
	Balanites wilsoniana	954
	Tribulus longipetalus	711
	Tribulus terrestris	711
Diospyrin, iso	Diospyros ismailii	1128
	Diospyros lotus	1128
	Diospyros maingayi	1128
	Diospyros morrisiana	604
	Diospyros siamang	1128
	Diospyros singaporensis	1128
	Diospyros sumatrana	1128
	Diospyros virginiana	170
	Diospyros wallichii	1128
Dopamine	Annona reticulata	326
Ecdysone, beta	Achyranthes fauriei	998
	Podocarpus elata	343
Eglandine-N-oxide, 10-methoxy	Ervatamia heyneana	380
Elemicin	Aniba species	668
	Croton nepetaefolius	232
	Dalbergia spruceana	223
	Monopteryx uaucu	370
Elipticine, 9-methoxy	Ochrosia acuminata	616
Ellagic acid	Aristotelia serrata	160
	Bridelia micrantha	807
	Diospyros cinnabarina	345
	Elaeocarpus hookerianus	162
	Eucalyptus camaldulensis	430
	Eucalyptus cypellocarpa	430
	Eucalyptus diversicolor	680
	Eucalyptus dives	430
	Eucalyptus pilularis	430

Table 10.6.3 (continued)

Chemical	Occurrence	Ref.
	Eucalyptus platypus	430
	Eucalyptus sideroxylon	430
	Rhus copallina	1121
	Rhus integrifolia	1121
	Rhus lanceolata	1121
	Rhus microphylla	1121
	Rhus oaxacana	1121
	Rhus pachyrrachis	1121
	Rhus rubifolia	1121
	Rhus schiedeana	1121
	Rhus terebinthifolia	1121
	Rhus typhina	1121
	Syncarpia glomulifera	872
	Syzygium cordatum	166
	Tamarix nilotica	725
	Terminalia arjuna	883
	Vochysia acuminata	253
Ellipticine	Bleekeria vitiensis	538
	Ochrosia acuminata	616
Ellipticine, 9-methoxy	Bleekeria vitiensis	538
Emodin	Cassia laevigata	939
	Cassia occidentalis	579
	Cassia sophera	651
	Juniperus formosana	582
	Polygonum cuspidatum	367
	Rhamnus sectipetala	719
	Rhamnus triquetra	22
	Sargentodoxa cuneata	1058
	Simaba multiflora	56
	Vatairea heteroptera	327
	Ventilago calyculata	405
Ephedrine	Ephedra equisetina	317
	Ephedra nevadensis	95
	Ephedra sinica	1
	Ephedra viridis	730
Epicatechin, (−)	Gossypium barbadense	895
	Lindera umbellata	701
	Lyonia ovalifolia	523
	Psorospermum febrifugum	137
	Uncaria elliptica	77
Erycristagallin	Erythrina crista-galli	693
Erythrabysin II	Erythrina abyssinica	514
Estragole	Croton zehntneri	231
	Monopteryx uaucu	370
	Pinus caribaea	948
	Pinus edulis	949
	Pinus elliottii	686
	Pinus kesiya	270
	Pinus monophylla	949

Table 10.6.3 (continued)

Chemical	Occurrence	Ref.
	Pinus patula	364
	Pinus serotina	372
	Pinus serotina	685
	Pinus sylvestris	411
	Pinus taeda	978
	Pinus tenuifolia	372
	Pinus tenuifolia	685
	Pinus yunnanensis	372
	Pinus yunnanensis	685
Eugenol	Aniba species	370
	Cinnamomum bodinieri	1023
	Melia azedarach	1050
	Nectandra polita	970
	Sassafras randaiense	193
Eugenol methyl ether	Croton nepetaefolius	232
	Croton zehntneri	231
	Dacrydium franklinii	808
	Nectandra polita	970
	Ocotea pretiosa	369
Eusiderin	Aniba species	267
	Virola guggenheimii	320
Euxanthone	Bonnetia stricta	576
	Calophyllum inophyllum	36
	Calophyllum ramiflorum	109
	Garcinia terpnophylla	81
	Mesua ferrea	212
	Mesua thwaitesii	82
	Rheedia gardneriana	142
	Vismia guaramirangae	256
Ferulic acid	Picea koraiensis	378
	Pterocarpus santalinus	1035
	Vitis vinifera	962
Flavone, 5-hydroxy-4'-7-dimethoxy	Rhus undulata	329
Formononetin	Dalbergia cearensis	266
	Dalbergia ecastophyllum	671
	Dalbergia inundata	607
	Dalbergia oliveri	278
	Dalbergia spruceana	223
	Dalbergia variabilis	588
	Diplotropis purpurea	143
	Ferreirea spectabilis	148
	Machaerium kuhlmannii	774
	Machaerium nictitans	774
	Machaerium vestitum	587
	Myroxylon balsamum	261
	Pericopsis species	460
	Thermopsis fabacea	61
	Vatairea heteroptera	327
	Wisteria floribunda	1001
	Zollernia paraensis	322

Table 10.6.3 (continued)

Chemical	Occurrence	Ref.
Forsythoside B	Forsythia koreana	295
Friedelin	Bridelia micrantha	807
	Callisthene major	226
	Calophyllum brasiliense	247
	Calophyllum cordato-oblongum	180
	Calophyllum tomentosum	521
	Caraipa densiflora	42
	Dalbergia volubilis	180
	Elaeocarpus hookerianus	162
	Grewia tiliaefolia	51
	Guazuma tomentosa	837
	Gyrinops walla	914
	Juniperus horizontalis	438
	Larix laricina	438
	Lithocarpus attenuata	452
	Lithocarpus cornea	452
	Spiraea formosana	210
	Syzygium cordatum	166
	Tripetaleia paniculata	1113
	Tsuga canadensis	438
	Uvaria angolensis	753
	Vanillosmopsis erythropappa	74
Gallic acid	Allanblackia floribunda	345
	Bridelia micrantha	807
	Caesalpinia sappan	964
	Dactylocladus stenostachys	4
	Dillenia indica	805
	Dillenia retusa	805
	Diospyros cinnabarina	345
	Diospyros fragrans	345
	Garcinia densivenia	345
	Garcinia mannii	345
	Glossocalyx longicuspus	345
	Paratecoma peroba	152
	Psidium guajava	688
	Rauvolfia vomitoria	652
	Rhus copallina	1121
	Rhus glabra	1121
	Rhus integrifolia	1121
	Rhus lanceolata	1121
	Rhus microphylla	1121
	Rhus oaxacana	1121
	Rhus pachyrrachis	1121
	Rhus rubifolia	1121
	Rhus schiedeana	1121
	Rhus terebinthifolia	1121
	Rhus typhina	1121
	Symphonia globulifera	345
	Syzygium cordatum	166
	Tamarix nilotica	725
	Vitis vinifera	962

Table 10.6.3 (continued)

Chemical	Occurrence	Ref.
Gardenoside	Gardenia jasminoides	462
Genistein	Lespedeza cyrtobotrya	695
	Prunus mahaleb	1040
	Pterocarpus angolensis	107
	Pueraria lobata	1000
	Pueraria pseudo-hirsuta	932
	Pueraria thomsonii	932
Gentisic acid	Pterocarpus santalinus	1035
Geranyl hydroquinone	Cordia elaeagnoides	656
Germichrysone	Cassia singueana	296
Ginsenoside RB-1	Panax ginseng	511
	Panax quinquefolius	539
Ginsenoside RC	Panax ginseng	797
	Panax pseudo-ginseng	903
	Panax quinquefolius	540
	Panax trifolius	637
Ginsenoside RD	Panax ginseng	539
	Panax pseudo-ginseng	903
	Panax quinquefolius	539
	Panax trifolius	637
Ginsenoside RE	Panax ginseng	1045
	Panax pseudo-ginseng	1045
	Panax pseudo-ginseng var. japonicus	1045
	Panax quinquefolius	1045
	Panax trifolius	605
Ginsenoside RF	Panax ginseng	797
	Panax pseudo-ginseng	540
	Panax quinquefolius	539
	Panax trifolius	605
Ginsenoside RG-1	Panax ginseng	712
	Panax pseudo-ginseng	1045
	Panax pseudo-ginseng var. japonicus	1045
	Panax quinquefolius	1045
Girinimbine	Murraya koenigii	854
Glaucarubolone	Soulamea soulameoides	401
Glycoperine	Haplophyllum ferganicum	105
Glycyrrhetinic acid	Glycyrrhiza glabra	878
Grandiflorenic acid	Lagascea rigida	124
	Montanoa tomentosa	298
	Verbesina oncophora	128
	Wedelia species	131
Harman	Mitragyna hirsuta	822
	Peganum harmala	352
	Zygophyllum fabago	639
Harringtonine	Cephalotaxus fortunei	642
	Cephalotaxus hainanensis	507

Table 10.6.3 (continued)

Chemical	Occurrence	Ref.
	Cephalotaxus harringtonia	255
	Cephalotaxus oliveri	786
	Cephalotaxus sinensis	642
Harringtonine, deoxy	Cephalotaxus harringtonia	255
Harringtonine, homo	Cephalotaxus fortunei	641
	Cephalotaxus hainanensis	444
	Cephalotaxus harringtonia	255
	Cephalotaxus sinensis	642
Harringtonine, iso	Cephalotaxus fortunei	1114
	Cephalotaxus hainanensis	444
	Cephalotaxus harringtonia	255
Hederagenin	Cephalaria kotschyi	33
	Cephalaria nachiczevanica	702
	Guettarda angelica	958
	Patrinia scabiosaefolia	1080
	Rhabdodendron macrophyllum	323
Hematoxylin	Haematoxylon brasiletto	842
Hentriacontane, N-	Acacia senegal	505
	Aesculus indica	704
	Aloe arborescens	432
	Boerhavia diffusa	689
	Colletia spinosissima	784
	Diospyros austro-africana	229
	Nothofagus fusca	625
	Nothofagus menziesii	625
	Nothofagus solanderi	625
	Nothofagus truncata	625
	Solanum platanifolium	845
	Zelkova oregoniana	733
Heyneatine, (−)	Ervatamia heyneana	380
Himachalol	Cedrus deodara	19
Himadarol	Cedrus deodara	803
Honokiol	Magnolia obvata	334
Hordenine	Alhagi pseudalhagi	350
	Sageretia gracilis	1136
	Zanthoxylum arborescens	375
	Zanthoxylum macrocarpum	139
	Zanthoxylum punctatum	168
Humulene, alpha	Abies nordmanniana	935
	Abies semenovii	536
	Larix olgensis	534
	Magnolia kobus	338
	Metasequoia glyptostroboides	337
	Picea orientalis	935
	Pinus edulis	949
	Pinus funebris	533
	Pinus koraiensis	404

Table 10.6.3 (continued)

Chemical	Occurrence	Ref.
	Pinus monophylla	949
	Syzygium cumini	233
	Vernonia polyanthes	125
	Wedelia grandiflora	127
Huratoxin	Hura crepitans	899
	Pimelea simplex	673
	Wikstroemia monticola	499
Hydroquinone	Decussocarpus wallichianus	163
Isoflavone, 7-hydroxy-4'-8-dimethoxyiso	Xanthocercis zambesiaca	408
Jacareubin	Calophyllum brasiliense	545
	Calophyllum cordato-oblongum	385
	Calophyllum inophyllum	482
	Calophyllum lankaensis	384
	Calophyllum ramiflorum	109
	Calophyllum tomentosum	521
	Calophyllum trapezifolium	373
Jacareubin, 6-deoxy	Calophyllum cuneifolium	381
	Calophyllum inophyllum	482
	Calophyllum lankaensis	384
	Calophyllum soulattri	381
	Calophyllum tomentosum	521
	Calophyllum trapezifolium	373
Jatrorrhizine	Berberis species	622
	Coptis chinensis	173
	Coptis deltoides	1106
	Coptis japonica	471
	Coscinium fenestratum	945
	Dioscoreophyllum cumminsii	341
	Fibraurea chloroleuca	943
	Jateorhiza palmata	173
	Mahonia aquifolium	568
	Mahonia repens	973
	Parabaena tuberculata	118
	Tinospora baenzigeri	118
	Tinospora capillipes	182
	Tinospora cordifolia	118
	Tinospora crispa	118
	Tinospora glabra	118
	Tinospora merrilliana	118
	Tinospora sagittata	118
	Tinospora sinensis	118
	Tinospora smilacina	118
Justicidin B	Boenninghausenia albiflora	885
	Haplophyllum dahuricum	93
	Haplophyllum obtusifolium	92
	Haplophyllum tuberculatum	929
Kuwanon G	Morus alba	488
	Morus bombycis	488
	Morus lhou	488

Table 10.6.3 (continued)

Chemical	Occurrence	Ref.
Kuwanon H	Morus alba	488
	Morus bombycis	488
	Morus lhou	488
Lapachol	Dolichandrone crispa	838
	Haplophragma adenophyllum	502
	Hibiscus tiliaceus	32
	Kigelia pinnata	504
	Macfadyena unguis-cati	503
	Markhamia platycalyx	503
	Millingtonia hortensis	836
	Paratecoma peroba	152
	Stereospermum suaveolens	859
	Tabebuia chrysantha	152
	Tabebuia donnell-smithii	152
	Tabebuia guayacan	657
	Tabebuia impetiginosa	151
	Tabebuia rosea	840
	Tecomella undulata	391
	Tectona grandis	664
	Zeyheria digitalis	248
Lapachone, beta	Haplophragma adenophyllum	502
	Markhamia platycalyx	503
	Markhamia stipulata	501
	Parataecoma peroba	152
	Tabebuia chrysantha	152
	Tabebuia donnell-smithii	152
	Tabebuia guayacan	657
	Tabebuia impetiginosa	151
	Tectona grandis	500
Leucodelphinidin	Alhagi pseudalhagi	472
	Plumbago rosea	407
Leucopelargonidin	Quercus robur	513
	Rourea santaloides	855
	Salacia chinensis	574
Liquiritigenin	Dalbergia retusa	658
	Diplotropis purpurea	143
	Glycyrrhiza glabra	931
	Hedysarum polybotrys	694
	Lespedeza cyrtobotrya	695
	Pterocarpus angolensis	107
	Pterocarpus marsupium	286
Liriodenine	Annona acuminata	1061
	Asimina triloba	1016
	Fissistigma glaucescens	633
	Goniothalamus amuyon	633
	Laurelia sempervirens	1030
	Limaciopsis sempervirens	175
	Liriodendron tulipifera	448
	Michelia champaca	371
	Monodora tenuifolia	960

Table 10.6.3 (continued)

Chemical	Occurrence	Ref.
	Pachygone ovata	246
	Rhigiocarya racemifera	288
	Rollinia papilionella	240
	Rollinia sericea	6
	Sinomenium acutum	455
Loganin	Hydrangea heteromalla	497
	Lonicera sempervirens	157
	Patrinia villosa	25
Lucidin omega-methyl ether	Morinda parvifolia	184
Lupanine	Cytisus purgans	75
	Cytisus scoparius	1074
	Genista tinctoria	843
Lupanine, 13-hydroxy	Cytisus scoparius	1074
Lupeol acetate	Cordia buddleioides	162
	Ectopopterys soejartoi	663
	Phoenix dactylifera	321
Lysicamine	Abuta rufescens	946
	Annona acuminata	135
	Chasmanthera dependens	756
	Illigera pentaphylla	882
	Liriodendron tulipifera	1148
	Rollinia papilionella	240
Magnoflorine	Berberis actinacantha	1067
	Berberis amurensis	1072
	Berberis darwinii	1029
	Berberis julianae	141
	Berberis koreana	567
	Berberis thunbergii	1072
	Berberis vulgaris	274
	Calophyllum robustum	1017
	Chasmanthera dependens	756
	Cocculus laurifolius	108
	Cyclea peltata	558
	Dioscoreophyllum cumminsii	341
	Magnolia grandiflora	858
	Mahonia aquifolium	568
	Monodora tenuifolia	1072
	Nandina domestica	458
	Pachygone ovata	292
	Phellodendron wilsonii	191
	Stephania hernandifolia	867
	Tiliacora racemosa	911
	Tinospora capillipes	182
	Tinospora cordifolia	782
	Zanthoxylum culantrillo	985
	Zanthoxylum dipetalum	325
	Zanthoxylum macrocarpum	139
	Zanthoxylum monophyllum	965
	Zanthoxylum nitidum	3
	Zanthoxylum punctatum	168

Table 10.6.3 (continued)

Chemical	Occurrence	Ref.
Magnolol	Magnolia obovata	334
	Sassafras randaiense	193
Mangiferin	Gyrinops walla	914
	Mangifera indica	900
Manniflavanone	Garcinia mannii	454
Marmesin	Afraegle paniculata	868
	Ruta pinnata	363
	Zanthoxylum arnottianum	467
	Zanthoxylum wutaiense	468
Matrine	Euchresta horsfieldii	759
	Sophora microphylla	547
	Sophora subprostrata	931
	Sophora tomentosa	188
Maytancyprine, nor-	Maytenus buchananii	951
	Putterlickia verrucosa	952
Maytanprine	Maytenus buchananii	584
	Maytenus confertiflora	1055
	Maytenus hookeri	1144
	Maytenus ilicifolia	23
	Putterlickia verrucosa	583
Maytansine	Maytenus acuminata	974
	Maytenus buchananii	584
	Maytenus confertiflora	1056
	Maytenus diversifolia	745
	Maytenus guangkiensis	846
	Maytenus heterophylla	974
	Maytenus hookeri	1144
	Maytenus mossambicensis	974
	Maytenus polycantha	974
	Maytenus senegalensis	974
	Putterlickia pyracantha	974
Menisperine	Magnolia grandiflora	858
	Nandina domestica	1072
	Rhigiocarya racemifera	288
	Tinospora capillipes	182
	Zanthoxylum nitidum	3
Menth-2-en-7-ol, para-	Ephedra sinica	976
Milliamine A	Euphorbia milii	666
Morin	Artocarpus integrifolia	800
	Artocarpus pithecogalla	703
	Maclura cochinchinensis	515
Morin	Maclura pomifera	1076
	Morus alba	265
	Morus indica	265
	Morus laevigata	265
	Morus rubra	265
	Morus serrata	265

Table 10.6.3 (continued)

Chemical	Occurrence	Ref.
Morroniside	Lonicera periclymenum	158
Myrcene	Abies concolor	1132
	Abies nordmanniana	935
	Abies semenovii	536
	Aralia cordata	1118
	Croton zehntneri	231
	Ephedra sinica	976
	Juniperus sabina	560
	Larix cajanderi	132
	Larix olgensis	534
	Larix sibirica	535
	Picea glauca	887
	Picea glehnii	592
	Picea orientalis	935
	Pinus contorta	659
	Pinus edulis	949
	Pinus eldarica	281
	Pinus funebris	533
	Pinus kesiya	270
	Pinus koraiensis	404
	Pinus merkusii	1068
	Pinus monophylla	949
	Pinus patula	364
	Pinus pityusa	281
	Pinus radiata	1127
	Pseudotsuga menziesii	1131
	Valeriana officinalis	1075
Myristic acid	Eleutherococcus spinosus	993
	Nothofagus fusca	625
	Nothofagus menziesii	625
	Nothofagus solanderi	625
	Nothofagus truncata	625
Naringenin	Ferreirea spectabilis	546
	Flemingia stricta	857
	Larix dahurica	598
	Prunus domestica	713
	Salix caprea	654
	Soymida febrifuga	860
Nicotine	Anthocercis frondosa	302
	Asclepias syriaca	665
	Duboisia hopwoodii	297
	Duboisia myoporoides	297
	Erythroxylum coca	1072
Nitidine	Zanthoxylum cuspidatum	469
	Zanthoxylum gillettii	1133
	Zanthoxylum inerme	469
	Zanthoxylum nitidum	237
	Zanthoxylum rubescens	700
	Zanthoxylum williamsii	965

Table 10.6.3 (continued)

Chemical	Occurrence	Ref.
Obaberine	Albertisia papuana	600
	Mahonia repens	972
	Pycnarrhena longifolia	944
Okanin	Abies pindrow	1015
	Albizia adianthifolia	165
	Albizia lebbek	390
Oleanolic acid	Adenanthera pavonina	181
	Aralia mandshurica	1041
	Fraxinus greggii	276
	Panax ginseng	1079
	Panax pseudo-ginseng var. japonicus	931
	Panax quinquefolius	539
	Patrinia intermedia	71
	Psidium guajava	688
	Rhabdodendron macrophyllum	323
	Tabebuia rosea	839
	Tetrapleura tetraptera	653
	Viscum album	481
Oleuropein	Olea europaea	927
Oxonantenine	Laurelia sempervirens	1030
Pericalline	Catharanthus lanceus	119
	Catharanthus roseus	981
	Ervatamia heyneana	380
	Muntfara sessilifolia	789
	Pandaca eusepala	850
	Pandaca ochrascens	225
Phenol	Gossypium mexicanum	1110
	Paeonia albiflora	696
Phyllanthoside	Phyllanthus acuminatus	818
	Phyllanthus brasiliensis	585
Phyllanthostatin 3	Phyllanthus acuminatus	818
Physcion	Abrus cantoniensis	1078
	Cassia occidentalis	579
	Cassia sophera	1043
	Picramnia parvifolia	831
	Rhamnus sectipetala	719
	Rheum species	1088
	Rubia cordifolia	1009
	Rubia tinctorum	1008
	Sargentodoxa cuneata	1058
	Ventilago calyculata	405
	Vochysia thyrsoidea	253
Picrasin B	Picrasma quassioides	428
	Quassia africana	1021
	Soulamea muelleri	829
	Soulamea soulameoides	403
Pimaric acid, iso	Abies semenovii	536
	Cedrus deodara	19

Table 10.6.3 (continued)

Chemical	Occurrence	Ref.
	Larix dahurica	936
	Larix olgensis	534
	Pinus ponderosa	333
Plaunol B	Croton sublyratus	552
Plaunol C	Croton joufra	752
	Croton sublyratus	552
Plaunol D	Croton sublyratus	552
Plaunol E	Croton oblongifolius	752
	Croton sublyratus	552
Plumieride coumarate glucoside	Allamanda cathartica	224
Prieurianin	Trichilia prieureana	638
Pristimerin	Catha edulis	235
	Crossopetalum uragoga	275
	Maytenus boaria	257
	Maytenus emarginata	1071
	Maytenus obtusifolia	257
	Plenckia populnea	358
	Salacia species	172
	Schaefferia cuneifolia	277
Procyanidin B-2	Juniperus sabina	801
	Psorospermum febrifugum	137
Procyanidin B-4	Juniperus sabina	801
	Rubus fruticosus	1010
	Rubus idaeus	1010
Prosopinine	Prosopis africana	207
Protocatechuic acid	Erica australis	169
	Picea koraiensis	378
	Rosa canina	729
Protopine	Nandina domestica	1138
Psoralen	Atalantia monophylla	581
	Ficus species	413
	Ruta pinnata	871
	Thamnosma texana	750
	Zanthoxylum arnottianum	466
	Zanthoxylum flavum	544
Puerarin	Pueraria lobata	707
	Pueraria mirifica	989
	Pueraria pseudo-hirsuta	241
	Pueraria tuberosa	116
Pyrazine, tetramethyl-	Ephedra sinica	976
Pyrogallol	Byrsonima intermedia	133
	Geum urbanum	844
	Rubus rigidus	889
Quercitrin	Arbutus unedo	249
	Cathaya argyrophylla	208

Table 10.6.3 (continued)

Chemical	Occurrence	Ref.
	Lyonia ovalifolia	524
	Ruta pinnata	870
Quinidine	Cinchona pubescens	705
Quinine	Cinchona ledgeriana	706
	Cinchona pubescens	705
	Picrolemma pseudocoffea	40
Reserpine	Catharanthus roseus	980
	Ervatamia coronaria	319
	Ervatamia hainanensis	319
	Ervatamia officinalis	319
	Evodia lepta	319
	Rauvolfia cambodiana	595
	Rauvolfia cubana	599
	Rauvolfia latifrons	319
	Rauvolfia mombasiana	479
	Rauvolfia oreogiton	29
	Rauvolfia serpentina	440
	Rauvolfia tiaolushanensis	440
	Rauvolfia verticillata	440
	Rauvolfia volkensii	29
	Rauvolfia vomitoria	1072
	Rauvolfia yunnanensis	1066
	Vinca minor	640
	Winchia calophylla	319
Resveratrol	Haplormosia monophylla	417
	Intsia species	459
	Morus alba	265
Retamine	Genista sphaerocarpa	1072
	Retama sphaerocarpa	75
Rhein	Cassia obovata	853
	Rheum species	1088
Robustine	Zanthoxylum arnottianum	466
	Zanthoxylum cuspidatum	469
	Zanthoxylum integrifoliolum	470
Rutacridone epoxide	Ruta graveolens	714
Rutaecarpine	Zanthoxylum integrifoliolum	470
Ryanodine	Ryania speciosa	1064
Ryanodine, 9-21-didehydro	Ryania speciosa	1064
Safrole	Aniba species	370
	Cinnamomum bodinieri	1023
	Cinnamomum camphora	336
	Cinnamomum camphora	819
	Cinnamomum parthenoxylon	147
	Croton zehntneri	231
	Licaria pichury-major	916
	Magnolia salicifolia	335

Table 10.6.3 (continued)

Chemical	Occurrence	Ref.
Safrole	Ocotea pretiosa	697
	Sassafras albidum	436
	Sassafras randaiense	193
Safrole, iso	Aniba species	370
Salsolinol	Annona reticulata	326
Sanggenone C	Morus alba	491
Sanggenone D	Morus alba	493
	Morus bombycis	493
	Morus lhou	493
Sanguinarine	Berberis chitria	394
	Berberis verruculosa	394
	Berberis vulgaris	394
	Chelidonium majus	1072
	Corydalis lutea	394
	Corydalis paniculigera	38
	Glaucium corniculatum	947
	Glaucium fimbrilligerum	518
	Glaucium flavum	475
	Sanguinaria canadensis	1014
Scoparone	Afraegle paniculata	868
	Hortia badinii	227
	Licaria armeniaca	24
	Zanthoxylum gillettii	544
Scopolamine	Anthocercis frondosa	302
	Datura meteloides	674
	Duboisia hopwoodii	635
	Duboisia leichhardtii	297
	Duboisia myoporoides	297
Scopoletin	Amyris pinnata	73
	Brunfelsia uniflora	480
	Diospyros maritima	272
	Diospyros quaesita	422
	Diospyros virginiana	170
	Diospyros walkeri	422
	Dipterocarpus glandulosus	386
	Dipterocarpus hispidus	386
	Dipterocarpus insignis	386
	Dipterocarpus zeylanicus	386
	Gossypium species	69
	Hopea brevipetiolaris	386
	Hopea jucunda	386
	Ledum decumbens	557
	Ledum hypoleucum	557
	Ledum macrophyllum	557
	Ledum palustre	557
	Nealchornea yapurensis	379
	Picrasma excelsa	1044
	Rhus copallina	1121
	Rhus glabra	1121

Table 10.6.3 (continued)

Chemical	Occurrence	Ref.
	Rhus integrifolia	1121
	Rhus lanceolata	1121
	Rhus microphylla	1121
	Rhus oaxacana	1121
	Rhus pachyrrachis	1121
	Rhus rubifolia	1121
	Rhus schiedeana	1121
	Rhus terebinthifolia	1121
	Rhus typhina	1121
	Ruta pinnata	870
	Shorea congestiflora	386
	Shorea lissophylla	386
	Shorea oblongifolia	386
	Shorea ovalifolia	386
	Shorea zeylanica	386
	Stemonoporus affinis	386
	Stemonoporus canaliculatus	386
	Stemonoporus cordifolius	386
	Stemonoporus oblongifolius	386
	Stemonoporus petiolaris	386
	Stemonoporus reticulatus	386
	Vatica affinis	386
Seneciphylline	Senecio douglasii	699
Sennoside B	Cassia angustifolia	940
	Rheum palmatum	1150
Sesamin, (+)	Cleistanthus collinus	50
	Cleistanthus patulus	910
	Gingko biloba	519
	Markhamia platycalyx	503
	Paulownia tomentosa	994
	Sassafras albidum	436
	Zanthoxylum integrifoliolum	470
	Zanthoxylum williamsii	965
Shikonin	Echium vulgare	315
	Lithospermum erythrorhizon	415
Sinapic acid	Erica australis	169
	Pterocarpus santalinus	1035
	Vitis vinifera	962
Singueanol I	Cassia singueana	296
Singueanol II	Cassia singueana	296
Skimmianine	Dictamnus albus	348
	Dictamnus angustifolius	975
	Esenbeckia flava	282
	Fagara mantshurica	368
	Flindersia maculosa	1072
	Geijera balansae	691
	Glycosmis bilocularis	140
	Glycosmis mauritiana	865
	Haplophyllum dauricum	104

Table 10.6.3 (continued)

Chemical	Occurrence	Ref.
	Haplophyllum foliosum	1072
	Haplophyllum latifolium	728
	Haplophyllum pedicellatum	1072
	Haplophyllum suaveolens	463
	Haplophyllum vulcanicum	804
	Hortia badinii	227
	Murraya paniculata	314
	Orixa japonica	1072
	Zanthoxylum ailanthoides	1072
	Zanthoxylum bungeanum	869
	Zanthoxylum culantrillo	985
	Zanthoxylum cuspidatum	469
	Zanthoxylum integrifoliolum	470
	Zanthoxylum nitidum	3
	Zanthoxylum nitidum	1052
	Zanthoxylum pluviatile	228
Sparteine	Cytisus multiflorus	1072
	Cytisus purgans	75
	Cytisus scoparius	1074
	Sophora secundiflora	189
Squalene	Baccharis calvescens	130
	Baccharis cassinaefolia	130
	Baccharis genistelloides	126
	Baccharis salzmannii	130
	Tilia vulgaris	619
Synephrine	Sinomenium acutum	906
	Zanthoxylum culantrillo	985
Syringaresinol, (+)	Gyrinops walla	914
	Passerina vulgaris	387
Syringin	Daphne giraldii	1053
	Eleutherococcus senticosus	933
Tabernulosine	Tabernaemontana glandulosa	11
Taraxerol	Bridelia micrantha	807
	Calophyllum tomentosum	521
	Canarium zeylanicum	78
	Clitoria ternatea	83
	Diospyros hirsuta	422
	Diospyros ismailii	1128
	Diospyros japonica	591
	Diospyros lotus	1119
	Diospyros maingayi	1128
	Diospyros siamang	1128
	Diospyros singaporensis	1128
	Diospyros sumatrana	1128
	Diospyros wallichii	1128
	Lithocarpus cornea	452
	Lithocarpus elizabethae	452
	Lithocarpus glabra	452
	Mimusops hexandra	690
	Myrica nagi	263
	Tripetaleia paniculata	1113

Table 10.6.3 (continued)

Chemical	Occurrence	Ref.
Taxifolin	Cedrus deodara	18
	Maclura cochinchinensis	709
	Prunus mahaleb	783
	Salix caprea	654
	Toxicodendron semialatum	1121
Tembetarine	Tinospora baenzigeri	118
	Tinospora cordifolia	118
	Tinospora crispa	118
	Tinospora glabra	118
	Tinospora merrilliana	118
	Tinospora sagittata	118
	Tinospora sinensis	118
	Zanthoxylum dipetalum	325
Terpinen-4-ol	Aralia cordata	1118
	Cupressus stephensonii	920
	Ephedra sinica	976
	Juniperus sabina	560
	Larix cajanderi	132
Terpinen-4-ol	Magnolia kobus	338
	Metasequoia glyptostroboides	337
	Picea glauca	887
	Picea glehnii	592
	Pinus edulis	949
	Pinus monophylla	949
	Pinus pumila	404
Terpineol, alpha	Cunninghamia konishii	203
	Cupressus dupreziana	824
	Larix cajanderi	132
	Larix sibirica	535
	Licaria puchury-major	916
	Picea glauca	887
	Santalum album	259
Thujaplicin, alpha	Thuja occidentalis	376
	Thuja plicata	26
Thujaplicin, beta	Cupressus benthami	1130
	Cupressus funebris	1130
	Cupressus glabra	1130
	Cupressus lindleyi	1130
	Cupressus macnabiana	1130
	Cupressus nevadensis	1130
	Cupressus pygmea	1130
	Cupressus sargentii	1130
	Cupressus sempervirens	299
	Cupressus stephensonii	920
	Juniperus californica	816
	Juniperus procera	816
	Thuja occidentalis	376
	Thuja orientalis	434
	Thuja plicata	301
Thujaplicin, gamma	Cupressus glabra	1130
	Cupressus lindleyi	1130

Table 10.6.3 (continued)

Chemical	Occurrence	Ref.
	Cupressus lusitanica	1130
	Cupressus sargentii	1130
	Cupressus stephensonii	920
	Juniperus procera	816
	Thuja occidentalis	376
	Thuja orientalis	434
	Thuja plicata	301
Tingenone	Crossopetalum uragoga	275
	Maytenus emarginata	1071
	Maytenus nemorosa	309
	Maytenus obtusifolia	257
	Salacia species	172
	Schaefferia cuneifolia	277
Torosachrysone	Cassia singueana	296
Triacontane, N-	Clerodendrum fragrans	941
	Eleutherococcus trifoliatus	194
	Gossypium hirsutum	835
	Keteleeria davidiana	204
	Nothofagus fusca	625
	Nothofagus menziesii	625
	Nothofagus solanderi	625
	Nothofagus truncata	625
	Rhododendron anthopogon	506
Tricin	Simaba multiflora	56
	Wikstroemia indica	602
Tripdiolide	Tripterygium wilfordii	287
Triptolide	Tripterygium wilfordii	205
Umbelliferone	Adina cordifolia	262
	Skimmia reevesiana	1086
Ursolic acid	Alhagi persarum	65
	Artocarpus integrifolia	252
	Boerhavia diffusa	689
	Catharanthus trichophyllus	918
	Corchorus capsularis	661
	Cotylelobium scabriusculum	386
	Diospyros ebenum	389
	Diospyros ismailii	1128
	Diospyros lotus	1128
	Diospyros maingayi	1128
	Diospyros melanoxylon	905
	Diospyros siamang	1128
	Diospyros singaporensis	1128
	Diospyros sumatrana	1128
	Diospyros wallichii	1128
	Dipterocarpus glandulosus	386
	Dipterocarpus hispidus	386
	Dipterocarpus insignis	386
	Dipterocarpus zeylanicus	386
	Doona congestiflora	79

Table 10.6.3 (continued)

Chemical	Occurrence	Ref.
	Doona macrophylla	79
	Grewia villosa	91
	Harpagophytum procumbens	1025
	Hopea brevipetiolaris	386
	Hopea cordifolia	386
	Hopea jucunda	386
	Shorea affinis	386
	Shorea congestiflora	386
	Shorea dyerii	386
	Shorea oblongifolia	386
	Shorea trapezifolia	386
	Shorea zeylanica	386
	Stemonoporus affinis	386
	Stemonoporus canaliculatus	386
	Stemonoporus cordifolius	386
	Stemonoporus oblongifolius	386
	Stemonoporus petiolaris	386
	Stemonoporus reticulatus	386
	Syncarpia glomulifera	872
	Tripetaleia paniculata	1113
	Wisteria floribunda	1001
	Xeromphis spinosa	504
Uvafzelin	Uvaria afzelia	450
Uzarigenin	Calotropis procera	300
Vanillic acid	Alnus japonica	742
	Elaeagnus pungens	990
	Erica australis	169
	Gossypium mexicanum	1110
	Melia azedarach	913
	Panax ginseng	398
	Paratecoma peroba	152
	Picea koraiensis	378
	Pterocarpus santalinus	1035
	Rosa canina	729
	Trachelospermum asiaticum	734
Vanillin	Abies sachalinensis	746
	Alnus japonica	741
	Caraipa densiflora	41
	Caraipa grandifolia	41
	Carissa edulis	13
	Geijera balansae	691
	Liquidambar styraciflua	928
	Melia azedarach	913
	Myroxylon peruiferum	662
	Picea ajanensis	609
	Picea koraiensis	609
	Picea obovata	609
	Zeyheria digitalis	248
Vasicine	Adhatoda vasica	114
	Peganum harmala	377

Table 10.6.3 (continued)

Chemical	Occurrence	Ref.
Vasicinone	Adhatoda vasica	484
Xanthone, 1-3-5-trihydroxy-2-methoxy	Calophyllum trapezifolium	726
	Kayea stylosa	382
	Pentadesma butyracea	383
Xanthone, 1-5-dihydroxy	Calophyllum tomentosum	521
	Garcinia echinocarpa	81
	Garcinia terpnophylla	81
	Mammea acuminata	80
	Mesua ferrea	1048
	Mesua thwaitesii	82
	Rheedia gardneriana	142
Xanthone, 3-8-dihydroxy-1-2-dimethoxy	Calophyllum tomentosum	521
	Garcinia echinocarpa	81
	Garcinia terpnophylla	81
	Mammea acuminata	80
	Mesua ferrea	1048
	Mesua thwaitesii	82
	Rheedia gardneriana	142
Xanthotoxin	Afraegle paniculata	868
	Atalantia ceylanica	21
	Feronia limonia	84
	Skimmia laureola	956
	Thamnosma texana	750
	Zanthoxylum flavum	544
Yohimbine	Amsonia elliptica	771
	Catharanthus lanceus	486
	Catharanthus roseus	796
	Pausinystalia johimbe	795
	Rauvolfia mombasiana	479
	Rauvolfia oreogiton	29
	Rauvolfia serpentina	1072
	Rauvolfia volkensii	29
	Rauvolfia vomitoria	1072
Yohimbine, alpha	Rauvolfia cumminsii	478
	Rauvolfia mombasiana	479
	Rauvolfia obscura	1013
	Rauvolfia oreogiton	29
	Rauvolfia vomitoria	477
Zeatin	Castanea species	1117
	Catharanthus roseus	250
	Spiraea arguta	891
Zeatin riboside	Spiraea arguta	891

References

1. Abdel-Wahab SM, Hashim FM, El-Keiv MA 1961 A pharmacognostical study of *Ephedra sinica* cultivated in Egypt. J Pharm Sci U A R 2: 59–89
2. Abdulla WA, Kadry H, Mahran SG, El-Raziky EH et al. 1977 Preliminary studies on the antischistosomal effect of *Ammi majus* L. Egypt J Bilharz 4: 19–26
3. Abe F, Mayumifurikawa G, Nonaka HO, Nishioka I 1973 Studies on *Xanthoxylum* spp. I. Constituents of the root of *Xanthoxylum piperitum*. Yakugaku Zasshi 93: 624–628
4. Abe Z, Minami K 1975 Extractives of hardwood. VI. Extractives from the wood of *Dactylocladus stenostachys*. Mokuzaki Gakkaishi 21: 194–197
5. Abeysekera BF, Abramowski Z, Towers GHN 1983 Genotoxicity of the natural furochromones, khellin and visnagin and the identification of a khellin-thymine photoadduct. Abstr 24th Ann. Meeting Am Soc Pharmacog., Univ. Mississippi, Oxford, July 24–28. American Society of Pharmacognosy, Dept of Pharmaceutical Sciences, Univ of Pittsburgh, Pittsburgh PA p. ABSTR-60
6. Abrash RM, Sneden AT 1983 Oxoaporphine alkaloids from *Rollinia sericea*. J Nat Prod 46: 437
7. Abtahi FS, Auletta FJ, Sadeghi D, Djahanguire B et al. 1978 Effect of sparteine sulfate on uterine prostaglandin F in the rat. Prostaglandins 16: 473–482
8. Abyshev AZ, Alieva SHA, Damirov IA, Denisenko PP et al. 1982 Coumarins of roots of *Doronicum macrophyllum* Fisch. ex Horten and their antiarrhythmic activity. Rast Resur 18: 249–252
9. Achenbach H, Raffelsberger B 1980a Alkaloids in *Tabernaemontana* Species. XI. Investigation of the alkaloids from *Tabernaemontana quadrangularis* (2Or)-20-Hydroxyibogamine, a new alkaloid from *T. quadrangularis*. Z Naturforsch Ser B 35: 219–225
10. Achenbach H, Rafflesberger B 1980b Alkaloids in *Tabernaemontana* species. XII. Investigation of alkaloids in *Tabernaemontana olivacea* Condylocarpine-N-oxide, a new alkaloid from *T. olivacea*. Z Naturforsch Ser B 35: 885–891
11. Achenbach H, Raffelsberger B 1980c Constituents of west African medicinal plants. Part 3. 19-ethoxycoronaridine, a novel alkaloid from *Tabernaemontana glandulosa*. Phytochemistry 19: 716–717
12. Achenbach H, Raffelsberger B, Addae-Mensah I 1982 Alkaloids in *Tabernaemontana* species, 8. Tabernulosine and 12-demethoxy-tabernulosine, two new alkaloids of the picrinine-type from *Tabernaemontana glandulosa*. Justus Liebigs Ann Chem 1982: 830–844
13. Achenbach H, Waibel R, Addae-Mensah I 1983 Constituents of west African medicinal plants. Part 12. Lignans and other constituents from *Carissa edulis*. Phytochemistry 22: 749–753
14. Achterrath-Tuckermann U, Kunde R, Flaskamp E, Isaac O et al. 1980 Pharmacological investigations with compounds of chamomile. V. Investigations on the spasmolytic effect of compounds of chamomile and kamillosan on the isolated guinea pig ileum. Planta Med 39: 38–50
15. Adams RJ, Seifen E 1978 The positive inotropic action of sanguinarine. Diss Abstr Int B 39: 2241–2242
16. Adesina SK 1985 Constituents of *Solanum dasyphyllum* fruit. J Nat Prod 48:147
17. Adesina SK, Ette EI 1982 The isolation and identification of anticonvulsant agents from *Clausena anisata* and *Afraegle paniculata*. Fitoterapia 53: 63–66
18. Agarwal PK, Agarwal SK, Rastogi RP 1980 Dihydroflavonols from *Cedrus deodara*. Phytochemistry 19: 893–896
19. Agarwal PK, Rastogi RP 1981 Terpenoids from *Cedrus deodara*. Phytochemistry 20: 1319–1321
20. Agostino D, Agostino N 1980 Enhancement of pulmonary metastases following the intravenous infusion of a suspension of ellagic acid in traumatized rats. Tumori 66: 285–293
21. Ahmad J, Shamsuddin KM, Zaman A 1984 A pyranocoumarin from *Atalantia ceylanica*. Phytochemistry 23: 2098–2099
22. Ahmad SA, Zaman A 1973 Chemical constituents of *Glochidion venulatum*, *Maesa indica*, and *Rhamnus triquetra*. Phytochemistry 12: 1826
23. Ahmed MS, Fong HHS, Soejarto DD, Dobberstein RH et al. 1981 High-performance liquid chromatographic separation and quantitation of maytansinoids in *Maytenus ilicifolia*. J Chromatogr 213: 340–344
24. Aiba CJ, Gottlieb OR, Maia JG, Pagliosa FM et al. 1978 Benzofuranoid neolignans from *Licaria armeniaca*. Phytochemistry 17: 2038–2039

25 Aimi N, Kohmoto T, Honma M, Sakai S et al. 1980 Constituents of *Nauclea racemosa*: Isolation of three glucosidic components and synthesis of noreugenin-7-O-β-D-glucoside. Shoyakugaku Zasshi 34: 151–154
26 Akers HA, Abrego VA, Garland E 1980 Thujaplicins from *Thuja plicata* as iron transport agents for *Salmonella typhimurium*. J Bacteriol 141: 164–168
27 Akhmedkhodzhaeva KS, Bessonova IA 1982 Some data on the relation between chemical structure and estrogen acitvity in a series of quinoline alkaloids. Dokl Akad Nauk Uzb Ssr. 34–36
28 Akhmedzhanova VI, Bessonova IA, Yunusov SY 1978 The roots of *Dictamnus angustifolius*. Khim Prir Soedin 14: 476–478
29 Akinloye BA, Court WE 1981a The alkaloids of *Rauvolfia oreogiton*. Planta Med 41: 69–71
30 Akinolye BA, Court WE 1981b The alkaloids of *Rauvolfia volkensii*. J Ethnopharmacol 4: 99–109
31 Alfenas AC, Hubbes M, Couto L 1982 Effect of phenolic compounds from *Eucalyptus* on the mycelial growth and conidial germination of *Cryphonectria cubensis*. Can J Bot 60: 2535–2541
32 Ali S, Singh P, Thomson RH 1980 Naturally occurring quinones. Part 28. Sesquiterpenoid quinones and related compounds from *Hibiscus tiliaceus*. J Chem Soc Perkin Trans I. 257–259
33 Aliev AM, Movsumov IS 1981 Biologically active substances from several representatives of the genus *Cephalaria*. Azerb Med Zh 58: 36–40
34 Aliev KHU, Akbarov MB 1972 Effect of α/β-allocryptopine on arrhythmia. Farmakol Alkaloidov Ikh Proizvod 133–136
35 Aliev KHU, Kamilov IK 1967 Pharmacology of the alkaloid protopine and its derivatives. Farmakol Alkaloidov Glikozidov pp. 176–178
36 Al-Jeboury FS, Locksley HD 1971 Xanthones in the heartwood of *Calophyllum inophyllum*: a geographical survey. Phytochemistry 10: 603–606
37 Al-Khayat MA, Blank G 1985 Phenolic spice components sporostatic to *Bacillus subtilis*. J Food Sci 50: 971–980
38 Alimova M, Israilov IA, Yunusov MS, Abdullaev ND et al. 1982 Alkaloids of *Corydalis paniculigera*. Khim Prir Soedin 18: 727–731
39 Alles GA, Ellis CH 1952 A comparative study of the pharmacology of certain *Cryptopine* alkaloids. J Pharmacol Exp Ther 104: 253–263
40 Altman RFA 1956 Chemical studies of Amazonian plants. I. Microchemical identification of *Cinchona* alkaloids. Bol Tec Inst Agron Norte Belem, Brazil, 9 p
41 Alves deLima R, Gottlieb OR, Lins Mesquita AA 1972 Chemistry of Brazilian Guttiferae. XXVIII. Xanthones from *Caraipa densiflora*. Phytochemistry 11: 2307–2309
42 Alves deLima R, Gottlieb OR, Lins Mesquita AA 1970 Chemistry of Brazilian Guttiferae. XXIV. Tetraoxygenated xanthones from *Caraipa densiflora*. An Acad Brasil Cienc 42: 133
43 Al-Yahya MA, Khafagy S, Shihata A, Kozlowski JF et al. 1984 Phytochemical and biological screening of Saudi medicinal plants, Part 6. Isolation of 2-Alpha-hydroxy-alantolactone, the antileukemic principle of *Francoeuria crispa*. J Nat Prod 47: 1013–1017
44 Andersen PH, Geisler A, Klysner R 1984 No change in rat cerebral cortex calmodulin content following chronic treatment with lithium, reserpine, imipramine, and lithium combined with reserpine or imipramine. Acta Pharmacol Toxicol 54: 394–399
45 Anderson LA, Harris A, Phillipson JD 1983 Production of cytotoxic canthin-6-one alkaloids by *Ailanthus altissima* plant cell cultures. J Nat Prod 46: 374–378
46 Anderson LA, Keene AT, Phillipson JD 1982 Alkaloid production by leaf organ, root organ and cell suspension cultures of *Cinchona ledgeriana*. Planta Med 46: 25–27
47 Andrejak M, Ward M, Schmitt H 1983 Cardiovascular effects of yohimbine in anaesthetized dogs. Eur J Pharmacol 94: 219–228
48 Andronescu E, Petcu P, Goina T, Radu A 1973 Antibiotic activity of the extract and the alkaloid isolate from *Berberis vulgaris*. Clujul Med 46: 627–631
49 Anjaneyulu AS, Murthy YL, Row LR 1978 Novel triterpenes from *Swietenia mahagoni* Linn: Part I-Isolation and structure determination of 9–19 cyclswietenol. Indian J Chem 16B: 650–654
50 Anjaneyulu ASR, Atchutaramaiah P, Ramachandrarow L, Venkat 1981 New lignans from the heartwood of *Cleistanthus collinus*. Tetrahedron 21: 3641–3652
51 Anjaneyuly B, Rao VB, Ganguly AK, Govindachari TR et al. 1965 Chemical investigation of some Indian plants. Indian J Chem 3: 237

52 Aojewole JAO, Adesina SK 1983 Mechanism of the hypotensive effect of scopoletin isolated from the fruit of *Tetrapleura tetraptera*. Planta Med 49: 46–50
53 Arens H, Borbe HO, Ulbrich B, Stockigt J 1982 Detection of pericine, a new CNS-active indole alkaloid from *Picralima nitida* cell suspension culture by opiate receptor binding studies. Planta Med 46: 210–214
54 Arichi H, Kimura Y, Okuda H, Baba K et al. 1982 Effects of stilbene components of the roots of *Polygonum cuspidatum* Sieb. et Zucc. on lipid metabolism. Chem Pharm Bull 30: 1766–1770
55 Arichi S, Hayashi T, Kubo M 1982 Antitumor agent. Canadian Patent 1,116,521
56 Arisawa M, Fujita A, Morita N, Kinghorn AD et al. 1985 Plant anticancer agents XXXV. Further constituents of *Simaba multiflora* (Simaroubaceae). Planta Med 49: 348–349
57 Arisawa M, Funayama S, Pezzuto JM, Kinghorn AD et al. 1984 Potential anticancer agents. XXXII. Hydroquinone from *Ipomopsis aggregata* (Polemoniaceae). J Nat Prod 46: 393–394
58 Arisawa M, Gunasekera SP, Cordell GA, Farnsworth NR 1981 Plant anticancer agents. XXI. Constituents of *Merrilliodendron megacarpum* (Icacinaceae). Planta Med 43: 404–407
59 Arisawa M, Kinghorn AD, Cordell GA, Farnsworth NR 1983a Plant anticancer agents. XXIII. 6-α-Senecioyl-oxy-chaparrin, a new antileukemic quassinoid from *Simaba multiflora*. J Nat Prod 46: 218–221
60 Arisawa M, Kinghorn AD, Cordell GA, Farnsworth NR 1983b Plant anticancer agents. XXIV. Alkaloid constituents of *Simaba multiflora*. J Nat Prod 46: 222–225
61 Arisawa M, Kyozuka Y, Hayashi T, Shimizm M et al. 1980 Studies on unutilized resources. Part IX. Isoflavonoids in the roots of *Thermopsis fabacea* D.C.(Leguminosae). Chem Pharm Bull 28: 3686–3688
62 Asahi KI, Ono I, Kusakabe H, Nakamura G et al. 1981 Studies on differentiation inducing substances of animal cells. 1. Differenol A, a differentiation inducing substance against mouse leukemia cells. J Antibiot 34: 919–920
63 Aschaffner-Sabba K, Schmidt-Ruppin KH, Wehrli W, Schuerch AR 1984 β-Lapachone: synthesis of derivatives and activities in tumor models. J Med Chem 27: 990–994
64 Asnis GM, Sachar EJ, Halbreich U, Ostrow LC et al. 1981 The prolactin-stimulating potency of reserpine in man. Psychiatry Res 5: 39–45
65 Asoeva EZ, Dauksha AD, Denisova EK 1963 Chemical composition of *Alhagi persarum*. Izv Akad Nauk Turkm SSR, Ser Biol Nauk 75–77
66 Atal CK 1980 Chemistry and pharmacology of vasicine - a new oxytocic and abortifacient. Regional Research Laboratory, Jammu, India. 155 p
67 Atal CK, Sharma ML, Khajuria A, Kaul A et al. 1982 Thrombopoietic activity of vasicine hydrochloride. Ind J Exp Biol 20: 704–709
68 Auf'Mkolk M, Kohrle J, Gumbinger H, Winterhoff H et al. 1984 Antihormonal effects of plants extracts: Iodothyronine deiodinase of rat liver is inhibited by extracts and secondary metabolites of plants. Horm Metabol Res 16: 188–192
69 Avazkhodzhaev MKH, Zhumaniyazov IZH 1980 Content of scopoletin and isohemigossypol in stems of cotton plants infected by wilt. Uzb Biol Zh. 10–11
70 Ayapbergenov EK, Baisheva SA 1969 Effect of leukodelphinoidine on the growth and radiosensitivity of transplanted tumors. Tr Kaz Nauch-Issled Instonkol Radiol 6: 151–154
71 Babash TA, Perel'Son ME 1982 Quantitative determination of the aglycone of the total patrinosides in the roots of *Patrinia intermedia*. Khim Prir Soedin 18: 621–624
72 Babbar OP, Chhatwal VK, Ray IB, Mehra MK 1982 Effect of berberine chloride eye drops on clinically positive trachoma patients. Indian J Med Res 76S: 83–88
73 Badawi MM, Seida AA, Kinghorn AD, Cordell GA et al. 1981 Potential anticancer agents. XVIII. Constituents of *Amyris pinnata* (Rutaceae). J Nat Prod 44: 331–334
74 Baker PM, Fortes CC, Fortes EG, Gazzinelli G et al. 1972 Chemoprophylactic agents in schistosomiasis: Eremanthine, costunolide, α-cyclocostunolide and bisabolol. J Pharm Pharmacol 24: 853–857
75 Balandrin MF, Robbins EF, Kinghorn AD 1982 Alkaloids of Papilionoideae. Part II. Alkaloid distribution in some species of the Papilionaceous tribes, Thermopsideae and Genisteae. Biochem Syst Ecol 10: 307–311
76 Balint G, Rabloczky G 1968 Antitussive effect of autonomic drugs. Acta Physiol Acad Sci Hung 33: 99–109
77 Balz JP, Das NP 1979 *Uncaria elliptica*, a major source of rutin. Planta Med 36: 174–177

78 Bandaranayake WM 1979 Terpenoids of *Canarium zeylanicum*. Phytochemistry 19: 255–257
79 Bandaranayake WM, Gunasekera SP, Karunanayake S, Sotheeswara 1975 Terpenes of *Dipterocarpus* and *Doona* species. Phytochemistry 14: 2043–2048
80 Bandaranayake WM, Karunanayake S, Sotheeswaran S, Sultanbawa 1980 Chemical investigation of Ceylonese plants. Part XXX. Xanthones and triterpenes of *Mammea acuminata* (Guttiferae). Indian J Chem Ser B 19: 463–467
81 Bandaranayake WM, Selliah SS, Sultanbawa MUS, Ollis WD 1975 Biflavonoids and xanthones of *Garcinia ternophylla* and *G. echinocarpa*. Phytochemistry 14: 1878–1880
82 Bandaranayake WM, Selliah SS, Sultanbawa MUS, Games DE 1975 Xanthones and 4-phenylcoumarins of *Mesua thwaitesii*. Phytochemistry 14: 265–269
83 Banerjee SK, Chakravarti RN 1963 Taraxerol from *Clitoria ternatea*. Bull Calcutta School Trop Med 11: 106–107
84 Banerji J, Ghoshal N, Sarkar S, Kumar M 1982 Studies on rutaceae: 2. Chemical investigations of the constituents of *Atalantia wightii, Limonia crenulata, Feronia limonia, Citrus limon* and synthesis of luvangetin, xanthyletin and marmin. Indian J Chem Ser B 21: 496–498
85 Banerji R, Nigam SK 1981 Studies On the heartwood of *Melia birmanica*. Fitoterapia 52: 3–4
86 Baranowski JD, Davidson PM, Nagel CW, Branen AL 1980 Inhibition of *Saccharomyces cerevisiae* by naturally occurring hydroxycinnamates. J Food Sci 45: 592–594
87 Bardyshev II, Zen'Ko RI, Pertsovskaya AL, Manukov EN et al. 1975 Presence of M-menthane series hydrocarbons in different turpentines from *Pinus silvestris*. Vestsi Akad Nauk B SSR Ser Khim Nauk. 107
88 Barnaulov OD, Manicheva OA, Shelyuto VL, Konopleva MM et al. 1985 Effect of flavonoids on development of experimental gastric dystrophies in mice. Khim Farm Zh 18: 935–941
89 Barnaulov OD, Manicheva OA, Zapesochnaya GG, Shelyuto VL et al. 1982 Effects of certain flavonoids on the ulcerogenic action of reserpine in mice. Khim Farm Zh 16: 300–303
90 Baser KHC, Bisset NG 1982 Alkaloids of Sri Lankan *Strychnos nux-vomica*. Phytochemistry 21: 1423–1429
91 Bashir AK, Turner TD, Ross MS 1982 Phytochemical investigation of *Grewia villosa* roots. Part 1. Fitoterapia 53: 67–70
92 Batirov EK, Matkarimov AD, Malikov VM 1981 Justidin-B and diphyllin from *Haplophyllum obtusifolium*. Khim Prir Soedin 17: 386–387
93 Batirov EKH, Matkarimov AD, Batsuren D, Malikov VM 1981 New coumarins and lignans from *Haplophyllum obtusifolium* and *H. dahuricum*. B. Atanasova (Ed) Proc 1st Int Conf Chem Biotechnol Biol Act Nat Prod., B 3: 120–123
94 Baumann J, Wurm G, Bruchhausen FV 1980 Prostaglandin synthetase inhibition by flavonoids and phenolic compounds in relation to their O_2^- scavenging properties. Arch Pharm (Weinheim) 313: 330–337
95 Beasley, JL, Jr, Harris LE 1942 A chemical study of Oklahoma plants. 5. *Ephedra nevadensis*. J Am Pharm Ass 31: 171–173
96 Bejar E, Enriquez R, Lozoya X 1984 The in vitro effect of grandiflorenic acid and zoapatle aqueous crude extract upon spontaneous contractility of the rat uterus during oestrus cycle. J Ethnopharmacol 11: 87–97
97 Beladi I, Mucsi I, Pusztai R, Bakay M et al. 1981 In vitro and in vivo antiviral effects of flavonoids. Stud Org Chem (Amsterdam) 11: 443–450
98 Beljanski M, Beljanski MS 1982 Selective inhibition of in vitro synthesis of cancer DNA by alkaloids of β-carboline class. Expl Cell Biol 50: 79–87
99 Beljanski M, Beljanski MS 1984 Three alkaloids as selective destroyers of the proliferative capacity of cancer cells. IRCS Med Sci 12: 587–588
100 Beretz A, Cazenave JP, Anton R 1982 Inhibition of aggregation and secretion of human platelets by quercetin and other flavonoids: Structure-activity relationships. Agents Actions 12: 382–387
101 Beretz A, Stoclet JC, Anton R 1980 Inhibition of isolated rat aorta contraction by flavonoids. Possible correlation with C-AMP phosphodiesterase inhibition. (Abstract). Planta Med 39: 236B
102 Bernard J, Riou G 1976 Biochemical action of ellipticine derivatives in *Trypanosoma cruzi*. In H. van den Bossche (ed) Biochem Parasites Host-parasites Relate. Proc Int Symp. Orth-Holland, Amsterdam p. 477
103 Beschia M, Leonte A, Oancea I 1982 Phenolic components with biological activity in vegetable extracts. Bul Univ Galati Fasc 6 (5): 59–63

104 Bessonova IA, Batsuren D, Yunusov SYU 1984 Alkaloids of *Haplophyllum dahuricum*. Khim Prir Soedin 20: 73–76
105 Bessonova IA, Yunusov SY 1982 Alkaloids of *Haployphyllum ferganicum*. Khim Prir Soedin 18: 530–531
106 Betina V, Nemec P, Barathova H, Durackova Z 1981 Isolation of the antibiotic repentin B and metabolites echinulin, erythroglaucin, flavoglaucin and physcion. Czech. Patent Cs 192,679
107 Bezuidenhoudt BCB, Brandt EV, Roux DG 1981 A novel α-hydroxydihydro-α-chalcone from the heartwood of *Pterocarpus angolensis* D.C.: Absolute configuration, synthesis, photochemical transformations. J Chem Soc Perkin Trans I. 263–269
108 Bhakuni DS 1984 Biosynthesis, aberrant biosynthesis and biogenetic type synthesis of some alkaloids of Indian medicinal plants. Proc Fifth Asian Symp Med Plants Spices, Aug 5. Nat Prod Res Inst, Seoul Natl Univ, Seoul, Korea. pp. 509–518
109 Bhanu S, Scheinmann F, Jefferson A 1975 Xanthones from the heartwood of *Calophyllum ramiflorum*. Phytochemistry 14: 298–299
110 Bhargav K, Varma KC 1978 Analgesic effect of petroleum ether, chloroform and ethanol extracts from *Ipomoea palmata*. Indian J Hosp Pharm 15: 81–82
111 Bhatnagar SS, Divekar PV 1953 Symposium on antibacterial substances from soil, plants and other sources. XIII. Pristimerin, the antibacterial principle from *Pristimeria indica*. Indian J Pharm 15: 307–308
112 Bhattacharya SK, Lal R, Basu K, Das PK 1970 Pharmacological studies on the roots of *Abroma augusta* (ulatkambal). J Res Indian Med 4: 176
113 Bhattacharya SK, Pilla CR, Mathur M, Sen P 1984 Effect of chloroquine and some other antimalarials on the immune mechanism in experimental animals. J Pharm Pharmacol 36: 268–269
114 Bhavsar GC, Pundarikashudu K, De S 1982 Studies on *Adhatoda vasica* (Nees). Indian J Pharm Sci Suppl 44: 17
115 Bhide MB, Naik PY 1980 Antiasthmatic potentiality of vasicinone-an alkaloid from *Adhatoda vasica*, Nees. (Abstract) Proc Fourth Asian Symp Med Plants Spices. Mahidol Univ, Bangkok, Thailand. p. 137
116 Bhutani SP, Chibber SS, Seshadri TR 1969 Components of the roots of *Pueraria tuberosa*: isolation of a new isoflavone C-glycoside (di-O-acetylpuerarin). Indian J Chem 7: 210
117 Bian R, Yang QH, Geng BQ, Xie QM et al. 1981 Antiasthmatic constituents in the essential oil of *Artemisia argyi* - Pharmacological study on terpinenol-4. Zhonghua Jiehe He Huxixi Jibing Zazhi 4: 203–206
118 Bisset NG, Nwaiwu J 1983 Quaternary alkaloids of *Tinospora* species. Planta Med 48: 275–279
119 Blomster RN, Martello RE, Farnsworth NR, Draus FJ 1964 Studies on *Catharanthus lanceus*. V. Preparation of root alkaloid fractions and the isolation of ajmalicine, yohimbine, pericalline, perimivine and cathalanceine. Lloydia 27: 480–485
120 Bodalski T, Kantoch M, Rzadkowska H 1957 Antiphage activity of some alkaloids of *Chelidonium majus*. Diss Pharm 9: 273–286
121 Bodalski T, Pelczarska H, Ujec M 1958 Action of some alkaloids of *Chelidonium majus* on *Trichomonas vaginalis* in vitro. Arch Immunol Terap Doswiadczalnej 6: 705–711
122 Bodalski T, Rzadkowska H 1957 Infusoriocidal action of some alkaloids of *Chelidonium majus*. Diss Pharm 9: 266–271
123 Bogutskii BV, Nikolaevskii VV, Eremenko AE, Tikhomirov AA et al. 1980 Mitosis stimulant. Russian Patent 950,392
124 Bohlmann F, Jakupovic J 1978 Novel chromenes and other constituents from *Lagascea rigida*. Phytochemistry 17: 1677–1678
125 Bohlmann F, Jakupovic J, Gupta RK, King RM et al. 1981 Naturally occurring terpene derivatives 296: Allenic germacranolides, bourbonene derived lactones and other constituents from *Vernonia species*. Phytochemistry 20: 473–480
126 Bohlmann F, Knauf W, King RM, Robinson H 1979 Naturally occurring terpene derivatives. 196. A new diterpene and additional constituents form *Baccharis* species. Phytochemistry 18: 1011–1044
127 Bohlmann F, Ngoleva N 1977 Naturally occurring terpene derivatives. 97. New kaurene derivatives from *Wedelia* species. Phytochemistry 16: 579–581

128 Bohlmann F, Zdero C 1976 Naturally occurring terpene derivatives. 72. New terpene constituents from *Verbesina* species. Phytochemistry 15: 1310–1311

129 Bohlmann F, Zdero C 1982 Naturally occurring terpene derivatives. Part 454. Sesquiterpene lactones and other constituents from *Tanacetum parthenium*. Phytochemistry 21: 2543–2549

130 Bohlmann F, Zdero C, Grenz M, Dhar AK et al. 1981 Naturally occurring terpene derivatives. Part 307. Five diterpenes and other constituents from nine *Baccharis* Species. Phytochemistry 20: 281–286

131 Bohlmann F, Zdero C, Jakupovic J, Ates N et al. 1983 Steiractinolides from *Aspilia* and *Wedelia* species. Justus Liebigs Ann Chem pp. 1257–1266

132 Bol'Shakova VI, Khan VA, Dubovenko ZV, Shmidt EN et al. 1980 Terpenoids of the oleoresin of the larch growing in Kamchatka. Khim Prir Soedin 16: 340–343

133 Bonzani daSilva J 1970 Gallic acid, pyrogallol, and pyrocatequine in *Brysonima intermedia latifolia*. Rev Farm Bioquim Univ Sao Paulo 8: 187–192

134 Boonchird C, Flegel TW 1982 In vitro antifungal activity of eugenol and vanillin against *Candida albicans* and *Cryptococcus neoformans*. Can J Microbiol 28: 1235–1241

135 Borup I, Kingston DGI 1981 Isolation and structure elucidation of cytotoxic aporphine alkaloids from *Annona acuminata*. (Abstract). Proc Joint Mt Am Soc Pharmacog and Soc for Economic Botany, Boston, July 13–18. American Society of Pharmacognosy, Dept of Pharmaceutical Sciences, University of Pittsburgh, Pittsburgh PA p. 31

136 Borup-Grochtmann I, Kingston DGI 1982 Aporphine alkaloids from *Annona acuminata*. J Nat Prod 45: 102

137 Botta B, DelleMonache F, DelleMonache G, Marini-Bettolo GB 1985 Prenylated bianthrones and vismione F from *Psorospermum febrifugum*. Phytochemistry 24: 827–830

138 Boulware RT, Southard GL, Walborn D, Senior SL et al. 1984 Salivary glycolysis testing of sanguinarine chloride, chlorhexidine and other antimicrobials. (Abstract) Proc Internat Assoc Dental Res. Internat Assoc Dental Res, Washington, D.C. pp. ABSTR-1274

139 Boulware RT, Stermitz FR 1981 Constituents of *Zanthoxylum*. Part 6. Some alkaloids and other constituents of *Zanthoxylum microcarpum* and *Z. procerum*. J Nat Prod 44: 200–205

140 Bowen IH, Perrera KPWC, Lewis JR 1980 Alkaloids from the stem of *Glycosmis bilocularis* (Rutaceae). Phytochemistry 19: 1566–1568

141 Brazdovicova B, Kostalova D, Tomko J 1975 Alkaloids of *Berberis julianae*. (Abstract) Proc Conf Med Plants (Marienbad) p. 11

142 Braz Filho R, DeMagalhaes GC, Gottlieb OR 1970 Xanthones of *Rheedia gardneriana*. Phytochemistry 9: 673

143 Braz Filho R, Gottlieb OR, Lameg Vieira Pinho S, Queiroz Monte et al. 1973 Flavonoids from Amazonian Leguminosae. Phytochemistry 12: 1184–1186

144 Briggs LH, White GW 1975 Constituents of the essential oil of *Araucaria araucana*. Tetrahedron 31: 1311–1314

145 Brissemoret A 1925 Note on "anti-opium" plants. C R Soc Biol 93: 1341–1343

146 Britton DR, Rivier C, Shier T, Bloom F et al. 1982 In vivo and in vitro effects of tetrahydroisoquinolines and other alkaloids on rat pituitary function. Biochem Pharmacol 31: 1208–1211

147 Brown E, Coomes TJ, Islip HT, Matthews WSA 1954 Oils of *Cinnamomum* species from Sarawak as a source of safrole. Colonial Plant Anim Prod(London) 4: 239

148 Brown PM, Thomson RH, Hausen BM 1974 Naturally occurring quinones. XXIV. Extractives from *Bowdichia nitida*. The first isoflavone quinone. Justus Liebigs Ann Chem pp. 1295

149 Brue F, Laborit H, Baron C 1961 Metabolic mechanism of the hypertensive activity of hydroquinone in rabbits. Agressologie 2: 275–280

150 Bulliard H, Ghali JD 1943 Action of yohimbine-HCL on the white rat. Compt Rend Soc Biol 137: 33–34

151 Burnett AR, Thomson RH 1967 Naturally occurring quinones. Part X. The quinonoid constituents of *Tabebuia avellanedae* (Bignoniaceae). J Chem Soc C. 2100–2104

152 Burnett AR, Thomson RH 1968 Naturally occurring quinones. Part XII. Extractives from *Tabebuia chrysantha* Nichols. and other Bignoniaceae. J Chem Soc C. 850–853

153 Cahn J 1981 Possible mechanism of psoralen-induced melanogenesis. Proc Int Symp Psoralens Cosmet Dermatol. Pergamon, Paris. 31–39

154 Cai DG, Wang YZ, Lu YQ 1983 Studies on the expectorant constituents of *Eupatorium fortunei*. Chung Yao T'Ung Pao 8: 30–31

155 Cai YY, Xu CL, Li SH, Liu Y 1983 Studies on sister chromatid exchanges induced by antitumor and antiparasitic drugs before or after activated with microsome enzyme S-9. Chung-Kuo I Hsueh K'O Hsueh Yuan Hsueh Pao 5: 161–164
156 Cala PM, Norby JG, Tosteson DC 1982 Effects of the plant alkaloid sanguinarine on cation transport by human red blood cells and lipid bilayer membranes. J Membrane Biol 64: 23–31
157 Calis I, Lahloub MF, Sticher O 1984 18. Loganin, loganic acid and periclymenoside, a new biosidic ester iridoid glucoside from *Lonicera periclymenum* L. (Caprifoliaceae). Helv Chim Acta 67: 160–165
158 Calis I, Sticher O 1984 Secoiridoid glucosides from *Lonicera periclymenum*. Phytochemistry 23: 2539–2540
159 Calle LM, Sullivan PD 1982 Screening of antioxidants and other compounds for antimutagenic properties towards benzo(a)pyrene-induced mutagenicity in strain Ta98 of *Salmonella typhimurium*. Mutat Res 101: 99–114
160 Cambie RC 1959 Wood extractives of *Aristotelia serrata*. N Z J Sci 2: 257–259
161 Cambie RC, Cox RE, Sidwell D 1984 Phenolic diterpenoids of *Podocarpus ferrugineus* and other podocarps. Phytochemistry 23: 333–336
162 Cambie RC, Parnell JC 1969 A New Zealand phytochemical survey. Part 7. Constituents of some dicotyledons. N Z J Sci 12: 453–465
163 Cambie RC, Sidwell D, Cheong CK 1984 Extractives of *Decussocarpus wallichianus*. J Nat Prod 47: 562
164 Campanelli AR, D'Alagni M, Marini-Bettolo GB 1980 Spectroscopic evidence for the interaction of tingenone with DNA. FEBS Lett 122: 256–260
165 Candy HA, Brookes KB, Bull JR, McGarry EJ et al. 1978 Flavonoids of *Albizia adianthifolia*. Phytochemistry 17: 1681–1682
166 Candy HA, McGarry EJ, Pegel KH 1968 Constituents of *Syzygium cordatum*. Phytochemistry 7: 889–890
167 Cao LX, Huang HP, Cao CP 1984 Semisynthetic deoxyharringtonine for treating leukemia in mice and humans. Int J Cell Cloning 2: 327–333
168 Caolo MA, Stermitz FR 1979 A new *Zanthoxylum* alkaloid structurally related to tetrahydrocannabinol. Tetrahedron 35: 1487–1492
169 Carballeira A 1980 Phenolic inhibitors in *Erica australis* L. and in associated soil. J Chem Ecol 6: 593–595
170 Carter FL, Garlo AM, Stanley JB 1978 Termiticidal components of wood extracts: 7-methyljuglone from *Diospyros virginiana*. J Agr Food Chem 26: 869
171 Cassady JM, Douros JD 1980 Anticancer agents based on natural product models. Academic Press, New York, NY
172 Castro V, Mojica E, Calzada J, Poveda L 1982 Preliminary study of *Salacia* sp. Ing Cienc Quim 6: 180
173 Cava MP, Reed TA, Beal JL 1965 An efficient separation of the common alkaloids of the berberine group: The isolation and characterization of columbamine. Lloydia 28: 73–83
174 Cave A, Debourges D, Lewin G, Moretti C et al. 1984 Alkaloids from Annonaceae: IV. Chemistry and pharmacology of *Cymbopetalum brasiliense*. Planta Med 50: 517–519
175 Cave A, LeBoeuf M, Hocquemiller R, Bouquet A et al. 1979 Alkaloids of *Limaciopsis loangensis*. Planta Med 35: 31–41
176 Cha L, Chien CC, Lu FH 1981 Antiarrhythmic effect of *Angelica sinensis* root, tetrandrine and *Sophora flavescens* root. Yao Hsueh T'Ung Pao 16: 53–54
177 Chakraborty DP 1980 Some aspects of the carbazole alkaloids. Planta Med 39: 97–111
178 Chalmers WT, Kutney JP, Salisbury PJ, Stuart KL et al. 1982 Tripdiolide, triptolide, and celastrol. United States Patent 4,328,309
179 Chan SYW, Tang LCH 1984 Effect of reserpine on fertilizing capacity of human spermatozoa. Contraception 30: 363–369
180 Chandler RF, Hooper SN 1979 Friedelin and associated triterpenoids. Phytochemistry 18: 711–724
181 Chandra S, Verma M, Saxena H 1982 Triterpenoids of *Adenanthera pavonina* root. Int J Crude Drug Res 20: 165–167
182 Chang HM, El-Fishawy AM, Slatkin DJ, Schiff PL Jr 1984 Quaternary alkaloids of *Tinospora capillipes*. Planta Med 50: 88–90

183 Chang IM, Hur SO, Lee ES 1985 Biological activities of iridoid glycosides I. Antimicrobial activities and inhibition of RNA and protein synthesis. Abstr Internat Res Cong Nat Prod. Coll Pharm, Univ N Carolina, Chapel Hill NC. p. ABSTR-192

184 Chang P, Lee KH 1984 Cytotoxic antileukemic anthraquinones from *Morinda parvifolia*. Phytochemistry 23: 1733–1736

185 Chang Z, Wang S, Liu F, Zhu M et al. 1982 No physical dependence of skimmianine in mice, rats and monkeys. Chung-Kuo Yao Li Hsueh Pao 3: 223–226

186 Chao CY, Tong LS, Ng WY 1980 Studies on the effects of ginsenosides Rg 1 and Rb 1 of *Panax ginseng* on mitosis. Proc Third Internat Ginseng Symp, pp. 121–129

187 Chatterjee DR 1951 Plant alkaloids. 1. *Berberis floribunda*. J Indian Chem Soc 28: 225–228

188 Chavez PI, Sullivan G 1982 A comparison of the alkaloids of the fasciated and non-fasciated stems in two species of *Sophora*. Diss Abstr Int B 42: 2626

189 Chavez PI, Sullivan G 1984 A qualitative and quantitative comparison of the quinolizidine alkaloids of the fasciated and normal stems of *Sophora secundiflora*. J Nat Prod 47: 735

190 Che CT, Fong HHS 1982 Phytochemical investigations of *Aristolochia indica* Linn. and *Caulophyllum thalictroides* (L.) Michaux. Ph D-Dissertation-Univ Illinois Medical Center Chicago, IL. 190 p

191 Chen AH 1981 Studies on the applications of alkaloids of *Phellodendron wilsonii* Hay. et Kaneh. I. Studies on the analysis of alkaloids of *Phellodendron wilsonii* Hay K'O Hseuh Fa Chan Yueh K'An 9: 398–411

192 Chen CC, Chen YL, Chen YP, Hsu HY 1983 A study on the constituents of *Millettia reticulata* Benth. Tai-Wan Yao Hsueh Tsa Chih 35: 89–93

193 Chen FC, Lee JS, Lin YM 1983 Biphenyls from the heartwood of Taiwan *Sassafras*. Phytochemistry 22: 616–617

194 Chen FC, Lin YM, Yu PL 1973 Constituents of *Acanthopanax trifoliatus*. Phytochemistry 12: 467

195 Chen PC, Wu SC 1976 Anisodamine in migraine type vascular headache Chung-hua Nei K'o Tsa Chih 1: 151

196 Chen QH, Liu CY, Qiu CH 1980 Studies on Chinese rhubarb. XII. Effect of anthraquinone derivatives on the respiration and glycolysis of Ehrlich Ascites Carcinoma Cell. Yao Hsueh Hsueh Pao 15: 65–70

197 Chen WM, Yan YP, Liang XT 1983 Alkaloids from roots of *Alstonia yunnanensis*. Planta Med 49: 62

198 Chen X, Huang QX, Zhou TJ, Dai HY 1980 Studies of *Citrus aurantium* and its hypertensive ingredients on the cardiac functions and hemodynamics in comparison with dopamine and dobutamine. Yao Hsueh Hsueh Pao 15: 71–77

199 Chen X, Wang Z, Yan Y 1982 Studies of reactivity of bronchial smooth muscle. 1. Effects of yiqi huozue (YH) and anisodamine on bronchial smooth muscle. Chung-kuo Yao Li hsueh Pao 4: 57–59

200 Chen YJ, Yu CH, Wang SS, Che J et al. 1981 Studies on the active principles in leaves of *Cannabis sativa*. Chung Ts'Ao Yao 12: 44

201 Chen ZF, Mu ZW, Zhou ZH 1984 The relationship between noradrenaline content and nerve growth factor level in mouse submandibular gland in the recovery course after reserpine. Sheng Li Hsueh Pao 36: 601–606

202 Cheng JT, Lin CW, Chen IS 1982 Pharmacological studies on skimmianine. Part I. Effect on the central nervous system of animals. Tai-Wan Yao Hsueh Tsa Chih 34: 138–145

203 Cheng YS, Lin CS 1979 Study of the extractive constituents from the wood of *Cunninghamia konishii* Hayata. J Chin Chem Soc(Taipei) 26: 169–172

204 Cheng YS, Lu SB 1978 The chemical constituents of the wood of *Keteleeria davidiana*. J Chin Chem Soc(Taipei) 25: 47–53

205 Cheng ZZ, Yang XQ, Wang LQ, Lu P 1984 Quantitative determination of *Triptolide* in common three-wing-nut (*Tripterygium wilfordii*). Chung Ts'Ao Yao 15: 339–340

206 Chih Wu Hsueh Pao, 1977 Correlation between phylogeny, chemical constituents and pharmaceutical aspects of plants and their applications in drug research (Part II). Chih Wu Hsueh Pao 19: 257–262

207 Chimique OSA 1966 Prosopine and prosopinine, alkaloids from *Prosopis africana*. French Patent 1,524,395 pp. 5PP

208 Chiu YP, Chen HC, Chuan TC, Huang SL 1981 Brief report on the study of flavonoids in *Cathaya argyophylla*. Chung Ts'Ao Yao 12: 7
209 Chou CJ, Liao C 1982 Phytochemical and pharmacological studies on the flower of *Melastoma candidum*. Ann Rep Natl Res Inst Chin Med pp. 69–132
210 Chou CJ, Wang CB, Lin LC 1977 Triterpenoids and some other constituents from *Spiraea formosana*. J Chin Chem Soc (taipei) 24: 195–198
211 Chow PYW, Holland MK, Suter DAI, White IG 1980 Evaluation of ten potential organic spermicides. Int J Fertil 25: 281–286
212 Chow YL, Quon HH 1968 Chemical constituents of the heartwood of *Mesua ferrea*. Phytochemistry 7: 1871–1874
213 Chuang CY, Chen CF, Lin MT, Teh GW et al. 1983 Pharmacological studies on the hypothermic constituents of the root of *Sophora subprostrata* (Leguminosae). Proc Natl Sci Counc Repub China Part B 7: 536–561
214 Ciccio JF, Herrera CH, Castro VH, Ralitsch M 1979 Isolation and characterization of alkaloids from *Stemmadenia glabra*. Rev Latinoamer Quim 10: 67–68
215 Cicotti I 1981 Pharmaceutical composition for treating gastro-duodenal ulcers. British Patent Appl 2,054,373
216 Clark AM, El-Feraly F, Li WS 1981 Antimicrobial activity of phenolic constituents of *Magnolia grandiflora* L. J Pharm Sci 70: 951–952
217 Cliffton EE, Girolami A, Agostino D 1966 The effect of ellagic acid on the thrombin-induced hemorrhagic syndrome in the rat. Blood 28: 253–257
218 Colomer LA 1945 Estrogenic action of yohimbine applications in gynecology. Obstet Ginecol 2: 390–395
219 Combes G, Escaut A 1982 Leaf extract of *Olea europea* rich in oleuropeine, products from it, their application as medicines and compositions containing them. French Patent Demande 2,507,477
220 Combes G, Gaignault JC 1984 On the coumarins of *Toddalia asiatica*. Fitoterapia 55: 161–170
221 Conner AH, Foster DO 1981 Triterpenes from Douglas fir sapwood. Phytochemistry 20: 2543–2546
222 Connolly JD, Harrison LJ, Rycroft DS 1984 The structure of tamariscol, a new pacifigorgiane sesquiterpenoid alcohol from the liverwort *Frullania tamarisci*. Tetra Lett 25: 1401–1402
223 Cook JT, Ollis WD, Sutherland IO, Gottlieb OR 1978 Pterocarpans from *Dalbergia spruceana*. Phytochemistry 17: 1419
224 Coppen JJW 1983 Iridoids with algicidal properties from *Allamanda cathartica*. Phytochemistry 22: 179–182
225 Cordell GA 1979 *Aspidosperma* alkaloids. Alkaloids 17: 199–384
226 Correa DB, Guerra LFB, DePadua AP, Gottlieb OR 1985 Ellagic acids from *Callisthene major*. Phytochemistry 24: 1860–1861
227 Correa DB, Gottlieb OR, DePadua AP 1979 Dihydrocinnamyl alcohols from *Hortia badinii*. Chemistry of Brazilian Rutaceae. III. Phytochemistry 18: 351
228 Corrie JET, Green GH, Ritchie E, Taylor WC 1970 The chemical constituents of Australian *Zanthoxylum* species. Aust J Chem 23: 133
229 Costa MAC, Paul MI, Alves AAC, VanderVijver LM 1978 Aliphatic and triterpenoid compounds of Ebenaceae species. Rev Port Farm 28: 171–174
230 Court WE 1983 Alkaloid distribution in some African *Rauvolfia* species. Planta Med 48: 228–233
231 Craveiro AA, Andrade CHS, Matos FJA, Alencar, JW 1978 Anise-like flavor of *Croton* aff. *zehntneri*. J Agr Food Chem 26: 772–773
232 Craveiro AA, Andrade HS, Matos FJA, Alencar JW et al. 1980 Fixed and volatile constituents of *Croton* aff. *nepetaefolius*. J Nat Prod 43: 756–757
233 Craveiro AA, Andrade CHS, Matos FJA, Alencar JW et al. 1983 Essential oil of *Eugenia jambolana* Lamk. J Nat Prod 46: 591–592
234 Cresteil T, Lesca P 1983 Enhancement of DNA binding, mutagenicity and carcinogenicity of polycylic aromatic hydrocarbons after induction of cytochrome P-450 by ellipticines. Chem Biol Int 472: 145–156
235 Crombie L 1980 The cathedulin alkaloids. Bull Nar 32: 37–50

236 Crout JR 1978 Apartene sulfate intramuscular injection and oxytocin citrate buccal tablets. Fed Reg 43: 58634–5863
237 Cushman M, Cheng L 1978 Total synthesis of nitidine chloride. J Org Chem 43: 286
238 D'Alagni M, DePetris M, Marini-Bettolo GB 1981 On the interaction between jatrophone and DNA. FEBS Lett 136: 175–179
239 D'Cruz JL, Nimbkar AY, Kokate CK 1980 Evaluation of fruits of *Piper longum* Linn. and leaves of *Adhatoda vasica* Nees for anthelmintic activity. Indian Drugs 17: 99–101
240 Dabrah TT, Sneden AT 1983 Oxoaporphine alkaloids from *Rollinia papilionella*. J Nat Prod 46: 436
241 Daigo K, Higuchi SH, Kiriyama K, Yamada S et al. 1972 *Pueraria* extract. Japanese Patent 72,07,471
242 Dallmeier K, Carlini EA 1981 Anesthetic, hypothermic, myorelaxant and anticonvulsant effects of synthetic eugenol derivatives and natural analogues. Pharmacology 22: 113–127
243 Dalvi RR, Peeples A 1981 In vivo effect of toxic alkaloids on drug metabolism. J Pharm Pharmacol 33: 51–53
244 Da Rocha RF, Lapa AJ, Ribeiro Do Vale J, Braz RF et al. 1981 Pharmacological activity of crude and purified extracts from *Acosmium dasycarpum* (Vog) Yakvol. Cienc Cult(Sao Paulo) 33: 158–162
245 Darwish FA, Evans FJ, Phillipson JD 1980 Bruceolides and dehydrobruceolides from Fijian *Brucea javanica*.(abstract). Planta Med 39: 232–233
246 DasGupta S, Ray AB, Bhattacharya SK, Bose R 1979 Constituents of *Pachygone ovata* and pharmacological action of its major leaf alkaloid. J Nat Prod 42: 399–406
247 Da Silva Pereira MO, Gottlieb OR, Taveira Magalhaes M 1966 Chemistry of Brazilian Guttiferae. IX. Xanthonic constituents of *Calophyllum brasiliense*. An Acad Brasil Cienc 38: 425–427
248 Da Silveira JC, Gottlieb OR, De Oliveira GG 1975 Zeyherol, a dilignol from *Zeyhera digitalis*. Phytochemistry 14: 1829–1830
249 Dauguet JC, Foucher JP, Dolley J 1982 The flavonols of *Arbutus unedo* L. (Ericaceae). Plant Med Phytother 16: 185–191
250 Davey JE, VanStaden J, DeLeeuw GTN 1981 Endogenous cytokinin levels and development of flower virescence in *Catharanthus roseus* infected with mycoplasmas. Physiol Plant Pathol 19: 193–200
251 Dawidar AAM, Fayez MBE 1969 Steroid sapogenins-XIII. The constituents of *Balanites aegyptiaca*. Phytochemistry 8: 261–265
252 Dayal R, Seshadri TR 1974 Colourless components of the roots of *Artocarpus heterophyllus*: isolation of a new compound, artoflavone. Indian J Chem 12: 895–896
253 DeBarros Correa D, Birchal E, Aguilar JEV, Gottlieb OR 1975 Ellagic acids from Vochysiaceae. Phytochemistry 14: 1138–1139
254 DelaForge M, Ioannides C, Parke DV 1982 Selective inhibition of the safrole-induced mixed function oxidase activities by 9-hydroxyellipticine. Chem Biol Interact 42: 279–289
255 Delfel NE 1980 Alkaloid distribution and catabolism in *Cephalotaxus harringtonia*. Phytochemistry 19: 403–408
256 DelleMonache F, Mac-Quhae MM, DelleMonache G, Marini-Betto, GB 1983 Chemistry of the *Vismia* genus. Part 9. Xanthones, xanthonolignoids and other constituents of the roots of *Vismia guaramirangae*. Phytochemistry 22: 227–232
257 DeLuca C, DellaMonache FD, Marini-Bettolo GB 1978 Triterpenoid quinones of *Maytenus obtusifolia* and *Maytenus boaria*. Rev Latinoamer Quim 9: 208–209
258 Deman E, Peeke HVS 1982 Dietary ferulic acid, biochanin A, and the inhibition of reproductive behavior in japanese quail (*Coturnix coturnix*). Pharmacol Biochem Behav 17: 405–411
259 Demole E, Demole C, Enggist P 1976 A chemical investigation of the violatile constituents of East Indian sandalwood oil (*Santalum album*). Helv Chim Acta 59: 737–747
260 Denisova SI, Men'Shikov GP, Utkin LM 1953 New alkaloid from the plant *Heliotropium supinum*. Dokl Akad Nauk SSSR 93: 59–61
261 De Oliveira AB, Iracema M, Madruga LM, Gottlieb OR 1978 Isoflavonoids from *Myroxylon balsamum*. Phytochemistry 17: 593–595
262 Desai PD, Dutia MD, Ganguly AK, Govindachari TR et al. 1967 Chemical study of some Indian plants. Part III. Indian J Chem 5: 523–524

263 Desai PD, Ganguly AK, Govindachari TR, Joshi BS et al. 1966 Chemical investigation of some Indian plants. Part II. Indian J Chem 4: 457–459
264 DeSantana CF, Lins LJP, Asfora JJ, Melo AM et al. 1981 Primary observations of the use of lapachol in human carriers of malignant neoplasms. Rev Inst Antibiot 20: 61–68
265 Deshpande VH, Srinivasan R, Ramarao AV 1975 Wood phenolics of *Morus* species. IV. Phenolics of the heartwood of five *Morus* species. Indian J Chem 13: 453–457
266 De Souza Guimaraes IS, Gottlieb OR, Andrade CHS, Taveir 1975 Flavonoids From *Dalbergia cearensis*. Phytochemistry 14: 1452
267 Dias SMC, Fernandes JB, Maia JGS, Gottlieb OR et al. 1982 The chemistry of Brazilian Lauraceae. 64. Neolignans from an *Aniba* species. Phytochemistry 21: 1737–1740
268 Didry N, Pinkas M, Torck M 1982 The chemical composition and antibacterial activity of the leaves of several species of *Grindelia*. Plant Med Phytother 16: 7–15
269 Didry N, Pinkas M, Torck M, DuBreuil L 1982 Components and activity of *Tussilago farfara*. Ann Pharm Fr 40: 75–80
270 Ding JK, Ding LS, Yi YF, Wu Y et al. 1983 The chemical constituents of the turpentine of *Pinus kesiya* var. *langbianensis*. Yun-Nan Chih Wu Yen Chiu 5: 224–226
271 Dixit R, Teel RW, Stoner GD 1984 Inhibition of DNA binding and metabolism of benzo(a)pyrene and benzo(a)pyrene-7,8-diol in cultured strain A mouse lung tissues by ellagic acid. Proc Am Assoc Cancer Res 75th Ann Mtg. Waverly Press, Baltimore, MD
272 Dixit RS, Perti SL, Agarwal RN 1965 New repellents. Labdev 3: 273
273 Dobbie GC, Langham ME 1961 Reaction of animal eyes to sanguinarine and argemone oil. Brit J Ophthalmol 45: 81–95
274 Domagalina E, Smajkiewicz A 1971 Isolation of magnoflorine and columbamine from *Berberis vulgaris*. Acta Pol Pharm 28: 81–87
275 Dominguez XA, Franco R, Alcorn JB, Garcia S et al. 1984 Medicinal plants from Mexico. L. Pristimerin and other components from *Crossopetalum uragoga* O. Ktze (Celastraceae). Rev Latinoamer Quim 15: 42
276 Dominguez XA, Franco R, Cano G, Castillo Osuna MS et al. 1980 Medicinal plants of Mexico. XIII. Terpenoids of "escobilla" *Fraxinus greggii* (Oleaceae) and heliettin from the roots. Rev Latinoamer Quim 11: 116–117
277 Dominguez XA, Franco R, Cano G, Garcia S et al. 1979 Mexican medicinal plants. Part 33. Triterpene quinone-methides from *Schaeferia* roots. Phytochemistry 18: 898–899
278 Donnelly DMX, Kavanagh PJ 1974 Isoflavanoids of *Dalbergia oliveri*. Phytochemistry 13: 2587–2591
279 Donnelly DMX, Nangle BJ, Prendergast JP, O'Sullivan AM 1968 *Dalbergia* species—5. Isolation of R-5-O-methyllatifolin from *Dalbergia cochinchinensis* Pierre. Phytochemistry 7: 647–649
280 Douros JD, Suffness M 1981 New antitumor substances of natural origin. (Review). Cancer Treat Rev 8: 63–87
281 Drebushchak TD, Shmidt EN, Khan VA, Dubovenko ZV et al. 1982 Terpenoids of the oleoresins of *Pinus pityusa* and *P. eldarica*. Khim Prir Soedin 18: 316–320
282 Dreyer DL 1980 Chemotaxonomy of the Rutaceae. Part XIV. Alkaloids, limonoids and furocoumarins from three Mexican *Esenbeckia* species. Phytochemistry 19: 941–944
283 Drost-Karbowska K, Hajdrych-Szaufer M, Kowalewski Z 1984 Search for alkaloid-type bases in *Lycium halimifolium*. Acta Pol Pharm 41: 127–129
284 Du BY, Zhang YX 1984 Experimental model for the screening of immunoactive drugs. Chung Ts'Ao Yao 15: 309–311
285 Du JZ, Li QF, Chen XG 1983 Protective effects of *Swertia mussotii* on liver damage induced by hypoxia. Yao Hsueh Hsueh Pao 18: 174–178
286 Duah FK, Schiff PL Jr, Slatkin D 1984 The isolation and characterization of alkaloids from *Monodora tenuifolia* Benth. Part II. Constituents of the heartwood of *Pterocarpus marsupium* Roxb. Diss Abstr Int B 45: 520–521
287 Dujack LW 1981 Production and mechanism of action of biologically active agents from *Tripterygium wilfordii*. Diss Abstr Int B 41: 3277–3278
288 Dwuma-Badu D, Ayim JSK, Withers SF, Agyemang NO et al. 1980 Constituents of west African medicinal plants. XXVII. Alkaloids of *Rhigiocarya racemifera* and *Stephania dinklagei*. J Nat Prod 43: 123–129

289 Dzhumagalieva FD, Seidakhanova TA 1971 Various aspects of the pharmacological action of preparations of polyphenol compounds. Akad Nauk Kaz SSR 16: 33–38
290 Dziedzic SZ, Hudson BJF, Barnes G 1985 Polyhydroxydihydrochalcones as antioxidants for lard. J Agr Food Chem 33: 244–246
291 Ekkin AL 1972 Role of cholinoreactive systems for some aspects of the action of rhodosin and *Eleutherococcus* extract. Lek Stedstra Dalnego Vostoku 11: 91–93
292 El-Kawi MA, Slatkin DJ, Schiff PL Jr, DasGupta S et al. 1984 Additional alkaloids of *Pachygone ovata*. J Nat Prod 47: 459–464
293 Elson LN 1937 The treatment of erythema multiforme. Urol Cutan Rev 41: 812–815
294 Elujoba AA, Sofowora EA 1977 Detection and estimation of total acid in the antisickling fraction of *Fagara* species. Planta Med 32: 54–59
295 Endo K, Takahashi K, Abe T, Hikino H 1982 Validity of Oriental medicines. 32. Structure of forsythoside B, an antibacterial principle of *Forsythia koreana* stems. Heterocycles 19: 261–264
296 Endo M, Naoki H 1980 Antimicrobial and antispasmodic tetrahydroanthracenes from *Cassia singueana*. Tetrahedron 36: 2449–2452
297 Endo T, Yamada Y 1985 Alkaloid production in cultured roots of three species of *Duboisia*. Phytochemistry 24: 1233–1236
298 Enriquez RG, Escobar LI, Romero ML, Chavez MA et al. 1983 Determination of grandiflorenic acid in organic and aqueous extracts of *Montanoa tomentosa* (Zoapatle) by reversed-phase high-performance liquid chromatography. J Chromatogr 258: 297–301
299 Enzell C, Erdtman H 1957 Chemistry of the natural order Cupressales. XIX. Occurrence of manool in *Cupressus sempervirens*. Acta Chem Scand 11: 902–903
300 Erdman MD 1983 Nutrient and cardenolide composition of unextracted and solvent-extracted *Calotropis procera*. J Agr Food Chem 31: 509–513
301 Erdtman H, Gripenberg J 1948 Antibiotic substances from the heart wood of *Thuja plicata*. II. The constitution of γ-Thujaplicin. Acta Chem Scand 2: 625
302 Evans WC, Hamsey KPA 1979 Alkaloids of *Anthocercis frondosa* (Solanaceae). J Pharm Pharmacol 31: 09P
303 Evdokimova NI, Polievtsev NP, Sultanov MB 1972 Synergism of skimmianine with hypnotics and narcotics and its antagonism to analeptics. Farmakol Alkaloidov Ikh Proizvod 47–55
304 Faddejeva MD, Bclyaeva TN, Novikov JP, Shalabi HG 1980 Possible intercalative binding of alkaloids sanguinarine and berberine to DNA. IRCS Med Sci-Biochem 8: 612
305 Fakhrutdinov SF 1972 Pharmacology of haplofoline, dubamine, robustine, and foliosine. Farmakol Alkaloidov Ikh Proizvod 64–67
306 Fan LL, O'Keefe DD, Powell WJ Jr 1984 Effect of puerarin on regional myocardial blood flow and cardiac hemodynamics in dogs with acute myocardial ischemia. Yao Hsueh Hsueh Pao 19: 801–807
307 Fan SF 1983 Chinese medicinal herbs used to treat heart diseases. Trends Pharm Sci 4: 208–210
308 Fan YJ, Zhou J, Li M 1981 Effect of nitidine chloride on the life cycle of Ehrlich Ascites carcinoma cells in mice. Chung-Kuo Yao Li Hsueh Pao 2: 46–49
309 Fang SD, Berry DE, Lynn DG, Hecht SM et al. 1984 The chemistry of toxic principles from *Maytenus nemorosa*. Phytochemistry 23: 631–633
310 Farnsworth NR 1984 How can the well be dry when it is filled with water. Econ Botany 38: 4–13
311 Farnsworth NR, Akerele O, Bingle AS, Soejarto DD et al. 1985 Medicinal plants in therapy. Bull WHO 6: 965–981
312 Farnsworth NR, Morris RW 1976 Higher plants - the sleeping giant of drug development. J Pharm 148: 46–52
313 Farnsworth NR, Soejarto DD 1985 Potential consequences of plant extinction in the United States on the current and future availability of prescription drugs. Econ Botany 39: 231–240
314 Fauvel MT, Gleye J, Moulis C, Fouraste I 1978 Alkaloids of *Murraya paniculata* (L.) Jack. Plant Med Phytother 12: 207–211
315 Fedoreev SA, Krivoshchekova OE, Denisenko VA, Gorovoi PG et al. 1979 Quinoid pigments of far eastern representatives of the family Boraginaceae. Khim Prir Soedin 15: 625–630
316 Fellous A, Luduena RF, Prasad V, Jordan MA et al. 1985 Effects of Tau and Map2 on interaction of maytansine with tubulin: inhibitory effect of maytansine on vinblastine-induced aggregation of tubulin. Cancer Res 45: 5004–5011

317 Feng CT 1972 A method of preparing pure ephedrine hydrochloride from *Ephedra equisetina* Bge. Chin J Physiol 1: 63–68
318 Feng XZ, Kan C, Potier P, Kan SK et al. 1982 Monomeric indole alkaloids from *Ervatamia hainanensis*. Planta Med 44: 212–214
319 Feng YX, Wang YP, Lou ZC 1981 Pharmacognostical study on several *Rauvolfia* roots and their adulterants. Yao Hsueh Hsueh Pao 16: 45–52
320 Fernandes JB, Nilce Des Ribeiro MN, Gottlieb OR, Gottlieb HE 1980 Chemistry of Brazilian Myristicaceae. Part 13. Eusiderins and 1,3-diarylpropanes from *Virola* species. Phytochemistry 19: 1523–1525
321 Fernandez MI, Pedro JR, Seoane E 1983 Constituents of a hexane extract of *Phoenix dactylifera*. Phytochemistry 22: 2087–2088
322 Ferrari F, Botta B, Alves deLima R 1983 Flavonoids And isoflavonoids from *Zollernia paraensis*. Phytochemistry 22: 1663–1664
323 Filho WW, Da Rocha AI, Voshida M, Gottlieb OR 1985 Ellagic acid derivatives from *Rhabdodendron macrophyllum*. Phytochemistry 24: 1991–1993
324 Filipski J, Kohn KW 1982 Ellipticine-induced protein-associated DNA breaks in isolated L1210 nuclei. Biochim Biophys Acta 698: 280–286
325 Fish F, Gray AI, Waterman PG 1975 Alkaloids, coumarins, triterpenes and a flavanone from the root of *Zanthoxylum dipetalum*. Phytochemistry 14: 2073–2076
326 Forgacs P, DesConclois JF, Mansard D, Provost J et al. 1981 Dopamine and tetrahydroisoquinoline alkaloids of *Annona reticulata* L. Annonaceae. Plant Med Phytother 15: 10–15
327 Formiga MD, Gottlieb OR, Mendes PH, Koketsu M et al. 1975 Constituents of Brazilian Leguminosae. Phytochemistry 14: 828
328 Forst AW 1943 Activity of *Arnica montana*. Arch Exp Pathol Pharmakol 201: 243–260
329 Fourie TG, Snyckers FO 1984 A flavone with antiinflammatory activity from the roots of *Rhus undulata*. J Nat Prod 47: 1057–1058
330 Franchi-Micheli S, Lavacchi L, Freidmann CA, Zilletti L 1983 The influence of rhein on the biosynthesis of prostaglandin-like substances in vitro. J Pharm Pharmacol 35: 262–264
331 Frencel I, Koscinski R 1960 Fungistatic action in vitro of some alkaloids from *Chelidonium majus*. Diss Pharm 12: 7–10
332 French Patent Demande 1980 Ajmalicine and its acid addition salts. French Patent Demande 2,442,238
333 Fujii R, Zinkel DF 1984 Minor components of ponderosa pine oleoresin. Phytochemistry 23: 875–878
334 Fujita M, Itokawa H, Sashida Y 1973 Studies on the components of *Magnolia obovata*. III. Occurrence of magnolol and honokiol in *M. obovata* and other allied plants. Yakugaku Zasshi 93: 429–434
335 Fujita SI, Fujita Y 1975 Comparative biochemical and chemo-taxonomical studies of the essential oils of *Magnolia salicifolia*. III. Chem Pharm Bull 23: 2443–2445
336 Fujita Y, Fujita S, Tanaka Y 1974 Biogenesis of the essential oils in camphor trees. XXXIV. Essential oil components of a residue of *Cinnamomum camphora* excavated at Konoyama, Moriyama, of district in Tokushima Prefecture. Nippon Nogei Kagaku Kaishi 48: 693–696
337 Fujita Y, Fujita SI, Iwamura JI, Nishida S 1975 Miscellaneous contributions to the essential oils of the plants from various territories. XXXVIII. On the components of the essential oils of *Metasequoia*. Yakugaku Zasshi 95: 349–351
338 Fujita Y, Kikuchi M, Fujita SI 1975 Miscellaneous contributions to the essential oils of the plants from various territories. XXXVII. On the components of the essential oils of *Magnolia*. Yakugaku Zasshi 95: 241–242
339 Fukada M 1980 Interaction between Sv40 DNA and camptothecin, an antitumor alkaloid. J Biochem 87: 1089–1096
340 Funayama S, Hikino H 1981 Hypotensive principles from plants. Heterocycles 15: 1239–1256
341 Furuya T, Yoshikawa T, Kiyohara H 1983 Alkaloid production in cultured cells of *Dioscoreophyllum cumminsii*. Phytochemistry 22: 1671–1673
342 Fuzellier MC, Mortier F, Lectard P 1982 Antifungic activity of *Cassia alata* L. Ann Pharm Fr 40: 357–363
343 Galbraith MN, Horn DHS 1966 Insect-moulting hormone from a plant. J Am Chem Soc Chem Commun 905–906

344 Galsky AG, Wilsey JP, Powell RG 1980 Crown gall tumor disc bioassay. A possible aid in the detection of compounds with antitumor activity. Plant Physiol 65: 184–185

345 Gartlan JS, McKey DB, Waterman PG, Mbi CN et al. 1980 A comparative study of the phytochemistry of two African rain forests. Biochem Syst Ecol 8: 401–422

346 Gautam CS, Sharma PL 1982 Potentiation of oxytocin evoked responses by (+) sotalol, deoxysotalol and vasicine HCL on isolated rat and rabbit uterus. Indian J Med Res 76: 107–114

347 Gershbein LL 1977 Regeneration of rat liver in the presence of essential oils and their components. Food Cosmet Toxicol 15: 173–182

348 Gertig H, Grabarczyk H 1961 Isolation of the alkaloids from the roots and the herb of *Dictamnus albus*. Acta Pol Pharm 18: 97

349 Ghosal S, Dutta SK 1971 Alkaloids of *Abrus precatorius*. Phytochemistry 10: 195

350 Ghosal S, Srivastava RS 1973 Chemical investigation of *Alhagi pseudalhagi* (Bieb.) Desv.: β-Phenethylamine and tetrahydroisopuinoline alkaloids. J Pharm Sci 62: 1555–1556

351 Ghosh A, Chakravarty D, Adhikari PC 1983 Action of isothymol (carvacrol) on *Vibrio cholerae* and *Vibrio parahaemolyticus*. J Inst Chem(India) 55: 88–90

352 Gill S, Raszeja W 1973 Chromatographic analysis of harman alkaloid derivatives in some plant raw materials. Rozpr Wydz 3:Nauk Mat-Przyr, Gdansk Tow Nauk 8: 137–143

353 Gillin FD, Reiner DS, Suffness M 1982 Bruceantin, a potent amoebicide from a plant, *Brucea antidysenterica*. Antimicrob Agents Chemother 22: 342–345

354 Giralt EG, Coelho MS 1970 Methoxy-9-ellipticine. II. Analysis in vitro of the mechanism of action. Rev Eur Etud Clin Biol 15: 539–541

355 Goijman SG, Turrens JF, Marini-Bettolo GB, Stoppani AOM 1985 Effect of tingenone, a quinonoid triterpene, on growth and macromolecule biosynthesis in *Trypanosoma cruzi*. Experientia 4: 646–648

356 Goina T, Petcu P, Pitea M 1979 Composition for treating hepatobiliar ailments. Romanian Patent 66,691 pp. 2PP

357 Goncalves deLima O, de Barros Coelho JS, D'Albuquerque, IL 1971 Antimicrobial and antitumor activity of lawsone in comparison with lapachol. Rev Inst Antibiot Univ Fed Pernambuco Recife 11: 21–26

358 Goncalves deLima O, Weigert E, Marini-Bettolo GB, Medeiros M et al 1972 Antimicrobial substances from higher plants. XXXVIII. Isolation and identification of maitenine & pristimerin from the roots of *Plenckia populnea* (Hippocrateaceae), from the cerrado region, Brasilia, D.F. Rev Inst Antibiot Univ Fed Pernambuco Recife 12: 13–18

359 Gonzalez AG, Agullo Martinez E, Martinez Iiniguez MA, Rodrigu EZ, Luis F 1971 New sources of natural coumarins. XVI. Products of the twigs of *Ruta pinnata* L. fil.: Structure of benahorin. An Quim 67: 441–451

360 Gonzalez AG, Cardona RJ, Lopez Dorta H, Medina JM et al. 1977 The chemistry of Rubiaceae. III. Anthraquinones of *Plocama pendula*. An Quim 73: 869–871

361 Gonzalez AG, DelaFuente G, Diaz R 1982 Four new diterpenoid alkaloids from *Delphinium pentagynum*. Phytochemistry 21: 1781–1782

362 Gonzalez AG, Estevez R, Jaraiz I 1972 New sources of natural coumarins. XX. Coumarins of *Ruta oreojasme* Webb. An Quim 68: 415–420

363 Gonzalez AG, Rodriguez FL 1971 Chemistry of the Rutaceae of Canary Isles. Herba Hung 10: 95–107

364 Goonetilleke LA, Jansz ER, Balachandran S, Vivekanandan K 1980 Studies on the *Pinus* species growing in Sri Lankan plantations. IV. Composition of oleoresin and turpentine of *Pinus patula* and *Pinus caribaea*. J Nat Sci Counc Sri Lanka 8: 161–165

365 Gopalakrishnan C, Shankaranarayanan D, Nazimudeen SK, Viswan 1980 Anti-inflammatory and C.N.S. depressant activities of xanthones from *Calophyllum inophyllum* and *Mesua ferrea*. Indian J Pharmacol 12: 181–191

366 Gorewit RC 1983 Pituitary and thyroid hormone responses of heifers after ferulic acid administration. J Dairy Sci 66: 624–629

367 Goris A, Crete L 1907 Purgative action of *Polygonum cuspidatum*. Bull Sci Pharmacol 14: 698–703

368 Goto R 1941 Constituents of Manchurian *Fagara mantschurica* Honda. Yakugaku Zasshi 61: 91–94

369. Gottlieb OR, Fineberg M, Magalhaes T 1962 Physiological varieties of *Ocotea pretiosa*. IV. Further data on nitrophenylethane containing specimens. Perfumery Essent Oil Rec 53: 299–301
370. Gottlieb OR, Koketsu M, Magalhaes MT, Guilhermesmaia J et al. 1981 Essential oils of Amazonia. VII. Acta Amazonica 11: 143–148
371. Govindachari TR, Joshi BS, Kamat VN 1965 Structure of parthenolide. Tetrahedron 21: 1509–1519
372. Govindachari TR, Nagarajan K, Pai BR, Rajappa S 1958 Chemical investigation of khet-papra. J Sci Ind Res B 17: 73–75
373. Govindachari TR, Subramaniam PS, Pai BR, Kalyanaraman PS et al. 1971 Heartwood constituents of *Calophyllum trapezifolium*. The isolation and structure of two new xanthones. Indian J Chem 9: 772–775
374. Govindachari TR, Viswanathan N 1972 Alkaloids of *Mappia foetida*. Phytochemistry 11: 3529–3531
375. Grina JA, Ratcliff MR, Stermitz FR 1982 Constituents of *Zanthoxylum*. Part 7. Old and new alkaloids from *Zanthoxylum arborescens*. J Org Chem 47: 2648–2651
376. Gripenberg J 1949 The constituents of wood of *Thuja occidentalis*. Acta Chem Scand 3: 782
377. Groger D, Mothes K 1960 Biogenesis of peganine. Arch Pharm(Weinheim) 293: 1049–1052
378. Gromova AS, Lutskii VI, Tyukavkina NA 1974 Phenolic acids and their glycosides from the phloem of *Picea jezoensis* and *P. koraiensis*. Khim Prir Soedin 10: 798–799
379. Gunasekera SP, Cordell GA, Farnsworth NR 1980a Constituents of *Nealchornea yapurensis* (Euphorbiaceae). J Nat Prod 43: 285–287
380. Gunasekera SP, Cordell GA, Farnsworth NR 1980b Potential anticancer agents. XV. Anticancer indole alkaloids of *Ervatamia heyneana*. Phytochemistry 19: 1213–1218
381. Gunasekera SP, Jayatilake GS, Selliah SS, Sultanbawa MUS 1977 Chemical investigation of Ceylonese plants. Part 27. Extractives of *Calophyllum cuneifolium* Thw. and *Calophyllum soulattri* Burm. f. (Guttiferae). J Chem Soc Perkin Trans I 1505–1511
382. Gunasekera SP, Selliah SS, Sultanbawa MUS 1975 Chemical investigation of Ceylonese plants. Part XV. Extractives of *Kayea stylosa* (Guttiferae). J Chem Soc Perkin Trans I 1539–1544
383. Gunasekera SP, Sivapalan K, Sultanbawa MUS, Ollis WD 1977 Chemical investigation of Ceylonese plants. Part 21. Extractives of *Pentadesma butyracea* Sabine (Guttiferae). J Chem Soc Perkin Trans I 11–14
384. Gunasekera SP, Sotheeswaran S, Sultanbawa MUS 1981 Two new xanthones, calozeyloxanthone and zeyloxanthonone, from *Calophyllum zeylanicum* (Guttiferae). J Chem Soc Perkin Trans I 1831–1835
385. Gunasekera SP, Sultanbawa MUS 1975 Chemical investigation of ceylonese plants. XVI. Extractives of *Calophyllum cordato-oblongum* (Guttiferae). J Chem Soc Perkin Trans I 2215
386. Gunawardana YAGP, Sultanbawa MUS, Balasubramanian S 1980 Chemical investigation of Ceylonese plants. Part 41. Distribution of some triterpenes and phenolic compounds in the extractives of endemic Dipterocarpaceae species of Sri Lanka. Phytochemistry 19: 1099–1102
387. Guo JX, Handa SS, Pezzuto JM, Kinghorn AD et al. 1983 Plant anticancer agents. XXXIII. Constituents of *Passerina vulgaris*. Planta Med 48: 264–265
388. Gupta MB, Nath R, Gupta GP, Bhargava KP 1981 Antiulcer activity of some plant triterpenoids. Indian J Med Res 73: 649–652
389. Gupta RK, Mahadevan V 1967 Chemical examination of the heartwood of *Diospyros ebenum*. Indian J Pharmacy 29: 289
390. Gupta SR, Malik KK, Seshadri TR 1966 Chemical components of *Albizzia lebbeck* heartwood. Indian J Chem 4: 139–141
391. Gupta SR, Malik KK, Seshadri TR 1969 Lapachol from the heartwood of *Tecoma undulata* and a note on its reactions. Int J Cancer 7: 457–459
392. Gurni AA, Wagner ML 1984 Proanthocyanidins from some Argentine species of *Ephedra*. Biochem Syst Ecol 12: 319–320
393. Hakim SAE 1954 Argemone oil, sanguinarine, and epidemic-dropsy glaucoma. Brit J Ophthalmol 38: 193–215
394. Hakim SAE, Mijovic V, Walker J 1961 Distribution of certain poppy-*Fumaria* alkaloids and a possible link with the incidence of glaucoma. Nature(London) 189: 198–201

395 Hall IH, Liou YF, Okano M, Lee KH 1982 Antitumor agents 46: In vitro effects of esters of brusatol, bisbrusatol, and related compounds on nucleic acid protein synthesis of P-388 Lymphocytic leukemia cells. J Pharm Sci 71: 345–348
396 Hall IH, Tagahara K, Lee KH 1982 Antitumor agents 53: The effects of daphnoretin on nucleic acid and protein synthesis of Ehrlich Ascites tumor cells. J Pharm Sci 71: 741–744
397 Hamsaveni GR, Mohan V, Purushothaman KK 1981 Activity of mangiferin on malarial parasite, bacteria and fungi. J Natl Integ Med Assoc 23: 109–112
398 Han BH, Park MH, Han YN 1981 Studies on the antioxidant components of Korean ginseng. III. Identification of phenolic acids. Arch Pharmacol Res 4: 53–58
399 Han CW, Lee SC 1980 An experimental study of the uterine contractility in response to various oxytocics on progesterone primed rabbit myometria. Koryo Taehakkyo Uikwa Taehak Chapchi 17: 85–98
400 Han HW, Wang JZ, Sun FL, Zeng GY 1981 Effect of DL-demethylcoclaurine on cultured rat heart cells. Chung-Kuo Yao Li Hsueh Pao 2: 111–114
401 Handa SS, Kinghorn AD, Cordell GA, Farnsworth NR 1981 Investigation of the antileukemic and cytotoxic constituents of *Soulamea soulameoides*. Abstr Joint Mt Am Soc Pharmacog And Soc For Economic Botany, Boston, July 13–18. American Society of Pharmacognosy, Dept of Pharmaceutical Sciences, Univ of Pittsburgh, Pittsburgh PA p. 27
402 Handa SS, Kinghorn AD, Cordell GA, Farnsworth NR 1983 Plant anticancer agents. XXVI. Constituents of *Peddiea fischeri*. J Nat Prod 46: 248–250
403 Handa SS, Kinghorn AD, Cordell GA, Farnsworth NR 1983 Plant anticancer agents. XXV. Constituents of *Soulamea soulameoides*. J Nat Prod 46: 359–364
404 Hang VA, Gatilov YV, Dubovenko ZV, Pentegova VA 1980 Mono-and sesquiterpenoids of the oleoresin of *Pinus koraiensis* and *P. pumila* crystal structure of 1β, 4α-H, 7α-H, 10β-H-Guaiane-5α,14-diol. Khim Prir Soedin 16: 361–365
405 Hanumaiah T, Rao BK, Rao KVJ 1983 Chemical examination of *Ventilago calyculata* Tulasne. Acta Cienc Indica Ser Chem 9: 209–211
406 Harats N, Ackerman Z, Shalit M 1984 Quinine-related hypoglycemia. N Engl J Med 310: 1331
407 Harborne JB 1967 Comparative biochemistry of the flavonoids IV. Correlations between chemistry, pollen morphology and systematics in the family Plumbaginaceae. Phytochemistry 6: 1415–1428
408 Harper SH, Shirley DB, Taylor DA 1976 Isoflavones from *Xanthocercis zambesiaca*. Phytochemistry 15: 1019–1023
409 Hartmann E, Renz B, Jung JA 1981 Studies on the bacterial and fungicidal inhibitors in higher plants. Isolation identification and antimicrobial activity of two resin acids from spruce Phytopathol Z 101: 31–42
410 Hartzell A, Wexler E 1946 Histological effects of sesamin on the brain and muscles of the housefly. Contrib Boyce Thompson Inst 14: 123–126
411 Hasselstrom T, Hampton BL 1938 Identification of methylcravicol in American gum spirits of turpentine. J Am Chem Soc 60: 3086
412 Hatfield GM, Keller WJ 1981 Structure of gramodendrine, a novel alkaloids from *Lupinus arbustus* subsp. *calcaratus*. (Abstract). Joint Mt Am Soc Pharmacog and Soc for Econ Botany, Boston, July 13–18. American Society of Pharmacognosy, Dept of Pharmaceutical Sciences, Univ of Pittsburgh, Pittsburgh PA p. 41
413 Hatsuda Y, Murao S, Terashima N, Yokota T 1960 Biochemical studies on soil sickness. 1. Toxic substance in fig roots. Nippon Nogei Kagaku Kaishi 34: 484–486
414 Hausen BM, Schmalle H 1981 The sensitizing capacity of naturally occurring quinones. Part VI. Quinonoid constituents as contact sensitisers in Australian blackwood (*Acacia melanoxylon* R. Br.). Brit J Ind Med 38: 105–109
415 Hayashi K, Isaka T 1950 Chemical identification of vegetable dyes used in ancient Japanese silk (a preliminary report). Shigen Kaghku Kenkyusho Iho 33
416 Hayashi T, Kubo M 1980 Antitumor compositions comprising saponins. British Patent UK Pat Appl 2,035,082
417 Hayashi Y, Sakurai K, Takahashi T, Kitao K 1974 Chemistry of wood extractives. XX. Heartwood constituents of *Haplormosia monophylla*. Mokuzaki Gakkaishi 20: 595–599

418 Hayat M, Mathe G, Janot MM, Potier P et al. 1974 Experimental screening of 3 forms and 19 derivatives or analogs of ellipticine: Oncostatic effect on L1210 leukemia and immunosuppressive effect of 4 of them. Biomedicine 21: 101
419 Hazanagy A 1970 Recent results with Plantaginis folium (plantain) leaves. Herba Hung 9: 57–63
420 He LY 1982 TLC separation and densitometric determination of tropine alkaloids Chung Ts'ao yao 132: 61–64
421 Herald PJ, Davidson PM 1983 Antibacterial activity of selected hydroxycinnamic acids. J Food Sci 48: 1378–1379
422 Herath WHMW, Rajasekera NDS, Sultanbawa MUS, Wannigama GP et al. 1978 Triterpenoid, coumarin and quinone constituents of eleven *Diospyros* species (Ebenaceae). Phytochemistry 17: 1007–1009
423 Herman E, Vick,J, Burka B 1971 The cardiovascular actions of ellipticine. Toxicol Appl Pharmacol 18: 743
424 Herman EH, Lee IP, Mhatre RM, Chadwick DP 1974 Prevention of hemolysis induced by ellipticine (NSC-71795) in rhesus monkeys. Cancer Chemother Rep Part 1 58: 637
425 Hikino H, Kiso Y, Kubota M, Hattori M et al. 1984 Antihepatotoxic principles of *Swertia japonica* herb. Shoyakugaku Zasshi 38: 359–360
426 Hikino H, Kiso Y, Wagner H, Fiebig M 1984 Antihepatotoxic actions of flavonolignans from *Silybum marianum* fruits. Planta Med 50: 248–250
427 Hikino H, Ogata K, Kasahara Y, Konno C 1985 Pharmacology of ephedroxanes. J Ethnopharmacol 13: 175–191
428 Hikino H, Ohta T, Takemoto T 1970 Stereostructure of picrasin B, simarubolide of *Picrasma quassioides*. Chem Pharm Bull 18: 219–220
429 Hikino H, Taguchi T, Fujimura H, Hiramatsu Y 1977 Antiinflammatory principles of *Caesalpinia sappan* wood and of *Haematoxylon campechianum* wood. Planta Med 31: 214–220
430 Hillis WE, Rozsa AN, Lau LS 1975 Rapid determination of ellagic acids by gas-liquid chromatography. J Chromatogr 109: 172–174
431 Hiranuma S, Hudlicky T 1982 Synthesis of homoharringtonine and its derivative by partial esterification of cephalotaxine. Tetra Lett 23: 3431–3434
432 Hirata T, Suga T 1977 Biologically active constituents of leaves and roots of *Aloe arborescens* var. *natalensis*. Z Naturforsch 32C: 731–734
433 Hirose Y, Iwama F 1984 Antioxidant isolated from grape seed optical isomer mixtures of catechin. Yakugaku Zasshi 33: 435–438
434 Hirose Y, Nakatsuka T 1958 Composition of the essential oil of *Biota orientalis* wood. Mokuzaki Gakkaishi 4: 26
435 Ho LX, Xu YX, Xue HZ, Wang WH et al. 1982 Studies on the antifertility constituents of huai jiao. I. Isolation of constituents I-XI and identification of I-IV and IX. Sheng Chih Yu Bi Yun 2: 23–27
436 Hoke M, Hansel R 1972 New investigation of sassafras root. Arch Pharm (Weinheim) 305: 33
437 Homma N, Kono M, Kadohira H, Yoshihara S et al. 1961 Influence of berberine on the intestinal flora of children. Arzneim-Forsch 11: 450–454
438 Hooper SN, Chandler RF 1984 Herbal remedies of the maritime Indians: phytosterols and triterpenes of 67 plants. J Ethnopharmacol 10: 181–194
439 Horiuchi I 1982 Aricine as a bactericide and fungicide. Japanese Patent Kokai Tokkyo Koho 82 75,906
440 Hsiao PK, Feng YS, Hsia KC, Feng SC et al. 1973 Search for biologically active substances in Chinese medicinal plants. New sources of five medicinal alkaloids. Chih Wu Hsueh Pao 15: 64–70
441 Hsu B 1980 Recent progress in antineoplastic drug research in China. (review). In: Marks, PA (ed.) Cancer Research In The People's Republic Of China And The United States of America - Epidemiology, Causation and Approaches to Therapy. Grune and Stratton, New York NY p. 235–250
442 Hsu B 1982 The influence of several anticancer agents on cell proliferation, differentiation and the cell cycle of murine erythroleukemia cells. Am J Chin Med 9: 268–276
443 Hsu B, Chen JT, Yang JL, Chang SY et al. 1980 New results in pharmacologic research of some anticancer agents. Proc U.S.-China Pharmacology Symp. National Academy of Science, Washington DC. p. 151–188

444 Hua Hsueh Hsueh Pao 1976 Studies on the antitumor constituents of *Cephalotaxus hainanensis*. Hua Hsueh Hsueh Pao 34: 283–293

445 Huang CC, Han CS, Yue XF, Shen CM et al. 1983 Cytotoxicity and sister chromatid exchanges induced in vitro by six anticancer drugs developed in the People's Republic of China. J Nat Cancer Inst 71: 841–847

446 Huang M-T, Wood AW, Newmark HL, Sayer JM et al. 1983 Inhibition of the mutagenicity of bay-region diol-epoxides of polycyclic aromatic hydrocarbons by phenolic plant flavonoids. Carcinogenesis 4: 1631–1637

447 Huang ZX, Li ZH 1980 Studies on the antitumor constituents of *Zanthoxylum nitidum* (Roxb.) DC. Hua Hsueh Hsueh Pao 38: 535–542

448 Hufford CD, Funderburk MJ, Morgan JM, Robertson LW 1975 Two antimicrobial alkaloids from heartwood of *Liriodendron tulipifera*. J Pharm Sci 64: 789

449 Hufford CD, Oguntimein BO, Baker JK 1981 New flavonoid and coumarin derivatives of *Uvaria afzelii*. J Org Chem 46: 3073–3078

450 Hufford CD, Oguntimein BO, VanEngen D, Muthard D et al. 1980 Vafzelin and uvafzelin, novel constituents of *Uvaria afzelii*. J Am Chem Soc 102: 7365–7367

451 Hufford CD, Sharma AS, Oguntimein BO 1980 Antibacterial And antifungal activity of liriodenine and related oxoaporphine alkaloids. J Pharm Sci 69: 1180–1182

452 Hui WH, Ko PDS, Lee YC, Li MM et al. 1975 Triterpenoids from ten *Lithocarpus* species of Hong Kong. Phytochemistry 14: 1063–1066

453 Hume WR 1984 Effect of eugenol on respiration and division in human pulp, mouse fibroblasts, and liver cells in vitro. J Dent Res 63: 1262–1265

454 Hussain RA, Waterman PG 1982 Lactones, flavonoids and benzophenones from *Garcinia conrauana* and *Garcinia mannii*. Phytochemistry 21: 1393–1396

455 Ichikawa K, Kinoshita T, Ital A, Litaka Y et al. 1984 Isolatin of (-)-stepholidine, an alkaloid of antiserotonergic-like activity from *Sinomenium acutum*. Heterocycles 22: 2071–2077

456 Iizuka C 1980 Antiviral substance. French Patent Demande Fr-2,485,373

457 Ikeyama S, Takeuchi M 1981 Antitubulin activities of ansamitocins and maytansinoids. Biochem Pharmacol 30: 2421–2425

458 Ikuta A, Itokawa H 1982 Studies on the alkaloids from tissue culture of *Nandina domestica*. In: Fujiwara, A. (ed) Proc 5th Int Congr Plant Tissue Cell Cult. Maruzen, Tokyo 315–316

459 Imamura H, Nomura K, Hibino Y, Ohashi H 1974 A new flavonol in the shake of merbau wood (*Intsia* sp.). Gifu Daigaku Nogakubu Kenkyu Hokoku pp. 93–101

460 Imamura H, Tanno Y, Takahashi T 1968 Isolation and identfication of four isoflavone derivatives from Eurasian teak (*Pericopsis* sp). Mokuzaki Gakkaishi 14: 295–297

461 Inamoto N, Masuda S, Shimamura O, Tsuyuki T 1961 4,5-Dimethoxycanthin-6-one and 2,6-dimethoxy-para-benzoquinone from *Picrasma ailanthoides*. Bull Chem Soc Jap 34: 888–889

462 Inouye H, Saito S, Taguchi H, Endo T 1969 Two new iridoid glucosides from *Gardenia jasminoides*: gardenoside and geniposide. Tetra Lett 2347–2350

463 Ionescu M, Vlassa M, Mester I, Vicol EC 1971 Alkaloids of *Haplophyllum suaveolens*. IV. Distribution of alkaloids in the organs of *Haplophyllum suaveolens*. Rev Roum Biochim 8: 123–127

464 Ishiguro K, Yamaki M, Takagi S, Ikeda Y et al. 1984 Studies on iridoid-related compounds: Antimicrobial and antitumor activity. Abstr 5th Symp Develop Appl Naturally Occurring Drug Materials pp. 58–60

465 Ishii H, Chen IS, Akaike M, Ishikawa T et al. 1982 Studies on the chemical constituents of rutaceous plants. 44. The chemical constituents of *Xanthoxylum integrifoliolum* (Merr.) Merr. (*Fagara integrifoliolum* Merr.) (1) The chemical constituents of the rootwood. Yakugaku Zasshi 102: 182–195

466 Ishii H, Hosoya K, Ishikawa T, Haginiwa J 1974 Studies on the chemical constituents of Rutaceous plants: XX. The chemical constituents of *Xanthoxylum arnottianum* Maxim.(1) Isolation of the chemical constituents of the xylem of stems. Yakugaku Zasshi 94: 309–321

467 Ishii H, Hosoya K, Ishikawa T, Ueda E et al. 1974 Studies on the chemical constituents of Rutaceous plants. XXI. The chemical constituents of *Xanthoxylum arnottianum*. Isolation of the chemical constituents of the xylem of roots. Yakugaku Zasshi 94: 322–331

468 Ishii H, Ishikawa T, Chen IS, Lu ST 1982 Structures and absolute configurations of some new optically active phenyl propanoids. Proc 25th Symp Nat Prod 25: 353–360

469 Ishii H, Ishikawa T, Tohojoh T, Murakami K et al. 1982 Studies on the chemical constituents of Rutaceous plants. Part 45. Novel phenyl propanoids: cuspidiol, boninenal, and methyl boninenalate. J Chem Soc Perkin Trans I. 2051–2058

470 Ishii H, Koyama K, Chen I-S, Lu ST et al. 1982 Studies on the chemical constituents of rutaceous plants. 46. The chemical constituents of *Xanthoxylum integrifoliolum* (Merr.) Merr. (*Fagara integrifoliolum* Merr.) 2. Structural establishment of integriquinolone: a new phenolic quinolone. Chem Pharm Bull 30: 1992–1997

471 Ishikawa O, Hashimoto T 1979 Separation of alkaloids of *Coptis japonica*. Japanese Patent - Jappan Kokai Tokkyo Koho 79 129,112

472 Islambekov SY, Mirzakhidov KA, Karimdzhanov AK, Ishbaev AI 1982 Catechins and proathocyanidins of *Alhagi pseudoalhagi*. Khim Prir Soedin 18: 653

473 Istudor V, Mantoiu M, Badescu I 1981 Study on the preparation of an "instant" type product (ulcovex) for gastric and duodenal ulcers. part I. Chemical study of calendulae flores products. Farmacia (Bucharest) 29: 41–48

474 Ito T, Hsuan CP, Chang HH 1981 Saponins as neoplasm inhibitors. Japanese Patent Kokai Tokkyo Koho 81 46,817

475 Ivanov V, Ivanova L 1958 Alkaloid content of *Glaucium flavum* var. *lejocarpum*. Farmatsiya(Sofia) 8: 27–33

476 Iwanishi K, Suzuki I 1982 Mechanism of the antishock effect of anisodamine. Int J Immunopharmacol 44: 332

477 Iwu MM, Court WE 1977 Root alkaloids of *Rauvolfia vomitoria*. Planta Med 32: 88–99

478 Iwu MM, Court WE 1977 Root alkaloids of *Rauvolfia cumminsii*. Planta Med 32: 158–161

479 Iwu MM, Court WE 1980 The alkaloids of *Rauvoifia mombasiana* roots. Planta Med 38: 260–263

480 Iyer RP, Brown JK, Chaubal MG, Malone MH 1977 *Brunfelsia hopeana* I: Hippocratic screening and antiinflammatory evaluation. Lloydia 40: 356–360

481 J Agr Chem Soc Japan 1941 The components of *Viscum album*. IV. The acid constituents of the resin wax of the woody portions. J Agr Chem Soc Japan 17: 1102

482 Jackson B, Locksley HD, Scheinmann F 1969 The isolation of 6-desoxyjacareubin,2-(3,3-dimethylallyl)-1,3,5,6-tetrahydroxy-xanthone and jacareubin from *Callophyllum inophyllum*. Phytochemistry 8: 927

483 Jahodar L, Kolb I, Leifertova I 1981 Unedoside in *Arctostaphylos uva-ursi* Roots. Pharmazie 36: 294–296

484 Jain MP, Sharma VK 1983 Phytochemical investigation of roots of *Adhatoda vasica*. Planta Med 46: 250

485 Jain SC, Lohiya NK 1981 Antifertility effects of hederagenin on male mice. Abstr Joint Mt Am Soc Pharmacog Soc For Econ Botany, Boston July 13–18. American Society of Pharmacognosy, Dept of Pharmaceutical Sciences, Univ of Pittsburgh, Pittsburgh PA p. 66

486 Janot MM, Lemen J, Hammouda Y 1936 Presence of yohimbine in the roots of *Lochnera lancea*. Ann Pharm Fr 14: 344

487 Januszewski P 1983 Medical composition for controlling the diurnal and nocturnal biological rhythm of humans and for treating human sexual problems. French Patent Demande 2,515,044

488 Japanese Patent Kokai Tokkyo Koho 1981a 8-Cyclohexenylflavones from mulberry species. Japanese Patent Kokai Tokkyo Koho 81 123,979

489 Japanese Patent Kokai Tokkyo Koho 1981b Ginsenosides for treatment of stress accompanied by autonomic imbalance. Japanese Patent Kokai Tokkyo Koho 81 43,214

490 Japanese Patent Kokai Tokkyo Koho 1981c Wound healing formulations containing allantoin. Japanese Patent Kokai Tokkyo Koho 81 131,517

491 Japanese Patent Kokai Tokkyo Koho 1982 Sanggenone C. Japanese Patent Kokai Tokkyo Koho 57 145,894

492 Japanese Patent Kokai Tokkyo Koho 1983a Pharmaceuticals containing quercitrin for protection against liver damage. Japanese Patent Kokai Tokkyo Koho 58 140,021

493 Japanese Patent Kokai Tokkyo Koho 1983b Sanggenone D. Japanese Patent Kokai Tokkyo Koho 58 41,894

494 Japanese Patent Kokai Tokkyo Koho 1984a Antimold for food. Japanese Patent Kokai Tokkyo Koho 84 140,868

495 Japanese Patent Kokai Tokkyo Koho 1984b Promotion of sex activity by syringin. Japanese Patent Kokai Tokkyo Koho 59 33 221
496 Japanese Patent Kokai Tokkyo Koho 1984c Trimethylsteroids as anticholesteremics. Japanese Patent Kokai Tokkyo Koho 59 27,824
497 Jensen SR, Nielsen BJ, Dahlgren R 1975 Iridoid compounds, their occurrence and systematic importance in the angiosperms. Bot Not 128: 148–180
498 Johnson CC, Johnson G, Poe CF 1952 Toxicity of alkaloids to certain bacteria. II. Berberine, physostigmine and sanguinarine. Acta Pharmacol Toxicol 8: 71–78
499 Jolad SD, Hoffmann JJ, Timmermann BN, Schram KH et al. 1983 Daphnane diterpenes from *Wikstroemia monticola*: Wikstrotoxins A-D, huratoxin and excoecariatoxin. J Nat Prod 46: 675–680
500 Joshi KC, Singh P, Pardasani RT 1977 Chemical components of the roots of *Tectona grandis* and *Gmelina arborea*. Planta Med 32: 71–75
501 Joshi KC, Singh P, Pardasani RT 1978 Chemical constituents of the stem heartwood of *Markhamia stipulata*. Planta Med 34: 219
502 Joshi KC, Singh P, Pardasani RT, Singh G 1979 Quinones and other constiuents from *Haplophragma adenophyllum*. Planta Med 37: 60–63
503 Joshi KC, Singh P, Sharma MC 1985 Quinones and other constituents of *Markhamia platycalyx* and *Bignonia unguiscati*. J Nat Prod 48: 145
504 Joshi KC, Singh P, Taneja S 1981 Crystalline components of the stem heartwood of *Randia dumetorum* and the root heartwood of *Kigelia pinnata*. J Indian Chem Soc 58: 825–826.
505 Joshi KC, Tholia MK, Sharma M 1975 Triterpenoids, wax, alcohols and other constituents from *Acacia senegal*. Indian J Chem 13: 638–639
506 Joshp YC, Dobhal MP, Joshp BC, Barar FSK 1981 Chemical investigation and biological screening of the stem of *Rhododendron anthopogon* (D.Don). Pharmazie 36: 381
507 K'O Hsueh T'Ung Pao 23: 1978 Separation and identification of harringtonine and epiharringtonine. K'O Hsueh T'Ung Pao 23: 696–699
508 Kabiev OK, Vermenichev SM 1966 Antitumor activity of leucoanthocyanidins and catechins. Vopr Onkol 12: 61–64
509 Kafi S, Gaillard JM 1980 Pre- and postsynaptic effect of yohimbine on rat paradoxical sleep. Sleep (Basel) 5: 292–293
510 Kageyama K, Nagasawa T, Kimura S, Kobayashi T et al. 1980 Cytotoxic activity of unsaturated fatty acids to lymphocytes. Can J Biochem 58: 504–508
511 Kaku T, Kawashima Y 1977a Studies on *Panax ginseng* C.A. Meyer. Part 1. Yamanouchi Seiyaku Kenkyu Hokoku 3: 22–29
512 Kaku T, Kawashima Y 1977b Studies on *Panax ginseng* C.A. Meyer. Part 2. Yamanouchi Seiyaku Kenkyu Hokoku 3: 30–44
513 Kalra VK, Kukla AS, Seshardi TR 1966 A chemical examination of *Quercus robur* and *Quercus incana*. Curr Sci 35: 204–205
514 Kamat VS, Chuo FY, Kubo I, Nakanishi K 1981 Antimicrobial agents from an East African medicinal plant *Erythrina abyssinica*. Heterocycles 15: 1163–1170
515 Kanchanapee P 1966 Studies on medicinal plants in Thailand. I. Phenolic constituents of *Cudrania javanensis* Trec. Shoyakugaku Zasshi 20: 63–66
516 Kaplanski O, Shemesh M, Berman A 1981 Effects of phytoestrogens on progesterone synthesis by isolated bovine granulosa cells. J Endocrinol 89: 343–348
517 Kar K, Mohanta PK, Popli SP, Dhawan BN 1981 Inhibition of passive cutaneous anaphylaxis by compounds of *Camellia sinensis*. Planta Med 42: 75–78
518 Karimova SU, Israilov IA, Vezhnik F, Yunusov MS et al. 1983 Alkaloids of *Glaucium fimbrilligerum*. II. Khim Prir Soedin 19: 493–496
519 Kariyone T, Imura H, Nakamura I 1958 Components of *Ginko biloba*. I. Isolation of bilbanone. Yakugaku Zasshi 78: 1152
520 Kartnig T 1981 Plants having activity on the endocrine system.(review). Osterr Apoth Ztg 35: 121–128
521 Karunanayake S, Sotheeswaran S, Uvais M, Sultanbawa MUS et al. 1981 Xanthones and triterpenes of *Calophyllum tomentosum*. Phytochemistry 20: 1303–1304
522 Kasai T, Okuda M, Sano H, Mochizuki H et al. 1982 Biological activity of the constituents in roots of ezo-no-gishigishi (*Rumex obtusifolius*). Agr Biol Chem 46: 2809–2813

523 Kato Y, Kato N, Baba M 1984 Phenolic constituents of *Lyonia ovalifolia*. Pharmazie 39: 425–426
524 Kato Y, Yasue M 1984 Studies on the constituents of *Lyonia ovalifolia* Drude var. *elliptica* Hand.-Mazz. XXI. The difference of flavonoid glycosides between individuals. Shoyakugaku Zasshi 37: 412–417
525 Kaumann AJ 1983 Yohimbine and rauwolscine inhibit 5-hydroxytryptamine-induced contraction of large coronary arteries of calf through blockade of 5 Ht2 receptors. Naunyn-Schmiedebergs Arch Exp Pathol Pharmakol 323: 149–154
526 Kavli G, Krokan H, Midelfart K, Volden G et al. 1983 Extraction, separation, quantification and evaluation of the phototoxic potency of furocoumarins in different parts of *Heracleum laciniatum*. Photobiochem Photobiophys 5: 159–168
527 Kawai A, Akimoto H, Kozai Y, Ootsu K et al. 1984 Chemical modification of ansamitocins. III. Synthesis and biological effects of 3-acyl esters of maytansinol. Chem Pharm Bull 32: 3441–3451
528 Ke MM, Zhang XD, Wu LZ, Zhao YI et al. 1982 Studies on the active principles of *Corydalis saxicola* Bunting. Chih Wu Hsueh Pao 24: 289–291
529 Kelentey B 1960 The pharmacology of chelidonine and sanguinarine. Arzneim-Forsch 10: 135–137
530 Kellis JT Jr, Vickery LE 1984 Inhibition of human estrogen synthetase (aromatase) by flavones. Science 225: 1032–1034
531 Kemp MS, Holloway PJ, Burden RS 1985 3-β,19-α-Dihydroxy-2-oxours-12-en-28-oic acid: a pentacyclic triterpene induced in the wood of *Malus pumila* Mill. infected with *Chondrostereum purpureum* (Pers. ex. Fr.) Pouzar, and a constituent of the cuticular wax of apple fruits. J Chem Res (S) 154–155
532 Khan KA, Shoer A 1985 A lignan from *Lonicera hypoleuca*. Phytochemistry 24: 628–630
533 Khan VA, Bol'Shakova VI, Grigorovich MI, Shmidt EN et al. 1984 Terpenoids of *Pinus funebris*. Khim Prir Soedin 20: 444–448
534 Khan VA, Bol'Shakova VI, Shmidt EN, Dubovenkoz V et al. 1983 Terpenoids of the oleoresin of *Larix olgensis*. Khim Prir Soedin 19: 110
535 Khan VA, Dubovenko ZV, Pentegova VA 1976 Monoterpenoids and sesquiterpenoids of the oleoresin of *Larix sibirica*. Khim Prir Soedin 11: 100
536 Khan VA, Pankrushina NA, Shmidt EN, Dobovenko ZHV et al. 1984 Terpenoids of the oleoresin of *Abies semenovii*. Khim Prir Soedin 20: 115–116
537 Kikvidze IM, Bessonova IA, Mudzhiri KS, Yunusov SY 1971 Alkaloids of *Dictamnus caucasicus*. Khim Prir Soedin 7: 675
538 Kilminster KN, Sainsbury M, Webb B 1972 Alkaloids and terpenoids of *Bleekeria vitiensis*. Phytochemistry 11: 389–392
539 Kim JY, Staba EJ 1973 Studies on the ginseng plants. I. Saponins and sapogenins from American ginseng plants. Korean J Pharmacog 4: 193–203
540 Kim MW, Lee JS, Choi KJ 1982 Comparative studies on the chemical components in ginseng. 1. The ginsenosides and the free sugars content of various ginseng plants. Korean J Ginseng Sci 6: 138–142
541 Kim YC, Kim EK 1980 Studies on the effect of ginseng extract on chick embryonic nerve and muscle cells. Yakhak Hoe Chi 24: 143–150
542 Kimura Y, Ohminami H, Okuda H, Baba K et al. 1983 Effects of stilbene components of roots of *Polygonum* ssp. on liver injury in peroxidized oil-fed rats. Planta Med 49: 51–54
543 King CL, Kong YC, Wong NS, Yeung HW et al. 1980 Uterotonic effect of *Evodia rutaecarpa* alkaloids. J Nat Prod 43: 577–582
544 King FE, Housley JR, King TJ 1954 The chemistry of extractives from hardwoods. XVI. Coumarin derivatives of *Fagara macrophylla*, *Zanthoxylum flavum* and *Chloroxylon swietenia*. J Chem Soc pp. 1392
545 King FE, King TJ, Manning LC 1953 Constitution of jacareubin, a pyranoxanthone from *Calophyllum brasiliense*. J Chem Soc pp. 3932
546 King FS, Grundon MF, Neill KG 1952 The chemistry of extractives from hardwoods. Part IX. Constituents of the heartwood of *Ferreirea spectabilis*. J Chem Soc pp. 4580–4584
547 Kinghorn AD, Balandrin MF, Lin LJ 1982 Alkaloids of the Papilionoideae. Part 1. Alkaloid distribution in some species of the Papilionaceous tribes Sophoreae, Dalbergieae, Loteae, Brongniartieae and Boissiaeeae. Phytochemistry 21: 2269–2275

548 Kinoshita T, Sankawa U, Takuma T, Asahi K 1984 Induction of differentiation of cultured tumor cells by flavonoids. Abstr 5th Symp Develop Appl Naturally Occurring Drug Materials. pp. 67–69

549 Kiso Y, Suzuki Y, Watanabe N, Oshima Y et al. 1983 Antihepatotoxic principles of *Curcuma longa* rhizomes. Planta Med 49: 185–187

550 Kiso Y, Tohkin M, Hikino H, Hattori M et al. 1984 Mechanism of antihepatotoxic activity of glycyrrhizin, I: Effect on free radical generation and lipid peroxidation. Planta Med 51: 298–302

551 Kita T, Hata T, Kawashima Y, Kaku T et al. 1981 Pharmacological actions of ginseng saponin in stress mice. J Pharm Dyn 4: 381–393

552 Kitazawa E, Sato A, Takahashi S, Kuwano H et al. 1980 Novel diterpenelactones with anti-peptic ulcer activity from *Croton sublyratus*. Chem Pharm Bull 28: 227–234

553 Kitts DD, Krishnamurti CR, Kitts WD 1980 Uterine weight changes and 3h-uridine uptake in rats treated with phytoestrogens. Can J Anim Sci 60: 531–534

554 Kitzman JV, Booth NH, Hatch RC, Wallner B 1982 Antagonism of xylazine sedation by 4-aminopyridine and yohimbine in cattle. Amer J Vet Res 43: 2165–2169

555 Kitzman JV, Wilson RC, Hatch RC, Booth NH 1984 Antagonism of xylazine and ketamine anesthesia by 4-aminopyridine and yohimbine in geldings. Am J Vet Res 45: 875–879

556 Klocke JA, Arisawa M, Handa SS, Kinghorn AD et al. 1985 Growth inhibitory, insecticidal and antifeedant effects of some antileukemic and cytotoxic quassinoids on two species of agricultural pests. Experientia 41: 379–382

557 Klokova MV, Seryhk EA, Berezovskaya TP 1982 Coumarins of the genus *Ledum*. Khim Prir Soedin 18: 517

558 Klughardt G, Zymalkowski F 1982 Magnoflorine and protoquercitol as contents of *Cyclea barbata* Miers. Arch Pharm(Weinheim) 315: 7–11

559 Kodaira H, Ishikawa M, Komoda Y, Nakajima T 1983 Isolation and identification of anti-platelet aggregation principles from the bark of *Fraxinus japonica* Blume. Chem Pharm Bull 31: 2262–2268

560 Koedam A, Looman A 1980 Comparison of isolation procedures for essential oils. Part V. Effect of pH during distillation on the composition of the volatile oil from *Juniperus sabina*. Planta Med Suppl 40: 22–28

561 Kohli JD, Mukerji B 1955 Comparative activity of reserpine and total alkaloids of *Rauvolfia*. Curr Sci 24: 198–199

562 Kohn KW, Ross WE, Glaubiger D 1979 Ellipticine. Antibiotics 5: 195–213

563 Koike K, Bevelle C, Talapatra SK, Cordell GA et al. 1980 Potential anticancer agents. V. Cardiac glycosides of *Asclepias albicans* (Asclepiadaceae). Chem Pharm Bull 28: 401–405

564 Kong YC, King CL 1980 Studies in Fructus Evodiae - a multidisciplinary exercise in ethnomedical research. Korean J Pharmacog 11: 212–218

565 Korneva EA, Guschin GV, Poltavchenko GM 1982 Influence of cholino- and adrenotropic drugs on immunogenesis. Ann 1st Super Sanita 18: 99–101

566 Koshihara Y, Neichi T, Murota SI, Lao AN et al. 1984 Caffeic acid is a selective inhibitor for leukotriene biosynthesis. Biochim Biophys Acta 792: 92–97

567 Kostalova D, Brazdovicova B, Hwang YJ 1982 Alkaloids from the aboveground parts of *Berberis koreana* Palib. Farm Obz 51: 213–216

568 Kostalova D, Brazdovicova B, Tomko J 1981 Isolation of quaternary alkaloids from *Mahonia aquifolium* (Pursh) Nutt. 1. Chem Zvesti 35: 279–283

569 Kosuge T, Ishida H 1985 Studies on active substances in the herbs used for oketsu ("stagnant blood") in Chinese medicine. IV. On the anticoagulative principle in Rhei Rhizoma. Chem Pharm Bull 33: 1503–1506

570 Kosuge T, Ishida H, Satoh T 1985 Studies on antihemorrhagic substances in herbs classified as hemostatics in Chinese medicine. V. On antihemorrhagic principle in *Biota orientalis* (L.). Chem Pharm Bull 33: 206–209

571 Kraemer RP, Hindorf H, Jha HC, Kallage J et al. 1984 Antifungal activity of soybean and chickpea isoflavones and their reduced derivatives. Phytochemistry 23: 2203–2205

572 Kreitmair H 1936 Pharmacological trials with some domestic plants. E Merck's Jahresber 50: 102–110

573 Krishnamurty HG, Parkash B, Parthasarathy MR, Seshadri TR 1974 Chemical components of the heartwood of *Dipterocarpus macrocarpus*. Isolation of two new sesquiterpenes, dipterolone and dipterol. Indian J Chem 12: 520–522

574 Krishnan V, Rangaswami S 1967 Proanthocyanidins of *Salacia chinensis* Linn. Tetra Lett 2441–2446

575 Krogsgaard-Larsen P, Christensen B, Koford H 1983 Natural Products and Drug Development. Proceedings Of The Alfred Benzon Symposium 20. Munksgaard, Copenhagen

576 Kubitzki K, Mesquita AAL, Gottlieb OR 1978 Chemosystematic implications of xanthones in *Bonnetia* and *Archytaea*. Biochem Syst Ecol 6: 185–187

577 Kubo I, Matsumoto A 1984 Molluscicides from olive *Olea europaea* and their efficient isolation by countercurrent chromatographies. J Agr Food Chem 32: 687–688

578 Kubo M, Kimura Y, Shin H, Haneda T et al. 1981 Studies on the antifungal substance of crude drugs (II) On the roots of *Polygonum cuspidatum* Sieb. and Zucc.(Polygonaceae). Shoyakugaku Zasshi 35: 58–61

579 Kudav NA, Kulkarni AB 1974 Chemical investigations on *Cassia occidentalis* Linn.: Part II. Isolation of islandicin, helminthosporin, xanthorin and NMR spectral studies of cassiollin and its derivatives. Int J Cancer 12: 1042–1044

580 Kuenzig W, Chau J, Norkus E, Holowaschenko H et al. 1984 Caffeic and ferulic acid as blockers of nitrosamine formation. Carcinogenesis 5: 309–313

581 Kulkarni GH, Haribal MM, Sabata BK 1980 Constituents of the heartwood of *Atalantia monophylla*. Indian J Chem Ser B 19: 424–425

582 Kuo YH, Wu TR, Lin YT 1982 A new lignan detetrahydroconidendrin from *Juniperus formosana* Hayata. J Chin Chem Soc (Taipei) 29: 213–215

583 Kupchan SM, Branfman AR, Sneden AT, Verma AK et al. 1975 Novel maytansinoids. Naturally occurring and synthetic antileukemic esters of maytansinol. J Am Chem Soc 97: 5294

584 Kupchan SM, Komoda Y, Thomas GJ, Hintz HPJ 1972 Maytanprine and maytanbutine, new antileukemic ansa macrolides from *Maytenus buchananii*. J Chem Soc Chem Commun 1065

585 Kupchan SM, Lavoie EJ, Branfman AR, Fei BY et al. 1978 Phyllanthocin, a novel bisabolane aglycone from the antileukemic glycoside phyllanthoside. J Am Chem Soc 99: 3199–3201

586 Kupchan SM, Sigel CW, Matz MJ, Saenz Renauld JA et al. 1970 Tumor inhibitors. LIX. Jatrophone, a novel macrocyclic diterpenoid tumor inhibitor from *Jatropha gossypiifolia*. J Am Chem Soc 92: 4476–4477

587 Kurosawa K, Ollis WD, Redman BT, Sutherland IO et al. 1978 Vestitol and vesticarpan, isoflavonoids from *Machaerium vestitum*. Phytochemistry 17: 1413

588 Kurosawa K, Ollis WD, Sutherland IO, Gottlieb OR 1978 Variabilin, a 6-hydroxypterocarpan from *Dalbergia variabilis*. Phytochemistry 17: 1417–1418

589 Kurosawa K, Ollis WD, Sutherland IO, Gottlieb OR et al. 1978 Mucronulatol, mucroquinone and mucronucarpan, isoflavonoids from *Machaerium mucronulatum* and *M. villosum*. Phytochemistry 17: 1405–1411

590 Kuroyanagi M, Ebihara T, Tsukamoto K, Fukushima S et al. 1984 Screening of antibacterial constituents of crude drugs and plants. Abstr 5th Symposium On Development And Application of Naturally Occurring Drug Materials pp. 55–57

591 Kuroyanagi M, Yoshihira K, Natori S 1971 Naphthoquinone derivatives from the Ebenaceae. III. Shinanolone from *Diospyros japonica*. Chem Pharm Bull 19: 2314–2317

592 Kurvyakov PN, Khan VA, Dubovenko ZV, Pentegova VA 1978 Mono- and sesquiterpenoids of the oleoresin of *Picea glehnii*. Khim Prir Soedin 14: 408–409

593 Kutney JP, Beale MH, Salisbury PJ, Stuart KL et al. 1981 Isolation and characterization of natural products from plant tissue cultures of *Maytenus buchananii*. Phytochemistry 20: 653–657

594 Kutney JP, Hewitt GM, Kurihara T, Salisbury PJ et al. 1981 Cytotoxic diterpenes tripolide, tripdiolide, and cytotoxic triterpenes from tissue cultures of *Tripterygium wilfordii*. Can J Chem 59: 2677–2683

595 Ky PT, Mau ND, Vinh PV, Sau VT, Hien, NT 1983 The *Rauvolfia cambodiana* Pierre ex. Pitard Apocynaceae plant. Tap Chi Duoc Hoc 11–14

596 Lagrota MHDC, Wigg MD, Pereira LOB, Fonseca MEF et al. 1983 Antiviral activity of lapachol. Rev Microbiol 14: 21–26

597 Lambev I, Pavlova N, Krushkov I, Manolov P 1980 Study of the antiinflammatory activity of mangiferin extracted from *Colladonia triquetra*. Probl Vutr Med 8: 109–115

598 Lapteva KI, Lutskii VI, Tyukavkina NA 1974 Some flavanones and flavanonols of the heartwood of *Larix dahurica*. Khim Prir Soedin 10: 97B

599 Lastra H, Palacios M, Menendez R, Larionova M et al. 1982 Chemical evaluation and pharmacology of *Rauvolfia tetraphylla* and *Rauvolfia cubana*. First Latin American and Caribbean Symp Pharmacologically Active Natural Products, Havana, Cuba, June 21–28. UNESCO, Paris. p. 172

600 LeBoeuf PM, Abouchacra ML, Sevenet T, Cave A 1982 Alkaloids of *Albertisia papuana* Becc., Menispermaceae. Plant Med Phytother 16: 280–291

601 Lee KH, Hayashi N, Okano M, Nozaki H et al. 1984 Antitumor agents. 65. Brusatol and cleomiscosin-A, antileukemic principles from *Brucea javanica*. J Nat Prod 47: 550–551

602 Lee KH, Tagahara K, Suzuki H, Wu RY et al. 1981 Antitumor agents. 49. Tricin, kaempferol-3-O-β-D-glucopyranoside and (+)-nortrachelogenin, antileukemic principles from *Wikstroemia indica*. J Nat Prod 44: 530–535

603 Lee SW, Kozukue N, Bae HW, Lee JH 1978 Studies on the polyphenols of ginseng. I. Comparison of polyphenols pattern of various ginseng products and *Acanthopanax* with gas chromatogram. Hanguk Sikp'Um Kwahakhoe Chi 10: 245–249

604 Lee TJ, Lin YM, Shih TS, Chen FC 1981 New isodiospyrin analogs from Morris Persimmon. Proc Int Conf Chem Biotechnol Biol Act Nat Prod 3: 290–292

605 Lee TM, Der Marderosian AH 1981 Two-dimensional TLC analysis of ginsenosides from root of dwarf ginseng (*Panax trifolius* L.) Araliaceae. J Pharm Sci 70: 89–91

606 Leet JE, Hussain SF, Minard RD, Shamma M 1982 Sindamine, punjabine and gilgitine: three new secobisbenzylisoquinoline alkaloids. Heterocycles 19: 2355–2360

607 Leite De Almeida ME, Gottlieb OR 1974 Iso- and neo-flavonoids from *Dalbergia inundata*. Phytochemistry 13: 751–752

608 Leite de Almeida ME, Gottlieb OR 1975 Further isoflavones from *Pterodon appariciol*. Phytochemistry 14: 2716

609 Leont'Eva VG, Modonova LD, Tyukavkina NA 1974 Lignans from the wood of *Picea koraiensis*. Khim Prir Soedin 10: 399–400

610 Lesca P 1983 Protective effects of ellagic acid and other plant phenols on benzo(A)pyrene-induced neoplasia in mice. Carcinogenesis 4: 1651–1653

611 Li S, Liu G, Zhang Y, Xu J 1982 Experimental study of antitussive effect of arbutin. Yao Hsueh T'Ung Pao 17: 720–722

612 Li YH, Guo SF, Zhou FY, Xu SZ et al. 1983 Combined harringtonine or homoharringtonine chemotherapy for acute nonlymphocytic leukemia in 25 children. Chung-Hua I Hsueh Tsa Chih (Engl Ed) 96: 303–305

613 Lie LH, Cowie CH 1974 Biochemical effects of ellipticine on leukemia L1210 cells. Biochim Biophys Acta 353: 375

614 Lieb WA, Scherf HJ 1956 Papaveraceae alkaloids and eye pressure. Klin Monatsbl Augenheilk 128: 686–705

615 Lin CM, Hamel E, Wolpert-Defillippes MK 1981 Binding of maytansine to tubulin: Competition with other mitotic inhibitors. Res Commun Chem Pathol Pharmacol 31: 443–451

616 Lin YM, Juichi M, Wu RY, LeE KH 1985 Antitumor agents LXIX. alkaloids of *Ochrosia acuminata* Planta Med 1985: 545–546

617 Lin ZB, Chai BL, Wang P, Guo QS et al. 1980 Studies on the antiinflammatory effect of chemical principles of zi-cao (*Arnebia euchroma* (Royle) Johnst). Pei-Ching I Hsueh Yuan Hsueh Pao 12: 101–106

618 Lindgren BO, Svahn CM 1968 Extractives of elm wood. Phytochemistry 7: 1407–1408

619 Lindgren BO, Svahn CM 1968 Extractives of linden wood (*Tilia vulgaris* Hayne). Phytochemistry 7: 669

620 Ling HC, King ML, Su MH, Chen GL et al. 1981 Study on antitumor plant *Gymnosporia trilocularis*. J Chin Chem Soc (Taipei) 28: 95–101

621 Liu GQ, Wang QJ, Yang HJ, Liao ZH 1983 Pharmacological study of gallic acid from *Ampelopsis brevipedunculata*. Nan-Ching Yao Hsueh Yuan Hsueh Pao pp. 43–47

622 Liu GS, Chen BZ, Song WZ, Xiao PG 1978 Comprehensive application of *Berberis* Poiret. II. Therapeutical value of berbamine and its content in 22 species of *Berberis*. Chih Wu Hsueh Pao 20: 255–259

623 Liu XJ, Yao MH, Fang TH, Song YY et al. 1982 Some cardiovascular effects of cyclovirobuxine D. Chung-Kuo Yao Li Hsueh Pao 3: 101–104

624 Liu YL, Xie FZ 1979 Alkaloids of *Physochlaina physaloides* G.Don Yao Huseh Huseh Pao 14: 497–501

625 Lloyd JA 1975 Extractives of New Zealand *Nothofagus* species. 1. petroleum ether solubles. N Z J Sci 18: 221

626 Lokhandwala MF, Sabouni MH, Jandhyala BS 1985 Cardiovascular actions of an experimental antitumor agent, homoharringtonine, in anesthetized dogs. Drug Dev Res 5: 157–163

627 Lomax NR, Narayanan VL 1981 Chemical structures of interest to the Division of Cancer Treatment. Compounds in development. Drugs with clinical activity. National Cancer Institute, Bethesda MD 31pp

628 Loub WD, Farnsworth NR, Soejarto DD, Quinn ML 1985 NAPRALERT: Computer-handling of natural product research data. J Chem Inf Computer Sci 25: 99–103

629 Lown JW, Chen HH 1980 Studies related to antitumor antibiotics. XIX. Studies on the effects of the antitumor agent camptothecin and derivatives on deoxy-ribonucleic acid. M Biochem Pharmacol 29: 905–915

630 Lown JW, Chen HH, Plambeck JA 1981 Studies related to antitumor antibiotics part XXI. Studies on the antitumor agent camptothecin and derivatives on DNA. Camptothecin potentiated cleavage of DNA by bleomycin in vitro. Chem Biol Interactions 35: 55–70

631 Lu IC, Hsing CP, Wei SL 1981 Studies on chemical constituents of *Rhododendron capitatum* Maxim. I. Yao Hsueh T'Ung Pao 16: 54

632 Lu ST, Su TL, Duh CY 1979 Studies on the alkaloids of Formosan Lauraceous Plants. XXIV. Alkaloids of *Litsea kawakamii* Hayata and *Litsea akoensis* Hayata. J Taiwan Pharm Assoc. 31: 23–27

633 Lu ST, Wu YC, Leou SP 1985 Alkaloids Of Formosan *Fissistigma* and *Goniothalamus* species. Phytochemistry 24: 1829–1834

634 Lu YC 1980 Studies on the constituents of the essential oil of *Rhododendron tsinghaiense* Ching. Hua Hsueh Hsueh Pao 38: 241–249

635 Luanratana O, Griffin WJ 1982 Alkaloids of *Duboisia hopwoodii*. Phytochemistry 21: 449–451

636 Luchnik NV 1960 Radiation injuries and protection. 4. Influence of various substances given to mice on the effect of irradiation. Tr Inst Biol Akad Nauk SSSR Ural'Sk Filial 46–75

637 Lui JHC, Staba EJ 1980 The ginsenosides of various ginseng plants and selected products. J Nat Prod 43: 340–346

638 Lukacova V, Polonsky J, Moretti C, Pettit GR et al. 1982 Isolation and structure of 14,15-β-epoxyprieurianin from the South American tree *Guarea guidona*. J Nat Prod 45: 288–294

639 Lutomski J, Malek B 1975 Pharmacological investigations on raw materials of the genus *Passiflora*. 4. The comparison of contents of alkaloids in some harman raw materials. Planta Med 27: 381–384

640 Lyapunova PM, Borisyuk YG 1961 Investigation of *Vinca minor*. I. Alkaloidal constituents. Farm Zh(Kiev) 16: 42–47

641 Ma GE, Sun GQ, Elsohly MA, Turner CE 1982 Studies on the alkaloids of *Cephalotaxus*. 3. 4-Hydroxycephalotaxine, a new alkaloid from *Cephalotaxus fortunei*. J Nat Prod 45: 585–589

642 Ma YQ, Guo MR, Zhu TP, Ma GE et al. 1984 The estimation of harringtonine and homoharringtonine of antileukemic alkaloids in *Cephalotaxus fortunei* Hook F. and *C. sinensis* (Rehd. et Wils.) Li. Chih Wu Hsueh Pao 26: 405–410

643 Mack DO, Curtis TA, Johnson BC 1980 Stimulation of vitamin K-1 protein carboxylation by several 1,4-naphthoquinones. In: Suttie JW (ed) Vitamin K Metab Vitamin K-dependent Proteins (Proc Steenbock Symp) University Park Press, Baltimore, 467–470

644 MaCrae WD, Towers GHN 1984 An ethnopharmacological examination of *Virola elongata* bark: A South American arrow poison. J Ethnopharmacol 12: 75–92

645 Mahajan VM, Sharma A, Rattan A 1982 Antimycotic activity of berberine sulfate: An alkaloid from an Indian medicinal herb. Sabouraudia 20: 79–81

646 Maiti M, Chaudhuri K 1981 Interaction of berberine chloride with naturally occurring deoxyribonucleic acids. Indian J Biochem Biophys 18: 245–250
647 Maiti M, Nandi R, Chaudhuri K 1984 Interaction of sanguinarine with natural and synthetic deoxyribonucleic acids. Indian J Biochem Biophys 21: 158–165
648 Majumder P, Saha S 1978 1,4-Bis(2'-hydroxy-5'-methylphenyl)- butan-1,4-dione, a biogenetically rare type of phenolic of *Berberis coriaria*. Phytochemistry 17: 1439–1440
649 Malan E, Roux DG 1975 Flavonoids and tannins of *Acacia* species. Phytochemistry 14: 1835–1841
650 Maldoni BE 1984 Alkaloids of two species of *Lycium*. Rev Latinoamer Quim 15: 83
651 Malhotra S, Misra K 1982 A new anthraquinone from *Cassia sophera* heartwood. Planta Med 46: 247–249
652 Malik A, Afza N 1983 Reserpic acid, gallic acid, and flavonoids from *Rauvolfia vomitoria*. J Nat Prod 46: 939
653 Mallavarapu GR, Muralikrishna E 1983 Maslinic lactone from the heartwood of *Terminalia alata*. J Nat Prod 46: 930–931
654 Malterud KE, Bremnes TE, Faegri A, Moe T et al. 1985 Flavonoids from the wood of *Salix caprea* as inhibitors of wood-destroying fungi. J Nat Prod 48: 559–563
655 Manabe Y, Manabe A 1981 Abortion during mid-pregnancy by rivanol-catheter supplemented with 2-α drip infusion or quinine hydrochloride. Contraception 23: 621–628
656 Manners GD 1983 The hydroquinone terpenoids of *Cordia elaeagnoides*. J Chem Soc Perkin Trans I 39–43
657 Manners GD, Jurd L 1976 A new naphthaquinone from *Tabebuia guayacan*. Phytochemistry 15: 225–226
658 Manners GD, Jurd L, Stevens KL 1974 Minor phenolic constituents of *Dalbergia retusa*. Phytochemistry 13: 292–293
659 Manning TDR, Hemmingson JA 1975 Bark and oleoresin monoterpene hydrocarbons of *Pinus contorta* grown in New Zealand. N Z J Sci 18: 115
660 Mansurov MM, Mansurova UM 1983 Action of elecampane and helenin on the retraction properties of blood and the coagulation of oxalate-treated plasma in vitro. Med Zh Uzb 51–54
661 Manzoor-I-Khuda M, Habermehl G 1979 Chemical constituents of *Corchorus capsularis* and *C. olitorius* (jute plant). Part IV. Isolation of corosolic acid, ursolic acid and oxo-corosin and correlation of corosin with tormentic acid. Z Naturforsch Ser B 34B: 1320–1325
662 Maranduba A, DeOliveira AB, DeOliveira GG, Depreis JE et al. 1979 Isoflavonoids from *Myroxylon peruiferum*. Phytochemistry 18: 815–817
663 Marcelle GB, Cordell GA, Farnsworth NR, Fonnegra R 1982 Preliminary investigation of *Ectopopterys soejartoi*. Planta Med 46: 190–191
664 Marini-Bettolo GB, Paoletti A, Nicoletti M 1981 Determination of components of exotic wood powders. Ann 1st Super Sanita 17: 189–198
665 Marion L 1939 The occurrence of L-nicotine in *Asclepias syriaca*. Can J Res 17B: 21–22
666 Marston A, Hecker E 1984 Isolation and biological activities of diterpene anthraniloyl and 3-hydroxyanthraniloyl peptide esters from *Euphorbia milii*. Progress In Tryptophan And Serotonin Research. Walter de Gruyter and Co., Berlin p. 789–792
667 Marston A, Msonthi JD, Hostettmann K 1984 Naphthoquinones of *Diospyros usambarensis*: Their molluscicidal and fungicidal activities. Planta Med 50: 279–280
668 Martinezv JC, Maia JGS, Yoshida M, Gottlieb OR 1980 Chemistry of Brazilian Lauraceae. 55. Neolignans from *Aniba* species. Phytochemistry 19: 474–476
669 Martins T, Valle JR, Porto A 1939 Endocrine control of motility of the male accessory sex organs studied in vitro on the seminal vesicles and *vas deferens* of normal and castrated Rhesus monkeys with and without hormonal treatment. Pfluegers Arch Gesamte Physiol Menschen Tiere 242: 155–167
670 Marvola M, Koponen A, Hiltunen R, Hieltala P 1981 The effect of raw material purity on the acute toxicity and laxative effect of sennosides. J Pharm Pharmacol 33: 108–109
671 Matos FJA, Gottlieb OR, Andrade CHS 1975 Flavonoids from *Dalbergia ecastophyllum*. Phytochemistry 14: 825–826
672 Mazur M, Polakowski P, Szadowska A 1966 Pharmacology of lupanine and 13-hydroxylupanine. Acta Physiol Polon 17: 299–300

673 McClure TJ, Mulley RC 1984 Some cytotoxic effects of mixtures of simplexin and huratoxin obtained from the desert rice flower, *Pimelea simplex*. Toxicon 22: 775–782
674 McGraw BA, Woolley JG 1982 Biosynthesis of tropane ester alkaloids in *Datura*. Phytochemistry 21: 2653–2657.
675 McPherson DD, Pezzuto JM 1983 Interaction of 9-aminoacridine, ethidum bromide and harman with DNA characterized by size exclusion high-performance liquid chromatography. J Chromatogr 281: 348–354
676 Mehta RC 1979 Inhibition of embryonic development in *Earias vittella* by terpenoids. Naturwissenschaften 66: 56–57
677 Meltzer HY, Simonovic M, Gudelsky GA 1983 Effect of yohimbine on rat prolactin secretion. J Pharmacol Exp Ther 224: 21–27
678 Mercier PF, Delphaut J, Blache P 1938 Comparison of the pharmacodynamic actions of papaverine, cryptopine and berberine. CR Soc Biol Marseilles 127: 1022–1028
679 Meyer WE, Coppola JA, Goldman L 1973 Alkaloid studies VIII: Isolation and characterization of alkaloids of *Tabernaemontana heyneana* and antifertility properties of coronaridine. J Pharm Sci 62: 1199–1201
680 Michael M, White DE 1955 Chemistry of western Australian plants. IX. Extractives from the timber of *Eucalyptus diversicolor*. Aust J Appl Sci 6: 359–364
681 Miller RW 1980 A brief survey of *Taxus* alkaloids and other taxane derivatives. (Review). J Nat Prod 43: 425–437
682 Miller RW, Powell RG, Smith CR Jr., Arnold E et al. 1981 Antileukemic alkaloids from *Taxus wallichiana* Zucc. J Org Chem 46: 1469–1474
683 Mirand C, Delaude C, Levy J, Lemen-Olivier J et al. 1979 Alkaloids of *Strychnos aculeata*. Plant Med Phytother 13: 84–86
684 Mironova VN, Kalashnikova LA 1982 Hypolipemic effect of β-sitosterol β-D-glucoside in rats. Farmakol Toksikol (Moscow) 45: 45–47
685 Mirov NT 1958 Composition of gum turpentine of pines. XXX. A report on *Pinus serotina*, *Pinus tenuifolia* and *Pinus yunnanensis*. J Am Pharm Assoc. Sci Ed 47: 410–413
686 Mirov NT, Frank E, Zavarin E 1965 Chemical composition of *P. elliottii* var *ellottii* turpentine and its possible relation to taxonomy of several pine species. Phytochemistry 4: 563–568
687 Mirvish SS, Salmasi S, Lawson TA, Pour P et al. 1985 Test of catechol, tannic acid, *Bidens pilosa*, croton oil, and phorbol for cocarcinogenesis of esophageal tumors induced in rats by methyl-N-amylnitrosamine. J Nat Cancer Inst 74: 1283–1289
688 Mishra CS, Misra K 1980 Chemical constituents of *Psidium guava* heartwood. J Indian Chem Soc 38: 201–202
689 Misra AN, Tiwari HP 1971 Constituents of roots of *Boerhaavia diffusa*. Phytochemistry 10: 3318–3319
690 Misra G, Mitra CR 1968 *Mimusops hexandra*—3. Constituents of root, leaves and mesocarp. Phytochemistry 7: 2173–2176
691 Mitaku S, Skaltsounis A-L, Tillequin F, Koch M et al. 1985 Plants of New Caledonia, XCVI. Alkaloids of *Geijera balansae*. J Nat Prod 48: 772–777
692 Mitchell JA, Hammer RE 1985 Effects of nicotine on oviductal blood flow and embryo development in the rat. J Reprod Fertil 74: 71–76
693 Mitscher LA, Ward JA, Drake S, Rao GS 1984 Antimicrobial agents from higher plants. Erycristagalin, a new pterocarpene from the roots of the Bolivian coral tree, *Erythrina cristagalli*. Heterocycles 22: 1673–1675
694 Miyase T, Fukushima S, Akiyama Y 1984 Studies on the constituents of *Hedysarum polybotrys* Hand.-Mazz. Chem Pharm Bull 32: 3267–3270
695 Miyase T, Ueno A, Noro T, Fukushima S 1980 Studies on the constituents of *Lespedeza cyrtobotrya* Miq. I. The structures of a new chalcone and two new isofla-3-ens. Chem Pharm Bull 28: 1172–1177
696 Miyazawa M, Maruyama H, Kameoka H 1984 Essential oil constituents of Paeoniae radix *Paeonia lactiflora* Pall. (*P. albiflora* Pall.). Agr Biol Chem 48: 2847–2849
697 Mollan TRM 1961 The essential oils of the sassafras laurels. III. The safrole-cinnamic aldehyde type of sassafras laurel. Perfum Essent Oil Rec 52: 411
698 Molokhova LG, Suslina ML, Datskovskii SB, Figurkin BA 1973 Antiinflammatory effect of fumitory alkaloids. Tr Permsk Gos Med Inst 118: 26–28

699 Molyneux RJ, Johnson AE, Boitman JN, Benson ME 1979 Chemistry of toxic range plants. Determination of pyrrolizidine alkaloid content and composition in *Senecio* species by nuclear magnetic resonance spectroscopy. J Agr Food Chem 27: 494

700 Moody JO, Sofowora EA 1984 Leaf alkaloids of *Zanthoxylum rubescens*. Planta Med 50: 101–103

701 Morimoto S, Nonaka G-I, Nishioka I, Ezaki N et al. 1985 Tannins and related compounds. XXIX. Seven new methyl derivatives of flavan-3-ols and a 1,3-diarylpropan-2-ol from *Cinnamomum cassia*, and *C. obtusifolium* Chem Pharm Bull 33: 2281–2286

702 Movsumov IS, Aliev AM, Kondratenko ES, Abubakirov NK 1975 Triterpene glycosides of *Cephalaria kotschyi* and *C. nachiczevanica*. Khim Prir Soedin 11: 519

703 Mu QZ, Li QX 1982 The isolation and identification of morin and morin-calcium chelate compound from *Artocarpus pithecogallus* C.Y.Wu and *Artocarpus heterophyllus* Lan. Chih Wu Hsueh Pao 24: 147–153

704 Mukherjee KS, Bhattacharya MK, Ghosh PK 1983 Phytochemical investigation of *Aesculus indica* Linn. and *Fragaria indica* Andr. J Indian Chem Soc 60: 507–508

705 Mulder-Krieger T, VerPoorte R, DeGraaf YP, VanderKreek M 1982 The effects of plant growth regulators and culture conditions on the growth and the alkaloid content of callus cultures of *Cinchona pubescens*. Planta Med 46: 15–18

706 Mulder-Krieger TH, VerPoorte R, DeWater A, VanGessel M et al. 1982 Identification of the alkaloids and anthraquinones in *Cinchona ledgeriana* callus cultures. Planta Med 46: 19–24

707 Murakami T, Nishikawa Y, Ando T 1960 Studies on the constituents of Japanese and Chinese crude drugs. IV. On the constituents of *Pueraria* root.(2). Chem Pharm Bull 8: 688–691

708 Murav'Ev MA, Vasilenko YK, Chyok KD, Lisevitskaya LI et al. 1983 Some evidence for the creative properties of Vietnamese plant *Lycium chinensis* Mill. Farmatsiya(Moscow) 32: 17–19

709 Murti VVS, Seshadri TR, Sivakumaran S 1972 Cudraniaxanthone and butyrospermol acetate from the roots of *Cudrania javanensis*. Phytochemistry 11: 2089–2092

710 Myers HB 1926 Comparative fungicidal action of certain volatile oils. J Pharmacol Exp Ther 27: 248

711 Nag TS, Mathur GS, Goval SC 1979 Phytochemical studies of *Tribulus alatus*, *T. terrestris* and *Agave wightii*: Contents of primary and secondary products. Comp Physiol Ecol 4: 157–160

712 Nagai Y, Tanaka O, Shibata S 1971 Chemical studies on the Oriental plant drugs. XXIV. Structure of ginsenoside Rg-1, a neutral saponin of ginseng root. Tetra Lett 881–892

713 Nagarajan GR, Parmar VS 1977 Chemical examination of the heartwood of *Prunus domestica*. Planta Med 31: 146–150

714 Nahrstedt A, Eilert U, Wolters B, Wray V 1981 Rutacridone-epoxide, a new acridone alkaloid from *Ruta graveolens*. Z Naturforsch Ser C 36: 200–203

715 Nakadate T, Yamamoto S, Aizu E, Kato R 1984 Effects of flavonoids and antioxidants on 12-O-tetradecanoyl-phorbol-13-acetate-caused epidermal ornithine decarboxylase induction and tumor promotion in relation to lipoxygenase inhibition by these compounds. Gann 75: 214–222

716 Nakajima S, Kawazu K 1980 Coumarin and euponin, two inhibitors for insect development from leaves of *Eupatorium japonicum*. Agr Biol Chem 44: 2893–2899

717 Nakamoto H, Miyamura S, Inada K, Nakamura N 1975 The study of the aqueous extract of Puerariae radix. I. The preparation and the components of the active extract. Yakugaku Zasshi 95: 1123–1127

718 Nakanishi K 1982 Recent studies on bioactive compounds from plants. J Nat Prod 45: 15–26

719 Nakashima T, Moreira EA, Brasilesilva GAA, Miguel OG 1982 Study of hydroxyanthraqinones in *Rhamnus sectipetala* Mart. Rhamnaceae. Trib Farm 37–60

720 Nakatani N, Inatani R 1981 Structure of rosmanol, a new antioxidant from rosemary (*Rosmarinus officinalis* L.). Agr Biol Chem 45: 2385–2386

721 Namba T, Hattori M, Tsunezuka M, Yamagishi T et al. 1982 Studies on dental caries prevention by traditional Chinese medicines. 3. In vitro susceptibility of a variety of bacteria to magnolol and honokiol. Shoyakugaku Zasshi 36: 222–227

722 Narasimhachari N, VonRudloff E 1973 Lyoniside and aucuparins from wood of North American *Sorbus* species. Phytochemistry 12: 2551–2552

723 Natl Med J China 1975 Pharmacologic effects of anisodamine Natl Med J China 55: 133–138

724 Natl Med J China 1979 The use of vasoactive drugs in the treatment of late experimental hemorrhagic shock Natl Med J China 59: 674–676

725 Nawwar MAM, Buddrus J, Bauer H 1982 Dimeric phenolic constituents from the roots of *Tamarix nilotica*. Phytochemistry 21: 1755–1758
726 Nazimuddin SK, Gopalakrishnan C, Shankarnarayan D, Nazeemuni 1982 Anti-inflammatory and CNS depressant activities of xanthones from *Calophyllum trapezifolium*. Bull Islamic Med 2: 500–507
727 Nelson L, Young MJ, Gardner ME 1980 Sperm motility and calcium transport: A neurochemically controlled process. Life Sci 26: 1739–1749
728 Nesmelova EF, Bessonova IA, Yunusov SY 1978 Alkaloids of *Haplophyllum latifolium*. The structure of haplatine. Khim Prir Soedin 14: 758–764
729 Netien G, Combet J 1971 Phenolic acids present in *in vitro* plant tissue cultures compared to the original plant. C R Acad Sci Ser D 272: 2491–2494
730 Nielsen C, McCausland H, Spruth HC 1927 The occurrence and alkaloidal content of various *Ephedra* species. J Am Pharm Ass 16: 288–294
731 Niikura H, Takizawa Y, Kim K, Hagiwara S et al. 1983 Effect of phyto-steryl-glucoside (glm-3401) on the function of human platelets. Ketsueki To Myakkan 14: 189–191
732 Nikaido T, Ohmoto T, Sankawa U, Tanaka O et al. 1984 Inhibitors of cyclic AMP phosphodiesterase in *Panax ginseng* C.A. Meyer and *Panax japonicus* C.A.Meyer. Chem Pharm Bull 32: 1477–1483
733 Niklas KJ, Giannasi DE 1977 Flavonoids and other chemical constituents of fossil Miocene *Zelkova*. Science 196: 877–878
734 Nishibe S, Okabe K, Hisada S 1981 Isolation of phenolic compounds and spectroscopic analysis of a new lignan from *Trachelospermum asiaticum* var. *intermedium*. Chem Pharm Bull 29: 2078–2082
735 Nishimoto N, Inque J, Ogawa S, Takemoto T 1980 Studies on the active constituents of Geranii herba. III. The relaxative active substance on intestines. Shoyakugaku Zasshi 34: 131–137
736 Nishimoto N, Inque J, Ogawa S, Takemoto T 1980 Studies on the active constituents of Geranii herba. I. The contractile active substances on intestines. Shoyakugaku Zasshi 34: 122–126
737 Nishimoto N, Kato R, Hayashi S 1980 Simplified quantitative analysis of alkaloids in *Scopolia* extract. Yakugaku Zasshi 100: 396–401
738 Nishino H, Kitagawa K, Iwashima A 1984 Antitumor-promoting activity of glycyrrhetic acid in mouse skin tumor formation induced by 7,12-dimethylbenz[A]anthracene. Carcinogenesis 5: 1529–1530
739 Nishino H, Nagao M, Fujiki H, Sugimura T 1983 Role of flavonoids in suppressing the enhancement of phospholipid metabolism by tumor promoters. Chem Lett 21: 1–8
740 Niu XY, Wang WC, Chin YC 1980 Further study on the dualistic central action of scopolamine and its effect on avoidance conditioned response in dogs. Proc U.S.-China Pharmacology Symposium, National Academy of Sciences, Washington, DC, p. 235–252
741 Nomura M, Tokoroyama T, Kubota T 1975 Further phenolic components from *Alnus japonica*. J Chem Soc Chem Commun pp. 316
742 Nomura M, Tokoroyama T, Kubota T 1981 Biarylheptanoids and other constituents from wood of *Alnus japonica*. Phytochemistry 20: 1097–1104
743 Nomura T, Fukai T 1980 Kuwanon G, a new flavone derivative from the root bark of the cultivated mulberry tree (*Morus alba* L.) Chem Pharm Bull 28: 2548–2552
744 Nomura T, Fukai T, Narita T 1980 Hypotensive constituent, kuwanon H, a new flavone derivative from the root bark of the cultivated mulberry tree (*Morus alba* L.). Heterocycles 14: 1943–1951
745 Nozaki H, Suzuki H, Hirayama T, Kasai R et al. 1986 Antitumour triterpenes of *Maytenus diversifolia*. Phytochemistry 25: 479–485
746 Numata A, Hokimoto K, Takemura T, Matsunaga S et al. 1983 Plant constituents biologically active to insects. 2. Juvabione analogs from *Abies sachalinensis* Mast. Chem Pharm Bull 31: 436–442
747 Odajima Y 1981 Effects of ginseng on cancer cells. In: H. Oura (Ed). Yakuyo Ninjin:Sono Kenkyu To Shinpo, Kyoritsu, Tokyo, p. 194–209
748 Odebiyi OO 1980 Antibacterial property of tetramethylpyrazine from the stem of *Jatropha podagrica*. Planta Med 38: 144–146
749 Odebiyi OO, Sofowora EA 1979 Antimicrobial alkaloids from a Nigerian chewing stick (*Fagara zanthoxyloides*). Planta Med 36: 204–207

750 Oertli EH, Beier RC, Ivie GW, Rowe LD 1984 Linear furocoumarins and other constituents from *Thamnosma texana*. Phytochemistry 23: 439–441
751 Offner E, Duhault J 1981 Medicine containing a biflavanone compound. European Patent Appl 39,265
752 Ogiso A, Kitazawa E, Mikuriya I, Promdej C 1981 Original plant of a Thai crude drug, plau-noi. Shoyakugaku Zasshi 35: 287–290
753 Oguntimein BO, Hufford CD 1981 Biologically active constituents of *Uvaria* species. Diss Abstr Int B 42: 577
754 Ogura M, Cordell GA, Farnsworth NR 1977 Quinquangulin, a new naphthopyrone from *Cassia quinquangulata* (Leguminosae). Lloydia 40: 347–351
755 Ohigashi H, Minami S, Fukui H, Koshimizu K et al. 1982 Flavanols, as plant growth inhibitors from roots of peach, *Prunus persica* Batsch, cv. 'hakuto'. Agr Biol Chem 46: 2555–2561
756 Ohiri FC, VerPoorte R, Baerheim-Svendsen A 1982 Alkaloids from *Chasmanthera dependens*. Planta Med 46: 228–230
757 Ohkuma T, Otagiri K, Tanaka S, Ikekawa T 1983 Intensification of host's immunity by squalene in sarcoma 180-bearing ICR mice. J Pharm Dyn 6: 148–151
758 Ohminami H, Kimura Y, Okuda H, Tani T et al. 1981 Effects of ginseng saponins on the actions of adrenaline, ACTH and insulin on lipolysis and lipogenesis in adipose tissue. Planta Med 41: 351–358
759 Ohmiya S, Higashiyama K, Otomasu H, Murakoshi I et al. 1979 (+)-5,α,9α-dihydroxymatrine, a new lupin alkaloid from *Euchresta horsfeldii*. Phytochemistry 18: 645–647
760 Ohta N, Mikumo K 1982 Effect of soybean isoflavones on lipase activity of soybean sprouts. Kumamoto Joshi Daigaku Gakujutsu Kiyo 34: 73–77
761 Ohta S, Furukawa M, Shinoda M 1984 Studies on chemical protectors against radiation. XXIII. Radioprotective activities of ferulic acid and its related compounds. Yakugaku Zasshi 104: 793–797
762 Ohta T, Watanabe K, Moriya M, Shirasu Y et al. 1983 Anti-mutagenic effects of coumarin and umbelliferone on mutagenesis induced by 4-nitroquinoline 1-oxide or UV irradiation in *E. coli*. Mutat Res 117: 135–138
763 Ojewole JAO 1983 Antibronchoconstrictor and antiarrhythmic effects of chemical compounds from Nigerian medicinal plants. Fitoterapia 54: 153–161
764 Ojewole JAO 1983 Blockade of autonomic transmission by scopoletin. Fitoterapia 54: 203–211
765 Ojewole JAO 1984 Effects of scopoletin on autonomic transmissions. Int J Crude Drug Res 22: 81–93
766 Ojewole JAO, Adesina SK 1983 Cardiovascular and neuromuscular actions of scopoletin from fruit of *Tetrapleura tetraptera*. Planta Med 49: 99–102
767 Ojewole JAO, Odebiyi OO 1981 Mechanism of the hypotensive effects of tetramethylpyrazine, an amide alkaloid from the stem of *Jatropha podagrica*. Planta Med 41: 281–287
768 Ojewole JAO, Odebiyi OO 1980 Neuromuscular and cardiovascular actions of tetramethylpyrazine from the stem of *Jatropha podagrica*. Planta Med 38: 332–338
769 Okano M, Fukamiya N, Kondo K, Fujita T et al. 1982 Picrasinoside-A, a novel quassinoid glucoside from *Picrasma ailanthoides*. Chem Lett pp. 1425–1426
770 Okano M, Lee KH, Hall IH, Boettner FE 1981 Antitumor agents. 39. Bruceantinoside-A and -B novel antileukemic quassinoid glucosides from *Brucea antidysenterica*. J Nat Prod 44: 470–473
771 Okigaki T 1958 Effect of yohimbine alkaloid on the testicular tissue in rats and mice. Yamashina Chorui Kenkyusho Kenku Hokoku 12: 25–28
772 Oksuz S, Ayyidiz H, Johansson C 1984 6-Methoxylated and c-glycosyl flavonoids from *Centaurea* species. J Nat Prod 47: 902–903
773 Okuda T, Yoshida T, Hatano T, Mori K et al. 1984 Structures and activites of tannins of crude drugs (Part 3). Abstr 5th Symp Devel Appl Naturally Occurring Drug Materials. pp. 46–48
774 Ollis WD, Redman BT, Roberts RJ, Sutherland IO et al. 1978 Neoflavonoids and the cin

777 Onda M, Takiguchi K, Hirakura M, Fukushima H et al. 1965 The constituents of *Macleaya cordata*. I. Nematocidal alkaloids. Nippon Nogei Kagaku Kaishi 39: 168–170

778 Otsuka H, Fujimura H, Sawada T, Goto M 1981 Studies on anti-inflammatory agents. 2. Anti-inflammatory constituents from rhizome of *Coptis japonica* Makino. Yakugaku Zasshi 101: 883–890

779 Otsuka H, Fujioka S, Komeya T, Mizuta E et al. 1982 Studies on anti-inflammatory agents. 6. Anti-inflammatory constituents of *Cinnamomum sieboldii* Meissn. Yakugaku Zasshi 102: 162–172

780 Otsuka H, Komiya T, Fujioka S, Goto M et al. 1981 Studies on anti-inflammatory agents. IV. Anti-inflammatory constituents from roots of *Panax ginseng* C.A.Meyer. Yakugaku Zasshi 101: 1113–1117

781 Paalzow GHM, Paalzow LK 1983 Yohimbine both increases and decreases nociceptive thresholds in rats: Evaluation of the dose-response relationship. Naunyn-Schmiedebergs Arch Exp Pathol Pharmakol 322: 193–197

782 Pachaly P, Schneider C 1981 Alkaloids from *Tinospora cordifolia* Miers. Arch Pharm (Weinheim) 314: 251–256

783 Pacheco H, Brachet AM 1960 Flavonic heterosides of the wood of *Prunus mahaleb*. Compt Rend F 60: 1106–1107

784 Pacheco P, Silva M, Sammes PG, Tyler TW 1973 Triterpenoids of *Colletia spinosissima*. Phytochemistry 12: 893–897

785 Palmeri M, Dominguez XA, Franco R, Cano G 1978 Mexican medicinal plants. XXXII. Terpenoids from ether extracts of the Celastraceae, *Mortonia gregii*. Rev Latinoamer Quim 9: 33–35

786 Pan WD, Mai LT, Li YJ 1983 Studies on the variation of harringtonine content in *Cephalotaxus oliveri* with seasons, parts and ages of the tree. Chung Yao T'Ung Pao 8: 15–16

787 Pan ZK, Chang ZY, Wang YC, Lee K et al. 1980 Studies on harringtonine as a new antitumor drug. In: Marks, PA (ed). Cancer Research In The People's Republic of China and the United States of America-Epidemiology, Causation and Approaches to Therapy. Grune & Stratton, New York, pp. 261–267

788 Pan ZK, Han R, Wang YC 1980 The cytokinetic effect of harringtonine on leukemia L1210 Cells. I. Autoradiographic studies. Sheng Wu Hua Hsueh Yu Sheng Wu Wu Li Hsueh Pao 12: 13–20

789 Panas JM, Richard B, Potron C, Razafindrambao RS et al. 1975 Alkaloids of *Muntafara sessilifolia*. Phytochemistry 14: 1120–1123

790 Panda RK, Sasmal BM, Panda HP, Joseph N et al. 1978 Isolation and structure of a compound from *Poecilocerus pictus*. Indian J Biochem Biophys 15: 483–486

791 Pandey VB, Dasgupta B, Bhattacharya SK, Lal R et al. 1971 Chemistry and pharmacology of the major alkaloid of *Fumaria indica* (Haussk) Pugsley. Curr Sci 40: 455–457

792 Paoletti,C, Auclair C, Lesca P, Tocanne JF et al., 1981 Ellipticine, 9-hydroxyellipticine, and 9-hydroxyellipticinium: some biochemical properties of possible pharmacologic significance Cancer Treat Rep 65: 107–118

793 Papageorgiou VP 1978 Wound healing properties of naphthaquinone pigments from *Alkanna tinctoria*. Experientia 34: 1499–1501

794 Parfent'Eva EP, Vasilenko YK, Lisevitskaya LI, Oganesyan ET 1980 Effect of ursolic acid on some indexes of lipid metabolism in experimental atherosclerosis. Vopr Med Khim 26: 174–179

795 Paris RR, Letouzey R 1960 Alkaloid distribution in *Pausinystalia yohimbe*. J Agr Trop Bot Appl 256–258

796 Paris RR, Moyse-Mignon H 1953 A sympatholytic Apocynaceae, *Vinca rosea*. C R Acad Sci 240: 1993–1995

797 Park H, Park-Lee Q, Lee CH 1980 Saponin pattern of *Panax ginseng* root in relation to stem color. Hanguk Nonghwa Hakhoe Chi 23: 222–227

798 Park YH, Kim YS, Cho BH 1983 A study on the diuretic action of buxuletin. Taehan Yakrihak Chapchi 19: 17–24

799 Parmar NS 1984 The gastric anti-ulcer activity of naringenin, a specific histidine decarboxylase inhibitor. Int J Tissue React 5: 415–420

800 Parthasarathy PC, Radhakrishnan PV, Rathi SS, Venkataraman K 1969 Colouring matters of the wood of *Artocarpus heterophyllus*: Part V. Cycloartocarpesin and oxydihydroartocarpesin, two new flavones. Indian J Chem 7: 101–102

801 Pashinina LT, Abil'Kaeva SA, Sheichenko BI 1982 Dimeric flavanols of *Juniperus sabina*. 1. Khim Prir Soedin 18: 307–312
802 Patil AD, Deshpande VH 1982 A new dimeric proanthocyanidin from *Cassia fistula* sapwood. Indian J Chem Ser B 21: 626–628
803 Patnaik GK, Puri VN, Kar K, Dhawan BN 1976 Spasmolytic activity of sesquiterpenes from *Cedrus deodara*. I.D.M.A. Bull 7: 238–242
804 Patra A, Valencia E, Minard RD, Shamma M et al. 1984 Furoquinoline alkaloids from *Haplophyllum vulcanicum*. Heterocycles 22: 2821–2825
805 Pavanasasivam G, Sultanbawa MUS 1975 Flavonoids of some Dilleniaceae species. Phytochemistry 14: 1127–1128
806 Peeples A, Dalvi RR 1982 Toxic alkaloids and their interaction with microsomal cytochrome P-450 in vitro. J Appl Toxicol 2: 300–302
807 Pegel KH, Rogers CB 1968 Constituents of *Bridelia micrantha*. Phytochemistry 7: 655–656
808 Penfold AR, Morrison FR 1952 Some Australian essential oils in insecticides and repellants. Soap Perfum Cosmet 25: 933–934
809 Peng JZ, Chen ZX, Chen XY 1982 Effects of anisodine on EEG and behavior of conscious cats. Chung-kuo Yao Li hsueh Pao 3: 78–81
810 Peng JZ, Jin LR, Chen XY, Chen ZX 1983 Central effects of anisodamine, atropine, anisodine and scopolamine after intraventricular injection. Chung-Kuo Yao Li Hsueh Pao 4: 81–87
811 Perera P, Kanjanapothy D, Sandberg F, VerPoorte R 1985 Muscle relaxant activity and hypotensive activity of some *Tabernaemontana* alkaloids. J Ethnopharmacol 13: 165–173
812 Perera PK 1985 Phytochemical and pharmacological studies on *Tabernaemontana dichotoma*. Acta Pharm Suecica 22: 64
813 Perez I, Sierra P 1980 Alkaloids of *Tabernaemontana amblyocarpa* Urb. Rev Latinoamer Quim 11: 132
814 Perez I, Sierra P 1983 Alkaloids of *Tabernaemontana amblyocarpa* Urb roots. Rev Latinoamer Quim 14: 31–33
815 Perry BD, U'Prichard DC 1981 (3H)Rauwolscine (α-yohimbine): A specific antagonist radioligand for brain alpha-2-adrenergic receptors. Eur J Pharmacol 76: 461–464
816 Pettersson E, Runeberg J 1961 The chemistry of the natural order Cupressales. XXXIV. Heartwood constituents of *Juniperus procera* Hochst. and *Juniperus californica* Carr. Acta Chem Scand Ser A 15: 713–720
817 Pettit GR, Barton DHR, Herald CL, Polonsky J et al. 1983 Antineoplastic agents. 97. Evaluation of limonoids against the murine P388 lymphocytic leukemia cell line. J Nat Prod 46: 379–390
818 Pettit GR, Cragg GM, Suffness MI, Gust D et al. 1984 Antineoplastic agents. 104. Isolation and structure of the *Phyllanthus acuminatus* Vahl (Euphorbiaceae) glycosides. J Org Chem 49: 4258–4266
819 Petyaev SI 1938 The Camphor tree as a source of different substances. Sov Subtrop (Moscow) 63–70
820 Pfyffer GE, Pfyffer BU, Towers GHN 1982 Monoaddition of dictamnine to synthetic double-stranded polydeoxyribonucleotides in UVA and the effect of photomodified DNA on template activity. Photochem Photobiol 35: 793–797
821 Phillipson J.D., Darwish FA 1981 Bruceolides from Fijian *Brucea javanica*. Planta Med 41: 211–220
822 Phillipson JD, Hemingway SR, Ridsdale CE 1982 The chemotaxonomic significance of alkaloids in the Naucleeae s.l. (Rubiaceae). J Nat Prod 45: 145–162
823 Pierre A, Robert-Gero M, Tempete C, Polonsky J 1980 Structural requirements of quassinoids for the inhibition of cell transformation. Biochem Biophys Res Commun 93: 7675–686
824 Piovetti L, Combaut G, Diara A 1980 Monoterpenes and oxygenated sesquiterpenes from *Cupressus dupreziana*. Phytochemistry 19: 2117–2120
825 Plouvier V 1956 A new cyclitol, L-pinitol, isolated from *Artemisia dracunculus*. Compt Rend 243: 1913–1915
826 Plouvier V 1968 Fraxin and coumarin heterosides occurring in various botanical groups. C R Acad Sci Ser D 267: 1883–1885
827 Polievtsev NP, Evdokinova NI, Sultanov MB 1972 Effect of *Haplophyllum* alkaloids on animal body temperatures. Farmakol Alkaloidov Ikh Proizvad pp. 41–47

828 Polonsky J, Lederer E 1959 Note on the isolation of 2,6-dimethoxybenzoquinone from the bark and wood of some Simarubaceae and Meliaceae. Bull Soc Chim France pp. 1157–1158

829 Polonsky J, Maivantri, Varon Z, Prange T et al. 1980 Quassinoids. Isolation from *Soulamea muelleri* and structures of 1,12-di-O- acetyl soulameanone and δ-2-picrasin B. X-Ray analysis of soulameanone. Tetrahedron 36: 2983–2988

830 Polonsky J, Varenne J, Prange T, Pascard C 1980 Antileukaemic quassinoids: Structure (X-ray analysis) of bruceine C and revised structure of bruceantinol. Tetra Lett 21: 1853–1856

831 Popinigis I, Moreira EA, Nakashima T, Krambeck R et al. 1980 Pharmacognostic study of *Picramnia parvifolia* Engler. Simaroubaceae. Trib Farm 48: 24–43

832 Popp FD, Veeraraghavan S 1982 Synthesis of potential antineoplastic agents. XXVII. (Reissert compound studies. XLIV.). The ellipticine Reissert compounds as an intermediate in the syntheses of ellipticine analogs. J Heterocycl Chem 19: 1275–1280

833 Powell RG, Mikolajczak KL, Weisleder D, Smith CR Jr 1972 Alkaloids of *Cephalotaxus wilsoniana*. Phytochemistry 11: 3317–3320

834 Powell RG, Miller RW, Smith CR Jr 1979 Cephalomannine: A new antitumor alkaloid from *Cephalotaxus mannii*. J Chem Soc. Chem Commun pp. 102–104

835 Power FB, Chesnut VK 1926 Non-volatile constituents of the cotton plant. J Am Chem Soc 48: 2721–2737

836 Prakash L, Garg G 1981 Chemical constituents of the roots of *Millingtonia hortensis* Linn. and *Acacia nilotica* (Linn.) Del. J Indian Chem Soc 58: 96–97

837 Prakash L, Garg G 1981 Chemical constituents from the stem heartwood of *Guazuma tomentosa*, Kunth (DC) and stem of *Clerodendron indicum* (Linn) Kuntze. J Indian Chem Soc 58: 726–727

838 Prakash L, Singh R 1980 Chemical constituents of the stem bark and stem heartwood of *Dolichandrone crispa* Seem. Pharmazie 35: 122–123

839 Prakash L, Singh R 1981 Chemical examination of the leaves and stem heartwood of *Tabebuia pentaphylla* (Linn) Hemsl (Bignoniaceae). J Indian Chem Soc 58: 1122–1123

840 Prakash L, Singh R 1980 Chemical constituents of stem bark and root heartwood of *Tabebuia pentaphylla* (Linn) Hemsl. (Bignoniaceae). Pharmazie 35: 12

841 Pratt DE, Miller EE 1984 A flavonoid antioxidant in Spanish peanuts (*Arachis hypogaea*). J Am Oil Chem Soc 61: 1064–1067

842 Pratt R, Yuzuriha Y 1959 Antibacterial activity of the heartwood of *Haematoxylon brasiletto*. J Am Pharm Assoc Sci Ed 48: 69–72

843 Przyborowska M, Soczewinski E, Waksmundzki A, Golkiewicz W 1967 Investigations on the alkaloids from *Genista tinctoria*. Diss Pharm Pharmacol 19: 289–295

844 Psenak M, Jindra A, Kovacs P, Dulovcova H 1970 Biochemical study on *Geum urbanum*. Planta Med 19: 154–159

845 Puri RK, Bhatnagar JK, Piatak DM, Totten CE 1976 *Solanum platanifolium*: Studies on the petroleum ether extract of the berries. Trans Ill State Acad Sci 69: 114–117

846 Qian XL, Yao SH 1982 Studies on the antileukemic principle of *Maytenus guangxiensis* Cheng Et Sha. II. In: Wang Y (ed) Chemistry of Natural Products-Proc Sino-American Chem Nat Prod. Science Press, Beijing, p. 263–264

847 Qian ZW, Xiao J, Fan YL, Chen ZB 1983 The effects of scopolamine and δ-9-tetrahydrocannabinol on respiration and blood pressure in rabbits. Shang-Hai Ti I Hsueh Yuan Hsueh Pao 10: 181–186

848 Qin GW, Yang XM, Gu WH, Wang BD et al. 1982 The structures of two new triterpene lactones from *Tripterygium wilfordii*—wilforlides A And B. Hua Hsueh Hsueh Pao 40: 637–647

849 Quinton RM 1963 The increase in the toxicity of yohimbine induced by imipramine and other drugs in mice. Brit J Pharmacol 21: 51–66

850 Quirin F, Debray MM, Sigaut C, Thepenier P et al. 1975 Alkaloids Of *Pandaca eusepala*. Phytochemistry 14: 812–813

851 Racz G, Fuzi I, Kisgyorgy Z, Ilies G 1961 Arbutin content in indigenous *Vaccinium vitis-idaea*. Farmacia(Bucharest) 9: 505–511

852 Raghavrao GS, Hanumaiah T, Jagannadharao KV 1980 New anthraquinones from the roots of *Ventilago maderaspatana* Gaertn.(Rhamnaceae) Indian J Chem Ser B 19: 97–100

853 Rai PP, Shok M 1983 Anthraquinone glycosides from plant parts of *Cassia occidentalis*. Indian J Pharm Sci 45: 87–88

854 Ramarao AV, Bhide KS, Mujumdar RB 1980 Mahanimbinol. Chem Ind(London) 697–698
855 Ramiah N, Prasad NBR, Abraham K 1976 Rapanone and leucopelargonidin from the roots of *Rourea santaloides*. J Inst Chem(Calcutta) 48: 196–197
856 Randerath K, Haglund RE, Phillips DH, Reddy MV 1983 Alkenylbenzene-DNA binding in vitro: 32p-postlabeling analysis of structure-activity relationships for safrole, estragole and congeners. Proc Amer Assoc Cancer Res 75th Ann Mtg. Waverly Press, Baltimore, MD. p. 85
857 Rao CP, Vemuri VSS, Jagannadharao KV 1982 Chemical examination of roots of *Flemingia stricta* (Leguminosae). Indian J Chem Ser B 21: 167–169
858 Rao KV, Davis TL 1982 Constituents of *Magnolia grandiflora*. 3. Toxic principle of the wood. J Nat Prod 45: 283–287
859 Rao KV, McBride TJ, Oleson JJ 1967 Lapachol as an antitumor agent: recognition and evaluation. Proc Amer Assoc Cancer Res 9: 55
860 Rao MM, Gupta PS, Krishna EM, Singh PP 1979 Constituents of the heartwood of *Soymida febrifuga*: Isolation of flavonoids. Indian J Chem Ser B 17: 178–180
861 Rao MM, Kingston DGI 1982 Plant anticancer agents. XII. Isolation and structure elucidation of new cytotoxic quinones from *Tabebuia cassinoides*. J Nat Prod 45: 600–604
862 Rao MNA, Mukherjee KC, Patnaik GK, Rastogi RP 1979 Chemical and pharmacological investigation of *Parkinsonia aculeata* L. Indian Drugs 17: 43–46
863 Rao MR, Liang MD 1980 Effects of protocatechuic acid on myocardial oxygen consumption and tolerance to anoxia in animals. Chung-Kuo Yao Li Hsueh Pao 1: 95–99
864 Rao RVK, Murty N, Rao JVLNs 1978 Occurrence of borneol and camphor and a new terpene in *Annona squamosa*. Indian J Pharm Sci 40: 170–171
865 Rastogi K, Kapil RS, Popli SP 1980 New alkaloids from *Glycosmis mauritiana*. Phytochemistry 19: 945–948
866 Ratnasooriya WD 1984 Effect of atropine on fertility of female rat and sperm motility. Indian J Exp Biol 22: 463–466
867 Ray AB, Chattopadhyay S, Tripathi RM, Gambhir SS et al. 1979 Isolation and pharmacological action of epistephanine, an alkaloid of *Stephania hernandifolia*. Planta Med 35: 167–173
868 Reisch J, Muller M, Mester I 1981 Constituents from the rootbark and heartwood of *Afraegle paniculata* (Rutaceae). Planta Med 43: 285–289
869 Ren LJ, Xie FZ 1981 Alkaloids from the root of *Zanthoxylum bungeanum* Maxim. Yao Hsueh Hsueh Pao 16: 672–677
870 Reyes RE, Gonzalez AG 1968 New sources of natural coumarins. IX. Coumarins from *Ruta pinnata* branches. An Quim 64: 641–646
871 Reyes RE, Gonzalez AG 1970 New sources of natural coumarins. Part XIV. Structure of pinnarin and furopinnarin, two new coumarins from the roots of *Ruta pinnata*. Phytochemistry 9: 833–840
872 Ritchie E, Taylor WC 1961 Chemical studies of the Myrtaceae. V. Constituents of the wood of *Syncarpia laurifolia*. Aust J Chem 14: 660
873 Ritschel WA, Alcorn GJ, Ritschel-Beurlin G 1982 Antipyretic and pharmacokinetic evaluation of coumarin in the rabbit after endotoxin administration. Methods Find Exp Clin Pharmacol 4: 407–411
874 Rizk AM, Hammouda FM, Ismail SI, Ghaleb HA et al. 1984 Alkaloids from *Senecio desfontainei* Druce (=*S. coronopifolius* Desf.) Fitoterapia 54: 115–122
875 Rizvi SH, Kapil RS, Shoeb A 1985 Alkaloids and coumarins Of *Casimiroa edulis*. J Nat Prod 48: 146
876 Rizvi SH, Shoeb A, Kapil RS, Popli SP 1980 Antidesmanol-A new pentacyclic triterpenoid from *Antidesma menasu* Miq. ex. Tul. Experientia 36: 146–147
877 Robinson GM 1937 Leucoanthocyanins. III. Formation of cyanidin chloride from a constituent of the gum of *Butea frondosa*. J Chem Soc pp. 1157–1160
878 Roder E 1982 Close effects of healing plants. Dtsch Apoth Ztg 122: 2081–2092
879 Rojas NM, Cuellar A 1982 *Catharanthus roseus* G.Don. I. Microbiological study of its alkaloids. First Latinamerican and Caribbean Symp Pharmacologically Active Natural Products, Havana, Cuba,June 21–25. UNESCO, Paris. p. 194
880 Roquebert J, DeMichel P 1984 Inhibition of the α-1- and α-2-adrenoceptor-mediated pressor response in pithed rats by raubasine, tetrahydroalstonine and akuammigine. Eur J Pharmacol 106: 203–205

881 Rosenthal S, Harris DT, Horton J, Glick JH 1980 Phase II study of maytansine in patients with advanced lymphomas: An Eastern Cooperative Oncology Group pilot study. Cancer Treat Rep 64: 1115–1117

882 Ross SA, Minard RD, Shamma M, Fagbule MO et al. 1985 Thaliporphinemethine: A new phenanthrene alkaloid from *Illigera pentaphylla*. J Nat Prod 48: 835–836

883 Row LR, Subbarao GSR 1962 Chemistry of *Terminalia* species. Part IV. Chemical examination of *T. arjuna*: Isolation of arjunolic acid saponin and (+) leucodelphinidin. J Indian Chem Soc 39: 89–92

884 Rowe JW, Conner AH 1979 Extractives in eastern hardwoods. US for Ser Gen Tech Rep FPL 18

885 Rozsa ZS, Reisch J, Mester I, Szendrei K 1981 Acridone alkaloids in *Boenninghausenia albiflora*. Fitoterapia 52: 37–43

886 von Rudloff E 1962 The isolation of guaiol and α-cadinol from the wood oil of *Neocallitropsis araucarioides*. Chem Ind (London) pp. 743–744

887 von Rudloff E 1975 Seasonal variation in the terpenes of the foliage of black spruce. Phytochemistry 14: 1695–1699

888 Rusakova SV, Glyzin VI, Kocherga SI, Anan'Eva AA et al. 1982 Mangiferin. French Patent Demande-2,486,941

889 Rwangabo PC, Dommisse R, Esmans E, Vlietinck A 1981 Isolation and identification of pyrogallol from the stems of *Rubus rigidus* Sm.(Rosaceae). Plant Med Phytother 15: 230–233

890 Ryan DE, Thomas PE, Levin W 1980 Properties of liver microsomal cytochrome P-450 from rats treated with isosafrole. In: Coon, MJ (ed) Microsomes Drug Oxid Chem Carcinog(Int Symp Microsomes Oxid) Academic Press, New York, p. 167–170

891 Rybicka H 1981 Zeatin and zeatin riboside in root exudate of *Spiraea arguta*. Acta Physiol Plant 3: 51–53

892 Sabir M 1981 Pharmacological evaluation of antiheparin and antitrachoma actions of *Berberis aristata*. Bull Islamic Med 1: 431–438

893 Sack RB, Froehlich JL 1982 Berberine inhibits intestinal secretory response of *Vibrio cholerae* And *Escherichia coli* enterotoxins. Infect Immun 35: 471–475

894 Sadritdinov FS 1980 Pharmacological properties of alkaloids from a *Berberis integerrima* Bge. plant. Med Zh Uzb pp. 54–55

895 Sadykov AS, Karimdzhanov AK, Ismailov AI, Islambekov SY 1968 Plant catechins. Fenol'Nye Soedin Ikh Biol Funkts, Mater Vses Simp, 208–212

896 Saeed SA, Butt NM, McDonald-Gibson WJ 1981 The effect of ellagic acid and tannic acid on prostaglandin synthase activity in bovine seminal-vesicle homogenates. Biochem Soc Trans 9: 443

897 Saito H 1980 Ginsenoside RBL and nerve growth factor. Proc Third International Ginseng Symp. pp. 181–185

898 Saito H 1983 Pharmacological studies of ginseng and crude drugs. Jap J Pharmacol Suppl 33: 9pp

899 Sakata K, Kawazu K, Mitsui T 1971 Studies on a piscicidal constituent of *Hura crepitans*. Part I. Isolation and characterization of huratoxin and its piscicidal activity. Agr Biol Chem 35: 1084–1091

900 Saleh NAM, El-Ansari MAI 1975 Polyphenolics of 20 local varieties of *Mangifera indica*. Planta Med 28: 124–130

901 Salem FS 1980 Evaluation of clove oil and some of its derivatives as trichomonacidal agents. J Drug Res(Egypt) 12: 115–119

902 Salgues R 1962 Botanical, chemical, and toxicological study of various species of *Atriplex*. Qual Plant Mater Veget 9: 71–102

903 Sanada S, Shoji J 1978 Comparative studies on the saponins of ginseng and related crude drugs. Shoyakugaku Zasshi 32: 96–99

904 Sandberg F 1958 The alkaloids of *Retama raetam* (*Genista raetam*). Pharm Weekblad 93: 8–10

905 Sankaram AVB, Sidhu GS 1964 Extractives from bark and wood of *Diospyros melanoxylon*. Indian J Chem 2: 467–468

906 Sankawa U 1980 Screening of bioactive compounds in Oriental medicinal drugs. Korean J Pharmacog 11: 125–132

907 Sankawa U, Ebizuka Y, Miyazaki T, Isomura Y et al. 1977 Antitumor activity of shikonin and its derivatives. Chem Pharm Bull 25: 2392–2395

908 Sano M, Terada M, Akira, Ishii I et al. 1981 Comparative pharmacology in parasitic helminths and host organs (1). Neuropharmacological action of alkaloids from *Stemona japonica* and *Sophora flavescens*. Jap J Pharmacol Suppl 31: 80P

909 Sasamori H, Reddy KS, Kirkup MP, Shabanowitz J et al. 1983 New cytotoxic principles from *Datisca glomerata*. J Chem Soc Perkin Trans I pp. 1333–1347

910 Sastry KV, Rao EV 1983 Isolation and structure of cleistanthoside A. Planta Med 47: 227–229

911 Saxena NK, Bhakuni DS 1980 Quaternary alkaloids of *Tiliacora racemosa* Colebr. J Indian Chem Soc 57: 773–774

912 Schimmer O 1984 Protective effects of some coumarin derivatives on the cytotoxic and mutagenic activity of 5-methoxypsoralen and UV-A irradiation in *Chlamydomonas reinhardii*. Planta Med 51: 316–319

913 Schulte KE, Rucker G, Matern HU 1979 Some constituents of the fruits and root of *Melia azedarach*. Planta Med 35: 76–83

914 Schun Y, Cordell GA 1985 Studies in the Thymelaeaceae III. Constituents of *Gyrinops walla*. J Nat Prod 48: 684–685

915 Schwenk E 1962 Tumor action of some quinonoid compounds in the cheek-pouch test. Arzneim-Forsch 12: 1145–1149

916 Seabra AP, Guimaraes EC, Mors WB 1967 Gas-liquid chromatography of "puzuri" essential oils. An Asoc Bras Quim 26: 73

917 Seeler AO, Dusenbery E, Malanga C 1943 The comparative activity of quinine, quinidine, cinchonine, cinchonidine and quinoidine against *Plasmodium lophurae* infections of Pekin ducklings. J Pharmacol Exp Ther 78: 159–163

918 Segelman AB, Farnsworth NR 1974 *Catharanthus* alkaloids. XXXI. Isolation of ajmalicine, pericalline, tetrahydroalstonine, vindolinine and ursolic acid from *Catharanthus trichophyllus* roots. J Pharm Sci 63: 1419–1422

919 Sekiya K, Okuda H, Arichi S 1982 Selective inhibition of platelet lipoxygenase by esculetin. Biochim Biophys Acta 713: 68–72

920 Senter P, Zavarin E, Zedler PH 1975 Volatile constituents of *Cupressus stephensonii* heartwood. Phytochemistry 14: 2233–2235

921 Seshadri TR, Vedantham TNC 1981 Betulaceae. Chemical examination of the barks and heartwoods of *Betula* species of American origin. Phytochemistry 10: 897–898

922 Seshadri TR, Vydeeswaran S 1971 Chemical constituents of *Daemia extensa* (roots). Curr Sci 40: 594–595

923 Sethi SC, Agarwal JS 1956 Stabilization of edible fats by spices: Part II. A new antioxidant from betel leaf. J Sci Ind Research(India) 15B: 34–36

924 Shah CS 1981 (-)Epicatechin as a cardiac drug. Indian Drugs 18: 302

925 Sharma RN, Cameron RG, Farber E, Griffin MJ et al. 1979 Multiplicity of induction patterns of rat liver microsomal mono-oxygenases and other polypeptides produced by administration of various xenobiotics. Biochem J 182: 317–327

926 Sharp PJ 1975 The effect of reserpine on plasma LH concentrations in intact and gonadectomised domestic fowl. Brit Poult Sci 16: 79–82

927 Shasha B, Leibowitz J 1961 On the oleuropein, the bitter principle of olives. J Org Chem 26: 1948–1954

928 Shen JY 1981 Cultivation of *Liquidambar styraciflua* L. and preliminary report on its application. Chung Ts'Ao Yao 12: 33–38

929 Sheriha GM, Abouamer KM 1984 Lignans of *Haplophyllum tuberculatum*. Phytochemistry 23: 151–153

930 Shi J, Miao Z, Zhang S, Zhou X et al. 1984 The protective effect of anisodamine (654–2) on the acute ischemic myocardium in anesthetized dogs. Tienchin I Yao 129: 551–553

931 Shibata S 1972 Some chemical studies on Chinese drugs. Some recent developments in the chemistry of natural products. Prentice-Hall of India, New Delhi, India, p. 1–23

932 Shibata S, Murakami T, Nishikawa Y, Harada M 1959 The constitutions of *Pueraria* Root. Chem Pharm Bull 7: 134–136

933 Shih CL 1981 Study on chemical constituents in *Acanthopanax senticosus* Harms. Yao Hsueh T'Ung Pao 16: 53

934 Shimizu M, Uchimaru F 1958 Isolation of alkaloids from *Vinca* (*Lochnera*) *rosea*. Chem Pharm Bull 6: 324

935 Shmidt EN, Khan VA, Isaeva ZA, Drebushchak TD et al. 1982 Terpenoids of the oleoresins of *Picea orientalis* and *Abies normanniana* growing in the Caucasus. Khim Prir Soedin 18: 189–194

936 Shmidt EN, Pentegova VA 1974 Chemical composition of the oleoresin of *Larix dahurica*. Khim Prir Soedin 10: 675

937 Shugart L, Kao J 1984 Effect of ellagic and caffeic acids on covalent binding of benzo[a]pyrene to epidermal DNA of mouse skin in organ culture. Int J Biochem 16: 571–573

938 Singh G, Kaur S 1981 Abscisic acid level in immature opened bolls of *Gossypium hirsutum* L. Plant Biochem J 8: 129–131

939 Singh J, Tiwari AR, Tiwari RD 1980 Anthraquinones and flavonoids of *Cassia laevigata* roots. Phytochemistry 19: 1253–1254

940 Singh P, Rao MM 1982 Optimum stage of harvest of leaflets and pods of senna, *Cassia angustifolia* Vahl, in relation to yield of crude drug and anthraquinone. Indian J Pharm Sci 44: 12–13

941 Singh P, Singh CL 1981 Chemical investigations of *Clerodendron fragrans*. J Indian Chem Soc 58: 626–627

942 Singh S, Bhagwat AW 1981 An analysis of the effect of estragole obtained from *Feronia limonia* on the smooth muscle of rat intestine. Indian J Pharmacol 13: 123

943 Siwon J, Thijs C, VerPoorte R, Baerheim-Svendsen A 1978 Studies on Indonesian medicinal plants. The alkaloids of *Fibraurea chloroleuca*. Pharm Weekbl 113: 1153–1156

944 Siwon J, VerPoorte R, VanBeek T, Meerburg H et al. 1981 Alkaloids from *Pycnarrhena longifolia*. Phytochemistry 20: 323–325

945 Siwon J, VerPoorte R, VanEssen GFA, Baerheim-Svendsen A 1980 Studies on Indonesian medicinal plants. III. The alkaloids of *Coscinium fenestratum*. Planta Med 38: 24–32

946 Skiles JW, Saa JM, Cava MP 1979 Splendidine, a new oxoaporphine alkaloid from *Abuta rufescens* Aublet. Can J Chem 57: 1642–1646

947 Slavik J, Slavikova L 1956 Alkaloids of the Papaveraceae. VIII. *Glaucium corniculatum* Curt. Chem Listy 50: 969–974

948 Smith RM 1975 Note on gum turpentine of *Pinus caribaea* from Fiji. N Z J Sci 18: 547–548

949 Snajberk K, Zavarin E 1975 Composition of turpentine from *Pinus edulis* wood oleoresin. Phytochemistry 14: 2025–2028

950 Sneden AT 1979 The development of bruceantin as a potential chemotherapeutic agent. (Review). Adv Med Oncol Res Educ Proc 12th Int Cancer Congr 5: 75–82

951 Sneden AT, Beemsterboer GL 1980 Normaytansine, a new antileukemic ansa macrolide from *Maytenus buchananii*. J Nat Prod 43: 637–640

952 Sneden AT, Sumner WC Jr, Kupchan SM 1982 Normaytancyprine, minor antileukemic ansa macrolide from *Putterlickia verrucosa*. J Nat Prod 45: 624–628

953 Soedigdo S, Manjang J, Cholies N, Soedigdo P 1980 Studies on the chemistry and pharmacology of some Indonesian medicinal plants. (Abstract). Abstr 4th Asian Symp Med Plants Spices. Mahidol Univ, Bangkok, Thailand. p. 112

954 Sofowora EA, Hardman R 1973 Steroids, phthalyl esters and hydrocarbons from *Balanites wilsoniana* stem bark. Phytochemistry 12: 403–406

955 Somanabandhu A, Wungchinda S, Wiwat C 1980 Chemical composition of *Combretum quadrangulare* Kurz.(abstract). Abstr 4th Asian Symp Med Plants Spices. Mahidol Univ, Bankgok, Thailand. p. 114

956 Sood S, Gupta BD, Banerjee SK 1978 Constituents of *Skimmia laureola*. Planta Med 34: 338–339.

957 Soto U 1929 Pharmacology of stimulants of bile secretion. Rev Soc Argentina Biol 5: 286

958 Sousa MP, Matos MEO, Machado MIL, Filho RB et al. 1984 Triterpenoids from *Guettarda angelica*. Phytochemistry 23: 2589–2592

959 Sparnins VL, Wattenberg LW 1981 Enhancement of glutathione S-transferase activity of the mouse forestomach by inhibitors of benzo[a]pyrene-induced neoplasia of the forestomach. J Nat Cancer Inst 66: 769–772

960 Spiff AI, Duah FK, Slatkin DJ, Schiff PL Jr 1982 Alkaloids of *Monodora tenuifolia*. Abstr 23rd Ann Mt Am Soc Pharmacog. American Society of Pharmacognosy, Dept of Pharmaceutical Sciences, Univ of Pittsburgh, Pittsburgh PA. p. ABSTR-26

961 Staba EJ, Chung AC 1981 Quinine and quinidine production by *Cinchona* leaf, root and unorganized cultures. Phytochemistry 20: 2495–2498
962 Stankova NV, Smirnova TA 1975 Phenocarboxylic acids of grape roots. Khim Prir Soedin 11: 508–509
963 Stegers CP, Fruhling A, Younes M 1983 Influence of dithiocarb, (dl)-catechin and silybine on halothane hepatotoxicity in the hypoxic rat model. Acta Pharmacol Toxicol 53: 125–129
964 Steinmetz EF 1960 *Caesalpinia sappan*. Acta Phytother 7: 115–117
965 Stermitz FR, Caolo MA, Swinehart JA 1980 Chemistry of *Zanthoxylum*. Part 5. Alkaloids and other constituents of *Zanthoxylum williamsii*, *Z. monophyllum* And *Z. fagara*. Phytochemistry 19: 1469–1472
966 Stich HF, Rosin MP, Bryson L 1982 Inhibition of mutagenicity of a model nitrosation reaction by naturally occurring phenolics, coffee and tea. Mutat Res 95: 119–128
967 Stich HF, Rosin MP, Wu CH, Powrie WD 1981 A comparative genotoxicity study of chlorogenic acid (3-O-caffeoylquinic acid). Mutat Res 90: 201–212
968 Strappaghetti G, Corsano S, Craveiro A, Proietti G 1982 Constituents of essential oil of *Boswellia frereana*. Phytochemistry 21: 2114–2115
969 Su JY, Hock CE, Lefer AM 1984 Beneficial effect of anisodamine in hemorrhagic shock. Naunyn-Schmiedebergs Arch Exp Pathol Pharmakol pp. 1297–1301
970 Suarez M, Bonilla J, Dediaz AMP, Achenbach H 1983 Studies of Colombian Lauraceae. 1. Dehydrodieugenols from *Nectandra polita*. Phytochemistry 22: 609–610
971 Subramanian SS 1981 (-) Epicatechin as an anti-diabetic drug. Indian Drugs 18: 259
972 Suess TR, Stermitz FR 1981 The constituents of *Castilleja rhexifolia*, *Mahonia repens* and *Oncidium cebolleta*. Diss Abstr Int B 42: 1025–1026
973 Suess TR, Stermitz FR 1981 Alkaloids of *Mahonia repens* with a brief review of previous work in the genus *Mahonia*. J Nat Prod 44: 680–687
974 Suffness M, Douros JD 1979 Drugs of plant origin. Methods Cancer Res 16: 73–126
975 Sultanov SA, Yunusov SY 1969 Alkaloids of *Dictamnus angustifolius*. Khim Prir Soedin 5: 195–196
976 Sun JY 1983 Novel active constituents of *Ephedra sinica*. Chung Ts'Ao Yao 14: 345–6-350
977 Susnik-Rybarski I, Mihelic F, Durakovic S 1983 Antioxidation properties of substances isolated from olive leaves. Hrana Ishrana 24: 11–15
978 Sutherland MD, Wells JW 1956 A re-examination of Indian and loblolly turpentines. J Org Chem 21: 1272–1275
979 Svejda J, Slavik J, Dvorak R, Adamek R 1969 Changes in L-cells after treatment with some (isochinoline) alkaloids. Scr Med(Brno) 42: 291–295
980 Svoboda GA 1969 The alkaloids of *Catharanthus roseus* G.Don (*Vinca rosea* L.) In: J.E. Gunckel (Ed.), Current Topics In Plant Science. Academic Press, New York, p. 303–335
981 Svoboda GH 1963 Alkaloids of *Vinca rosea*. (*Catharanthus roseus*) XVIII. Root alkaloids. J Pharm Sci 52: 407–408
982 Svoboda GH, Poore GA, Montfort MI 1968 Alkaloids of *Ochrosia maculata* (*Ochrosia borbonica*). Isolation of the alkaloids and study of the antitumor properties of 9-methoxyellipticine. J Pharm Sci 57: 120-X
983 Swanbeck G 1966 Interaction between deoxyribonucleic acid and some anthracene and anthraquinone derivatives. Biochim Biophys Acta 123: 630–633
984 Swarnalakshmi T, Gomathi K, Sulochana N, Baskar EA et al. 1981 anti-inflammatory activity of (-)-Epicatechin, a bioflavonoid isolated from *Anacardium occidentale* Linn. Indian J Pharm Sci 43: 205–208
985 Swinehart JA, Stermitz FR 1980 Constituents of *Zanthoxylum*. Part 4. Bishordeninyl terpene alkaloids and other constituents of *Zanthoxylum culantrillo* and *Z. coriaceum*. Phytochemistry 19: 1219–1223
986 Syrov VN, Mel'Nikova EV, Sultanov MB 1981 Effect of the phytoecdysteroid ecdysterone on the course of heliotrine-induced toxic hepatitis in rats. Dokl Akad Nauk Uzb SSR. 36–38
987 Szadowska A 1962 Pharmacology of galenic preparations and flavonoids isolated from *Helichrysum arenarium*. Acta Polon Pharm 19: 465–479
988 Tabata M, Tsukada M, Fukui H 1982 Antimicrobial activity of quinone derivatives from *Echium lycopsis* callus cultures. Planta Med 44: 234–236

989 Tabnilanidhi B, Kalyaisarasena D, 1957 Constituents of the tuberous roots of *Pueraria mirifica*. Proc Pacific Sci Congr 5. Pacific Sci Assoc, Honolulu, Hawaii. p. 41

990 Tagahara K, Suzuta Y, Kiniwa M, Unemi N 1982 Constituents of the genus *Elaeagnus* (2). On the constituents of the stems of *Elaeagnus pungens* Thunb., and the antiinflammatory effect of β-sitosterol-β-D-glucoside therein. Shoyakugaku Zasshi 36: 370

991 Takagi K, Park EH, Kato H 1980 Anti-inflammatory activities of hederagenin and crude saponin isolated from *Sapindus mukorossi* Gaertn. Chem Pharm Bull 23: 1183–1188

992 Takagi S, Yamaki M, Inoue K 1983 Antimicrobial agents from *Bletilla striata*. Phytochemistry 22: 1011–1015

993 Takahashi M, Osawa K, Banba K 1982 Components of *Acanthopanax spinosus* Miq. Annu Rep Tohoku Coll Pharm Tohoko Coll Pharm., Sendai, Japan 77–80

994 Takahashi K, Nakagawa T 1966 Constituents of medicinal plants. VIII. The stereochemistry of paulowin and isopaulowin. Chem Pharm Bull 14: 641–647

995 Takasugi M, Anetai M, Masamune T, Shirata A et al. 1980 Broussonins A and B, new phytoalexins from diseased paper mulberry. Chem Lett 339–340

996 Takasugi M, Okinaka S, Katsui N, Masamune T et al. 1985 Isolation and structure of lettucenin a, a novel guaianolide phytoalexin from *Lactuca sativa* var. *capitata* (Compositae). J Chem Soc Chem Commun 621–622

997 Takeda S, Yasua K, Endo T, Aburada M 1980 Pharmacological studies on iridoid compounds, II. Relationship between structures and choleretic actions of iridoid compounds. J Psychedelic Drugs 3: 485–492

998 Takemoto T, Ogawa S, Nishimoto N, Hirayama H et al. 1968 Constituents of Achyranthis radix. VII. The insect-molting substances in *Achyranthes* and *Cyathula* genera. Yakugaku Zasshi Suppl. 88: 1293–1297

999 Takemura Y, Ohnuma T, Chou TC, Okano T et al. 1984 Effects of harringtonine (ht) on various human leukemia-lymphoma cell lines and concentration X exposure time studies. Proc Am Assoc Cancer Res 75Th Ann Mtg. Waverly Press, Baltimore, MD. p. 343

1000 Takeya K, Itokawa H 1982 Isoflavonoids and the other constituents in callus tissues of *Pueraria lobata*. Chem Pharm Bull 30: 1496–1499

1001 Tanaka I, Ohsaki K, Takahashi K 1975 Studies on constituents of medicinal plants. XV. The constituents of the bark and wood of *Wisteria floribunda*. Yakugaku Zasshi 95: 1388–1390

1002 Tanaka Y, Odani T, Kanaya T 1974 Pharmacodynamic study of shiunko. II. Extractive condition of shikon in making shiunko. Shoyakugaku Zasshi 28: 173–178

1003 Tandon R, Jain GK, Khanna NM 1982 Ecdysterone from *Forrestia mollissima* Blume. Indian J Chem Ser B 21: 265–266

1004 Taniguchi S 1960 Anthelmintic constituents of *Melia azedarach*. Osaka Shiritsu Daigaku Igaku Zasshi 9: 445–465

1005 Taraskina KV, Chumbalov TK, Ushakova MT, Anisimova AD 1966 Leukoephdine and ephdine from *Ephedra equisetina* and the study of their P-vitamin activity. Med Prom SSSR 20: 27–29

1006 Taylor WC 1981 The constituents of *Eupomatia laurina*. Proc Fourth Asian Symp Med Plants Spices. Mahidol Univ, Bangkok, Thailand. p. 150–160

1007 Terada M, Sano M, Ishii AI, Kino H et al. 1982 Studies on chemotherapy of parasitic helminths (4). Effects of alkaloids from *Sophora flavescens* on the motility of parasitic helminths and isolated host tissues. Folia Pharmacol Japon 79: 105–111

1008 Tessier AM, Champion B, Delaveau P 1981 New anthraquinones in *Rubia cordifolia* root. Planta Med 41: 337–343

1009 Tessier AM, Delaveau P, Champion B 1980 New anthraquinones from the roots of *Rubia cordifolia*. (Abstract). Planta Med 39: 279–280

1010 Thompson RS, Jacques D, Haslam E, Tanner RJN 1972 Plant proanthocyanidins. Part 1. Introduction: The isolation, structure, and distribution in nature of plant procyanidins. J Chem Soc Perkin Trans I 1387–1399

1011 Thoms H, Dambergis C 1930 Constituents of *Dictamnus albus*. Arch Pharm(Weinheim) 268: 39–48

1012 Timmins P, Court WE 1976 Stem alkaloids of *Rauvolfia obscura*. Phytochemistry 15: 733–735

1013 Timmins P, Court WE 1976 Further alkaloids from the roots of *Rauvolfia obscura*. Planta Med 29: 283–288

1014 Tin-Wa M, Farnsworth NR, Trojanek J 1970 Biological and phytochemical evaluation of plants. VIII. Isolation of a new alkaloid from *Sanguinaria canadensis*. Lloydia 33: 267–269

1015 Tiwari KP, Minocha PK 1980 A chalcone glycoside from *Abies pindrow*. Planta 19: 2501–2503

1016 Tomita M, Mutsuo K 1965 Alkaloids of *Asimina triloba*. Yakugaku Zasshi 85: 77–82

1017 Tomita M, Takahashi T 1958 Alkaloids of Berberidaceous plants. XVIII. Alkaloids of *Caulophyllum robustum*. Yakugaku Zasshi 78: 680–681

1018 Toyota A, Ninomiya R, Kobayashi H, Kawanishi K et al. 1979 Studies of Brazilian crude drugs. 1. Muirapuama. Shoyakugaku Zasshi 33: 57–64

1019 Traganos F, Staiano-Coico L, Darzynkiewicz Z, Melamed MR 1980 Effects of ellipticine on cell survival and cell cycle progression in cultured mammalian cells. Cancer Res 40: 2390–2399

1020 Trager W, Polonsky J 1981 Antimalarial activity of quassinoids against chloroquine-resistant *Plasmodium falciparum* in vitro. Am J Trop Med Hyg 30: 531–537

1021 Tresca JP, Alais L, Polonsky J 1971 Bitter principles from *Quassia africana* (Simaroubaceae). Simalikalactones A,B,C,D and Simalikahemiacetal A. C R Acad Sci Ser C 273: 601–604

1022 Trowell OA 1960 The cytocidal action of mitotic poisons on lymphocytes in vitro. Biochem Pharmacol 5: 53–63

1023 Tsai Y, Ting CK, Nieh SL 1964 Essential oils of the family Lauraceae from Yunnan. I. Chemical constituents of the essential oils of *Cinnamomum glanduliferum* and *Cinnamomum bodinieri*. Yao Hsueh Hsueh Pao 11: 801–808

1024 Tsuruo T, Lida H, Kitatani Y, Yokota K et al. 1984 Effects of quinidine and related compounds on cytotoxicity and cellular accumulation of vincristine and adriamycin in drug-resistant tumor cells. Cancer Res 44: 4303–4307

1025 Tunmann P, Bauersfeld HJ 1975 Some constituents from the roots of *Harpagophytum procumbens*. Arch Pharm(Weinheim) 308: 655–657

1026 Turova AD, Konovalov MN, Leskov AI 1964 Berberine, an effective cholagogue. Med Prom SSSR 18: 59–60

1027 Uchizumi S 1957 Pharmacological action of berberine. Nippon Yakurigaku Zasshi 53: 63–74

1028 Ulrichova J, Walterova D, Preininger V, Slavik J et al. 1983 Isolation, chemistry and biology of alkaloids from plants of the Papaveraceae. Part 86. Inhibition of acetylcholinesterase activity by some isoquinoline alkaloids. Planta Med 48: 111–115

1029 Urzua Moll A, Torres R, Villarroel L, Fajardo V 1984 Secondary metabolites of *Berberis darwinii*. Rev Latinoamer Quim 15: 27–29

1030 Urzua Moll A 1981 Alkaloids from the wood of *Laurelia sempervirens*. Contrib Cient Tecnol(Univ Tec Estado Santiago) 11: 41–47

1031 VandenBroucke CO, Lemli JA 1980 Antispasmodic activity of *Origanum compactum*. Planta Med 38: 317–331

1032 Varshney IP, Vyas P 1982 Saponin and sapogenin contents of *Balanites roxburghii*. Int J Crude Drug Res 20: 3–7

1033 Vasil'Eva SV, Davinchenko LA 1983 Selective effect of para-aminobenzoic acid on mutagenesis: Phenotypic analysis of Arg+ revertants induced by n-nitroso-n-methylures in *Escherichia coli* K-12AB1157. Genetika (Moscow) 19: 1916–1920

1034 Vasudevan TN, Luckner M 1968 Alkaloids from *Ruta angustifolia*, *Ruta chalepensis*, *Ruta graveolens* and *Ruta montana*. Pharmazie 23: 520–521

1035 Venkataramaiah V, Prasad SV, Rao GR, Swamy PM 1980 Levels of phenolic acids in *Pterocarpus santalinus* L. Indian J Exp Biol 18: 887–889

1036 Vermel EM 1964 Search for antitumor substances of plant origin. Acta Unio Intern Contra Cancrum 20: 211–213

1037 Vichkanova SA, Adgina VV 1971 Antifungal properties of sanguinarine. Antibiotiki(Moscow) 16: 609–612

1038 Vieira PC, deAlvarenga MA, Gottlieb OR, deNazare M et al. 1980 Chemistry of Brazilian Rutaceae. Part IV. Structural confirmation of dihydrocinnamic acids from *Adiscanthus fusciflorus* by ^{13}C NMR. Phytochemistry 19: 472–473

1039 Vilarem MF, Charcosset JY, Primaux F, Gras MP et al. 1985 Differential effects of ellipticine and aza-analogue derivatives on cell cycle progression and survival of balb/C 3t3 cells released from serum starvation. Cancer Res 458: 3906–3911

1040 Ville A, Ville R, Pacheco H 1965 Incorporation of phenylpyruvate-3-C into the flavones of *Prunus mahaleb*. Compt Rend 260: 206–208

1041 Voskanyan VL, Vasilenko YK, Ponomarev VD 1983 Hypolipidemic properties of saparal and oleanolic acid. Khim Farm Zh 17: 177–180

1042 Waage SK 1984 Plant flavonoids: Isolation, characterization, and biological evaluation. Diss Abstr Int B 45: 879

1043 Wagner H, El-Sayyad SM, Seligmann O, Chari VM 1978 Chemical consituents of *Cassia siamea*. I. 2-methyl-5-acetonyl-7-hydroxy-chromone (cassiachromone). Planta Med 33: 259–261

1044 Wagner H, Nestler T, Neszmelyi A 1979 New constituents from *Picrasma excelsa*. Planta Med 36: 113–118

1045 Wagner H, Wurmbock A 1977 Chemistry, pharmacology and thin-layer chromatography of ginseng and *Eleutherococcus* drugs. Dtsch Apoth-Ztg 117: 743–748

1046 Wakhloo RL, Kaul G, Gupta OP, Atal CK 1980 Safety evaluation in human subjects of vasicine hydrochloride, a promising oxytocic and abortifacient agent. Indian J Pharmacol 12: 58A

1047 Waki I, Kyo H, Yasuda M, Kimura M 1982 Effects of a hypoglycemic component of Ginseng radix on insulin biosynthesis in normal and diabetic animals. J Pharm Dyn 5: 547–554

1048 Walia S, Mukerjee SK 1984 Ferrxanthone, a 1,3,5,6-tetraoxygenated xanthone from *Mesua ferrea*. Phytochemistry 23: 1816–1817

1049 Wall ME, Wani MC, Cook CE, Palmer KH et al. 1966 Plant antitumor agents. I. The isolation and structure of camptothecin, a novel alkaloidal leukemia and tumor inhibitor from *Camptotheca acuminata*. J Am Chem Soc 88: 3888–3890

1050 Wang CP, Kameoka H 1978 The constituents of the steam volatile oil from *Melia azedarach* var *japonica*. Part I. Nippon Nogei Kagaku Kaishi 52: 297–299

1051 Wang FVL 1950 In vitro antibacterial activity of some common Chinese herbs on *Mycobacterium tuberculosis*. Chin Med J 68: 169–172

1052 Wang MH 1981 Isolation of antitumor alkaloids from *Zanthoxylum nitidum* and structural study of its alkaloid C. Yao Hsueh T'Ung Pao 16: 48

1053 Wang MS 1980 Studies on the chemical constituents of *Daphne giraldii* Nitsche. III. Identification of syringin, a styptic principle. Chung Ts'Ao Yao 11: 389–390

1054 Wang X, Shao H, Pan S, Tang Y et al. 1982 Metabolic fate of cyclovirobuxine D in rats. Yao Hsueh T'Ung Pao 17: 198–202

1055 Wang XF, Chen JY, Wei RF, Djiang DT 1982 Isolation and characterization of the antitumor principles maytansine and maytanprine from the stems of *Maytenus confertiflora* Luo et Chen. In: Wang Y (ed). Chem Nat Prod-Proc Sino-American Chem Nat Prod. Science Press, Beijing p. 267

1056 Wang XF, Chen JY, Wei RF, Jiang DQ 1981 Studies on the antitumor constituents of *Maytenus confertiflora* Luo et Chen (Celastraceae). II. Isolation and characterization of maytansine and maytanprine from the stems. Yao Hsueh Hsueh Pao 16: 628–630

1057 Wang YZ, Zhou SL 1983 Structure-activity relationships of sesquiterpenoid antitumor agents using molecular orbital method. Yao Hsueh Hsueh Pao 18: 25–32

1058 Wang ZQ, Yang ZH, Wang XR 1982 Studies on the chemical constituents of hongteng (*Sargentodoxa cuneata*). Chung Ts'Ao Yao 13: 7–9

1059 Wang W 1985 Vasodilatory effects of four atropine-like drugs on rat footpad. Chung-kuo Yao Lik Hsueh Pao 61: 26–29

1060 Wani MC, Ronman PE, Lindley JT, Wall ME 1980 Plant antitumor agents. 18. Synthesis and biological activity of camptothecin analogues. J Med Chem 23: 554–560

1061 Warthen D, Gooden EL, Jacobson M 1969 Tumor inhibitors: Liriodenine, a cytotoxic alkaloid from *Annona glabra*. J Pharm Sci 58: 637

1062 Watanabe K, Goto Y, Hara N, Kanaoka S 1981 Neuronal factors involved in the pathogenesis of experimental ulceration and antiulcer effect of magnolol, an active component of *Magnolia* bark. Wakanyaku Shinpojumi(Kiroku) 14: 1–6

1063 Watanabe K, Watanabe H, Goto Y, Yamaguchi M et al. 1983 Pharmacological properties of magnolol and honokiol extracted from *Magnolia officinalis*: Central depressant effects. Planta Med 49: 103–105

1064 Waterhouse AL, Holden I, Casida JE 1984 9,21-Didehydroryanodine: a new principal toxic constituent of the botanical insecticide *Ryania*. J. Chem Soc Chem Commun pp. 1265–1266

1065 Wattenberg LW, Coccia JB, Lam LKT 1980 Inhibitory effects of phenolic compounds of benzo(A)pyrene-induced neoplasia. Cancer Res 400: 2820–2823

1066 Wei CH 1965 The constituents of the roots of *Rauvolfia yunnanensis*. I. Isolation of ajmalicine and reserpine. Yao Hsueh Hsueh Pao 12: 429–434

1067 Weiss I, Valencia E, Freyer AJ, Shamma M et al. 1985 Andesine: an alkaloidal naphthalenopyrone. Heterocycles 23: 301–303

1068 Weissmann G, Simatupang MH 1974 Indonesian rosin of *Pinus merkusii*. Resin with unusual properties. Farbe Lack 80: 932–936

1069 Wennstrom J, Lindhe J 1984 Clinical effectiveness of a sanguinarine mouthrinse on plaque and gingivitis. Abstr Internat Assoc Dental Res. Internat Assoc Dent Res, Washington, D.C. p. ABSTR-481

1070 Wick MM 1978 Dopamine a novel antitumor agent active against B-16 melanoma in vivo. J Invest Dermatol 71: 163–164

1071 Wijeratne DBT, Kumar V, Sultanbawa MUS, Balasubramaniam S 1982 Chemical investigation of Ceylonese plants. 46. Triterpenes from *Gymnosporia emarginata*. Phytochemistry 21: 2422–2423

1072 Willaman JJ, Schubert BG 1961 Alkaloid bearing plants and their contained alkaloids. US Dep Agr ARS Tech Bull 1234

1073 Williams RH, Martin FB, Henley ED, Swanson HE 1959 Inhibitors of insulin degradation. Metabolism 8: 99–113

1074 Wink M, Witte L, Hartmann T 1981 Quinolizidine alkaloid composition of plants and of photomixotropic cell suspension cultures of *Sarothamnus* and *Orobanche rapum-genistae*. Planta Med 43: 342–352

1075 Witek S, Krepinsky J 1966 Terpenes. CLXXVII. The composition of valerian oil (*Valeriana officinalis*). Collect Czech Chem Commun 31: 1113–1123

1076 Wolfrom ML, Bhat HB 1965 Osage orange pigments. XVII. 1,3,6,7-tetrahydroxyxanthone from the heartwood. Phytochemistry 4: 765–768

1077 Wolters B, Eilert U 1981 Antimicrobial substances in callus cultures of *Ruta graveolens*. Planta Med 43: 166–174

1078 Wong SM, Chang HM 1982 Hydroxyanthraquinones and triterpenes from *Abrus cantonensis* Hance. (Abstract). Abstr 23rd Ann Mt Am Soc Pharmacog. American Society of Pharmacognsy. Dept of Pharmaceutical Sciences, Univ of Pittsburgh, Pittsburgh, PA

1079 Woo LK, Han BH, Park DS, Lah WL 1973 Species differences of dammarane aglycones of ginsengs. Korean J Pharmacog 4: 181–184

1080 Woo WS, Choi JS, Seligmann O, Wagner H 1983 Sterol and triterpenoid glycosides from the roots of *Patrinia scabiosaefolia*. Phytochemistry 22: 1045–1047

1081 Woo WS, Lee CK, Shin KH 1982 Studies on crude drugs acting on drug metabolizing enzymes. 4. Isolation of drug metabolism modifiers from roots of *Angelica koreana*. Planta Med 45: 234–236

1082 Wood AW, Huang MT, Chang RL, Newmark HL et al. 1982 Inhibition of the mutagenicity of bay-region diol epoxides of polycyclic aromatic hydrocarbons by naturally occurring plant phenols: Exceptional activity of ellagic acid. Proc Nat Acad Sci (USA) 79: 5513–5517

1083 Woynarowski JM, Konopa J 1980 Inhibition of DNA biosynthesis in HeLa cells by cytotoxic and antitumor sesquiterpene lactones. Mol Pharmacol 19: 97–102

1084 Wu GY, Fang F, Zuo J, Han R et al. 1983 Epiharringtonine induced enhancement of harringtonine inhibition on protein and DNA synthesis in tumor cells. Chung-Kuo I Hsueh K'O Hsueh Yuan Hsueh Pao 5: 157–160

1085 Wu LS, Yang GD, Jiang JL, Dong C 1984 Effects of henbane drugs on the phagocytosis of colloidal gold-198 by liver and spleen in mice Chung-kuo Yao Li Hsueh Pao 52: 140–142

1086 Wu TS 1985 The constituents of *Skimmia reevesiana*. Abstr Internat Res Cong Nat Prod. Coll Pharm, Univ N Carolina, Chapel Hill NC. p. ABSTR-75

1087 Wu TS, Tien HJ, Yeh MY 1980 Studies on the constituents of Formosan folk medicine. VII. Constituents of the flowers of *Vanilla somai* Hayata and the roots of *Camptotheca acuminata*. Ch'Eng-Kung Ta Hsuen Hsuen Pao 15: 65–67

1088 Xiao PG, Chen BZ, Wang LW, Ho LY et al. 1980 A preliminary study of the correlation between phylogeny, chemical constituents and therapeutic effects of *Rheum* species. Yao Hsueh Hsueh Pao 15: 33–39

1089 Xiao PG, He LY 1982 *Przewalskia tangutica* - a tropane alkaloid-containing plant. Planta Med 45: 112–115

1090 Xie PS, Yang ZX 1980 Isolation and identification of a hemostatic constituent from Chinese drug Jiu bi ying, the bark of *Ilex rotunda* Thunb. Yao Hsueh Hsueh Pao 15: 303–305

1091 Xiu SB, Ma HL, Wang XS, Shi JS et al. 1982 Active constituents of Xiao xuan cao (*Hemerocallis minor*). Chung Ts'Ao Yao 13: 49–52

1092 Xu LN, Rong OY, Yin ZZ, Zhang LY et al. 1981 The effect of Dang-gui (*Angelica sinensis*) and its constituent ferulic acid on phagocytosis in mice. Yao Hsueh Hsueh Pao 16: 411–414

1093 Xu QC 1983 A distributive survey of antimalaria constituents in plants. Chung Ts'Ao Yao 14: 93–95

1094 Xu YT, Du CZ 1981 Effect of harringtonine and its allied alkaloids on DNA synthesis in mice bearing leukemias P388, L615 and normal mice. Chung-Kuo Yao Li Hsueh Pao 2: 252–256

1095 Xu YT, Du CZ, Zhang FR, Ji XJ 1981 The effect of harringtonine and its allied alkaloids on the incorporation of labeled amino acids into proteins of cells of transplantable leukemias L-615 and P-388. Yao Hsueh Hsueh Pao 16: 661–666

1096 Xu ZW, Bao SL, Zhao JJ, Wang Q 1983 Study on hemostatic constituents in *Biota orientalis* (L.) Endl. leaves. Chung Yao T'Ung Pao 8: 30–32

1097 Xue SB, Xu P, Li SW, Hu YY et al. 1984 Effects of harringtonine and homoharringtonine on the traverse of cell cycle of P-388 leukemic cells by flow cytometry. Chung-Kuo I Hsueh K'O Hsueh Yuan Hsueh Pao 6: 28–31

1098 Yamahara J, Kobayashi M, Miki K, Kozuka M et al. 1985 Cholagogic and antiulcer effect of Saussureae radix and its active components. Chem Pharm Bull 33: 1285–1288

1099 Yamahara J, Kobayashi M, Saiki T, Sawada T et al. 1983 Biologically active principles of crude drugs. Pharmacological evaluation of cholagogue substances in clove and its properties. J Pharm Dyn 6: 281–286

1100 Yamahara J, Matsuda H, Watanabe H, Sawada T et al. 1980 Biologically active principles of crude drugs. Analgesic and anti-inflammatory effects of "Keigai" (*Schizonepeta tenuifolia* Briq.). Yakugaku Zasshi 100: 713–717

1101 Yamahara J, Mibu H, Sawada T, Fujimura H et al. 1981 Biologically active principles of crude drugs. Antidiabetic principles of Corni fructus in experimental diabetes induced by streptozotocin. Yakugaku Zasshi 101: 86–90

1102 Yamahara J, Miki S, Matsuda H, Fujimura H 1986 Screening test for calcium antagonists in natural products. The active principles of *Magnolia obovata* Yakugaku Zasshi 106: 888–893

1103 Yamamoto H, Ogura N, Kobayashi M, Iso R et al. 1979 Studies on filariasis: Effects of insect hormones and theophylline on the development of filarial larvae of *Brugia pahangi* in the mosquitoes. Dokkyo J Med Sci 6: 24–30

1104 Yamasaki H, Weinstein IB, VanDuuren BL 1981 Induction of erythroleukemia cell adhesion by plant diterpene tumour promoters: A quantitative study and correlation with in vivo activities. Carcinogenesis 2: 537–543

1105 Yamazaki M, Arai A 1984 Antiulcerogenic activity of oxymatrine. Abstr 5th Symp Develop Naturally Occurring Drug Materials pp. 4–6

1106 Yang MK, Liao RH, Huang ZM, Ran CX 1983 Investigation on the alkaloids of yalian (*Coptis deltoidea*). Chung Ts'ao Yao 14: 151–153

1107 Yang ML, Shi Y 1983 Effects of anesthesia with scopolamine on the kidney of rabbits in hemorrhagic shock. An optical and electron microscopic study. Zhonghua Mazuixue Zazhi 3: 136–137

1108 Yang PF, Pratt DE 1984 Antithiamin activity of polyphenolic antioxidants. J Food Sci 49: 489–492

1108 Yang SR, Fang FD, Wu GY 1982 Effects of harringtonine on nucleotide metabolism in tumor cells. Yao Hsueh Hsueh Pao 17: 721–727

1110 Yanishevskaya EN, Pulatov BK, Abduazimov KA 1983 Lignin of the cotton plant of variety Tashkent 6. Khim Prir Soedin 19: 516–517

1111 Yao Hsueh Hsueh Pao 1980 Studies on the antiasthmatic principles of *Gnetum parvifolium*. Yao Hsueh Hsueh Pao 15: 434–436

1112 Yao T, Xiao YF, Sun XY, Tong HH et al., 1984 Effects of anisodamine on cardiovascular activities in endotoxin-shocked dogs. Chin Med J pp. 871–876

1113 Yasue M, Sakakibara J, Ina H, Yanagisawa I 1974 Studies on the constituents of *Tripetaleia paniculata*. V. On the constituents of the flowers and of the wood. Yakugaku Zasshi 94: 1634–1638

1114 Ye CQ, Fan JH, Ren ML 1983 Glass capillary gas chromatographyy of *Cephalotaxus* alkaloids. Yao Hsueh Hsueh Pao 18: 934–937

1115 Yegnanarayan R, Shrotri DS 1982 Comparison of antidiarrhoeal activity of some drugs in experimental diarrhoea. Indian J Pharmacol 14: 293–299

1116 Yin ZZ, Zhang LY, Xu LN 1980 The effect of Dang-gui (*Angelica sinensis*) and its ingredient ferulic acid on rat platelet aggregation and release of 5-Ht. Yao Hsueh Hsueh Pao 15: 321–326

1117 Yokota T, Takahashi N 1980 Cytokinins in shoots of the chestnut tree. Phytochemistry 19: 2367–2370

1118 Yoshihara K, Hirose Y 1973 Terpenes from *Aralia* species. Phytochemistry 12: 468

1119 Yoshihira K, Tezuka M, Natori S 1971 Naphthaquinone derivatives from the Ebenaceae: II. Isodiosyprin, bisisodiosyprin and mamegakinone from *Diospyros lotus* and *D. morrisiana*. Chem Pharm Bull 19: 2308

1120 Yoshikawa-Fukada M, Yoshikawa K, Notake K 1980 Effects of camptothecin on leukemia cells. Aichi Ika Daigaku Igakkai Zasshi 8: 40–47

1121 Young DA 1979 Heartwood flavonoids and the infrageneric relationships of *Rhus* (Anacardiaceae). Am J Bot 66: 502–510

1122 Yu RS 1983 Studies on the constituents of *Swertia davidii*. Yao Hsueh T'Ung Pao 18: 434–435

1123 Yu SR, Ma JW 1982 Expectorant effect of anisodamine (654). Chung-kuo I Huseh K'o Hsueh Yuan hsueh Pao 44: 258–260

1124 Yu X, Luo Z, You J, Luo H 1985 Effect of anisodamine (compound 654–2) on the cells from bronchoalveolar lavage in experimental respiratory distress syndrome. Hu-nan I Hsueh Yuan Hsueh Pao 101: 13–18

1125 Yue KL 1981 Analgesic and antispastic effects of protopine. Chung-Kuo Yao Li Hsueh Pao 2: 16–18

1126 Yurkow EJ, Lee E, Laskin DL, Gallo MA et al. 1983 Phototoxicity of psoralens in mouse and human tumor cells in culture. Pharmacologist 25: ABSTR-572

1127 Zabkiewicz JA, Allan PA 1975 Monoterpenes of young cortical tissue of *Pinus radiata*. Phytochemistry 14: 211–212

1128 Zakaria M, Jeffreys JAD, Waterman PG, Zhong SM 1984 Naphthoquinones and triterpenes from some Asian *Diospyros* species. Phytochemistry 23: 1481–1484

1129 Zaki AI, Zentmyer GA, Pettus J, Sims JJ et al. 1980 Borbonol from *Persea* spp. Chemical properties and antifungal activity against *Phytophthora cinnamomi*. Physiol Plant Pathol 16: 205–212

1130 Zavarin E, Smith LV, Bicho JG 1967 Tropolones of Cupressaceae. III. Phytochemistry 6: 1387–1394

1131 Zavarin E, Snajberk K, Fisher J 1975 Geographic variability of monoterpenes from cortex of *Abies concolor*. Biochem Syst Ecol 3: 191–203

1132 Zavarin E, Snajberk K 1975 *Pseudotsuga menziesii* chemical races of California and Oregon. Biochem Syst Ecol 2: 121–129

1133 Zee-Cheng KY, Cheng CC 1973 Synthesis of 5,6-dihydro-6-methoxy-nitidine and a practical preparation of nitidine chloride. J Heterocycl Chem 10: 85–88

1134 Zeitune MG, Bazerque PM 1983 Reserpine inhibits non-selectively the contraction of rat uterus in vitro. Commun Biol 1: 245–251

1135 Zeng GY, Xie SY, You SQ, Jin YC 1980 Effects of reserpine on the development of neuropsychogenic hypertension in dogs. Chung-Kuo K'O Hsueh 23: 796–802

1136 Zhang MY, Lan ZS, Chou J, Yang TR et al. 1980 Chemical and pharmacological studies of *Sageretia gracilis* Dunn et Spr., an antitumor plant in Yunnan (China). Yun-Nan Chih Wu Yen Chiu 2: 62–66

1137 Zhang ZY, Han PF, Liang MD, Rao MR 1980 Experimental studies on relationship of chemical structure of protocatechuic acid derivatives and cardiac oxygen consumption and coronary flow. Yao Hsueh Hsueh Pao 15: 641–647

1138 Zhao Y, Zheng JJ, Huang SY, Li XJ et al. 1981 Studies on the antimalarial activity of protopine derivatives. Yao Hsueh T'Ung Pao 16: 7–10

1139 Zheng XF, Chen LF, Bian RL 1983 Effect of α-terpineol on rabbit pulmonary artery. Zhejiang Yike Daxue Xuebao 16: 293–295, 285

1140 Zheng XZ, Wu FH 1981 Antiasthmatic effect of the active principle from Maimateng (*Gnetum parvifolium*). Chung Ts'Ao Yao 12: 30–32

1141 Zheng YF, Bao GM 1984 Effects of anisodine on the electrophysiology of cortical neurons in rats. K'o Huseh T'ung Pao pp. 1273–1276
1142 Zhou JM, Yu TC, Zhou Y, Zhou BC 1981 Active principles in Shehanqi (*Corydalis sheareri*). Chung Ts'Ao Yao 12: 3–4,11
1143 Zhou YL, Huang LY, Zhou QR, Jiang FX et al. 1981 Studies on the active principles of *Maytenus hookeri* Loes. I. Isolation and characterization of two antitumor principles-maytansine and maytanprine. Hua Hsueh Hsueh Pao 39: 427–432
1144 Zhou YL, Yang YM, Huang LY, Liu BN 1981 Studies on the active principles of *Maytenus hookeri* Loes. 2. Characterization of the antitumor principle maytanbutine by mass spectrometry. Hua Hsueh Hsueh Pao 39: 933–936
1145 Zhu B, Ahrens F 1983 Antisecretory effects of berberine with morphine, clonidine, L-phenylephrine, yohimbine or neostigmine in pig jejunum. Eur J Pharmacol 96: 11–19
1146 Zhu ZP, Xie RM, Zhou PF, Miao AR et al. 1982 Pharmacology of active components of *Thermopsis lanceolata*. Comparison of effects of (+)-sparteine and (-)-sparteine sulfate. Shan-Hsi Hsin I Yao 11: 53–54
1147 Zipper J, Bruzzone ME, Angelo S, Munoz V et al. 1982 Effect of topically applied adrenergic blockers on fertility. Int J Fertil 27: 242–245
1148 Ziyaev R, Abdusamatov A, Yunusov SY 1974 (+)-Isolaureline, a new alkaloid from *Liriodendron tulipifera*. Khim Prir Soedin 10: 685A
1149 Zutshi U, Rao PG, Soni A, Atal CK 1978 Absorption, distribution and excretion of scoparone: A potent hypotensive agent. Indian J Exp Biol 16: 836–838
1150 Zwaving JH 1974 Investigation of the qualitative and quantitative determination of anthracene derivatives in *Rheum palmatum*. I. Anthracene derivatives in the roots of *Rheum palmatum*. Pharm Weekbl 109: 1117–1125
1151 Zwelling LA, Michaels S, Kerrigan D, Pommier Y et al. 1982 Protein-associated deoxyribonucleic acid strand breaks produced in mouse leukemia L1210 cells by ellipticine and 2-methyl-0-hydroxy-ellipticinium. Biochem Pharmacol 31: 3261–3267

Chapter 11

The Future of Wood Extractives

H. L. HERGERT

11.1 Introduction

For many years visionaries have been forecasting a world in which a substantial part of the chemicals, fibers, plastics, and liquid fuels would come from the forest. Integrated factories featuring the production of cellulose for fibers and thermoplastics, terpenes for fine chemicals, phenolic polymers from bark for dispersants and adhesives, lignin pyrolysis for aromatic hydrocarbons, and so on, are readily imagined. Abundant raw material appears to be available. Bark, for example, which contains waxes, flavonoids, phenolic polymers, and a whole host of other compounds as described in preceding chapters, is estimated to be available worldwide in amounts exceeding one hundred million tons (dry basis) annually at wood products mill sites (13). Despite raw material availability and the ample technology reviewed in these pages, we are farther from the goal of effectively utilizing wood extractives than forty years ago when the wood-based chemical industry reached its peak (41).

Two examples are illustrative: Shortly after the turn of the century, one of the ten largest companies in the United States was totally dependant upon extractives for its raw material (23). The Central Leather Company, successor to the United States Leather Company, had tanneries scattered all over the state of Pennsylvania, each of which extracted tannin from the bark of the eastern hemlock (*Tsuga canadensis*). Today not only is it the only one of the ten companies no longer in existence, but there is not a single producer of native tannin in North America.

The gum naval stores industry showed a similar pattern. During 1908, the peak year of production in the United States, more than 120000 tons of spirits of turpentine and 600000 tons of rosin were produced. Due to competition from wood naval stores after World War I and sulfate turpentine and tall oil after World War II, gum naval stores production declined to less than 4000 tons in 1985. The United States government, recognizing the inevitable demise of this product, completed the liquidation of its stocks in 1972, terminated the Naval Stores Conservation Program, and closed the Agricultural Research Service's Olustee (Florida) Research Laboratory in 1973 (44).

If we were limited to these two examples, it might be concluded that the commercial future of wood-based extractive industries is doubtful. On the other hand, natural rubber production has continued as a major activity in a number of countries in spite of all the attempts to replace it with synthetic substitutes. Carbohydrate gums and mucilages continue to be collected, refined, and utilized in applications unimagined only a few years ago. Almost all of the world's tall oil and sulfate turpentine that can be conveniently captured is sold to fractionators who convert it to a myriad of saleable products and intermediates (Chap. 10.1).

In trying to forecast the industrial future of wood extractives, it is vitally important to identify reasons why some major segments have died, some have prospered and grown, and others with substantial potential did not even reach initial success in spite of substantial financial and technical backing. This should help us to focus our research on projects that are likely to bear fruit. Even more important, there is a good rationale for continuing and expanding wood extractives science quite apart from potential commercial exploitation of extractives. As pointed out in Chap. 9, wood extractives play a profound role in the pulpability of wood, the quality of pulp and paper products, the stability of lumber towards its environment, and the utilization of wood for decorative purposes. Indeed, the life and death of the forest is controlled by the trace chemical constituents that signal growth and senescence. There is, therefore, compelling reason to improve our understanding of not just the chemistry of extractives but their distribution, function, and possible manipulation through external factors. Therein lies the future direction that extractives research ought to take. To defend this thesis, elaboration will be made on the business requirements for successful extractives-based enterprises, reasons for some failed ventures, prognosis for existing extractives-based industries, and some new directions for both industrial-oriented and basic research.

11.2 Requirements for Future Wood Extractives Business Ventures

At the risk of stating the patently obvious to an industrial scientist, there are a number of requirements that must be met for any new extractives utilization project if it is to receive serious consideration by business. Since research and development budgets are usually tightly limited, management asks the project initiator to think through the probable consequences of the work even before the first major experiment is conducted. If the business venture resulting from the project meets a number of requirements, and there is a reasonable probability that the research can be successfully carried out within the time and equipment restraints of a given research facility, then, and only then, is the project likely to be funded. The wise researcher would do well, therefore, to keep the following objectives in mind when trying to gain support for a wood extractives project:

11.2.1 Low Investment Risk

In times of high interest rates, volatile currency exchange rates, and risk of continued inflation, no corporate executive is going to be interested in investing his money in a project that does not yield any better return than the purchase of securities or placing money in the bank. A current example of ignoring that principle is involved in the extensive efforts during the past decade that North American government and university researchers have devoted to wood utilization schemes for simultaneous production of furfural, benzene, ethylene and/or ethanol, and energy. These are to be generated from a waste wood chemical plant by some combination of hydrolysis, solvent pulping, fermentation, and high-temperature cleavage at a forecasted return on investment (ROI) significantly lower than the current prime interest rate. Such projects are doomed to failure when it comes time to attract a potential investor in the scheme. In the United States a 35% before tax ROI appears to be generally accepted minimum target. A 25% ROI would be the absolute minimum and would require meeting all of the other requirements listed below.

If there are objections by some to the setting of reasonable ROI targets for research projects, it should be kept in mind that virtually all forest products corporations have strong limitations on their capital. (It is assumed that the forest product companies are those who are most likely to sponsor extractives from wood research since they control the raw material base.) The extractives project will have to compete head-on with other projects in which a modest investment in an existing plant will increase production rates, improve reliability, or meet new environmental regulations. All other things being equal, utilization schemes requiring low capital will, therefore, be more attractive.

There is a particular situation in which the high ROI requirement can be set aside. If a wood-using manufacturing plant is faced with the requirement to eliminate or significantly reduce a waste stream, one of the alternatives may be utilization. Assuming that a reliable process and adequate product outlets are available, the projected ROI could even be negative as long as it was better than the nonutilization treatment method such as incineration, bioponds, etc. An example of such a situation might be a pulpmill that discharges effluent containing small quantities of resin acids and terpenes. Certain of the resin acids, even at parts-per-million concentration, are highly toxic to fish. A scheme to extract and sell these compounds from the wood before it is processed may be less expensive than investment in exotic treatment methods to remove the offending compounds from the effluent. Financially, the utilization alternative might be viewed as the lesser of two evils but from a philosophical point of view, production of a marketable item would seem to be a much more attractive way to spend time and effort than simply to destroy the unwanted waste stream.

11.2.2 Good Sales Potential

The combining of a new technology with a new or untested market represents a risk that few managements of larger corporations are willing to take unless the profit and sales potential are so large that the risk can be justified. Thus, aiming for known markets appears to be a vital necessity. The size of the market and the chances for displacement of competitive products can be usually be forecast with reasonable certainty. Unfortunately this objective has been ignored in all too much of wood extractives research during the past two or three decades. The usual project has involved a lengthy, detailed search for exotic compounds in wood or bark, followed by development of elegant schemes for isolation, and lastly, a search for possible uses of the isolated substance. With few exceptions, these projects have resulted in good science and zero utilization. It would seem to make the most sense to identify good, continuing markets, followed by identification of a potentially wood-derived product that might fit into that market rather than the other way around. For example, the production and use of non-caloric sweeteners represent a rapidly growing worldwide market. If one chooses to enter and serve this market, the whole field of wood extractives would be examined for potential candidates. Hydrochalcone or diterpene glucosides with known sweetening capacity might then be selected for possible modification and tested for product safety and cost competitiveness compared to the compounds currently approved for this end-use. This approach to the marketplace would require the wood utilization researcher to acquaint himself with the properties and economics of the selected end-use and then determine how he might isolate and modify wood extractives to supply the particular market.

11.2.3 Inexpensive Raw Material

More often than not the yield of a desirable wood extractive is quite small based on the volume of wood or bark that has to be processed. To assure reasonable manufacturing costs of the desired endproduct, the cost of the raw material must be kept low unless, of course, the final product is a relatively rare pharmaceutical that can command a high price. Since large volumes of wood and bark residues are generated at forest products manufacturing plants such as sawmills, pulpmills, etc., there is a common misconception that so-called wood wastes represent a "free" raw material. The cost of the raw material is then ignored in product cost calculations. In fact, there is no thing as free raw material. Collection and transportation of wastes to provide a steady supply at the utilization site may well be more expensive than the chemical processing costs.

One way to assure low raw material costs is to have multiple use of the material. If wood chips, sawdust, or bark is to be extracted, the residue should be useful for another purpose. Examples of this approach are: 1) Isolation of arabinogalactan gum from western larch (*Larix occidentalis*) heartwood chips followed by kraft pulping of the extracted chips; 2) isolation of mimosa tannins from bark of trees in which the peeled wood (*Acacia* sp.) serves as raw material for chemical cellulose production (South Africa); and 3) tall oil separation from the concentrated black liquor derived from kraft pulping of southern pines. In each of these cases the brunt of the raw material cost is carried by some product other than the extractive. Assurance of low raw material costs virtually dictates extractive isolation and processing ancillary to a forest products processing plant.

It is also important to keep in mind that just because wood is a renewable raw material does not mean that it is independent of oil economics. As the price of oil goes up, so will the value of wood residues as fuel. Most wood processing plants, especially energy-intensive facilities such as chemical pulpmills, have recently invested in energy-generating plants that consume all the waste wood and bark generated within an economic radius of the plant. These plants are now so locked into the use of these residues that the residues are no longer available for alternate wood utilization schemes. If the residues are to be extracted, they must not end up saturated with water thereby rendering them unsuitable for combustion. Extraction with a volatile organic solvent followed by steaming or heating to remove solvent residues, such as is practiced in the wood naval stores industry, is preferable to aqueous extraction in terms of using the residue for heat recovery.

It almost goes without saying that there may be a product that can be made only from a given wood or bark substrate. In that event, the substrate is the preferred raw material, and the financial risk from substitute raw materials is very low. While there are relatively few utilization schemes that fall into this category, some important examples are natural rubber, the naval stores industry, production of carbohydrate gums from acacia trees, essential oils from foliage, and so on.

11.2.4 Shared Capital Expense

A greenfield plant requires installation of transportation, energy generation facilities, and environmental treatment systems in addition to the primary manufacturing facility. When the cost of these is added to those of the extraction plant, production costs are likely to be so high as to totally destroy the feasibility of a project. This is another reason for locating an extraction plant, especially if it is small, adjacent to an existing wood-using facility. The project will only have to carry the capital costs of the extractives processing plant or, at worst, a pro rata share of the energy generating and treatment facility capital costs.

11.2.5 National Priority

Some types of raw materials are considered to be vital for national defense. When this occurs, government policy will generally override individual economics either through direct subsidy or through the erection of protective tariffs. An example of this situation is natural rubber that has been considered vital for transport during times of war or international stress. Stockpiling of a 1- to 2-year supply and/or development of an indigenous source have been common responses by government. Secondly, governments may wish to favor a natural product industry such as hevea rubber in Liberia or gum rosin in the People's Republic of China, because it represents a substantial local source of employment or a potential source of hard currency.

11.2.6 Realistic Research, Development, and Engineering (RD&E)

There always seems to be a terrible temptation for natural products scientists to get involved in challenging but overly detailed structural studies. While this may satisfy the psyche, it tends to unduly prolong the time and expense of a utilization-oriented project. If the marketing objectives are well in hand and the type of information that needs to be generated to get from the raw material to the final product have been well thought out, the RD&E most likely will be economic and successful. With few exceptions the type of inputs needed for such realistic RD&E can only be generated within the framework of the organization that will ultimately be carrying out the business venture. It is hard to visualize an extractive business venture that is not site-, raw material-, and market-specific. The chances that universities or government laboratories can do this type of pragmatic RD&E are problematic, and they could better focus their efforts on identifying opportunities or doing the basic researches needed (see Sect. 11.6).

11.3 Prospects for Existing Extractives-Based Industries

In terms of volume and value, the most significant wood-based extractives industries are natural rubber, resin and terpenes, carbohydrate gums, and tannins. The reasons for survival, growth, or decline in the future for each of these industries are quite different and are illustrative of the business principles described in the preceding section.

11.3.1 Natural Rubber

Less than two decades ago, the natural rubber industry was faced with prices about equal to production costs and the prospect of intense competition from synthetic *cis*-polyisoprene. This was, of course, not the first threat to the future of natural rubber.

Up to the advent of World War II, practically all of the rubber consumed was of the natural variety, mostly derived from *Hevea brasiliensis* plantations in the Far East. With the entrance of Japan into the war, both the United States and western Europe were cut off from the main source. At this point one of the most massive organic chemistry research and development efforts in history was launched, which ultimately resulted in the development of synthetic elastomers such as Buna N, Neoprene, Thiokol, and styrene-butadiene. While these were not equivalent to the natural product in some respects, product improvement continued so that by 1970 worldwide synthetic rubber production was 3.8 million metric tons compared to 2.7 million of the natural product. Styrene-butadiene rubber sold for about the same price as natural rubber. Although national defense interest dictated a policy of independence from foreign raw materials, conventional bias-belted tires still required 15% natural rubber for optimum performance.

What could not be seen in the rather gloomy natural rubber consumption forecast of the late 1960s was the radical change in tire technology that was to take place in the following decade. Original-equipment passenger-car tires have largely changed from bias to radial construction.

The latter uses an average of 2.34 kg of natural and 3.61 kg of synthetic rubber compared to 0.70 and 4.08 for bias tires (43). Radial truck tires, which are just now growing in importance, use an even larger ratio of natural to synthetic rubber. The net result is that natural rubber production and consumption have continued to grow more than 2% per year over the last decade. Production is now estimated to grow about 1.5% per year while consumption is expected to increase to 3%, leading to possible shortages. However, eastern European countries are currently mounting major efforts to produce synthetic *cis*-polyisoprene to free themselves from dependence on the import product. Furthermore, there is still room for productivity improvement from *Hevea*. The use of chemical stimulants, the earliest of which were 2,4-D and 2,4,5-T, has progressed as has higher-yielding and faster-maturing clones. Continued research in these two areas can be expected to further improve productivity and keep costs competitive with synthetic products.

Another aspect of natural rubber economics is the strategy for dealing with mature trees. Although *Hevea* can be productive for more than 50 years, maintenance of high yield plantations may require replacement of trees with improved growth stock. Prior to the last decade, this represented an expense. During the 1970s, investigations were conducted to use chipped rubber tree wood for paper-making, thus converting a nuisance into an asset (54). Chemical and fiber physical properties showed that good quality printing and writing paper (38) or dissolving pulp (37) could be produced but there were problems with latex particles and dark specks resulting from blue stain attack on extractives following felling of the trees. Approximately 0.5 million tons of rubber tree chips are exported to Japan per year, but use seems to be limited to production of corrugating medium because of the aforementioned problems. Larger quantities of rubber tree wood are potentially available as a result of plantation recycles planned for the next decade in Malaysia and Indonesia. Practical solutions to the latex problems in pulp, such as chemical debarking, inexpensive flotation additives, or mechanical treatment could open up new routes to utilization of this resource and help to defray the cost of replanting.

Growth of *Hevea* is limited to a region about 15° north and south of the equator. Research efforts designed to expand the geographic range of rubber production have shown that rubber can also be obtained from a number of other plants, the most interesting of which is the desert shrub, guayule (*Parthenium argentatum*) (7). The United States Congress passed legislation allocating $30 million for guayule research in 1978. A prototype guayule rubber processing plant is to be built in Gila River Indian land in Arizona (8, 9). Current experiments show that rubber output can be doubled by the use of the plant growth regulator, DCPTA (2-(3,4-dichlorophenoxy)-triethylamine). This treatment causes the plant to synthesize rubber during the summer months, which it does not do in the untreated state, and opens up regions like southern Texas to commercial growth of guayule rubber (9). Guayule rubber is probably at least 10 years away from commercial viability. Development will continue because success could free the United States from dependence upon imported natural rubber and eliminate the need for a government stockpile. Alternatively, guayule could be left unharvested until needed, thus providing a "living stockpile" (43).

Since it seems unlikely that synthetic *cis*-polyisoprene will become economically attractive in the next 5 to 10 years, and excellent markets persist for natural rubber, there is every reason to expect that the natural rubber industry will remain viable into the next century. The substantial potential for increasing yields and lowering costs through chemical treatment and genetic studies should encourage continued research in this field. According to a recent estimate (10), *Hevea* produces 95 tons of dry matter per acre each year. Even under the best of conditions only about 5% of this is biochemically converted into rubber by the tree, but the genetic potential exists for ultimately increasing this to 20%. Yield improvement of this magnitude coupled with various structural modifications potentially achievable through biological systems or chemical reactions (the "ene" reaction using N-phenyl carbamoyl azoformate to alter natural rubber resilience and hardness to desired levels is an example) might well reverse the inroads made by synthetics into the world's rubber industry.

Late in 1986 an important new modification of natural rubber, ENR (epoxidized natural rubber), was announced by British and Malaysian scientists at the Tun Abdul Razak laboratory in Hartford, England (58). ENR is made by treating natural rubber with peracetic acid. The new material absorbs impacts in a different way, making it suitable for engine mounting and automobile suspension systems. Adhesion to polyvinylchloride is much improved over natural rubber and opens up new composite applications such as conveyer belts and shoe soles. The biggest potential breakthrough is improved grip of tires under wet conditions. A blend of ENR and synthetic polybutadiene is visualized as providing a contribution of improved wet traction, good wearability, and low energy absorption while rolling.

ENR is currently being produced on a pilot scale in Malaysia at a rate of 1 ton per day. Plans are to scale up the process to 5000 to 10000 tons per year and then to build a hydrogen peroxide production plant to supply the ENR process locally. Although Malaysia's natural rubber is more expensive to produce than that from Indonesia or Thailand, the new ENR could provide it with new markets. If the experimental program continues to be successful, a doubling of the world market for natural rubber is envisioned.

11.3.2 Rosin and Terpenes from Pine

The production of resin acids and terpenes from pine stump extraction and the distillation of tapped gum (oleoresin), sulfate turpentine, and kraft tall oil are equal to or greater than the existing markets, most of which continue to show marginal declines. The net results are depressed pricing and a somewhat cloudy future, particularly for tall oil rosin, which is faced with changing technology in one of its major end-uses – paper-sizing additives.

Despite the electronic communications revolution, paper production continues to grow worldwide at a rate of 2% to 3% annually. Much of the incremental increase is based on pine because no other long-fibered softwood offers such high yield rates per hectare coupled with a short growth cycle. New kraft pine pulpmills and most existing mills will continue to practice tall oil recovery because non-removal of tall oil soap from kraft black liquor adversely affects evaporator and recovery boiler capacity (14). Increased tall oil production has been accompanied by new fractionator capacity in Europe and incremental expansion in the United States. Since the paper industry is the largest user of rosin size, it might be expected that the increased paper production would absorb the increased availability of tall oil rosin.

Producers of rosin size have developed modified emulsions that permit a lower net addition of rosin derivatives per ton of paper. Furthermore, there is a rapidly accelerating shift to alkaline sizing because of substantial cost saving and improved longevity of paper. Alkaline sizing is carried out by the use of alkyd ketene dimer and alkenyl succinic anhydride. Thus far, the decrease in tall oil rosin usage in papermaking, at least in Europe, has been offset by increased use in adhesives and emulsifiers. Harrop (in 31) notes that the real loser in Europe has been gum rosin, where current demand is only 25% of that used in 1965. All of this leads to the conclusion that naval stores producers must develop new materials and uses for rosin and its derivatives if demand is to catch up to supply. This is an increasingly acute problem for the tall oil fractionator who is faced with good, long-term markets for tall oil fatty acids but increasingly poor markets for rosin, the other half of his output. Tall oil fatty acid markets are cost sensitive so reasonably inexpensive outlets for rosin are imperative for the well-being of the fatty acid part of the business.

In his 1978 assessment of the future for rosin products, Zinkel (64) offered the opinion that "the key to expansion of rosin as a prime chemical raw material is the development of a large-scale dependable new source such as offered by the induced lightwood technology." This opinion was likely influenced by the somewhat erratic price and supply of rosin in the United States during the middle 1970s. Since that time tall oil rosin output has increased in the southeastern United States due to improved soap separation methods at pulpmills, increased fractionator capacity, and the installation of a number of large new kraft pulpmills using pine as their primary raw material. Furthermore, plentiful supplies of gum rosin have been made available in world markets through expanded production capacity in countries such as Portugal and especially the People's Republic of China (56). Hercules has expanded its Pinex process technology, assuring the supply of lightered stumps for wood rosin extraction. Thus, current pricing of rosin is weak and there is increasing worry that the market is becoming glutted.

The only answer, therefore, seems to be development of new markets. Gonick (16) points out that the rosin business amounts to approximately a $0.5 billion annually, yet there is probably less than 1% of sales spent on research and development. Since the profitability of the rosin business is being strongly impacted by the inroads of alternate materials from the chemical industry, he believes three courses of action are necessary: 1) significant increases in research and development must be made to lower processing costs and improve products; 2) new growth markets must be sought and cultivated; and 3) markets for rosin-based products must be diversified into more specialized end-uses that will be less prone to entry by alternate materials.

Notwithstanding problems with existing markets, there are good reasons for the tall oil producers – i.e., pulp and paper companies – to join the rosin processors in expanding the research and development effort needed on rosin. If the current "lightwood" programs, such as the Hercules Pinex process, are proven to be economically viable, the pine wood destined for pulp and paper mills will contain substantially greater quantities of rosin acids and turpentine. Tall oil and sulfate turpentine yields will be correspondingly greater. The increased byproduct yield will be of no value to the paper producer unless the tall oil and turpentine processors, in turn, have suitable markets for fractions and derivatives. Even though there seems to be opportunity to breed pines for higher extractives content and to find stimulants more effective than paraquat in generating lightwood (see Sect. 11.5), work devoted to such objectives will be in vain if it is not carried out parallel with market-oriented research.

The situation with turpentine is similar to rosin except that much more R&D continues to be expended on terpene-derived products. Many turpentine processors are more closely allied with the chemical industry, which understands the need for adequate R&D, than tall oil processors, who usually are owned and managed as subsidiaries of the forest products industry. Quantitative data on the current markets for turpentine are difficult to obtain, but the overall market in the United States currently seems to be about half for synthetic pine oil (terpineol) and the remainder for flavors, perfumes, pinene- and dipentene-derived resins, insecticides, and a host of small diverse uses (33). Insecticide production, primarily of the toxaphene-type, has declined radically in the last several years due to governmental pressure to eliminate chlorinated hydrocarbons from the environment. Markets for pine oil, terpene resins, and terpene-derived flavors and fragrances appear to be good and likely to continue. Not more than 10% to 20% of the total current terpene derivative markets are weak and need replacement in the future. Since 90% of the raw material in the United States is sulfate turpentine, there is some possibility that production of the raw material may decline and keep supply and demand in balance. One reason for this is the increasing use of continuous digestors in kraft pulpmills; continuous digestors are generally not as effective in capturing the terpenes as batch digestors. Another reason is the switch to offsite chipping. Terpene losses are unavoidable once the wood is chipped. With longer times and distance between chipping and pulping, yields of sulfate turpentine are proportionately lower. Both of these trends appear to be irreversible, so sulfate turpentine production would be downward except for the new sources from new pulp production coming on stream. The overall future supply, therefore, is likely to be flat unless the lightwood and genetic developments, mentioned above, come into substantial commercial practice.

11.3.3 Carbohydrate Gums

The only carbohydrate gums produced by the extraction of wood in the United States are *arabino*-galactans isolated from western larch (*Larix occidentalis*) chips by hot water extraction (25). The products are marketed under the trade name Stractan, by the St. Regis Paper Co. (now part of Champion International). They have found a highly specific end-use in the pharmaceutical industry, so their production is likely to be continued unless a cheaper or more effective substitute should happen to be found. Most of the other gums, such as gum arabic, etc., come from Third World countries, which desperately need hard currency and which have very low labor rates. This keeps prices relatively low, and the supply-demand balance is maintained. Almost without exception they could be substituted by the new generation of polymers produced by fermentation of glucose or lactic acid. Production costs of these products are gradually being lowered, especially by the introduction of fixed enzyme technology. There does not seem to be a particularly bright picture for most of the wood-derived carbohydrate gums in the light of these developments, especially in view of the high degree of labor intensity in their production.

11.3.4 Tannins

Most of the natural tannins still used in the production of leather are of the "condensed" variety and are obtained by aqueous extraction of the heartwood of the quebracho tree (*Schinopsis lorentzii* and *S. balansae*) in central South America, mangrove (*Rhizophora* spp.) in Australia, or the bark of wattle (*Acacia mollissima*) in South Africa. After a long slow decline in the use of these extracts, supply and demand now seem to be in balance insofar as they are used for leather manufacture. Quebracho tannin was used quite extensively as an oil-well drilling fluid viscosity control agent in the 1950s and 1960s, but it was only suitable for shallow wells where there was minimal salt contamination. Most of the wells drilled today are deep and make use of modified lignosulfonates. At one time there was fear that the quebracho forests of Paraguay would be depleted. Reduced demand for oil-well use and improved husbandry of the resource suggest the probable continuation of quebracho availability for tanning. Low-cost labor and the need for hard currency are further incentives for continued quebracho tannin production. Changes in leather technology may alter the picture, however.

Fifteen to twenty years ago there were rather extensive forecasts that many uses of leather would be replaced by poromeric polymer compositions. This and environmental concerns with the use of chromium salts, now categorized as "hazardous wastes" in the United States, caused some technologists to forecast a dim future for the industry. The industry has applied technology to its problems; substitutes for chromium, such as Montedison's aluminum polychloride complex, have recently been found. New uses and properties of leather are expected to counter the forecast demise. An example of the latter is dry-cleanable leather based on the addition of polyacrylates or polymethacrylates to leather during the dyeing or fat liquoring stage of manufacture. Synthetic poromers have not offered the same degree of comfort in footwear as leather and, in some instances, have been more expensive than the material for which they were intended to substitute. Advancing technology is not necessarily a blessing for the vegetable tannins. In the United States, where all tannins are imported, there continues to be substitution by syntans such as sulfonated phenol or naphthol condensed with formaldehyde.

The producers of wattle tannins have recognized the possible problems associated with the use of vegetable tannins in leathermaking and have developed alternative uses through an active research program in South Africa (47). Approximately 70% of the industrial phenolic adhesives used in the composite wood industry are based on wattle. Urea-formaldehyde tannin formulations have replaced resorcinol in strengthening starch adhesives used in corrugated cardboard packaging, and ethanolamine-formaldehyde tannin condensation products have proven useful in municipal water clarification systems. Many other possible uses are continuing to be uncovered so the future of the wattle tannin industry seems to be reasonably secure.

11.4 Failed Wood Extractives Ventures

An analysis of the reasons for termination of some North American extractives ventures should be helpful in illustrating the importance of the principles proposed in Sect. 11.2 with regard to the potential for future extractives-based businesses. During and following World War II, there developed a widespread recognition that the forest products industry of the western United States generated greatly excessive quantities of waste wood and bark residues that were becoming a problem in disposal. Some successful attempts were made near large urban areas to use the material for power generation, but usually it was just set aside in large piles or burned in waste "tepee" burners.

Kurth carried out extensive work on the extractives of redwood bark at the Institute of Paper Chemistry and subsequently extended this work to Douglas-fir bark at Oregon State University after the War. He was a strong advocate of sequential extraction with increasingly polar solvents to yield wax, flavanones, water-soluble tannins, and bark phenolic acids. Pilot experiments were carried out at the Oregon Forest Products Laboratory (now part of Oregon State University) and the Oregon wood hydrolysis plant at Springfield. In 1957, the M. W. Kellogg Company (now part of Kellogg-Pullman) licensed Kurth's patents and built a substantial pilot plant to produce wax by countercurrent extraction with hydrocarbon solvents. An excellent wax product, competitive in properties with candelilla or carnauba waxes, could be produced when bark was carefully obtained from fresh, mature logs. Mill-run bark, usually contaminated with considerable wood because of the use of hydraulic barkers, yielded wax that was soft and dark green in color. The product was contaminated from resinous extractives from the wood and iron from the debarking equipment, thus necessitating solvent fractionation to remove the resin and fatty contaminants and treatment with acid or bleaching agents to remove the colored iron-phenolic complexes. After extensive market surveys and cost estimates, it was concluded that the yield of a single saleable product, wax, was too low and financial margins were, therefore, insufficient to justify building a major plant. Multiple products, as originally envisioned by Kurth, ran into competition with bark phenolic extractives simultaneously being developed, promoted, and sold by the Pacific Lumber Company, Weyerhaeuser Corporation, and Rayonier, Inc. (now ITT Rayonier).

Weyerhaeuser developed an alternative approach to wax recovery (19). Although exact details of their processes were never published, patents granted to Weyerhaeuser suggest that they used aqueous alkaline extraction to remove a mixture of dispersed wax, dihydroquercetin, tannins, and phenolic acid, which was subsequently resolved by acidification and solvent fractionation. High hopes were held for dihydroquercetin in pharmaceutical and antioxidant applications (20), but they did not materialize. The price of natural wax was depressed by the increasing use of aqueous dispersions of synthetic polymers that did not require buffing to produce a shiny surface. The phenolic polymers were promoted for oil-well drilling additives but were not particularly successful, most likely as a result of the presence of trace residues of fatty acids (from the waxes), which generated foam during pumping of the fluid.

The Pacific Lumber Company built a plant at Scotia, California, in the early 1950s to produce tannins by alkaline extraction of redwood (*Sequoia sempervirens*) bark fractions. The products, marketed as Palcotan and Sodium Palconate, were promoted for use in tanning, mineral dispersants, water treatment, and oil-well drilling. Expanding upon the work of Gardner and coworkers at the University of British Columbia, Rayonier designed processes and built plants in 1954, 1956, and 1967 in Hoquiam, Washington, and Vancouver, British Columbia, to produce polyphenolic extracts from western hemlock (*Tsuga heterophylla*) bark by high-temperature extraction with aqueous sodium bisulfite, sodium hydroxide, or ammonium hydroxide. Rayonier's products were used in oil-well drilling muds, boiler water additives, correction of micronutrient deficiencies in agriculture, chemical grouting systems, and components of phenol-formaldehyde adhesives for exterior-type plywood, particleboards, and laminated timbers (24). When these plants were built, there was a substantial market for leather-tanning agents in North America, but neither the redwood nor hemlock bark product could effectively compete in price or color with imported tannins. Fortunately, the boom in oil-well drilling provided markets for the products until the middle 1960s when additives made from metal complexes of spent sulfite liquor, an inexpensive byproduct of sulfite pulp mills, became more cost-effective in the increasing numbers of deep holes that were being drilled. Pacific Lumber shut their plant down in the early 1960s and Rayonier discontinued production in 1976, having converted their Hoquiam plant to the manufacture of lignin-based chemicals.

Reasons for the discontinuance of these products have not been publicly disclosed, but it was not for want of effort or investment in research time. More than 30 patents and papers were published by ITT Rayonier during the course of the 23 years that they were active in producing and promoting phenolic bark extractives. Manufacturing costs were, unfortunately, substantial because it was a stand-alone product that ultimately had to compete with byproduct lignins, which were available in very large supply at a cost no greater than their net heat value. If end-uses for these bark polyphenolic extractives could have been found that were sufficiently unique to justify their manufacturing cost, no doubt they would still be made today.

In light of the difficulties in commercializing Douglas-fir bark wax and bark fractions in the 1950s and 1960s, it was surprising that Bohemia, Inc. (Eugene, Ore.) decided to build a $4.25 million plant capable of processing 40 000 tons per year of Douglas-fir bark to extract wax. The plant came on-stream in 1975 and was based on a process developed by Trocino (59). Hammermilled bark was dried in a rotary steam dryer and extracted in a Rotocel countercurrent extractor, a modification of an apparatus used to extract soybean oil, with a solvent described only as being composed of paraffins and aromatic compounds. Evaporation of the solvent yielded a light green wax and deodorized ground bark. The wax was sold for $0.22 per pound into various applications such as furniture polish, carbon paper, cosmetics, and an internal lubricant in plastics. The ground bark was sold as a plywood adhesive extender at $75 per ton. Anticipated markets did not materialize (12) and the value of the wax produced in any given year did not exceed $300 000 (6). In the early 1980s the solvent extraction unit was shut down, but ground bark continues to be sold in small quantities.

Trocino's concept of total utilization of the raw material, Douglas-fir bark, to produce several salable products was good, and earned Bohemia the 1976 Environmental Award from the American Paper Institute and the American Forest Products Institute. Unfortunately a certain amount of solvent losses is inevitable. Thus, efficient solvent extraction and recovery of solvent to obtain the primary product in a 3% yield, based on bark, could only be expected to be cost effective if the product sold in the "dollars per kilogram" range, such as carnauba wax imported from Brazil or Mexico. Unrefined Douglas-fir wax is soft because of the presence of terpenes, unsaturated fats, etc., and is subject to discoloration by iron salts because of the presence of ferulate esters, which promote the formation of complexes. As in the case of the polyphenolic extractives from redwood and hemlock bark, the product end-use was not sufficiently unique to ultimately justify a price that would support production and operating costs, and generate a reasonable profit.

11.5 Future Directions for Industrially Oriented Extractives Research

Much of the industrially sponsored research effort on wood extractives during the past several decades has been directed toward finding and developing processes for salable products from wood, bark, or pulping liquor residues. The work was frequently frustrating because the products, though effective, could not compete in price or versatility with similar products based on petroleum. The oil crisis temporarily breathed new life into these endeavors, but the possibility for producing wood-based chemicals with a lower price than petrochemicals has been pre-empted by use of woody raw materials for energy generation. Thus, at least in North America, we do not seem to have made any progress at all during the last decade in developing new extractives-based industries.

Does this mean that the future is bleak for carrying out extractives research in the industrial environment? On the contrary, we believe that there are still many opportunities that will generate a significant economic payout. These opportunities do not seem to lie in a continuation of classical organic extractives research – i.e., substance isolation, structural identification, and confirmation of structure by synthesis – even though this type of work is certainly essential to the basic research arena (see Sect. 11.6). A significant change in focus needs to be made. The approach that we have in mind needs to forge the disciplines of structural organic chemistry, wood anatomy, cell biology, and tree physiology into a unit that will find ways to control extractives formation and transposition by external stimuli or genetic manipulation. Our rationale for this change in direction and the possible benefit that it could provide are based on two premises: a) the site and amount of extractives in the tree are affected by internal or external chemical signals, and, more importantly, b) the rate of cell division, cell wall thickness, heartwood formation (senescence), etc. are under the control of trace constituents – auxins and the like – as well as the genetic makeup of the species. Consider some of the consequences if we could manipulate (a) or (b).

11.5.1 Control of Extractives Deposition

Anyone who has examined the chapters leading up to this one, even in the most cursory fashion, must be well aware of the role of extractives in utilizing wood. Intensity and uniformity of color affect the selection and value of fine veneers for furniture manufacture. Absence of extractives permits the invasion of microorganisms that promote sap-stain or actual wood destruction. On a positive note, absence of extractives permits pulpability by the sulfite process. Extractives harm bleached wood pulp absorptivity when used in sanitary applications such as disposable diapers. Traces of extractives in pulpmill effluents threaten the life of important marine organisms. Extractives impede pressurized wood treatment. The list could go on and on. One thing that is particularly evident about all the industrially important aspects of these effects of extractives is that they are accepted by the user simply as "a fact of nature" and nothing much is done about them other than the simplest sorts of physical or chemical treatment.

The amount of money consumed in forest products industries to cope with the presence or absence of extractives dwarfs any and all current business that involves the collection and sale of extractives. One simple illustration should suffice to establish this point. Nearly every chemical pulpmill throughout the world has problems with foam on the unbleached stock washers and many have problems with cavitation of pulp slurries in pumps or poor bleached pulp washing from the same cause. These problems can be traced to soaps formed from wood extractives that carry through the pulping stage into the bleach plant. Standard procedure for dealing with these problems is the application of defoamers, frequently in quantities as high as several kilograms per ton of pulp. The same family of extractives may cause pitch specks in paper, poor absorption in fluffing pulp, rayon strength problems, and poor acetate color stability from dissolving pulp. Prescription in this instance is the liberal application of nonionic detergents. The total cost of detergents and defoamers to control the undesirable properties of wood-derived fats and resins during pulp and paper processing in North America probably exceeds the value of byproduct tall oil and turpentine produced by a factor of two. Add to this the inventory costs of roundwood or chip storage for extractive control in most mills north of 40° latitude, purchase of additives such as talc for pitch speck control, defoamers in waste-treatment systems, etc., and one can readily see that negative costs of extractives greatly outweigh value gained from byproduct sale.

The time is overdue to control, collect, or eliminate these materials at their source, the tree, rather than inefficiently dealing with them during processing. For the most part polyphenol, fat, and resin generation seems to be confined to parenchyma and epithelial cells within the wood. Extractives are formed as a result of a chemical message. If we could intercept or overrule this message at a time shortly before a tree was to be harvested, we could have a tree with induced heartwood with all of its coloration, preservative effects, etc., or one in which the cells self-consumed the reserve carbohydrates and fats to yield a tree essentially devoid of these extractives. The pulp and paper manufacturer could then use the wood from the extractive-free tree without spending money for detergents, defoamers, etc. If the onset of heartwood could be delayed, many more species of wood would be suitable for sulfite pulping. Avoidance of pit membrane closure, which is the usual byproduct of heartwood formation, would facilitate penetration of chemicals during the pulping operation or the even distribution of preservatives during wood treating.

Hillis has long recognized the potential for inducement of extractive formation in wood (26, 29). The amount of extractives in sapwood is usually small and is much larger in heartwood. The conversion of sapwood to heartwood is generally accompanied by the closure of pits and the generation of a whole variety of secondary metabolites, many of which seem to be totally absent in the sapwood. The mechanism that triggers this process has naturally been of considerable interest to a number of investigators. A variety of experiments carried out during the last ten years has shown that small amounts of ethylene or an ethylene-releasing compound can initiate the production of rubber, polyphenols (27, 28), tannins, and coloring bodies (39, 40), and oleoresins (62). Most of this work has been done to determine possible mechanisms of extractives formation rather than commercial exploitation of the phenomenon. The specific exception, of course, is the use of paraquat to induce oleoresin formation in the butt of pine trees (Hercules Pinex process and others) so that the stumps can be processed for naval stores extraction.

Oleoresin formation seems to be primarily a function of the epithelial cells surrounding resin canals, so the use of paraquat to stimulate resin production replicates only part of the processes involved in pine heartwood formation. We still do not know which cells do what, but certainly ray parenchyma and bordered pit tori must be major sites where heartwood constituents are elaborated. Programs for studying the systematic administration of various combinations of natural or synthetic auxins and ethylene-releasing or -inducing compounds just prior to harvest might ultimately result in producing all "heartwood" stems with the attendant benefits already mentioned. It may be possible to achieve the same end result by the use of other types of compounds that would internally trigger the formation of ethylene. Another possibility is the use of microorganisms, but this seems to be so fraught with danger of infecting nearby trees that the overall cost of overseeing the process would be prohibitive.

Means for preventing extractive deposition or heartwood formation appear to be much more difficult to formulate until we have a much more precise idea of the mechanism(s) that initiate the process in the first place. We do not even know, for example, whether the onset of heartwood formation is under genetic control or whether it is precipitated by a combination of environmental factors such as growing season characteristics, moisture availability, temperature cycles, and so on. There are some suggestions that heartwood formation is clone-, species-, or genus-dependent, but data are inadequate. Perhaps there would be more support for data gathering if the ultimate goal of the work were enunciated as being manipulation of heartwood formation and its attendant value.

11.5.2 Manipulation of Wood Growth by Chemicals

In the preceding section, further research was proposed on the use of chemical stimulants to induce cells to produce secondary metabolites in greater quantities than normal, or not to produce them at all. While the payout for successfully meeting this objective would be substantial, it is small compared to the rewards obtainable if chemicals (and a low-cost method for administering them to the tree) could be found that would radically alter wood-growth characteristics. If we could increase the rate of cell division over the course of a growing season, the yields of wood per hectare per year would be increased and would lower its cost. If tracheid length could be increased, trees could grow taller within a given time frame and could give improved paper properties when pulped. If ratios of vessels to fibers and fibers to parenchyma cells could be altered, wood with improved machinability could be produced. If ratios of summerwood to springwood could be changed, or if wall thickness of tracheids could be increased or decreased, depending upon a given species, dimension lumber strength properties could be improved, penetrability of preservatives into crossties or poles could be facilitated, and paper burst-tear-tensile strength relationships could be optimized.

As far as can be determined, research with these objectives has been primarily limited to genetic selection. It ought to be possible to bypass the lengthy time needed to breed for given wood traits such as higher density, faster growth, etc., by some judicious use of auxins and/or chemical change in the growing environment. If the question be raised as to why such a proposal is made in a book focused on extraneous substances of woody plants, the answer is that a variety of such growth hormones must be already present in trees. Research is needed to identify them, produce them synthetically or generate even more effective compounds, and then test them for efficiency. Another question that might be asked is whether there is a precedent for possible success in this type of work. The answer from recent work is quite positive.

Longleaf pine (*Pinus palustris*) is one of the important forest species in the southeastern United States. Although the tree has many excellent qualities for forest products and for papermaking, it is gradually being replaced in plantations by slash pine (*Pinus elliottii*) or loblolly pine (*Pinus taeda*). The reason for this is the delayed seedling height growth of longleaf, which results in a prolonged grass stage. Because of their slow growth, seedlings often die from weed competition or from brown spot needle blight. In work carried out at the Southern Forest Experiment Station, Hare (18) showed that treatment of longleaf pine seedlings with cytokinin-like substances in spring and early summer stimulated seedling height growth. This effect was enhanced by gibberellic acid and an experimental triazinone derivative. Synthetic cytokinin benzyladenine derivatives also stimulated growth but had deleterious side effects – bud proliferation and death; a seaweed extract with cytokinin proved effective without deleterious side effects. Thus, some combination of auxins could be expected to stimulate growth and improve the economics of species such as longleaf pine.

A different type of example of cell change in response to an external stimulus is the formation of compartments (52) or barriers (5) in response to injury or infection. Zones of extractives-loaded parenchyma cells and swollen ray parenchyma cells are formed between infected xylem tissue and the healthy cambium. The zones act as a barrier allowing formation of new healthy xylem and protecting it from further damage by the pathogen of its metabolites. This reaction does not always take place. Elms (*Ulmus* spp.) susceptible to Dutch elm disease do not form a barrier zone. Possibly they do not possess a gene for elaborating a hormone that triggers the process. In any event, there is a variety of physiological effects such as tyloses formation, ring shakes, etc., accompanied by extractive deposition which show that woody cells amplify or change their function in response to chemical stimuli.

Recent work by Albersheim and his coworkers (60) is helping to elucidate the pathway by which cell morphology is or might be altered by the influence of plant hormones. They have shown that there is a new class of regulatory molecules in plants in addition to the five major hormones already identified: auxin, abscisic acid, cytokinin, ethylene, and gibberellin. This new category has been named oligosaccharins, and each of them appears to be able to deliver a message regulating a particular plant function. The first of these compounds to be identified is a heptaglucoside in which two D-glucose molecules are joined as side chains to the second and fourth D-glucose unit of a backbone of five β-linked D-glucose units. This substance is reposited in the cell wall and can be released by an auxin or an enzyme. Once released, it acts on a set of cell genes to initiate antibiotic formation, a variety of morphological changes, etc., depending on the circumstances. Albersheim believes that it may be possible at some time in the future to spray specific oligosaccharins on plants or to manipulate the genes controlling the release and metabolism of oligosaccharins (possibly by the use of as-yet-undiscovered hormones), in order to tell plants to flower, to form seeds and fruits, to become resistant to a disease, to grow faster, or to change their shape. Functional elaboration has already been accomplished with cell cultures, and there is a good reason to believe that this experimental approach will eventually be very useful in the field of forestry, provided the extensive research work still needed is undergirded with adequate financial support.

11.5.3 New Techniques for Extractive Isolation

Outside of the tall oil-turpentine naval stores industry and of the natural rubber industry, the most promising extractives-based product group for potential exploitation are the polyflavanoids. This family of polymers comprises 30% to 50% of all the bark generated as a by product of the lumber and pulpmills throughout the world. Enough work has been done to prove conclusively that these polymers can be used as a substantial replacement for phenol or resorcinol in a multiplicity of adhesive formulations. The question remains as to why the use of these materials has not developed to any extent, and the answer has to be that we have not found a way to separate them from bark at a low-enough cost and in a sufficiently undegraded form. Furthermore, the isolation procedures currently used give polyphenolic polymers so contaminated with other bark constituents, usually pectins or related carbohydrates, that an expensive technique such as membrane dialysis is required to render them usable in adhesive formulations.

The major problem seems to be that much of the polyphenolic polymers in the bark are primarily insoluble in simple neutral solvents such as water or alcohol under non-pressurized conditions. When they are extracted with dilute alkali, they are strongly subject to chain degradation, oxidation to quinones, ring-opening, etc., and the solvent also removes carbohydrates, low-molecular weight lignins, and hydrophobic waxes that can subsequently interfere with adhesive bonds. We cannot suggest which low-cost isolation techniques might be more useful; all we can do is point out that there is a very large, low-cost raw material source and a very large potential market — so large that it would certainly seem to justify further industrial investigation. There may be a temptation to respond to this proposal by saying if we knew more about the structure of these compounds, we might find better ways to isolate them. It can readily be conceded that there is always more to learn about the structure of natural polymers. In this particular instance, additional research could better be directed towards improved, lower-cost isolation techniques than structural studies in an environment of limited financial support.

A second area, in which new isolation techniques could be very interesting, follows from the considerations in Sect. 11.5.1. Much of the North American and northern European chemical pulp industry is based on conifers that contain resin canals — i.e., pine, spruce, and Douglas-fir. These trees ought to be capable of resin and terpene inducement through preharvest treatment of these trees with an auxin that functions like paraquat but would have improved capability. It has already been noted that pulping is inhibited in resin-saturated chips because of the consumption of alkaline pulping chemicals and the difficulty of penetration of aqueous cooking liquor into hydrophobic chips. One way to overcome this would be pre-extraction of chips with a low-boiling organic solvent, such as methanol or hexane, which could be subsequentially distilled out of the chips as they are being brought up to temperature with kraft cooking liquor. Ideally the extraction should take place in a separate vessel from that used in the pulping reaction so that the terpenes and resins would not be contaminated with the volatile sulfur-containing byproducts generated during the pulping process.

Some of the advantages that can be visualized for a process of this type are: a large, low-cost source of clean terpenes and resin acids; improved uniformity of kraft cooking that would not be inhibited by the presence of resinous constituents; no fatty or resin acid soaps that create foaming on brown stock washers and require additions of defoamers for control; and, most importantly of all, no toxic fatty or resin acids in the bio-basin effluent discharge. The inherently more expensive northern wood resource, compared to subtropic plantation eucalyptus or other hardwoods, would be enhanced by the co-production of products, which would help to provide a competitive balance.

There are some potential disadvantages to such a utilization scheme: Are there markets that could absorb a very large increase in availability of resin acids and terpenes? New end-uses would certainly need to be developed. Another major unanswered question has to do with fats. Would they also be produced in the auxin pretreatment process? Will they be present in the wood as free fatty acids or as glycerides? If the latter, what will be their effect on processing? These are the types of questions that need to be answered in an industrially-oriented program that combines increased resin and terpene generation with new isolation schemes.

A technique that initially held promise for replacing whole plants, or parts of plants, for production of useful secondary metabolites is plant cell culture. After about ten years of intensive industrially-supported effort, many technical problems have been solved in scale-up. Unfortunately none of the cells derived from commercially important pharmaceutical-yielding plants have yielded the desired compounds in anything close to commercial quantities (3).

One of the problems is that compounds are not accumulated in cells but are released into the culture medium where they are lost, if volatile, or may interfere biologically with subsequent cell growth and division. This problem has been dealt within cell cultures of *Thuja occidentalis* (eastern white cedar) by Berlin and co-workers (4) by adding an insoluble solvent "trap" to collect terpenes and tropolones released into the medium. The yield of thujaplicins was nearly tripled over the duration of the fermentation by this technique. Nevertheless, the secondary metabolite generated by the *Thuja* cultures did not replicate what might be expected from any specific part of the tree.

Flores (11) attributes the failures of plant cell fermentation thus far to the fact that cell cultures typically consist of undifferentiated cells while the desired compounds are often produced by specialized cells in the mature plant. Thus, some methods of forcing the cultured plant cells to act as though they were parenchyma or epithelial cells, root cells, etc., is needed. Berlin (3) sees this activity as taking a great deal more work, possibly requiring the techniques of genetic engineering. We might add that if more were known as to what prompts a particular type of cell within a plant to generate secondary metabolites, it might be possible to stimulate cultured cells to undergo the same expression. This presumes, of course, that cells can be externally stimulated by a chemical signal to change their metabolism and their function.

In the meantime, we expect plant cell culture experimentation to be continued even though the initial promise has not yet been fulfilled. More important than the production of compounds may be the use of cultures in biotransformation reactions. The cells may elaborate enzymes that can cause an organic reaction to take place that could be next to impossible to carry out synthetically. An example of this is the 12-β hydroxylation of digitoxin to digoxin by cell lines of *Digitalis lanata*. Thus, cells from certain woody species might be put to work as part of an overall synthesis of valuable pharmaceuticals. For this to happen, much more work is needed on biotransformations taking place within woody plants.

11.6 Areas of Needed Basic Research

In spite of the large volume of research already conducted on wood extractives, as attested to by the preceding pages, there is need for much additional basic research, particularly if we are to attain the practical objectives proposed in Sect. 11.5. The following suggestions for future basic work do not pretend to be exhaustive. Rather they should be viewed as representative of the types of information that ought to be generated in the future. Regardless of the particular thrust of future work, we think that it is important for future investigators to frame their work in a broader context than much of what has been done in the past. Invariably, this will require something more than a narrow disciplinary approach. For example, in earlier times a chemist might have been content to survey the identity of phenolic constituents of a particular species of wood. In the future, it is hoped that the same chemist might carry out his examination on serial sections from the center of the tree to the cambium or at least separately from sapwood and heartwood. The investigator should also check his findings on a number of trees to gain some idea of whether the compounds discovered are common throughout the species or whether there are chemical varieties of the species. This would also help to uncover possible contamination of a wood sample by microorganisms, which are, of course, capable of generating unique compounds not typical of the wood species. The work then might branch out into examination of closely allied species, or it might take a biochemical tack in which a search is made for probable biosynthetic routes. And once the constituents are identified with certainty in vitro, examination of wood sections should be made microscopically for the probable location (see Sect. 11.6.6 for rationale) of the site at which the compounds are formed or to which they are translocated. An excellent example of this approach is the work of Nobuchi and coworkers (42) who have coupled measurements of various *Cryptomeria* polyphenols formed in the tree with ultrastructural changes in parenchyma cells associated with heartwood formation. Complimentary work by Kai and coworkers (30) on the distribution of polymeric polyphenol fractions is helping to establish a good understanding of the composition, distribution, and biosynthesis of the phenolic extractives in sugi (*Cryptomeria japonica*), one of the most important forest trees in Japan.

Other needs for broader perspectives are outlined in the following.

11.6.1 Cambial Constituents – Growth Regulators

Reference has already been made to treatment of pine with paraquat or an ethylene generator such as Ethrel, 2-chloroethylphosphonic acid, to induce the formation of oleoresin (35, 62). This work was originally based on experimental treatment of trees with a variety of growth hormones and/or herbicides to stimulate oleoresin exudation, and it just happened that these two compounds were efficacious in inducing oleoresin soaking of the wood. To make better progress in this area of endeavor, we need to have information on the hormone systems naturally operating in trees. There is some speculation that certain compounds such as gibberellins, abscisic acid, cytokinins, etc., are operative, as they are in the agricultural plants that have been investigated. Detailed analysis of soluble cambial constituents and their seasonal variability need to be made.

Good starts in this direction have recently been reported. Savidge and Wareing (50) have studied the level of indol-3-yl acetic acid and (*S*)-abscisic acid in the cambial cells of lodgepole pine (*Pinus contorta*) and related it to xylem development. Are these growth-promoting compounds translocated to these cells, or is their formation within the cell triggered by a message compound or a physical stimulus such as the temperature change, change in sap pressure, etc.? These questions are not answerable at present, but related work by Sandberg (48) on *Pinus sylvestris* needles has shown that L-tryptophan and tryptamine are converted endogenously to indole-3-ethanol, which is subsequently transformed to indole-3-acetic acid. Whether this is then transported to the cambium in the phloem sap, or whether the circumstances that prompt biosynthesis in the needles are similarly operative in the cambium remains to be determined. Radiometric tracing of labeled indoleacetic acid in needles, phloem, and cambium would be helpful. As the composition of cambial extractives or those in the sap are being investigated, fractions should be checked for physiological activity. One method of doing this would be to treat a tissue culture of callus cells from the species being investigated and examine it for morphological differentiation or extractive generation. Yamaguchi and coworkers (63) have shown significant changes in the extractives produced by *Eucalyptus robusta* callus when the medium was inoculated with kinetin and 2,4-D. Before inoculation, the callus cells contained polyphenols of the same composition and roughly the same concentration as the sapwood. Following administration of phytohormones, flavanols and polyphenols characteristic of the bark were formed.

11.6.2 Root Constituents – Role of Mycorrhizae

The growth of many forest trees is dependent upon a symbiosis between mycorrhizal fungi and the plant. The fungi attach themselves to the roots and send out hyphae that increase the volume of soil tapped by the tree. This allows improved efficiency in the uptake of phosphate, greater tolerance to drought and resistance to certain pathogens. Abelson in a plea for greater study of plant-fungal symbiosis (1), points out that techniques are now available to grow mycelia of fungi specifically adapted to given trees from pure culture and that these are being used to treat a large fraction of the pine seedlings now being planted in the United States. Benefits include better performance of the seedling in the nursery and better survival and growth in the field.

One of the anomalies of wood extractives research is the paucity of information on the composition and biogenesis of extractives in the roots of trees compared to stem or foliage extractives. In particular, we ought to have substantial investigations on the role of the mycorrhizae, assuming that the symbiotic relationship may actually change the composition of compounds in roots compared to the stem. Work of this type may be helpful in optimizing combinations of trees and fungi to adapt to hostile environments. The worldwide removal of forests in the tropics accompanied by laterization of soil must ultimately be corrected by reforestation. The latter will require all the skills that biologists can muster; the more we know about the symbiosis of mycorrhizae and tree roots, the more likely we will be to succeed.

11.6.3 Environmental Relationships

The slow demise or even death of conifers, especially at higher elevation, attributed to atmospheric conditions (so called acid rain) has become a well known phenomenon during the last few years. As part of the documentation of the effects of air pollution, we would expect future research to be needed on the interaction of the external environment with extractive deposition and vice versa. At this time, higher-than-normal concentrations of ozone, sulfuric acid, sulfur dioxide, and nitric acid have been blamed for forest injury in North America and Europe. The mechanism has not been established, but it could well be imagined that there are interactions with tree extractives. For example, under normal summertime conditions, coniferous forests emit substantial quantities of terpenes into the atmosphere, so much in fact that they are responsible for a visible blue haze. Obviously this behavior has most likely been occurring since these trees evolved, but now some interaction of terpenes with air containing high contents of pollutants from automobile emissions and from power plant usage of high sulfur coal could be visualized as resulting in terpene epoxides, nitrates, etc., which might return to the trees during rainfall. Studies of the emission, interaction with ozone, etc., and effect on tree growth of atmospherically altered terpenes would establish whether this phenomenon has important consequences for the forest.

Another environmental relationship that needs to be studied is the array of possible effects of external pH on secondary metabolite formation. Does acid rain impinging on leaves or needles change the ratio or amounts of phytohormones, the messenger chemicals that control tree growth? Or might acid rain affect the availability of metallic ions in the soil, either by selectively leaching certain acid-soluble ions or by increasing concentration? A recent study of the presence of trace elements in tree rings of shortleaf pine (2) showed a correlation between suppressed growth and atmospheric iron and sulfur dioxide associated with nearby smelting activity. No information on effect of contaminants on extractive content or concentration was reported.

This raises a broader question about the effect of trace elements in the soil upon extractive formation. Recent work with pine, spruce, and Douglas-fir seedlings (51) under controlled conditions showed that tree seedlings took up a variety of elements including gold, which was formerly thought to be too insoluble or resistant to chemical reactions to be extracted from mineral deposits by trees. Effect on extractive generation, tree growth, etc., was not reported, probably because the intent of the work was to demonstrate the use of trace metal occurrence in trees as a means for detection of buried ore deposits.

The physical effects of drought on many types of plants are well established, but drought resistance mechanisms are not so well known. Work by Pizzi and Cameron (45) has now shown that flavonoid tannins are involved in South African species such as *Myrothamnus flabellifolia*, where they protect the cell wall from cracking after it has been subjected to drought and subsequent rehydration. The helicoidal tridimensional structure of the tannins function as springs preventing cracking. Tannins undoubtedly have other important effects that are of importance under environmental duress. When the tree is under stress, tannins may be generated by various cells in a tree to help protect it against microorganisms or insects while the tree is in a weak condition. On the other hand, plants injured by atmospheric influence may emit volatile oils such as terpenes or diterpenes that act as signals to initiate insect attack. Elegant techniques for trapping and identifying volatile compounds, such as that reported by Kratzl and coworkers (34), should facilitate comparative biochemical studies of plants injured by atmospheric influences, fungi, or insects, or some combination of the three.

Ecological effects are often extremely complex, so care must be given to avoid oversimplification. Some plant secondary constituents may be generally toxic to predators but they can become a problem for the plant if the predator adapts the use of the chemical to its own benefit. Smiley and coworkers (53) have shown, for example, that salicin, a toxic phenol glycoside, is used by the larvae of a certain beetle as a substrate for producing defensive secretions against other insect predators. Thus, Sierra willow (*Salix orestora*) with a high content of salicin, instead of being protected by the compound, is more readily attacked by this specific beetle species that has adapted salicin to its own benefit.

11.6.4 Pharmacologically Active Compounds

It is hard to imagine a field of endeavor in extractive chemistry that would be easier to justify than the search for pharmacologically active compounds. Yet the fact remains that the surface has hardly been scratched. Probably not more than 5% of the woody species throughout the world have been chemically examined in sufficient detail to determine whether they contain organic compounds with novel structures. Of these only part have been examined for pharmacological activity. Although primary interest so far has been for antitumor agents (46), it seems likely that additional searches will reveal new types of tranquilizers, antibacterials, termiticides, and so on. On the negative side, we need to continue work to identify harmful constituents in wood, especially carcinogens (36).

Since the extractive constituents of many north temperate zone trees have been studied, at least in some detail, future work should be primarily directed to subtropical and tropical forests such as those of Indonesia, New Guinea, Central Africa, Central America, and the northern half of South America. Many of these areas have up to 200 different species of trees in as little as a hectare of land. With the growing concern for the displacement of tropical forests by agriculture or homogeneous forest plantations, the need for an extractives inventory becomes increasingly pressing. Since many species are becoming rare or even extinct, a newly identified attractive compound could be chemically synthesized.

It might be argued that the finding of a potentially useful drug in a rare species might not be particularly helpful because of the difficulty in finding sufficient quantities of the tree or the time required to grow additional numbers of trees. It should be kept in mind that plant cell culture techniques are advancing rapidly (11). In the past, researchers have frequently been unable to obtain even trace quantities of a desired product from plant cells growing in fermentors, but this was frequently caused by undifferentiated cells in the culture. As means are perfected for causing cultured cells to perform chemically as roots, ray cells, phloem parenchyma, etc., we may be able to preserve the synthesizing capability of a given species or variety for medically useful compounds.

The search for pharmacologically-active compounds can also be assisted by the findings generated in the field of chemical systematics – i.e., identification of chemical constituents unique to a species, genus, or family. This activity grew enormously in the middle 1960s, thanks to blossoming microanalytical techniques such as GLC-MS, TLC, gel electrophoresis, and so on. Activity seems to have waned recently, not so much for want of sound results as for the difficulty in securing financial support to do the work. Some combination of chemical systematics with a research objective of finding new sources of useful compounds might help to stimulate support of systematics research.

There are still some questions as to the type of compounds and part of the tree to be analyzed in studying the systematics of wood species. Most work so far has been done on terpenes and sesquiterpenes, most likely a consequence of the ease in doing the chemical analysis. Fats, phytosterols, simple sugars, simple phenols (vanillin, coniferyl aldehyde quaiacol, etc.), and flavanols (catechin, flavan dimers, etc.) seem to be too widely distributed and not sufficiently species-unique to be of much help for generic distinction. Thus far, the more complex polyphenols (lignans, flavanones, flavones, etc.) and alkaloids seem to be the most useful. Di- and triterpenes are also of interest, particularly when their differentiation is coupled with a polyphenol inventory. Heartwood and leaf or needle secondary constituents are generally the most useful; exudates must be used with great care since their composition can vary with weathering, autooxidation, and the types of tissue and circumstance producing the response.

11.6.5 Phenolic Polymers

One of the paradoxes of wood extractives research, especially the work dealing with heartwood constituents, is the meticulous attention paid to the elucidation of trace substituents while simultaneously ignoring the composition of some of the major components – i.e., carbohydrate and phenolic polymers. While it is true that techniques for polymer separation and structural analysis have only become well-developed in recent years, this hardly accounts for ignoring a whole family of extractives that comprise up to 75% or more of the organic solvent soluble portion of heartwood extractives. There is not even an agreed-upon name for this category of extractive constituents, but it has been variously called native lignin, Braun's native lignin, secondary lignin, polylignan, or lignin-like materials. There is even disagreement as to whether these substances are secondary metabolites or are degradation products of lignin resulting from acidic conditions present in heartwood (22).

Polar solvent extraction of sapwood, heartwood, or rhytidome (outer bark) yields a pale tan- to brown-colored solid, which is insoluble in water and in less polar solvents such as diethyl ether or dichloromethane. Proximate analysis, methoxyl content, color reactions, and infrared spectra (32) are so close to those of lignin to have lead F. E. Brauns and others in the 1940s to believe that this material was a low molecular weight fraction of lignin. Later on, Hergert (21) showed that this material was present in relatively large amounts in heartwood and only in very small quantities in sapwood. It must, therefore, be formed biosynthetically in the same way as other heartwood extractives since solubility characteristics alone would preclude translocation from the sapwood or the cambium.

One of the troubling aspects of these polymers is that they appear to have lignans which are characteristic of the species from which they are isolated, incorporated into the polymeric chain. Heartwood lignans are invariably optically active, suggesting that they are biosynthesized by a different mechanism than lignin. The latter is believed to involve laccase or peroxidase-like enzymes acting on coniferyl alcohol to produce dimer intermediates via a quinonemethide intermediate. These dimers are structurally related to heartwood lignans, but because of the reaction mechanism they are racemic. The dimers undergo further polymerization to form lignin. It can be hypothesized that the heartwood lignin-like polymers involve enzymatic coupling of coniferyl alcohol and species-specific lignans. Proof of this would require cleavage of these heartwood lignins to find out whether their monomeric or dimeric structural units are optically active. (Interestingly enough, experiments have never been done to definitively establish the absence of optical activity in the dimeric units of cell wall lignin. Generally the amount of any dimer isolated in lignin cleavage studies is so small that measurement of optical activity cannot be conveniently carried out.)

Co-occurring with heartwood solvent-soluble lignin-like polymers are polyphenolic polymers based on flavan-3,4 diol or gallic acid derivatives. These polymers have only been studied in detail in the case of a few commercial tannin products [quebracho heartwood (*Schinopsis lorentzii* and *S. balansae*), wattle bark (*Acacia mollisima*) extracts] and from a few coniferous barks, mainly pine (*Pinus* spp.). Further complicating the picture is the fact that substantial quantities of all three types of these polymers are deposited in heartwood and bark in a form that renders them insoluble in neutral solvents. By definition they are secondary metabolites but in analytical procedures they show up as cell-wall constituents, usually in the Klason lignin fraction. At the moment there are no good methods for isolation of these cell-bound polymers without structural alteration or degradation.

Suffice it to say, this particular area of wood extractives chemistry needs clarification. In particular some sound work needs to be done on polymer structure of the solvent-soluble "lignins" from several representative species of angiosperms and gymnosperms. Sites of deposition need to be established. It might be added that this subject is of more than academic interest. It is quite likely that the co-occurring polyflavanoid and lignin-like extractives, even though they are present in small quantities, can interfere with normal pulping reactions in the acid bisulfite process (17). They may also play a role in organosolv, NSSC, or CTMP processes, but this is pure speculation at the moment.

11.6.6 Sites and Control Mechanisms of Biosynthesis

The many secondary constituents identified as being present in wood can be visualized in a time-dependent mode. Their presence and location in the tree is fixed as a consequence of the circumstances under which the experimenter isolates and identifies them. If it were practicable to do so, extractive composition and location should be identified in the living tree. Short of that, extractives should be examined immediately upon harvesting before any post-mortem changes take place. If we are studying compounds as they actually exist in the living tree, or at least as we deduce they exist, compounds must fall into one or more of the following categories: a) Messenger compounds sent from one part of the tree to another to trigger a chemical reaction – e.g., auxins; b) intermediates in biosynthesis; c) intermediates in respiration; d) energy reserves – e.g., fats and starch; e) compounds generated in response to an external stimulus such as a physical wound or a pathogenic invader; f) end products of metabolism; g) degradation products or altered compounds in dead parts of the tree resulting from oxidative, acidic, or other processes; and h) compounds that seem to have no function.

If the only objective of one's research is to catalogue the presence or absence of secondary constituents, precise determination of physiological location will be of little interest. If, on the other hand, the goal of our work is to understand the role played in the plant by the various secondary compounds, site and function must be ascertained, especially if the ultimate objective of the work is to stimulate the yield of useful materials or minimize the effect of harmful substances as outlined in Sect. 11.5.1. All too frequently, extractives chemists treat wood as if it were a homogeneous material. Sapwood, heartwood, cambium, wound tissue, knots, compression wood, etc., are all ground together into one sample and extracted without regard to the differences in the various parts of the tree, age of the tree, time elapsed since the harvest of the tree, and so on. Hopefully, future generations of extractives chemists will be sufficiently sensitive to these issues so that they carefully document the origins of their raw material and separate it into the type of fractions that will permit assessment of function.

While heartwood and outer bark fractions are certainly the most interesting parts of the tree from the standpoint of variety and content of many types of organic compounds, we certainly need to develop a much better understanding of the physiologically active zones of the tree. More emphasis needs to be placed on the water-soluble constituents transported down the phloem from the site of biosynthesis and centripetally from the cambium to the rhytidome and to the heartwood via the rays. We need more information on resin canal or axial parenchyma strand constituents as opposed to contents of ray parenchyma. Serial section analysis from the center of the tree to the outer bark will be helpful. Finally, this work must be buttressed by in vivo labeled compound experiments to determine if our biosynthetic speculations follow the same course in nature as they do in vitro.

As an example of the type of clarification of function needed for key secondary constituents, we cite a specific instance: Our understanding of the biosynthesis and structure of lignin was largely based on the in vitro experiments of the Freudenberg school in which coniferyl alcohol alone or in mixtures with coumaryl and sinapyl alcohol was dehydrogenatively polymerized to products that are nearly identical with milled wood lignins isolated from wood. Coniferyl alcohol was not detected in its free state in trees, but the glucoside of coniferyl alcohol, coniferin, had been found in cambial extracts. Freudenberg envisaged the translocation of coniferin from the cambial zone into lignifying tissues (15) where it encountered β-glucosidase and the liberated coniferyl alcohol was polymerized into lignin. This simplistic picture was modified by Higuchi and others who showed that a single lignifying cell contains all the necessary enzymes to transform glucose to lignin. Nevertheless, coniferin in the sap of conifers is still generally considered to act as a reservoir (49) to augment the precursor supply of lignifying cells.

The latter hypothesis may need to be altered or at least augmented by new studies being conducted by Terazawa and Miyake (57) on cambial constituents.

It now appears that there are many other phenolic glucosides present in the cambial zone. Some of them have structures that suggest that they could be incorporated into lignin in their free phenolic form. Some very careful work will need to be done to ascertain the precise location of these materials. Perhaps they and coniferin have nothing to do with lignification. The availability of analytical techniques that involve silylation/GLC or HPLC will greatly assist extractive site location and will help to unravel the mystery of the relationship of cambial phenolic glucosides to lignification.

A useful method for relating extractives composition to synthesis or storage site is illustrated by the recent work of White and Nilsson (61). By comparing the number of and size of certain physiological features versus yield and composition of extractives, their location can be deduced. For example, an absolute correlation was established between resin canal frequency and amounts of monoterpenes and sesquiterpenes in *Pinus contorta* needles. This showed that resin storage and synthesis is largely compartmentalized in resin canals in needles. Quantitative extractive studies on a variety of wood samples that had been subjected to quantitative analysis of their content of vertical and horizontal parenchyma content, ray cell volume, resin canal frequency, etc., would help to establish sites of extractives storage. Combined expertise in organic chemistry and wood physiology will be necessary in such an approach.

Another technique that will be of increasing help in our understanding of extractives biosynthesis are detailed studies of the location of various enzymes within the tree. A recent example of such work is that of Stich and Ebermann (55) who showed that isoenzymes of peroxidase and polyphenol oxidase were identical in the sapwood and heartwood of oak. This work shows that there is a potential for polymerization of monomers to lignin in unlignified parenchyma cells at the sapwood/heartwood boundary as well as in the zone of active lignification close to the cambium. It could also help to explain the formation of the heartwood lignin-like polymers mentioned in Section 11.6.5.

Regardless of the technique used, we believe that a much better understanding of the function and the potential for control of extractive formation will be forthcoming when we have detailed information of the precise location of various types of extractives in the tree.

11.7 Conclusion

The availability of large quantities of bark and wood residues as a byproduct from the pulp and wood products industries, a large body of technology just waiting to be utilized, and a world in which renewable resources must ultimately displace the use of oil and other fossil fuel ought to augur well for the future of wood extractives. The fact is, however, that there is a rapidly declining base of financial support for work in this field. In view of the promise of wood science, and that of progressive development of knowledge of secondary components in particular, what can those of us who are optimistic about future prospects do to reverse this trend? For without financial support for further research and development, there is little likelihood of providing the results that will surely be needed in the near future.

As Hillis has recently so aptly summarized the situation (29), "Perhaps the biggest challenge facing wood science research, and particularly for that concerning secondary products, is not the development of the discipline but the advancement of a widespread realization of its importance for the development of forest products everywhere." To that, we would add the admonition that each scientist who is active in this field must be sure that the ultimate goals of his work are not only consistent with advancing the knowledge base of the discipline, but are clearly stated in each piece of published work. In that way the reader and, more importantly, the sponsor will be able to understand how the work is advancing environmental and material needs of present and future generations.

References

1 Abelson P H 1985 Plant-fungal symbiosis. Science 229:617
2 Baes C F, McLaughlin S B 1984 Trace elements in tree rings (as) evidence of recent and historical air pollution. Science 224:494–497
3 Berlin J 1984 Plant cell cultures – a further source of natural products? Endeavour New Series (8)1:5–8
4 Berlin J, Witte L, Schubert W, Wray V 1984 Determination and quantification of monoterpenoids secreted into the medium of cell cultures of *Thuja occidentalis*. Phytochemistry 26:1277–1279
5 Bonsen K J M, Scheffer R J, Elgersma D M 1985 Barrier zone formation as a resistance mechanism of elms to Dutch elm disease. IAWA Bull 6(1):71–77
6 Bratt L C 1979 Wood-derived chemicals: Trends in production in the U. S. Pulp Pap 6:102–108
7 Campos-Lopez E, Roman-Alemany A 1980 Organic chemicals from the Chihuahuan desert. J Agr Food Chem 28:171–183
8 Chemical and Engineering News 1983 Test plant for guayule rubber to be built. Chem Eng News 16:7–8
9 Chemical and Engineering News 1983 Bioregulator doubles guayules's rubber output. Chem Eng News 25:7–8
10 Chemical and Engineering News 1985 Science, technology role stressed in outlook for developing countries. Chem Eng News 14:48–52
11 Chemical and Engineering News 1985 Biotech research for agriculture advances on several fronts. Chem Eng News 30:71–72
12 Chemical Week 1986 They're not 'biting' at this bark. Chem Week 22:39
13 Corder S E 1976 Properties and uses of bark as an energy source. Oreg State Univ For Res Lab Paper 31
14 Drew J 1984 Editorial: Let's face it. Naval Stores Rev 94(4):4
15 Freudenberg K 1964 In: Zimmermann M H (ed) The formation of wood in forest trees. Academic Press New York, 203
16 Gonick E 1984 The future of rosin. Naval Stores Rev 94(6):5
17 Gray R L, Rickey R G, Hergert H L 1983 The influence of sapwood-heartwood conversion of bordered pit tori in western hemlock on bisulfite pulping. Wood Fiber Sci 15:251–262
18 Hare R C 1984 Stimulation of early height growth in longleaf pine with growth regulators. Can J For Res 14:459–462
19 Hemingway R W 1981 Bark: Its chemistry and prospects for chemical utilization. In: Goldstein I S (ed) Organic chemicals from biomass. CRC Press Boca Raton, 190–235
20 Hergert H L 1962 Economic importance of flavonoid compounds: Wood and bark. In: Geissman T A (ed) The chemistry of flavonoid compounds. Macmillan New York, 553–592

21 Hergert H L 1971 Infrared spectra. In: Sarkanen K V, Ludwig C H (eds) Lignins: Occurrence, formation, structure, and reactions. Wiley-Interscience New York, 267–298
22 Hergert H L 1977 Secondary lignification in conifer trees. In: Arthur J C (ed) Cellulose chemistry and technology. Am Chem Soc Symp Series 48:227–243
23 Hergert H L 1983 Tannins. In: Davis R (ed) Encyclopedia of American forest and conservation history. Macmillan New York, 2:631–632
24 Herrick F W 1980 Chemistry and utilization of western hemlock bark extractives. J Agr Food Chem 28:228–237
25 Herrick F W, Hergert H L 1977 Utilization of chemicals from wood: Retrospect and prospect. In: Loewus F A, Runeckles V C (eds) Recent advances in phytochemistry, vol II. Structure, biosynthesis, and degradation of wood. Plenum Press New York, 443–515
26 Hillis W E 1962 Factors controlling the amount of polyphenols formed (in heartwood). In: Hillis W E (ed) Wood extractives and their significance to the pulp and paper industry. Academic Press New York, 118
27 Hillis W E 1975 Ethylene and extraneous material formation in woody tissues. Phytochemistry 14:2559–2562
28 Hillis W E 1977 Secondary changes in wood. In: Loewus F A, Runeckles V C (eds) Recent advances in phytochemistry, vol II. Structure, biosynthesis, and degradation of wood. Plenum Press New York, 247–309
29 Hillis W E 1986 Forever amber – A story of the secondary wood components. Wood Sci Technol 20:203–227
30 Kai Y, Teratani F 1979 Studies on the color of the heartwood of sugi (*Cryptomeria japonica*). II. Radial distribution of heartwood pigment. Mokuzai, Gakkaishi 25:77–81
31 Krumbein J P 1984 Review of technical papers at naval stores meeting. Naval Stores Rev 94(6):6
32 Lai Y Z, Sarkanen K 1971 (Lignin) Occurrence and structure. In: Sarkanen K V, Ludwig C H (eds) Lignins: Occurrence, formation, structure, and reactions. Wiley-Interscience New York, 165–240
33 Mattson R H 1984 Turpentine, a by-product of the kraft paper industry. Naval Stores Rev 94(4):10–16
34 Mayr M, Hausmann B, Zelman N, Kratzl K 1984 Identification of the volatile leaf oils of common spruce by the stripping method and composition of the nonvolatile extractives in the bark of *Picea* sp. Proc 1984 TAPPI Res Development Conf Atlanta, 165–168
35 McReynolds R O, Kossuth S V 1984 CEPA in sulfuric acid paste increases oleoresin yields. S J Appl For 8:168–172
36 Morton J F 1980 Search for carcinogenic principles. In: Swain T, Kleiman R (eds) Recent advances in phytochemistry, vol 14. The resource potential in phytochemistry. Plenum Press New York, 53–73
37 Nagoshi A 1970 Dissolving sulphite pulp from rubberwood. Indonesian Pulp Paper Tech Assoc 7:223–230
38 Nakayama K, Usui T, Hiraishi S, Onosato K 1972 Pulping of para rubber tree. Jpn Tappi 26:323–326
39 Nelson N D 1978 Xylem ethylene, phenol-oxidizing enzymes and nitrogen and heartwood formation in walnut and cherry. Can J Bot 56:626–634
40 Nelson N D, Hillis W E 1978 Genetic and biochemical aspects of kino vein formation on eucalyptus. Aust For Res 8:75–91
41 National Air and Space Administration 1978 Bioconversion study conducted by the Jet Propulsion Laboratory. JPL Publication 79-9. JPL Pasadena, 115 pp
42 Nobuchi T, Matsuno H, Harada H 1985 Radial distribution of heartwood phenols and the related cytological structure in the fresh wood of sugi (*Cryptomeria japonica* D. Don). Mokuzai Gakkaishi 31:711–718
43 Oosterhof D, Bakker J, Kamateri O 1983 Natural rubber. In: Chemical economics handbook. SRI Section 525, 2001A. Stanford Research International Stanford
44 Perry P 1983 Naval stores. In: Davis R (ed) Encyclopedia of American forest and conservation history. Macmillan New York, 2:471–479
45 Pizzi A, Cameron F A 1986 Flavonoid tannins – structural wood components for drought-resistance mechanisms of plants. Wood Sci Technol 20:119–124

46 Powell R G, Smith C R 1980 Antitumor agents from higher plants. In: Swain T, Kleiman R (eds) Recent advances in phytochemistry, vol 14. The resource potential in phytochemistry. Plenum Press New York, 23–51
47 Roux D G, Ferreira D, Botha J J 1980 Structural considerations in predicting the utilization of tannins. J Agr Food Chem 28:216–222
Chem Week 22:39
48 Sandberg G 1984 Biosynthesis and metabolism of indole-3-ethanol and indole-3-acetic acid by *Pinus sylcerstris* L. needles. Planta 161:398–403
49 Sarkanen K V 1971 Precursors and their polymerization. In: Sarkanen K V, Ludwig C H (eds) Lignins: Occurrence, formation, structure, and reactions. Wiley-Interscience New York, 110
50 Savidge R A, Wareing P F 1984 Seasonal ambial activity and xylem development in *Pinus contorta* in relation to endogenous indol-3-yl-acetic and (S)-abscisic acid levels. Can J For Res 14:676–682
51 Science News 1985 Tracking down ores with trees. Sci News 128:191
52 Shigo A L 1985 Compartmentalization of decay in trees. Sci Am 252(4):96–103
53 Smiley J T, Horn J M, Rank N E 1985 Ecological effects of salicin at three trophic levels: New problems from old adaptions. Science 229:649–651
54 Stacy D L 1971 Practical experiences in the use of rubberwood for the production of pulp and paper. UNIDO Document ID/WG102/15. United Nations Industrial Development Organization New York, 30
55 Stich K, Ebermann R 1984 Peroxidase and polyphenol oxidase isoenzymes in sapwood and heartwood of oak. Holzforschung 38:239–242
56 Su Z-a 1984 Naval stores industry in China. Naval Stores Rev 8:13
57 Terazawa M, Miyake M 1984 Phenolic compounds in living tissue of woods. 2. Seasonal variations of phenolic glycosides in cambial sap of woods. J Jpn Wood Res Soc 30:329–334
58 The Economist 1986 Natural rubber new attire. The Economist (London) 15:107
59 Trocino F S 1974 Method of extracting wax from bark. US Patent 3789058
60 Van K T T, Toubart P, Cousson A, Dravill A G, Gollin D J, Chelf P, Albersheim P 1985 Manipulation of the morphogenetic pathways of tobacco explants by oligosaccharins. Nature 314:615–617
61 White E E, Nilsson J E 1984 Genetic variation in resin canal frequency and relationship to terpene production in foliage of *Pinus contorta*. Silvae Genet 33:79–84
62 Wolter K E, Zinkel D F 1984 Observation on the physiological mechanism and chemical constituents of induced oleoresin synthesis in *Pinus resinosa*. Can J For Res 14:452–458
63 Yamaguchi T, Fukuzumi T, Yoshi T 1986 Effects of phytohormones on phenolic accumulations in callus of *Eucalyptus robusta*. Mokuzai Gakkaishi 32:209–212
64 Zinkel D F 1981 Turpentine, rosin and fatty acids from conifers. In: Goldstein I S (ed) Organic chemicals from biomass. CRC Press Boca Raton, 163–187

Index of Plant Genera and Species

There are a number of synonyms in the index; we have not tried to remove them. Where synonymous names exist for a species, all synonyms should be checked.

Abies   165, 539, 558, 674, 852, 884, 889, 895, 1003, 1057
Abies alba   159–161, 264, 319, 821
Abies amabilis   327, 331, 557, 1053
Abies balsamea   319, 323, 344, 717, 1051, 1054
Abies concolor   309–310, 327, 331, 333, 557, 703, 1109
Abies firma   657, 673, 857
Abies grandis   310, 319
Abies lasiocarpa   319
Abies magnifica   310, 737, 825
Abies nephrolepis   442
Abies nordmanniana   1104, 1109
Abies pectinata   704
Abies pindrow   1110
Abies procera   319, 858
Abies sachalinensis   673, 717, 857–858, 1118
Abies semenovii   1104, 1109–1110
Abies sibirica   302
Abies webbiana   576
Abroma augusta   1096
Abrus cantoniensis   1110
Abrus precatorius   1096
Abuta rufescens   1107
Acacia   5, 164, 168, 171, 283, 374, 521, 526, 539, 544, 547, 563–564, 590, 592, 612, 617, 670–672, 675–676, 683, 979, 983, 986, 1057, 1168
Acacia auriculiformis   604
Acacia baileyana   619, 660
Acacia catechu   9, 587, 997
Acacia cultriformis   606–607
Acacia cyclops   168
Acacia decurrens   563
Acacia elata   168
Acacia excelsa   607, 672
Acacia farnesiana   407
Acacia fasciculifera   604
Acacia filicifolia   168
Acacia georginae   268
Acacia giraffae   1096
Acacia glogulifera   283
Acacia harpophylla   607
Acacia hebeclada   282
Acacia implexa   168
Acacia kempeana   539, 545
Acacia longifolia   168
Acacia luderitzii   613–614
Acacia mearnsii   6, 168, 539, 563, 596, 600, 604, 606, 613, 615–616, 619–620, 652, 660, 925, 993–995, 997–998, 1002, 1012
Acacia melanoxylon   586, 604, 607, 619, 621, 679, 935, 1095
Acacia mimosa   545
Acacia mollissima   161, 543, 1175, 1190
Acacia nigrescens   545, 604
Acacia nilotica   5, 993
Acacia obtusifolia   606
Acacia ovites   606
Acacia podalyriaefolia   168
Acacia pycnantha   168, 171, 563
Acacia raddiana   320
Acacia rhodoxylon   544
Acacia salinga   168
Acacia saxatilis   606–607
Acacia senegal   5, 168, 171, 320, 322, 326, 345, 1104
Acacia seyal   316, 321
Acacia sieberiana   282
Acacia spargitalia   606
Acacia villosa   940
Acanthopanax chiisanensis   446
Acanthopanax koreanum   446
Acer   163, 187, 404, 429, 431, 857, 895, 924
Acer campestre   160, 162, 408, 413
Acer ginnale   408
Acer griseum   328
Acer mono   858
Acer platanoides   408, 413
Acer pseudoplatanus   157–158, 184, 187, 193, 328
Acer rubrum   413, 605, 865
Acer saccharinum   408, 414
Acer saccharum   156, 163, 166, 260, 267, 320
Acer tartaricum   408
Achras sapota   671, 777, 1049
Achyranthes fauriei   1099
Aconitum   248
Aconium lindleyi   407
Acorus calamus   320, 322, 345, 447, 718–719, 722, 724, 745
Acritopappus longifolius   741
Acronychia baueri   211, 782
Acronychia muelleri   445
Actinidia   675

Actinidia chinensis 674
Actinocheita filicina 561
Actinodaphne madraspatane 536
Adansonia digitata 267, 320
Adenanthera pavonina 545, 1110
Adenia lobata 283
Adenia volkensii 284
Adhatoda vasica 1065, 1118–1119
Adina cordifolia 1117
Adiscanthus fusciflorus 1098
Aegle marmelos 1098
Aesculus 300, 626, 675
Aesculus carnea 662
Aesculus flava 268
Aesculus hippocastanum 267–268, 311, 320, 622, 625, 666, 674
Aesculus indica 1104
Aesculus persea 624
Afraegle paniculata 1098, 1108, 1113, 1119
Afrormosia elata 550, 925
Afzelia 541–542, 590
Agalinis purpurea 79
Agathis alba 4, 852
Agathis australis 4, 319, 519, 673, 756, 759
Agathis microstachya 751, 758
Agave 331
Agave americana 308, 318, 331, 345,
Agrimonia pilosa 421, 522
Ailanthus altissima 227, 322
Ailanthus excelsa 227
Ailanthus glandulosa 308, 312
Ailanthus malabarica 227, 783–784
Akimmia reevesiana 1117
Alangium 224
Alangium lamarckii 224
Albertisia papuana 1110
Albizia 168, 539, 823
Albizia adianthifolia 545, 547, 1110
Albizia amara 823
Albizia brownei 168
Albizia glaberrima 168
Albizia julibrissin 823, 835
Albizia lebbeck 260, 672, 1110
Albizia procera 823
Albizia sericocephela 168
Albizia zygia 168
Aldina heterophylla 551
Aleurites cordata 264
Alhagi camelorum 9, 158
Alhagi persarum 1117
Alhagi pseudalhagi 1104, 1106
Alkanna tinctoria 1093
Allamanda cathartica 1111
Allanblackia florigunda 1102
Allium cepa 333
Alluaudia 540, 542
Alluaudia ascendens 540

Alluaudia dumosa 540
Alluaudia humbertii 540
Alnus 524, 895
Alnus firidis 514
Alnus firma 514, 522
Alnus glutinosa 445, 782
Alnus hirsuta 522
Alnus japonica 522, 822, 826, 1118
Alnus rubra 311, 522, 524
Alnus sieboldiana 422, 514
Aloe arborescens 1104
Alpinia nutans 705
Alstonia yunnanensis 1094
Amanoa oblongifolia 445
Amaroria soulameoides 825
Amorpha fruticosa 735
Ampelopsis hederaceae 264
Amphipterigium 547, 562
Amphipterygium adstrigens 561–562
Amsonia elliptica 1119
Amyris pinnata 1094, 1113
Anacardium 170, 576
Anacardium excelsum 561
Anacardium occidentale 170, 577
Anacridium melanorhodon 683
Anacyclus pyrethrum 446
Ananas comosus 345
Andenochlaena siamensis 753
Andira parviflora 549, 1098
Angelica archangelica 697
Aniba 371, 444, 536, 548, 558, 1099, 1101, 1112–1113
Aniba affinis 443
Aniba burchellii 443
Aniba canella 373
Aniba citrifolia 443
Aniba ferrea 443
Aniba guianensis 443
Aniba lancifolia 443
Aniba megaphylla 443
Aniba riparia 539, 544–545, 547
Aniba rosaeodora 544, 547–548, 702
Aniba simulans 443
Aniba terminalis 444
Anisodus tanguticus 1094
Anisoptera 853, 924
Annona acuminata 1094, 1106–1107
Annona reticulata 1099, 1113
Annona squamosa 1094
Anogeissus 164, 170
Anogeissus latifolia 169, 264, 986
Anogeissus leiocarpus 169
Anopterus 248
Anthemis cotula 745
Anthemis cretica 723
Anthemis montana 702
Anthocercis frondosa 1109, 1113

Anthriscus nemerosa 446
Anthriscus sylvestris 446
Antiaris 855
Antiaris africana 854
Aptenia cordifolia 181
Aptosimum spinescens 447
Apuleia 563–564
Apuleia leiocarpa 534, 539–540, 542, 559, 563
Aquilaria 853
Aquilaria agallocha 728–729
Arachis 624
Arachis hypogea 622, 990, 998, 1008–1009, 1013–1014
Aralia cordata 1109, 1116
Aralia mandshurica 1110
Araucaria 3, 576, 751, 852
Araucaria angustifolia 442, 980
Araucaria araucana 761, 1095
Araucaria bidwillii 171
Araucaria klinkii 171
Arbutus unedo 264, 1111
Arctium lappa 446
Arctium leiospermum 446
Arctostaphylos glandulosa 684
Arctostaphylos uva-ursi 403, 407, 413, 780, 1094
Areca catechu 625–627, 990
Arenga 163
Aristolochia 444
Aristolochia albida 444
Aristolochia argentina 280
Aristolochia debilis 744
Aristolochia indica 444, 729, 743
Aristolochia taliscana 444
Aristolochia triangularis 444
Aristotelia serrata 1099
Arseodaphne 216
Artemisia 934
Artemisia absinthium 446
Artemisia fragrans 708
Artemisia pallens 745
Artocarpus 7, 536–538, 556, 558–559, 901
Artocarpus integrifolia 544, 1108, 1117
Artocarpus nobilis 537
Artocarpus pithecogalla 1108
Asarum sieboldi 444, 708
Asarum taitonense 444
Ascidia nigra 45
Asclepias syriaca 1109
Ascodichaena rugosa 348
Asimina triloba 1106
Aspidosperma 230–231
Aspidosperma marcgravianum 1094
Aspidosperma olivaceum 231
Aspidosperma peroba 10
Aspidosperma quebracho blanco 161, 229

Aspidosperma subincanum 231
Aster acaber 817
Astragalus 164, 169, 985
Astragalus gummifer 5, 169
Astragalus microcephalus 169, 172
Atalantia ceylanica 1119
Atalantia monophylla 1111
Athrotaxis selaginoides 517, 519
Atractylodes lancea 729
Atriplex halimus 1096
Atriplex hastata 1096
Atriplex littoralis 1096
Atropa 203
Austrobaileya scandens 443, 1094
Azadirachta 171
Azadirachta indica 17, 171
Azima tetracantha 321

Baccharis 752
Baccharis calvescens 1115
Baccharis cassinaefolia 1115
Baccharis genistelloides 1115
Baccharis salzmannii 1115
Baikiaea plurijuga 671, 925
Balanites aegyptiaca 322, 836, 1099
Balanites orbicularis 836
Balanites pedicellaris 836
Balanites roxburghii 1099
Balanites wilsoniana 836, 1099
Balfourodendron riedelianum 210
Baliospermum montanum 766
Baphia 527
Baphia nitida 7, 527
Barteria fistulosa 284
Bauhinia purpurea 260, 264
Begonia 268
Berberis 216, 1095, 1105
Berberis actinacantha 1107
Berberis amurensis 1107
Berberis asiatica 1093
Berberis buxifolia 445
Berberis chilensis 445
Berberis chitria 1096, 1113
Berberis coriaria 1093
Berberis darwinii 1107
Berberis floribunda 1093
Berberis julianae 1107
Berberis koreana 1107
Berberis lycium 1093
Berberis thunbergii 1093, 1107
Berberis verruculosa 1096, 1113
Berberis vulgaris 218, 1093, 1096, 1107, 1113
Bergenia 590, 592
Bergenia cordifolia 407–408
Bergenia crassifolia 407–408
Betula 160, 163, 300, 392, 822, 857, 887, 890, 894, 936

Betula alba 156–157, 392
Betula alleghaniensis 835, 1098
Betula lenta 835, 1098
Betula manshurica 897
Betula maximowicziana 858
Betula papyrifera 320
Betula pendula 157–159, 311, 320, 328, 343, 780
Betula platyphylla 311, 328, 522, 777
Betula populifolia 159, 320
Betula pubescens 159
Betula verrucosa 157, 302, 323, 328, 787
Biota 557
Blechnum niponicum 837
Bleekeria vitiensis 1100
Blumea balsamifera 710
Boehmeria biloba 671
Boehmeria tricuspis 445
Boenninghausenia albiflora 1105
Boerhavia diffusa 1104, 1117
Bombyx mori 37–38
Bonnetia stricta 1101
Boswellia 7, 170, 1051
Boswellia carteri 170, 764
Boswellia cartevii 709
Boswellia frereana 1098
Boswellia papyrifera 170
Boswellia serrata 703, 708
Bothriochloa intermedia 727
Bowdichia nitida 550, 845
Brabejum stellatifolium 171
Brachychiton diversifolium 171
Brachyostegia spiciformis 347
Brassica 819
Brassica napobrassica 308
Brassica napus 989
Brassica oleracea 313–316
Brassica rapa 816
Brickellia guatemaliensis 716
Bridelia micrantha 1099, 1102, 1115
Broussonetia 590
Broussonetia papyrifera 909
Brucea javanica 1095
Bruguiera 6
Bruguiera conjugata 293–294
Bruguiera cylindrica 293–294
Bruguiera gymnorrhiza 321
Brunfelsia uniflora 1113
Brya ebenus 847
Buddleja davedii 447, 739
Bunesia sarmienti 730
Bupleurum fruticescens 446
Bursera ariensis 446
Bursera coalifera 320
Bursera delpechiana 702
Bursera fagaroides 446
Bursera graveolens 705

Bursera microphylla 446
Bursera morelensis 446
Butea 540
Butea frondosa 9, 605
Butyrospermum parkii 817
Buxus microphylla 1098
Buxus sempervirens 250
Byrsonima intermedia 1111

Cacalia 729
Cacalia decomposita 729
Caesalpinia 554
Caesalpinia braziliensis 847
Caesalpinia brevifolia 406, 408, 410–411, 419, 427
Caesalpinia coriaria 406, 411
Caesalpinia echinata 7, 554
Caesalpinia sappan 6, 846–847, 1095, 1102
Caesalpinia spinosa 406, 414
Callisthene major 1102
Callitris 730
Callitris collumellaris 442, 724, 727, 731
Callitris drummondi 442
Callitris glauca 702
Callitris quadrivalvis 704
Callitropsis araucarioides 725
Calluna 403
Calluna vulgaris 633
Calocedrus decurrens, see Libocedrus decurrens 888
Calocedrus formosana 442
Caloncoba echinata 285
Calophyllum 293, 525, 554, 894
Calophyllum brasiliense 1102, 1105
Calophyllum cordato-oblongum 1102, 1105
Calophyllum cuneifolium 1105
Calophyllum inophyllum 1101, 1105
Calophyllum lankaensis 293, 1105
Calophyllum ramiflorum 1101, 1105
Calophyllum robustum 1107
Calophyllum thwaitesii 293
Calophyllum tomentosum 1102, 1105, 1115, 1119
Calophyllum trapezifolium 1105, 1119
Calophyllum vexans 858
Calophyllum walkeri 293
Calophyllum zeylanicum 293
Calotropis procera 1118
Calycanthus 141, 225
Calycanthus floridus 225–226
Calycanthus glaucus 225
Calycanthus occidentalis 225
Calygonium squamulosum 940
Camellia 590–591, 817
Camellia japonica 422
Camellia sinensis 260, 267, 403–404, 407, 940, 989, 991

Index of Plant Genera and Species 1201

Camponotus 280
Camptotheca acuminata 244–245, 1095, 1098
Canarium 853
Canarium commune 773
Canarium indicum 858
Canarium luzonicum 724, 773
Canarium samonense 744
Canarium strictum 780
Canarium zeylanicum 1115
Capraria biflora 764
Caraipa densiflora 1102, 1118
Caraipa grandifolia 1118
Carapa procera 777
Cardiospermum hirsutum 282
Carica papaya 344
Carissa edulis 447, 1118
Carya 539, 935
Carya illinoensis 267, 998, 1008–1009, 1013–1014
Carya ovata 525
Casearia esculenta 671
Casimiroa 558
Casimiroa edulis 210, 536, 538
Cassia 526, 857
Cassia angustifolia 1063, 1114
Cassia fistula 613–614, 672, 1096
Cassia laevigata 1100
Cassia obovata 1112
Cassia obtusifolia 267
Cassia occidentalis 1100, 1110
Cassia quinquangulata 1097
Cassia reniger 536, 544
Cassia sieberiana 672
Cassia singueiana 1103, 1114, 1117
Cassia sophera 1100, 1110
Cassine matabelica 728
Cassinopsis ilicifolia 224
Cassipourea verticillata 672
Castanea 406, 927, 993–994, 997, 999, 1119
Castanea crenata 407
Castanea sativa 6, 326, 328, 408, 422, 993
Castanopsis cuspidata 858
Castanospermum australe 537
Castilla 8
Castilloa elastica 161
Casuarina equisetifolia 264
Casuarina stricta 422
Catalpa 300
Catalpa ovata 525
Catalpa speciosa 20
Catha 213
Catha edulis 213, 940, 1096, 1111
Catharanthus lanceus 1110, 1119
Catharanthus roseus 1093–1094, 1110, 1112, 1119
Catharanthus trichophyllus 1117
Cathaya argyrophylla 1111

Ceanothus americanus 264, 267
Ceanothus velutinus 536, 824
Cedrela 853
Cedrela toona 734–736, 750, 776
Cedrelopsis grefei 373
Cedrus 558
Cedrus atlantica 557–558
Cedrus deodara 442, 545–546, 556–558, 717, 740–741, 852, 1055, 1093, 1096, 1104, 1110, 1116
Cedrus libani 2, 10, 302, 558
Ceiba samauma 855
Celastrus scandens 1096
Celtis 854–855
Celtis kajewskii 858
Celtis reticulosa 853
Centella asiatica 780
Centrolobium 514, 522–523
Centrolobium robustum 523
Cephaelis 223
Cephaelis ipecacuanha 223
Cephalanthus spathelliferus 545
Cephalaria kotschyi 1104
Cephalaria nachiczevanica 1104
Cephalotaxus 222–223, 576, 1064
Cephalotaxus fortunei 1103–1104
Cephalotaxus hainanensis 1103–1104
Cephalotaxus harringtonia 222, 1096, 1104
Cephalotaxus mannii 1094
Cephalotaxus oliveri 1104
Cephalotaxus sinensis 1104
Cephalotaxus wilsoniana 1096
Cerasus vulgaris 312
Ceratonia siliqua 408, 672, 680, 989
Cercidiphyllum 427, 429
Cereus peruvianus 171
Cetinus obovatus 561
Chaenomeles 658, 663
Chaenomeles chinensis 625, 655, 657–658, 662, 674, 680
Chaenomeles speciosa 675
Chaerophyllum maculatum 446
Chamaecyparis 852, 870
Chamaecyparis formosensis 709, 727
Chamaecyparis lawsoniana 310, 702–703, 710, 717, 821, 858
Chamaecyparis nootkatensis 327, 331, 673, 709, 711, 729, 761, 820, 857–858, 929
Chamaecyparis obtusa 442, 519, 673, 857–858
Chamaecyparis obtusa formosana 704
Chamaecyparis pisifera 673, 858
Chamaecyparis taiwanensis 702, 934
Chamaecyparis thyoides 720
Chasmanthera dependens 1107
Chelidonium majus 269, 1113
Chimosanthus fragrans 225

1202 Index of Plant Genera and Species

Chlorophora excelsa 514, 517, 925
Chlorophora rigia 925
Chlorophora tinctoria 7, 539, 542, 544, 846, 848
Chloroxylon 170
Chondodendron 217
Chondodendron candicans 217
Chondodendron platyphyllum 217
Chondodendron tomentosum 217
Chrysanthemum cinerariaefolium 721
Chrysothamnus nauseosa 705
Cinchona 9
Cinchona calisaya 9
Cinchona ledgeriana 9, 1097, 1112
Cinchona officinalis 9
Cinchona pubescens 1097, 1112
Cinchona succirubra 237, 593
Cinnamomum 444, 590, 626–627, 670, 853
Cinnamomum bodinieri 1101, 1112
Cinnamomum camphora 10, 214, 293, 300, 444, 710, 858, 1112
Cinnamomum cassia 592, 990
Cinnamomum fragrans 744
Cinnamomum japonicum 704
Cinnamomum laubattii 214
Cinnamomum parthenoxylon 1112
Cinnamomum zeylanicum 7, 622, 625, 990
Cinnamosma madagascarinensis 443
Cissampelos pareira 162
Cissus pallida 514
Cistus 752
Citrus 267, 317
Citrus aurantium 343
Citrus decumana 268
Citrus junos 743
Citrus limon 170, 321
Citrus maxima 170
Citrus natsudaidai 700
Citrus paradisi 264, 317–318, 345, 729,
Citrus reticulata 321
Citrus sinensis 170, 260, 264, 267
Citrus unshiu 264
Cladrastis 850
Cladrastis platycarpa 547, 550, 845
Cleistanthus collinus 445, 1114
Cleistanthus patulus 445, 1114
Cleistanthus schlechteri 756
Clereospermum kunthianum 447
Clerodendron, see Clerodendrum
Clerodendron inerme 447
Clerodendron infortunatum 752
Clerodendrum 817, 832
Clerodendrum fragrans 1098, 1117
Clinostemon 279
Clinostemon mahuba 275
Clitoria ternatea 1115
Cneorum tricoccon 777
Cocculus 216

Cocculus laurifolia 216, 222
Cochlospermum gossypium 171
Cochlospermum kunth 171
Cocos nucifera 989
Coffea 989
Coleus 181, 758
Colletia spinosissima 1104
Colophospermum 590
Colophospermum mopane 592, 613, 615–616, 671
Combretum 164
Combretum elliottii 170
Combretum hartmannianum 170
Combretum leonense 170
Combretum micranthum 264
Combretum quadrangulare 1098
Combretum salicifolium 170
Commiphora 7
Commiphora incisa 446
Commiphora mukul 701
Commiphora myrrha 170
Conium maculatum 932
Connarus monocarpus 672
Conocarpus erectus 445
Conopharingia 230
Copaifera copallifera 4
Copaifera demeusi 4
Copaifera officinalis 752
Coptis chinensis 1105
Coptis deltoides 1105
Coptis japonica 1105
Corchorus capsularis 374, 1117
Corchorus olitorius 374
Cordia 10
Cordia buddleioides 1107
Cordia elaeagnoides 1103
Cordyline terminalis 321
Coreopsis 547
Cornus 157, 267, 300, 422
Cornus florida 37
Cornus mas 260
Cornus sericea 267
Cornus stolonifera 311
Coronilla 990
Coronilla varia 989
Corydalis lutea 1113
Corydalis paniculigera 1113
Corynanthe yohimbe 236
Coscinium fenestratum 1105
Cotinus 431
Cotinus coggygria 408, 413, 671
Cotoneaster 674
Cotylelobium 853, 858
Cotylelobium scabriusculum 1117
Couroupita guianensis 260
Crataegus monogyna 989
Crataegus oxyacantha 662, 671, 674, 681, 992

Cratoxylon 894
Crossopetalum uragoga 1111, 1117
Croton 216
Croton diasii 754
Croton eluteria 293
Croton joufra 1111
Croton macrostachys 288
Croton nepetaefolius 1099, 1101
Croton oblongifolius 1111
Croton sublyratus 750, 1068, 1111
Croton zehntneri 1100–1101, 1109, 1112
Crudia amazonica 1094
Cryptocarya 901
Cryptocarya amygdalina 292
Cryptocarya massoia 280
Cryptocarya pleurosperma 10
Cryptomeria 1185
Cryptomeria japonica 519, 673, 726, 761, 852, 857–858, 899–900, 909, 920, 1185
Cryptostegia madagascariensis 933
Cucurbita 186
Cucurbita maxima 817–818
Cunninghamia 824, 852
Cunninghamia konishii 1116
Cupressus 852, 870
Cupressus benthami 1116
Cupressus dupreziana 1116
Cupressus funebris 1116
Cupressus glabra 1116
Cupressus leylandii 327
Cupressus lindleyi 1116
Cupressus lusitanica 708, 1117
Cupressus macnabiana 1116
Cupressus nevadensis 1116
Cupressus pygma 1116
Cupressus sargentii 1116–1117
Cupressus sempervirens 704, 750, 1116
Cupressus stephensonii 1116–1117
Cupressus torulosa 750
Curcuma 522
Curcuma aromatica 717
Curcuma zedoaria 722, 724
Cussonia spicata 171
Cyanthus stercoreus 348
Cycas revoluta 576
Cyclea 216
Cyclea burmanni 162
Cyclea peltata 1107
Cyclolobium 547
Cyclolobium clausseni 552
Cydonia oblonga 659, 662, 674–675, 989
Cylicodiscus gabunensis 547–548
Cyperus rotundus 732
Cytisus laburnum 209
Cytisus multiflorus 1115
Cytisus purgans 1107, 1115
Cytisus scoparius 209, 1094, 1107, 1115

Dacrydium bidwillii 756
Dacrydium biforme 750
Dacrydium colensoi 751, 754, 756
Dacrydium dupressinum 758
Dacrydium franklinii 1101
Dacrydium intermedium 442, 837
Dacrydium kirkii 750
Dactylocladus stenostachys 1102
Daemonorops 9
Daemonorops draco 527
Dalbergia 10, 383, 526, 550, 554, 562–564, 566, 845–846, 849, 857, 859
Dalbergia baroni 565
Dalbergia barretoana 565
Dalbergia bugra 565
Dalbergia cearensis 565, 1101
Dalbergia cochinchinensis 565, 1094
Dalbergia decipularis 565
Dalbergia ecastophyllum 545, 565, 1098, 1101
Dalbergia frutescens 565
Dalbergia inundata 1101
Dalbergia lanceolaria 565
Dalbergia latifolia 554, 565
Dalbergia melanoxylon 554, 565
Dalbergia miscolobium 565
Dalbergia nigra 554
Dalbergia nitidula 554, 564, 619
Dalbergia obtusa 565
Dalbergia oliveri 552, 1101
Dalbergia paniculata 553-554
Dalbergia parviflora 716
Dalbergia retusa 544, 1106
Dalbergia riparia 550
Dalbergia sericea 544
Dalbergia sissoo 554, 565, 716, 941
Dalbergia spruceana 565, 1094, 1099, 1101
Dalbergia stevensonii 549
Dalbergia variabilis 551, 1101
Dalbergia villosa 565
Dalbergia volubilis 565, 1102
Daniellia 4, 853
Daniellia oliveri 751
Daphnandra 216
Daphne 268
Daphne giraldii 1115
Daphne mezereum 766
Daphne oleoides 445
Daphne tangutica 445
Daphniphyllum 249
Daphniphyllum macropodum 249
Datisca glomerata 1098
Datura 203
Datura meteloides 1113
Daucus carota 72, 308, 348, 721
Davidia involucrata 429
Davilla flexuosa 671
Decarya madagascariensis 540

Decodon 207
Decussocarpus wallichianus 1105
Deidamia clematoides 283
Delonix regia 824
Delphinium 248
Dendroctonus 865
Dendroctonus ponderosae 865
Dendropanax trifidus 292
Derris 553
Derris rariflore 514
Dialyanthera otoba 321
Dichapetalum barteri 268
Dichapetalum cymosum 264, 268-269
Dichapetalum toxicarium 268–269
Dichrostachys cinerea 824
Dicranum 572
Dictamnus albus 1096, 1099, 1114
Dictamnus angustifolius 1099, 1114
Dictamnus caucasicus 1099
Dictyopteris 293
Diderea madagascariensis 539
Dillenia indica 1102
Dillenia pulchella 910
Dillenia retusa 1102
Dioscoreophyllum cumminsii 1105, 1107
Diospyros 2, 844, 846–847, 857, 859, 935, 946
Diospyros austro-africana 1104
Diospyros celebica 10
Diospyros cinnabarina 1099, 1102
Diospyros ebenum 780, 1117
Diospyros fragrans 1102
Diospyros hirsuta 1115
Diospyros ismailii 1099, 1115, 1117
Diospyros japonica 1115
Diospyros kaki 445, 989, 992
Diospyros lotus 1099, 1115, 1117
Diospyros maingayi 1099, 1115, 1117
Diospyros maritima 1113
Diospyros melanoxylon 1117
Diospyros morrisiana 1099
Diospyros peregrina 671
Diospyros quaesita 1113
Diospyros siamang 1099, 1115, 1117
Diospyros singaporensis 1099, 1115, 1117
Diospyros sumatrana 1099, 1115, 1117
Diospyros virginiana 525, 940, 1099, 1113
Diospyros walkeri 1113
Diospyros wallichii 1099, 1115, 1117
Diphylleia grayi 445
Diplotropis purpurea 544, 1101, 1106
Dipterocarpus 730, 739, 743, 853, 857–858, 897, 901
Dipterocarpus glandulosus 1097, 1113, 1117
Dipterocarpus hispidus 1097, 1113, 1117
Dipterocarpus insignis 1097, 1113, 1117
Dipterocarpus macrocarpus 1097
Dipterocarpus zeylanicus 1097, 1113, 1117

Dirca occidentalis 445
Distemonanthus 563
Distemonanthus benthamianus 539–540, 563, 937
Distylium racemosum 858
Docyniopsis tschonoski 537
Dolichandrone crispa 447, 1106
Doliocarpus guianensis 671
Doliocarpus macrocarpus 671
Doona 853
Doona congestiflora 1117
Doona macrophylla 1118
Doryphora 216
Dracaena cinnabri 9
Dracontomelon edule 561
Drimys winteri 744
Dryas octopetala 671
Dryobalanops 853, 857, 920, 922
Dryobalanops aromatica 9
Dryobalanops camphora 710
Duboisia hopwoodii 1109, 1113
Duboisia leichhardtii 1113
Duboisia myoporoides 1113
Ducampopinus krempfii 556
Dyera 855, 857, 894
Dysoxylum muelleri 10

Echium vulgare 1114
Ectopepterys soejartoi 1107
Elaeagnus pungens 1118
Elaeocarpus 204–205
Elaeocarpus altisectus 204
Elaeocarpus densifolius 204
Elaeocarpus dolichostylis 204–205
Elaeocarpus hookerianus 1099, 1102
Elaeocarpus polydactus 204–205
Elaeocarpus sphaericus 204–205
Eleusine coracana 989
Eleutherococcus senticosus 1115
Eleutherococcus spinosus 1109
Eleutherococcus trifoliatus 1117
Encephalartos longifolius 171
Endiandra introrsa 291
Engelhardtia formosana 545
Entandrophragma condollei 321
Entandrophragma cylindricum 773
Ephedra 137
Ephedra alata 442
Ephedra andina 1098
Ephedra equisentina 1100
Ephedra frustillata 1098
Ephedra nevadensis 1100
Ephedra sinica 1096, 1100, 1108–1109, 1111, 1116
Ephedra viridis 1100
Epilobium angustifolium 408
Epimedium grandiflorum 445

Eremophila freelingii 716
Eremophila mitchelli 728
Erica 403
Erica australis 1095, 1111, 1114, 1118
Erica multiflora 264
Eriophyllum staechadifolium 740
Ervatamia 231
Ervatamia coronaria 1112
Ervatamia hainanensis 1097, 1112
Ervatamia heyneana 245, 1097, 1099, 1104, 1110
Ervatamia officinalis 1112
Ervatamia orientalis 231
Erythrina 221-222
Erythrina abyssinica 1093, 1100
Erythrina blakei 309, 311
Erythrina crista-galli 221, 549, 553, 1100
Erythrina stricta 309, 312
Erythrina suberosa 824
Erythrina variegata 309, 550
Erythrobalanus 875
Erythrophleum chlorostachya 249
Erythrophleum guineense 248
Erythrophleum letestui 10
Erythroxylum coca 202-203, 1109
Erythroxylum ellipticum 203
Erythroxylum novogranatens 203
Esenbeckia flava 1099, 1114
Esenbeckia litoralis 1095, 1099
Espeletiopsis guacharaca 737
Eucalyptus 157, 181, 267, 513-514, 558, 590, 675, 683, 743, 868, 883, 896, 898, 912, 993
Eucalyptus calophylla 672
Eucalyptus camaldulensis 9, 887, 1099
Eucalyptus cypellocarpa 1099
Eucalyptus deglupta 858
Eucalyptus delegatensis 886, 897
Eucalyptus diversicolor 672, 1099
Eucalyptus dives 1099
Eucalyptus elaeophora 321
Eucalyptus fastigata 321, 333
Eucalyptus gigantea 897
Eucalyptus globulus 912
Eucalyptus hemiphloia 445
Eucalyptus maculata 544, 672, 922
Eucalyptus manna 157
Eucalyptus marginata 321, 345, 882, 888
Eucalyptus microcorys 875
Eucalyptus obliqua 321, 333, 897
Eucalyptus pilularis 1099
Eucalyptus platypus 1100
Eucalyptus regnans 157, 159-161, 875, 886, 897, 922
Eucalyptus robusta 1186
Eucalyptus sideroxylon 536, 874, 1100
Eucalyptus stjohnii 321, 333
Eucalyptus tereticornis 312

Eucalyptus torelliana 536
Eucalyptus triantha 875
Eucalyptus wandoo 6
Euchresta horsfieldii 1108
Eucommia ulmoides 260, 267, 446
Eugenia capensis 321
Eugenia caryophyllata 422
Euglena gracilis 313
Euonymus 213
Euonymus alatus 320, 328
Euonymus europaeus 213
Euonymus sieboldiana 213
Eupatorium rugosum 933
Eupatorium triplinerve 940
Euphorbia 407, 427, 429, 684, 765, 809, 813
Euphorbia hermentiana 936
Euphorbia jolkini 758, 765
Euphorbia kansui 766
Euphorbia milii 1108
Euphorbia pulcherrima 816, 826
Eupomatia larina 443
Euryops brevipapposus 739
Eusideroxylon zwageri 444
Evodia lepta 1112
Evodia rutaecarpa 212-213
Evodia xanthoxyloides 211
Exostemma caribaerum 554

Fagara 777
Fagara boninensis 445
Fagara macrophylla 1094
Fagara mantshurica 1114
Fagara xanthoxyloides 445, 1094, 1096
Fagus 855, 895
Fagus crenata 857-858
Fagus grandifolia 320
Fagus sylvatica 159, 167, 267, 308, 311, 320, 329, 347, 822-823
Feronia elephantum 171
Feronia limonia 1119
Ferreirea spectabilis 1009, 1101
Ferula communis 744
Ferula foetida 744
Ferula jaeschkeana 721
Fibraurea chloroleuca 1105
Ficus 671, 829, 1111
Ficus bengalensis 822
Ficus glomerata 933
Ficus racemosa 605
Filipendula ulmaria 421, 431
Fissistigma glaucescens 1106
Fitzroya cupressoides 442
Flemingia stricta 1109
Flindersia 522
Flindersia dissosperma 1099
Flindersia laevicarpa 824
Flindersia maculosa 1099, 1114

Flourensia heterolepis 740
Fluggea microcarpa 824
Forsythia arten 446
Forsythia koreana 447, 1102
Forsythia suspensa 446
Forsythia vercissima 447
Fouquieria diguetii 320
Fouquieria splendens 320
Fragaria ananassa 989
Fragaria chiloensis 321
Fraxinus 157, 160, 163
Fraxinus excelsior 329
Fraxinus graggii 1110
Fraxinus japonica 447
Fraxinus mandshurica 447
Fraxinus ornus 9, 160
Fraxinus pennsylvanica 321
Frullania 937
Fuchsia 408, 422, 426
Funtumia 250

Gaillardia aristata 704
Galbulimima belgraveana 246
Garcinia 572, 576, 583, 894
Garcinia conrauana 281
Garcinia densivenia 1102
Garcinia echinocarpa 1119
Garcinia gambogia 267
Garcinia mannii 279, 281, 1102, 1108
Garcinia multiflora 536
Garcinia terpnophylla 1101, 1119
Gardenia jasminoides 1103
Gardneria multiflora 243
Gardneria nutans 243
Garrya veatchii 248
Garuga 583
Gastrolobium grandiflorum 268
Geijera balansae 1099, 1114, 1118
Gelonium multiflorum 782
Gelsemium 241, 243
Gelsemium elegans 242
Gelsemium rankinii 243
Gelsemium sempervirens 241–242
Genista sphaerocarpa 1112
Genista tinctoria 1107
Geranium 404, 427, 429
Geranium macrorhizum 722
Geranium robertanium 408
Geum 431
Geum japonicum 421–422
Geum rivale 421
Geum urbanum 1111
Ginkgo 572, 583
Ginkgo biloba 21, 264, 327, 603, 717, 759, 817, 822, 1114
Glaucium fimbrilligerum 1113
Glaucium flavum 1113

Glaucium paniculigera 1113
Gleditsia 540
Gleditsia japonica 606, 671
Gleditsia triacanthos 321, 539
Gliricidia 544
Gliricidia sepum 545
Globularia alypum 447
Glossocalyx longicuspus 1102
Gluta 902–904, 925
Glycosmis bilocularis 1099, 1114
Glycosmis citrifolia 211
Glycosmis mauritiana 1114
Glycosmis pentaphylla 211
Glycyrrhiza glabra 778, 1103, 1106
Gmelina arborea 447, 822
Gmelina asiatica 447
Gmelina leichhardtii 447
Gnetum 137
Gnetum parvifolium 1097
Gnidia 268
Gnidia latifolia 445
Goniothalamus 281, 894
Goniothalamus amuyon 1106
Gonolobus condurango 162
Gonystylus 854
Gonystylus bancanus 855, 942
Gossweilerodendron balsamiferum 752
Gossypium 1113
Gossypium barbadense 1100
Gossypium hirsutum 325, 329, 732, 992–993, 1117
Gossypium mexicanum 1110, 1118
Grevillea leucopteris 157
Grevillea robusta 171, 657
Grevillea striata 522
Grewia tiliaefolia 1102
Grewia villosa 1118
Grias cauliflora 320
Griffonia simplicifolia 286
Grindelia robusta 751
Guaiacum officinale 446
Guazuma tomentosa 824, 1102
Guettarda angelica 1104
Guettarda eximia 238
Guettarda heterosepala 239
Guibourtia 671
Guibourtia coleosperma 606, 613
Gymnosporia montana 825
Gynocardia odorata 283
Gyrinops walla 1095, 1102, 1108, 1115

Haematoxylon brasiletto 1095, 1104
Haematoxylon campechianum 7, 554, 846, 848
Hakea acicularis 171
Hakea suaveolens 321
Hamamelis 264, 993

Hamamelis mollis 413
Hamamelis virginiana 408
Haplophragma 822
Haplophragma adenophyllum 1106
Haplophyllum dauricum 445, 1105, 1114
Haplophyllum ferganicum 1103
Haplophyllum foliosum 1115
Haplophyllum hispanicum 445
Haplophyllum latifolium 1115
Haplophyllum obtusifolium 445, 1105
Haplophyllum pedicellatum 1115
Haplophyllum perforatum 445
Haplophyllum popovii 445
Haplophyllum suaveolens 1099, 1115
Haplophyllum tuberculatum 445, 1105
Haplophyllum versicolor 445
Haplophyllum vulcanicum 445, 1115
Haplophyton cimicidum 161
Haplormosia monophylla 1094, 1112
Hardwickia pinnata 740, 752
Harpagophytum procumbens 1118
Harpullia pendula 825
Hedycarya angustifolia 724
Hedyotis lawsoniae 447
Hedysarum polybotrya 1106
Heimia 207
Helenium autumnale 731
Helichrysum 764
Helichrysum bracteatum 446
Helinus ovatus 260
Heliopsis buphthalmoides 446
Heliopsis scabra 446
Heliothis zea 683
Hernandia cordigera 444
Hernandia ovigera 444
Hernandia peltata 444, 853
Herpestospermum caudigerium 445
Heterophragma 822
Heterophragma adenophylum 825
Heterotropa takahoi 444
Hevea 8, 1171
Hevea brasiliensis 8, 162, 287, 320, 348, 788, 901, 1028, 1170
Hibiscus abelmoschus 716
Hibiscus elatus 735
Hibiscus rosa-sinensis 312
Hibiscus tiliaceus 735, 1106
Himantandra baccata 443
Himantandra belgraveana 443
Hippomane mancinella 941–942
Holarrhena 250
Holarrhena antidisenterica 826
Hopea 4, 514, 901
Hopea bravipetiolaris 1097, 1113, 1118
Hopea cordifolia 1118
Hopea jucunda 1097, 1113, 1118
Hopea odorata 773

Hopea pierrei 858
Hordeum 989, 991
Hordeum vulgare 313, 344–345, 584, 630–631, 659, 679–680
Horsfieldia iryaghedi 443
Hortia badinii 1113, 1115
Hoya carnosa 320
Humboldtia 590
Humboldtia laurifolia 592
Hura crepitans 1105
Hydnocarpus anthelminthica 285
Hydrangea 513
Hydrangea heteromall 1107
Hydrangea particulata 166
Hyoscyamus 203
Hyptis pectinata 280

Iboza riparia 280
Ilex 1057
Ilex crenata 782
Ilex opaca 933
Ilex paraguaiensis 940
Ilex varitoria 940
Illicium religiosum 264, 266
Illicium verum 990
Illigera pentaphylla 1107
Impatiens balsamina 672
Inga saman 823
Intsia 901, 1112
Intsia bijuga 12
Inula 248
Inula royleana 758
Inula salicina 705
Ipomoea 63
Ipomoea batatas 326, 331–332, 342
Iresine celosioides 744
Iris pallida 706
Iryanthera grandis 443
Isocoma wrightii 740

Jacaranda acutifolia 539, 541
Jateorhiza palmata 753, 1105
Jatropha curcas 1057
Jatropha gossypifolia 445, 766
Jatropha macrorhiza 765
Jatrorhiza palmata 704
Juglans 300, 422, 684, 927, 935
Juglans hindsii 347
Juglans nigra 189, 320, 344, 525, 544, 874–876
Juglans regia 311, 320, 347, 544
Juglans sieboldiana 858
Julbernadia globiflora 613–614, 671
Juniperus 576, 583, 852, 870, 980
Juniperus bermudiana 442
Juniperus californica 1096, 1116

Juniperus chinensis 673, 1052
Juniperus communis 264, 621, 741–742, 990
Juniperus conferta 742
Juniperus formosana 442, 1100
Juniperus horizontalis 310, 1102
Juniperus macropoda 559
Juniperus mexicana 1052
Juniperus occidentalis 319, 344, 704
Juniperus phoenica 442
Juniperus procera 1096, 1116–1117
Juniperus rigida 718–719
Juniperus sabina 442, 1109, 1111, 1116
Juniperus virginiana 719, 938, 1052
Justicia extensa 447
Justicia flava 447
Justicia hayatay 447
Justicia procumbens 447
Justicia prostata 447
Justicia simplex 447

Kadsura coccinea 443
Kadsura japonica 443
Kadsura longipedunculata 443
Kalanchoe diagramontiana 817
Kayea stylosa 1119
Kerria japonica 671
Keteleeria 558
Keteleeria davidiana 1117
Khaya 936
Khaya anthotheca 935–937
Khaya grandifolia 171
Khaya grandifoliola 775–776
Khaya senegalensis 170–171
Kielmeyera 520
Kielmeyera coriacea 311, 329
Kigelia pinnata 447, 1106
Kopsia 591
Kopsia dasyrachis 593
Krameria 993
Krameria cistisoides 446
Krameria ixinia 940
Krameria ramosissima 446
Krameria triandra 446, 940

Laburnum alpinum 514, 549
Laburnum anagyroides 329, 549
Laburnum vossii 321
Lagascea rigida 1103
Lagerstroemia 206
Lagerstroemia faurier 207
Lagerstroemia indica 206
Lagerstroemia subcostata 207
Lancea tibetica 447
Lannea coromandelica 170
Lannea humilis 170
Lannea schimperi 170
Lansium anamalayanum 717

Lappa major 446
Lappa minor 446
Lappa tomentosa 446
Larix 165–166, 264, 545, 557–558, 821, 852, 889, 895, 978–979, 994
Larix cajanderi 1098, 1109, 1116
Larix dahurica 544, 1109, 1111
Larix decidua 267, 319, 322, 345, 442, 820–821, 824, 924
Larix gmelini 165, 571, 669, 673, 857
Larix laricina 319, 344, 979, 1102
Larix leptolepis 384, 442, 673, 820–821, 857–858, 920
Larix occidentalis 165, 979, 1168, 1174
Larix olgensis 1104, 1109, 1111
Larix russica 302
Larix sibirica 159, 165, 267, 302, 757, 1098, 1109, 1116
Larrea cuneifolia 446
Larrea divaricata 446
Larrea nitida 446
Larrea tridentata 322, 345
Laserpitium latifolia 721
Lathyrus 568
Laurelia novae-zelandiae 443
Laurelia sempervirens 1106
Laurus nobilis 300, 320, 345–346, 990
Ledum decumbens 1113
Ledum hypoleucum 1113
Ledum macrophyllum 1113
Ledum palustre 743, 1113
Legnephora moorii 162
Lens culinaris 990
Leptospermum scoparium 309, 312
Leptosphaeria coniothyrium 348
Lespedeza cyrtobotrya 550, 552, 1098, 1103, 1106
Lespedeza sericea 79, 989
Lethospermum officinale 286
Lethospermum purpureocaeruleum 286
Libanotis transcaucasica 739
Libocedrus 557, 870
Libocedrus bidwillii 319, 442
Libocedrus decurrens 310, 852, 874, 888, 980, see Calocedrus decurrens
Libocedrus yateensis 371, 442, 519
Licaria 444
Licaria aritu 444
Licaria armeniaca 444, 1113
Licaria canella 444
Licaria chrysophylla 444
Licaria macrophylla 444
Licaria pichury-major 1112, 1116
Licaria rigida 444
Ligularia fischeri 728, 730
Ligularia hodgsoni 722, 729
Ligularia sibirica 731

Ligustrum japonicum 447
Limaciopsis 216
Limaciopsis loangensis 1095
Limaciopsis sempervirens 1106
Lindera 670
Lindera obtusiloba 278
Lindera umbellata 853, 1100
Linum usitatissimum 282–283
Liquidambar orientalis 3, 777, 853
Liquidambar styraciflua 300, 302, 865, 1118
Liriodendron tulipifera 321, 443, 817, 832, 938, 945, 1106–1107
Litchia chinensis 264, 276–268
Lithocarpus attenuata 1102
Lithocarpus cornea 1102, 1115
Lithocarpus elizabethae 1115
Lithospermum erythrorhizon 1114
Litsea 894
Litsea citrata 702
Litsea gracilipes 444
Litsea grandis 444
Litsea japonica 275
Litsea kawakamii 1096
Litsea turfosa 444
Lobadium 562
Lolium temulentum 346
Lonicera hypoleuca 447
Lonicera periclymenum 1109
Lonicera sempervirens 1107
Lophopetalum rigidum 824
Lotus 568, 989–990
Lotus corniculatus 181
Lotus pedunculatus 676
Lotus tenuis 672
Lovoa klaineana 942
Loxopterygium sagottii 561
Lunasia amara 211
Lupinus luteus 777
Lycium chilense 1096
Lycium tenuispinum 1096
Lycopersicon esculentum 187, 190, 261, 317, 332, 343, 345
Lyonia ovalifolia 446, 1100, 1112

Maackia amurensis 513–514
Macaranga peltata 547
Macfadyena ungius-cati 1106
Machaerium 10, 526, 550–551, 554, 562-564, 566, 845-846
Machaerium acutifolium 565
Machaerium kuhlmannii 554, 565, 1101
Machaerium mucronulatum 552, 565
Machaerium nictitans 554, 565, 1101
Machaerium opacum 565
Machaerium pedicellatum 554
Machaerium scleroxylon 565, 935

Machaerium vestitum 565, 1101
Machaerium villosum 547, 549, 565, 1098
Machilus 216
Machilus edulis 444
Machilus japonica 444
Machilus thunbergii 444
Machilus zuihoensis 444
Maclura cochinchinensis 1116
Maclura pomifera 525, 874, 941, 1108
Macropiper excelsum 444
Magnolia 932
Magnolia acuminata 443
Magnolia denudata 443
Magnolia fargesii 443
Magnolia grandiflora 218, 344, 443, 1066, 1094, 1107-1108
Magnolia kachirachirai 443
Magnolia kobus 853, 1098, 1104, 1116
Magnolia liliflora 443
Magnolia obovata 1104, 1108
Magnolia officinale 443
Magnolia salicifolia 443, 1098, 1112
Magnolia stellata 443
Mahonia 216
Mahonia aquifolium 1105, 1107
Mahonia repens 1093, 1105, 1110
Mallotus japonicus 429
Malosma laurina 561
Malouetia 250
Malphigia coccigera 672
Malus domestica 159
Malus pumila 321, 329, 335, 340, 671, 674–675, 989, 991
Malus sylvestris 1094
Mammea 554
Mammea acuminata 1119
Mammea africana 554
Mammea americana 264
Mangifera altissima 561
Mangifera indica 267, 539, 559, 941, 1108
Manihot 282
Manihot cartheginensis 282
Manihot esculenta 282, 325, 990
Manihot utilissima 990
Manilkara zapota 972
Maniltoa 901
Mansonia altissima 10, 526, 735, 845, 848, 920, 935,
Markhamia platycalyx 1106, 1114
Marsdenia volubilis 824
Maruba deiko 221
Matthiola incana 679
Mauritia 163
Maytenus 213
Maytenus acuminata 1108
Maytenus boaria 1111
Maytenus buchananii 1108

Maytenus confertiflora 1098, 1108
Maytenus dispermus 758
Maytenus diversifolia 1098, 1108
Maytenus emarginata 1111, 1117
Maytenus guangkiensis 1108
Maytenus heterophylla 1108
Maytenus hookeri 1108
Maytenus ilicifolia 1108
Maytenus mossambicensis 1108
Maytenus nemorosa 1117
Maytenus obtusifolia 1111, 1117
Maytenus ovata 213–214
Maytenus polycantha 1108
Maytenus rigida 672
Maytenus senegalensis 825, 1108
Medicago 568
Medicago sative 817
Melaleuca quinquenervia 321
Melampodium longipilum 723
Melanorrhoea 559, 902–904, 925
Melanorrhoea aptera 547
Melanorrhoea usitata 5
Melia azedarach 264, 773, 776, 1096, 1098, 1101, 1118
Melia birmanica 835, 1098
Melicope 540, 542, 558
Melicope broadbentiana 541
Melicope mantellii 540
Melicope octandra 822, 824
Melicope sarcocolla 544
Melicope simplex 541
Melicope ternata 541
Meliosma oldhamii 672
Menispermum 216
Menispermum cocculus 737
Menispermum dauricum 286
Mentha piperita 735
Mentha spicata 700
Merrilliodendron megacarpum 245, 1095
Mesua 554
Mesua ferrea 576, 1101, 1119
Mesua thwaitesii 1101, 1119
Metasequoia 576
Metasequoia glyptostroboides 371, 519, 852, 1098, 1104, 1116
Metopium brownei 561
Metopium toxiferum 561
Metroxylon 163
Metroxylon sagu 990
Mezilaurus 279
Mezilaurus itauba 444
Mezilaurus synandra 279
Michelia champaca 1106
Michelia 216
Michelia figo 344
Mildbraediodendron excelsa 550

Millettia 540, 857, 926
Millettia laurentii 935
Millettia pendula 551
Millingtonia 822
Millingtonia hortensis 825, 1106
Mimosa 545
Mimusops balata 1049
Mimusops hexandra 1115
Mimusops littoralis 823
Mitragyna hirsuta 1103
Mitragyna parvifolia 235
Monochamus alternatus 856
Monodora tenuifolia 1094, 1106–1107
Monopteryx uaucu 1099, 1100
Montanoa tomentosa 1103
Moquinea velutina 726
Mora excelsa 777
Morinda parvifolia 1093–1094, 1107
Moringa 172
Mortonia gregii 1096
Mortonia palmeri 1096
Morus 514, 537–539, 541, 555, 558
Morus alba 517, 544, 1105–1106, 1108, 1112–1113
Morus bombycis 1105–1106, 1113
Morus indica 267, 1108
Morus laevigata 514, 1108
Morus lhou 1105–1106, 1113
Morus rubra 1108
Morus serrata 1108
Mundulea sericea 550
Muntfara sessilifolia 1097, 1110
Murraya euchretifolia 213
Murraya koenigii 213, 1103
Murraya paniculata 1115
Musa sapientum 816, 989
Myoporum platycarpum 9
Myosotis scorpioides 65
Myrica 542
Myrica nagi 522, 524, 630–631, 1115
Myrica rubra 267, 541, 630–631, 680
Myristica cagayanensis 443
Myristica fragrans 443, 990
Myristica malabarica 443
Myristica otoba 443
Myristica simarum 443
Myrothamnus flabellifolia 1187
Myroxylon balsamum 3, 1101
Myroxylon pereirae 716
Myroxylon peruiferum 1118

Nandina domestica 260, 1097, 1107–1108, 1111
Narcissus incomparibilis 269
Nardostachys jatamansi 729, 744
Nauclea undulata 901
Nealchornea yapurensis 1113

Nectandra 444
Nectandra elaiophora 702, 704
Nectandra miranda 444
Nectandra polita 444, 822, 1101
Nectandra puberula 444
Nectandra rigida 444
Nectandra rubra 278
Nelia meyeri 622
Nemuaron humboldtii 384
Neocallitropsis araucarioides 702, 1095
Neochamaelea pulverulenta 777
Neolitsea sericea 822–823
Neonauclea 901
Neorautanenia amboensis 606, 671
Nepeta cataria 706
Nerium odorum 838
Nerium oleander 933
Neviusia alabamensis 671
Nicotiana glutinosa 344
Nothapodytes foetida 245, 1095
Nothofagus 539, 544–545, 590
Nothofagus fusca 1104, 1109, 1117
Nothofagus menziesii 320, 1109, 1117
Nothofagus solanderi 1104, 1109, 1117
Nothofagus truncata 1104, 1109, 1117
Nuphar japonicum 408

Ochna 577
Ochroma lagopus 857
Ochrosia elliptica 231
Ochrosia sandwicensis 231
Ochrosia silvatica 231
Ochrosia viellardii 231
Ocotea 444
Ocotea aciphylla 444
Ocotea caparrapi 716, 744
Ocotea catharinensis 444
Ocotea cymbarum 444
Ocotea petalanthera 1098
Ocotea porosa 444
Ocotea pretiosa 373, 853, 1101, 1113
Ocotea usambarensis 703–704
Ocotea veraguensis 444
Octomeles sumatrana 853–855, 858
Odina wodier 170
Olea africana 447
Olea capensis 447
Olea cunninghamii 447
Olea europaea 9, 47–48, 260, 281, 447, 1110
Olearia paniculata 777, 780
Onobrychis 990
Onobrychis viciifolia 657, 675, 989
Ononis spinosa 785
Onopordon alexandrinum 446
Ophiorrhiza mungos 245
Opuntia 171
Opuntia aurantiaca 171

Opuntia ficus–indica 171
Opuntia fulgida 171
Opuntia monacantha 171
Opuntia nopalea–coccinillifera 171
Orixa japonica 211, 1115
Ormosia excelsa 549
Oroxylum indicum 536
Oryza sativa 345, 621
Osteophloeum 559
Osteophloeum platyspermum 443
Ostrya japonica 522
Otanthus maritimus 740
Ouratea 621
Oxalis 268
Oxystigma manii 736

Pachycormus discolor 320
Pachysandra terminalis 345
Paeonia albiflora 709
Paeonia officinalis 413
Palaquium 857
Palicourea marcgravii 268
Paliurus aculeatus 268
Panax ginseng 773
Pangium edule 283
Parabenzoin praecox 373
Parabenzoin trilobum 444
Parartocarpus 894
Parashorea malaanonan 857
Paratecoma 870
Paratecoma peroba 10, 935
Parrottia persica 413
Parthenium 936
Parthenium argentatum 936
Passerina 268
Passiflora adenododa 283
Passiflora allardii 283
Passiflora altocaerulea 285
Passiflora biflora 284
Passiflora caerulea 283, 285
Passiflora capsularis 285
Passiflora coccinea 285
Passiflora incarnata 283
Passiflora suberosa 283
Passiflora talamancensis 284
Passiflora trifascia 284
Paulownia tomentosa 447
Pausinystalia yohimbe 236
Pelargonium 269, 413, 431
Pennisetum americanum 345
Pentacme contorta 857
Perezia 717
Pericopsis montana 550
Perilla frutescens 700
Persea 675
Persea americana 674
Persea gratissima 622

Persea nan-mu 2
Petasites hybridus 311
Petasites japonica 729
Peucedanum japonicum 697
Phaeanthus 216
Phaseolus lunatus 287
Phaseolus vulgaris 344, 817
Phellodendron amurense 218, 777
Philodendron pertusum 325
Photinia 675
Photinia glabrescens 631–632, 662
Phyllanthus 208
Phyllanthus discoides 208
Phyllanthus emblica 777
Phyllanthus engleri 782
Phyllanthus niruri 445
Phyllanthus simplex 260
Phyllocladus 590
Phyllocladus alpinus 321, 593
Phyllocladus glaucus 309, 312
Phyllogeiton 577
Physostigma venenosum 830
Picea 165, 300, 373, 514, 539, 674, 785, 852, 884, 887, 889, 894–895
Picea abies 157–161, 167, 264, 300, 302, 308–310, 322, 345, 347, 370, 373, 384, 442, 810, 821, 857, 864, 866, 894, 924
Picea ajanensis 442
Picea engelmannii 684
Picea excelsa 302, 442
Picea glauca 163, 310, 319, 323, 675, 866
Picea glehnii 384, 673
Picea jezoensis 442, 673, 857–858
Picea obovata 442
Picea pungens 373
Picea sitchensis 187, 319, 327, 331, 346, 557–558, 785, 821
Picea vulgaris 442
Picrasma quassioides 226
Picrorhiza kurrooa 706
Pimelea flava 268
Pinca 891
Pinus 161, 165, 181, 264, 300, 373, 513–514, 516, 534, 536, 539–540, 542–545, 557–558, 673–674, 701, 741–742, 757, 785, 821, 852, 887, 889, 891, 895, 912, 925, 929
Pinus banksiana 163, 184, 308–310, 513–514, 821, 896
Pinus brutia 319, 673–674, 955
Pinus caribaea 310
Pinus clausa 545
Pinus contorta 308, 310, 384, 555, 557–558, 674, 703, 824
Pinus densiflora 327, 333, 674, 857–858, 909
Pinus echinata 300, 319, 629, 955
Pinus eldarica 703

Pinus elliottii 182, 300, 629, 674, 708, 755, 955
Pinus excelsa 514, 539, 545, 547, 705
Pinus frempfii 544
Pinus griffithii 539, 556–557
Pinus halepensis 161, 310, 319, 955
Pinus insularis 955
Pinus jeffreyi 246, 301
Pinus kesiya 955
Pinus koraiensis 857
Pinus lambertiana 161
Pinus laricio 442
Pinus lumholtzii 703
Pinus massoniana 373, 442, 955
Pinus monticola 308–310, 331, 824–825, 834
Pinus nigra 160–161, 319, 821, 865
Pinus palustris 300, 319, 629, 631, 674, 703, 755, 758, 955
Pinus patula 674
Pinus pentaphylla 673
Pinus pinaster 166–167, 327, 955
Pinus pinceana 703
Pinus pinea 319, 703, 817
Pinus ponderosa 310, 540, 557, 684–685, 701, 703, 858, 922
Pinus quadrifolia 701
Pinus radiata 160, 167, 184, 309–310, 319, 327, 371, 374, 514, 516, 621, 657, 667, 673–674, 676, 679, 681, 684–685, 824–825, 865, 884, 890, 892–893, 896
Pinus reflexa 703
Pinus resinosa 319, 370, 384, 865
Pinus roxburghii 311, 708, 742
Pinus sabiniana 246, 301
Pinus serotina 319
Pinus sibirica 442, 955
Pinus strobus 180, 319, 536, 545, 676, 821
Pinus subg. Diploxylon 557
Pinus subg. Haploxylon 544–545, 557
Pinus succinifera 953
Pinus sylvestris 2, 157, 159–161, 165, 167, 302, 309, 311, 319, 347, 370–371, 373, 384, 558, 629–630, 667, 673–674, 682, 701, 703–704, 734, 736, 741, 755, 757, 821, 857–858, 876, 887, 955
Pinus taeda 167, 182, 300, 311, 319, 327, 347, 514, 596, 626, 629, 631, 633, 635, 666, 674, 864–865
Pinus thunbergii 331, 674, 857–858
Pinus torreyana 246
Pinus virginiana 158, 300, 327
Pinus yunnanensis 955
Piper brachystachyum 288, 444
Piper clusii 444
Piper cubeba 444
Piper futokadsura 288, 445

Piper guineense 445
Piper hookeri 288
Piper lacunosum 445
Piper peepuloides 445
Piper sylvaticum 445
Piptadenia macrocarpa 835
Piptanthus nepalensis 549
Piptocalix moorei 443
Piqueria trinervia 706
Piscidia erythrina 550
Pistacia 545
Pistacia chinensis 539
Pistacia lentiscus 4, 773
Pistacia texana 561
Pistacia vera 191, 559
Pisum 568
Pisum sativum 314-315, 817
Pithecellobium dulce 312
Pithecellobium saman 823
Planchonella 854-855
Planchonella thysoidea 858
Platanus americana 941
Platanus hybrida 312
Platanus vulgaris 545
Plathymenia reticulata 544, 547, 753
Platycodon grandiflorum 63
Platymiscium praecox 539, 545
Playlopsis falcisepala 447
Plectranthus 758
Plumeria rubra 706
Podanthus ovatifolius 723
Podocarpus 539, 556, 822, 837, 870
Podocarpus elatus 837
Podocarpus ferrugineus 758
Podocarpus lambertii 822, 824-825
Podocarpus macrophyllus 673, 837
Podocarpus spicatus 556-557, 559, 761
Podocarpus totara 673
Podophyllum emodi 445
Podophyllum hexandrum 445
Podophyllum peltatum 445
Podophyllum sikkimensis 445
Pogonopus tubulosum 224
Pogostemon patchouli 732
Polyalthia 281
Polygala chinensis 446
Polygala polygama 446
Polygonum 590, 624
Polygonum multiflorum 593, 622, 670, 680
Polygonum reynoutria 267
Polypodium 591
Polypodium vulgare 593, 837
Polystichum acrostichoides 817
Pongamia glabra 537, 539-541
Populus 193, 289, 300, 370-371, 853-854, 890-891, 895
Populus balsamifera 312, 373

Populus deltoides 545
Populus gelrica 343
Populus grandidentata 163, 290, 321
Populus maximowiczii 857
Populus nigra 312
Populus tremula 329, 886
Populus tremuloides 166-167, 312, 321, 331, 822, 825,
Populus trichocarpa 264, 290, 321, 374
Potentilla 421, 422, 431
Potentilla erecta 622
Pouchea odorata 705
Pouteria 942
Primula 562
Prinsepia utilis 445
Prionostema aspera 672
Pristimera indica 782
Prochloron 128
Prosopis 164, 168, 590
Prosopis chilensis 168
Prosopis glandulosa 168, 602, 682
Prosopis juliflora 168
Prosopis spicigera 321
Prunus 267, 374, 534, 536-537, 543-545, 547, 550, 555, 558-559
Prunus aequinoctialis 537
Prunus amygdalus 169, 282
Prunus armeniaca 169, 282
Prunus avium 545, 555-556
Prunus cerasus 169, 260, 264, 267, 544, 547, 674
Prunus cornuta 312
Prunus domestica 169, 260, 373, 539, 545, 864
Prunus emarginata 541
Prunus insitia 169
Prunus jamasakura 864
Prunus mume 539, 541
Prunus persica 169, 282, 321, 347, 544, 621, 685
Prunus puddum 312, 537, 545
Prunus serotina 371
Prunus serrata 537
Prunus spinosa 169
Prunus ssiori 536-537
Prunus virginiana 169
Prunus yedoensis 371, 378
Pseudosindora palustris 902-904, 906
Pseudotsuga 300, 539, 545, 557
Pseudotsuga japonica 673
Pseudotsuga menziesii 158, 185-186, 188, 191, 194, 264, 309, 311, 319, 323, 326-327, 331, 333, 335-336, 555, 558, 673, 679, 821, 852, 857-858, 885, 887, 889, 896, 924
Pseudowintera colorata 743
Psidium guajava 264, 267, 422
Ptaeroxylon obliquum 696

1214   Index of Plant Genera and Species

Ptelea trifoliata 211
Pteris vittata 442
Pterocarpus 7, 514, 527, 562-563, 566-567, 846, 848
Pterocarpus angolensis 550, 942
Pterocarpus indicus 552
Pterocarpus marsupium 9
Pterocarpus santalinus 7, 527
Pterocymbium beccarii 854-855, 858
Punica granatum 205, 267, 429
Puya 172
Pycnarrhena 216
Pygeum acuminatum 445
Pyrus 267
Pyrus communis 347, 370, 520
Pyrus malus 260

Qualea labouriauana 544-545
Qualea paraensis 544
Quassia africana 1110
Quassia amara 777
Quercus 162-163, 275, 406, 422, 542, 558, 855, 868, 927, 989, 993, 998
Quercus acuta 857-858
Quercus acutissima 329
Quercus aegilops 6, 406
Quercus alba 320, 822, 873, 875, 923
Quercus borealis 408
Quercus crispula 333, 857-858
Quercus discolor 7
Quercus falcata 320, 923, 940
Quercus ilex 325, 329
Quercus infectoria 406, 408, 414
Quercus lusitanica 414
Quercus pedunculata 264
Quercus phellus 555
Quercus robur 167, 320, 329, 344-345, 406, 411, 602, 822, 1106
Quercus rubra 163, 320, 370, 384, 408, 940
Quercus sessiflora 422
Quercus stenophylla 264, 593
Quercus suber 162, 304, 311, 316, 320, 323, 329, 331, 336, 781
Quercus tinctoria 534, 541, 555
Quercus valonea 423
Quercus velutina 320
Quillaja saponaria 1098
Quintinnia serrata 545

Ranunculus acris 333
Raphanus sativa 332
Ratanhia radix 446
Rauvolfia 227, 1057
Rauvolfia cambodiana 1112
Rauvolfia canescens 227
Rauvolfia cubana 1112
Rauvolfia cumminsii 1119

Rauvolfia heterophylla 227
Rauvolfia hirsuta 227
Rauvolfia latifrons 1112
Rauvolfia mombasiana 1094, 1112, 1119
Rauvolfia obscura 1094, 1119
Rauvolfia oreogiton 1094, 1112, 1119
Rauvolfia serpentina 227-228, 1112, 1119
Rauvolfia tetraphylla 227
Rauvolfia tiaolushanensis 1112
Rauvolfia verticillata 227, 1112
Rauvolfia volkensii 1093-1094, 1112, 1119
Rauvolfia vomitoria 227-228, 1093-1094, 1102, 1112, 1119
Rauvolfia yunnanensis 1112
Retama sphaerocarpa 1112
Rhabdodendron macrophyllum 1104, 1110
Rhamnella franguloides 268
Rhamnus 267-268, 541, 558
Rhamnus alaternus 541
Rhamnus cathartica 541
Rhamnus sectipetala 1097, 1100, 1110
Rhamnus triquetra 1100
Rhaphiolepis 590
Rhaphiolepis umbellata 593, 670
Rheedia gardneriana 1101, 1119
Rheum 590-591, 624, 1110, 1112
Rheum officinale 408, 410
Rheum palmatum 680, 1114
Rheum rhaponticum 512
Rheum rhizoma 407, 622
Rhigiocarya racemifera 1107-1108
Rhizobium meliloti 82
Rhizophora 6, 1175
Rhizophora mangle 172, 321
Rhizophora stylosa 669
Rhododendron 392, 403
Rhododendron anthopogon 1117
Rhododendron arboreum 777
Rhododendron chrysanthum 392
Rhododendron maximum 320
Rhus 264, 300, 431, 537, 545, 547, 559-560, 562, 576, 613, 616, 845, 935
Rhus chondroloma 561
Rhus copallina 561, 1100, 1102, 1113
Rhus coriaria 406, 408, 413
Rhus cotinus 7, 543, 850
Rhus glabra 561, 1102, 1113
Rhus integrifolia 1100, 1102, 1114
Rhus javanica 539, 547
Rhus lancea 617, 671
Rhus lanceolata 1100, 1102, 1114
Rhus leptodictya 615
Rhus microphylla 561, 1100, 1102, 1114
Rhus oaxacana 1100, 1102, 1114
Rhus pachyrrachis 1100, 1102, 1114
Rhus rubifolia 561, 1102, 1114
Rhus schiedeana 1100, 1102, 1114

Rhus semialata 406, 408, 412-413
Rhus succedeana 559
Rhus terebinthifolia 1100, 1102, 1114
Rhus typhina 320, 404, 406, 408, 413, 561, 1100, 1102, 1114
Rhus undulata 1101
Rhus vernicifera 5, 520
Rhus virens 561
Ribes 260, 267-268, 623, 676
Ribes americanum 330
Ribes aureum 264
Ribes davidii 330
Ribes futurum 330
Ribes grossularia 330
Ribes houghtonianum 330
Ribes nigram 330
Ribes sanguineum 630, 657
Rizoclonia corcorium 348
Robinia 540, 544, 937, 990
Robinia fertilis 672
Robinia pseudoacacia 187, 191, 193, 302, 321, 539, 542, 544-545, 547, 606, 671, 869, 874
Rollinia papilionella 1107
Rollinia sericea 1107
Rosa 421-423, 431
Rosa canina 408, 421, 1111, 1118
Rosa glauca 167
Rosa rugosa 423
Rosmarinus 264
Rourea santaloides 1106
Rubia cordifolia 1110
Rubia tinctorum 1110
Rubus 267, 421, 431, 989
Rubus crataegifolius 321, 344
Rubus fruticosus 260, 408, 421, 1111
Rubus idaeus 329, 336, 348, 408, 421, 663, 681, 989, 1111
Rubus monogyna 989
Rubus rigidus 1111
Rumex 268
Ruta chalepensis 1099
Ruta graveolens 445, 1112
Ruta microcarpa 445
Ruta oreojasme 1095
Ruta pinnata 445, 1093, 1095, 1098, 1111-1112, 1114
Rutanilla 216
Ryania speciosa 1112

Saccharum officinarum 261
Sageretia gracilis 1104
Salacia 1111, 1117
Salacia chinensis 1106
Salacia macrosperma 782
Salix 290, 371, 373, 544, 547, 621

Salix alba 167
Salix caprea 629-630, 1096, 1109, 1116
Salix orestora 1188
Salix purpurea 290
Salix sieboldiana 593, 622
Salix triandra 373, 384
Salvia hypoleuca 767
Salvia officinalis 700
Samanea saman 823, 826
Sambucus nigra 264, 267, 328
Sanguinaria canadensis 1113
Sanguisorba 590, 624
Sanguisorba officinalis 421, 592, 622
Santalum 8
Santalum album 8, 710, 720, 853, 1051, 1116
Santalum lanceolatum 717
Santalum spicatum 8, 716
Sapium japonicum 264
Sapium sebiferum 264, 782
Sapota achras 172
Saraca indica 312, 824
Sarcocephalus didderrichii 161
Sarcopetalum 216
Sargentodoxa cuneata 1100, 1110
Sassafras 938
Sassafras albidium 940, 822, 1113-1114
Sassafras randaiense 444, 1101, 1108, 1113
Saururus cernuus 444
Saussurea lappa 706, 722
Schaefferia cuneifolia 1111, 1117
Schinopsis 539, 548, 559, 606, 613, 616, 652, 660, 671, 993-994, 997, 999
Schinopsis balansae 6, 615, 1175, 1190
Schinopsis lorentsii 539, 1175, 1190
Schinopsis quebracho-colorada 559
Schinus molle 161, 311
Schinus terebinthifolius 936
Schisandra 443
Schisandra chinensis 443
Schisandra rubriflora 443
Schisandra sphenanthera 443
Sciadopitys verticillata 673, 752, 765, 852, 858,
Sclerocarya 170
Scopolia 203, 1094
Scorodocarpus borneensis 853
Secale cereale 344-345
Securinega 208
Securinega suffruticosa 208
Securinega virosa 208
Segeretia minutifolia 268
Selinum carvifolia 706
Semecarpus 576, 583
Semecarpus anacardium 572, 576
Senecio crassissimus 739
Senecio deltoideus 740
Senecio douglasii 1114

Sequoia 557, 925, 929
Sequoia sempervirens 161, 320, 371,
    519–520, 852, 858, 874, 924, 942, 1177
Sequoiadendron giganteum 371, 519, 545
Sesamum angolense 447
Sesamum indicum 187
Shorea 4, 853–855, 858, 882, 891, 897,
    901–902, 924, 1012
Shorea affinis 1097, 1118
Shorea congestiflora 1097, 1114, 1118
Shorea dyerii 1097, 1118
Shorea laevifolia 897
Shorea lissophylla 1097, 1114
Shorea maranti 705, 708
Shorea negrosensis 857, 897
Shorea oblongifolia 1097, 1114, 1118
Shorea ovalifolia 1097, 1114
Shorea stipularis 514, 1097
Shorea trapezifolia 1097, 1118
Shorea vulgaris 773
Shorea zeylanica 1097, 1114, 1118
Sideritis canariensis 447
Silphium perfoliatum 740
Simaba multiflora 1096–1098, 1100, 1117
Simarouba amara 227
Simmondsia californica 286
Sinomenium acutum 217, 1107, 1115
Siphonodon australe 781
Skimmia japonica 211
Skimmia laureola 1119
Skimmia wallichii 824
Solanum dulcamara 816
Solanum macrocarpum 836
Solanum platanifolium 1104
Solanum tuberosum 308, 316, 318, 325–326,
    331–333, 335–337, 341–342, 346–348,
    992
Solidago 752
Sophora 264
Sophora microphylla 1108
Sophora secundiflora 1115
Sophora subprostrata 1108
Sophora tomentosa 1108
Sorbus aucuparia 280, 529
Sorbus sylvestris 1094
Sorghum 675, 989
Sorghum bicolor 345
Sorghum vulgare 940
Soulamea muelleri 1110
Soulamea soulameoides 1095–1097, 1103,
    1110
Soymida febrifuga 539–540, 545, 1109
Spinacia oleracea 314, 339, 817
Spiraea arguta 1119
Spiraea formosana 1102
Spiraea japonica 248
Spodoptera exempta 288

Spondianthus preussii 268
Spondias 170, 855
Spondias cytheria 170
Spondias dulcis 858
Spondias mombin 561
Stachyurus praecox 422–423
Steganotaenia araliacea 446
Stellera chamaejasme 268, 445, 572, 577
Stelletta 785
Stemmadenia 230
Stemmadenia glabra 1097
Stemonoporus affinis 1097, 1114, 1118
Stemonoporus canaliculatus 1097, 1114, 1118
Stemonoporus cordifolius 1097, 1114, 1118
Stemonoporus oblongifolius 1097, 1114, 1118
Stemonoporus petiolaris 1097, 1114, 1118
Stemonoporus reticulatus 1097, 1114, 1118
Stephania 216
Stephania hernandifolia 1107
Sterculia 171
Sterculia caudata 171
Sterculia foetida 539
Sterculia platanifolia 267
Sterculia setigera 170–171
Sterculia urens 171, 984
Stereospermum kunthianum 447
Stereospermum suaveolens 1106
Stevia rhombifolia 741
Stevia serrata 741
Strophanthus 250
Strychnos 217, 240
Strychnos nux-vomica 239, 1095
Strychnos pseudoquina 576
Strychnos toxifera 240
Styrax 3
Styrax benzoin 853
Swartzia leiocalycina 551
Swertia caroliniensis 707
Swietenia 857
Swietenia macrophylla 935
Swietenia mahagoni 1098
Swintonia luzonensis 561
Symphonia globulifera 1102
Symplocos 590–591
Symplocos uniflora 593, 670
Syncarpia glomulifera 1100, 1118
Synchlisia 216
Syringa vulgaris 157, 312
Syzygium cordatum 1100, 1102
Syzygium dumini 1105

Tabebuia 870
Tabebuia chrysantha 1106
Tabebuia guayacan 1106
Tabebuia impetiginosa 1094, 1106
Tabebuia rosea 825, 1106, 1110
Tabernaemontana 230

Index of Plant Genera and Species 1217

Tabernaemontana amblyocarpa 1097
Tabernaemontana glandulosa 1115
Tabernaemontana olivacea 1097
Tabernaemontana quadrangularis 1097
Tabernanthe iboga 230
Taiwania cryptomerioides 442, 734, 920
Talauma 216
Talauma hodgsoni 443
Tamarix aphylla 322
Tamarix nilotica 1100, 1102
Tamarix pentandra 312
Tapirira guanensis 561
Taraxacum koksaghz 1028
Taraxacum officinale 722, 777, 782
Taxillus kaempferi 679
Taxodium 539, 557, 932
Taxodium dubicum 519
Taxus 300, 539–540, 765, 837
Taxus baccata 264, 442, 557, 765, 822, 1094
Taxus brevifolia 822
Taxus cuspidata 442, 673, 822, 858
Taxus floridana 822
Taxus mairei 920
Taxus wallichiana 442
Tecomella undulata 822, 1106
Tectona grandis 2, 10, 322, 849, 857, 874, 920, 923, 935, 1106
Tellima grandiflora 407, 422
Tephrosia maxima 696
Terminalia 164, 170, 267, 993, 995, 997
Terminalia arjuna 778, 1100
Terminalia chebula 406, 408, 410–411, 429
Tetraclinis articulata 4
Tetradenia fruticosa 280
Tetramerista glabra 910
Tetrapathaea tetrandra 284
Tetrapleura tetraptera 1110
Thalictrum 216
Thalictrum dasycarpum 286
Thalictrum rugosum 286
Thamnosma texana 1095, 1111, 1119
Thea 817
Theobroma cacao 267, 674, 989, 991
Thermopsis fabacea 1101
Thomasiniana theobaldi 348
Thuja 557, 852, 870, 874, 1184
Thuja occidentalis 727, 938, 1054, 1116–1117, 1184
Thuja orientalis 720, 1116–1117
Thuja plicata 10, 309, 311, 328, 336, 442, 708, 858, 889, 924, 934, 937–938, 941–942, 944, 1052, 1054–1055, 1057, 1116–1117
Thuja standishi 442, 673, 858, 942
Thujopsis dolabrata 673, 679, 702, 852, 858, 720
Thymelaea 268

Thymus longiflorus 447
Tibouchina pulchra 933
Tieghemella heckelii 935
Tilia 157, 163, 300, 890, 895
Tilia americana 167, 322
Tilia cordata 157, 300, 302, 322
Tilia fulgaris 1115
Tilia japonica 857–858, 924
Tilia platyfolia 545
Tiliacora acuminata 162
Tiliacora racemosa 1107
Tinospora baenzigeri 1105, 1116
Tinospora capillipes 1105, 1107, 1108
Tinospora cordifolia 445, 1105, 1107, 1116
Tinospora crispa 1105, 1116
Tinospora glabra 1095, 1105, 1116
Tinospora malabarica 537
Tinospora merrilliana 1105, 1116
Tinospora sagittata 1105, 1116
Tinospora sinensis 1105, 1116
Tinospora smilacina 1105
Toddalia asiatica 1095
Tolmiea menziesii 407
Torreya nucifera 716–717, 852
Totomita 525
Toxicodendron 576, 935–936
Toxicodendron diversilobum 561
Toxicodendron semialatum 1116
Toxicodendron succedaneum 545
Toxicodendron vernix 936
Trachelospermum asiaticum 447, 1118
Trachelospermum foetidum 447
Trachelospermum liukiuense 447
Trachylobium 548
Trachylobium verrucosum 547
Trapa japonica 422
Tribulus longipetalus 1099
Tribulus terrestris 1099
Trichilia prieureana 1111
Trichilia roka 75
Trichoscypha arborea 561
Triclisia 216
Trifolium 568, 989–990
Trifolium arvense 675
Trifolium repens 662, 990
Trigonella 568
Trimenia papuana 443
Tripetaleia paniculata 1102, 1115, 1118
Tripterygium wilfordii 758, 1096, 1098, 1117
Tristania 894
Tristania conferta 321
Triticum aestivum 345, 347
Tsuga 300, 539, 558, 674, 884, 889, 932, 993, 998
Tsuga canadensis 319, 323, 344, 821, 1102, 1165
Tsuga chinensis 442

Tsuga heterophylla 311, 320, 327, 331, 442, 557–558, 631, 673, 857–858, 869, 896, 998, 1003, 1008, 1011–1012, 1053, 1177
Tsuga mertensiana 311, 328, 331, 685, 821
Tsuga sieboldii 442, 673, 857–858
Tulipa kaufmanniana 333
Turnera ulmifolia 284–286

Ulmus 163, 526, 591, 735, 845, 941, 1181
Ulmus americana 322, 593, 822
Ulmus fulva 157, 166
Ulmus glabra 1098
Ulmus parvifolia 735
Ulmus rubra 526, 822, 834, 849
Ulmus thomasii 445, 822
Uncaria 235
Uncaria elliptica 1100
Uncaria florida 235
Uncaria formosa 235
Uncaria gambir 232–233, 599, 619, 670, 672
Uncaria rhynchophylla 233–235
Uragoga ipecacuanha 268
Urbanodendron verrucosum 444
Urtica dioica 671
Uvaria afzelii 1098, 1118
Uvaria angolensis 1094, 1102
Uvaria catocarpa 287–288
Uvaria ferruginea 288–289
Uvaria purpurea 288–289
Uvaria rufas 536
Uvaria zeylanica 288–289

Vaccinium 675
Vaccinium corymbosum 267, 320, 657
Vaccinium macrocarpon 280
Vaccinium membranaceum 989
Vaccinium myrtillus 267
Vaccinium oxycoccus 675, 989
Vaccinium vitis-idaea 622, 1094
Valeriana officinalis 697, 706, 729, 1109
Valeriana wallichii 720, 744
Vanilla planifolia 316, 321
Vanillosmopsis erythropappa 1097, 1102
Vatairea heteroptera 1100–1101
Vateria copallifera 514
Vatica 4, 858, 901, 909
Vatica affinis 514, 1114
Ventilago calyculata 1097, 1100, 1110
Ventilago maderaspatana 1097
Verbesina oncophora 1103
Vernonia polyanthes 1105
Vernonica hymenolepis 724
Vetiveria zizanioides 720
Viburnum 300
Viburnum nudum 267
Vicia 989–990
Vicia faba 325, 337, 339

Vicia sativa 666
Vigna unguiculata 989
Viguiera 590
Viguiera quinqueradiata 592
Vinca 1057
Vinca minor 1112
Virgilia divaricata 169
Virola 550, 559, 855
Virola caducifolia 550
Virola calophylloidea 443
Virola cuspidata 443
Virola elongata 443
Virola guggenheimii 443, 1101
Virola multinerva 550
Virola pavonis 443
Virola sebifera 443
Virola surinamensis 443
Viscum album 446, 1110
Vismia guaramirangae 1101
Vitex lucens 537, 541
Vitex negundo 537, 541
Vitex pubescens 672
Vitis 624, 672, 675, 681, 989
Vitis glauca 672
Vitis vinifera 260, 264, 267–269, 322, 344, 676, 680, 1095–1096, 1101–1102, 1114
Vochysia acuminata 1100
Vochysia thyrsoidea 1095, 1110
Vouacapoua 514
Voucapoua macropetela 756

Warburgia ugandensis 774
Watsonia ardnerei 655, 662
Watsonia pyramidata 657
Wedelia 1103
Wedelia grandiflora 1105
Welwitschia 137
Wethia 559
Widdringtonia 720
Wigginsia erinacea 171
Wikstroemia 268
Wikstroemia foetida 445
Wikstroemia indica 445, 1098, 1117
Wikstroemia monticola 1105
Wikstroemia uva-ursi 445
Winchia calophylla 1112
Wisteria 623
Wisteria floribunda 553, 1101, 1118
Wisteria sinensis 622

Xanthoceras sorbifolia 1093
Xanthocercis zambesiaca 1105
Xeromphis spinosa 1118
Xylia dolabriformis 756

Zanthoxylum 446
Zanthoxylum acanthopodium 446

Zanthoxylum ailanthoides 219, 446, 1099, 1115
Zanthoxylum alatum 446
Zanthoxylum americanum 219
Zanthoxylum argorescens 1104
Zanthoxylum arnottianum 446, 1099, 1108, 1111–1112
Zanthoxylum bungeanum 1115
Zanthoxylum capense 446
Zanthoxylum chalybeum 446
Zanthoxylum conspersipunctatum 824
Zanthoxylum coriaceum 1096
Zanthoxylum culantrillo 1107, 1115
Zanthoxylum cuspidatum 384, 1099, 1112, 1115
Zanthoxylum dinklagei 446
Zanthoxylum dipetalum 1095–1096, 1107, 1116
Zanthoxylum elephantiasis 1095
Zanthoxylum fagara 72
Zanthoxylum flavum 1111, 1119
Zanthoxylum gillettii 1109, 1113
Zanthoxylum inerme 219, 446, 1099, 1109
Zanthoxylum integrifoliolum 1093, 1099, 1112, 1114–1115
Zanthoxylum lemairei 1095
Zanthoxylum leprieurii 1095
Zanthoxylum macrocarpum 1104, 1107
Zanthoxylum monophyllum 1095–1096, 1107
Zanthoxylum nitidum 219, 1107–1109, 1115
Zanthoxylum oxophyllum 446
Zanthoxylum piperitum 446
Zanthoxylum planispium 218
Zanthoxylum pluviatile 446, 1115
Zanthoxylum podocarpum 446
Zanthoxylum punctatum 1104, 1107
Zanthoxylum rubescens 1095, 1109
Zanthoxylum senegalense 446
Zanthoxylum tsihanimposa 219
Zanthoxylum viride 1095
Zanthoxylum williamsii 1096, 1109, 1114
Zanthoxylum wutaiense 1097, 1099, 1108
Zea mays 317–318, 331, 344–345, 834
Zelkova oregoniana 1104
Zelkova serrata 920, 541–542
Zeyheria digitalis 1106, 1118
Zingiber officinale 325, 717
Ziziphus 216, 267
Ziziphus jujuba 267
Ziziphus lotus 268
Zollernia paraensis 514, 1101
Zygophyllum fabago 1103

Index of Organic Compounds

abacunone   777
abiesin   576
abietadiene   757–758
abietic acid   757–758, 869, 891–892, 936, 943, 967–968
abietinal   757–758
abietinol   757–758
abscisic acid   745–746, 789, 1182, 1185
abyssinone   1073, 1075, 1090, 1093
acacipetalin   282–283
acamelin   935
acanthine, oxy   1073, 1093
acanthoside   463
acer tannin   400, 408, 410
acetic acid   854–855
acetovanillone   373
acetyl shikonin   1082, 1093
acetyldigoxin   1070
aconitic acid   260–261
acoradiene   718, 720
acoradin   477
acorenol   718, 720
acoric acid   719–720
acorone   718–720
acronycine   211–212
acronycine, de-N-methyl   211-212
acronylin   390
acrydone, α,3-dimethosy-N-methyl   211
acrylic acid, β,β-dimethyl   22
aculeatin   378
acuminatin   494
adoniside   1070
aegeline   376
aescin   1070
aesculetin   378, 1070, 1073, 1075–1077, 1079–1080, 1093
africanal   466
afrormosin   549–550
afzelechin   542, 590, 685
afzelin   1078, 1089
agarofuran   727–728, 853
agarospirol   729–730, 853
agatharesinol   517–519, 926
agathic acid   751
agathisflavone   576, 579
agbanindiol   752–753
agrimolide   522–523
agrimoniin   421

agrimophol   1070
ajmalicine   227–228, 238, 1062, 1070
ajmalicine, 19-epi   238
ajmaline   227–228
akuammidine   242–243
akuammigine   235–236
alanine   265
alaskene   718
alizarin-1-methyl ether   1079, 1084, 1093
alkannin   1074, 1093
allantoin   1070, 1074, 1080, 1093
alliodorin   391
allocryptopine   1083, 1086–1087, 1091, 1093
allocryptopine, α   1073, 1081
allodevadarool   756
alloevodionol   385
allohimachalol   741–742, 852, 1078, 1093
alloptaeroxylin   385
allosecurinine   208
alloxanthoxyletin   382
alloyohimbine   236–237
allyl isothiocyanate   1070
allysulfide   851
alnusdiol   524
alnusenone   781–782
alnusiin   422
aloe emodin   1075, 1093
alpinone   545, 555–556
alstonine   1084, 1086, 1093
alstonine, tetrahydro   235–236, 238, 1086, 1094
altantone   717
altholactone   281
amber   3, 954, 1051
amentoflavone   572, 574–577
ammodendrine   1083, 1094
ammoidin   1072
amorphene   735
amygdalinic acid   291
amyrenone   780
amyrin   309, 767–777, 779–780, 809
anabasine   1070
anagyrine   209
andeographolide   1070
androst-1,4-dien-3-one   833
androsten-3,17-dione   976
anethole   853, 961, 965
angelic acid   697

angeloylgomisin 481
angolensin 552
anibine 548
anisaldehyde 965
anisodamine 1064, 1070, 1072–1073, 1075–1078, 1080, 1082–1083, 1086–1087, 1089–1090, 1093–1094
anisodine 1064, 1070, 1094
anonaine 1084, 1094
anosidine 1083, 1093
ansamacrolide 1065
anthocyanidin 131, 608–609, 660, 681
anthocyanin 608–609, 679, 681, 844, 851
anthothecol 936
anthraquinone, 1,8-dihydroxy 1070
anthraquinone, 2-hydroxymethyl 1079, 1084, 1094
anthricin 457, 468
anticopalic acid 967–968
anwulignan 475
aphanamol 83
apigenin 536–537, 541, 543, 576, 583, 1073, 1085–1088, 1091–1092, 1094
apioside 593
apopolygamain 472
aptosimol 463
aptosimone 463
apuleidin 540, 566
apulein 540, 566
apuleirin 540, 566
apuleisin 540, 566
apuleitrin 540, 566
arabinogalactan 924, 978–979, 1168
arachadic acid 323
araliangine 473
araucarolone 755–756
arboreol 462
arbutin 371, 1080, 1088, 1094
arctigenin 455
arctiin 455
arecoline 1070
argentilactone 280
aricine 1073, 1094
ariensin 453
aristolene 744
arjunolic acid 777
armenin 487
armepavine methosalt 219–220
arnottianin 381
aromadendrene 743–744, 853
aromadendrin 545, 677
artemisia ketone 710
artemone 746
artobilichromene 537
artocarpanone 544
artocarpesin 537–538
artocarpetin 536

artocarpin 538
asadanin 522, 524
asaraldehyde 389
asarinin 460, 465
asatone 504–505
ascidiacyclamide 70
ascorbic acid 268–269
asiatic acid 779–780
asiatic acid, 11-oxo 780
asiaticoside 1070
aspidospermidine 231
aspidospermidine, 1,2-dehydro 231
aspidospermidine, N_a-methyl 231
aspidospermine 229–231
aspidospermine, deacetyl 229–231
aspidospermine, N_a-methyldeacetyl 231
astringenin 514
astringin 514
athosterol, 24-methylene 817
athrotaxin 517–519
athrotaxin, hydroxy 518–519
atisine 761–762
atisirene 761–762
atlantone 718
atropine 1061, 1080–1081, 1092, 1094
aucuparin 520–521, 1075, 1094
aucuparin, hydroxy 520–521
aucuparin, methoxy 520–521
aurein 485, 502
aureusidin 535, 547
aurone 851
austrobailignan 467, 475–476, 478, 1079, 1084, 1094
avenasterol 816–818, 829
avicine 219
ayanin 563, 566, 937
azadirachtin 17–18

baccatin 1084, 1094
bakkenolide 729
balata 1029, 1048–1049
balfourodine 210
balfourodinium salt, O-methyl 210
baliospermin 766
barterin 284
bathorhodopsin 100–101
bauerenol 782
bayin 537
behenic acid 323
benihiol 852
benzaldehyde 371
benzaldehyde, hydroxy 336, 1073, 1094
benzoic acid 371, 853, 1074, 1082, 1090, 1094
benzoic acid, p-hydroxy 1077
benzophenone 281
benzoquinone 921

benzoquinone, 2,6-dimethoxy-1,4- 935, 943, 1084, 1095
benzoylgomisin 481
benzyl benzoate 371, 1070
benzylalcohol 853
berbamine 1089, 1095
berberine 218, 1062, 1070, 1073−1078, 1080, 1082−1083, 1085−1089, 1091−1092
bergamotane 720
bergapten 379, 1078, 1080−1081, 1084−1085, 1087, 1091, 1095
bergenin 388, 1070
betacyanin 141
betalin 844
betaluprenol 787
betaxanthin 141
betuligenol 392
betulin 309, 777, 779, 976
betulinic acid 777, 779
betuloside 392
beyerol 761−762
bhirra gum 170
biapigenin 576
bicyclogermacrene 742
biflavonoid 281
biflorin 764−765
bile acids 808, 811
bilobanone 717−718
bilobetin 574
biluteolin 572, 575
binetinidin 611
biocahnin 549−550, 553
bisabolene 717−718, 720
boehmenan 473
boivinose 285
borane 698−699
borbonol 1075, 1095
borneol 710, 852−853, 933, 1070
bornesitol 161
bornyl acetate 852−853
bornyl alcohol 852
boronolide 280−281
boswellic acid 779−780
bowdichinone 845
bowdichione 550−551
brassicasterol 816, 830
brassinolide 830
brayleyanin 378
braylin 382
brazilein 847
brazilin 553−554, 847, 1074, 1095
brevetoxin 67−68
brevifolin carboxylic acid 415−416
brevilagin 419, 427
bromelain 1070
brosiparin 377

brosiparin, O-prenyl 378
brosiprenin 377
broussinol 592
bruceantin 1068, 1073, 1077, 1079, 1086, 1091, 1095
bruceantinol 1079, 1095
bruceantinoside 1079
brucein 1079, 1085, 1090, 1095
brucine 239−240, 1083, 1095
brugierol 293−294
bryaquinone 847
buddledin 739
buddlenol 473−474
bufacienolides 837
bulgarene 735
bunesol 730, 732
burchellin 440, 485, 501
butan-2-ol, 4-(p-hydroxyphenyl)- 392
butanolide, 2-dodec-ω-enyl-3-hydroxy-4-methyl 279
butanolide, 2-hydroxy-4-(heptaced-8'-enyl) 279
butein 546−547, 560−562, 904, 908
butin 543−544, 904
butyric acid 854−855
butyrospermal 774

cacalol 729
cadalenal, 7-hydroxy 849
cadalenal, 7-hydroxy-3-methoxy 849
cadalene 735, 853, 903
cadambine 234-235
cadambine, 3α-dihydro 234−235
cadambine, 3β-isodihydro 234−235
cadaverine 209
cadinene 732, 734−735, 852−853
cadinol 734−735, 852−853, 1086, 1095
caffeic acid 401−402, 404, 1055, 1074−1075, 1077−1078, 1080, 1082−1083, 1085−1086, 1090, 1095
caffeine 667−668, 1070
cagayanin 479
calacone 735, 746
calacorene 736, 852−853
calamenene 735−736, 852−853
calamenene, 7-hydroxy 391
calarene 853
callitrin 724
callitrisin 727
calocendrin 458
calopiptin 476
calycanthidine 225−226
calycanthine 225−226
calycosin 550, 567
campest-4,6-dien-3-one 825
campest-4-en-3-one 825
campest-4-en-3-one, 6β-hydroxy 826
campestanol 820

Index of Organic Compounds 1223

campesterol 816, 819–822, 824
camphene 708, 852–853, 856, 960, 964
camphor 9–10, 700, 710, 851, 853, 933–934, 964, 1057, 1070
camptothecin 225, 244–246, 1064, 1079, 1082, 1084–1086, 1091, 1095
camptothecin, 9-methoxy 244–245, 1085, 1095
camptothecin, 10-hydroxy 245
camptothecin, 10-methoxy 244–245
camptothecin, 11-hydroxy 244–245
camptothecin, 11-methoxy 244–245
camptothecin, 20-deoxy 244–245
canarone 853
canellin 487, 490–491
canthin-6-one 226–227, 1084, 1095
canthin-6-one, 1-methoxy 226–227
canthin-6-one, 4,5-dimethoxy 226–227
canthin-6-one, 5-hydroxy 226–227
canthin-6-one, 5-methoxy 226
canthin-6-one, 8-hydroxy 226
canthin-6-one, 11-hydroxy 1084, 1096
canthin-6-one, 5-methoxy 227
canthin-6-one-3-oxide 226–227
canthron 1070
caparrapi oxide 744–745
caparrapidiol 717
caparrapitriol 717
caphnoretin 1091
caprilic acid 855
caproic acid 854–855
caracurine 240
carane 698–699
carapanaubine 228–229
carapin 776–777
carbazole 213
carboline, 1-acetyl-4-methoxy-β- 227
carboline, 1-acetyl-β- 227
carboline, 1-methoxycarbonyl-β- 227
carboline, 4,8-dimethoxy-1-vinyl-β- 227
carboline, 4-methoxy-1-vinyl-β- 227
carboline-1-carboxamide 227
cardenolides 837, 933
cardiospermin 282–283
carene 708, 852, 960, 965–966
carinatanol 485
carinatidin 495
carinatidiol 497
carinatin 495
carinatol 493
carinatone 440, 494
carinol 453
carissanol 457
carnosol 1077, 1096
caronaridine 1097
carotene 128, 130, 786
carotenoid 695, 812, 844

carozeylanic acid 293
carpanone 480
carquejol, 2-isobutyryloxy 706
carreic acid 374
carvacrol 393, 704, 852, 1074, 1077–1078, 1081, 1096
carvacrol methyl ether 704, 852
carvacrol, p-methoxy 852
carvane 965
carvone 705
carvotagetone, angeloyloxy 705
caryatin 539, 542
caryophyllene 711, 737, 739–740, 852–853, 961
caryophyllene alcohol 740
caryophyllene oxide 739
casbene 765–766
cascarillic acid 293
cascarillin 753
casegravol 378
cassaine 248–249, 756
cassameridine 1078, 1080, 1096
cassinin 727–728
castalagin 422, 424–425, 431
castalin 422, 425
castanaguyone 72
castanaguyone, diacetyloctahydro 72
castavalonenic acid 423
casuarictin 420–425
casuariin 422, 425
casuarinin 422, 425
catalpalactone 386
catalposide, 6-O-veratryl 389
catechin 400, 535, 571–635, 653–654, 656–657, 659–661, 663–665, 670, 678–681, 684, 896, 898–899, 989, 991–992, 995, 1018, 1068, 1070, 1073–1075, 1090–1091, 1096, 1189
catechin-3-gallate 407
catechinic acid 598, 635
catechol 370
cathenamine 238
cathidine 213–214
cedeodarin 545, 556–557
cedrelone 776
cedrelopsin 377
cedrene 719–720, 851–852
cedrenol 852
cedrin 545, 556–557
cedrinoside 556–557
cedrol 719–720, 852, 1052–1053, 1056
celastrol 1085, 1096
celphinidin 1098
cembranolide 764
cembrene 764–765, 960
centdarol 852, 1078, 1096
centdarol, iso 1078

centrolobine 522–523
centrolobol 522–523
cephaeline 223–224, 1062
cephalotaxine 222, 1084, 1096
cerasinone 544
cerberin 838
cerin 781
cerotic acid 323
ceylantin 382
chagual gum 172
chalcomoracin 516–517
chalconaringenin 571
chalcone 556, 851
chalepin 381
chamaecynenol, dehydro 727
chamaecyparin 582
chamaejasmin 577, 581
chamic acid 708–709
chaminic acid 709
chanootin 711
chaparrinone 1077, 1079, 1085, 1090, 1096
chaparrinone, 6-α-senecioyl-oxy 1080, 1096
chaplashin 537–538
chebulagic acid 427–429
chebulic acid 415–416
chebulinic acid 400, 427–428
chelerythrine 220, 1073, 1078–1080, 1084, 1096
chelidonic acid 268, 269
chelrythrine 219
chettaphanin 753–754
chicle 787, 972, 1028–1050
chimonanthine 225–226
chinensin 469
chinensinaphthol 469–470
Chinese gallotannin 412
chitosenine 243–244
chlionasterol, dehydro 817
chlorellagic acid 415
chlorogenic acid 375, 402, 1074, 1077, 1079–1080, 1083, 1086, 1096
chlorophorin 514, 517, 925
chlorophyll 16, 128, 130, 179, 695–696, 750, 788–789
cholest-4,6-dien-3-one 825
cholestanol 808, 810–811, 815, 817
cholesterol 808–839
cholesterol, 7-dehydro 815, 827
cholesterol, 24-methyl 820
cholesterol, 24-methylene 829
cholesterol, methyl 819
cholic acid 60–61, 809–810
choline 1086–1088, 1091, 1096
chondrillasterol 828
chondrillasterol, dehydro 817, 830
chromomycin 42, 47
chrysanthemic acid 710, 966

chrysanthemyl alcohol 710
chrysarobin 941
chrysin 536–537, 555–557, 1086, 1096
chrysophanic acid 1085, 1097
chrysophyllon 495–496
chymopapain 1070
ciasin 754
cinchonain 625
cinchonamine 239
cinchonidine 237, 1077, 1097
cinchonine 237, 1077, 1097
cineole 705, 853
cinnamate, 3,4,5-trimethoxy 203
cinnamic acid 374
cinnamic acid, 3,4,5-trihydroxy 401
cinnamic acid, 3,4-dimethoxy 331
cinnamic acid, 3,5-dimethoxy-4-hydroxy 331
cinnamic aldehyde 853
cinnamic aldehyde, o-methoxy 852
cinnamonol 460
cinnamyl cinnamate 853
cinnamylphenol 565
cissampeline 1070
citral 702, 853, 963–964
citric acid 260–261
citronella 966
citronellal 702, 853, 962–964
citronellic acid 702–703
citronellol 962–964
citrostadienol 822, 829
cleistanthin 470
cleistanthol 756
cleistanthoside 471
cleomiscosin 383, 1084, 1097
clerodin 752–753
clerosterol 817–818
clerosterol, dehydro 817
clionasterol 819, 828
clusin 457
cneorane 775
cneorin 776–777
cneoroids 774
cocaine 203–204, 1070
cocaine, cinnamoyl 203
coclaurine 216
coclaurine, N-demethyl 1073, 1078, 1081–1083, 1087, 1091, 1097
codeine 1070
codisterol 818
colchiceine amide 1070
colchicine 1070
colensenone 754
collinin 378
collumellarin 731–732
columbin 753
colvane diol 740
communic acid 751, 967–969

compactin 81, 83
conduritol 162
conessine 250
conidendral 466
conidendrin 470–472, 866, 868, 1053–1054
coniferin 384
coniferyl alcohol 384
coniferyl aldehyde 384, 1189
conocarpin 494
conocarpol 475
conrauana lactone 281
convallatoxin 1070
copaborneol 736–737
copacamphor 737
copaene 736, 852–853
copal 1051
copalliferol 514–515
copalol 750–751
coprostanol 808, 811
coptisine 1076, 1084, 1097
cordigerin 457
corilagin 417–418, 428–429
coronaridine 1084, 1087, 1091
cortisol 830–831, 833
cortisone 20
corynanthe 239
corynantheal 237–238
corynantheine 233–234, 237
corynantheinealdehyde, 4,21-dehydro 238
corynantheol 239
corynanthine 236–237
corynoxeine 235
costic acid 726
costunolide 722–723, 1075, 1080, 1082, 1085–1086, 1097
costunolide, dihydro-β-cyclo 726
cotonefuran 520–521
coumaranone 546, 548
coumaranone, 4-methoxy-6-hydroxy 904
coumaranone, hydroxy 904
coumaric acid 309, 342, 374, 401–402, 1075, 1097
coumarin 219, 378, 552, 554, 1078, 1082, 1089, 1098
coumaroylferuloylmethane 523
coumaryl alcohol 384
coumestan 548, 552
coumestrol 933
coumurrayin 377
crenatidine 227
crenatine 227
cresol 852
cretanin 407
crombenin 526–527
crotepoxide 288
crotocaudin 753

crusdecdysone 836
cryogenine 206–207
cryotimerin 582
cryptocaryic acid 291
cryptofauronol 729–730
cryptomeridiol 726
cryptophenol 900
cryptostrobin 544, 555, 557
cubebane 735
cubebin 457
cubebinin 457
cubebinolide 456
cubebinone 456
cumic acid 371
cuminaldehyde 705
cuparene 720–721, 852
cuparenic acid 720–721
cuparenone 720–721
cupressic acid 750–751
cupressuflavone 576, 578
curare 240, 931
curarine 241
curcumene 717–718
curcumin 522–523, 1070
curdione 722–723
curzerenone 724
cutch 9
cyanidin 535, 608, 652, 656, 681
cyanogenic glycosides 933
cycloartanol 826
cycloartanol, 24-methylene 828–829
cycloartenol 309, 812–814, 826, 828, 832, 1076, 1098
cycloartocarpesin 537–538
cycloartocarpin 537–538
cyclobuxine 250
cyclocopacemphenic acid 736
cycloeucalenol 828–829
cycloheterophyllin 537–538
cyclohexanediol 290
cyclolaudenol 828
cyclomicophyllin 250
cyclomulberrin 537–538
cyclomulberrochromene 537, 541
cycloolivil 466
cyclopentenone, 5-dodecanyl-4-hydroxy-4-methyl-2- 293
cyclosativane 733
cyclosativene 736–737
cyclovirobuxine 1076, 1081, 1083, 1088, 1098
cymene 393, 703, 852–853, 1082, 1086, 1098
cystine 46
cytisine 209
cytisine, methyl 209
cytokinin 34, 789

Index of Organic Compounds

daidamin   283
daidzein   549–550, 553, 1075, 1077, 1086, 1089, 1098
dalbergin   378, 383, 526, 553–554, 564
dalbergioidin   550–551, 553-554, 564, 568
dalbergione, 3,4-dimethoxy   935
dalbergione, 4'-hydroxy-4-methoxy   849
dalbergione, 4,4'-dimethoxy   849
dalbergione, 4-methoxy   849
dalbergiquinol   554, 564
dalgergin   565
dambonitol   161
dammar   3-4
dammarendiol   773
daniellic acid   751
daphnetoxin   765–766
daphniphylline   249
daphnoretin   382, 1079, 1085, 1090, 1098
daphylloside, 10-caffeoyldeacetyl   376
dasycarponilide   286
dasycarponin   286
daucol   721–722
daucosterol   1075–1076, 1081, 1085, 1087, 1092, 1098
daurinol   471
davidiin   429
debilone   744
decamine   206
decenoic acid   293
decinine   206
decompostin   729
decynoic acid   312
dehydroabietic acid   893, 967–970
dehydroabietylamine   973
dehydrocostus lactone   731–732
dehydrodicatechin   601
dehydrodieugenol   440
dehydrodigallic acid   415–416, 421
dehydroferruginol   899
dehydrogallic acid   415
dehydrogeranic acid   702–703
dehydroheliobuphthalmin   453
dehydroheliobuphthalminlactone   458
dehydrohexahydroxydiphenoyl esters   419
dehydromoracin   516–517
dehydroneryl isovalerate   702
dehydroquinic acid   266, 404
dehydroshikimic acid   266
dehydrotodomatuic acid   717–718
dehyotol   474
deidaclin   283–286
delphinidin   608, 656, 1077
demecolcine   1070
dendrolasin   852
denudatin   493
denudatone   492
deodarin   546

deodarone   717–718
deoxyharringtonine   1090, 1104
deoxyisodiphyllin   471
deoxyjacareubin   525, 1105
deoxylapachol   936
deoxyschisandrin   481, 483
deserpidine   1062, 1070
deslanoside   1070
desmethoxymatteucinol   544
desmosterol   815, 821, 827
devadarool   756
diaboline   240–241
diaboline, desacetyl   240
diaeudesmin   462
diasesartemin   463
diasyringaresinol   463
dibenzofuran   521
dicatechin   598, 601
dicoumaroylmethane   522–523
dictamnine   210–211, 1074–1075, 1085, 1098
dictyopterene   293
didehydroryanodine   1088, 1112
digitalin   1070
digitogenin   835
digitonin   835
digitoxigenin   838–839
digitoxin   838, 1070, 1184
digoxin   1070, 1184
dihydroabietyl alcohol   936
dihydroagarofuran   853
dihydroagathic acid   967
dihydrobenzofuran   565
dihydrobrassicasterol   816, 819–822, 828, 830
dihydrocarinatin   495
dihydrocarinatinal   495
dihydrocarveol   1073, 1087, 1092, 1096
dihydroclusin   453
dihydroconiferyl alcohol   384
dihydrocorynantheine   233–234
dihydrocubebin   453
dihydrodaidzein   550–551
dihydrodieugenol   496, 498
dihydroflavonol   676–677, 679
dihydroguaiaretic acid   440, 475, 479
dihydrohinokiflavone   582
dihydrokaempferol   545–546, 555–557, 571
dihydrokawain   387
dihydromahuba lactone   275, 278
dihydromahubenolide   276
dihydromelanoxetin   545
dihydromethysticin   388
dihydromyricetin   545–546, 555–557, 603, 679
dihydropyranoquinoline   210
dihydroquercetin – see taxifolin   533, 543, 545–546, 555–558, 569, 603, 608, 678–679, 888–889, 896, 1053–1054, 1176

Index of Organic Compounds 1227

dihydrorobinetin 545, 559, 869
dihydrosciadipitysin 575
dihydrosecurinine 208
dihydrosesamin 459
dihydrosesartemin 459
dihydrospinasterol 816–817, 824
dihydrouroquinoline 210
dihydroverticillatine 206
dihydrowogonin 544
dihydroyan-gambin 459
dihydroyashabushiketol 522–523
dihydrozymosterol 827
diisoeugenol, 5-methoxydehydro 495
dillapiol 390
dimethoxybenzoquinone 933
dimethylpinocembrin 556
dimoracin 516–517
dioonflavone 575–576
diosgenin 835–836, 1087, 1099
diospyrin 1075
dipentene 703, 852, 961–962, 965, 1174
diphenic acid, hexahydroxy 401
diphyllin 470
diphyllinin 471
distemonanthin 937
dityrosine 186
djalonensin 386
docosanedioic acid 325
dopa 1070
dopamine 1079, 1086, 1099
dracogenin 836
dracorubin 527
dregamine 232
drimanolide 744
drimenol 744–745
duartin 551
dulcitol 160

eburnamenine 230–231
ecdysone 136, 808, 830, 836, 1074–1076, 1083, 1099
ecdysone, 20-hydroxy 837
ecdysteroid 37–38, 836
ecdysterone, 20-hydroxy 836
ecgonine, benzoyl 203
ecgonine, methyl 203
edgeworthin 382
edulein 210
edulitine 210
eglandine-N-oxide, 10-methoxy 1084, 1099
eicosanoic acid 973
eicosanyl ferulate 375
elaeocarpidine 203, 205
elaeocarpiline 203, 205
elaeocarpine 203, 205
elaidic acid 973
elcarvone 708

elemenal 852
elemene 724, 852–853
elemicin 384, 1085, 1099
elemol 724, 852–853
elemolic acid 773–774
elenolide 281
eleutheroside 464
ellagic acid 371, 401, 520–521, 868, 873, 878, 898, 911, 940, 1072, 1075–1076, 1078–1079, 1081, 1085, 1087, 1089–1090, 1092, 1099
ellagitannin 403, 406, 868–869, 872–875, 878, 895–899, 989, 911
elliotinic acid – see communic acid 751
ellipticine 231–232, 1065, 1077–1079, 1082, 1084–1088, 1100
ellipticine, 9-methoxy 1078–1079, 1084, 1086, 1088, 1091, 1099, 1100
emetine 223–224, 1062, 1070
emodin 1074–1075, 1082, 1085–1086, 1100
endiandric acid 291–292
enshicine 478
enterolactone 440–441
ephedrine 1062, 1070, 1072–1073, 1076, 1078, 1080–1081, 1083, 1087, 1091–1092, 1100
epiafzelechin 571, 590, 592, 600, 610, 613–614, 619, 622–623, 632, 679–680
epiasadanol 522, 524
epiaschantin 462
epicaryophyllene 739
epicaryophyllene oxide 739
epicatechin 400, 571–635, 653–659, 661, 664–665, 670, 674, 676, 678, 680–681, 684, 989–991, 1017, 1073, 1076–1077, 1088, 1100
epicatechin-3,5-digallate 407
epicatechin-3-gallate 407
epiergosterol 819
epieudesmin 462
epiexcelsin 462
epifisetinidol 590, 592, 595, 610, 615
epigallocatechin 591, 592, 629–630, 654–659, 670, 679, 681, 989, 991–993
epigallocatechin-3,5-digallate 407
epigallocatechin-3-gallate 407
epigomisin 482
epigynocardin 283
epilupeol 777
epimagnolin 463
epioleanolic acid 777
epipassibiflorin 284–285
epipassicoccin 285
epipinoresinol 461
epipodophyllotoxin glycoside 1063
epiproacacipetalin 283
episerratenediol 785

Index of Organic Compounds

episesamin 460
episesartemin 463
episyringaresinol 463
epitetraphyllin 284–286
epoxycyclolignan 466
epoxylignan 459
equol 551
eremoligenol 728
eremophilone 728
ergonine 203
ergosterol 819, 830–831
eriodictyol 535
erioflorin 723
erthrocentaurine 387
ervatamine 231–232
ervatamine, 20-epi 231–232
ervatamine, dehydro 231–232
erycristagallin 1073, 1100
erythoxytriol 756
erythrabysin 1090, 1100
erythrabysin II 1074
erythraline 221
erythratine 221
erythrinine 221
eserine 1071
estradiol 830
estragole 965, 1085, 1087–1088, 1091, 1100
ethane, 1-nitro-2-phenyl 373
ethylphenol 373, 853
etoposide 1070
eucalyptin 536
eucarvone 707
eudalene 725–726
eudesmin 462
eudesmol 725–726, 852
eugeniin 420, 422
eugenol 384, 440, 739, 853, 1074, 1077–1078, 1081–1083, 1087, 1089, 1091, 1101
eugenol methyl ether 853, 1083, 1085, 1087, 1091, 1101
eugenol, methoxy 964
euphol 773, 809
eupodienone 484–485
eupomatenoid 494–495
euponatene 494
eusiderin 497–498, 1092, 1101
euxanthone 1076, 1081, 1083, 1101
evodiamine 212
evodiamine, 7-carboxy 212
evodiamine, dehydro 212–213
evonine 213–214
evoxanthidine 211–212
evoxanthine 211–212
excelsin 462
exostemin 553–554

fagaramid 376

fagarol 460
fargesin 463
farnesane 853
farnesene 305, 852
farnesiferol 744–745
farnesol 692, 712, 716, 768
farnesyl pyrophosphate 812
fascicuferol 521
fenchane 698–699
fenchol 710
fenchone 934
ferrearin 489
ferrudiol 288–289
ferruginol 758, 869–870, 898–900, 908, 920
ferruol 380
ferulate, octacosanyl acetoxy 375
ferulene 744
ferulic acid 309, 311, 331, 342, 374, 402, 869, 922, 1055, 1074–1075, 1077, 1079, 1081–1082, 1090-1091, 1101, 1178
ferulol 706
fibraurine 1071
fisetin 533, 542, 559–563, 849
fisetinidin 604
fisetinidol 590, 592, 595–596, 600, 610, 615–617, 660
flavan-3,4-diol 677, 851, 1190
flavan-3-ol 404, 656, 677, 679–680, 682, 851, 991–992
flavogallol 415
flavokawin 547
flavone, 5-hydroxy-4'-7-dimethoxy 1101
floribin 377
fluoroacetic acid 268–269
fluorocapric acid 269
fluorocurine 240–241
fluoromyristic acids 269
fluorooleic acid 269
fluoropalmitic acid 269
fluoropyruvic acid 270
folicanthine 225–226
formic acid 855
formononetin 551, 553, 567, 1075, 1088–1089, 1101
formosanine 235
formosanol 467
forsythoside 1074, 1102
fragilin 372
fragranol 707–708
frankincense 7, 170
fraxidin 377
fraxinol 377
freelingyne 716–717
friedelin 309, 767, 781, 1085, 1102
fucosterol 829
fukegetin 580
fukugiside 576, 580

Index of Organic Compounds

fumaric acid 260, 262
furanoeremophilanolide 729
furoguaiacin 477
furoquinoline 210
fustin 543, 545, 559–561
futoenone 440, 492, 501–502
futoquinol 492
futoxide 288

gadain 472
galactinol 161
galactitol 160
galangin 539, 555, 557
galbacin 476
galbegin 476
galbulin 477, 479
galcatin 477, 479
galgravin 476
gallagic acid 415
gallic acid 371, 399, 405, 684, 868–869, 873, 889, 897–898, 1073–1074, 1076–1077, 1079, 1080–1082, 1088, 1091, 1102, 1190
gallocatechin 591–592, 601–602, 619–621, 629–630, 654–657, 659, 680, 991
gallotannin 868, 896–897, 911
galloyloxyphenyl-β-D-glucoside 407
galloylquinol-β-D-glucoside 407
gambir 233
gambirin 599, 619, 622
gambirine 233
gambirtannine 233
gantisyl alcohol 371
garbanzol 545, 559–560
gardenoside 1074, 1090–1091, 1103
gardfloramine 243–244
gardmultine 243–244
gardmultine, demethoxy 243–244
gardneramine 243–244
gardneramine, 18-demethyl 243–244
gardnerine 243–244
gardnerine, hydroxy 242
gardnutine 243–244
gardnutine, hydroxy 243–244
garryine 248, 761–762
gascardic acid 766–767
geijerin 377
geissoschizine 238–239
geissoschizine, 4,21-dehydro 238–239
geissoschizol, 10-hydroxy 230
geissoschizol, 10-methoxy 230
gelsedine 241–242
gelsedine, 14β-hydroxy 241–242
gelsemicine 241–242
gelsemicine, 14-hydroxy 241–242
gelsemine 225, 241–243
gelsemine, 21-oxo 241
gelsenicine 242

gelsevirine 241–243
gelsevirine, 19-hydroxydihydro 242
gelsevirine, 21-oxo 243
gemin 421
genistein 549–550, 557, 567, 933, 1073–1074, 1082, 1088–1089, 1103
genkwanin 537
gentiodelphin 63
gentiopicroside 707
gentisic acid 371, 1073, 1080, 1103
gentisyl alcohol 373
geranial 964
geraniin 419, 427–429
geraniol 692, 700, 702, 853, 963–964
geranyl hydroquinone 1074, 1081, 1103
geranyl-linalool 750
geranylfarnesol 767, 787
geranylgeraniol 692, 750
geranylgeraniol, 19-hydroxy 750
germacranolide 723
germacrene 725
germacrolide 722–723
germacrone 721, 722
germichrysone 1074, 1103
gibberellic acid 763
gibberellin 34, 695, 746, 760, 762, 789, 1185
ginkgetin 574, 583
ginkgolide 21–22, 759, 1068
ginsenoside 773, 1072, 1074, 1077–1078, 1080–1083, 1085–1086, 1088–1089, 1103
girinimbine 213, 1073, 1103
gitalin 1070
gitoxigenin 838
glaucarubin 1070
glaucarubolide 1063
glaucarubolone 1073, 1080, 1085, 1090, 1103
glaucine 1057, 1070
glaziovine 1071
gleditsin 585, 606
glucitol 156, 159
glucogallin 402, 406, 408, 410
gluconic acid 269
glucose 923
glutamine 262
glyceric acid 264–265
glycerol-1-O-gallate 407
glycine, cyclopentenyl 285
glycitein 549–550
glycocitrine 211–212
glycolaldehyde 265
glycolic acid 264
glycoperine 1086, 1103
glycyrrhetic acid 779, 1071
glycyrrhetinic acid 1076, 1082, 1103
glycyrrhizin 778, 1063, 1071
glycyrrhizin acid 778
glyoxylic acid 260, 263, 265

gmelinol 462
gnidifolin 455
gomisin 418, 480–484
goniothalamin 281
gossypol 732, 734, 1071
gramisterol 828–829
granatin 427–429
grandidentatin 290, 375
grandidentin 290
grandidentoside 290, 375
grandiflorenic acid 1092, 1103
grandifloric acid 761–762
grandifloroside 376
grandifoliolenone 775
grandisin 476
grayanotoxin 762
grevillol 391
griffonilide 286
griffonin 286
grindellic acid 751
guaiacin 476–477
guaiacin, 3'-demethoxy 478
guaiacol 384, 1189
guaiacyl
glycerol 384
guaiaretic acid 475
guaiol 730, 732, 852
guayule 1031
guayulin 936
guettardine 239
guianin 440, 489, 492
guibourtacacidin 585, 604, 606, 608, 670
guibourtacacidol 600, 610, 613–614, 632
gum arabic 5, 978, 983
gum ghatti 169, 978, 986
gum jeol 170
gum karaya 978, 984
gum tragacanth 169, 978, 985
gummadiol 461
gurjunene 730, 732, 743–744, 853
gutta 787, 1028–1050
gynocardin 283

haematein 848
haematoxylin 553–554, 848
haemaxylin 1056
haginin 552
halfordin 379
hamameli tannin 400, 408, 410
hardwickiic acid 752–753
harman 1085, 1087, 1103
harringtonine 222, 1064, 1077, 1079, 1082,
    1084, 1086, 1088, 1090–1091, 1103
harringtonine, deoxy 222, 1079, 1086
hederagenin 777, 1074, 1076, 1104
hedycaryol 724
hedyotisol 474

heliangolide 722–723
heliobuphthalmin 453
heliobuphthalmin lactone 458
heliobuphthalminlactone, dihydro 471
helioxanthin 472
hematin 7
hematoxylin 7, 1074, 1104
hemitoxiferine 240–241
hemsleyadin 1071
hentriacontane 1076, 1085, 1104
herclavin 376
hernandian 468
herpetrione 473
hesperidin 1071
heterophyllin 537–538
heterotropan 477
heterotropanone 498
heterotropatrione 498
heveaflavone 575
hexacosyl ferulate 375
hexadecanoic acid, dihydroxy 339
hexahydroxydiphenic acid 406, 411, 415–416
heyneatine 1084, 1104
hibaene 760–761
hibalactone 457
hibiscone 735–736
hibiscoquinone 735, 736
himachalene 740, 742
himachalol 742, 1078, 1104
himachalonolide 741
himadarol 852, 1078, 1104
himandridine 247
himasecolone 391
himbacine 247
hinesol 729–730
hinokiflavone 577, 582
hinokiic acid 720–721
hinokinin 454
hinokiresinol 517–519, 926
hinokitin 934
hinokitiol 934
hirsutanonol 522–523
hirsuteine 233–235
hirsutenone 522–523
hirsutine 233–235
holarrhimine 250
homobutein 547
homoerythrinan alkaloids 222
homoharringtonine 1064, 1078–1079,
    1081–1082, 1087–1088, 1090–1091, 1104
homopterocarpin 550–551
honokiol 496, 1066, 1074–1075, 1078–1079,
    1081, 1083, 1091, 1104
hopane 782–783
hopanone, hydroxy 782, 784
hopeaphenol 514–515
hordenine 1084, 1104

Index of Organic Compounds 1231

humantendine 242
humantenine 242
humantenirine 242–243
humulene 711, 737–739, 852, 1080, 1104
humulene epoxide 739
huratoxin 1084, 1105
hydrangeic acid 513
hydrastine 1062, 1071
hydrocortisone 833
hydroquinone 370, 1073, 1079, 1084–1086, 1090, 1105
hydrothymoquinone 870
hydroxyacetophenone 373
hydroxybenzaldehyde 331, 341
hydroxybenzoic acid 371, 402
hydroxycadalenal 526
hydroxychalcone 587
hydroxycinnamic acid 374, 378
hydroxycitronellal 964
hydroxydalbergione 845
hydroxydammarenone 901
hydroxydocosanoic acid 325
hydroxylupanine 1092, 1107
hydroxymatairesinol 455, 864, 866, 869, 1053
hydroxymetasequirin 518–519
hydroxyotobain 480
hydroxyoxootobain 480
hydroxyphenylglycerol 384
hydroxysesamin 463
hydroxysugiresinol 518–520, 942
hygrine 204
hyoscyamine 1061, 1071
hypericin 937
hypophyllanthin 466, 472
hyptolide 280

ibogaine 230–231
idonic acid 269
imbricatadiol 751
imbricataloic acid 967–968
imbricatolal 751
imbricatolic acid 751
imonene 965
imperatorin 379, 696
incensole oxide 764–765
indoleacetic acid 1185–1186
indolenine 228–229
ingenane 936
ingenol, deoxy 766
inositol 160–161
integrin 537
intermedeol 727
ionone 66, 706, 963–964
ionone, methyl 964
ipecoside, desacetyl 223
iresin 745
iridane 698

iridoid glycoside 47
iridoids 706
irisolidone 567
iriyelliptin 388
irone 706
ishwarane 729–730
ishwarone 729–730
isoanthricin 455
isoasadanol 522, 524
isoasatone 440, 498
isobalfourodine 210
isobornyl acetate 964, 966
isobornyl thiocyanoacetate 966
isobrucein 1085, 1090, 1095
isobrugierol 293–294
isobutyric acid 854
isocamphane 698–699
isocarapanaubine 229
isocentdarol 852, 1096
isochamaejasmin 581
isochlorogenic acid 402
isochorismic acid 289
isocitric acid 260–261, 263
isocryptomerin 582
isocupressic acid 967–968
isocycloheterophyllin 537–538
isodihydromahubanolide 278
isodiospyrin 935, 1099
isodiphyllin 469
isodityrosine 186
isoelemicin 390
isoeugenol 390, 440
isoferulol 706
isoflavene 926–927
isoflavone, 7-hydroxy-4'-8-dimethoxyiso 1088, 1090, 1105
isoformosanine 235
isofraxidin 377
isofucosterol 816, 829
isofucosterol, dehydro 830
isofutoquinol 452, 493
isogalcatin 477, 479
isoginkgetin 574
isoguaiacin 477
isohalfordin 379
isoharringtonine 1084, 1086, 1090, 1104
isoheterotropatrione 498
isohibaene 760–761
isokadsuranin 481
isokaurene 305
isolariciresinol 465
isoleucine 283
isoliquiritigenin 547, 845, 850
isomagnolol 498
isomahubanolide 278
isomelacacidin 604, 607, 609, 619
isomitraphylline 236

Index of Organic Compounds

isomyristicin 390
isoobtusilactone 278
isookanin 544
isoolivil 466, 864, 868
isootobain 477
isootobaphenol 477
isopelletierine 205
isophyllocladene 305
isopicropodophyllone 468
isopimaric acid 967–969, 1073, 1084–1085, 1110
isopimpinellin 379
isoprene 384, 696
isoprene alcohol 696
isoprunetin 549
isopteropodine 235–236
isopulegol 853, 962
isoreserpiline 228–229
isoreserpiline-φ-indoxyl 229
isorhynchophylline 234–235
isosafrole 1078, 1084, 1088–1089, 1113
isosakuranetin 544
isosalipurposide 547
isoscopoletin 377
isosequiric acid 942
isosequirin 520
isosparteine 209
isotaxiresinol 465, 920, 922
isoteracacidin 606
isoterchebin 422
isothujone 708–709
isovalencenic acid 728
isovaleric acid 854–855
isovalolaginic acid 423
isovitexin 537, 541
isoyatein 456
izalpinin 555–557

jacareubin 1076, 1080–1081, 1083, 1105
jacareubin, 6-deoxy 1080, 1083
jaeschkeanadiol 721–722
jamaicin 549–550
jatrophatrione 765–766
jatropholone 766
jatrophone 1068, 1079, 1084–1085
jatrorrhizine 1076, 1105
jeediflavanone 575
jolkinol 765–766
jolkinolide 758
juglone 525, 684, 933, 935
junenol 853
juniperic acid 323
juniperol 741
juruenolide 388
justicidin 469, 471, 1084, 1105
justicinol 472
juvabione 717–718, 1053–1054

kachirachirol 494
kadayo gum 171
kadsuranin 480, 484
kadsurenone 439–440, 493
kadsurin 482, 493, 503, 505
kadsutherin 481
kaempferol 401, 539, 541–542, 558
kaerophyllin 456
kainic acid 1071
kanugin, demethoxy 540
karaya 171
katilo gum 171
kauranol 760–761, 763
kaurene 305, 760–761, 763
kaurenoic acid 763
kaurenol 763
kaurenolide 761
kauri gum 4
kawain 1063
kawaina 1071
kayaflavone 575
ketha gum 171
ketoglutaric acid 259–262
keyakinin 542
khellin 1071
khusinol 734–735
kigeliol 464
kobusin 464
koenigicine 213
koenimbine 213
kokusaginine 210
kolaflavanone 580
kolavenol 752–753
konyanin 455
kopsirachin 591, 593
koumidine 242–243
koumine 242
koumine, 11-methoxy 242
kulinone 773–774
kullo gum 171
kusunokinin 455
kusunokinol 457
kuteera gum 171
kuwanon 1068, 1087, 1105–1106

lactic acid 266–267
laevojunenol 726
lagerine 206
lagerine, methyl 206
lagerstroemine 206
lambertianic acid 967–968
lamentoflavone 584
lanceol 717–718
lancifolin 440, 492
lancilin 498
lanosterol 774, 809, 812–813, 832
lantaside 1071

lantibeside 465
lapachol 849, 937, 943, 1066, 1074, 1079–1080, 1084, 1090, 1106
lapachol, dehydroxy 920
lapachone 525, 1079, 1106
lapachonone 870, 935
lappaol 473–474
lariciresinol 458
larixinol 571, 577
laserpitin 721–722
lasubine 207
lathosterol 815, 818, 824, 827
lathosterol, 24α-ethyl 816
lathosterol, methyl 824, 828
latifolin 553–554
laucine 265
lauric acid, hydroxy 331
lectin 181–183, 186–187, 193
ledol 743–744
leiocalycin 551
leiocarpin 563–564
lemmasterone 837
leptolepisol 474
leucoanthocyanin 669
leucocordiachrome 391
leucocyanidin 585, 595, 604–605, 609, 611, 679
leucodelphinidin 1076, 1078–1079, 1082, 1091–1092, 1106
leucofisetinidin 585, 604, 606, 609
leucopelargonidin 605, 1079, 1106
leucorobinetinidin 595, 604, 606, 609, 925
leucorobinetinidol 619
leucotriene 945
leukocristine 1072
levopimaric acid 758, 891, 967, 969
licarin 494
lignanolide 456
lignoceric acid 323
ligstroside 48
ligstroside, hydroxy 389
liguloxide 730, 732
liliflodione 493
liliflone 493
limonene 700, 703, 852–853, 856
limonin 777
limonoids 774, 810
linalool 700, 702, 853, 963–964
linalool oxide 702
linamarin 283, 287
linderane 722–723
lindleyin 407
lingularinon, 8β-(angeloyloxy) 722
linmein 282
linoleic acid 973–974
linolenic acid 973, 975
lintetralin 466

linustatin 283
liovil 457
lipecoside, desacetyl 224
liquiritigenin 543–544, 1086, 1106
liriodendrin 464
liriodenine 1073, 1075, 1084, 1106
lirioresinol 460, 462–463, 470
lithospermoside 286–287
litsenolide 270, 278–279
lobeline 1071
loganin 700–701, 1074, 1079, 1082, 1107
longiborneol 741–742
longicyclene 742
longifolene 742, 852
longipilin 723
longipinene 741–742
lophenol 826
lophenol, 24-methylene 822
lotaustralin 282–283
lucenin 537
lucidin ω-methyl ether 1084, 1107
lunacrine 210
lupanine 209, 1087, 1092, 1107
lupanine, 13-hydroxy 1087
lupeol 777, 779, 1076, 1087
lupeol acetate 1076, 1080, 1107
lupinine 209
luteolin 535–537, 541
luvangetin 382
luvangetin, demethyl 382
lycopene 786, 787
lyoniresinol 465
lysicamine 1074–1075, 1078, 1080, 1107
lythridine 206
lythrine 206–207

maackiasin 513, 515, 568
maaliol 744
maalioxide 744
macassar 847
machilin 475, 494, 497
machilusin 476
maclurin 848
macluroxanthone 525
macrophyllin 493
maculosidine 210
macusine 240–241
maglifloenone 492
magnoflorine 218, 1081, 1083, 1087, 1092, 1107
magnolin 463
magnolol 496, 1066, 1074, 1078–1081, 1083, 1091, 1108
magnosalicin 496
magnosalin 477
magnostellin 459, 477
mahuba lactone 275, 279

1234 Index of Organic Compounds

mahubenolide 277
makisterone 837
malabaricol 783–784
malanic acid 265
maliacane 775
malic acid 260, 262–263
mallotinic acid 429
mallotusonic acid 429
malonic acid 264
maltose 158–159
mamegaki-quinone 847
mammea 380, 553–554
mammeisin 381
manassantin 498
mandschurin 377
mangiferin 1075–1076, 1080, 1088, 1108
manisdaurin 287
manngashinoro 924
manniflavanone 581, 1076, 1108
mannitol 9, 157, 160, 923
mannose 108
manool 750–751
manoyl oxide 751
mansone 735, 736
mansonone 526, 735–736, 845, 848, 920
mappicine 244–245
marmesin 381, 1075, 1108
marmin 378
massoilactone 280
massoniresinol 459
mastic 3, 4
masticadienonic acid 773–774
masuaferrone 578
matairesinol 440, 454, 868–869
matrine 1072, 1080, 1087, 1089, 1091–1092, 1108
maturinone 729
mavacurine 240–241
maytanprine 1079, 1108
maytansine 214, 1065, 1077, 1079, 1089, 1092, 1108
maytenoquinone 758–759
maytine 213–214
maytoline 213–214
mayurone 720–721
medicagol 552
medicarpin 550–551, 568
medioresinol 461
megaphone 485, 501
megaphyllone 485
melacacidin 585, 603–604, 607–609, 670
melacacidol 619
melacacinidin 619
melampolide 722–723
melannein 553–554
melanoxetin 539
melezitose 158

melicopicine 211–212
melicopidine 211–212
melicopine 211–212
melisimplexin 541
melisimplin 541
meliternatin 540, 542
meliternin 541
menisdaurilide 287
menisdaurin 286
menisperine 1087, 1108
menth-2-en-7-ol 1073, 1087, 1093, 1108
menthadiene 852
menthane 698–699, 703
menthol 84, 962–963, 966, 1071
menthone 705
mercusic acid 967
mesuferrone 576
metasequirin 518–519
methoxybilobetin 572, 575
methoxydalbergione 554
methoxypsoralen 1063, 1072
methoxythymol 870
methoxyyatein 456
methyl angolensate 776
methyl benzoate 853
methyl chavicol 961, 965
methyl ferulate 904
methyl p-coumarate 903–904, 906, 908
methyl reserpate 229
methyl salicylate 851, 1071
methyl thujate 852, 1053, 1055
methylcatechol 370
methylchavicol 853
methyldalbergin 553–554
methylgrandifloroside 376
methyljuglone 525
methyllatifolin 553
mexolide 383
mexoticin 378
micromelin 381
milliamine 1082, 1108
mimosine 932
mirandin 493
mitomycin 95
mitraphylline 236
modhephene 740
mollisacacidin 585, 595–596, 602, 604, 606, 608–611, 619, 670
monocrotaline 1071
montanic acid 323
morachalcone 516–517
moretenol 784
morin 539, 542, 556, 1078–1080, 1086, 1090–1091, 1108
morolic acid 777, 779
morphine 215, 1071
morroniside 1082, 1109

Index of Organic Compounds 1235

mucara gum 171
mucroquinone 551–552
mukoeic acid 213
mukulol 764–765
mulberranol 537
mulberrin 537–538
mulberrochromene 537–538
multiflorenol 781–782
mumenin 541
mundulone 549–550
murrayanine 213
muurolene 734–735
muurolol 734–735
muzigadial 745
myrcane 698, 701, 852–853, 856, 963, 1073, 1087, 1093, 1109
myricanol 522, 524
myricanone 522, 524
myricetin 136, 401–403, 407, 540–542, 555, 557–558, 603
myricitrin 542
myristic acid 1091, 1109
myristicin 384, 390
myrrh 7, 170
myrtenal 709
myrtenic acid 709
myrtenol 709

naphthoquinone, dimethallyl-1,4- 935
narcotine 1071
naringenin 407, 543–544, 1075, 1078, 1080, 1086, 1090–1091, 1109
naringenochalcone 535
nectandrin 476
neoabietic acid 758, 891, 967–968
neoandrographolide 1071
neochamaejasmin 581
neocryptomerin 582
neodiosgenin 835
neoflavene 554, 564, 851
neohesperidoside 541
neoisostegane 473
neolinustatin 283
neoolivil 458
neoplathymenin 547
neoruscogenin 836
neosa uranin 547
neotorreyol 716–717
nepetalactone 706–707
neral 963–964, 700, 702, 963–964
nerolidol 716
neurosporene 786–787
nezukone 852
nicotine 1071, 1076, 1084, 1087, 1092, 1109
nigakinone 226
nimbin 776
niranthin 453

nirtetralin 466
nitidine 219–220, 1064, 1079, 1086, 1091, 1109
nodakenetin, 8-prenyl 381
nonacosan-15-one 315
nonacosane 315
nonacosanol 315
nonynoic acid 312
nootkatin 711, 870
nootkatone 728
noracronycine 211–212
noracronycine, de-N-methyl 211–212
norartocarpin 537
norartocarpetin 536
nordihydroguaiaretic acid 475, 1071
nordihydrotoxiferine 240–241
norephedrine 43
norerythrostachaldine 249
norisoguaiacin 477
normacusine 230
normaytancyprine 1079, 1084, 1108
normaytansine 1079, 1084
norpseudoephedrine 213, 1062, 1071
norpurpeline 228, 230
norseredamine 228, 230
nortetraphyllicine 228, 230
nortetraphyllicine, 10-hydroxy 230
nortrachelogenin 455
noscapine 1071
novacine 239
nuciferal 717–718, 852

obaberine 1076, 1110
obacunone 776–777
obtusifoliol 829
obtusilactone 277–279
occidentalol 727, 852, 1053–1055
occidentoside 577, 582
occidol 727, 852
ochnaflavone 577, 582
ocimene 701, 964
ocosanic acid, hydroxy 331
ocotillone 773
octacosanyl ferulate 375
octadecadiene-12,14-diyne-1,16-diol 9, 17, 292
octadecadiene-12,14-diynoic acid 9, 17, 292
octadecanedioic acid, 9,10-dihydroxy 340
octadecanoic acid, 9,10,18-trihydroxy 325–326, 340
octadecanoic acid, 9,10-epoxy-18-hydroxy 325, 339–340
octadecanoic acid, hydroxy 331, 339
octadecanoic acid, tri-hydroxy 339
okanin 546–547, 1077, 1110
oleandrigenin 839
oleanolic acid 309, 777, 779, 1073, 1076–1077, 1087, 1110

1236 Index of Organic Compounds

oleic acid   973, 975
oleuropein   48, 1076, 1089, 1110
olivacine   231–232
olivil   458
olivil-glucopyranoside   459
onocerin   785
ophiobolin   766–767
oregonin   522–524
orixine   210–211
ornithine   46
orobol   535, 554
orosunol   472
oroxylon   536
osajaxanthone   525
osthol   377
otobaene   479
otobain   480
otobanone   480
otobaphenol   477
oubain   1071
oxalic acid   267–269, 1057
oxaloacetic acid   259–260, 263
oxidohimachalene   740, 742
oxogambirtannine   233
oxonantenine   1074–1075, 1079, 1081, 1110
oxyayanin   540, 563, 566, 937
oxydihydroartocarpesin   537
oxynitidine   219–220

pachycarpine   1061, 1071
paeoniflorin   709
pallidol   514–515
palmatine   1071
palmitic acid   323, 973, 975
palustric acid   891, 967, 969
papain   1071
papaverine   1071
paracotoin, 4-methoxy   387
parasorbic acid   280
parbenzlactone   454
parthein   936
parthenin   936
passibiflorin   284–285
passicapsin   285
passicoccin   285
patagonaldehyde   390
paulownin   460
pectin   683
pectolinarigenin   536
pedunculagin   418–420, 422, 424–425
peganine   1072
pelargonidin   656
pelletierine   205–206
pelletierine, N-methyl   205
peltatin   468
peltogynoids   604
peltogynol   526–527

pentadecylcatechol   935–936
penta-O-galloyl-D-glucose   406, 410–414, 419–420, 423, 426–427, 431, 433
pentagalloyl glucose   668
pentalenolactone   72–73
perezone   717–718, 935
pericalline   1073, 1084, 1089, 1110
perillaldehyde   700, 705, 853
persicogenin   544
pessitrifasciatin   285
phellandrene   703, 852
phellogenic acid   323
phellonic acid   323
phenol   370, 853, 1084–1085, 1110
phenylacetic acid   373
phenylacetic acid, 4-hydroxy-3-methoxy   373
phenylalanine   262
phenylchroman   548
phenylcoumarin   548
phenylethanol   853
phillygenol   462
phlobatannin   635, 661
phloionic acid   323
phloionolic acid   323
phloroglucinol   370–371
photomethylquercetin   526–527
phycobilin   128
phyllanthin   453
phyllanthine   208
phyllanthol   782
phyllanthoside   1068, 1079, 1084, 1110
phyllanthostatin   1068, 1079, 1110
phyllocladanol   761–762
phyllocladene   305, 761
phyllodulcin   1071
phylloflavan   590, 593
phylloquinone   788
phyltetralin   466
physcion   1074, 1110
physostigmine   1071
phytane   747
phytoene   695, 786–787
phytofluene   786–787
phytol   750
piceatannol   513–514
picein   373
picrasin   1083, 1110
picropodophyllin   468
picropolygamain   470
picroside   706–707
picrotin   737
picrotoxin   737, 1071
picrotoxinin   737
pilocarpine   1071
pimaradiene   749, 755
pimaric acid   755, 967–969
pimarinal   755

Index of Organic Compounds 1237

pinane 698–699
pinene 246, 701, 708, 851–853, 856, 865, 869, 960–964, 966, 1174
pinidine 246–247
pinitol 161, 1071
pinobanksin 545–546, 555–557, 896
pinocarveol 709
pinocembrin 543–544, 548, 555–558
pinomyricetin 540, 555, 557
pinoquercetin 540, 542, 555
pinoresinol 461, 868–869
pinostrobin 544, 555
pinosylvin 513–514, 516–517, 864–865, 873, 876, 888, 925
pinosylvin dimethyl ether 513–514, 517
pinosylvin monomethyl ether 514, 516, 864, 873, 888
pinosylvin, dihydro 517
piperenone 493
piperideine 206
piperitol 464
piperitone 705
piperlongumin 392
piperonylic acid 389
pipoxide 288
plastoquinone 788–789
plathymenin 544
plathyterpol 753
platyconin 63
platydesminium salt, N-methyl 210
platyphyllin 522
platyphyllonol 522–523
plaunol 1068, 1080, 1084, 1111
plicatic acid 465, 934, 937–938, 943, 945, 1053–1055
plicatinaphthol 469
plumericin 707
plumieride coumarate glucoside 1072, 1111
pluviatilol 461, 464–465
pluviatolide 454
podecdysone 837
podocarpic acid 758–759
podocarpusflavone 574, 584
podophyllotoxin 439–440, 467–468, 933, 1063, 1071
podophyllotoxone 468
podospicatin 557
pollinastanol 826–827
polygamain 470
polygamatin 472
polyprenol 695, 787
pongaglabrone 537
populin 290, 372
populoside 372
poriferasterol 829
poriferasterol, dehydro 817–818, 830
porosin 486, 501

potentillin 420–423
praecoxin 423
pregn-1,4-dien-3-one, 20-carboxy 833
pregn-4-en-3,20-dione 833
pregn-5-en-20-one, 3β-hydroxy 833
pregnenolone 833
pregomisin 476
preisocalamendiol 722–723
prenyl alcohol 696
prenyletin 696
presilphiperfolane, hydroxy 740
prestegane 440, 456–457
prezizaene 720
prieurianin 1085, 1111
primeveroside 526
prinsepiol 464
pristimerin 781–782, 1074, 1111
pristine 305
proacacipetalin 283
proanthocyanidin 400, 403–404, 608–609, 651–690, 886, 888, 897–899, 940, 989–992
procyanidin 401, 571–651, 651–690, 898, 945, 989–992, 1015–1016, 1018, 1076, 1111
prodelphinidin 6, 401–403, 584, 603, 612, 619, 629–631, 634, 652, 654–655, 657, 660–661, 670, 673–674, 679–680, 896, 898–899, 989–991, 1016
profisetinidin 584, 600, 611, 613, 615–616, 618–619, 632, 652, 660, 663, 670–671, 676
progesterone 830, 833
proguibourtacacidin 612
proguibourtinidin 584, 613–614, 652, 670–671
proline 46, 181
proluteolinidin 989
promelacacidin 604, 610, 619, 621
propacin 383
propelargonidin 584, 619, 621, 652, 654, 661, 670–671, 680, 685
propionic acid 854
propiovanillone 384
prorobinetinidin 584, 619–620, 652, 663, 670–671
prosopinine 1072–1073, 1082, 1111
prostalidin 472
proteracacidin 619
protocatechuic acid 371, 402, 404, 1073–1075, 1080, 1089–1090, 1111
protocatechuic aldehyde 904
protoemetine 223, 224
protopine 1072, 1075–1078, 1081–1083, 1087–1088, 1090–1092, 1111
protoveratrine 1071
prunasin 373–374

1238   Index of Organic Compounds

prunetin 550, 554
prunin 544
pseudo-γ-schisandrin 483
pseudobaptigenin 550, 567
pseudoephedrine 1062, 1071
pseudoionone 964
pseudopelletierine 205-206
pseudotropine 203
pseudoyohimbine 236-237
psicose 74
psoralen 379, 1087, 1091, 1111
psoralen, 5-methoxy 937
psychotrin, O-methyl 223
psychotrine 223, 224
ptaeroxylin 386
pteleatinium salt 210
pterecortine 210-211
pterocarpan 548, 550, 563-565, 568
pterocarpane 851
pterocarpdiolone 726
pterofuran 552
pteropodine 235-236
pterostilbene 514
puearin 1111
puerarin 1081, 1083, 1087, 1089
pulegone 705
pungenin 373
purifolidine 231
purpeline 228, 230
purpurein 290
purpurogallin 601
pygeoside 466
pyranojacareubin 525
pyranoquinoline, dihydro 210
pyrazine, tetramethyl 1073, 1078, 1081, 1083, 1087, 1091, 1111
pyrethrosin 721-722
pyridine 851, 853
pyrocatechol 290, 929
pyrogallol 370-371, 411, 921, 1075, 1077, 1080, 1088, 1111
pyrrolidine, methyl 851, 853
pyrrolizidine alkaloids 65
pyrufuran 520-521
pyruvic acid 264-265

quassane 775
quassin 226, 776-777
quassinoids 774, 810
quebrachamine 230-231
quebrachitol 161-162
quebracho 400
quercetin 401, 407, 535, 539, 541-542, 555, 557-558, 568-569, 583, 869, 896, 941, 1080
quercitol 162
quercitrin 542, 1075, 1077, 1080, 1086, 1111

quercus lactone 275
quibourtacacidin 592, 602
quinic acid 162, 264, 266, 404
quinidine 237, 1063, 1071, 1077, 1079, 1085, 1112
quinine 9, 225, 237-239, 1063, 1071-1072, 1077, 1087-1088, 1092, 1112
quinoline 210
quinolizidine 1061
quinolone, 4-methoxy-1-2- 210
quinolone, 4-methoxy-1-methyl-2- 219
quinones 933-935, 944
quisqualic acid 1071

raffinose 157-158
randainol 496
rankinidine 242-243
rastevione 741-742
ratanhiaphenol 494-495
rauvoxine 229
rengasin 559-561, 906, 908, 925-926
rescinnamine 1062, 1071
resene 902-903
reserpiline 228-229
reserpiline, 19,20-dehydro 229
reserpine 227-228, 1062, 1071, 1076, 1078, 1081, 1083, 1086-1090, 1092, 1112
resorcinol 370
resorcylic acid 402
resveratrol 512-516, 873, 1073, 1075, 1082, 1088, 1112
retamine 1092, 1112
reticuline 214-215
retinal 45, 101
retrochinensin 472
retusin 513
rhein 1075, 1085-1086, 1090, 1112
rhetsinine 212
rhizobactin 80-82
rhododendrin 392
rhododendrol 392-393
rhodopsin 100-101
rhomitoxin 1071
rhusflavanone 579
rhusflavone 579
rhynchophine 234
rhynchophylline 234-235
ribose 159
robinetin 539, 542, 559-560, 869
robinetinidin 585, 602, 608
robinetinidol 590, 592, 620
robtein 547
robtin 544
robustaflavone 575-576
robustine 1081, 1112
robustol 522, 524
rorifone 1071

Index of Organic Compounds   1239

rotenoid   552
rotenone   1071
rotunidine   1071
roxburghine   233
royleanone   758
rubber   8–9, 11, 695, 787, 970, 976, 1028–1050, 1165, 1169–1170, 1180
rubraflavones   537, 541
rubranine   546–548
rubrenolide   278–280
rubrynolide   278–280
rugosin   421, 423–424, 431
rutacridone epoxide   1074, 1079, 1112
rutaecarpine   212, 1092, 1112
rutaecarpine, 13β,14-dihydro   212–213
rutin   569, 1071
ryanodine   1088, 1112

safrole   384, 851, 938, 1074, 1083–1085, 1087, 1089, 1112–1113
sakuranetin   544
salaspermic acid   781, 782
salicin   290, 371–373, 1071, 1188
salicortin   290, 372
salicyl alcohol   371–373
salicylic acid   290
salicyloylsalicin   372
salicyloyltremuloidin   372
salidroside   373, 389
salireposide   372–373
salsolinol   1086, 1089, 1113
salvileucolide   767
sandarac   3–4
sandaracopimaradiene   755–756
sandaracopimaric acid   755–756, 967–968
sanggenone   1076, 1113
sanguin   421, 431
sanguinarine   1062, 1071–1076, 1078, 1080–1090, 1092–1093, 1113
sanguisorbic acid   415, 417, 421
santal   567
santalene   720–721, 853
santalin   527, 848
santalol   721, 853
santanin   726
santarubin   527
santenol   853
santenone   853
santonin   725, 1071
sapelin   773–774
sapogenin   812
saponin   37, 45, 808, 835, 933
sapote gum   172
sarmentogenin   838
sarpagine   230
sativene   736–737

saucerneol   498
saussurea lactone   723–724
savinin   454
schinol   936
schisandrin   481
schisandrol   481
schisandrone   478
schisanhenol   482
schisantherin   483–484
schizandrin   417–418, 482
schottenol   816
sciadin   751–752
sciadopitysin   574, 583
scillarin   1071
scoparone   377, 1074, 1083, 1087, 1097, 1113
scopolamine   183, 203–204, 1061, 1071, 1081, 1083–1084, 1091, 1113
scopoletin   378, 864, 1073–1074, 1081, 1083–1085, 1087, 1089, 1091–1092, 1113
secodaphniphylline   249
secodehydroabietic acid   969
secoisolariciresinol   453
secologanin   223–225, 238
securinine   208
securinol   208
securitinine   208
selinene   852
sempervirine   242
senecioic acid   697
seneciphylline   1087, 1091, 1114
seneol   287–289
senepoxide   287–289
sennoside   1063, 1071, 1088, 1114
sequirin   518–520, 925–926
sequoiaflavone   574, 583
sequoyitol   161
serpentine   227–228
serrantendiol   309, 785
sesamin   460, 1088, 1114
sesartemin   464
seselin   382
sesquifenchene   720–721
sesquineolignan   505
shahjan gum   172
shikimic acid   264, 266, 404
shikonin   1074, 1079, 1114
shikonin, acetyl   1076
shonanic acid   707–708
shyobunone   724
siderin   380
sikkimotoxin   469
silicolin   468
silphinene   740
silphiperfol-6-ene   740
silymarin   1071
simmondsin   286
simplexoside   464

Index of Organic Compounds

sinapaldehyde 939
sinapic acid 374, 402, 1074–1075, 1081, 1114
singueanol 1074, 1078, 1114
sinine 206
sinomenine 217
sitost-3,5-dien-7-one 825
sitost-4,6-dien-3-one 825
sitost-4-en-3-one 825
sitost-4-en-3-one, 6β-hydroxy 826
sitostanol 821–822
sitostanone 826
sitosterol 808–839, 907, 913, 969, 976
sitosterol-β-D-glucoside 835, 904, 907–908
skatole 853–853, 855
skimmianine 210–211, 1972, 1074, 1078, 1081, 1083, 1087, 1114
skimmin 377
skytanthine 593
sobrerol 704
solanesol 787
sorbifolin 385
sorbitol 159–160
sotetsuflavone 574
sparteine 209, 1061, 1071, 1087–1088, 1090, 1092, 1115
specionin 20
speciophylline 235–236
specioside 376, 580
spinasterol 817–818, 823–824, 828, 830, 835
spinasterone 826
spinescin 464
spiradine 248
spironolactone 976
squalene 695, 768, 771, 1080, 1087, 1115
squalene oxide 809–810, 813–814, 826, 832
stachyose 157–159
stachyurin 422, 425
starch 978
stearic acid 323
steganolide 472
stemmadenine N-oxide 231–232
stemonoporol 514–515
stevioside 761–762, 1071
stigmasterol 816, 818–822, 824–826, 830, 832–833
stilbene(s) 133, 136, 512, 514, 851, 864, 868–874, 896, 898, 911, 926, 943
stilbene, 2,4,3'-trihydroxy 514
stilbene, 3,3'-dimethoxy-4,4'-dihydroxy 896
stilbene, 3,3',4,5' tetrahydroxy 925, 927
stilbene, 3,4,3',3-tetrahydroxy 513
stilbene, 3,4,3',4'-tetrahydroxy 513
stilbene, 3,4,3',5'-tetrahydroxy 513–514, 613
stilbene, 3,4,5,3',5'-pentahydroxy 514
stilbene, 3,5,4'-trihydroxy 514
stilbene, 3,5-dihydroxy 514
stilbene, 3,5-dimethoxy-4'-hydroxy 514

stilbene, 3,5-dimethoxy-4-isopentenyl 514
stilbene, 3-methoxy-4,5-dihydroxy 514
stilbene, 4-hydroxy 514
storax 3, 1057
striatene 62, 64
striatol 522, 524
strictosamide 234, 246
strictosidine 225, 228, 238, 246
strobane 966
strobobanksin 545, 555–557
strobochrysin 536, 556
strobopinin 544, 555–557
strophanthidine 935
strychnine 225, 239, 1071
strychnobiflavone 575
styrax, see storax 1057
styrene, 2,4,5-trimethoxy 390
subcosine 207
suberinic acid 323
suberosin 377
suberosin, 7-demethyl 377
succedaneaflavone 576, 579
succinic acid 260, 262
suchilactone 456
sucrose 923
sugiresinol 517–519, 942
sulferalin 731–732
sulfuretin or sulphuretin 546–548, 559–561, 845, 849, 905, 908
surinamensin 496
suvanine 69
sventenin 454–455
swietenocoumarin 379
swietenol 377
swietenone 379
symplocoside 593
symplocosigenol 461
synephrine 1078, 1088, 1092, 1115
syringaldehyde 331–332, 336
syringaresinol 462–463, 1115
syringetin 540, 542
syringic acid 371
syringin 384, 1078, 1081, 1086, 1115

tabernaemontanine 232
tabernanthine 230–231
tabernulosine 1076, 1115
taiwaniaflavone 576, 580
taiwanin 453, 469, 920
talaromycin 77–78
tanghinin 838
tannic acid 412, 923
tannin 5, 6, 9, 11, 37, 132, 226, 229, 233, 571–635, 864, 873, 875, 899, 922–924, 927, 939–942, 944, 988–1018, 1057, 1165, 1168, 1170, 1175–1177, 1080, 1087, 1190
taraxasterol 777, 779

taraxerol 781–782, 1080, 1086, 1115
taraxinic acid 722
tartaric acid 267–269
taxifolin – see dihydroquercetin 373, 535, 543, 545, 679, 869, 920, 922, 1075, 1077, 1090, 1092, 1116
taxol 1056
tectochrysin 536, 555, 557
tectol 920
tectoquinone 849
tectorigenin 549–550
tellimagrandin 420–422
tembamide 389
tembetarine 1081, 1083, 1088, 1116
teracacidin 585, 595, 603–604, 606–609, 670
terchebin 427, 429
terebinthone 936
teresantalal 710
teresantalic acid 710
teresantalol 710
terpin 704
terpin hydrate 961
terpinen-4-ol 1072, 1074
terpinene 703, 852
terpinenol 1087
terpineol 700, 703–704, 852–853, 961, 1073, 1082, 1093, 1116, 1174
terpinin 1080, 1086, 1116
terpinolene 703, 852, 856
terpinyl acetate 852
testosterone 830, 833
tetol 756
tetracosyl ferulate 375
tetrahydrocannabinol 1072
tetrahydroiangonin 387
tetrahydropalmatine 1071–1072
tetrahydrorobustaflavone 575
tetramethylpyrazine 1066
tetrandrine 1072
tetraphyllin 284
thalictoside 373
thapsic acid 323
theaflavin 601
thearubigin 601–602
theobromine 1072
theogallin 402, 407
theophylline 1072
thevetin 838
thioxanthene oxide 43
thomasic acid 465
threonine 46
thronic acid 269
thujane 698–699
thujaplicatin 454–455
thujaplicin 711, 869–870, 889, 929, 934, 1053, 1055, 1082, 1116, 1184
thujene 708

thujic acid 707–708, 1055
thujone 708–709, 933–934
thujopsene 720–721, 852
thunbergene 764
thymohydroquinone dimethylether, isobutyryloxy 704–705
thymol 704, 1072
thymol, p-methoxy 852
tiglic acid 697
tigloylgomisin 481
tingenone 1078, 1085–1086, 1090–1091, 1117
tingtanoxide 288–289
tirucallol 773, 809
tocopherol 788–789
tocotrienol 789
tocotrienolquinone 788
todomatuic acid 717–718
todomatuic acid, 7-hydroxy 718
torosachrysone 1074, 1117
torreyal 716–717
torulosal 751
torulosol 751
totarol 758–759, 870
toxaphene 966
toxiferine 240–241
trachelogenin 455
trachylobane 759–762
tragacanth 171
trapain 422
tremetone 933
tremulacin 290, 372
tremuloidin 290, 372
triacontane 1076
triandrin 384
trichilin 75–76
trichocarpin 372–373
trichocarposide 372
trichosanthin 1072
tricin 1117
tricontane 1117
tricyclene 698–699, 708
trilloic acid 415, 417
trilobine 216
tripdiolide 1068, 1077, 1079, 1084, 1086, 1090–1091, 1117
triptolide 758–759, 1068, 1077
tropane 1061
tropine 203
tropine benzoate 203
tropine, 3,4,5-trimethoxy 204
tropine, benzoyl 204
tropinone 204
tropolone(s) 711, 868, 870, 872–873, 888–889, 929
tryptamine, N$_b$-methyl 226
tsugacetal 467

Index of Organic Compounds

tubocurarine 217, 1072
tubulosine 224
tubulosine, deoxy 224
tunichrome 45
turriane 522, 524
tyrosine 262

ubiquinone 58, 788–789
ugandensolide 744–745
umbelliferone 378, 1073, 1077–1078, 1082, 1090–1091
umbellulone 708–709
umuravumbolide 280
uncarine 235–236
ursolic acid 309, 767, 779–780, 1076, 1080, 1083, 1087, 1093
urushiol 5, 520–521
uvafzelin 1074, 1079
uzarigenin 839, 1084

valepotriates 1072
valeranone 729–730
valerianine 706–707
valerianol 728–729
valeric acid 854–855
valerosidate 706–707
valine 265, 283
vallesiachotamine 234
valolaginic acid 423
valoneic acid 415, 417, 421, 424
valtrate 706–707
vanillic acid 309, 371, 374, 920, 1072
vanillin 331, 336, 341, 373, 853, 1075
vanillyl alcohol 373
variabilin 551
vascalin 422
vasicine 1065, 1072, 1088, 1091–1092
vasicinone 1072, 1081
vaticaffinol 514
vaucapenyl acetate 756
veatchine 248
velutin 536
veraguensin 476
verbascose 157–158
verbenol 709
verbenone 709
vernolepin 724
verrucosin 476
versicoside 464
verticillatine, dihydro 206
verticillol 765
vertine 206–207
vescalagin 422, 424–425, 431
vescalin 425
vetivane 730
vetivone 728–729
vicenin 537

vinblastine 1072
vincaleukoblastine 1072
vincamine 1072
vincoside lactam 234
vincristine 1072
viniferin 514–515
violaxanthin 745
viroallosecurinine 208
virolaflorine 391
virolin 497
virosecurinine 208
vitamin K 788–789
vitexin 537, 541
voacarpine, 16-epi 242
voachalotine, 16-epi 243
vobasine 232, 242
volkenin 284
volkensiflavone 576, 580
vomicine 239

wharangin 541
widdrol 720–721, 852
wighteone 549–550
wikstromol 456
winterin 744–745
wodeshiol 460
wuweizizu 481

xanthevodine 211–212
xanthoarnol 381
xanthochymusside 581
xanthone 281, 851
xanthone, 1-3-5-trihydroxy-2-methoxy 1076, 1081
xanthone, 1-5-dihydroxy 1077
xanthone, 3-8-dihydroxy-1-2-dimethoxy 1077, 1081
xanthone, tetrahydroxy 848
xanthoperol 899
xanthophyll 130, 787
xanthotoxin 379, 1063, 1072, 1077, 1079, 1084–1085, 1091–1092
xanthoxyletin 382
xanthyletin 382
xylosmacin 372
xylotenin 379

yabunikkeol 704
yamogenin 835–836
yangambin 459–460
yashabushiketol 522–523
yatein 456
yateresinol 518
ylangene 736
yohimbine(s) 225, 228–229, 236–237, 1072–1073, 1078, 1080–1081, 1086, 1089, 1090–1092
yohimbine, 18-hydroxy 228–229

yuanhuacine 1072
yuanhuadine 1072
yuzurimine 249

zanthoxylol 390
zapotin 538
zapotinin 538
zeatin 789−790, 1080

zeatin riboside 1080
zehyerin 577
zeylena 289
zeylenol 288−289
zingiberene 717−718
zizaene 720
zuihonin 453
zurumbone 739

Printed and bound by PG in the USA